MULTIDIMENSIONAL SOLID-STATE NMR AND POLYMERS

MULTIDIMENSIONAL SOLID-STATE NMR AND POLYMERS

K. Schmidt-Rohr
and
H. W. Spiess
Max-Planck-Institut für Polymerforschung
Postfach 3148, D-55021 Mainz, Germany

ACADEMIC PRESS
Harcourt Brace & Company, Publishers
LONDON SAN DIEGO NEW YORK BOSTON
SYDNEY TOKYO TORONTO

ACADEMIC PRESS LIMITED
24–28 Oval Road
London NW1 7DX

US edition published by
ACADEMIC PRESS INC
San Diego, CA 92101

This book is printed on acid-free paper

A catalogue record for this book is available from the British Library

ISBN 0-12-626630-1

Typeset by Mathematical Composition Setters Ltd, Salisbury, Wiltshire
Printed in Great Britain by Hartnolls Ltd, Bodmin, Cornwall

Contents

Preface

Another book on NMR spectroscopy: why? As in other fields of science, NMR of polymers is divided into two major areas. From the chemist's point of view, high-resolution NMR provides detailed information about the chain microstructure in solution and is an indispensable tool of polymer chemistry. From the materials science point of view, many of the macroscopic properties of polymers arise from the organization of macromolecules in the bulk. Therefore, experimental techniques like solid-state NMR are needed that can characterize the behaviour of polymers on a molecular level and on mesoscopic scales in order to relate it to the macroscopic properties. However, contrary to the situation in liquids, where the detailed information about chain microstructure as available from high-resolution NMR is unique, solid-state NMR competes with well-established techniques, such as light, X-ray, or neutron scattering, electron microscopy, and dielectric or mechanical relaxation. In order to compete favourably with these techniques, the potential of solid-state NMR has to be fully exploited.

This calls for multidimensional methods, which increase the spectral resolution and, therefore, the selectivity, but also provide routes to new information not accessible from 1D spectra even in the limit of high resolution. Whereas excellent monographs exist on high-resolution NMR in liquids and in solids as well as on two-dimensional NMR in general, a book on multidimensional solid-state NMR is not available. Substantial progress has been achieved in this area in recent years, in particular as far as the elucidation of molecular dynamics of polymers is concerned. This provided the impetus for our endeavour. When thinking of an outline, it quickly became apparent that the book will address a heterogeneous audience with backgrounds in polymers and spectroscopy. Therefore, we expanded the scope to include basic and advanced material of solid-state NMR (Chapters 2 and 3), Fourier theory (Chapter 4), and polymer science (Chapter 5), before we systematically discuss multidimensional separation and correlation experiments in Chapter 6. The primary aim of our monograph is molecular dynamics and multidimensional NMR techniques designed to unravel its different aspects; these are treated in the central Chapters 7–11. The book is concluded by addressing two other important aspects of macromolecular organization in the solid state, namely chain alignment in partially ordered systems (Chapter 12) and domain sizes in heterogeneous polymers (Chapter 13).

Although we restrict ourselves to discussing applications of multidimensional NMR to synthetic polymers, it is obvious that the techniques described here are equally applicable to many problems in solid-state physics, chemistry, materials science and notably, biophysics. The reader will quickly find out that different chapters are written on rather different levels, descriptive and technical or mathematical ones. Our intention was to generate a text that, first of all, is useful for many. Applying sophisticated spectroscopic techniques in an interdisciplinary area like polymer science inevitably involves use of concepts from different fields of science. More specifically, rather subtle aspects of quantum mechanics, statistical mechanics, signal processing, and chemistry are involved. It is our experience that newcomers to the field often are 'overqualified' concerning some but have severe difficulties in other aspects. It is for this reason that we have included details which will seem trivial to many but will hopefully be appreciated by others. For the same reason we sometimes use simplification which the specialists might find hard to accept, but will hopefully help readers with different backgrounds to follow. In all cases we emphasize those aspects which are most important for our specific applications at the expense of generality. Thus, we tried to give as many details as possible in order to provide a self-contained text which can be read without constantly being forced to consult the literature cited. However, this does not imply that our text should be taken as 'the last word' on the subjects treated.

The references are by no means intended to provide an exhaustive compilation of the relevant literature. In a rapidly developing area like multidimensional solid-state NMR this would be quickly outdated. Likewise, we often took the liberty of ignoring aspects of priority and did not cite pioneering work, if adequate descriptions are readily available in books or review articles. We apologize to all the authors affected by this procedure. For the reader's convenience we have included short titles of the papers in the citation, in order to indicate which aspect is actually treated in a particular reference. Moreover, at the beginning of each chapter keywords are listed as a guidance for the reader.

Our work on NMR of polymers has substantially benefited from the close contact and the advice of Professor Ulrich Haeberlen at the Max-Planck-Institute at Heidelberg and Professor Hans Sillescu at the University of Mainz. Therefore, special thanks are due to them for their continuous support and specifically for their helpful comments concerning individual chapters.

This book is based on the efforts of our research group during several years. It is our pleasure to acknowledge the contributions of a large number of co-workers, who participated in our program as staff members, PhD students, postdoctoral fellows or as visiting scientists, providing ideas, samples, hardware or software: Bernhard Blümich, Christine Boeffel, Ralf Born, Lothar Brombacher, Brad Chmelka, Joachim Clauss, Norbert Egger, Stefan Féaux de Lacroix, Ewald Günther, Alfred Hagemeyer, Michael Hansen, Gerard Harbison, Manfred Hehn, Jérôme Hirschinger, Sigmar Kaufmann, Julia Kornfield, Andrzej Kulik, Johannes Leisen, Volker Macho, Angela Ohlemacher, Ute Pawelzik, Claudia Schmidt, Jeremy Titman, Der-Lii Tzou, Stephan Wefing, Michael Werth, Manfred Wilhelm, Yuliang Yang, Karen Zemke. Two of our co-workers deserve special thanks: Haskell Beckham closely followed the actual writing and checked the manuscript at various stages. His advice on scientific and, not least, linguistic aspects was invaluable. Dieter Schaefer carefully proofread the whole work and removed inconsistencies both in the text and in the mathematics. Various sections have been improved after discussions with the Pines group in Berkeley, in particular Mei Hong thoroughly checked several chapters.

The contact with colleagues in the international community is an indispensable aspect of science in general and of interdisciplinary work in particular. We would like to specifically thank Professors Richard Ernst, Erhard Fischer, Robert Griffin, Alan Jones, Zeev Luz, Michael Mehring, Alex Pines, Helmut Ringsdorf, Gerhard Scheler, John Waugh, Ian Ward, Gerhard Wegner, and the members of their research groups.

K. S.-R.
H. W. S.

Acknowledgements

We would like to thank the following authors and publishers for permission to reproduce figures.

G. Bodenhausen, Fig. 10.10
W. T. Dixon, Fig. 6.13
H. T. Edzes, 13.12
R. R. Ernst, Figs. 1.7, 6.27, 6.28, 13.6, 13.15, 13.16
R. Freeman, Figs. 2.24, 3.10
Z. Gan, Fig. 6.9
K. K. Gleason, Fig. 3.22
D. M. Grant, Figs. 6.23, 6.24
R. G. Griffin, Figs. 1.2, 1.3, 2.6, 13.17
T. Hashimoto, Fig. 5.5
J. Herzfeld, Fig. 3.13
A. A. Jones, Fig. 8.6
F. F. Karasz, Fig. 6.26
G. Kothe, Fig. 6.32
G. E. Maciel, Figs. 1.2, 6.6, 6.8, 7.28
S. J. Opella, Figs. 6.16, 13.14
A. Pines, Figs. 2.6, 3.17, 3.18, 3.19, 3.20, 3.21, 6.10, 6.31
J. Schaefer, Figs. 6.17, 6.18
H. Sillescu, Fig. 9.2
T. Terao, Figs. 1.3, 6.19, 6.29, 6.30
R. Tycko, Figs. 6.11, 6.20, 13.13
W. S. Veeman, Fig. 10.9

Academic Press

from *Advances in Magnetic Resonance*: Figs. 7.17, 10.9, 12.12
from the *Journal of Magnetic Resonance*: Figs. 1.2, 2.24, 3.10, 3.20, 6.6, 6.8, 6.11, 6.14, 6.17, 6.23, 6.24

American Chemical Society

from *Accounts of Chemical Research*: Figs. 3.19, 6.31
from *Biochemistry*: Fig. 13.17
from the *Journal of the American Chemical Society*: Figs. 1.3, 3.17, 3.18, 3.21, 6.9, 6.16, 7.28, 13.12, 13.13, 13.14
from *Macromolecules*: Figs. 1.4, 1.5, 1.6, 1.7, 1.8, 3.22, 5.5, 6.18, 6.21, 6.22, 6.26, 7.26, 7.27, 7.29, 7.30, 7.31, 8.2, 8.6, 8.11, 8.12, 12.19, 12.20, 12.25, 12.26, 13.5, 13.6

American Institute of Physics

from the *Journal of Chemical Physics*: Figs. 1.6, 2.6, 3.13, 6.10, 6.13, 6.27, 6.28, 6.29, 6.30, 6.32, 8.1, 8.4, 8.5, 8.7, 9.2, 10.9, 10.10, 11.3, 12.4, 12.9, 12.10, 12.11, 12.13, 12.14, 12.15, 12.16, 12.17, 12.23, 12.24
from *Physical Review Letters*: Figs. 8.9, 8.10

Butterworth Heinemann Ltd

from *Polymer*: Fig. 12.2

Elsevier Science Publishers

from *Chemical Physics Letters*: Figs. 1.2, 6.20, 11.6, 13.15, 13.16

VCH Verlagsgesellschaft

from *Acta Polymerica*: Figs. 13.1, 13.3, 13.7, 13.8, 13.9, 13.10

J. Wiley & Sons Ltd

from *Magnetic Resonance in Chemistry*: Figs. 1.3, 6.19, 6.29

NMR Acronyms and Abbreviations

List of NMR acronyms and abbreviations used in the text. Polymer acronyms are explained in Section 5.7.

Acronym	Full wording	Specific discussion: section(s)
ADC	Analog-to-Digital Converter	4.5
BR-24	pulse sequence by Burum and Rhim	3.3.1
CHF	Coupled Hartree-Fock technique	2.4.6
CORD	Correlation of ORder and Dynamics	12.6
CP	Cross polarization	3.2, 3.4.3
CRAMPS	Combined Rotation And Multiple-Pulse Spectroscopy	3.2
CS	Chemical Shift	2.4.1
CSA	Chemical-Shift Anisotropy	2.4.2
DANTE	Delays Alternating with Nutation for Tailored Excitation	3.8
DAS	Dynamic-Angle Spinning	3.10
DBM	Double-Balance Mixer	4.5.1
DD	Dipolar decoupling	6.2
DECODER	Direction Exchange with Correlation for Orientation-Distribution Evaluation and Reconstruction	12.7
deTOSS	time-reversed TOSS	3.9.4
DF	Director Frame	12.2
DFT	Discrete Fourier Transformation	4.2
DOR	DOuble Rotation	3.10
DRAMA	Dipolar Recovery At the Magic Angle	6.7.4
FFT	Fast Fourier Transformation	4.2
FID	Free Induction Decay	2.5.7
FT	Fourier Transformation	Chapter 4
fwhm	full width at half maximum	12.3.1
GIAO	Gauge-Invariant Atomic Orbitals	2.4.6
IGLO	Individual Gauge for Localized Orbitals	2.4.6
KWW	stretched exponential according to Kohlrausch, Williams, and Watts	8.1
LORG	Localized Orbitals/local oRiGin	2.4.6
LC	Liquid Crystal(line)	5.6
LF	Laboratory Frame	2.4.2

continued

Continued

Acronym	Full wording	Specific discussion: section(s)
MAS	Magic-Angle Spinning	2.4.4, 3.9
MF	Molecular Frame	12.2
MQ	Multiple-Quantum	3.12
NOE	Nuclear Overhauser Effect	3.13.2
OMAS	Off-Magic-Angle Spinning	2.4.4
PAS	Principal-Axes System	2.4.2, D.3
PASS	Phase-Altered Spinning Sidebands	6.5.3
REDOR	Rotational-Echo DOuble Resonance	6.6.5
RF	Rotor Frame	3.9, 12.2
rf	radio frequency	2.6
R^2	Rotational Resonance	13.5.7
SATRAS	SAtellite-TRansition Spectroscopy	3.10
SASS	Switching-Angle Sample Spinning	6.4.3, 6.6, 6.7.4
SAXS	Small-Angle-X-ray Scattering	13.3.4, 5.2
SELDOM	SELectivity by Destruction Of Magnetization	3.8
SF	Sample(-fixed) Frame	2.4.2
SLF	Separated-Local-Field	6.6.1
S/N	Signal-to-Noise ratio	4.9
sync-MAS	rotation-synchronized MAS	12.4
T_g	Glass-transition temperature	5.3, 8.1
TOSS	TOtal Suppression of (Spinning) Sidebands	3.9.3
TPPI	Time-Proportional Phase Incrementation	4.5.1
VACSY	Variable-Angle Correlation SpectroscopY	6.4.3
VFT	temperature dependence according to Vogel, Fulcher, and Tammann	8.1
WHH-4	pulse sequence by Waugh, Huber, and Haeberlen	3.3.1
WAXS	Wide-Angle X-ray Scattering	12.7.3
WIM	Windowless Isotropic Mixing	3.4.3
WISE	WIdeline SEparations	6.6.2
WLF	temperature dependence according to Williams, Landel, and Ferry	8.1

CHAPTER ONE

Introduction

1.1 Overview

It is one of the key goals of materials science to establish structure–property relationships, in order to improve known and design new materials. This holds in particular for synthetic polymers, whose materials properties depend on both the molecular structure and the organization of the macromolecules in the solid state: their phase structure, morphology, molecular order and molecular dynamics (Flory, 1953; Kroschwitz, 1990). Macroscopic as well as microscopic parameters are influenced by the processing that follows the chemical synthesis. This calls for powerful analytical tools that can probe these aspects in the material as it is used, predominantly in the solid state. The structural aspects are studied mostly by scattering techniques or by microscopy. Information about dynamic aspects are deduced mainly from scattering or relaxation experiments (McCrum *et al.*, 1967). Various spectroscopic techniques are also applied to solid polymers, in particular for determining their electronic structure.

Nuclear magnetic resonance (NMR) (Abragam, 1961; Slichter, 1980) is well established in structural characterization of liquids or compounds in solution, but much less so of solids (Fyfe, 1983; Bovey, 1988). Moreover, NMR offers numerous ways to study dynamic aspects over a large range of characteristic motional rates. The main advantage of NMR is its unprecedented selectivity. Thus it is desirable to exploit this technique for studying structure and dynamics of solid polymers (Komoroski, 1986). However, due to the presence of angular-dependent anisotropic interactions, the spectral resolution of solid-state NMR spectra is orders of magnitude lower than that of high-resolution NMR in liquids. Important improvements were achieved in the 1970s by combining high-speed mechanical rotation of the sample with ingenious manipulations of the nuclear spins such as multiple-pulse irradiation, high-power decoupling and cross polarization (Mehring, 1983). Moreover, two-dimensional and higher-dimensional NMR techniques (Jeener, 1971; Müller *et al.*, 1975; Aue *et al.*, 1976; Ernst *et al.*, 1987) have been introduced that offer fundamental advantages. First, as in liquids, the introduction of a new frequency dimension provides a means to increase the spectral resolution. Even more important, multidimensional spectroscopy also provides routes to new information, unavailable from 1D spectra even in the limit of high resolution. Experiments can be designed which correlate different spin interactions providing different structural information, or relate various states taken up by the molecular unit during different time periods by exchange and in this way probe dynamic processes in real time.

Progress in multidimensional solid-state NMR has been hampered by experimental and conceptual difficulties, which have now been overcome. This introductory chapter briefly outlines some of the main concepts and illustrates the information available about polymer structure and dynamics through selected experimental examples. Chapters 2–4 review the principles of Fourier transform solid-state NMR. The purpose of these chapters is twofold. They should enable the newcomer to follow the book without being forced to work through other monographs on NMR beforehand. For readers with a strong NMR background, these chapters emphasize the concepts that are particularly relevant for the experiments described later. These chapters also contain important details that should be appreciated by those who perform (or plan to perform) such experiments.

The advances in multidimensional NMR can be exploited in many fields of physics, chemistry, materials science, and, last but not least, biology. This book emphasizes applications of multidimensional solid-state NMR to synthetic polymers. This reflects the specific interest of the authors, the need for improved analytical tools in order to adequately characterize these complex materials, and the fact that ample experimental examples are already available. The rather short Chapter 5 reviews some background information on structure and dynamics of polymers that is needed to appreciate the remainder of the text. It is followed by an overview of experiments for separation and correlation of spectral patterns in Chapter 6, which also points out differences between the main objectives of multidimensional liquid- and solid-state NMR. Experiments are discussed which offer means to determine bond lengths and orientations within the molecular framework; others correlate molecular structure and mobility.

Polymer dynamics is of central interest in our studies. Of the great variety of molecular motions possible in polymers (e.g. translations, rotations, vibrations), rotations have the most pronounced effects on NMR lineshapes and relaxation parameters. Thus, multidimensional NMR provides especially unique information about rotational motions. Their timescales are followed in real time over many orders of magnitude, covering in particular the regime of slow motions which govern the mechanical properties of polymers. Moreover, the higher-order correlation functions provided by multidimensional NMR yield previously inaccessible model-independent information about the geometry of rotational motions, the orientational memory of molecular units involved in complex dynamics, and the nature of nonexponential relaxation in disordered systems. This information collected for a variety of polymers should eventually lead to a better understanding of their mechanical and rheological behaviour, which is of interest not only for conventional but also for new advanced polymeric materials. The corresponding NMR techniques and examples concerning chain dynamics in crystalline and amorphous polymers are treated in Chapters 7–11 at various levels of sophistication. Chapters 7 and 8 address in particular readers with interest in polymers by giving a descriptive introduction to exchange NMR techniques, indicating their scope and presenting results on polymers, especially above the glass transition. In Chapter 9 the NMR techniques are explicitly compared to well-established scattering and relaxation experiments. Complex motions whose time dependencies cannot be described by single correlation times are treated in detail. Finally, the techniques for recording 2D exchange NMR spectra of solids as well as the algorithms developed for simulation and quantitative analysis of multidimensional NMR spectra are described in Chapters 10 and 11. Through numerous examples, these chapters should provide sufficient information not only to appreciate the unique information content of such spectra but also to actually perform the experiments.

Two other aspects are particularly important for establishing structure–property relationships for polymer materials, namely chain alignment in partially ordered systems and domain sizes in heterogeneous polymer materials. The orientation of macromolecular units is used to improve the properties of polymers for such different applications as high-tensile-strength fibres and nonlinear optical materials for information technology. Advanced polymer materials almost invariably consist of more than one component, which often leads to phase separation. Careful design of the molar ratios as well as size, composition, and morphology of the different phases offers a means to control the mechanical, electrical, and optical properties. Small domains that extend over only a few nanometers, as well as interfacial regions between the different phases, are particularly difficult to characterize. As described in Chapters 12 and 13, major advances have been achieved in these areas by introducing the concepts of multidimensional spectroscopy. The new solid-state NMR techniques nicely supplement well-established scattering and microscopic methods as demonstrated by various experimental examples and by explicit comparison.

The book addresses a heterogeneous audience with backgrounds in polymers and in spectroscopy. Therefore, most of its material is presented in two ways: a descriptive and a mathematical one. The text is self-contained in so far as most treatments start from obvious facts or first principles. More complex equations are followed by hints regarding the proofs or are at least justified in a heuristic fashion. Appendices are provided in which the mathematics and the physics needed to follow the derivations are collected. Also included is theoretical background which is readily available in principle but in practice often has been found to represent a problem to the beginner in solid-state NMR. Wherever possible, the discussion is based on geometric visualizations and principles rather than detailed analytical formulas. This applies for the various topics that are treated: the 'spin-operator gymnastics', the orientation dependencies of the NMR interactions, as well as the analysis of molecular dynamics, order, and structure. In many instances, analogies between multidimensional NMR spectroscopy and other techniques will be pointed out, and the information content of the corresponding spectra will be compared. For the study of dynamics, reference to relaxation techniques such as dielectric, dynamic–mechanical, or NMR relaxation is obvious. More surprisingly, 2D NMR exhibits a variety of analogies to scattering techniques such as quasielastic neutron scattering or wide- and small-angle scattering of neutrons and X-rays. It is hoped that these analogies will help the appreciation of the high information content of multidimensional solid-state NMR spectra.

1.2 Examples

Before starting the systematic treatment, a number of experimental examples are presented which address the various aspects and advantages of multidimensional NMR. Details of the experiments designed to record multidimensional NMR spectra are not given here, but are left to the later chapters. However, basic knowledge of the concept of two-dimensional (2D) Fourier spectroscopy (Aue *et al.*, 1976; Ernst *et al.*, 1987) is needed to read this section: A 2D NMR spectrum is generated by recording a two-dimensional data set following pulsed irradiation as a function of two time variables t_1 and t_2 and subsequent double Fourier transformation, as indicated in

Figure 1.1. The development of the nuclear spin system in the *evolution period* with the incremented time t_1 at the beginning of the pulse sequence provides the basis for the first frequency dimension ω_1. The NMR signal is detected in the *detection period* with time t_2 at the end of the pulse sequence, providing the basis for the second frequency dimension ω_2. In many 2D experiments, a *mixing period* separates the evolution and detection times. This mixing period of duration t_m is particularly important in exchange experiments, allowing for dynamic processes to take place between the measurement of the correlated frequencies ω_1 and ω_2 (for details see Chapters 6 and 7). The concept of 2D spectroscopy is readily extended to three and higher dimensions by inserting additional evolution and mixing times.

Let us first consider experiments which increase the spectral resolution of solid-state NMR spectra in order to characterize polymer structure in terms of chemical moieties on the basis of isotropic chemical shifts. The information from isotropic chemical shifts represents the most important feature of standard NMR techniques. The corresponding spectra make up one or two dimensions of many multidimensional experiments in which the chemical structure is correlated with other molecular properties such as mobility or order. In the solid state, the tensorial nature of the chemical shift makes the NMR frequency also depend on the orientation of the molecular unit under study with respect to the applied magnetic $\vec{\mathbf{B}}_0$ field. Since the orientational dependence of the NMR frequency is comparable to or even larger than the variation of the isotropic chemical shifts of different structural units, the powder patterns resulting from the anisotropic chemical shifts severely overlap in all but the simplest polymer structures. Thus, in many cases a quantitative analysis is virtually impossible. The effect of the anisotropy can be removed by rapidly spinning the sample about an axis inclined at the 'magic angle' of 54.7°(magic-angle spinning, MAS). However, this is accompanied by a loss of the information about the molecular orientation, which is the basis of structural and dynamic information typical of the solid state. Thus, experiments are desired that retain this information without sacrificing spectral resolution.

This can be achieved by extension to a second frequency dimension. For instance, in a 2D spinning experiment the angle of the rotor axis is set to 54.7° in the detection and to 90° in the evolution time (Bax *et al.*, 1983). As displayed in Figure 1.2(a), the powder patterns for the chemically distinct sites are then recorded along the frequency dimension ω_1 and separated according to their respective isotropic chemical shifts, recorded along the frequency dimension ω_2. In standard MAS experiments the spinner axis is kept at the magic angle. For moderate spinning speeds in the range of a few kHz, the anisotropic chemical shift is not 'spun out', but leads

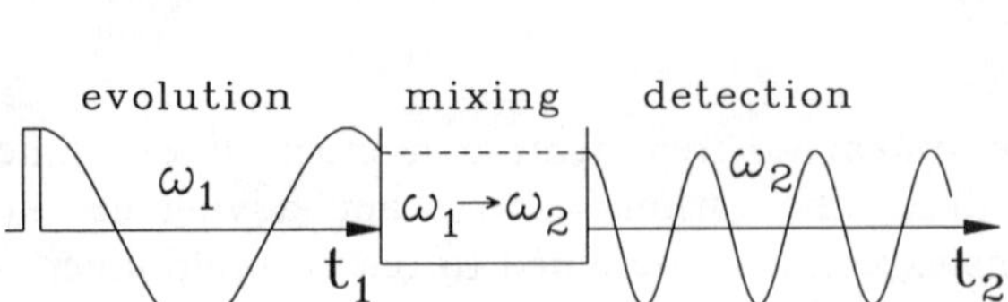

Figure 1.1 Schematic representation of the generic two-dimensional NMR pulse sequence, with evolution time t_1, mixing period t_m, and detection time t_2. The frequencies ω_1 and ω_2 are correlated in the corresponding two-dimensional spectrum obtained by Fourier transformation with respect to t_1 and t_2.

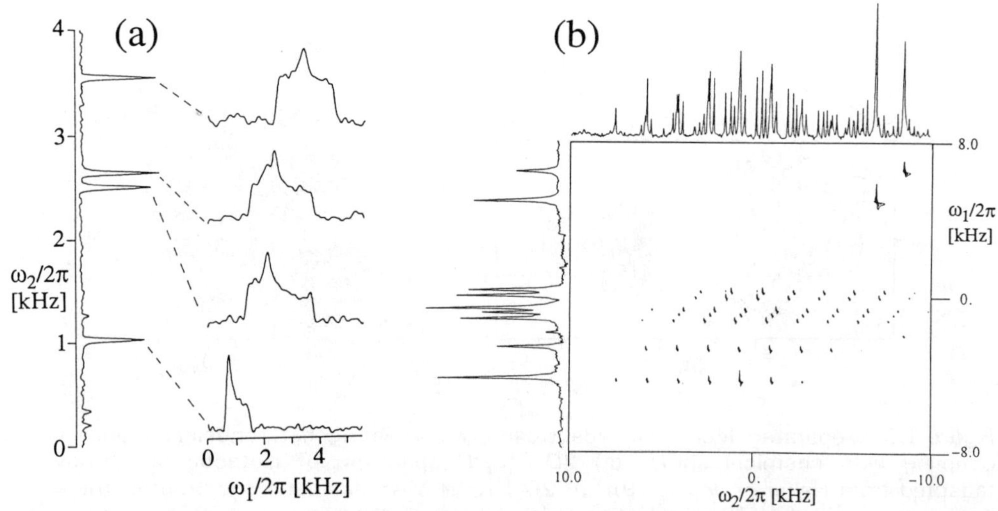

Figure 1.2 Separation of anisotropic-chemical-shift patterns by the isotropic chemical shifts. (a) 2D ^{13}C NMR spectrum of *p*-dimethoxybenzene, recorded with flipping of the spinner axis (adapted from Bax *et al.*, 1983), (b) 2D ^{13}C sideband MAS spectrum of the amino acid Tyr-HCl (adapted from Kolbert and Griffin, 1990).

to sideband patterns which may again overlap considerably and hamper a quantitative analysis. By ingenious spin manipulations through multiple-pulse sequences these sideband patterns can be removed from the spectrum in one dimension and retained in the other (Kolbert and Griffin, 1990). As shown in Figure 1.2(b), the crowded sideband patterns can nicely be separated in a second frequency dimension and can then be exploited to provide structural and dynamic information.

The dipole–dipole coupling, for instance between ^1H and ^{13}C, or ^1H and ^{15}N, provides valuable structural information. Since the C—H bond lengths are known, the measurements of dipolar splittings can be interpreted in terms of angles between individual bonds and the applied magnetic fields. For proteins, measurements of N—H bond lengths are of considerable interest because they vary due to hydrogen bonding and therefore contain important structural information. In order to be useful, the dipolar patterns have to be separated, however, according to the chemically distinct sites in a molecule or monomer unit as identified by their ^{13}C or ^{15}N chemical shifts. Since the dipolar couplings correspond to local fields, such experiments are often named 'separated local field' (SLF) experiments. Figure 1.3 displays two examples of SLF spectroscopy, both obtained by sample spinning techniques. In Figure 1.3(a), ^{13}C chemical shift/$^{13}C-^1H$ dipolar powder patterns for the three carbons of isotactic polypropylene are correlated in the 2D plane (Nakai *et al.*, 1989). From a quantitative analysis of such patterns, bond lengths and bond angles, as well as the orientations of the principal axes of the chemical-shift tensor in relation to structural units such as CH_2 groups can be determined. Such information is needed for a quantitative analysis of other experiments exploiting anisotropic chemical shifts. As shown in Figure 1.3(b), similar patterns can be generated based on MAS sidebands reflecting the dipolar couplings.

As stated above, multidimensional NMR provides especially interesting information about polymer dynamics. A long-standing method for a qualitative characterization of

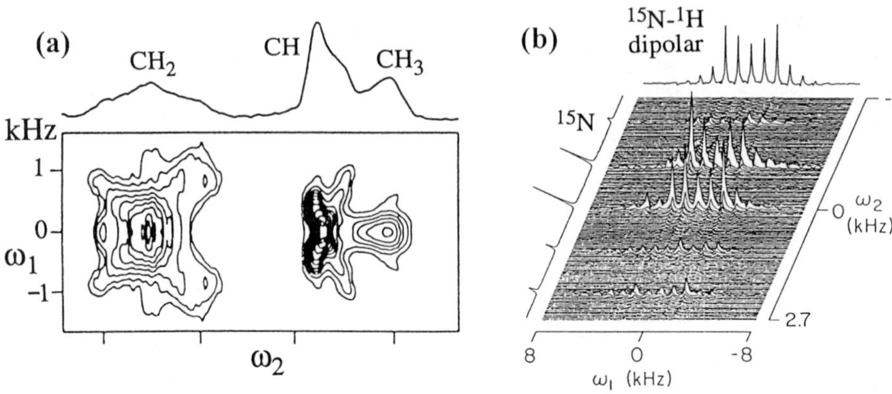

Figure 1.3 Separated local field spectroscopy correlating heteronuclear dipole–dipole coupling with chemical shifts. (a) 2D $^{13}C/^{1}H$ spectrum of isotactic poly(propylene) (adapted from Nakai *et al.*, 1989), (b) 2D $^{15}N/^{1}H$ MAS sideband spectrum of the amino acid [^{15}N]acetylvaline (adapted from Roberts *et al.*, 1987).

molecular mobility is ^{1}H wideline NMR spectroscopy. There, large-amplitude motions are detected through the reduction of the dipolar line width. However, 1D proton lineshapes leave many questions open, since they typically represent superpositions of broad and narrow lines and their relation to different structural units is often not obvious. In a straightforward combination of ^{1}H wideline NMR, cross polarization (CP), and ^{13}C MAS spectroscopy in a 2D experiment (Schmidt-Rohr *et al.*, 1992), it is possible to separate the dipolar patterns for the different structural units (wideline separation, WISE). This is demonstrated in Figure 1.4, where a WISE NMR spectrum of a 50:50 wt% blend of poly(styrene) (PS) and poly(vinyl methylether) (PVME) is presented. The ^{1}H wideline spectrum, Figure 1.4(a), consists of a rather featureless superposition of components with different dipolar linewidths, which are nicely separated in the second frequency dimension (Figure 1.4(b)) and related to structural units according to their ^{13}C chemical shifts. Substantial motional heterogeneities, PVME being more mobile than PS, are detected despite the fact that the spectrum is recorded about 60 K above the caloric glass transition of this blend, which appears homogeneous by most classical techniques. Such information about the mobility of the different structural units is highly valuable for many practical applications.

For a thorough understanding of the chain motions in polymers, qualitative information provided by ^{1}H wideline spectra is not sufficient. In order to relate chain motions to the structure of the polymer itself or to the packing of the macromolecular chains, knowledge is required concerning the geometry of the motion, for example the angles about which a molecular unit rotates during individual motional steps. This information, which is hard to get otherwise, is indeed provided by 2D exchange NMR as applied to rotational motions. In simple cases the 2D spectra exhibit elliptical ridge patterns from which the angle about which the molecules have rotated can be directly read off with a ruler (Schmidt *et al.*, 1986). As a specific example, Figure 1.5(a) displays such a ^{2}H 2D exchange spectrum for poly(vinylidene fluoride) (PVF_2), a polymer of considerable technological interest because of its electric properties. The geometric information from 2D NMR, together with knowledge of the dipole-moment change generated by the motion, allowed the identification of the conformational

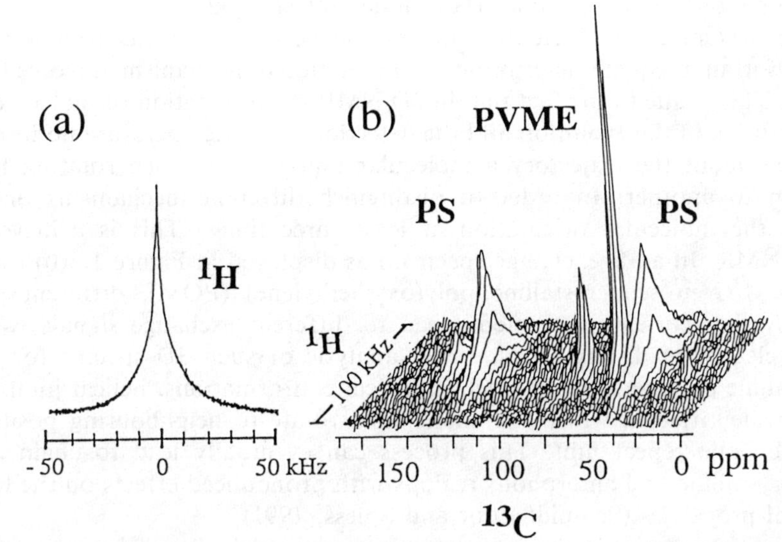

Figure 1.4 2D WISE NMR separating ^1H wideline spectra for different structural units according to their ^{13}C chemical shifts. (a) Conventional ^1H wideline spectrum of a blend of poly(styrene) (PS) and poly(vinyl methylether) (PVME), (b) 2D $^1H/^{13}C$ WISE NMR spectrum indicating different mobilities of the two components (adapted from Schmidt-Rohr *et al.*, 1992).

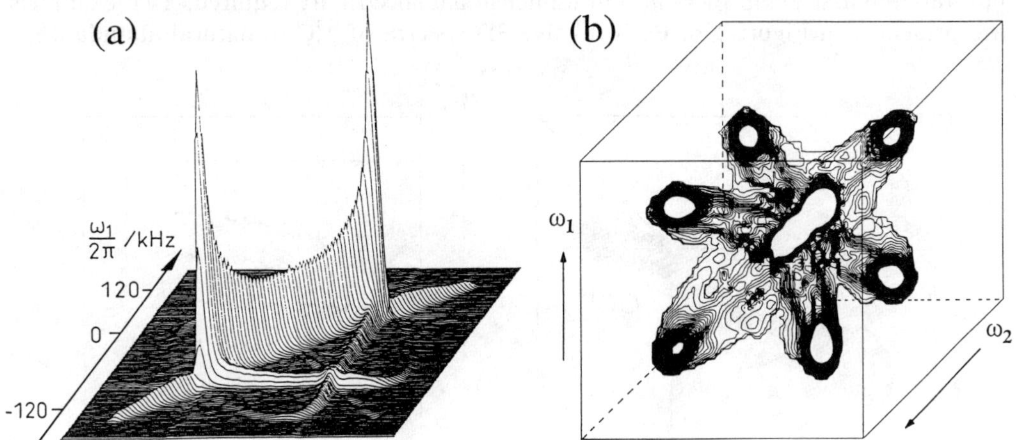

Figure 1.5 Multidimensional exchange NMR spectra elucidating slow molecular motions. (a) 2D ^2H NMR spectrum of crystalline poly(vinylidene fluoride) reflecting chain motions through defect diffusion (adapted from Hirschinger *et al.*, 1991), (b) 3D ^{13}C exchange spectrum of oriented poly(oxymethylene) reflecting helical jumps in the crystalline regions.

defect of the chain responsible for the mechanical and dielectric relaxation in the crystalline regions of this polymer (Hirschinger *et al.*, 1991).

For complex motions, even the information accessible by 2D techniques is not sufficient for an adequate description of the motional mechanisms involved in chain dynamics. This is due to the fact that in 2D NMR the orientation of molecules is only measured twice, in the evolution and the detection periods. Therefore no information is provided about the trajectory a molecular unit follows when rotating from one orientation to another. In order to distinguish different mechanisms one has to determine the molecular orientation at least three times. This is achieved in 3D exchange NMR. In a 3D exchange spectrum as displayed in Figure 1.5(b) for natural abundance ^{13}C in semicrystalline poly(oxymethylene) (POM), different pathways pursued by the reorienting segments lead to different exchange signals, which can therefore clearly be distinguished. From analysis of such 3D spectra for different semicrystalline polymers, which pack in helical conformations, helical jump motions have been identified in which the chain units rotate to neighbouring positions and translate by one repeat unit. This process can eventually lead to chain diffusion between crystalline and amorphous regions with pronounced effects on the long-term mechanical properties (Schmidt-Rohr and Spiess, 1991).

Structural characterization of polymers often requires the determination of the alignment of macromolecular chains. Orientation in polymers is frequently induced by the production process and has strong effects on product properties. Through the angle-dependent NMR interactions, orientation is amenable to NMR experiments. As opposed to most classical techniques, order of both crystalline and amorphous components can be studied. For ^{2}H- or ^{13}C-enriched samples, one-dimensional experiments are sufficient to obtain orientation distributions in terms of a single angle. In order to reconstruct two-dimensional orientation distributions, or to resolve overlapping patterns in natural-abundance ^{13}C spectra, multidimensional spectra are required. Two examples are presented in Figure 1.6. Both involve 3D spectra of ^{13}C in natural abundance.

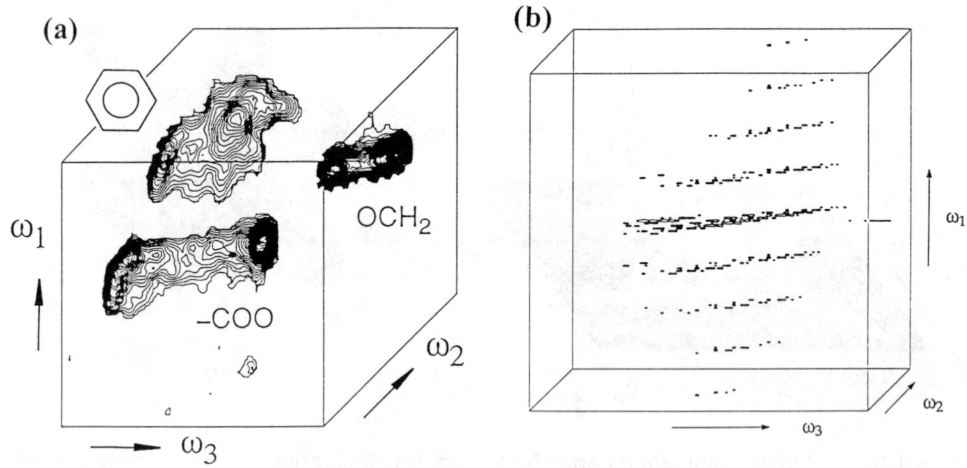

Figure 1.6 3D NMR spectra for determining molecular order. (a) 3D ^{13}C DECODER NMR spectrum of biaxially drawn poly(ethylene terephthalate) (adapted from Chmelka *et al.*, 1993), (b) 3D ^{13}C MAS spectrum of a liquid-crystalline side-group polymer (adapted from Titman *et al.*, 1993).

In the case of a biaxially drawn film of poly(ethylene terephthalate) (PET), the orientational distribution was mapped out by flipping the sample between different orientations with respect to the magnetic field (direction exchange with correlation for orientation distribution evaluation and reconstruction, DECODER). The NMR signals of the different structural units of PET are resolved in the 3D cube of Figure 1.6(a) and their orientation distributions thus separately determined (Chmelka *et al.*, 1993). In the case of a liquid-crystalline side-group polymer (Figure 1.6(b)), the extension to a third dimension is performed in order to achieve the necessary spectral resolution (Titman *et al.*, 1993). Rotor-synchronized MAS is applied to map out the degree of molecular order along ω_1. The sidebands are separated by their order along ω_2, and the chemical structure is probed via the isotropic chemical shift along ω_3. Each dot in the 3D spectrum represents a single sideband of a carbon position in the repeat unit with resolved ^{13}C chemical shift in the MAS spectrum. This allows the molecular order of relatively complex polymers to be analysed quantitatively, even for liquid-crystalline polymers, where different moieties of the repeat unit exhibit substantially different degrees of alignment (McArdle, 1989).

In heterogeneous polymers, domain sizes, or structures on the scale of up to a few hundred nanometres, can be investigated. In NMR the proximity of molecular units is probed by spin diffusion (Abragam, 1961), which is most effective among protons. Thus, NMR is particularly suited for characterizing small domains, nanoheterogeneities or concentration fluctuations on length scales of a few nm where other

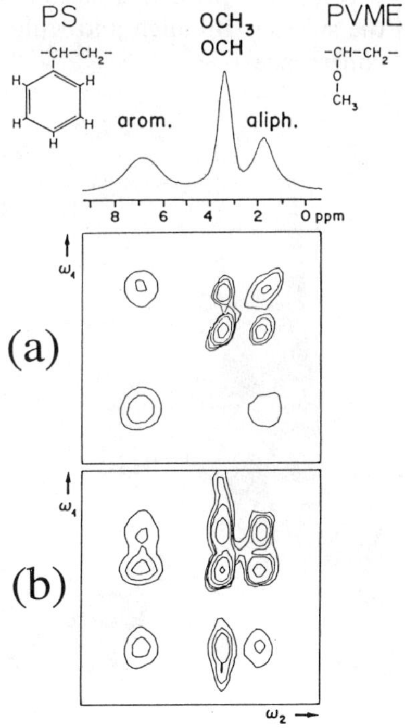

Figure 1.7 2D ^1H CRAMPS spectra of PS/PVME blends indicating different degrees of mixing. (a) Blend cast from chloroform, heterogeneous blend, (b) cast from toluene, homogeneous blend (adapted from Caravatti *et al.*, 1985).

methods often fail due to limitations in resolution or contrast. An advanced approach exploiting ^{1}H spin diffusion with high sensitivity ^{1}H detection was introduced by Ernst and co-workers, employing a combination of rotation and multiple-pulse spectroscopy (CRAMPS) in a 2D exchange experiment (Caravatti *et al.*, 1985). As displayed in Figure 1.7, the cross-peaks indicating proximity between the aromatic protons of PS and the ether groups of PVME are missing in the heterogeneous blend cast from chloroform (Figure 1.7(a)), but are clearly detected in the homogeneous blend cast from toluene (Figure 1.7(b)). However, these results should not be taken as evidence for mixing on a molecular level: recent studies combining WISE NMR with spin diffusion have shown that the system is neither homogeneous nor phase-separated, but nano-heterogeneous (Schmidt-Rohr *et al.*, 1992) (see Chapter 13).

Eventually one would like to obtain the information about molecular structure and dynamics accessible by solid-state NMR not only for the sample as a whole but spatially resolved. For instance, one would like to distinguish the molecular parameters in regions close to the surface from those in the bulk. In fact, by applying pulsed magnetic field gradients the concepts of multidimensional NMR can also be used to generate a spatial dimension in 2D-spectra through Fourier imaging (Ernst *et al.*, 1987). As a first example a spatially resolved ^{13}C NMR spectrum of a drawn poly(propylene) sample is displayed in Figure 1.8 (Günther *et al.*, 1992). It reflects differences of density and chain alignment between skin and core due to the processing of the material and demonstrates that spectroscopic NMR imaging is indeed possible. This exciting field is still in its infancy and considerable progress is expected in the near future. Thus, rather than describing the present situation here we refer the interested reader to a recent book on the subject (Blümich and Kuhn, 1992) which is based on lectures at an international conference.

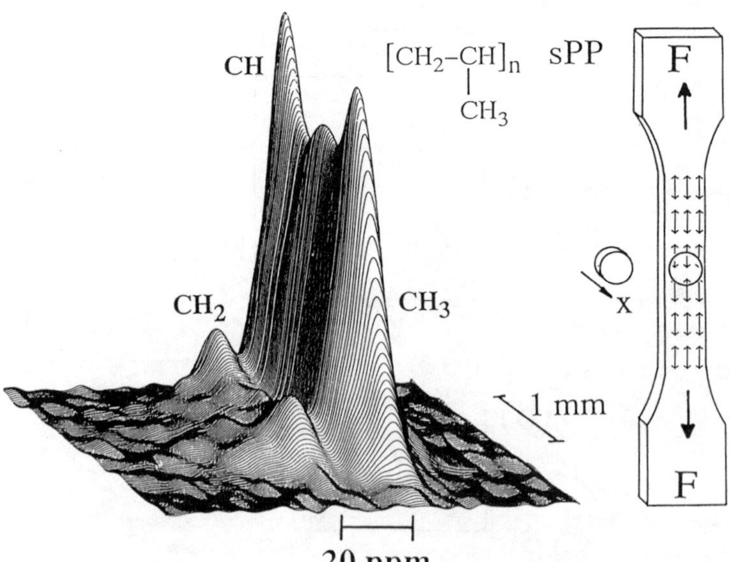

Figure 1.8 Spatially resolved 2D ^{13}C spectrum of drawn poly(propylene) (adapted from Günther *et al.*, 1992).

As these examples indicate, a wealth of information about polymer structure and dynamics is available through advanced multidimensional solid-state NMR techniques. Profound correlations between macroscopic and molecular behaviour are emerging from such studies. Examples are helical jumps or defect diffusion in crystallites, specific local motions in amorphous polymers and conformational transitions coupled to relaxations of larger chain units in highly viscous polymer melts. Such information has already successfully been used in the design of polymers with improved properties. Thus, it can be anticipated that the importance of multidimensional solid-state NMR in materials science will substantially increase in the future.

1.3 References

Abragam, A. (1961). *The Principles of Nuclear Magnetism*. Oxford University Press, London.

Aue, W. P., Bartholdi, E. and Ernst, R. R. (1976). 2D spectroscopy. Application to NMR. *J. Chem. Phys.*, **64**, 2229.

Bax, A., Szeverenyi, N. M. and Maciel, G. E. (1983). Chemical Shift Anisotropy in Powders Studied by 2D FT NMR with Flipping of the Spinning Axis. *J. Magn. Reson.*, **55**, 494.

Blümich, B. and Kuhn, W. (eds) (1992). *Magnetic Resonance Microscopy*. VCH-Verlagsgesellschaft, Weinheim.

Bovey, F. A. (1988). *Nuclear Magnetic Resonance Spectroscopy*. Academic Press, San Diego.

Caravatti, P., Neuenschwander, P. and Ernst, R. R. (1985). Characterization of Heterogeneous Polymer Blends by 2D ^{1}H Spin Diffusion Spectroscopy. *Macromolecules*, **18**, 119.

Chmelka, B. F., Schmidt-Rohr, K. and Spiess, H. W. (1993). Orientation Distributions in PET Films and Fibers from Multidimensional DECODER NMR. *Macromolecules*, **26**, 2282.

Kroschwitz, J. I., Ed. (1990). *Concise Encyclopedia of Polymer Science and Technology*. Wiley, New York.

Ernst, R. R., Bodenhausen, G. and Wokaun, A. (1987). *Principles of Nuclear Magnetic Resonance in One and Two Dimensions*. Clarendon Press, Oxford.

Flory, P. J. (1953). *Principles of Polymer Chemistry*, Cornell University Press, Ithaca.

Fyfe, C. A. (1983). *Solid State NMR for Chemists*. C.F.C. Press, Guelph.

Günther, E., Blümich, B. and Spiess, H. W. (1992). Molecular Order and Density in Drawn Polypropylene: Spectroscopic ^{13}C NMR Imaging. *Macromolecules*, **25**, 3315.

Hirschinger, J., Schaefer, D., Spiess, H. W. and Lovinger, A. J. (1991). Chain Dynamics in the Crystalline α-Phase of Poly(vinylidene fluoride) by 2D Exchange ^{2}H NMR. *Macromolecules*, **24**, 2428.

Jeener, J. (1971). *Proc. Ampere International Summer School II, Basko Polje, Yugoslavia*.

Kolbert, A. C. and Griffin, R. G. (1990). 2D Resolution of Isotropic and Anisotropic Chemical Shifts in MAS NMR. *Chem. Phys. Letters*, **166**, 87.

Komoroski, R. A., Ed. (1986). *High Resolution NMR Spectroscopy of Synthetic Polymers in Bulk*. VCH Publ., Deerfield Beach, Florida.

McArdle, C. B., Ed. (1989). *Side Chain Liquid Crystal Polymers*. Blackie, Glasgow.

McCrum, N. G., Read, B. E. and Williams, G. (1967). *Anelastic and Dielectric Effects in Polymeric Solids*. Dover, New York.

Mehring, M. (1983). *Principles of High Resolution NMR in Solids*. Springer-Verlag, Berlin.

Müller, L., Kumar, A. and Ernst, R. R. (1975). 2D ^{13}C NMR. *J. Chem. Phys.*, **63**, 5490.

Nakai, T., Ashida, J. and Terao, T. (1989). Determination of ^{13}C Chemical Shift Tensors in Isotactic Polypropylene. *Magn. Reson. Chem.*, **27**, 666.

Roberts, J. E., Harbison, G. S., Munowitz, M. G., Herzfeld, J. and Griffin, R. G. (1987). Measurement of Heteronuclear Bond Distances in Polycrystalline Solids by Solid-State NMR Techniques. *J. Am. Chem. Soc.*, **109**, 4163.

Schmidt, C., Wefing, S., Blümich, B. and Spiess, H. W. (1986). Molecular Reorientations: Determination of Rotational Angles from 2D NMR of Powders. *Chem. Phys. Lett.*, **130**, 84.

Schmidt-Rohr, K., Clauss, J. and Spiess, H. W. (1992). Correlation of Structure, Mobility, and Morphology by 2D Wideline-Separation NMR. *Macromolecules*, **25**, 3273.
Schmidt-Rohr, K. and Spiess, H. W. (1991). Chain Diffusion in Polyethylene Detected by ^{13}C NMR. *Macromolecules*, **24**, 5288.
Slichter, C. P. (1980). *Principles of Magnetic Resonance*. Springer-Verlag, Berlin.
Titman, J. J., Féaux de Lacroix, S. and Spiess, H. W. (1993). Structure and Order in Partially Oriented Solids by 3D MAS NMR. *J. Chem. Phys.*, **98**.

CHAPTER TWO

Principles of NMR of Organic Solids

Keywords: *precession, interactions and Hamiltonians, angle-dependent frequencies, powder patterns, motional averaging, MAS and OMAS; magnetization and density operator, time-evolution formulas, quadrupolar and dipolar evolutions, rotating frames, pulses and direction of rotation, echoes, spectral excitation.*

2.1 Introduction

In this chapter, the basics of solid-state NMR in diamagnetic organic solids are described in an essentially self-contained manner. The chapter starts with a short description of useful semiclassical concepts concerning the behaviour of magnetization in magnetic fields. This is followed by a short discussion of NMR in terms of basic quantum-mechanical notions. The full quantum-mechanical treatment starts with the discussion of the nuclear magnetic interactions in the solid state. It continues immediately with translating the interactions into NMR spectra, because this relation is indispensable for the understanding of solid-state NMR experiments. It is demonstrated how solid-state NMR spectra can probe orientations of molecular segments relative to the magnetic field. Typical lineshapes and effects of fast molecular motion or sample rotation are derived.

The second part of the chapter treats the quantum-mechanical equation of motion of the density operator and the corresponding magnetization. For several relevant examples it is demonstrated how the time-dependence of the magnetization subject to the NMR interactions can be derived by fully quantum-mechanical calculations involving just a few matrix multiplications. The action of resonant radio-frequency pulses, including the generation of spin echoes, is treated in the final sections of this chapter.

2.2 Semiclassical description of pulsed NMR

The semiclassical description of magnetization in a magnetic field already contains the ingredients necessary to characterize most magnetic-resonance properties of spin-$\frac{1}{2}$ nuclei not subject to couplings to other spins. If transverse (T_2) and longitudinal (T_1) relaxation are included phenomenologically, one can describe the essential features of many multidimensional NMR experiments on spin-$\frac{1}{2}$ nuclei under dipolar decoupling, including spin echoes.

A magnetic moment $\vec{\mathbf{M}}$ in a homogeneous field $\vec{\mathbf{B}}$ is subject to a torque $\vec{\mathbf{N}}$:

$$\vec{\mathbf{N}} = \vec{\mathbf{M}} \times \vec{\mathbf{B}} \tag{2.1}$$

In quantum mechanics, nuclear magnetic moment $\vec{\mathbf{M}}$ and angular momentum $\vec{\mathcal{I}} = \hbar \vec{\mathbf{I}}$ are related by:

$$\vec{\mathbf{M}} = \gamma \vec{\mathcal{I}} \tag{2.2}$$

where γ is the magnetogyric ratio, which is positive for the most commonly observed nuclei ^{1}H, ^{2}H, ^{13}C, ^{19}F, and ^{31}P; it is negative for ^{29}Si and ^{15}N.

The basic equation of rotational motion, relating torque $\vec{\mathbf{N}}$ and angular momentum $\vec{\mathcal{I}}$:

$$\vec{\mathbf{N}} = \frac{\mathrm{d}}{\mathrm{d}t} \vec{\mathcal{I}} \left(\text{analogous to Newton's } \vec{\mathbf{F}} = \frac{\mathrm{d}}{\mathrm{d}t} \vec{\mathbf{p}}\right) \tag{2.3}$$

together with Equations 2.1 and 2.2 yields the equation of motion for the magnetization:

$$\frac{\mathrm{d}}{\mathrm{d}t} \vec{\mathbf{M}} = \gamma \vec{\mathbf{M}} \times \vec{\mathbf{B}}. \tag{2.4}$$

Setting $\vec{\mathbf{B}} = (0, 0, B_0)$ leads to the Bloch equations without relaxation:

$$\begin{aligned} \frac{\mathrm{d}M_x(t)}{\mathrm{d}t} &= M_y(t)\gamma B_0 \\ \frac{\mathrm{d}M_y(t)}{\mathrm{d}t} &= -M_x(t)\gamma B_0 \\ \frac{\mathrm{d}M_z(t)}{\mathrm{d}t} &= 0 \end{aligned} \tag{2.5}$$

which are solved by the magnetization:

$$\vec{\mathbf{M}}(t) = \begin{bmatrix} M_x(t) \\ M_y(t) \\ M_z(t) \end{bmatrix} := \begin{bmatrix} M_x^0 \cos \omega_0 t - M_y^0 \sin \omega_0 t \\ M_y^0 \cos \omega_0 t + M_x^0 \sin \omega_0 t \\ M_z^0 \end{bmatrix} \tag{2.6}$$

precessing around $\vec{\mathbf{B}}$ with the Larmor frequency:

$$\omega_0 = -\gamma B_0 \tag{2.7}$$

(for the minus sign see Section 2.6.2). Including exponential T_1 relaxation of M_z towards the equilibrium value M_z^∞, as well as T_2 relaxation of M_x and M_y, the time-dependence is given by:

$$\vec{\mathbf{M}}(t) = \begin{bmatrix} (M_x^0 \cos \omega_0 t - M_y^0 \sin \omega_0 t)\, \mathrm{e}^{-t/T_2} \\ (M_y^0 \cos \omega_0 t + M_x^0 \sin \omega_0 t)\, \mathrm{e}^{-t/T_2} \\ M_z^0 + (1 - \mathrm{e}^{-t/T_1})\,(M_z^\infty - M_z^0) \end{bmatrix}. \tag{2.8}$$

A transition into a rotating frame of reference, precessing with the magnetization around $\vec{\mathbf{B}}_0$, is convenient in NMR. Obviously, in this frame the magnetization appears static, that is, the effect of the $\vec{\mathbf{B}}_0$ field has apparently been removed. Therefore, even if $|\vec{\mathbf{B}}_1| \ll |\vec{\mathbf{B}}_0|$, a transverse $\vec{\mathbf{B}}_1$ field that is static in the rotating frame makes a magnetization vector precess around the $\vec{\mathbf{B}}_1$ direction. The requirement for $\vec{\mathbf{B}}_1$ to be static in the rotating frame means that in the frame of the sample coil (i.e. the laboratory frame), $\vec{\mathbf{B}}_1$ rotates with frequency ω_0. This 'resonance frequency' is located in the radio-frequency (rf) region for $\vec{\mathbf{B}}_0$ fields of several tesla. Application of a $\vec{\mathbf{B}}_1$ field oscillating with frequency ω_0 is therefore also referred to as 'rf irradiation'.

In the rotating frame, a flip of the magnetization is effected by the precession of the magnetization around $\vec{\mathbf{B}}_1$, with a frequency $|\omega_1| = \gamma|\vec{\mathbf{B}}_1|$. To have the commonly used flip angle of 90°, the $\vec{\mathbf{B}}_1$ field must be applied for a duration t_{90} that fulfils the condition:

$$\frac{\pi}{2} = |\omega_1|\, t_{90} = \gamma B_1 t_{90}. \tag{2.9}$$

The duration t_{90} of this rf pulse is referred to as the '90° pulse length'; t_{90} values of 2–5 μs are typical in solid-state NMR. If the $\vec{\mathbf{B}}_1$ field is oriented along the y axis of the rotating frame, the rf pulse is termed a 90° 'y pulse'; x, $-y$, and $-x$ pulses are defined correspondingly. The sign of ω_1 and the sense of precession will be discussed in Section 2.6.2.

A 90° x pulse can serve to flip the magnetization from the thermal equilibrium along the z axis into the transverse plane. In the rf coil, this transverse magnetization $\vec{\mathbf{M}}(t)$, precessing with ω_0, produces an oscillating magnetic flux density ('B field') $B_x(t) = \mu_0 M_x(t) = \mu_0 M_0 \cos \omega_0 t$ along the coil axis and thus a voltage $U = -\mathrm{d}\Phi/\mathrm{d}t$ with the flux $\Phi(t) = \int \vec{\mathbf{B}}(t)\, \mathrm{d}\vec{\mathbf{A}}$. This results in:

$$U(t) \sim -\frac{\mathrm{d}}{\mathrm{d}t} M_x(t) = \omega_0 M_0 \sin \omega_0 t. \tag{2.10}$$

Note that $\mathrm{d}M_x(t)/\mathrm{d}t$ is measured instead of $M_x(t)$, but the differences are only the prefactor ω_0 and the phase shift of $\frac{1}{2}\pi$ since $\mathrm{d} \cos \omega_0 t/\mathrm{d}t = -\sin \omega_0 t = \cos(\omega_0 t + \pi/2)$; this corresponds to an angle of $\pi/2$ between the vectors $\mathrm{d}\vec{\mathbf{M}}/\mathrm{d}t$ *and* $\vec{\mathbf{M}}$ in the rotating frame.

A *Hahn spin echo* (Hahn, 1950) can be generated by application of a second pulse, usually a 180° (or 'π') pulse. The echo occurs by refocusing the dephasing of magnetization components precessing with different frequencies (different 'isochromats'). The dephasing results in the decrease of the sum magnetization of the isochromats. Even with negligible T_2 relaxation, the various isochromats become less parallel the longer they precess and ultimately, when uniformly spread over the transverse plane, they produce a vanishing vector sum. The action of the 180° pulse on several isochromats that have been precessing for a time $t = t_e$ is shown in Figure 2.1. Because the isochromats that have precessed by the largest angle during t_e also return fastest, at a time t_e after the pulse, all isochromats are refocused along their common direction at the very start of the precession ($t = 0$).

Precession, spin echoes, and the effects of rf pulses are the essential components of various one-dimensional and multidimensional techniques. The relation between the NMR precession frequencies and molecular properties, to be discussed in Section 2.4,

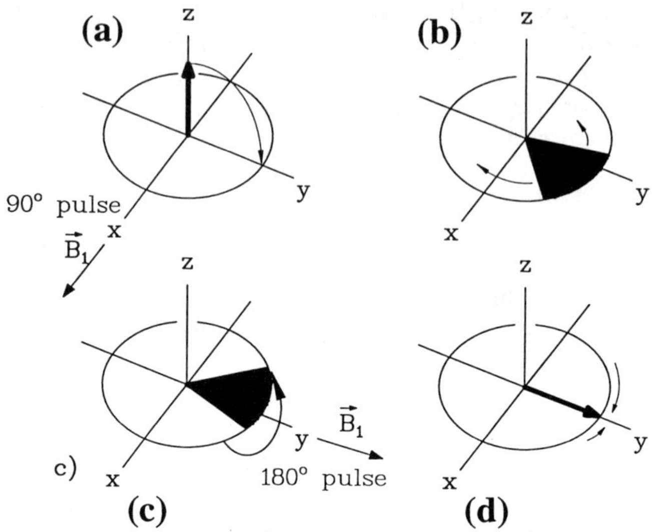

Figure 2.1 Principle of the Hahn spin echo. After having been flipped from the z axis by a 90° pulse around the x axis (a), the magnetization starts its precession on the y axis. (b) Seen in a system that rotates around the z axis (approximately) with the average precession frequency, the isochromats with their different precession frequencies $\omega^{(j)}$ gradually spread (dephase). (c) A 180° pulse around the y axis inverts the precessional phases $\omega^{(j)}t_e$ of the isochromats to $-\omega^{(j)}t_e$. This phase inversion is compensated by the phase $\omega^{(j)}t_e$ newly acquired at a time t_e after the pulse (d). Here, all magnetization vectors are again parallel, so a maximum in the signal (echo maximum) is observed. In the figure, the spread of the magnetization vectors is intentionally drawn asymmetric with respect to the y axis, in order to indicate that such a symmetry is by no means required for the echo to occur.

is the only further essential ingredient for understanding the scope of solid-state NMR spectra.

2.3 Concepts of NMR in terms of textbook quantum mechanics

Systems made up of isolated spins represent some of the simplest quantum-mechanical systems found in nature. Below, their behaviour will be discussed in the density-operator formalism, which is most appropriate for the purpose and closely related to the very useful concept of magnetization but not extensively treated in most textbooks on quantum mechanics. To bridge the gap somewhat, we shall give a short evaluation of some basic features of NMR in terms of perturbation theory, energy levels, and populations.

For the systems to be considered, the Zeeman interaction:

$$\hat{H} = -\vec{\mathbf{M}} \cdot \vec{\mathbf{B}}_0 = -\gamma \hbar \hat{I}_z B_0 \tag{2.11}$$

is by far the strongest interaction. Other contributions to the Hamiltonian, such as dipolar, quadrupolar or electron-cloud couplings which can be termed 'local fields', lead to small energy shifts that can be calculated in terms of (time-independent) first-order perturbation theory (Sakurai, 1985). On the basis of these energy eigenvalues ('levels') and the $\hat{I}_z$ eigenfunctions, *full* time-dependent solutions of the Schrödinger equation are determined for isolated spins or pairs of coupled spins. As demonstrated, for instance, in the Feynman Lectures (Feynman *et al.*, 1965), these two- and few-level systems can be treated exactly by means of a matrix representation of the Schrödinger equation, without recourse to time-dependent perturbation theory. A basic but conceptually most important result noted there is:

$$\omega_{pr} = \Delta E/\hbar \tag{2.12}$$

which represents the *equality of the precession or oscillation frequency of magnetization around the $\vec{B}_0$ field, to the energy difference ΔE between the energy levels divided by $\hbar$* (Figure 2.2). This relation will be derived for many-level systems in Section 2.5.4. Note that the 'transition energy' ΔE appears here even though the precession characterized by ω_{pr} does not involve any measurable transitions between the energy levels: the population difference corresponds to z magnetization:

$$\langle \hat{I}_z \rangle = \tfrac{1}{2}(|c_\uparrow|^2 - |c_\downarrow|^2) \tag{2.13}$$

(in the case of the two-level or $I = \frac{1}{2}$ system), and the z magnetization remains unchanged during precession around the z axis.

Transitions between the energy levels are induced only under the action of a time-dependent $\vec{\mathbf{B}}_1$ field which oscillates with a frequency close to the resonance frequency ω_{pr}. The transitions are equivalent to a precession of the magnetization around $\vec{\mathbf{B}}_0 + \vec{\mathbf{B}}_1(t)$ since the projection of the magnetization onto the z axis is proportional to the population difference of the energy levels. It should be noted, however, that the

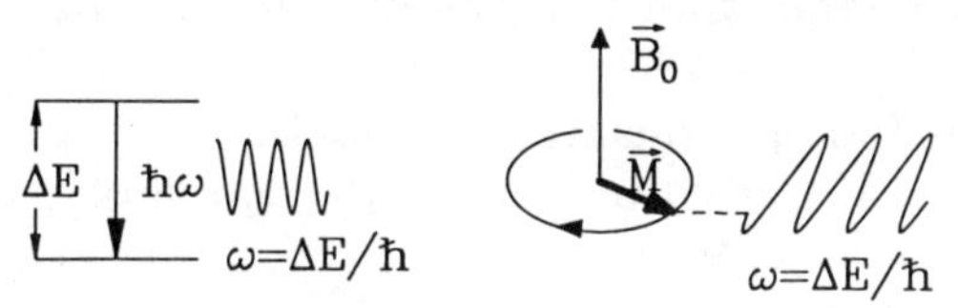

Figure 2.2 Schematic representation of the identity of the energy-level difference $\Delta E/\hbar$, the resonant transition frequency, and the NMR precession frequency.

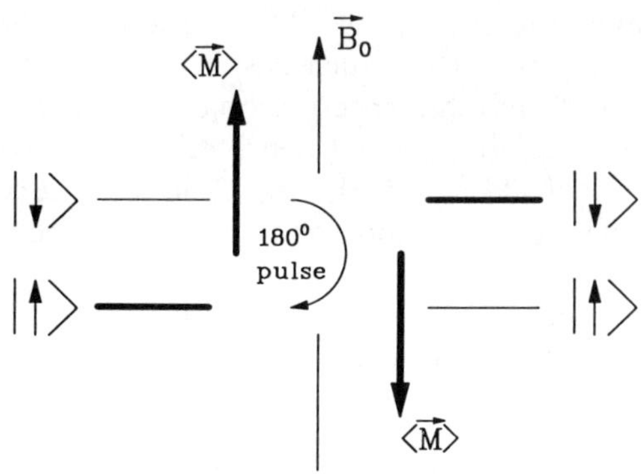

Figure 2.3 Population inversion by a 180° pulse: Comparison of magnetization and energy-level pictures. The thickness of the lines marking the energy levels indicates the populations (number of spins) in these levels.

common superficial picture of transitions in a two-level system tends to lead to some erroneous conceptions. For instance, it is often falsely concluded that the balance of transitions down and up leads to an asymptotic approach to equal population of both states. From this, the population inversion by a 180° pulse (Figure 2.3) as well as the effects of other large-angle pulses are not properly predicted. A related discrepancy involves the transition rates in standard time-dependent perturbation theory (Fermi's golden rule, see Section 3.6) and the exact on-resonance solution $M_z(t) = M_0\{1 - \cos(\gamma \vec{\mathbf{B}}_1 t)\}$: in solid-state NMR, the pulses are usually so short and strong that Fermi's golden rule is not applicable; they rather produce an oscillatory behaviour directly analogous to the Rabi oscillations (see Section 3.6).

2.4 Hamiltonians and spectra

Instead of further indulging in the quantum-mechanical derivation of the time evolution of spin systems, in this section we rather concentrate on the important relations between the NMR frequencies, molecular parameters and resulting spectra, in particular the dependence of the frequency on the segmental orientation. These are absolutely necessary for understanding and analysing the multidimensional spectra discussed below, while many other aspects of those experiments can be described in terms of the simple semiclassical picture of precessing isochromats. The quantum-mechanical verification of this picture and the treatment of more complex spin systems is therefore deferred to Sections 2.5 and 2.6.

2.4.1 Important NMR interactions and their Hamiltonians

In the semiclassical and simple quantum-mechanical description given above, the behaviour of the spins was treated only under the action of the Zeeman interaction (2.11):

$$\hat{H}_0 = -\gamma B_0 \hat{I}_z = -\sum_j \gamma^j B_0 \hat{I}_z^j$$

(from here on, we will work with Hamiltonians in frequency units, substituting $\hat{H}/\hbar$ by $\hat{H}$ as is convenient and common in the NMR literature). The Zeeman interaction usually dominates the behaviour of the nuclear spin system, but contains no relevant structural information itself. For the spectroscopic applications of NMR, the local fields that the nuclear spins feel are crucial (Mehring, 1983). In solid polymers, dipolar couplings between spins, shielding of the $\vec{\mathbf{B}}_0$ field by the electron clouds, and interactions of nuclear electric quadrupole moments with electric field gradients are particularly relevant.

The Hamiltonian of dipolar interactions between nuclear spins is constructed according to the Correspondence Principle:

$$\hat{H}_\mathrm{D} = -\frac{\mu_0}{4\pi}\hbar \sum\sum_{\text{all } j,k \text{ pairs}} \gamma_j\gamma_k \frac{3(\hat{\vec{\mathbf{I}}}^j \cdot \vec{\mathbf{r}}_{jk}/r_{jk})(\hat{\vec{\mathbf{I}}}^k \cdot \vec{\mathbf{r}}_{jk}/r_{jk}) - \hat{\vec{\mathbf{I}}}^j \cdot \hat{\vec{\mathbf{I}}}^k}{(r_{jk})^3}$$

$$=: \sum\sum_{\text{all } j,k \text{ pairs}} \hat{\vec{\mathbf{I}}}^j \mathbf{D}^{jk} \hat{\vec{\mathbf{I}}}^k \qquad (2.14a)$$

$$(\mathbf{D}^{jk})_{mn} = -\frac{\mu_0}{4\pi}\hbar\gamma_j\gamma_k \frac{3(\vec{\mathbf{e}}_m\cdot\vec{\mathbf{r}}_{jk}/r_{jk})(\vec{\mathbf{e}}_n\cdot\vec{\mathbf{r}}_{jk}/r_{jk}) - \vec{\mathbf{e}}_m\cdot\vec{\mathbf{e}}_n}{(r_{jk})^3} \tag{2.14b}$$

where $\vec{\mathbf{r}}_{jk}$ designates the vector from nucleus j to nucleus k, and we define $r_{jk} := |\,\vec{\mathbf{r}}_{jk}\,|$. In equation (2.14b), $\vec{\mathbf{e}}_1$, $\vec{\mathbf{e}}_2$ and $\vec{\mathbf{e}}_3$ are the basis vectors of the coordinate system in which the $\mathbf{D}^{jk}$ tensor is represented. The values of the prefactors in equation (2.14) can easily be calculated using:

$$\omega_\mathrm{d} := \frac{\mu_0}{4\pi}\hbar\frac{\gamma_1\gamma_2}{(r_{1,2})^3} = 2\pi\ \mathbf{122\ kHz}\ \frac{\gamma_1}{\gamma^{^1\mathrm{H}}}\frac{\gamma_2}{\gamma^{^1\mathrm{H}}}\frac{1}{(r_{1,2}/\mathbf{1\ \mathring{A}})^3}. \tag{2.15}$$

For example, we find that for a $^{13}C{-}^1H$ spin pair ($\gamma^{^{13}\mathrm{C}}/\gamma^{^1\mathrm{H}} = \omega_0^{^{13}\mathrm{C}}/\omega_0^{^1\mathrm{H}} \cong 75$ MHz/300 MHz $= \frac{1}{4}$) with a distance of 1.1 Å (0.11 nm), the coupling strength is $\omega_\mathrm{d}/2\pi = 122\ \mathrm{kHz} \times \frac{1}{4} \times 1 \times 0.751 = 23$ kHz.

Under the influence of the $\vec{\mathbf{B}}_0$ field, the electron cloud also generates an additional local field. Diamagnetic currents in the electron orbitals and partial unquenching of their paramagnetism (see Section 2.4.6) generate local fields $\vec{\mathbf{B}}_\mathrm{s}$ that scale with the $\vec{\mathbf{B}}_0$ field, $\vec{\mathbf{B}}_\mathrm{s} = \boldsymbol{\sigma}\vec{\mathbf{B}}_0$, such that the Hamiltonian of the chemical shift for a given spin reads:

$$\hat{H}_\mathrm{cs} = \gamma\,\hat{\vec{\mathbf{I}}}\boldsymbol{\sigma}\vec{\mathbf{B}}_0 = \gamma(\hat{I}_x\sigma_{xz}^\mathrm{LF} + \hat{I}_y\sigma_{yz}^\mathrm{LF} + \hat{I}_z\sigma_{zz}^\mathrm{LF})B_0. \tag{2.16}$$

Here, the $\sigma_{\alpha\beta}^\mathrm{LF}$ are elements of the laboratory-frame representation of the chemical-shift tensor. (To simplify the notation, we use the symbol $\hat{I}_z$ both for the isolated-spin operator and for the sum over all spins, $\hat{I}_z = \Sigma_j\hat{I}_z^j$, of which the single-spin case actually represents a special example; which of these related operators is concerned can usually be deduced from the context. Also, in writing $\hat{\vec{\mathbf{I}}}\boldsymbol{\sigma}\vec{\mathbf{B}}_0$ for the bilinear form $\hat{\vec{\mathbf{I}}}^\mathrm{T}\boldsymbol{\sigma}\vec{\mathbf{B}}_0$, we have dropped the 'T'; here and below it is implied that the vector to the left of the matrix is transposed, i.e. a row vector, while to the right we have a column vector.) The distinction of different chemical units in terms of their chemical shifts, to be discussed in Section 2.4.6, is the basis of the majority of standard NMR experiments, and also represents an important component of many multidimensional techniques.

The indirect or J coupling of nuclear spins, whose scalar part is of great importance in solution NMR, plays no relevant role in solid-state NMR of polymers. It has the same spin dependence as the dipolar coupling and since it amounts only to a few hundreds of Hz or less, it is usually hidden under the dominating and partially indistinguishable dipolar effects. Only the nonvanishing isotropic component ('scalar coupling') can sometimes be observed.

The $I = \frac{1}{2}$ nuclei can only have magnetic dipole moments, since their higher (static) multipole moments, magnetic as well as electric, vanish (the proof is based on the Wigner–Eckart Theorem, see Sakurai, 1985). Therefore, all their interactions are dipolar in nature, and the Hamiltonians are linear in $\hat{\vec{\mathbf{I}}}^j$. Nuclei with $I \geqslant 1$ possess a quadrupole moment Q, which, subject to an electric field gradient tensor (second spatial derivatives of the electric potential Φ):

$$V_{\alpha\beta}^j = \frac{\partial E_\alpha}{\partial r_\beta} = \frac{\partial^2\Phi}{\partial r_\alpha \partial r_\beta} \qquad (\alpha, \beta = x, y, z) \tag{2.17}$$

produces an interaction that can be described by the following Hamiltonian (for a single spin):

$$\hat{H}_{\text{Q}} = \frac{eQ}{2I(2I-1)\hbar}\,\hat{\mathbf{I}}\mathbf{V}\hat{\mathbf{I}}. \tag{2.18}$$

Here, e is the elementary charge.

As noted above, in fields of several tesla the local fields felt by ^{1}H, ^{2}H, ^{13}C, ^{15}N, ^{19}F, ^{29}Si, or ^{31}P nuclei are small compared with B_0. Therefore, the shifts of the energy levels of the local interactions can be treated in first-order perturbation theory with respect to $\hat{H}_0 = \omega_0 \hat{I}_z$. As the energy shifts are given by the diagonal elements of the perturbing Hamiltonians in a basis of $\hat{I}_z$ eigenfunctions, $E_m^{(1)} = \langle m \mid \hat{H}_1 \mid m \rangle$, only those parts of the local-field operators that are diagonal in this $\mid m\rangle$ basis need to be retained. These 'secular' parts of the Hamiltonians are the components that commute with I_z (operators that are diagonal in the same basis have this basis as their common eigenfunctions and commute). After this 'truncation' of the local-field Hamiltonians, only terms containing 1, $\hat{I}_z$, $\hat{\mathbf{I}}^j \cdot \hat{\mathbf{I}}^k$ and their products remain. The operator relations that represent a more profound basis of the truncation procedure are discussed in Appendix G.

The truncated chemical-shift Hamiltonian for a single spin is:

$$\hat{H}_{\text{cs}} = \gamma \hat{I}_z \sigma_{zz}^{\text{LF}} B_0. \tag{2.19}$$

The index LF indicates that σ_{zz}^{LF} is the zz element of the $\boldsymbol{\sigma}$ matrix in its laboratory-frame (LF) representation.

The truncated heteronuclear dipolar couplings are:

$$\begin{aligned} \hat{H}_{\text{D}}^{S,I} &= \sum_j \sum_k \hat{I}_z^j (\mathbf{D}_{\text{LF}}^{jk})_{zz}\, \hat{S}_z^k \\ &= -\frac{\mu_0}{4\pi}\hbar \sum_j \sum_k \frac{\gamma^I \gamma^S}{r_{jk}^3}\, \tfrac{1}{2}(3\cos^2(\theta_{jk}) - 1)\, 2\hat{I}_z^j \hat{S}_z^k. \end{aligned} \tag{2.20}$$

The truncated Hamiltonian of homonuclear dipolar couplings is given by:

$$\begin{aligned} \hat{H}_{\text{D}}^{II} &= \sum_j \sum_{k<j} \tfrac{1}{2}(\mathbf{D}_{\text{LF}}^{jk})_{zz}\, (3\hat{I}_z^j \hat{I}_z^k - \hat{\mathbf{I}}^j \cdot \hat{\mathbf{I}}^k) \\ &= -\frac{\mu_0}{4\pi}\hbar \sum_j \sum_{k<j} \frac{\gamma^2}{r_{jk}^3}\, \tfrac{1}{2}(3\cos^2(\theta_{jk}) - 1)(3\hat{I}_z^j \hat{I}_z^k - \hat{\mathbf{I}}^j \cdot \hat{\mathbf{I}}^k). \end{aligned} \tag{2.21}$$

In these expressions, θ_{jk} is the angle between the $\vec{\mathbf{B}}_0$ field and the axis $\vec{\mathbf{r}}_{jk}$ which connects the nuclei j and k. In NMR of organic solids, usually the abundant spins (protons or ^{19}F) are designated as the I spins, while the rare spins (^{13}C, ^{29}Si, or ^{15}N) are the S spins.

The quadrupolar Hamiltonian (2.18) for a single spin reads after truncation:

$$\hat{H}_{\text{Q}} = \frac{eQ}{2I(2I-1)\hbar}\, V_{zz}^{\text{LF}}\, \tfrac{1}{2}(3\hat{I}_z \hat{I}_z - \hat{\mathbf{I}} \cdot \hat{\mathbf{I}}). \tag{2.22}$$

In the truncation of $\hat{H}_D^{II}$ and $\hat{H}_Q$, not only the $\hat{I}_z^i$ contribute to the secular components, but homonuclear scalar-product terms $\hat{\mathbf{I}}^j \cdot \hat{\mathbf{I}}^k \equiv \hat{I}_x^j \hat{I}_x^k + \hat{I}_y^j \hat{I}_y^k + \hat{I}_z^j \hat{I}_z^k$ do, too. To prove that these are in fact secular, consider:

$$[\hat{\mathbf{I}}^j \cdot \hat{\mathbf{I}}^k, \hat{I}_z] = [\hat{\mathbf{I}}^j \cdot \hat{\mathbf{I}}^k, \sum_n \hat{I}_z^n]$$

$$= [\hat{\mathbf{I}}^j \cdot \hat{\mathbf{I}}^k, \hat{I}_z^1 + \hat{I}_z^2 + \cdots + \hat{I}_z^j + \cdots + \hat{I}_z^k + \cdots] = [\hat{\mathbf{I}}^j \cdot \hat{\mathbf{I}}^k, \hat{I}_z^j + \hat{I}_z^k]$$

where we take into account that $\hat{I}_z$ represents the sum of all z spin components, $\hat{I}_z = \Sigma_n \hat{I}_z^n$, and that for different spins $n \neq k$ the spin operators commute $[\hat{I}_\alpha^n, \hat{I}_\beta^k] = i\hat{I}_\gamma^n\, \delta_{nk}$. Employing the fundamental commutator relations of components of a spin operator:

$$\begin{aligned} [\hat{I}_x^j, \hat{I}_y^j] &= i\hat{I}_z^j \\ [\hat{I}_y^j, \hat{I}_z^j] &= i\hat{I}_x^j \\ [\hat{I}_z^j, \hat{I}_x^j] &= i\hat{I}_y^j \\ [\hat{B}, \hat{A}] &= -[\hat{A}, \hat{B}] \end{aligned} \tag{2.23}$$

it is finally found that indeed:

$$[\hat{\mathbf{I}}^j \cdot \hat{\mathbf{I}}^k, \hat{I}_z] = [\hat{\mathbf{I}}^j \cdot \hat{\mathbf{I}}^k, \hat{I}_z^j + \hat{I}_z^k] = [\hat{I}_x^j \hat{I}_x^k + \hat{I}_y^j \hat{I}_y^k + \hat{I}_z^j \hat{I}_z^k, \hat{I}_z^j + \hat{I}_z^k] = 0 \tag{2.24}$$

(the vanishing commutator $[\hat{\mathbf{I}}^j \cdot \hat{\mathbf{I}}^k, \hat{I}_z^j + \hat{I}_z^k] = 0$ has its analogue in the case of L–S or spin–orbit coupling in atomic physics, where the z component of the total spin $\hat{J}_z = \hat{L}_z + \hat{S}_z$ commutes with $\hat{\mathbf{L}} \cdot \hat{\mathbf{S}}$). By contrast, the heteronuclear scalar product does not commute with the Zeeman Hamiltonian $-\gamma_I B_0 \hat{I}_z - \gamma_S B_0 \hat{S}_z$:

$$[\hat{\mathbf{I}}^j \cdot \hat{\mathbf{S}}^k, \gamma_I \hat{I}_z^j + \gamma_S \hat{S}_z^k] \neq 0.$$

The different spin dependencies (2.20) and (2.21) of heteronuclear and homonuclear dipolar interactions lead to important differences in their decoupling behaviour, as will be discussed in Chapter 3.

2.4.2 Segmental-orientation dependence of NMR frequencies

The orientation dependence of NMR interactions in solids is an important feature of solid-state NMR. It permits the measurement of molecular orientations and reorientations of segments. The information content is particularly high when these angle-dependent frequencies are exploited several times for each segment in multidimensional experiments. These angular dependencies, or anisotropies, result from the tensorial nature of the interactions, and can be understood in terms of straightforward classical geometric considerations. For the understanding of solid-state NMR spectra, it is crucial to realize that the anisotropic NMR frequencies do not reflect the orientations of nuclear spins, but rather the orientations of molecular segments, to which the interaction tensors are generally fixed.

As the simplest example, let us consider the chemical shift of the Larmor frequency of a certain spin. According to equations (2.11) and (2.19), using $\omega_0 = -\gamma B_0$, the relevant Hamiltonian reads:

$$\hat{H}_0 + \hat{H}_{cs} = \omega_0 (1 - \sigma_{zz}^{LF}) \hat{I}_z.$$

It leads to a slow precession of the magnetization with a frequency:

$$\omega_{cs} = -\omega_0 \, \sigma_{zz}^{\mathrm{LF}} \tag{2.25}$$

relative to the Larmor frequency ω_0 of the unshielded nucleus.

Now $\sigma_{zz}^{\mathrm{LF}}$, the zz element of the shielding tensor expressed in the lab frame, depends on the orientation of the molecular segment relative to the $\vec{\mathbf{B}}_0$ field because the $\boldsymbol{\sigma}$ tensor is a molecular property and is therefore fixed relative to the segment. We have $\vec{\mathbf{B}}_0 = B_0(0, 0, 1)$ in the lab frame, and with the unit vector $\vec{\mathbf{b}}_0 = \vec{\mathbf{B}}_0 / B_0$ along the field direction we have the identity:

$$\sigma_{zz}^{\mathrm{LF}} = (0, 0, 1)\; \boldsymbol{\sigma}^{\mathrm{LF}} \begin{bmatrix} 0 \\ 0 \\ 1 \end{bmatrix} = \vec{\mathbf{b}}_0^{s}\; \boldsymbol{\sigma}^{s}\; \vec{\mathbf{b}}_0^{s}. \tag{2.26}$$

This is valid with both $\vec{\mathbf{b}}_0$ and $\boldsymbol{\sigma}$ expressed in any coordinate system s, since the bilinear form on the right-hand side of equation (2.26) is coordinate independent. For the further treatment, we decompose $\boldsymbol{\sigma}$ into its symmetric and its antisymmetric part:

$$\boldsymbol{\sigma} = \boldsymbol{\sigma}_{\mathrm{sy}} + \boldsymbol{\sigma}_{\mathrm{asy}}.$$

The antisymmetric portions of $\boldsymbol{\sigma}$:

$$\boldsymbol{\sigma}_{\mathrm{asy}} = \tfrac{1}{2}(\boldsymbol{\sigma} - \boldsymbol{\sigma}^{\mathrm{T}}) \qquad (\boldsymbol{\sigma}_{\mathrm{asy}})_{ij} = \tfrac{1}{2}(\sigma_{ij} - \sigma_{ji})$$

produce no contribution to the frequency (2.25), since:

$$\begin{aligned} \vec{\mathbf{B}}^{\mathrm{T}}\; \boldsymbol{\sigma}_{\mathrm{asy}} \vec{\mathbf{B}} &= \sum_{i=1}^{3} \sum_{j=1}^{3} B_i \tfrac{1}{2}(\sigma_{ij} - \sigma_{ji}) B_j \\ &= \tfrac{1}{2} \sum_{i=1}^{3} \sum_{j=1}^{3} B_i \sigma_{ij} B_j - \tfrac{1}{2} \sum_{j=1}^{3} \sum_{i=1}^{3} B_j \sigma_{ji} B_i = 0 \end{aligned} \tag{2.27}$$

(in the bottom line, an interchange of the arbitrary indices i and j in the second term shows that this term is exactly identical to the first term). Therefore, we can completely disregard the antisymmetric part of $\boldsymbol{\sigma}$ in the following.

A symmetric tensor like the symmetric part of $\boldsymbol{\sigma}$ is characterized most conveniently in the coordinate system in which it is diagonal; this is the 'principal axes system' (PAS) (for some mathematical background, see Appendix D). Thus, equations (2.25) and (2.26) produce:

$$\omega_{cs} = -\omega_0\; \vec{\mathbf{b}}_0^{s}\; \boldsymbol{\sigma}^{s}\; \vec{\mathbf{b}}_0^{s} = -\omega_0\; \vec{\mathbf{b}}_0^{\mathrm{PAS}}\; \boldsymbol{\sigma}^{\mathrm{PAS}}\; \vec{\mathbf{b}}_0^{\mathrm{PAS}}. \tag{2.28}$$

On the right-hand side, both the $\vec{\mathbf{B}}_0$ field direction $\vec{\mathbf{b}}_0$ and the tensor $\boldsymbol{\sigma}$ are expressed by their coordinates in the PAS. With $\vec{\mathbf{b}}_0$ expressed in terms of its polar coordinates (θ,ϕ) in the PAS and the principal values (eigenvalues) $\sigma_{xx}^{\mathrm{PAS}}$, $\sigma_{yy}^{\mathrm{PAS}}$, and $\sigma_{zz}^{\mathrm{PAS}}$ of $\boldsymbol{\sigma}$, equation (2.28) can be written as:

$$\begin{aligned} \omega_{cs} &= -\omega_0 \sum_{\alpha} (b_0^{\mathrm{PAS}})_{\alpha} \sigma_{\alpha\alpha}^{\mathrm{PAS}} (b_0^{\mathrm{PAS}})_{\alpha} \\ &= \omega_0 (\sigma_{xx}^{\mathrm{PAS}} (\cos\phi \sin\theta)^2 + \sigma_{yy}^{\mathrm{PAS}} (\sin\phi \sin\theta)^2 + \sigma_{zz}^{\mathrm{PAS}} (\cos\theta)^2). \end{aligned} \tag{2.29}$$

To extract the traceless part of the tensor, the isotropic chemical shift:

$$\sigma_{\text{iso}} = \tfrac{1}{3}(\sigma_{xx}^{\text{PAS}} + \sigma_{yy}^{\text{PAS}} + \sigma_{zz}^{\text{PAS}}) \tag{2.30}$$

is subtracted from each principal value to define:

$$\begin{aligned} \sigma_x &:= \sigma_{xx}^{\text{PAS}} - \sigma_{\text{iso}} \\ \sigma_y &:= \sigma_{yy}^{\text{PAS}} - \sigma_{\text{iso}} \\ \sigma_z &:= \sigma_{zz}^{\text{PAS}} - \sigma_{\text{iso}}. \end{aligned} \tag{2.31}$$

The corresponding frequencies for $\vec{\mathbf{B}}_0$ field orientations parallel to a principal axis α will be designated as ω_α:

$$\omega_\alpha := -\omega_0 \sigma_\alpha.$$

The frequency formula (2.29) can be simplified even further. If $\sigma_x = \sigma_y$ (axial symmetry of the tensors with respect to the principal z axis), one has $\sigma_z = -2\sigma_x = -2\sigma_y$ from equations (2.30) and (2.31), and after application of simple trigonometric relations (see Appendix A) to the frequency formula (2.29), one finds:

$$\omega_{\text{cs,aniso}} = -\omega_0 \sigma_z \tfrac{1}{2}(3\cos^2\theta - 1). \tag{2.32}$$

The relevance of this relation is indicated in Figure 2.4. The curve $\omega(\theta) \sim \frac{1}{2}(3\cos^2\theta - 1)$, which is the second Legendre polynomial, is shown (inverted) at the top of Figure 2.7.

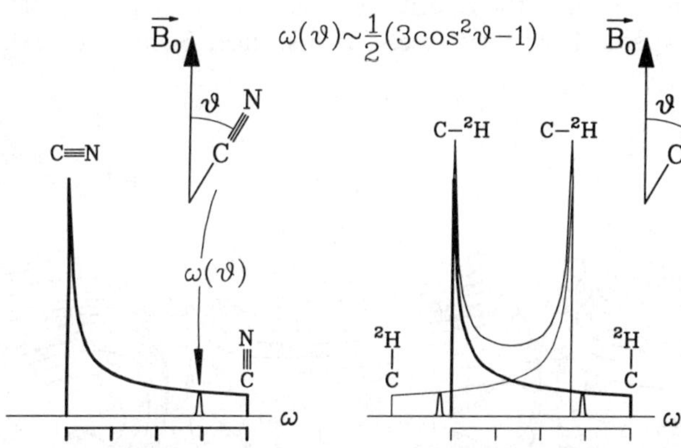

Figure 2.4 Dependence of the NMR frequency on the relative orientation of a molecular segment and the $\vec{\mathbf{B}}_0$ field (note that this has nothing to do with spin orientation!). The simple case of an axially symmetric interaction ($\eta = 0$) is shown. ω depends only on the angle θ between $\vec{\mathbf{B}}_0$ and the unique axis of the interaction tensor which is the C≡N vector for the ^{13}C chemical shift, the C—^{2}H vector for the ^{2}H quadrupolar coupling, and the C—H internuclear vector for the dipolar-pair coupling. In 'powder' samples, all possible orientations are present with equal probabilities, generating a well-defined inhomogeneously broadened 'powder' spectrum. The 'second transition' of dipolar and quadrupolar couplings gives rise to a second powder pattern, shown with thin lines here, which is superimposed on the first.

For the general case, it is useful to introduce the

$$\text{asymmetry parameter} \qquad \eta := \frac{\sigma_y - \sigma_x}{\sigma_z} = \frac{\omega_y - \omega_x}{\omega_z} \tag{2.33}$$

$$\text{anisotropy parameter} \qquad \delta := \omega_z \tag{2.34}$$

(except for $\omega_z \neq 0$, no conventions about the relative sizes of the principal values are required at this point). With the parameters η and δ, equation (2.29) can be transformed (using $\sigma_y = \sigma_x + (\sigma_y - \sigma_x)$, $2\sigma_x = -\sigma_z - \sigma_y + \sigma_x \Rightarrow \sigma_x = (-1-\eta)\sigma_z/2$, and $2\sin^2\phi - 1 = -\cos(2\phi)$) into the most important expression for the anisotropic frequency:

$$\omega(\theta, \phi) = \delta \tfrac{1}{2}(3\cos^2\theta - 1 - \eta \sin^2\theta \cos(2\phi)). \tag{2.35}$$

The orientation dependence of the frequency can thus be expressed quite conveniently in terms of the polar angles (θ, ϕ) of the $\vec{\mathbf{B}}_0$ field in the PAS. These are also the Euler angles (ϕ, θ, arbitr.) specifying the Euler transformation of the PAS into the laboratory ($\vec{\mathbf{B}}_0$) frame (see equation B.8 in Appendix B). The third Euler angle, corresponding to a rotation around the $\vec{\mathbf{B}}_0$ axis, is not relevant for the frequency, since the secular interactions are invariant under a rotation around $\vec{\mathbf{B}}_0$.

The geometrical meaning of equation (2.35) can be displayed most conveniently in terms of lines of equal frequency on the unit sphere (Haeberlen, 1976), as shown in Figure 2.5. On this unit sphere, orientation distributions can also be visualized relatively easily (in terms of $\vec{\mathbf{B}}_0$ vector distributions); for instance, conical distributions can be represented as circles on the unit sphere; this is an important advantage over the 'interaction ellipsoid' (see Appendix D.3) which is sometimes introduced. In several two-dimensional experiments, the relative course of equifrequency lines of two tensors is relevant for the spectral features (see, for instance, Section 6.7.3).

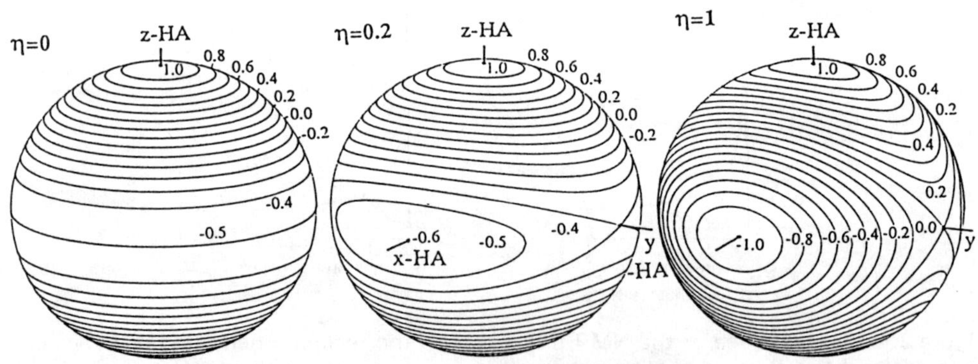

Figure 2.5 Lines of equal frequency ('equifrequency lines') on the unit sphere provide a useful visualization of the orientation dependence of NMR frequencies. Every point represents a direction of the $\vec{\mathbf{B}}_0$ field in the unified PAS; the corresponding frequency can be read off the equifrequency line running through this point. In the figure, the frequencies are given in units of δ, with increments of $\delta/10$. For all η values, a low density of equifrequency lines, and a weak angular dependence, is observed close to the principal axes. For the saddle point at the y axis, this results in a spectral singularity. The patterns for $\eta = 0.2$ and $\eta = 0$ are obviously similar except for the regions close to the x and y axes.

Rules for the attachment of the three principal values of the interaction tensor to the x, y and z axes of its principal-axes system have not been specified up to this point. In most texts, the discussion of the angular dependencies begins with conventions that attribute the principal values to the axes according to their signs and absolute values. The derivation of the frequency formula (2.35), however, shows that this relation can be obtained without such restrictions (except for requiring $\omega_z \neq 0$). One can interchange the principal values in the equations above, if only δ and η are changed accordingly. A convention convenient for general line shape analysis (Haeberlen, 1976; Spiess, 1978) requires:

$$|\omega_y| \leqslant |\omega_x| \leqslant |\omega_z| \tag{2.36}$$

This is a mathematical prescription which ensures that ω_y is located between ω_x and ω_z, and that ω_x is the principal value closer to ω_y (see Figure 2.8). This ordering of the principal values results in $0 \leqslant \eta \leqslant 1$ (see equation (2.33)). In particular, for $\eta = 0$ this convention always identifies the unique principal axis as the z axis, irrespective of the sign of δ.

Another convention has proved useful in a complementary way, namely for characterizing chemical-shift tensors within a given class of segments such as methylene or carboxyl groups. The principal-value frequencies, denoted as $\omega_{11}, \omega_{22}, \omega_{33}$ in this convention, are simply ordered with ω_{11} as the downfield (leftmost) and ω_{33} as the upfield (rightmost) principal-value frequency. The advantage consists in the rule that often the ω_{nn}, $n = 1, 2, 3$ values for similar segments in different chemical environments correspond to equivalent principal axes (e.g. ω_{33} corresponds to the principal axis normal to the plane of carboxyl groups). In contrast, in the $\omega_x, \omega_y, \omega_z$ convention the labelling of the ω_x and ω_z axes switches when the central principal value ($\omega_{22} = \omega_y$) shifts across the centre of the spectrum.

The chemical-shift tensor is determined by the electronic structure around the nuclear spin concerned. Therefore, the principal-axes system is fixed relative to the molecular segment. In Section 2.4.6, several typical ^{13}C shift tensor orientations in important molecular groups (obtained from single-crystal rotation patterns or from multidimensional experiments to be described below) are shown and discussed.

The equation (2.35) for the orientation dependent frequency $\omega(\theta, \phi)$ is valid also for interactions other than the chemical shift; only the expressions for δ have to be changed. This is so because the orientational dependencies of the truncated dipolar and quadrupolar interactions are also governed by lab-frame zz elements of second-rank interaction tensors (see equations (2.21)–(2.23)). In the same fashion as was demonstrated here for $\sigma_{zz}^{\mathrm{LF}}$, these $(D_{\mathrm{LF}}^{ij})_{zz}$ and V_{zz}^{LF} elements can be related to the principal values and the orientation (θ, ϕ) of the $\vec{\mathbf{B}}_0$ direction in the respective principal-axes systems.

For the dipolar coupling, we have found already in equations (2.20) and (2.21) that the angular dependence is $\frac{1}{2}(3\cos^2\theta - 1)$, hence $\eta_{\mathrm{D}} = 0$ (for isolated spin pairs). The corresponding anisotropy parameters are:

Heteronuclear interaction:

$$\delta_{\mathrm{D}}^{IS} = \omega_{\mathrm{d}} = \tfrac{1}{2} D_{zz}^{\mathrm{PAS}} = \frac{\mu_0}{4\pi} \hbar \frac{\gamma_1 \gamma_2}{r_{1,2}^3} = \frac{2\pi\ \mathbf{122\ kHz}}{(r_{1,2}/\mathbf{1\ \AA})^3} \frac{\gamma_1}{\gamma^{^1\mathrm{H}}} \frac{\gamma_2}{\gamma^{^1\mathrm{H}}}. \tag{2.37a}$$

Homonuclear interaction:

$$\delta_{\rm D}^{II} = \tfrac{3}{2}\omega_{\rm d}. \tag{2.37b}$$

The relation $\omega_{\rm d} = \frac{1}{2} D_{zz}^{\rm PAS}$ and the prefactor of 3/2 for the homonuclear interaction are due to the spin parts of the Hamiltonians; they will be proved below (see equations (2.91) and (3.43), respectively). The unique axis or z axis of the principal-axes system of the dipolar interaction of an isolated spin pair is along the internuclear vector of the interacting nuclei (as long as the structure is rigid).

For the field-gradient tensor element $V_{zz}^{\rm LF}$, in exact analogy to the preceding treatment for $\sigma_{zz}^{\rm LF}$ one finds:

$$V_{zz}^{\rm LF} = V_{zz}^{\rm PAS}\,\tfrac{1}{2}(3\cos^2\theta - 1 - \eta\sin^2\theta\cos 2\phi).$$

In the case of deuteron quadrupolar couplings ($I = 1$) one consequently obtains an angular-dependent frequency of the form of equation (2.35), with an anisotropy parameter of:

$$\delta_{\rm Q} = 3 \times \frac{1}{4}\frac{eQeq}{\hbar} \quad (\cong 2\pi \times 125 \text{ kHz for } {}^2\text{H in aliphatic C—}{}^2\text{H}). \tag{2.38}$$

$V_{zz}^{\rm PAS} = \mathbf{e}\cdot\mathbf{q}$ is the customary expression for the bond-structure-dependent principal value $V_{zz}^{\rm PAS}$ of the field-gradient tensor. Since $V_{zz} = \partial^2\Phi/\partial z^2$, the units of $\mathbf{e}\cdot\mathbf{q}$ are V/m^2 (where V $\triangleq$ Volt), while the units of the quadrupole moment Q are m^2. Planck's constant $\hbar$ can be expressed in terms of eV·s (eV = electron Volt = 1.6×10^{-19} Coulomb·Volt), so the units of δ_Q are indeed 1/s. The prefactor of 3 in equation (2.38) cannot be justified at this point; it is due to the spin $I = 1$ and the spin part of the Hamiltonian (2.22), as will be demonstrated in Section 2.5.5. Instead of δ_Q, quite often the 'quadrupole coupling constant':

$$\frac{eQeq}{\hbar} = \tfrac{4}{3}\delta_Q (\cong 2\pi \times 170 \text{ kHz for aliphatic C—}{}^2\text{H}) \tag{2.39}$$

is used to characterize the strength of the quadrupolar coupling. The asymmetry parameter $\eta_{\rm Q}$ of the electric field gradient tensor in aliphatic C—^{2}H bonds is usually negligible ($\eta_Q \cong 0$). This is due to the approximate uniaxiality of the electron density in the $\boldsymbol{\sigma}$ bonds, which are most relevant for the charge density distribution and resulting field-gradient tensor close to the ^{2}H nucleus. This cylindrical symmetry also causes the unique z axis of the PAS of the electric field gradient tensor to coincide with the C—^{2}H bond direction (more precisely, the C—^{2}H internuclear vector) within 3° or less. This well-defined tensor orientation is of great value in the interpretation of quadrupolar ^{2}H NMR spectra, both in one and two dimensions.

In Table 2.1, we have gone back and also expressed the Hamiltonians in terms of δ, η, θ, and ϕ, for chemical shift and quadrupolar coupling of a single site, as well as for dipolar interactions of spin pairs, to stress the eventual similarity of the spatial factors of these interactions after truncation.

For deuterons or isolated pairs of dipolar-coupled spins, the time evolution of the magnetization under the anisotropic quadrupolar or dipolar interaction is found to be given by $\cos(\omega(\theta)t) = \frac{1}{2}\exp\{i\omega(\theta)t\} + \frac{1}{2}\exp\{-i\omega(\theta)t\}$ (see Sections 2.5.5 and 2.5.6). Thus, for every $\omega(\theta)$ a pair of spectral lines, at $+\omega(\theta)$ and $-\omega(\theta)$, appear. The

Table 2.1

chemical shift:

$$\hat{H}_{cs} = \{\sigma_{iso}\gamma B_0 + \tfrac{1}{2}\delta_{cs}[3\cos^2\theta - 1 - \eta_{cs}\sin^2\theta\cos(2\phi)]\}\hat{I}_z \quad (2.40)$$

homonuclear dipolar pair coupling:

$$\hat{H}_D^{II} = -\frac{\mu_0}{4\pi}\hbar\frac{\gamma^2}{r^3}\tfrac{1}{2}(3\cos^2\theta - 1)(3\hat{I}_z\hat{J}_z - \hat{\mathbf{I}}\cdot\hat{\mathbf{J}}) \quad (2.41)$$

heteronuclear dipolar pair coupling:

$$\hat{H}_D^{SI} = -\frac{\mu_0}{4\pi}\hbar\frac{\gamma^I\gamma^S}{r^3}\tfrac{1}{2}(3\cos^2\theta - 1)2\hat{I}_z\hat{S}_z \quad (2.42)$$

(first-order) quadrupolar coupling:

$$\hat{H}_Q = \frac{eQeq}{2I(2I-1)\hbar}\tfrac{1}{2}[3\cos^2\theta - 1 - \eta_Q\sin^2\theta\cos(2\phi)]\tfrac{1}{2}[3\hat{I}_z^2 - I(I+1)] \quad (2.43)$$

spectrum is therefore symmetrical with respect to $\omega(\theta) = 0$. These relations are derived in detail further below.

In single crystals, the segmental orientations are well defined and few so that spectra with sharp lines result. From rotation patterns (i.e. the dependence of the line positions on the orientation of the crystal) the orientation of the interaction tensor's PAS relative to the crystal can be established. With the rotation axis perpendicular to $\vec{\mathbf{B}}_0$, the frequency for every site depends on the rotation angle ϕ according to:

$$\begin{aligned}\omega(\phi,\eta)/\delta &= A + \tilde{C}_2\cos(2\phi + 2\gamma) + \tilde{S}_2\sin(2\phi + 2\gamma)\\ &= A + B\cos(2\phi - \phi_0).\end{aligned} \quad (2.44a)$$

Note that this simple formula applies for all η values. The parameters $\tilde{C}_2$, $\tilde{S}_2$ and B are related via $B = (\tilde{C}_2^2 + \tilde{S}_2^2)^{1/2}$ and $\tan(\phi_0 + 2\gamma) = \tilde{S}_2/\tilde{C}_2$ according to relation (A.31). In order to visualize equation (2.44a), an experimental ^{13}C rotation pattern (Pines *et al.*, 1974) is shown in Figure 2.6. The equation can be derived on the basis of the frequency formula (2.28) written in a sample-fixed frame SF with a direction perpendicular to $\vec{\mathbf{B}}_0$ (such as the rf-coil axis) as its z axis. With a suitably chosen x axis, the $\vec{\mathbf{B}}_0$ direction in this frame is $\vec{\mathbf{b}}_0^{SF} = (\cos\phi, \sin\phi, 0)$. Thus:

$$\begin{aligned}\omega(\phi,\eta)/(\gamma B_0) &= \vec{\mathbf{b}}_0^{SF}\,\boldsymbol{\sigma}^{SF}\,\vec{\mathbf{b}}_0^{SF}\\ &= \tfrac{1}{2}(\sigma_{11}^{SF} + \sigma_{22}^{SF}) + \tfrac{1}{2}(\sigma_{11}^{SF} - \sigma_{22}^{SF})\cos(2\phi + 2\gamma) + \sigma_{12}^{SF}\sin(2\phi + 2\gamma).\end{aligned} \quad (2.44b)$$

The parameters A, $\tilde{C}_2$, and $\tilde{S}_2$ are proportional to δ and depend on η as well as on the Euler angles (α, β, γ) which relate the PAS with the sample-fixed frame SF. Via $\boldsymbol{\sigma}^{SF} = \tilde{\mathbf{R}}(\alpha,\beta,0)\,\boldsymbol{\sigma}^{PAS}\,\tilde{\mathbf{R}}^{-1}(\alpha,\beta,0)$, A, $\tilde{C}_2$ and $\tilde{S}_2$ can be determined explicitly:

$$\begin{aligned}A &= -\frac{1}{2}\frac{\delta}{2}(3\cos^2\beta - 1 - \eta\sin^2\beta\cos 2\alpha)\\ \tilde{C}_2 &= \tfrac{3}{2}\delta[\tfrac{1}{2}\sin^2\beta - \tfrac{1}{6}\eta(1 + \cos^2\beta)\cos 2\alpha]\\ \tilde{S}_2 &= \tfrac{1}{2}\delta\eta\cos\beta\sin 2\alpha.\end{aligned} \quad (2.44c)$$

This is analogous to the derivation of the frequency under magic-angle spinning in Section 3.9.1.

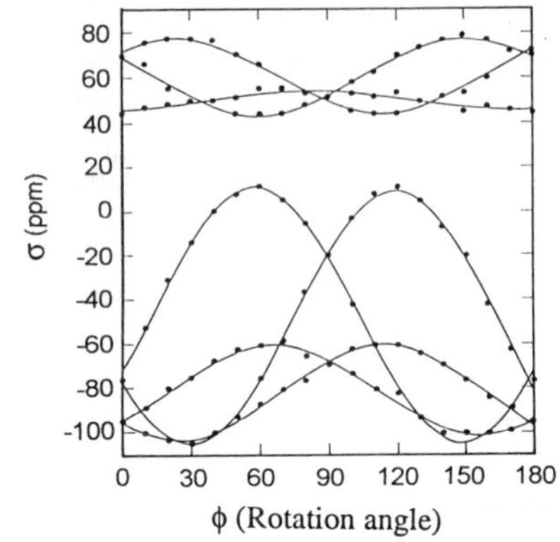

Figure 2.6 Example of a ^{13}C rotation pattern (adapted from Pines *et al.*, 1974) for a single crystal of ammonium D-tartrate. The positions of the spectral lines are plotted as a function of the crystal orientation with respect to the external $\vec{\mathbf{B}}_0$ field. The theoretical lines are all simple sinusoidal functions (see equation (2.44a)).

If the molecular structure is known, the tensor orientation in the molecular segment can be established from a sufficiently large set of A, B, and ϕ_0 values obtained from rotation patterns around different axes of the crystal. This is the basis for most of the available data on tensor orientations. In polymers, for highly oriented (and sufficiently simple) systems, similar studies can be performed to yield the relative orientation of the chain axis and the PAS of the interaction tensor. In polyethylene, the ^{13}C shift tensor orientation has been established in this way (VanderHart, 1976).

For large molecules, the one-dimensional spectra contain so many lines that the rotation patterns become hard to unravel. Here, a two-dimensional chemical-shift correlation experiment (described in Section 6.7.1) provides valuable correlations between corresponding lines at different orientations.

2.4.3 Powder lineshapes

From the anisotropic frequency shifts and the resulting inhomogeneously broadened lineshapes, valuable information about orientations and dynamics can be extracted by means of one-dimensional and multidimensional techniques. In many other applications, however, the same spread of frequencies is considered as an undesirable line-broadening that is often removed by magic-angle spinning. In unoriented samples, the orientation-dependent interactions as discussed above produce a 'powder' spectrum spread over the entire internal $[\omega_x \ldots \omega_z]$ (note that the term 'powder' in solid-state NMR usually refers to any isotropic sample of sufficiently rigid material, including, for example, unoriented blocks of polymer material).

For $\eta = 0$, the calculation of the powder spectrum is particularly simple due to the elimination of the ϕ dependence of the frequency. Every θ interval is to be transformed into the spectrum such that the integral intensity of corresponding intervals in θ and ω is equal (Figure E.1). Thus, the spectral intensity $S(\omega)$ and the angular distribution $P(\theta)$ are related by:

$$S(\omega(\theta)) \,|\, \mathrm{d}\omega \,| = P(\theta) \,|\, \mathrm{d}\theta \,| \,. \tag{2.45}$$

This is the fundamental 'principle of differential conservation of the integral' that is analysed in more detail in Appendix E. According to differential calculus, we can 'divide by $|d\beta|$' to transform equation (2.45) into:

$$S(\omega(\theta)) = P(\theta)/|d\omega/d\theta| = P(\theta)/|3\delta \sin\theta \cos\theta| \tag{2.46}$$

For powders, $P(\theta)$ is determined solely by the size of the surface elements, $P(\theta) = \sin\theta$ (normalization $\int_{0^\circ}^{90^\circ} P(\theta)\, d\theta = 1$); this results in:

$$\begin{aligned} S_{\text{pul}}(\omega(\theta)) &= \frac{1}{3\delta|\cos\theta|} \qquad |\cos\theta| = \frac{\sqrt{2}}{\sqrt{3}}\sqrt{\omega/\delta + \tfrac{1}{2}} \\ S_{\text{pul}}(\omega) &= \frac{1}{\sqrt{6\delta}}\frac{1}{\sqrt{\omega + \frac{1}{2}\delta}} \qquad -\delta/2 \leqslant \omega \leqslant \delta. \end{aligned} \tag{2.47}$$

There is an (integrable) square-root singularity at $\omega = -\frac{1}{2}\delta = \omega_x = \omega_y$, that is, at one edge of the spectrum (see Figure 2.7). It becomes the characteristic maximum of the spectrum if some homogeneous line-broadening of the individual spectral components is introduced. (At $\omega = \delta = \omega_z$ (i.e. $\sin\theta = 0$), there is also a $1/(\sin\theta = 0)$ singularity of the angular dependence $1/(d\omega/d\theta)$ in Equation (2.46), but it is cancelled by the vanishingly small contribution of the north-pole, $\theta = 0$, surface element, $P(0^\circ) = \sin(0^\circ) = 0$, to yield the finite value of $1/(3\delta)$. This is visualized in Figure E.1. The centre of the range covered by the $\eta = 0$ spectral pattern (for an isolated spin-$\frac{1}{2}$ as discussed so far) is at $\omega(45^\circ) = \delta/4$, the centre of gravity at $\omega(54.74^\circ) = 0$.

The symmetrical Pake powder pattern (Pake, 1948) in Figure 2.7 results for the cases of deuterons bonded to aliphatic carbons ($\eta \cong 0$), and of isolated pairs of dipolar-coupled spin-$\frac{1}{2}$ nuclei ($\eta = 0$), where due to the superposition of powder patterns for 'two transitions' (see Sections 2.5.5 and 2.5.6) the spectrum is symmetrical with respect to $\omega(\theta) = 0$.

The more complicated case of the powder spectrum for $\eta \neq 0$ has been treated by Bloembergen and Rowland (1955) and by Haeberlen (1976). The decisive step is the transformation of the integration variables from the angles $(\phi, \cos\theta)$ to $(\phi, \omega(\phi, \cos\theta))$. The distortion of the corresponding surface intervals can be calculated on the basis of the Jacobian determinant (See Appendix E):

$$\begin{aligned} \frac{\partial(\phi, \omega)}{\partial(\phi, \cos\theta)} &:= \begin{vmatrix} \dfrac{\partial\phi}{\partial\phi} & \dfrac{\partial\omega}{\partial\phi} \\ \dfrac{\partial\phi}{\partial\cos\theta} & \dfrac{\partial\omega}{\partial\cos\theta} \end{vmatrix} = 1\,\frac{\partial\omega}{\partial\cos\theta} - 0\,\frac{\partial\omega}{\partial\phi} \\ &= \delta\cos\theta(3 + \eta\cos 2\phi). \end{aligned} \tag{2.48}$$

In a shorthand notation, according to Appendix E the differential conservation of integral intensity under the transformation can be written as:

$$I(\phi, \omega) = P(\phi, \cos\theta)\,\frac{\partial(\phi, \cos\theta)}{\partial(\phi, \omega)} \tag{2.49}$$

where a corresponding transformation of the integration limits is implied. According to differential calculus:

$$\frac{\partial(\phi, \cos\theta)}{\partial(\phi, \omega)} = \left(\frac{\partial(\phi, \omega)}{\partial(\phi, \cos\theta)}\right)^{-1}. \tag{2.50}$$

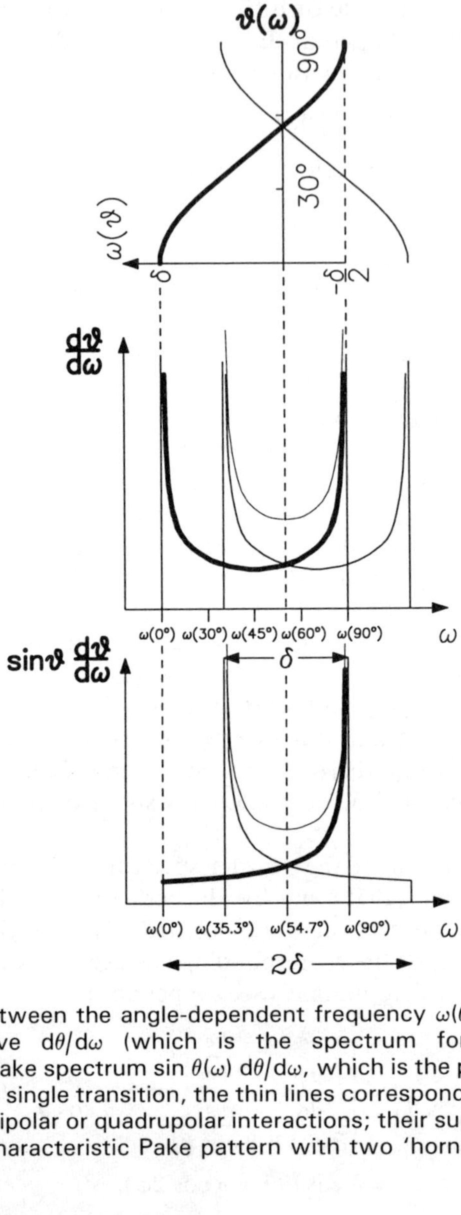

Figure 2.7 Relation between the angle-dependent frequency $\omega(\theta)$, its inversion $\theta(\omega)$, the corresponding derivative $d\theta/d\omega$ (which is the spectrum for a planar distribution $P(\theta) = \text{const.}$), and the Pake spectrum $\sin\theta(\omega)\, d\theta/d\omega$, which is the powder pattern for $\eta = 0$. The thick lines refer to a single transition, the thin lines correspond to the mirror-symmetric 'second transition' for dipolar or quadrupolar interactions; their superposition produces the characteristic Pake pattern with two 'horns'.

With $P(\cos\theta, \phi) = 1/4\pi$ in powders, $S(\omega)\, d\omega = d\omega \int I(\phi, \omega)\, d\phi$ with equations (2.49) and (2.50) inserted leads to:

$$S(\omega) = \frac{1}{4\pi} \int_{\phi_1(\omega)}^{\phi_2(\omega)} \left(\frac{\partial(\phi, \omega)}{\partial(\phi, \cos\theta)} \right)^{-1} d\phi$$

$$= \frac{1}{4\pi} \int_{\phi_1(\omega)}^{\phi_2(\omega)} 1/\{\delta \cos\theta(\omega, \phi)(3 + \eta \cos 2\phi)\}\, d\phi \tag{2.51}$$

where:

$$\cos\theta(\omega,\phi)=\sqrt{\frac{2\omega/\delta+1+\eta\cos 2\phi}{3+\eta\cos 2\phi}}.$$

These integrals can be evaluated in terms of the elliptical-integral functions; this is discussed in detail by Haeberlen (1976).

The resulting spectra span the range between ω_x and ω_z, or between $-\frac{1}{2}\delta(1+\eta)$ and $+\delta$; this can in fact be deduced directly from the frequency formula (2.35). The intensity distribution has sharp edges at ω_x and ω_z (Figure 2.8). At ω_y a singularity occurs; in real life, due to homogeneous line-broadening, it is a maximum of the spectral intensity, which decreases in relative height with increasing η. Generally, the actual spectrum is a convolution (for definition see equation (4.10)) of the theoretical spectral distribution function with the line-broadening function of the individual resonance lines.

The lineshapes can also be calculated numerically by scanning the angles θ and ϕ on the unit sphere, calculating $\omega(\theta,\phi)$, and weighting the intensity at $\omega=\omega(\theta,\phi)$ by $\cos\theta$, the size of the surface element represented by (θ,ϕ). Similar approaches to calculate multidimensional spectra will be discussed in chapter 10.

For partially oriented samples, the lineshapes obtained are often characteristic of the degree of orientation. Their discussion requires some more involved transitions between coordinate systems and is therefore deferred to Chapter 12 and Section 7.8. At this point, it is only to be mentioned that the spectral patterns for a highly oriented uniaxial sample are particularly sharp if the sample director $\vec{\mathbf{Z}}_D$ (e.g. the draw direction of fibres) is aligned along the $\vec{\mathbf{B}}_0$ field. It can be shown (see Section 7.8) that in simple systems with perfect uniaxial order, the director $\vec{\mathbf{Z}}_D$ has the same polar coordinates (θ,ϕ) in all principal-axes systems of equivalent sites (e.g. all CH_2 groups in the

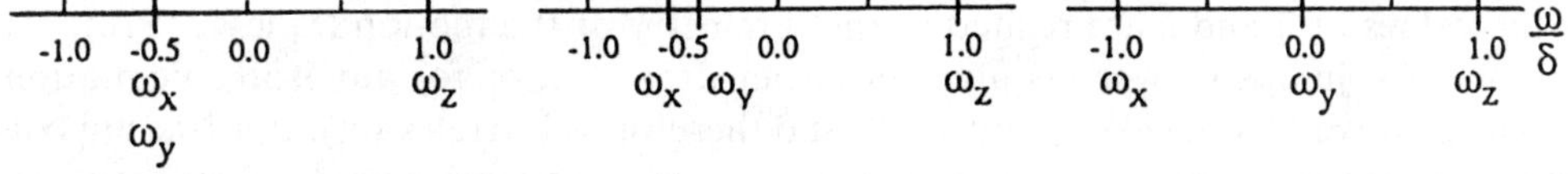

Figure 2.8 Powder spectra for various asymmetry parameters η. The isotropic shift is set to $\omega=0$. With the convention used in the text, the ω_z edge of the spectrum is always located at $\omega=\delta$. The ω_x edge shifts linear with η between $-\frac{1}{2}\delta$ and $-\delta$ (i.e. $\omega_x=-\frac{1}{2}\delta-\frac{1}{2}\delta\eta$), the ω_y singularity between $-\frac{1}{2}\delta$ and 0 (i.e. $\omega_y=-\frac{1}{2}\delta+\frac{1}{2}\delta\eta$). The spectra are normalized to equal integral (area). Near the ω_z edge of the spectrum, the spectral lineshape hardly changes with η. In fact, for $\eta\leqslant 0.3$ the spectral features of the $\eta=0$ spectrum can often be regarded as a good approximation. The distribution of the spectral intensity can be understood in terms of equifrequency lines as shown in Figure 2.5: the spectral intensity over the interval $[\omega\ldots\omega+d\omega]$ is proportional to the unit-sphere surface area between equifrequency lines at ω and $\omega+d\omega$.

polymer backbone). Since $\vec{\mathbf{B}}_0$ is parallel to $\vec{\mathbf{Z}}_D$ for the specifically chosen sample orientation, in all the principal-axes systems the $\vec{\mathbf{B}}_0$ field has the same coordinates (θ, ϕ). Thus, the same frequency $\omega(\theta, \phi)$ results for all the equivalent sites.

2.4.4 *Motional averaging of interaction tensors and NMR spectra*

Due to the dependence of the NMR frequencies on the segmental orientation, molecular reorientations usually have strong effects on the powder lineshape when the motional rate exceeds the width of the powder spectrum. While the description of the 'intermediate motional regime' ($\tau^{-1} \cong \delta$) is rather complicated (see Section 2.4.7), the situation becomes simple again in the 'fast-motion limit' (Abragam, 1961) when the motion is so fast that many frequency changes occur during one period of signal modulation by the interaction. Then, the effective precession frequency is the average:

$$\bar{\omega} = \sum_{j=1}^{N} p_j \omega_j = \gamma B_0 \sum_{j=1}^{N} p_j (\boldsymbol{\sigma}_j^{\mathrm{LF}})_{zz} =: \gamma B_0 (\bar{\boldsymbol{\sigma}})_{zz} \tag{2.52}$$

for reorientations that involve N different segmental orientations which are taken with occupation probabilities p_j. The $\boldsymbol{\sigma}_j^{\mathrm{LF}}$ are the interaction tensors for the various orientations expressed in the lab frame; the averaged tensor $\bar{\boldsymbol{\sigma}}$ is obviously given by:

$$\bar{\boldsymbol{\sigma}} = \sum_{j=1}^{N} p_j \boldsymbol{\sigma}_j. \tag{2.53}$$

This sum of second-rank tensors naturally is itself a second-rank tensor. Thus, $\bar{\boldsymbol{\sigma}}$ can be characterized by the anisotropy parameter $\bar{\delta}$ and the asymmetry parameter $\bar{\eta}$, which after diagonalization of the sum matrix $\bar{\boldsymbol{\sigma}}$ are obtained from the eigenvalues analogous to equations (2.33) and (2.34). (Note that in spite of the average bar, $\bar{\delta}$ is usually not equal to the average $\Sigma p_j \delta_j$, and $\bar{\eta} \neq \Sigma\, p_j \eta_j$). In terms of the parameters of the averaged tensor, the averaged frequency is given by:

$$\bar{\omega}(\theta_a, \phi_a) = \bar{\delta}\tfrac{1}{2}(3\cos^2\theta_a - 1 - \bar{\eta}\sin^2\theta_a \cos 2\phi_a) \tag{2.54}$$

where (θ_a, ϕ_a) denote the polar coordinates of the $\vec{\mathbf{B}}_0$ field in the PAS of the averaged tensor $\bar{\boldsymbol{\sigma}}$. The orientation of the averaged principal axes system in the molecule as well as the values of $\bar{\eta}$ and $\bar{\delta}$ are related to the symmetry of the motion (Spiess, 1978). For instance, for jumps between equivalent tetrahedral sites, or for any isotropic motion, the anisotropy is averaged out and $\bar{\delta} = 0$ (and therefore $\bar{\eta}$ is irrelevant). For free uniaxial rotations, but also for jumps between the N sites in systems with C_N symmetry and $N \geqslant 3$, the averaged tensor is axially symmetric ($\bar{\eta} = 0$) and has its principal z axis along the symmetry axis. To deduce these relations, the averaged tensor can be calculated as the sum:

$$\bar{\boldsymbol{\sigma}} = \sum_{j=1}^{N} \mathbf{R}^{-1}\left(\alpha_P, \beta_P, j\,\frac{360^\circ}{N}\right) \boldsymbol{\sigma}^{\mathrm{PAS}}\, \mathbf{R}\left(\alpha_P, \beta_P, j\,\frac{360^\circ}{N}\right)$$

where all $\boldsymbol{\sigma}_j$ are expressed in a frame in which the rotation axis (i.e. the C_N symmetry axis) is the z axis; β_P and α_P specify its orientation in the original (non-averaged) principal-axes system. The rotation matrices $\mathbf{R}^{-1} = \tilde{\mathbf{R}}(\alpha, \beta, \gamma)$ are given in equation (B.14). Under these conditions, the averaged anisotropy parameter:

$$\bar{\delta} = \tfrac{1}{2}\delta(3\cos^2\beta_P - 1 - \eta\sin^2\beta_P\cos 2\alpha_P) \tag{2.55}$$

is derived by a simple argument: $\bar{\delta} = -\omega_0\bar{\sigma}_z$ is the frequency observed if the $\vec{\mathbf{B}}_0$ field is along the z axis of the averaged tensor's PAS, which is the rotation axis. For $\vec{\mathbf{B}}_0$ along the rotation axis, the frequency can be calculated by the formula without motion since this $\vec{\mathbf{B}}_0$ orientation and the frequency are not affected by the rotation; therefore, $\bar{\delta} = -\omega_0\bar{\sigma}_z = \omega(\beta_P, \alpha_P)$. On the basis of the same invariance argument, the principal values for averaging by two-fold jumps between equally populated sites can be determined. In the case of $\eta = 0$ (see Figure 2.9), the nth principal value of the averaged tensor is given by:

$$\bar{\omega}_n = \omega(\theta_n) = \tfrac{1}{2}\delta(3\cos^2\theta_n - 1) \tag{2.56}$$

where $\cos^2\theta_n$ is equal to the square of the cosine of the angle between either of the

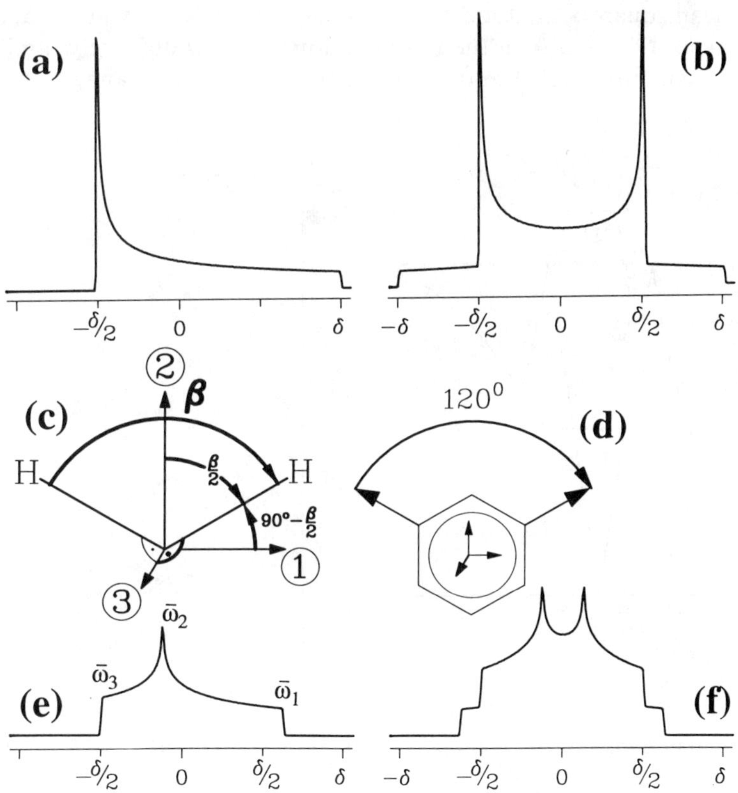

Figure 2.9 Motionally averaged interaction tensor for two-site jumps of C—H bonds (quadrupolar or C–H dipolar coupling). The original interaction is axially symmetric ($\eta = 0$). (a) Spectrum of single transition, (b) quadrupolar ^{2}H or dipolar-pair spectrum. (c) Orientation of the averaged principal axes 1, 2, and 3. (d) Geometry of the phenylene ring flip ($\beta = 120^\circ$). (e) Motionally averaged single-transition spectrum for $\beta = 120^\circ$ ($\bar{\eta} = 0.6$). (f) Phenylene-ring flip ^{2}H spectrum. The pattern is obtained from the spectrum in (e) by superimposing the mirror-symmetric powder pattern arising from the second transition.

C—^{2}H bonds and the averaged tensor's principal axis labelled by n (the actual angles of this axis to the C—^{2}H bonds may be θ_n and $180^\circ - \theta_n$). The corresponding principal-axes system is displayed in Figure 2.9(c). With an angle β between the $\vec{\mathbf{Z}}_P$ axes (C—^{2}H bonds), the relevant angles are $\theta_1 = 90^\circ - \frac{1}{2}\beta$, $\theta_2 = \frac{1}{2}\beta$, and $\theta_3 = 90^\circ$, so that we obtain:

$$\begin{aligned} \bar{\omega}_1 &= \delta\tfrac{1}{2}(3\sin^2\tfrac{1}{2}\beta - 1) && \text{(in } ^2\text{H—C—}^2\text{H plane} \perp \text{ to the bisector)} \\ \bar{\omega}_2 &= \delta\tfrac{1}{2}(3\cos^2\tfrac{1}{2}\beta - 1) && \text{(bisector between the C—}^2\text{H bonds)} \\ \bar{\omega}_3 &= -\tfrac{1}{2}\delta && (\perp \text{ to the } ^2\text{H—C—}^2\text{H plane).} \end{aligned} \tag{2.57}$$

Applied to phenylene-ring flips with $\beta = 120^\circ$ (Figure 2.9(d)), this yields:

$$\bar{\omega}_1 = \bar{\delta} = \tfrac{5}{8}\delta \quad \bar{\omega}_2 = -\tfrac{1}{8}\delta \quad \bar{\omega}_3 = -\tfrac{1}{2}\delta \tag{2.58}$$

$$\Rightarrow \bar{\eta} = \frac{3/8}{5/8} = 0.6.$$

The corresponding powder pattern is shown in Figure 2.9(e), and the symmetric ^{2}H pattern in Figure 2.9(f). The dependence of $\bar{\eta}$ and $\bar{\delta}$ on β for the two-site jump is plotted in Figure 2.10(a), while Figure 2.10(b) displays $\bar{\eta}$ and $\bar{\delta}$ for uniaxial libration with a root-mean-square amplitude σ_Φ on a cone with half-apex angle β_P.

The dependence of $\bar{\eta}$ and $\bar{\delta}$ on the reorientation angles is definitely nonlinear, as is exemplified by equation (2.57) and Figure 2.10. On the one hand, this means that

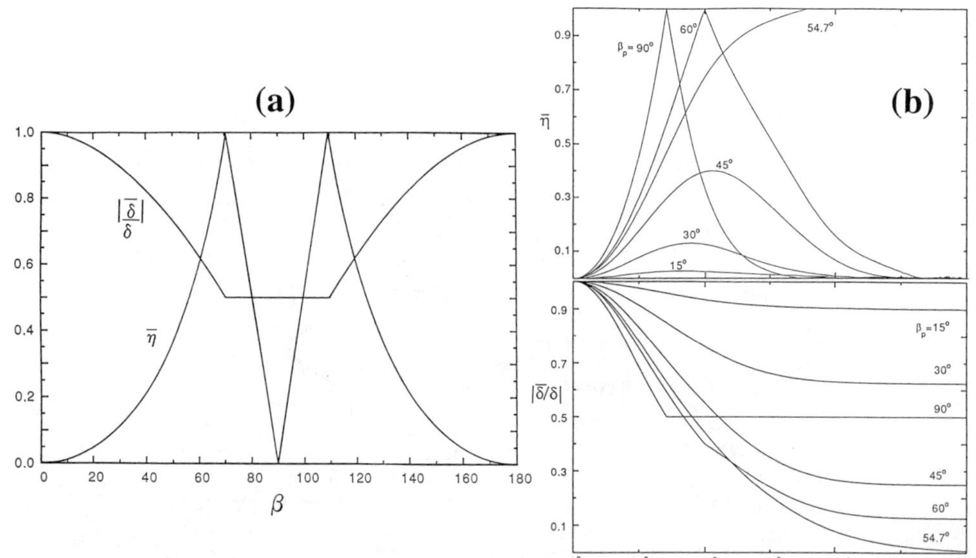

Figure 2.10 Dependence of the averaged interaction parameters $\bar{\eta}$ and $\bar{\delta}$ on the geometry of molecular motion (for $\eta = 0$, i.e., the original interaction is uniaxial). (a) Two-site jump with geometry as in Figure 2.9c. Sudden changes in the curves are observed where the asymmetry parameter attains a value of unity. At those points, the labelling of the x and z principal axes is interchanged to comply with the conventions (2.36). $|\bar{\delta}|/\delta$ remains constant at 0.5 between 70.5° and 109.5° since the principal value corresponding to the principal axis perpendicular to the two original unique axes is always equal to -0.5 and has the largest absolute value in the said angular range. (b) Uniaxial liberation with root-mean-square amplitude σ_Φ on a cone with half-apex angle β_P, for various values of β_P. The value of $|\bar{\delta}|/\delta$ in the unixial limit $\sigma_\Phi \to \infty$ is equal to $|\frac{1}{2}(3\cos^2\beta_P - 1)|$.

attention must be paid to the resulting error margins in regions where $\bar{\eta}$ and $\bar{\delta}$ are relatively insensitive to the angles (e.g. at $\beta \cong 0^\circ$ or $\beta \cong 180^\circ$ in equation (2.57)). On the other hand, small deviations of molecular bond angles or tensor orientations from the ideal values can sometimes have astonishingly large effects on the lineshape (e.g. between $\beta \cong 90^\circ$ and $\beta = 109^\circ$ in equation (2.57)).

The information content of motionally averaged lineshapes is restricted to only two parameters: $\bar{\eta}$ and $\bar{\delta}/\delta$. Quite different types of motions, both diffusive processes and well-defined jumps, can produce identical lineshapes. Likewise, variation of the occupation probabilities and of the reorientation angles can have indistinguishable effects. By contrast, in 2D exchange NMR (discussed in Chapter 7), the geometric and occupation effects are nearly completely separated, with the geometry determining the signal locations and patterns in the 2D plane and the occupation probabilities influencing the relative intensities of the patterns.

2.4.5 Basics of sample spinning (MAS and OMAS)

Averaging of tensors, sometimes occurring naturally by reorientations of molecular segments as described in the preceding section, can also be achieved by macroscopic sample rotation (Andrew *et al.*, 1958; Lowe, 1959; Schaefer and Stejskal, 1976). Thus, one can eliminate the anisotropic broadenings in order to obtain the structural information contained in the isotropic chemical shifts. Magic-angle spinning (MAS) of the sample is performed about an axis that makes the 'magic angle' of $\theta_m := \arccos(\sqrt{1/3}) = 54.74^\circ$ (the null point of $3\cos^2\theta - 1$) with the $\vec{\mathbf{B}}_0$ field. This angle is half the tetrahedral angle of 109.5°, or the angle between the (1,1,1) direction and any coordinate axis. MAS averages out all anisotropic interactions that can be described by second-rank tensors, if the rotation frequency exceeds the largest coupling of the spin species considered (strong interactions can 'protect' weak ones, as discussed in Section 3.9.5). This averaging will be rationalized in the following within the framework of motional averaging of interaction tensors.

Combining equations (2.54) and (2.55), for fast uniaxial rotation ($\Rightarrow \bar{\eta}_{\text{mol}} = 0$) we can write the averaged precession frequency as:

$$\omega = \overbrace{\tfrac{1}{2}(3\cos^2\beta_P - 1 - \eta\sin^2\beta_P\cos 2\alpha_P)\,\delta}^{\bar{\delta}_{\text{mol}}}\;\underbrace{\tfrac{1}{2}[3\cos^2\theta_r - 1]}_{\bar{\delta}_{\text{OMAS}}} \tag{2.59}$$

where θ_r is the angle between the rotation axis and the $\vec{\mathbf{B}}_0$ field, while (β_P, α_P) denote the polar coordinates of the rotation axis in the PAS. While for a single site or crystallite the frequency is given by equation (2.59) both for fast sample rotation and for fast molecular rotation (Figure 2.11(a)), the powder averages must be taken in different ways: in the case of uniaxial molecular rotations, the relative orientation of the original PAS to the rotation axis is fixed by the molecular structure, and the (molecule-fixed) rotation axes are distributed randomly relative to the $\vec{\mathbf{B}}_0$ field in a powder (Figure 2.11(b)). In contrast, for sample-rotation experiments, the angle θ_r between the (macroscopic) rotation axis and $\vec{\mathbf{B}}_0$ is fixed while the angles (β_P, α_P) of the relative orientations of the original principal axes to the rotation axis are distributed randomly. According to equation (2.59), under fast off-magic-angle spinning (OMAS) ($\theta_r \neq \theta_m$) a similar angular dependence of the frequency results as

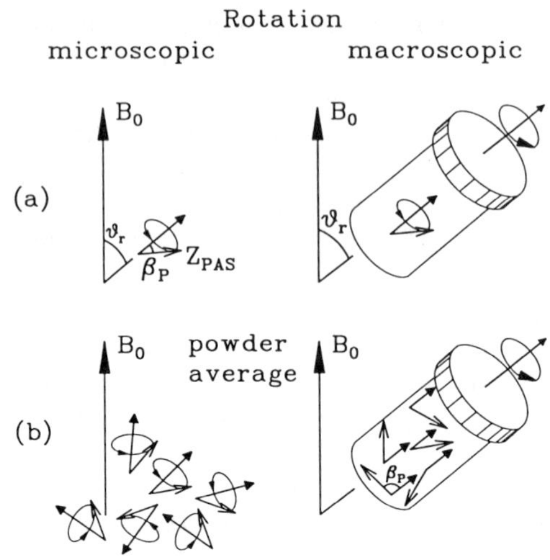

Figure 2.11 Microscopic vs. macroscopic uniaxial (continuous) rotation of NMR interaction tensors. (a) For a single crystallite, the microscopic and macroscopic rotations shown are completely equivalent, and identical spectra would result from both situations. (b) In the powder average, the two cases differ: For the macroscopic rotation, all principal-axes systems are rotated around the same (macroscopic) axis, while for the microscopic rotation, the rotation axes, each being fixed in the corresponding molecular segment, are distributed randomly.

for the static sample, equation (2.35). In fact, the asymmetry parameter η is the same as for the static spectrum, and the original δ parameter is only scaled:

$$\bar{\delta}_{\mathrm{OMAS}} = \tfrac{1}{2}(3\cos^2\theta_r - 1)\delta. \tag{2.60}$$

As a difference, it should be noted that according to equation (2.39) the individual line positions in the OMAS spectrum depend on the orientation (β_P, α_P) of the rotor axis in the PAS, while in a static sample the orientation (θ, ϕ) of the $\vec{\mathbf{B}}_0$ field in the PAS determines the frequencies.

The scaling factor (2.60) in equation (2.59) vanishes for every segment under MAS where $\theta_r = \theta_m = 54.74^\circ$. The discussion shows that, contrary to the frequently encountered opinion, the factor $(3\cos^2\theta_r - 1)$ in equation (2.59) is only indirectly related to the term $(3\cos^2\theta - 1)$ in the angular-dependent frequency $\omega_{\mathrm{aniso}} = \frac{1}{2}\delta(3\cos^2\theta - 1 - \eta\sin^2\theta\cos 2\phi)$ (equation (2.35)) in the static sample. The angles θ_r and θ are quite different: θ_r is fixed while θ varies between 0° and 180° in a powder.

The MAS spinning speed is often not sufficiently high to average out the anisotropies. Proton–proton and proton–carbon dipolar couplings in polymers are usually stronger than 20 kHz and are therefore *not* averaged away by conventional MAS. Rather, they are suppressed by high-power rf irradiation (see Section 3.4.1). The removal or reduction of chemical shift anisotropies is the main purpose of sample spinning applied to organic solids. Even though the corresponding inhomogeneous line-broadenings often exceed the spinning speed, the desired resolution of isotropic shifts can still be achieved. This is due to the splitting of the anisotropic chemical-shift broadening into spinning sidebands displaced by multiples of the rotation frequency ω_r from the centreband, which is located at the isotropic chemical shift (see Figure 2.12).

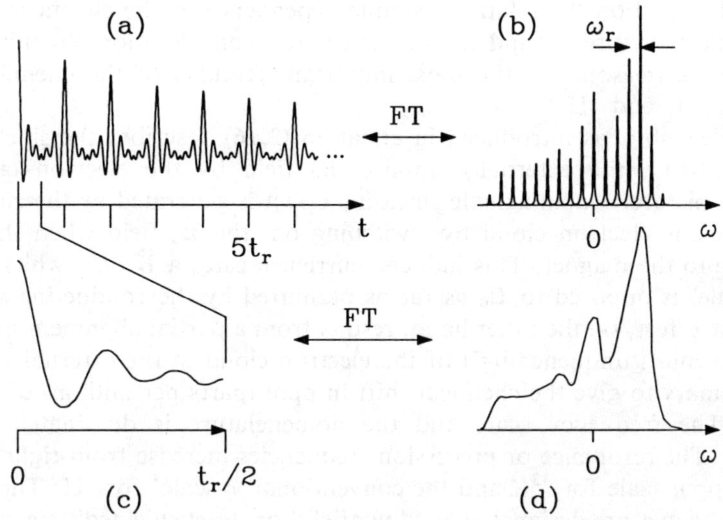

Figure 2.12 Magic-angle spinning. (a) Time signal (on resonance) with rotational echoes at multiples of t_r. (b) Corresponding spectrum with spinning sidebands at multiples of $\omega_r = 2\pi/t_r$. In ^{13}C MAS NMR of systems with several ^{13}C sites, the rotational echoes are often not so pronounced, because they are modulated by the isotropic chemical shifts, and differently for the various ^{13}C sites. (c) Expanded view of the time signal in the first half rotation period. Its Fourier transform (d) is an envelope to the sideband spectrum in (b). In the limit of very slow spinning $\omega_r \ll \delta$, this spectrum becomes identical to the pattern of a static sample.

Sharp chemical-shift sidebands occur because the chemical-shift Hamiltonian (in contrast to dipolar Hamiltonians which involve many spins) commutes with itself at different orientations so that the chemical-shift anisotropy is averaged out after each full rotor period (Maricq and Waugh, 1979) (see Section 3.9.5). As a result, the anisotropy component of the time signal is periodic with the rotor period t_r, and spectral intensity appears at the 'harmonics' $n(2\pi/t_r) = n\omega_r$.

In other words, the anisotropy effects of Hamiltonians self-commuting at different orientations are refocused at times $t = mt_r$ (if $\theta_r = \theta_m$) so that 'rotor echoes' appear (see Figure 2.12) which are modulated only by the isotropic shifts and the off-resonance effect. When sampled at $t = mt_r$, even for slow spinning the magnetization appears to evolve under the isotropic and the simply scaled anisotropic portions (equation (2.59)) of these Hamiltonians. Special techniques for sideband suppression like the TOSS (total suppression of sidebands) pulse sequences will be discussed in Section 3.9, where a full analysis of the behaviour of the magnetization under magic-angle spinning conditions will be provided.

2.4.6 ^{13}C and ^{1}H chemical shifts in theory and practice

Originally encountered as a bothersome shift of the resonance frequency which prevented the precise measurement of magnetogyric ratios γ of nuclear spins in condensed matter, the chemical shift soon became one of the most important foundations of the spectroscopic power of NMR. The large potential for structural elucidation is based on the relatively simple dependence of the chemical shift on the local chemical structure around the nucleus under consideration. In this section we want to summarize some of the most important features of the chemical shift, in particular of ^{13}C and ^{1}H.

The chemical shift as introduced in equation (2.16) describes the shielding of the nuclear spin from the externally applied $\vec{\mathbf{B}}_0$ field by the electron clouds. In a semiclassical picture, a diamagnetic shielding effect is generated by the current which is induced in the electron cloud by 'switching on' the $\vec{\mathbf{B}}_0$ field when the sample is introduced into the magnet. This induced current creates a $\vec{\mathbf{B}}$ field which, according to Lenz's rule, is opposed to $\vec{\mathbf{B}}_0$ as far as permitted by the conducting structure. A paramagnetic effect, on the other hand, results from a partial alignment of the orbital angular momenta ('unquenching') of the electron cloud in the external field.

It is customary to give the chemical shift in ppm (parts per million) of the Larmor frequency. The frequency scale and the nomenclature is dominated by historic conventions. The resonance or precession frequencies increase from right to left, and so does the ppm scale for ^{13}C and the conventional 'δ-scale' for ^{1}H. The resonances at large ppm values are designated as 'downfield' or 'least shielded', since they would resonate at the lowest field in a fixed-frequency cw spectrometer where the $\vec{\mathbf{B}}_0$ field strength was swept. Consistently, the 'upfield' or 'most shielded' region corresponds to smaller frequency (or ppm) values.

The chemical shift has a direct relationship to the chemical structure: Figure 2.13 displays a rough sketch of ^{13}C and ^{1}H isotropic-shift ranges relevant in organic polymeric solids, as measured by MAS NMR; more detailed diagrams and spectra can be found in the literature (Bovey, 1988, p. 94; Pham *et al.*, 1983). Overall, the tendencies for ^{13}C and ^{1}H shifts are found to be quite similar: aliphatics resonate at small ppm values, aromatics are found downfield, proximity of oxygen to a segment shifts its resonance downfield.

The anisotropies of the ^{13}C chemical shifts, reviewed by Veeman (1984), also follow a general pattern. Carbons in aliphatic environments have anisotropies of the order of 40 ppm. These relatively small values can be ascribed to the sp_3 hybridization which gives rise to electronic structures that are relatively close to the tetrahedral symmetry, for which the anisotropy must vanish due to the second-rank tensorial nature of the interaction (see below). With sp_2 hybridization, the anisotropies are larger: carbonyl anisotropies span about 150 ppm, those of aromatic carbons about 200 ppm. Consistently, C≡N groups with their sp hybridization have a particularly large anisotropy of 250 ppm. The anisotropies for ^{1}H (and ^{2}H) chemical shifts are of the order of only 6 ppm. Some ^{13}C anisotropies relevant in polymers are schematically summarized in Figure 2.14.

For ^{13}C nuclei, the chemical-shift tensor orientations, that is, the orientations of the principal axes, are often dominated by the local electronic structure around the site under consideration. As a consequence, local structural symmetries are reflected in the tensor orientations: if the electronic structure is nearly invariant under a certain symmetry operation, the same will hold for the chemical-shift tensor. In particular, if there exists a local plane of reflection, one principal axis must be perpendicular to it. This is the case for phenylene rings and carbonyl groups, where the most shielded axis is perpendicular to the planar structure. For X—CH_2—Y groups, the X–C–Y plane is a reflection plane, and one principal axis, usually relatively weakly shielded, is orthogonal to the plane. The H–C–H plane is a further reflection plane if the

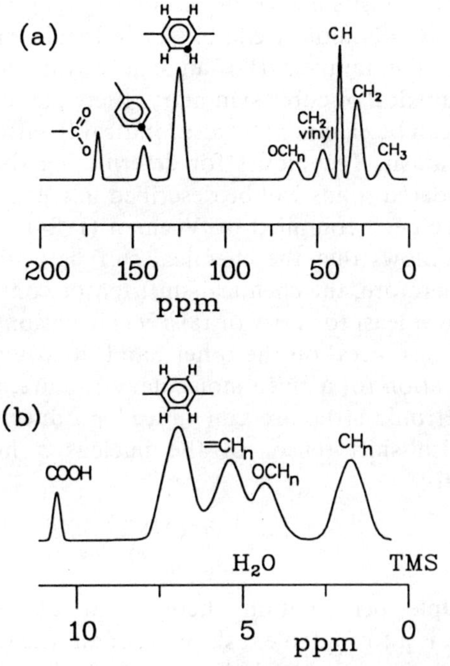

Figure 2.13 Schematic overview over isotropic shifts in organic polymers. (a) ^{13}C spectra, (b) ^{1}H spectra. Diagrams with the isotropic-shift ranges for many moieties can be found in the literature.

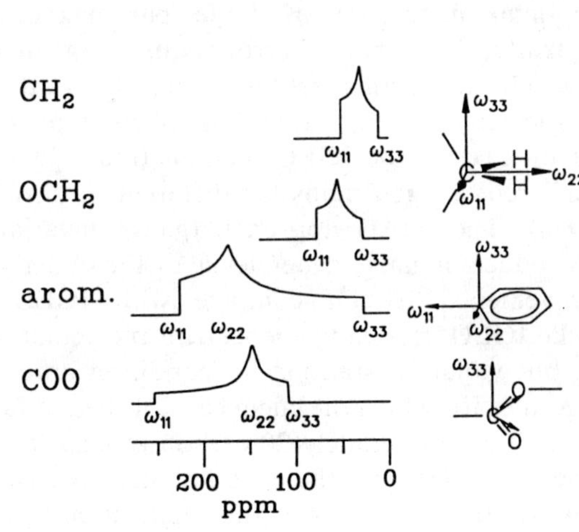

Figure 2.14 Schematic plot of typical ^{13}C chemical-shift anisotropies and tensor orientations in polymers. The ω_{33} axis is often perpendicular to the major symmetry plane of the segment. In carboxyl groups, the orientation of the ω_{33} axis is accordingly perpendicular to the O–C–O plane, while the orientation of the ω_{11} and ω_{22} axes is less well defined.

neighbouring groups X and Y are identical (for instance, other CH_2 groups in PE, or oxygens in POM) and the most shielded direction is perpendicular to it (i.e. along the local chain axis). The —C≡N moiety exhibits cylindrical symmetry around the triple bond, which is therefore the unique axis of an axially symmetric interaction tensor. In an environment of tetrahedral or cubic symmetry, there are more reflection planes and other symmetries than can be exhibited by a second-rank tensor, so that the anisotropy has no choice but to vanish. Techniques for determining shift-tensor orientations in single crystals and powdered solids will be described in Chapter 6. Measured ^{13}C shift tensor orientations have been compiled by Veeman (1984).

From the above it follows that the chemical shift is a direct consequence of the electronic structure. Therefore, the chemical-shift tensor contains information on that structure, and can serve at least to verify or falsify calculations of the binding structure of molecules. In many instances, on the other hand, it would be useful to be able to obtain the tensor orientation for a given molecular structure, and a sufficiently precise calculation of the electronic structure can indeed produce these values. In such a calculation, the chemical-shift tensor of the nucleus n in the $\vec{\mathbf{B}}_0$ field must be determined according to:

$$\sigma_{ij}^{(n)} = \frac{\partial^2 \langle \hat{H} \rangle}{\partial B_i \partial \mu_j^{(n)}}$$

in second-order multiple perturbation theory. An *ab initio* (i.e. full quantum mechanical) calculation is a formidable task. First of all, the wavefunctions with which to start the perturbation theory are only known approximately since the Schrödinger equation of the molecule cannot be solved exactly even for zero field and vanishing magnetic moments $\vec{\mu}_n$. Special approaches, such as the coupled Hartree–Fock (CHF) technique, must be used to deal with the shortcomings of the initial wave functions. In conventional CHF with a common gauge (Jackson, 1975) for all orbitals, the chemical shift is a small difference of large paramagnetic and diamagnetic contributions. Consequently, the numerical error is quite large unless fairly large basis sets are employed, which limits this rather straightforward approach to small molecules. More sophisticated techniques abandon the common gauge for all orbitals. The first such variant of CHF is the IGLO method (individual gauge for localized orbitals) which uses different gauge origins for different localized molecular orbitals (Kutzelnigg *et al.*, 1990). The GIAO approach (gauge invariant atomic orbitals), in spite of its name, introduces a gauge dependence of the underlying atomic orbitals. Due to recent improvements, it can compete with IGLO. LORG (localized orbitals/local origin) like IGLO uses gauge-dependent molecular orbitals. The LORG equations are simpler but violate a stationarity condition automatically fulfilled in IGLO. The largest molecules for which the chemical shift has so far been characterized by these techniques contain approximately 20 carbon atoms. It is promising that the differences between the theoretical and the experimental shift values are of the same order of magnitude as solvent shifts. For a thorough discussion of the virtues and drawbacks of the programs the reader is referred to the literature (Kutzelnigg *et al.*, 1990).

Equation (2.16) describes a linear dependence of the chemical shift on $|\vec{\mathbf{B}}_0|$. In liquids, it is empirically found that this linearity holds very precisely, although higher-order terms containing higher powers of B_0 do occur in the full theory. In solids, small nonlinear field-dependent shifts have been reported, which are related to heteronuclear dipolar couplings and become negligible at field strengths exceeding 5 tesla (VanderHart, 1986).

Conformational effects (Tonelli, 1989; Komoroski, 1986) play an important role in the chemical shifts in solids: they can contain some information, but also produce the dominant line-broadening in amorphous materials that even an increase of the B_0 field strength cannot overcome, since the broadening grows with the field. The best-documented and particularly useful conformational change of isotropic chemical shifts is the γ-gauche effect for methylene groups in aliphatic structures. The γ (third-neighbour) position from the $^{13}CH_2$ group under consideration is at the end of the β–γ bond that can be trans or gauche to the ^{13}C–α bond. For a gauche conformation, the signal is shifted upfield by about 5 ppm compared to the trans conformation. In a linear chain, there are two γ positions, and their γ-gauche effects add up, so that shifts by about 10 ppm or more can occur. Packing effects, in contrast, are found empirically to amount to only 1 ppm in aliphatics.

2.4.7 Intermediate motional regime

In the discussion on the effect of motion on the lineshape given in Section 2.4.4 it was assumed that the motion occurs so quickly that only averaged frequencies are measured. Between this fast-motion limit and the rigid limit we have an intermediate motional regime. It produces the most pronounced spectral features if just a few reorientations and concomitant frequency changes occur while the time signal is detected, as sketched in Figure 2.15(b), or equivalently if the exchange rate is comparable to the inhomogeneous width of the spectrum. This intermediate motional regime is more complicated to describe than the fast limit, while producing more

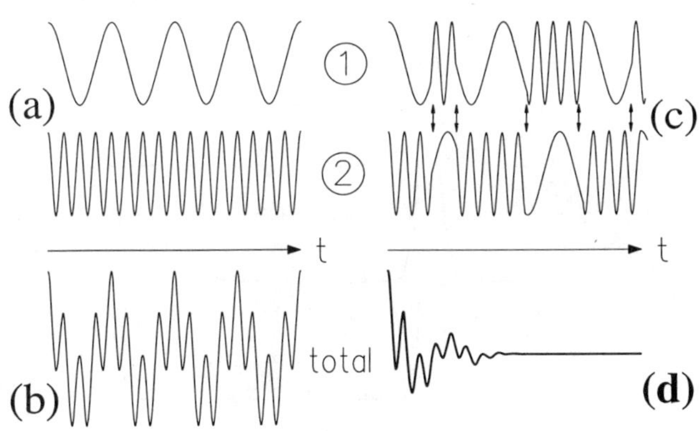

Figure 2.15 Schematic representation of the drastic effects of stochastic frequency changes on the time signal of spins exchanging between two sites with different NMR frequencies. For reference, (a) shows the individual time signals of the two sites 1 and 2 without motion and (b) their sum. (c) Signals of a specific pair of exchanging sites when the correlation time is of the order of the inverse of the two sites' frequency difference. The frequencies change stochastically. The experimentally observed time signal (d) is the weighted superposition of the immense number of different signals that arise when the frequency changes occur at other times. Due to destructive interference of the individual signals, the overall signal decays fast, and the corresponding spectral lines are broadened.

characteristic lineshapes. For demonstration, Figure 2.16 displays a series of calculated spectra for the process of methyl-group rotational three-site jumps as a function of the jump-rate k in units of the anisotropy parameter δ.

During the precession of the magnetization isochromats, the molecular orientation changes and thus the precession frequency is altered stochastically (see Figure 2.15(b)), which of course affects the spectrum drastically. For a simple analysis of these spectral changes, we consider the spin system in terms of magnetization isochromats $\mathcal{M}^j$ expressed by complex numbers where the real part represents the x component and the imaginary part the y component of the jth magnetization isochromat vector. The change with time of each magnetization isochromat $\mathcal{M}^j$ is due to two effects: the precession around the $\vec{\mathbf{B}}_0$ field:

$$\frac{\mathrm{d}\mathcal{M}^j}{\mathrm{d}t} = \mathrm{i}\omega_j \mathcal{M}^j \tag{2.61}$$

and the exchange between different isochromats:

$$\frac{\mathrm{d}\mathcal{M}^j}{\mathrm{d}t} = \sum_k \Pi_{jk} \mathcal{M}^k. \tag{2.62}$$

The Π_{jk} with $j \neq k$ are the transition rates from state k to state j, while Π_{jj} is the total transition rate from state j to all other states, with a negative sign. Exchange between

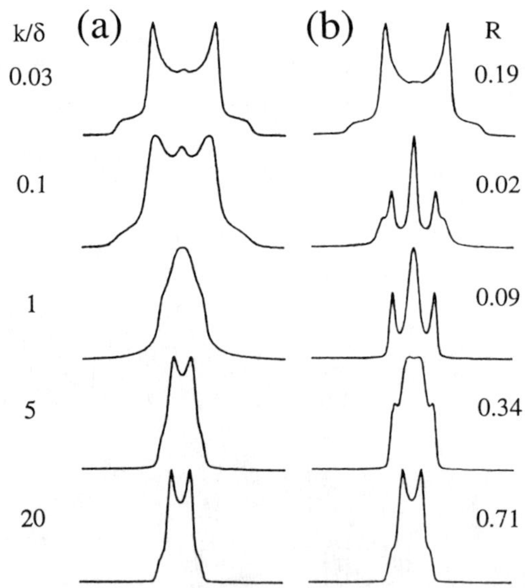

Figure 2.16 (a) ^{2}H quadrupolar line shapes for a three-site rotational jump of a methyl group as a function of the jump rate k in units of the anisotropy parameter δ. For very fast jumps, a Pake pattern results which is scaled by $-\frac{1}{3}$ compared to the rigid-limit spectrum (see Section 2.4.4). (b) Corresponding spectra calculated for data acquisition with the quadrupolar-echo pulse sequence (see Section 2.6.4). A reduction of the echo height arises due to incomplete refocusing of the echo. The reduction factors R are given next to the spectra. The range of rates which produce lineshape changes is increased compared with (a). The maxima of the spectra in the intermediate dynamic regime are located at frequencies that change little during the echo sequence so that their isochromats are refocused better than others.

N different sites can be characterized by such an $N \times N$ exchange matrix $\boldsymbol{\Pi}$ (for details see Section 9.4; if the dynamics is a Markov process, see Section 9.3). The two contributions (2.61) and (2.62) to $\mathrm{d}\mathcal{M}^j/\mathrm{d}t$ can be combined and written in vector and matrix format as:

$$\frac{\mathrm{d}\vec{\mathcal{M}}}{\mathrm{d}t} = (\mathrm{i}\boldsymbol{\omega} + \boldsymbol{\Pi})\vec{\boldsymbol{\mathcal{M}}} \tag{2.63}$$

where $\boldsymbol{\omega}$ is a diagonal matrix with $\omega_{jk} = \delta_{jk}\omega_j$. ($\vec{\boldsymbol{\mathcal{M}}}$ is written in script type since it is not a normal magnetization vector, but rather a large set of complex numbers $\mathcal{M}^j$, each of which represents an isochromat vector in the xy plane). The total magnetization $M(t)$ is the sum over all isochromats, $M = \Sigma_j\ \mathcal{M}^j$, which can be expressed as $\Sigma_j\ \mathcal{M}^j = \vec{\mathbf{1}}\cdot\vec{\boldsymbol{\mathcal{M}}}$, where $\vec{\mathbf{1}} = (1, 1, \ldots, 1)$. Thus, we obtain the formal solution:

$$M(t) = \vec{\mathbf{1}}\cdot[\mathrm{e}^{(\mathrm{i}\boldsymbol{\omega}+\boldsymbol{\Pi})t}\ \vec{\boldsymbol{\mathcal{M}}}(0)]\,. \tag{2.64}$$

The exponential matrix in this expression can be made tractable by means of diagonalization as described in Appendix D. The traditional more formal derivation of these relations has been described by Abragam (1961) and extended for the description of many motional processes in solid-state NMR by Spiess (1978). The exchange matrices for several motional processes are derived in Section 11.4 below.

2.5 Quantum-mechanical calculations of NMR time evolutions

So far we have discussed general Hamiltonians and the spectra that result for single-spin or two-spin interactions. The relationship between Hamiltonians and frequencies was introduced only on the basis of the semiclassical precession picture. In this section, we shall describe its quantum-mechanical foundation. The density-operator description of spin systems, which combines quantum mechanics and statistical mechanics, will be discussed, and it will be shown how the quantum-mechanical time evolutions can be calculated in practice.

2.5.1 Density operator $\hat{\rho}$ and magnetization

In elementary quantum mechanics, it is customary to characterize quantum-mechanical systems in terms of wave functions and spin functions (spinors, '$|\uparrow\rangle$' '$|\downarrow\rangle$') (Feynman *et al.*, 1965; Sakurai, 1985; Gerstein and Dybowski, 1985). For spins in magnetic fields and other few-level systems, it is, however, more convenient to describe the system in terms of the density operator $\hat{\rho}\ (t)$ (Sakurai, 1985).

In fact, for isolated $I = \frac{1}{2}$ spins, *the density operator behaves exactly like the magnetization* so that everything essential is already known. For $I \geqslant 1$, $\hat{\rho}$ can be regarded as an extended notion of magnetization: some of its components are observable magnetization, while others are unobservable terms.

The density operator represents a valid synthesis of quantum mechanics with statistical mechanics. Spin systems that are not completely polarized, and nuclear spins in NMR certainly rank with these, cannot be described in terms of simple spin-function solutions of the Schrödinger equation. All superpositions of such spinor functions represent perfectly polarized systems with only the spin orientation being different.

The density operator, in contrast, contains incoherent superpositions of such states which can account for any polarization or more complex spin orders.

All this is contained in the *ansatz* (to be justified below) that in thermal equilibrium at a temperature T and with the Hamiltonian $\hat{H}$ of the system, the density operator of the spin system is analogous to the classical Boltzmann distribution:

$$\hat{\rho}_{\text{eq}} = \frac{1}{Z}\,\mathrm{e}^{-\hat{H}/kT}$$
$$Z = \mathrm{Tr}(\mathrm{e}^{-\hat{H}/kT}). \tag{2.65a}$$

The exponential operator is defined by its series expansion according to:

$$\mathrm{e}^{\hat{A}} := \sum_{n=0}^{\infty} \frac{\hat{A}^n}{n!} = \hat{1} + \frac{\hat{A}}{1} + \frac{\hat{A}\cdot\hat{A}}{1\cdot 2} + \frac{\hat{A}\cdot\hat{A}\cdot\hat{A}}{1\cdot 2\cdot 3} + \cdots. \tag{2.65b}$$

For the nuclei of relevance to us, in B_0 fields of 1 tesla and greater, the dominant contribution to the spin Hamiltonian $\hat{H}$ is the Zeeman interaction (2.11):

$$\hat{H}_{\text{ze}} = -\vec{\mathbf{M}}\cdot\vec{\mathbf{B}}_0 = -\gamma\hbar\hat{I}_z B_0 = \hbar\omega_0\hat{I}_z.$$

(In order to obtain the unitless ratio $\hbar\gamma B_0/kT$ here, we reintroduce $\hbar$ into $\hat{H}$ for a moment. Nevertheless, for the calculations of quantum-mechanical time-dependences below, we shall return to the convenient convention of expressing $\hat{H}$ in frequency units.)

At temperatures above 1 K in the fields currently available, $|\hbar\gamma B_0/kT| \ll 1$ so that in the expansion of the exponential operator of equation (2.65) the quadratic and all higher terms are negligible compared with the linear term ('high temperature approximation'), hence:

$$\hat{\rho}_{\text{eq}} \sim \left(\hat{1} + \frac{\hbar\gamma B_0}{kT}\,\hat{I}_z\right). \tag{2.66}$$

The unity operator $\hat{1}$ commutes with all operators and is irrelevant in most cases. Therefore, we will often consider the partial density operator $\hbar\gamma B_0\hat{I}_z/kT$ in equation (2.66) simply as *the* density operator as such.

According to the basic density-operator theory outlined in many texts (Buckley *et al.*, 1975; Slichter, 1980; Gerstein and Dybowski, 1985; Sakurai, 1985; Ernst *et al.*, 1987), the physical observables, or the thermodynamic averages of the quantum-mechanical expectation values of operators, are obtained from the density operator by taking the following trace:

$$\langle\hat{A}\rangle = \mathrm{Tr}(\boldsymbol{\rho}\mathbf{A}) := \sum_i (\boldsymbol{\rho}\mathbf{A})_{ii} = \sum_i \sum_j \rho_{ij}A_{ji}. \tag{2.67}$$

ρ_{ij} and A_{ij} can be the elements of any (consistent) matrix representation $\boldsymbol{\rho}$ and $\mathbf{A}$ of the operators $\hat{\rho}$ and $\hat{A}$, since the trace (which is the sum of the diagonal elements of the matrix) is always independent of the representation. With the energy eigenstates $|\varepsilon_j\rangle$ as the basis, we have:

$$A_{ij} = \langle\varepsilon_i|A|\varepsilon_j\rangle. \tag{2.68}$$

In this particular basis, the representation **H** of the Hamiltonian is diagonal and has the energy eigenvalues ε_j as its diagonal elements. Therefore, the matrix representation of the density operator of equation (2.65) simplifies to:

$$(\rho_{\mathrm{eq}})_{ij} = \frac{1}{Z} \langle \varepsilon_i \,|\, \mathrm{e}^{-\hat{H}/kT} \,|\, \varepsilon_j \rangle = \frac{\mathrm{e}^{-\varepsilon_j/kT}\,\delta_{ij}}{\Sigma_k\, \mathrm{e}^{-\varepsilon_k/kT}} \tag{2.69}$$

(see Appendix D.4). With (2.68) and (2.69), equation (2.67) takes the appealing form:

$$\langle \hat{A} \rangle = \frac{\Sigma_j \langle \varepsilon_j \,|\, \hat{A} \,|\, \varepsilon_j \rangle\, \mathrm{e}^{-\varepsilon_j/kT}}{\Sigma_k\, \mathrm{e}^{-\varepsilon_k/kT}} \tag{2.70}$$

where the quantum-mechanical expectation values $\langle \varepsilon_j \,|\, \hat{A} \,|\, \varepsilon_j \rangle$ for the energy eigenstates are weighted with the thermodynamical Boltzmann factors $\exp(-\varepsilon_j/kT)$. This justifies the Boltzmann *ansatz* (2.65) for the density operator.

In NMR, the only observables of interest are magnetizations like:

$$\gamma\hbar \langle \hat{\mathbf{I}} \rangle := \gamma\hbar \begin{bmatrix} \langle \hat{I}_x \rangle \\ \langle \hat{I}_y \rangle \\ \langle \hat{I}_z \rangle \end{bmatrix}.$$

Taking the trace (2.67) (e.g., using the Pauli matrices as representations $\mathbf{I}_\alpha$ of the spin operators in the case of $I = \frac{1}{2}$; see Section 2.5.3), one can prove that a density operator $\hat{\rho}^{(z)} = c\hat{I}_z$ corresponds to magnetization along the z axis, $\hat{\rho}^{(x)} = c\hat{I}_x$ to x magnetization, etc. (The trace (2.67) of $\mathbf{I}_\alpha$ with terms of $\boldsymbol{\rho}$ that contain products of several spin operators in general vanishes, unless they 'collapse' like $\hat{I}_x\,(\hat{I}_x^2 + \hat{I}_y^2 + \hat{I}_z^2) = \hat{I}_x I(I+1)$, and therefore these product terms do not represent observable magnetization.)

2.5.2 The action of Hamiltonians on $\hat{\rho}$: time evolution

The density operator $\hat{\rho}(t)$ describes the state of the spin system, while the Hamiltonians represent the 'forces', or interactions, that try to change the state of the system. They are connected by the von Neumann equation (Sakurai, 1985):

$$\frac{\mathrm{d}}{\mathrm{dt}}\, \hat{\rho} = -\mathrm{i}\,[\hat{H}, \hat{\rho}] \tag{2.71}$$

which represents a generalization of the Schrödinger equation, which with our conventions takes a similar form:

$$\frac{\mathrm{d}}{\mathrm{d}t}\, \psi = -\mathrm{i}\hat{H}\psi. \tag{2.72}$$

If $\hat{H}$ and $\hat{\rho}$ commute, according to equation (2.71) the system characterized by $\hat{\rho}$ does not change with time. For instance, the first term of the density operator in equation (2.66), $\sim \hat{1}$, commutes with any $\hat{H}$, is therefore time invariant, and without further relevance.

For a Hamiltonian $\hat{H}$ acting constantly without a time dependence of its own ($\partial \hat{H}/\partial t = 0$), a formal solution of the von Neumann equation is given by:

$$\hat{\rho}(t) = \mathrm{e}^{-\mathrm{i}\hat{H}t}\hat{\rho}(0)\, \mathrm{e}^{\mathrm{i}\hat{H}t}. \tag{2.73}$$

This can be confirmed by inserting expression (2.73) into the von Neumann equation (2.71) (use the product rule for the derivative, without changing the order of the operators). The operator:

$$\hat{U}(t) = e^{-i\hat{H}t} \tag{2.74}$$

which 'pushes the density operator ahead in time' according to equation (2.73), is often termed a 'propagator'.

To evaluate the formal solution (2.73), one makes use of the Baker–Hausdorff relation ('time-evolution series') (Sakurai, 1985; Gerstein and Dybowski, 1985):

$$\begin{aligned}\hat{\rho}(t) &= e^{-i\hat{H}t}\hat{\rho}_0\, e^{i\hat{H}t}\\ &= \hat{\rho}_0 + [-i\hat{H}t, \hat{\rho}_0] + \tfrac{1}{2}[-i\hat{H}t, [-i\hat{H}t, \hat{\rho}_0]] + \tfrac{1}{6}[-i\hat{H}_t, [-i\hat{H}t, \hat{\rho}_0]]] + \cdots\\ &= \hat{\rho}_0 - \frac{it}{1!}\underbrace{[\hat{H}, \hat{\rho}_0]}_{=:\, r_1} + \frac{(it)^2}{2!}\underbrace{[\hat{H}, \hat{r}_1]}_{=:\, \hat{r}_2} - \frac{(it)^3}{3!}\underbrace{[\hat{H}, \hat{r}_2]}_{=:\, \hat{r}_3} + \cdots\end{aligned} \tag{2.75}$$

(the proof follows from expansion of the exponential operators according to (2.65b), and combination of terms with equal powers of t in the product). The last line of (2.75) shows that the cumulation of the commutators in the middle line can be avoided by calculating the commutators sequentially, exploiting $\hat{r}_{n+1} = [\hat{H}, \hat{r}_n]$. The infinite series of commutators in (2.75), although hardly tractable for complicated spin systems, proves to be quite harmless when:

$$\hat{r}_2 = \omega^2\hat{\rho}_0 \tag{2.76}$$

where ω^2 is a positive constant; this condition (2.76) holds in several cases of practical importance. In this case, one has $\hat{r}_3 = [\hat{H}, \hat{\rho}_0\omega^2] = \hat{r}_1\omega^2$, and the other multiple commutators also produce only terms proportional to $\hat{\rho}_0$ or to $\hat{r}_1$. This results in:

$$\begin{aligned}\hat{\rho}(t) &= \hat{\rho}_0\left(1 + \frac{(i\omega t)^2}{2!} + \frac{(i\omega t)^4}{4!} + \cdots\right) - \frac{\hat{r}_1}{\omega}\left(\frac{(i\omega t)^1}{1!} + \frac{(i\omega t)^3}{3!} + \cdots\right)\\ &= \hat{\rho}_0 \cos(\omega t) - i\,\frac{\hat{r}_1}{\omega}\sin(\omega t)\end{aligned} \tag{2.77}$$

(see the cosine and sine series (A.26) and (A.27)). Since $\hat{\rho}(t)$ is generated by complex exponential expressions $\exp(-i\hat{H}t)...\exp(i\hat{H}t)$ in (2.73) and recalling $\exp(-i\omega t) = \cos(\omega t) - i\sin(\omega t)$, the result (2.77) is actually not too surprising. Such an oscillatory behaviour results for magnetization under the action of strong $\vec{\mathbf{B}}$ fields (this is the well-known precession) as well as for the time evolution of $I = 1$ spins under the quadrupolar interaction and of isolated pairs of $I = \frac{1}{2}$ dipoles.

For instance, with $\hat{\rho}(0) = \hat{\rho}_0 = \hat{I}_x$ we can calculate the action of the $\vec{\mathbf{B}}_0$ field and the chemical shift with their Hamiltonian of:

$$\begin{aligned}\hat{H} = \hat{H}_0 + \hat{H}_{cs} = \omega_0(1 - \sigma_{zz}^{LF})\hat{I}_z = \omega_L\hat{I}_z\\ \omega_L := \omega_0(1 - \sigma_{zz}^{LF}).\end{aligned} \tag{2.78}$$

To evaluate the time-evolution series (2.75), we need the commutators:

$$\hat{r}_1 = [\hat{H}, \hat{\rho}_0] = \mathrm{i}\omega_L \hat{I}_y$$

$$\hat{r}_2 = [\hat{H}, \hat{r}_1] = \omega_L^2 \hat{I}_x$$

(see the commutators of equation (2.23)). Since $\hat{r}_2 = \omega_L^2 \hat{\rho}_0$ fulfils condition (2.76), we can use relation (2.77) to obtain:

$$\hat{\rho}(t) = \hat{I}_x \cos(\omega_L t) + \hat{I}_y \sin(\omega_L t). \tag{2.79}$$

This represents a quantum-mechanical derivation of the precession of the magnetization around the $\vec{\mathbf{B}}_0$ field, with the effective Larmor frequency ω_L. Derived in a fully quantum-mechanical calculation, equation (2.79) represents the valid basis for the semiclassical picture of the precessing nuclear magnets as introduced in Section 2.2. In the following sections, further examples for the calculation of the quantum-mechanical time evolution, under quadrupolar or two-spin dipolar couplings, will be given. These will serve as the basis for many other time signals and spectral features discussed in the context of one-dimensional and multidimensional experiments.

2.5.3 *Matrix representations of spin operators*

To calculate properties of simple spin systems, it is often convenient to express the spin operators $\hat{I}^i_\alpha$ in terms of appropriate matrices. For $I = \frac{1}{2}$, these are essentially the Pauli spin matrices:

$$\mathbf{I}_z = \tfrac{1}{2}\begin{bmatrix} 1 & 0 \\ 0 & -1 \end{bmatrix} \qquad \mathbf{I}_x = \tfrac{1}{2}\begin{bmatrix} 0 & 1 \\ 1 & 0 \end{bmatrix} \qquad \mathbf{I}_y = \tfrac{1}{2}\begin{bmatrix} 0 & -\mathrm{i} \\ \mathrm{i} & 0 \end{bmatrix} \tag{2.80a}$$

$$\mathbf{I}_x^2 = \mathbf{I}_y^2 = \mathbf{I}_z^2 = \tfrac{1}{4}\begin{bmatrix} 1 & 0 \\ 0 & 1 \end{bmatrix} =: \tfrac{1}{4}\mathbf{1} \tag{2.80b}$$

$$\mathbf{I}_x\mathbf{I}_y = \tfrac{1}{2}\mathrm{i}\mathbf{I}_z \qquad \mathbf{I}_y\mathbf{I}_x = -\tfrac{1}{2}\mathrm{i}\mathbf{I}_z \qquad \mathbf{I}_y\mathbf{I}_z = \tfrac{1}{2}\mathrm{i}\mathbf{I}_x \text{ etc.} \tag{2.80c}$$

For general spin I, the elements $(\mathbf{I}_\alpha)_{m',m} = \langle m' \,|\, \hat{I}_\alpha \,|\, m \rangle$ of the $(2I+1) \times (2I+1)$ matrices in the basis of $\hat{I}_z$ eigenstates $|\, m\rangle$ can be derived from basic bracket relations:

$$(\mathbf{I}_z)_{m',m} = \langle m' \,|\, \hat{I}_z \,|\, m\rangle = \langle m' \,|\, m \,|\, m\rangle = m\delta_{m',m} \tag{2.81a}$$

$$(\mathbf{I}_x \pm \mathrm{i}\mathbf{I}_y)_{m',m} = \langle m' \,|\, \hat{I}^{\pm} \,|\, m\rangle = \sqrt{I(I+1) - m(m \pm 1)}\; \delta_{m',m\pm 1}$$

$$\Rightarrow (\mathbf{I}_x)_{m',m} = (\pm\, \mathrm{i}\mathbf{I}_y)_{m',m} = \tfrac{1}{2}\sqrt{I(I+1) - m(m \pm 1)}\; \delta_{m',m\pm 1} \tag{2.81b}$$

where the orthonormality $\langle m \,|\, m'\rangle = \delta_{m,m'}$ was used. (Note that the indices m, m' used here for labelling the matrix elements are the eigenvalues of $\hat{I}_z$ and can therefore be negative or half-integer; for instance, for spin-$\frac{1}{2}$ they are $-\frac{1}{2}$ and $\frac{1}{2}$). The matrix representations of spin operators for $I = 1$ and $I = \frac{3}{2}$ will be explicitly given in Sections 2.5.5 and 3.10.

For a pair of coupled spins, one can work with representations in terms of the product basis $|\, m_1 m_2\rangle$. The matrix elements for a spin operator $\hat{I}^j_\alpha$ of spin j are then obtained as $\langle \tilde{m}_1 \tilde{m}_2 \,|\, \hat{I}^j_\alpha \,|\, m_1\ m_2\rangle$ analogous to the relations (2.81). However, these matrices are relatively large (4×4 for the simplest case of two spin-$\frac{1}{2}$). Below, for coupled spins we shall instead work with products of single-spin operators and matrices, which are more convenient to write down and produce more general results in certain cases (e.g. independent of the spin-quantum number of one spin).

2.5.4 Time evolution of the density matrix

By means of the matrix representations of spin operators as introduced in the previous section, we can express Hamiltonians and the density operator in terms of matrices. (It should be noted that these depend on the spin-quantum numbers of the nuclei concerned, so the calculations are less general than those expressed only in terms of spin operators on the basis of equations (2.75)–(2.77).) Using these matrices, the time evolution of spin systems can be calculated according to various strategies:

(a) Evaluate the commutators of the time-evolution series (2.75) in terms of the matrices. This is straightforward but sometimes not very elegant.
(b) Calculate with general $\hat{I}^i_\alpha$ operators according to relation (2.75) as far as possible, and make use of matrix properties, such as those listed in (2.80c) for $I = \frac{1}{2}$, when specific expressions are required. This will be demonstrated in Section 2.5.6 for the dipolar coupling between two spins.
(c) Calculate directly on the basis of the fundamental time-evolution equation (2.73) written in terms of matrices:

$$\boldsymbol{\rho}(t) = \mathrm{e}^{-\mathrm{i}\mathbf{H}t}\boldsymbol{\rho}(0)\mathrm{e}^{\mathrm{i}\mathbf{H}t}. \tag{2.82}$$

All truncated Hamiltonians are diagonal matrices in the basis of $\hat{I}_z$ eigenfunctions, since the truncation or first-order perturbation theory represents a selection of exactly those elements. Therefore, in the absence of rf fields (which are nondiagonal in this representation), the relevant Hamiltonian is diagonal. This considerably simplifies the calculation of the matrix product (2.82). According to Appendix D.4, the exponential of a diagonal matrix is a diagonal matrix of exponentials:

$$(\mathrm{e}^{-\mathrm{i}\mathbf{H}t})_{m,m'} = \mathrm{e}^{-\mathrm{i}(\mathbf{H})_{mm}t}\,\delta_{m,m'}.$$

On these grounds, the matrix product (2.82) can be easily evaluated:

$$\{\boldsymbol{\rho}(t)\}_{m,n} = \sum_j \sum_k \mathrm{e}^{-\mathrm{i}(\mathbf{H})_{mm}t}\,\delta_{mj}\{\boldsymbol{\rho}(0)\}_{jk}\,\mathrm{e}^{\mathrm{i}(\mathbf{H})_{nn}t}\,\delta_{kn}.$$

This produces the important density-matrix time-evolution formula:

$$\{\boldsymbol{\rho}(t)\}_{m,n} = \{\boldsymbol{\rho}(0)\}_{m,n}\,\mathrm{e}^{-\mathrm{i}\{(\mathbf{H})_{mm}-(\mathbf{H})_{nn}\}t}. \tag{2.83}$$

This formula proves a fundamental fact of quantum-mechanics: the oscillation frequencies $\omega_{mn} = (\mathbf{H})_{mm} - (\mathbf{H})_{nn}$ of transverse magnetization, etc., are equal to the energy-level differences (transition energies) $\langle m|\hat{H}|m\} - \langle n|\hat{H}|n\rangle = (\mathbf{H})_{mm} - (\mathbf{H})_{nn}$. This relation was already alluded to in Section 2.3 and is often tacitly used in the discussion of NMR spectra in terms of energy-level diagrams.

As a first example, we consider $I=\frac{1}{2}$, $\mathbf{H}=\omega_0\mathbf{I}_z$: $(\mathbf{H})_{\frac{1}{2},\frac{1}{2}}=\frac{1}{2}\omega_0$, $(\mathbf{H})_{-\frac{1}{2},\frac{1}{2}}=-\frac{1}{2}\omega_0$ and thus according to (2.83):

$$\boldsymbol{\rho}(0)=\mathbf{I}_x=\tfrac{1}{2}\begin{bmatrix}0 & 1\\ 1 & 0\end{bmatrix} \xrightarrow{\mathbf{H}=\begin{bmatrix}\frac{1}{2}\omega_0 & 0\\ 0 & -\frac{1}{2}\omega_0\end{bmatrix}}$$

$$\boldsymbol{\rho}(t)=\tfrac{1}{2}\begin{bmatrix}0 & 1\cdot e^{-i\omega_0 t}\\ 1\cdot e^{i\omega_0 t} & 0\end{bmatrix}=\mathbf{I}_x\cos\omega_0 t+\mathbf{I}_y\sin\omega_0 t.$$

This produces again the precession of magnetization. For many purposes, in particular for writing down multidimensional time signals, it is convenient to make a transformation of the following kind between the magnetization and the observed signal:

$$\hat{\rho}(t)=\hat{I}_x\cos\omega_0 t+\hat{I}_y\sin\omega_0 t \quad\rightarrow\quad f(t)=\cos\omega_0 t+i\sin\omega_0 t=\exp(i\omega_0 t). \qquad (2.84)$$

This is a special case of the formal relation between $\hat{\rho}(t)$ and the complex NMR signal $f(t)$, which is obtained by taking the trace of $\boldsymbol{\rho}(t)$ with $\mathbf{I}^+=\mathbf{I}_x+i\mathbf{I}_y$:

$$f(t)\sim \mathrm{tr}\{\boldsymbol{\rho}(t)\mathbf{I}^+\}. \qquad (2.85)$$

The NMR spectrum can be generated by Fourier transformation (Ernst and Anderson, 1966) of the superposition $\langle\exp(i\omega_L t)\rangle$ of the complex time signals generated by the various isochromats in the sample. Details of Fourier transformation in one and two dimensions are treated in Chapter 4.

2.5.5 ^{2}H quadrupolar interaction

The quadrupolar interaction of ^{2}H spins plays an important role in solid-state NMR of polymers. The corresponding evolution of x magnetization (possibly generated by a $-y$ pulse from the Boltzmann z magnetization) can be calculated conveniently by means of the theoretical relations of Sections 2.5.2 or 2.5.4.

The time dependence of the density matrix can be formulated quite elegantly in terms of spherical tensor operators as listed in Appendix C. However, we will rather derive it by means of a matrix representation here, since that calculation can be performed with a minimum of theoretical background and shows much more clearly how the extra prefactor of 3 in the quadrupolar anisotropy parameter δ_Q arises [cf. equation (2.38)].

For $I=1$, the representations of the $\hat{I}_\alpha$ operators in a basis of $\hat{I}_z$ eigenstates $|1\rangle, |0\rangle$, and $|-1\rangle$ are:

$$\mathbf{I}_z=\begin{bmatrix}1 & 0 & 0\\ 0 & 0 & 0\\ 0 & 0 & -1\end{bmatrix} \quad \mathbf{I}_x=\sqrt{\tfrac{1}{2}}\begin{bmatrix}0 & 1 & 0\\ 1 & 0 & 1\\ 0 & 1 & 0\end{bmatrix} \quad \mathbf{I}_y=\sqrt{\tfrac{1}{2}}\begin{bmatrix}0 & -i & 0\\ i & 0 & -i\\ 0 & i & 0\end{bmatrix} \qquad (2.86)$$

according to equation (2.81). Note the structural similarity of these $I=1$ matrices to the Pauli 2×2 matrices (2.80) (which represent the spin operators for spin $I=\frac{1}{2}$ with its two basis states $|\frac{1}{2}\rangle$ and $|-\frac{1}{2}\rangle$).

From these matrices, all the other spin operators for $I = 1$ can be generated by straightforward matrix multiplications, employing the relations between the spin operators (see Table C.1). In particular, the quadrupolar Hamiltonian matrix reads:

$$\mathbf{H}_Q = \frac{\omega_Q}{3}(3\mathbf{I}_z\mathbf{I}_z - I(I+1)\mathbf{1}) = \frac{\omega_Q}{3}\left(3\begin{bmatrix}1&0&0\\0&0&0\\0&0&1\end{bmatrix} - 2\begin{bmatrix}1&0&0\\0&1&0\\0&0&1\end{bmatrix}\right)$$

$$= \frac{\omega_Q}{3}\begin{bmatrix}1&0&0\\0&-2&0\\0&0&1\end{bmatrix} \tag{2.87}$$

where the prefactor of the matrix is given by:

$$\frac{\omega_Q}{3} = \frac{eQeq}{4\hbar}\,\tfrac{1}{2}(3\cos^2\theta - 1 - \eta_Q \sin^2\theta \cos 2\phi)$$

(see equations (2.22) and (2.43)). According to (2.83), the possible oscillation or precession frequencies of the magnetization are the differences between the diagonal elements of $\mathbf{H}_Q$ in equation (2.87), or equivalently between the corresponding energy levels. In the Hamiltonian matrix (2.87), these differences are: ω_Q for $|1\rangle \to |0\rangle$, $-\omega_Q$ for $|0\rangle \to |-1\rangle$, and 0 for $|1\rangle \to |-1\rangle$ (evolution with this last $\omega = 0$ will not appear for the magnetization since the corresponding matrix elements $\langle -1|\hat{I}_\alpha|1\rangle$ and $\langle 1|\hat{I}_\alpha|-1\rangle$ vanish). The full time dependence is obtained by the calculation according to the density-matrix evolution formula (2.83) for an initial density matrix $\boldsymbol{\rho}(0) = \mathbf{I}_x$:

$$\boldsymbol{\rho}(0) = \mathbf{I}_x = \sqrt{\tfrac{1}{2}}\begin{bmatrix}0&1&0\\1&0&1\\0&1&0\end{bmatrix} \xrightarrow{\mathbf{H}_Q}$$

$$\boldsymbol{\rho}(t) = \sqrt{\tfrac{1}{2}}\begin{bmatrix}0 & \exp(-\mathrm{i}\omega_Q t) & 0\\ \exp(\mathrm{i}\omega_Q t) & 0 & \exp(\mathrm{i}\omega_Q t)\\ 0 & \exp(-\mathrm{i}\omega_Q t) & 0\end{bmatrix}. \tag{2.88}$$

To obtain $\boldsymbol{\rho}(t)$, each element $\{\rho(0)\}_{mn}$ of the initial density matrix has simply been multiplied by $\exp\{-\mathrm{i}(\mathbf{H}_{mm} - \mathbf{H}_{nn})t\}$. By taking the trace of $\boldsymbol{\rho}(t)\mathbf{I}_x$ and $\boldsymbol{\rho}(t)\mathbf{I}_y$, it is found that the y signal vanishes for all times, while the x signal oscillates with $\cos(\omega_Q t)$. The full time evolution can be better appreciated if we rewrite the density matrix (2.88) (after splitting the matrix elements according to $\exp(\mathrm{i}\omega_Q t) = \cos(\omega_Q t) + \mathrm{i}\sin(\omega_Q t)$:

$$\boldsymbol{\rho}(t) = \mathbf{I}_x \cos\omega_Q t - \mathrm{i}\,\frac{\mathbf{r}_1}{\omega_Q}\sin\omega_Q t \tag{2.89}$$

where $\mathbf{r}_1$ is identical to the first commutator appearing in the time-evolution series (2.75):

$$\mathbf{r}_1 = [\mathbf{H}_Q, \mathbf{I}_x] = \mathbf{H}_Q\mathbf{I}_x - \mathbf{I}_x\mathbf{H}_Q = \omega_Q\sqrt{\tfrac{1}{2}}\begin{bmatrix}0&1&0\\-1&0&-1\\0&1&0\end{bmatrix}.$$

The term $\mathbf{r}_1$ is not observable, $\mathrm{tr}(\mathbf{I}_\alpha \mathbf{r}_1) = 0$ (check by summing up the diagonal elements of the $\mathbf{I}_\alpha \mathbf{r}_1$ product matrix). Thus, it is clear that the observable signal oscillates as:

$$\cos(\omega_Q t) = \tfrac{1}{2}\exp(\mathrm{i}\omega_Q t) + \tfrac{1}{2}\exp(-\mathrm{i}\omega_Q t) \overset{\mathrm{FT}}{\rightarrow} \tfrac{1}{2}\delta(\omega - \omega_Q) + \tfrac{1}{2}\delta(\omega + \omega_Q) \qquad (2.90)$$

(Figure 2.17). The corresponding spectrum consists of two lines, at ω_Q and $-\omega_Q$ (relative to the Larmor frequency). These two lines are commonly referred to as the two 'transitions' of the ^{2}H quadrupolar interaction, although strictly speaking the time-domain quadrupolar evolution does not involve any transitions between energy levels ($E = -\hbar\gamma B_0 \langle \hat{I}_z \rangle = \mathrm{const.}$). The terminology originates from CW NMR where actually two transitions are observed at $+\omega_Q$ and $-\omega_Q$ relative to the Larmor frequency. The equality of these transition frequencies and the oscillation frequencies is proved by equation (2.83).

According to equation (2.89), the quadrupolar Hamiltonian does not produce a precession of the observable ^{2}H magnetization, but rather an oscillation of the length of the observable magnetization along its initial direction. This is an important distinction from the precessional behaviour observed under the action of Hamiltonians linear in $\hat{I}_z$ (with only one $\hat{I}_z$ factor). Nevertheless, it is correct to state that the observable ^{2}H magnetization (including the invariant z component) of equation (2.88) behaves just like the xz projection vector of a virtual magnetization precessing with angular frequency ω_Q around the z axis; a y component (i.e. perpendicular to $\vec{\mathbf{B}}_0$ and to the direction of the initial magnetization) is not detectable. These relations are indicated in Figure 2.17.

The angular dependence $\omega_Q(\theta)$ and its potential for studying orientation and dynamics of molecular segments was discussed in Section 2.4 and has been exploited in multidimensional exchange experiments to be described below. The treatment given in this section describes the basis of the appearance of mirror-symmetric signals (two 'transitions') which are characteristic for both one-dimensional and two-dimensional ^{2}H quadrupolar spectra.

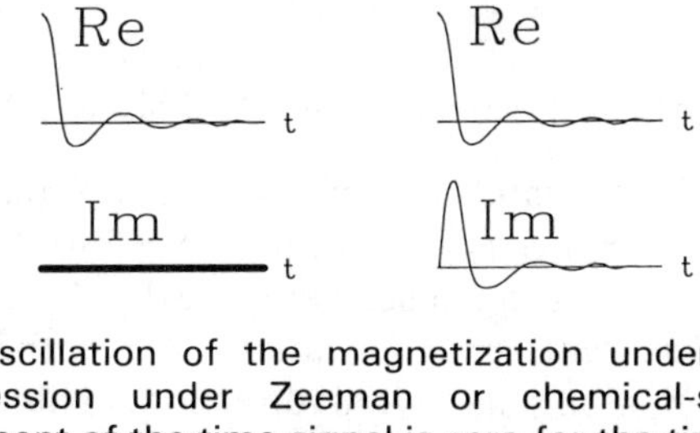

Figure 2.17 Harmonic oscillation of the magnetization under quadrupolar or dipolar couplings, versus precession under Zeeman or chemical-shift interaction. As a consequence, one component of the time signal is zero for the time signals evolving under pure quadrupolar or dipolar interactions. Sometimes this is explained in terms of two counter-rotating vectors, which would correspond to the two lines ('transitions') in the spectrum.

Equation (2.89) describes the undisturbed evolution of the ^{2}H magnetization under the action of the quadrupolar Hamiltonian. If rf pulses are applied at certain times, the density matrix has to be recalculated. It should be noted that the $\mathbf{r}_1$ term, though originally 'invisible', must not be neglected in that case, since a pulse followed by quadrupolar interaction can transform it into detectable magnetization. The most widely used pulse sequence, generating a solid echo (or quadrupole echo), will be analysed in Section 2.6.4 where we again exploit the matrix representation of the spin operators. At a certain point of complexity of the pulse sequence, however, the treatment in terms of spherical tensor operators becomes more convenient, in particular when the action of rf pulses on the terms appearing in the density operator is to be determined. The necessary relations are compiled in Appendix C; examples for their applications are provided by the derivation of two-dimensional time-domain signals in Section 10.2.4.

2.5.6 Heteronuclear dipolar couplings

Heteronuclear dipolar coupling of two spins I and S is based on the Hamiltonian:

$$\hat{H}_{IS} = 2\omega_{IS}\hat{I}_z\hat{S}_z = \hat{I}_z D_{zz}\hat{S}_z \tag{2.91a}$$

with a coupling frequency:

$$\omega_{IS} = -\frac{\mu_0\hbar}{4\pi}\gamma_I\gamma_S\,\tfrac{1}{2}(3\cos^2\theta_{IS} - 1)/r_{IS}^3 \tag{2.91b}$$

(see equation (2.42)). Starting from Boltzmann equilibrium, after a 90° $-y$ pulse on the S spin, the relevant part of the density operator is $\hat{\rho}(0) = c_S\hat{S}_x + c_I\hat{I}_z$. Because of $[\hat{H}_{IS}, \hat{I}_z] = 0$, $\hat{I}_z$ can be disregarded in the density operator. Generally, with this form of $\hat{H}$, the density operator of the I spins does not affect the behaviour of the S spins, and vice versa. Hamiltonians that do cause a transfer of magnetization between spins, e.g. cross polarization or spin diffusion, will be discussed in section 3.4.3. The commutators required for evaluating the time evolution formula (2.75) are:

$$\begin{aligned}\hat{r}_1 &= [\hat{H}_{IS}, \hat{S}_x] = \mathrm{i}2\omega_{IS}\hat{I}_z\hat{S}_y \\ \hat{r}_2 &= [\hat{H}_{IS}, \hat{r}_1] = 4\omega_{IS}^2\hat{I}_z^2\hat{S}_x.\end{aligned} \tag{2.92}$$

With $I = \frac{1}{2}$, according to (2.80b) we have $\mathbf{I}_z^2 = \frac{1}{4}$ and thus $\hat{r}_2 = \omega_{IS}^2\hat{S}_x$. This fulfils the condition (2.76) for applying the simplified time-evolution formula (2.77), which produces:

$$\hat{\rho}(t) = \hat{S}_x\cos(\omega_{IS}t) + \hat{S}_y\mathbf{I}_z\,2\sin(\omega_{IS}t). \tag{2.93}$$

This expression is obviously valid for an arbitrary S quantum number. Due to $\mathrm{tr}(\mathbf{I}_z) = 0$, it is found that $\mathrm{tr}(\mathbf{S}_\alpha\mathbf{I}_z\mathbf{S}_y) = \mathrm{tr}(\mathbf{I}_z)\,\mathrm{tr}(\mathbf{S}_\alpha\mathbf{S}_y) = 0$ so that only the $\hat{S}_x$ component represents observable magnetization. Therefore, the dipolar coupling does not produce a precession of the observable magnetization, but rather an oscillation of the observable magnetization along its initial direction. This behaviour is completely analogous to that of ^{2}H magnetization under the action of the quadrupolar Hamiltonian as treated in the previous section. Therefore, the visualization given in Figure 2.17 can also be applied to the magnetization of a heteronuclear spin pair subject to dipolar coupling. The corresponding spectrum for a single orientation is the Fourier transform of the time signal:

$$\cos(\omega_{IS}t) = \tfrac{1}{2}\exp(\mathrm{i}\omega_{IS}t) + \tfrac{1}{2}\exp(-\mathrm{i}\omega_{IS}t)$$

and therefore consists of two lines, at ω_{IS} and $-\omega_{IS}$.

In the majority of cases, the study of heterodipolar couplings in polymers does not involve fairly isolated C–H spin pairs but rather CH_2 and CH_3 groups, in which two or three protons are coupled to the same carbon (the ^{1}H–^{1}H couplings are interrupted by homonuclear decoupling). For a CH_2 group with protons A and B, the Hamiltonian reads:

$$\hat{H}_{SI} = \underbrace{\hat{S}_z\, 2\omega_A \hat{I}_z^A}_{\hat{H}^A} + \underbrace{\hat{S}_z\, 2\omega_B \hat{I}_z^B}_{\hat{H}^B} \tag{2.94}$$

where:

$$\omega_A = \tfrac{1}{2}D_{zz}^A = -\frac{\mu_0\hbar}{4\pi}\frac{\gamma^I\gamma^S}{r_A^3}\tfrac{1}{2}(3\cos^2\theta_A - 1) \tag{2.95}$$

according to equation (2.91); ω_B is defined correspondingly. The evolution of transverse S magnetization under this Hamiltonian is found by a calculation based on the results derived so far in this section. The operators $\hat{H}^A = \hat{S}_z\, 2\omega_A\hat{I}_z^A$ and $\hat{H}^B = \hat{S}_z\, 2\omega_B\hat{I}_z^B$ commute; therefore, we can split the propagator $\exp\{\mathrm{i}(\hat{H}^A + \hat{H}^B)t\} = \exp(\mathrm{i}\hat{H}^A t)\exp(\mathrm{i}\hat{H}^B t)$, so that:

$$\hat{\rho}(t) = e^{-\mathrm{i}\hat{H}^B t}\,\underbrace{e^{-\mathrm{i}\hat{H}^A t}\,\hat{\rho}(0)\,e^{\mathrm{i}\hat{H}^A t}}\,e^{\mathrm{i}\hat{H}^B t}. \tag{2.96}$$

The central part of the operator product can be evaluated as shown above for equation (2.93):

$$e^{-\mathrm{i}\hat{H}^A t}\,\hat{\rho}(0)\,e^{\mathrm{i}\hat{H}^A t} = \hat{S}_x\cos(\omega_A t) + \hat{S}_y\mathbf{I}_z^A\, 2\sin(\omega_A t). \tag{2.97}$$

Inserting (2.97) into (2.96) yields:

$$\begin{aligned}\hat{\rho}(t) &= e^{-\mathrm{i}\hat{H}^B t}\{\hat{S}_x\cos(\omega_A t) + 2\hat{S}_y\mathbf{I}_z^A\sin(\omega_A t)\}\,e^{\mathrm{i}\hat{H}^B t}\\ &\overset{(2.93)}{=} \cos(\omega_A t)\{\hat{S}_x\cos(\omega_B t) + 2\hat{S}_y\mathbf{I}_z^B\sin(\omega_B t)\} + \mathbf{I}_z^A(\ldots).\end{aligned} \tag{2.98}$$

Only the term oscillating with $\cos(\omega_A t)\cos(\omega_B t)$ represents observable S magnetization $\mathrm{tr}\{\hat{\rho}(t)\hat{S}^+\}$ (recall that $\mathrm{tr}(\mathbf{I}_\alpha^A) = 0$ and therefore $\mathrm{tr}\{\mathbf{I}_\alpha^A\, f(\hat{S}, \mathbf{I}_\beta^B)\hat{S}^+\} = 0$). This time-dependence can be rewritten as:

$$\begin{aligned}\cos(\omega_A t)\cos(\omega_B t) &= \tfrac{1}{2}(e^{\mathrm{i}\omega_A t} + e^{-\mathrm{i}\omega_A t})\tfrac{1}{2}(e^{\mathrm{i}\omega_B t} + e^{-\mathrm{i}\omega_B t})\\ &= \tfrac{1}{4}(e^{\mathrm{i}(\omega_A+\omega_B)t} + e^{\mathrm{i}(-\omega_A-\omega_B)t} + e^{\mathrm{i}(\omega_A-\omega_B)t} + e^{\mathrm{i}(-\omega_A+\omega_B)t}).\end{aligned} \tag{2.99}$$

Thus, for a single CH_2 group orientation a four-line spectrum results with two symmetric pairs of signals, at:

$$\pm(\omega_A + \omega_B) \text{ and } \pm(\omega_A - \omega_B).$$

relative to the Larmor frequency. The pairs of complex exponentials in equation (2.99) can be combined into cosine functions, so that we can also write:

$$\cos(\omega_A t)\cos(\omega_B t) = \tfrac{1}{2}\cos[(\omega_A + \omega_B)t] + \tfrac{1}{2}\cos[(\omega_A - \omega_B)t]$$

(see equation (A.7), which is actually proven analogous to equation (2.99)).

By superimposing the signals at $\pm(\omega_A + \omega_B)$ and $\pm(\omega_A - \omega_B)$, with ω_A and ω_B calculated according to equation (2.95) for all CH_2 orientations present in the sample, the total spectral pattern can be generated. However, the shape of the powder spectrum can be determined more directly after rewriting the frequencies in terms of the dipole tensor elements in the laboratory frame:

$$\omega_A + \omega_B = \tfrac{1}{2}D^A_{zz} + \tfrac{1}{2}D^B_{zz} = \tfrac{1}{2}(D^A + D^B)_{zz} \tag{2.100a}$$

$$\omega_A - \omega_B = \tfrac{1}{2}D^A_{zz} - \tfrac{1}{2}D^B_{zz} = \tfrac{1}{2}(D^A - D^B)_{zz}. \tag{2.100b}$$

Consequently, the powder spectrum can be viewed as the superposition of powder spectra governed by the second-rank tensors $\mathbf{D}^A + \mathbf{D}^B$ and $\mathbf{D}^A - \mathbf{D}^B$ which can be characterized in terms of their principal values obtained by diagonalization.

With $\eta = 0$ for the individual C–H dipole couplings, the diagonalization of the 'sum tensor' $\mathbf{D}^A + \mathbf{D}^B$ for two C—H bonds of equal lengths is mathematically similar to the calculation of the averaged tensor for two-site jumps and $\eta = 0$ as described in Section 2.4.4. The principal values are given by equation (2.57), where β is the angle between the C—H^A and C—H^B bonds. In particular, $\eta = 1$ results for the tetrahedral angle $\beta = 109^\circ$ which is typically encountered in CH_2 groups. The principal axes and spectrum of the sum tensor for $\beta = 109^\circ$ are displayed in Figure 2.18(a). For the difference tensor $\mathbf{D}^A - \mathbf{D}^B$, the principal axes also correspond to $\vec{\mathbf{B}}_0$ field directions for which the frequency is not changed by an exchange of A and B. This applies for the normal of the H–C–H plane, characterized by $\omega^d = \pm 0$. The other two axes must therefore be located in the H–C–H plane. The invariance under A–B exchange can be fulfilled only if $\omega_1 = -\omega_3$, that is, if $\eta = 1$ for any β, and the corresponding principal axes are disposed symmetrically $\pm 45^\circ$ from the H–C–H bisector (Terao *et al.*, 1986) (Figure 2.18(b)).

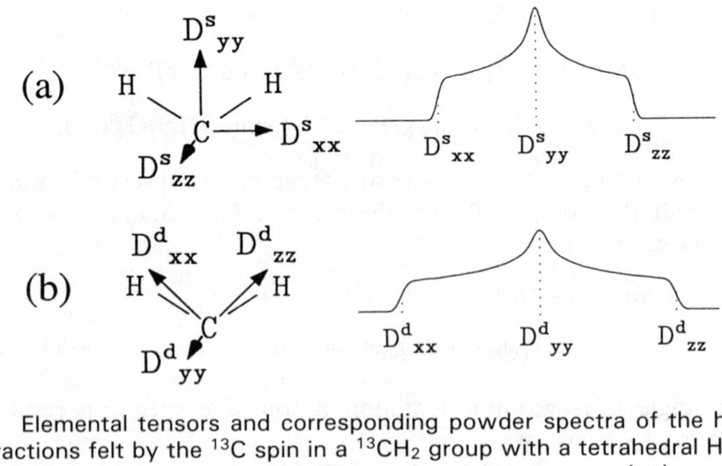

Figure 2.18 Elemental tensors and corresponding powder spectra of the heteronuclear dipolar interactions felt by the ^{13}C spin in a $^{13}CH_2$ group with a tetrahedral H–C–H angle. (a) 'Sum tensor', (b) 'difference tensor'. The total powder spectrum is the sum of the two elemental spectra shown.

The effect of the dipolar coupling of more than two (not necessarily bonded) protons on a certain ^{13}C spin can be determined by a straightforward extension of the treatment given so far. The ^{13}C signal under the action of four ^{1}H spins A, B, C and D, for instance, is:

$$\cos(\omega_A t)\cos(\omega_B t)\cos(\omega_C t)\cos(\omega_D t)$$

resulting in $2^4 = 16$ signals for a single orientation, namely at:

$$\pm(\omega_A + \omega_B + \omega_C + \omega_D) \qquad \pm(-\omega_A + \omega_B + \omega_C + \omega_D)$$
$$\pm(\omega_A + \omega_B + \omega_C - \omega_D) \qquad \pm(\omega_A + \omega_B - \omega_C - \omega_D)$$
$$\pm(\omega_A + \omega_B - \omega_C + \omega_D) \qquad \pm(\omega_A - \omega_B - \omega_C + \omega_D)$$
$$\pm(\omega_A - \omega_B + \omega_C + \omega_D) \qquad \pm(-\omega_A + \omega_B + \omega_C - \omega_D)$$

The spectrum for the simultaneous action of chemical shift and dipolar C–H coupling can be calculated in an analogous fashion. For simplicity, we consider a ^{13}C–^{1}H spin pair. The Hamiltonians commute, therefore the time-dependencies multiply (as for $\hat{H}^A$ and $\hat{H}^B$ in equations (2.96)–(2.98)). The resulting ^{13}C signal is given by $\cos(\omega_{IS}t)\exp(i\omega_{cs}t)$, so signals appear at $\omega_{cs} \pm \omega_{IS}$. Again, for determining the powder spectrum it is helpful to write:

$$\omega_{cs} + \omega_{IS} = -\omega_0\sigma_{zz} + \tfrac{1}{2}D_{zz} = (-\omega_0\boldsymbol{\sigma} + \tfrac{1}{2}\mathbf{D})_{zz} \tag{2.101a}$$

$$\omega_{cs} - \omega_{IS} = -\omega_0\sigma_{zz} - \tfrac{1}{2}D_{zz} = (-\omega_0\boldsymbol{\sigma} - \tfrac{1}{2}\mathbf{D})_{zz}. \tag{2.101b}$$

Thus, the powder spectrum can be characterized as a superposition of the powder patterns (Figure 2.19) of the second-rank tensors $-\omega_0\boldsymbol{\sigma} + \frac{1}{2}\mathbf{D}$ and $-\omega_0\boldsymbol{\sigma} - \frac{1}{2}\mathbf{D}$, which can be characterized by their respective principal values or the corresponding η and δ parameters (equations (2.34) and (2.35)). As is apparent in Figure 2.19, the spectrum is not symmetric in general, but the centre of gravity of either constituent powder pattern is at the isotropic chemical shift. The shapes of the powder spectra contain information on four parameters, namely the η and δ values for $-\omega_0\boldsymbol{\sigma} + \frac{1}{2}\mathbf{D}$ and $-\omega_0\boldsymbol{\sigma} - \frac{1}{2}\mathbf{D}$. Three unknowns, the C—H bond length and the polar coordinates (β, α) of the C—H bond in the chemical-shift PAS, are to be determined from the spectra. In practice it is found that the 1D lineshapes are not particularly characteristic of the α and β values, so that the broadening by couplings to other protons prevents a reliable analysis. In Section 6.7.3 we will compare these 1D spectra with two-dimensional separated-local field spectra, which due to a high redundancy of the information provide much more reliable values for α and β.

2.5.7 *^{1}H dipolar wideline spectra*

The coupling of a large number of spins leads to a multitude of lines, even in single crystals. In the preceding section, the number of lines for a single orientation and configuration was seen to rise exponentially ($\sim 2^N$) rather than linearly with the number N of coupled spins, even though many interactions were neglected. With the different configurations and orientations present in a powder, it is clear that a wide featureless 'line' results as the spectrum. Still, the width reflects the strength of the dipolar coupling; its reduction by fast molecular motions has been exploited in ^{1}H wideline NMR for a long time. This motional narrowing is most pronounced if the

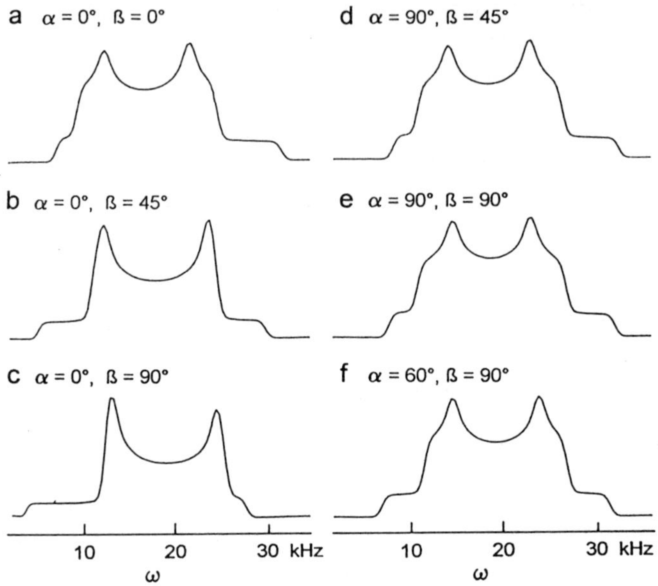

Figure 2.19 Typical powder patterns from simultaneous action of C–H dipole-pair coupling and ^{13}C chemical shift (adapted from Linder *et al.*, 1980). Each spectrum consists of the superposition of two powder patterns with the same centre of gravity but (in general) different δ and η parameters.

motion is isotropic and fast enough; under these circumstances, all the dipolar couplings are reduced to their common isotropic average, which is zero. An often used quantity related to the linewidth is the second moment $\int d\omega\, \omega^2 S(\omega)$ of the spectrum, which for a Gaussian $S(\omega)$ is the standard deviation σ, which can be calculated analytically for sufficiently simple crystal structures (Abragam, 1961). For organic polymers, the full-width-at-half-maximum (fwhm) is typically $2\pi \times 40$ kHz (according to equation (2.37), the proton pair interaction in an isolated CH_2 group would produce a Pake pattern with a splitting of the maxima of $2\pi \times 37$ kHz). In the crystalline portions of polyethylene, the fwhm of the ^1H line amounts to $2\pi \times 70$ kHz. The time signal, the free-induction decay (FID) or 'Bloch' decay of the proton magnetization after a single pulse, is correspondingly short (~ 150 μs) so that a deadtime of more than 10 μs leads to significant distortions in the ^1H wideline shape.

The multispin dipolar broadening of the ^1H wideline spectra is homogeneous, but coherent (see Section 3.13.3), which means that it can be eliminated by appropriate pulse sequences. These will be a central theme of Chapter 3.

2.5.8 Rotating frames

For the Zeeman interaction $\hat{H} = \omega_0 \hat{I}_z$, the time-evolution formula (2.73) for a magnetization $\hat{\rho}(0) = \hat{I}_\alpha$ reads:

$$\hat{\rho}(t) = e^{-i\omega_0 t \hat{I}_z}\, \hat{I}_\alpha\, e^{i\omega_0 t \hat{I}_z}. \tag{2.102}$$

This represents a rotation of $\hat{I}_\alpha$ by an angle $\omega_0 t$ about the z axis, as was shown from first principles in Sections 2.5.2 and 2.5.4. The effect of this precession is usually eliminated by a transition to a frame that rotates with a frequency $\omega_R \cong \omega_0$. The time evolution of the magnetization (density operator) is usually visualized in such a rotating frame. Formally, the density operator (2.102) seen from the frame rotating with ω_R can be written as:

$$\hat{\rho}_R(t) = \exp(i\omega_R t\hat{I}_z)\ \hat{\rho}(t)\ \exp(-i\omega_R t\hat{I}_z) \tag{2.103a}$$

$$\hat{\rho}_R(t) = \exp(i(\omega_R - \omega_0)t\hat{I}_z)\ \hat{I}_\alpha\ \exp(-i(\omega_R - \omega_0)t\hat{I}_z). \tag{2.103b}$$

Usually, the 'off-resonance operator' $(\omega_R - \omega_0)\hat{I}_z$ is combined with the truncated chemical shift. Indeed, both cannot be separated in practice because the resonance frequency ω_0 of the bare nucleus, without the electron cloud and its chemical shift, is usually unknown. In practice, the NMR spectrometer provides phase-sensitive detection of the nuclear spin induction signal by mixing with a reference signal of frequency $\omega_R \cong \omega_0$ which produces the difference $\omega_R - \omega_0$. According to equation (2.103), this represents essentially a transition into a frame rotating with an angular frequency ω_R.

Seen from the rotating frame, parts of the Hamiltonians of the local fields (before truncation), which are sample-fixed, become harmonically time-dependent with the frequency ω_R. They oscillate so fast that they do not affect the magnetization to first order. Only those terms that are not affected by the rotation operator $\exp(i\omega_R\hat{I}_z)$ remain because they commute with it. These are exactly those that commute with $\hat{I}_z$, that is, just the secular terms which also remained in the first order of time-independent perturbation theory as discussed in Section 2.4.1.

In the formal treatment of the 'interaction representation', equation (2.103a), these heuristic considerations are confirmed. The density operator in this representation is:

$$\hat{\rho}_R(t) = e^{i\hat{H}_0 t}\ \hat{\rho}(t)\ e^{-i\hat{H}_0 t}. \tag{2.104}$$

Here, the signs of the $e^{i\hat{H}_0 t}$ operators are reversed compared with the time-evolution formula (2.73): it is the purpose of the transition into the interaction frame to transform away the direct effects of the propagator $e^{-i\hat{H}_0 t}$ (2.74) on the density operator.

For a Hamiltonian $\hat{H} = \hat{H}_0 + \hat{H}_1$, the von Neumann equation in the $\hat{H}_0$-interaction representation reads:

$$\dot{\hat{\rho}}_R(t) = -i[\hat{H}_{1,R}(t), \hat{\rho}_R(t)]. \tag{2.105}$$

$\hat{H}_0$ does not appear as such in this equation: its direct effect has been transformed away. Indirectly, however, $\hat{H}_0$ plays an important role in the transformation of the other Hamiltonians:

$$\hat{H}_{1,R}(t) = e^{i\hat{H}_0 t}\hat{H}_1(t)\ e^{-i\hat{H}_0 t}. \tag{2.106}$$

This is the same kind of transformation as was applied to $\hat{\rho}$ in equation (2.104), which is reasonable since both $\hat{H}_{1,R}$ and $\hat{\rho}_R$ describe the appearance of laboratory-frame operators, $\hat{H}_1$ and $\hat{\rho}$, seen from the rotating frame. The equivalence of the interaction-representation (2.104)–(2.106) and the results of the original von Neumann equation of motion (2.71) can be proved by inserting the expressions for $\hat{\rho}_R$ and $\hat{H}_R$ into the von Neumann equation (2.105) in the rotating frame, which yields equation (2.71). With the exception of zero-field experiments, below (and actually already above) the relevant frame of reference for the density operator was and will be the rotating frame, and therefore the index R is dropped from $\hat{\rho}_R$.

2.6 The effects of radio-frequency (rf) pulses

Radio-frequency fields are of central importance in pulsed NMR, and essential for multidimensional NMR experiments. In this section, we discuss the effect of resonant pulses on the density operator, resulting in rotations of magnetization vectors in the simplest case. The Hahn spin echo and the solid echo as generated by suitable pulses will be derived in a quantum-mechanical description. Effects of finite pulse lengths and their partial compensation by means of composite pulses will be discussed briefly at the end of this section. The use of rf pulses to manipulate the effective Hamiltonians of the local fields will be treated in Chapter 3.

2.6.1 Rf pulses and the rotating frame

The rf field is generated by the probe coil in a suitable resonance circuit. The components relevant in NMR are those perpendicular to the $\vec{\mathbf{B}}_0$ field, $B_1^{\mathrm{LF}}(t) = 2B_1 \cos(\omega_{\mathrm{rf}} t + \phi)$. According to $\cos(\omega_{\mathrm{rf}} t + \phi) = \frac{1}{2}\exp(\mathrm{i}\omega_{\mathrm{rf}} t + \phi) + \frac{1}{2}\exp(-\mathrm{i}\omega_{\mathrm{rf}} t + \phi)$, the harmonic oscillation can be decomposed into two rotating fields with angular frequencies $+\omega_{\mathrm{rf}}$ and $-\omega_{\mathrm{rf}}$.

Choosing $\omega_{\mathrm{rf}} \cong \omega_{\mathrm{L}}$, the field rotating with $-\omega_{\mathrm{rf}}$ can be neglected for most practical purposes since the nuclear magnetization is influenced appreciably only by fields rotating with an angular frequency close to the nuclear Larmor frequency ω_{L}. In order to make the $\vec{\mathbf{B}}_1$ field appear static, it is customary to transform to the system rotating with ω_{rf}. Usually, spectrometers work with $\omega_{\mathrm{rf}} = \omega_{\mathrm{R}}$, where ω_{R} is the angular frequency of the rotating detection frame (i.e. irradiation and demodulation are performed with the same radio frequency). This frame is usually referred to as the 'rotating frame' in NMR. Remember that this means that the rotating frame usually does not rotate exactly with the Larmor frequency ω_{L}. In fact, due to the chemical shift of the effective Larmor frequencies, even for a single type of isotope, such as ^{13}C, there is usually no rotating frame in which all magnetization isochromats appear to be static.

In the rotating frame, the field acting on the magnetization is the vector sum of two components: the first, designated as *the* $\vec{\mathbf{B}}_1$ field in the following, is the portion of the radio-frequency (rf) field that is now static and perpendicular to the $\vec{\mathbf{B}}_0$ field (this applies independently of the orientation of the coil: components of the rf field parallel to $\vec{\mathbf{B}}_0$ oscillate rapidly even in the rotating frame (along its z axis) and can therefore be neglected like the nonsecular terms of the Hamiltonians in Section 2.4.1). The second field is the remainder $\vec{\mathbf{B}}_0(\omega_{\mathrm{L}} - \omega_{\mathrm{rf}})/\omega_{\mathrm{L}}$ of the $\vec{\mathbf{B}}_0$ field that has not been transformed away by the transition into the rotating system. In the following it will usually be assumed that this off-resonance field contribution is negligible since the B_1 field strength is great and the difference $(\omega_{\mathrm{L}} - \omega_{\mathrm{rf}})$ is relatively small. Then, the magnetization precesses around the $\vec{\mathbf{B}}_1$ field. (In Section 2.6.5, the effect of limited B_1 field strength and large off-resonance fields will also be discussed.)

The overall precession or flip angle can be set directly by adjusting the duration of the irradiation of the pulse. In most experiments, 90° pulses are employed, with:

$$\frac{\pi}{2} = \gamma B_1 t_{\mathrm{pulse}}$$

on the basis of $\phi_{\mathrm{prec}} = |\omega| t$ and $\omega = -\gamma B$ (see equation (2.7)). A 90° x pulse rotates the magnetization by 90° with respect to the x axis, which can in the first place be chosen arbitrarily, but must then remain in a fixed relation with respect to the other axes. (Note that in the laboratory frame, the orientation of the applied rf field of course cannot change, being fixed by the geometry of the coil. An x pulse and a $-y$ pulse in the rotating frame differ by 90° in the phase ϕ of the rf field in the laboratory frame, $2B_1 \cos(\omega_{rf}t + \phi)$.)

The Hamiltonian describing the effect of the $\vec{\mathbf{B}}_1$ field in the rotating frame is $\hat{H} = -\vec{\mathbf{M}} \cdot \vec{\mathbf{B}}_1$, which yields $-\gamma B_1 \hat{I}_x$ for an x pulse, $-\gamma B_1(-\hat{I}_x)$ for a $-x$ pulse, and analogous expressions for y and $-y$ pulses. Since in the rotating frame these Hamiltonians are time-independent, their effects on the density operator can be described by a propagator according to equation (2.74); for a $\vec{\mathbf{B}}_1$ field along the x axis of the rotating frame, it reads:

$$\hat{P}_x(t) = e^{-i\hat{H}_x t} = e^{i\gamma B_1 \hat{I}_x t}. \tag{2.107}$$

By inserting this propagator into $\hat{P}_x \hat{I}_\beta \hat{P}_x^{-1}$ and evaluating it according to equations (2.75)–(2.77), we find that this propagator produces a rotation of any spin operator $\hat{I}_\beta$ by an angle $-\gamma B_1 t$ around the x axis (see also Appendix B). This result is actually the quantum-mechanical basis of the precession picture used above.

2.6.2 *Pulse propagators and the sense of precession*

The definition of the sense of rotation of the rf pulses leads to a certain dilemma: for nuclei with $\gamma > 0$, such as ^1H, ^2H, ^{13}C, ^{19}F, or ^{31}P, the magnetization precesses with $\omega = -\gamma|\vec{\mathbf{B}}|$, that is, in the mathematically negative sense ('left-handed': see Figure 2.20 and Appendix B) around the $\vec{\mathbf{B}}$ field (this could be $\vec{\mathbf{B}}_0$ or $\vec{\mathbf{B}}_1$). This follows directly from first principles: namely, $\hat{H} = -\hat{\vec{\mathbf{M}}} \cdot \vec{\mathbf{B}}$, $\hat{\vec{\mathbf{M}}} = \gamma\hat{\vec{\mathbf{I}}}$, and the von Neumann equation with its solutions (2.73) and (2.75) which require commutator relations like

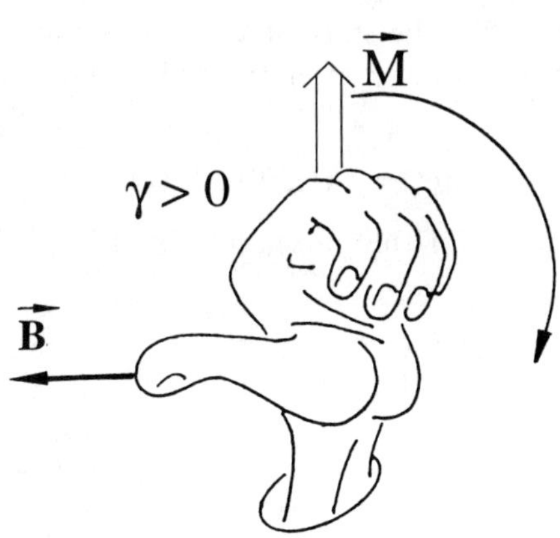

Figure 2.20 Visualization of the left-handedness of precession of magnetization from nuclear spins with $\gamma > 0$ around a $\vec{\mathbf{B}}$ field, arising from first principles and the sign conventions of electrodynamics. If the thumb of the left hand points along the $\vec{\mathbf{B}}$ field direction, the fingers indicate the direction of precession of the magnetization.

$[\hat{I}_x, \hat{I}_y] = \mathrm{i}\hat{I}_z$. These relations produce explicit formulas for the action of a $\vec{\mathbf{B}}_1$ field: for a $\vec{\mathbf{B}}_1$ field along the x axis:

$$\hat{\rho}(0) = \hat{I}_z \xrightarrow{\hat{H} = -\gamma B_1 \hat{I}_x} \hat{I}_z \cos(\gamma B_1 t) + \hat{I}_y \sin(\gamma B_1 t)$$

which is indeed identified as a left-handed rotation of initial z magnetization around the x axis, that is, around the $\vec{\mathbf{B}}_1$ field.

On the other hand, it is sometimes formally more convenient if the operator $\hat{P}_x(\phi)$ of a $+\phi$ x pulse corresponds to an active, mathematically positive ('right-handed') rotation by ϕ around the x axis (Ernst *et al.*, 1987, p. 27). In that case, the $\vec{\mathbf{B}}_1$ field producing the rotation is directed along the $-x$ direction, that is, antiparallel to the rotation axis.

Though the convention of the sense of rotation does not change anything about the physical action of a $\vec{\mathbf{B}}_1$ field, in the definitions of the $\hat{P}_\alpha$ operators different signs arise, and therefore different (but always equivalent) results. This can make the comparison of results in the literature confusing since the sense of rotation is not always clearly specified. In the right-handed, mathematically positive, convention, a 90° x pulse rotates a magnetization from the $+z$ direction into the $-y$ direction. In the left-handed convention, for a 90° $+x$ pulse the $\vec{\mathbf{B}}_1$ field is along the $+x$ direction and rotates $+z$ magnetization onto the $+y$ axis. The corresponding propagators read:

$$\text{'mathematical' (right-handed):} \quad \hat{P}_x(t) = \mathrm{e}^{-\mathrm{i}(\gamma B_1 t)\hat{I}_x} \tag{2.108}$$

$$\text{'physical' (left-handed):} \quad \hat{P}_x(t) = \mathrm{e}^{-\mathrm{i}(-\gamma B_1 t)\hat{I}_x} = \mathrm{e}^{\mathrm{i}\gamma B_1 t \hat{I}_x}. \tag{2.109}$$

In the tradition of Haeberlen and others, we adopt the 'physical' convention: with the thumb of the left hand pointing along the $\vec{\mathbf{B}}_1$ field or pulse direction, the fingers indicate the sense of rotation.

2.6.3 Calculating the Hahn spin echo

The Hahn spin echo can be generated for interactions linear in $\hat{S}_z^i$ (chemical shift and resonance offset, heteronuclear dipolar coupling) by application of a 180° pulse. If the pulse is applied at a time τ after the start of the precession in the xy plane, the echo occurs with its maximum at a time τ after the pulse. This can be derived quantum-mechanically via:

$$\hat{\rho}(t) = \hat{U}(t)\hat{\rho}(0)\hat{U}^{-1}(t).$$

The propagator at a time t_e after the pulse, that is, at $2t_e$ after the start of the precession, is given by:

$$\begin{aligned}
\hat{U}(2t_e) &= \underset{\text{precess}}{\mathrm{e}^{-\mathrm{i}\omega\hat{S}_z t_e}}\ \underset{\text{pulse}}{\mathrm{e}^{\mathrm{i}\pi\hat{S}_x}}\ \underset{\text{precess}}{\mathrm{e}^{-\mathrm{i}\omega\hat{S}_z t_e}} \\
&= \mathrm{e}^{-\mathrm{i}\omega\hat{S}_z t_e}\ \underbrace{\mathrm{e}^{\mathrm{i}\pi\hat{S}_x}\ \mathrm{e}^{-\mathrm{i}\omega\hat{S}_z t_e}\ \exp(-\mathrm{i}\pi\hat{S}_x)}_{\exp(-\mathrm{i}\omega\{-\hat{S}_z\}t_e)}\ \exp(\mathrm{i}\pi\hat{S}_x) \quad (\overbrace{\exp(-\mathrm{i}\pi\hat{S}_x)\exp(\mathrm{i}\pi\hat{S}_x)}^{\hat{1}\ \text{inserted}}) \\
&= \underbrace{\mathrm{e}^{-\mathrm{i}\omega\hat{S}_z t_e}\ \mathrm{e}^{\mathrm{i}\omega\hat{S}_z t_e}}_{=\hat{1}}\ \mathrm{e}^{\mathrm{i}\pi\hat{S}_x} = \mathrm{e}^{\mathrm{i}\pi\hat{S}_x}
\end{aligned} \tag{2.110}$$

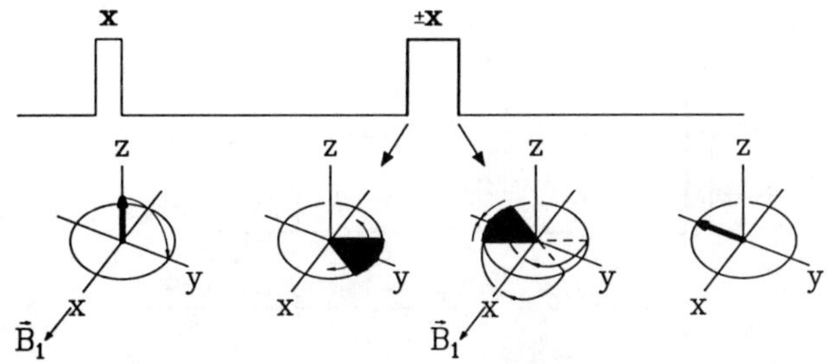

Figure 2.21 Negative Hahn echo for a $90°_x - t_e - 180°_{\pm x}$ pulse sequence. Both signs of the phase of the $180°$ pulse produce the same result.

where for $e^{i\pi\hat{S}_x} e^{i\omega\hat{S}_z t_e} e^{-i\pi\hat{S}_x} = e^{-i\omega(-\hat{S}_z)t_e}$ we have used the relation:

$$\hat{P}\exp(\hat{A})\hat{P}^{-1} = \exp(\hat{P}\hat{A}\hat{P}^{-1}) \tag{2.111}$$

(for a proof see equation (D.9)) with $\hat{A} = -i\omega\hat{S}_z t_e$, and the operator $\hat{P} = e^{i\pi\hat{S}_x}$ which generates a $180°$ inversion of $\hat{S}_z$.

For initial x magnetization, $\hat{\rho}(0) = \hat{S}_x$, we thus find that, independent of the precession frequency:

$$\hat{\rho}(2t_e) = e^{i\pi\hat{S}_x}\hat{S}_x e^{-i\pi\hat{S}_x} = \hat{S}_x = \hat{\rho}(0)$$

($e^{i\pi\hat{S}_x}\hat{S}_x e^{-i\pi\hat{S}_x}$ represents a rotation of $\hat{S}_x$ around the x axis, which leaves $\hat{S}_x$ invariant). Because of $\hat{\rho}(2t_e) = \hat{\rho}(0)$, the time dependence around the echo is:

$$\hat{\rho}(2\tau + t) = \hat{\rho}(t).$$

That is, at $2t_e$ the time evolution seems to start again. If the initial magnetization is $90°$ out of phase with the $180°$ pulse, in our case $\hat{\rho}(0) = \hat{S}_y$, we find from (2.110) that $\hat{\rho}(2t_e) = -\hat{S}_y$ and $\hat{\rho}(2t_e + t) = -\hat{\rho}(t)$: the magnetization appears inverted at the echo. This is quite easily confirmed in the picture of precessing isochromats as discussed above (Figure 2.21).

2.6.4 The solid (quadrupolar) echo

The solid echo serves to refocus the time evolution of spins under Hamiltonians bilinear in the same type of spins, such as dipolar (Powles and Strange, 1963) or quadrupolar (Davis *et al.*, 1976) couplings $(3\hat{I}_z^a\hat{I}_z^b - \hat{\mathbf{I}}^a \cdot \hat{\mathbf{I}}^b)$. It is generated by a $90°$ pulse applied at a time t_e after the $90°$ excitation pulse. The echo maximum is observed at a time t_e after the second pulse (Figure 2.22). The two $90°$-flip-angle pulses must be $90°$ out of phase, for example, an x pulse and a y pulse. (A Hahn $180°$ pulse cannot refocus homonuclear spin-bilinear interactions because the $180°$ inversion produces $\{3(-\hat{I}_z^a)(-\hat{I}_z^b) - (-\hat{\mathbf{I}}^a)\cdot(-\hat{\mathbf{I}}^b)\}$; therefore, the Hamiltonian remains unchanged and does not cancel itself in the overall propagator.)

For the solid echo, at a time $2t_e$ after the excitation x pulse we have the following propagator:

$$\hat{U}(2\tau) = \underbrace{e^{-i\omega_\theta(3\hat{I}_z^a\hat{I}_z^b - \hat{\mathbf{I}}^a\cdot\hat{\mathbf{I}}^b)t_e}}_{\text{free evolution}} \underbrace{e^{i\frac{1}{2}\pi\hat{I}_y}}_{90^\circ\text{ pulse}} \underbrace{e^{-i\omega_\theta(3\hat{I}_z^a\hat{I}_z^b - \hat{\mathbf{I}}^a\cdot\hat{\mathbf{I}}^b)t_e}}_{\text{free evolution}} \tag{2.112}$$

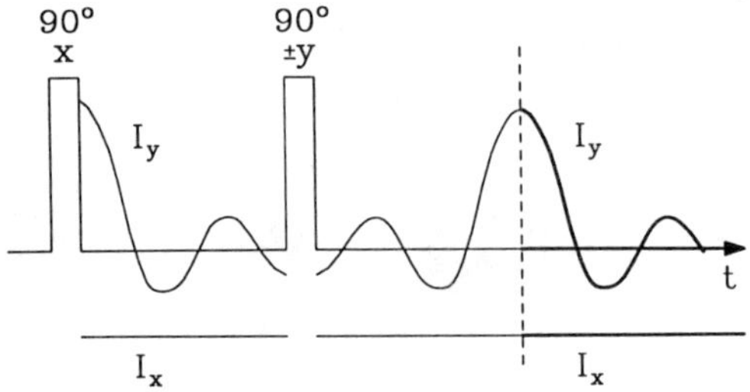

Figure 2.22 Solid-echo pulse sequence. The phase cycle indicated removes spurious signal distributions that are generated if the flip angle of the second pulse deviates from 90°.

with $\hat{\mathbf{I}}^a = \hat{\mathbf{I}}^b =: \hat{\mathbf{I}}$ for quadrupolar, and $\hat{\mathbf{I}} = \hat{\mathbf{I}}^a + \hat{\mathbf{I}}^b$ for dipolar coupling. Multiplying from the right with $\hat{1} = \exp(-i\frac{1}{2}\pi\hat{I}_y)\exp(i\frac{1}{2}\pi\hat{I}_y)$ analogously to the procedure in (2.110) and applying the rotation ($\hat{I}_z \to \hat{I}_x$) to the exponent according to (2.111), we obtain:

$$\hat{U}(2t_e) = e^{-i\omega_\theta(3\hat{I}_z^a\hat{I}_z^b - \hat{\mathbf{I}}^a\cdot\hat{\mathbf{I}}^b)t_e}\, e^{-i\omega_\theta(3\hat{I}_x^a\hat{I}_x^b - \hat{\mathbf{I}}^a\cdot\hat{\mathbf{I}}^b)t_e}\, e^{i\frac{1}{2}\pi\hat{I}_y}. \tag{2.113}$$

The exponents in equation (2.113) can be combined if $\hat{I}_z^a\hat{I}_z^b$ and $\hat{I}_x^a\hat{I}_x^b$ commute (prove for $I = \frac{1}{2}$ and $I = 1$ by spin-matrix multiplication with the $I = \frac{1}{2}$ matrices (2.80) or the $I = 1$ matrices of (2.86)). Using the 'magic zero':

$$3\hat{I}_z^a\hat{I}_z^b - \hat{\mathbf{I}}^a\cdot\hat{\mathbf{I}}^b + 3\hat{I}_y^a\hat{I}_y^b - \hat{\mathbf{I}}^a\cdot\hat{\mathbf{I}}^b + 3\hat{I}_x^a\hat{I}_x^b - \hat{\mathbf{I}}^a\cdot\hat{\mathbf{I}}^b = 0 \tag{2.114}$$

which will be of great relevance in the following chapter (prove (2.114) by using the definition of the scalar product $\hat{\mathbf{I}}^a\cdot\hat{\mathbf{I}}^b := \hat{I}_z^a\hat{I}_z^b + \hat{I}_y^a\hat{I}_y^b + \hat{I}_x^a\hat{I}_x^b$), we obtain:

$$\hat{U}(2t_e) = e^{+i\omega_\theta(3\hat{I}_y^a\hat{I}_y^b - \hat{\mathbf{I}}^a\cdot\hat{\mathbf{I}}^b)t_e}\, e^{i\frac{1}{2}\pi\hat{I}_y}. \tag{2.115}$$

For $\hat{\rho}(0) = \hat{I}_y$, created by an x pulse from z magnetization, we thus find the echo condition $\hat{\rho}(2t_e) = \hat{U}(2t_e)\hat{I}_y\hat{U}^{-1}(2t_e) = \hat{I}_y$ (on the basis of $[3\hat{I}_y^a\hat{I}_y^b - \hat{\mathbf{I}}^a\hat{\mathbf{I}}^b\hat{I}_y] = 0$, which is analogous to the Zeeman truncation). Repeating the calculation with $e^{-i\frac{1}{2}\pi\hat{I}_y}$ for the pulse propagator, it is found that the solid echo is independent of the sign of the (second) pulse. This is relevant for the quadrupolar echo if the first pulse is shortened, for instance, to improve the excitation profile (see Section 2.6.5). In that case, there is residual z magnetization after the first pulse, which is flipped into the xy plane by the second pulse, creating spurious signal contributions. By changing the sign of the second pulse, these undesired magnetization components are inverted. Thus, in a phase cycle with alternating signs for the second pulse the spurious signals cancel while the quadrupolar-echo signals add up.

For many-spin dipolar coupling, with the necessary sum over all pairs of spins in the Hamiltonian, the calculation can be performed in an analogous manner. However, the exponents after the rotation as in (2.113) do not commute in this case. Nevertheless, for sufficiently small times t_e, the linear approximation of the exponential operators:

$$e^{\hat{A}}e^{\hat{B}} \cong (1 + \hat{A})(1 + \hat{B}) \cong 1 + (\hat{A} + \hat{B}) \cong e^{\hat{A} + \hat{B}}$$

can be used to combine the exponents. For the dipolar solid echo to form, the echo delay t_e should therefore be smaller than the inverse coupling strength ($t_e < 1/50$ kHz $= 20$ μs).

The quadrupolar-echo sequence can be exploited to study molecular dynamics with correlation times τ in the range of approximately $1/\delta < \tau < 10t_e$. Due to the frequency changes during the quadrupolar evolution, the refocusing is incomplete and the echo intensity is reduced by a certain, t_e-dependent reduction factor R. Since the frequency changes are different for isochromats corresponding to different C—^{2}H bond orientations, the Fourier transform of the time signal after the echo maximum exhibits a lineshape characteristic of the geometry and rate of the motion. Examples are shown in Figure 2.16(b). In particular, signals corresponding to C—^{2}H orientations invariant in the motion, for instance along the rotation axis, are completely refocused and thus produce characteristic maxima in the spectrum. The evaluation of the position of these maxima in terms of the motional geometry is laid out in Section 6.8, where it is also discussed how the quadrupolar-echo experiment can be turned into a 2D experiment by incrementing and Fourier transforming with respect to t_e.

2.6.5 *Finite pulse lengths and their excitation effects*

For simple considerations, pulse strengths are assumed to exceed all internal NMR interactions, so that the magnetization of all spins is flipped in the same well-defined manner without any interference from the other interactions. In reality, pulse strengths are usually limited to less than $|\omega_1| = \gamma B_1 = 2\pi \times 100$ kHz (corresponding to a 360° pulse of $t_{2\pi} = (100\ \text{kHz})^{-1} = 10\ \mu\text{s}$, or a 90° pulse of $t_{\pi/2} = t_{2\pi}/4 = 2.5\ \mu\text{s}$ or more). This only slightly exceeds typical dipolar couplings (30–70 kHz) and is comparable to quadrupolar couplings of rigid deuterons ($\cong 125$ kHz).

If the $\vec{\mathbf{B}}_1$ field does not dominate over the local fields, decreased amplitudes and phase distortions result for spectral features appearing more than $\sim \gamma B_1$ off the carrier frequency ω_{rf}, which is taken as $\omega = 0$ in the following (rotating frame). These excitation effects, resulting from the simultaneous action of pulse and local fields, can be treated approximately in terms of an excitation function multiplying the spectral intensity. It has its maximum at $\omega = 0$ (vanishing local fields) and decreases towards the wings of the spectrum; in addition, there is also a frequency-dependent phase shift. It is customary, and a useful approximation, to derive the spectral excitation function as the Fourier transform of the pulse envelope. For a square pulse of amplitude $\omega_1 = \gamma B_1$, according to equation (4.27) this produces:

$$E(\omega) = \omega_1 t_p \left(\frac{\sin(\omega t_p)}{\omega t_p} + \mathrm{i}\, \frac{\cos(\omega t_p) - 1}{\omega t_p} \right). \tag{2.116}$$

The $\omega_1 t_p\ (\sin(\omega t_p)/\omega t_p)$ function is displayed in Figure 2.23 (thin line). Its first zero crossing is at $\omega = 2\omega_1$ for an on-resonance 90° pulse (i.e. with $\omega_1 t_p = \pi/2$). The FT-approach is based on linear-response theory: the ideal time signal $s(t)$ is considered as the δ-pulse response:

$$s(t) = \int_0^t \mathrm{d}t'\ s(t - t')\ \delta(t'). \tag{2.117}$$

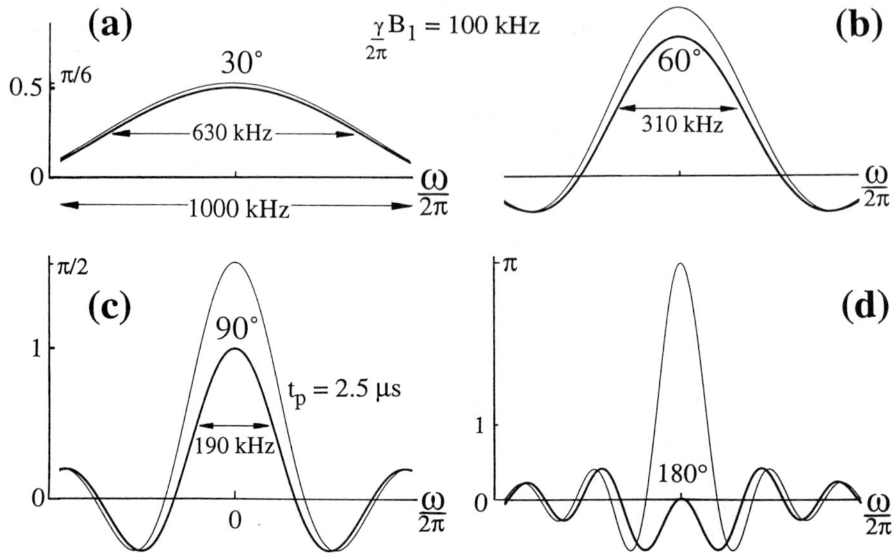

Figure 2.23 Excitation profiles for pulses generated by a B_1 field with a strength of $\gamma B_1 = 2\pi \times 100$ kHz, applied for durations of (a) 0.833 μs, (b) 1.667 μs, (c) 2.5 μs and (d) 5 μs, corresponding to on-resonance flip angles of 30°, 60°, 90°, and 180°, respectively. Thick lines: correct precessional profiles $X(\omega)$ according to equation (2.121a). Thin lines: linear-response approximation (real part of equation (2.116)). Note that these functions also apply for ^{2}H NMR, but with the ω axis scaled by a factor of two (wider excitation).

The finite-pulse time signal is then the superposition of δ-pulse responses starting at times t' with the shifted time-dependence $s'(t) = s(t - t')$ and the amplitude $e(t')$ proportional to the excitation profile at time t':

$$\tilde{s}(t) = \int_0^{t(\infty)} \mathrm{d}t' s(t - t')\, e(t'). \tag{2.118}$$

If the pulse has ended before the time t (i.e. $e(t') = 0$ for $t' > t$) the integration limit t can be replaced by ∞. According to the FT convolution theorem (4.11), the Fourier transform of equation (2.118) is easily found:

$$\tilde{s}(t) \overset{\mathrm{FT}}{\leftrightarrow} \tilde{S}(\omega) = S(\omega)E(\omega). \tag{2.119}$$

This approach is valid only in the linear region, that is, in the limit that the xy component $M_0 \sin(\omega_1 t_p)$ for on-resonance magnetization grows linear in $\omega_1 t_p$, which is true only for small flip angles $\omega_1 t_p$. For 90° pulses, this approach is no longer justified.

The exact excitation effect with a chemical-shift or off-resonance frequency offset $\omega := \omega_L - \omega_R$ can be determined by considering the precession around the effective field $\vec{\mathbf{B}}_{\mathrm{eff}} = (0, B_1, B_0\, \omega/\omega_0)$ during the y pulse. The precession frequency is:

$$\omega_{\mathrm{eff}} = -\gamma \mid \vec{\mathbf{B}}_{\mathrm{eff}} \mid = -\sqrt{\omega_1^2 + \omega^2}. \tag{2.120}$$

$X(\omega)$ and $Y(\omega)$, the x and y components (projections on the x and y axes), respectively, of each isochromat of frequency ω, are found as follows: take $\vec{\mathbf{B}}_{\mathrm{eff}}$ as the x_3 axis, the x axis as the x_1 axis, ζ as the angle between $\vec{\mathbf{B}}_{\mathrm{eff}}$ and the initial

z magnetization, such that $\sin\zeta = B_1/B_{\text{eff}} = \omega_1/\omega_{\text{eff}}$, and express the precessing magnetization in the (x_1, x_2, x_3) system as $\vec{\mathbf{M}} = (\sin(\omega_{\text{eff}}t_p)\sin\zeta, -\cos(\omega_{\text{eff}}t_p)\sin\zeta, \cos\zeta)$. The x, or x_1, component of fthe magnetization is:

$$X(\omega) = \omega_1 t_p \frac{\sin(\omega_{\text{eff}}t_p)}{\omega_{\text{eff}}t_p} \tag{2.121a}$$

the y component is $\vec{\mathbf{M}} \cdot (0, \cos\zeta, \sin\zeta)$, with $\cos\zeta = -\omega/\omega_{\text{eff}}$:

$$Y(\omega) = \frac{\omega}{\omega_{\text{eff}}}\,\omega_1 t_p \frac{\cos(\omega_{\text{eff}}t_p) - 1}{\omega_{\text{eff}}t_p}. \tag{2.121b}$$

The $X(\omega)$ function is plotted in Figure 2.23 (thick line). Its first null point occurs at $\omega = \sqrt{3}\omega_1$ for an on-resonance 90° pulse.

The effect on the spectrum must be considered with some care. The excitation effect is *not* simply obtained from multiplication of the absorptive or real part of the spectrum by $X(\omega)$. Rather, the distortion of a signal at ω is determined by the length of the projection $L(\omega) = \sqrt{X^2(\omega) + Y^2(\omega)}$ of the corresponding isochromat in the xy plane, and by the initial phase $\Phi(\omega)$ of the xy plane component, $\cos\{\Phi(\omega)\} = X(\omega)/L(\omega)$ and $\sin\{\Phi(\omega)\} = Y(\omega)/L(\omega)$. The spectrum $\tilde{S}(\omega)$ detected after the pulse is consequently:

$$\tilde{S}(\omega) = L(\omega)\, e^{i\Phi(\omega)} S(\omega) = \{X(\omega) + iY(\omega)\} S(\omega) \tag{2.122}$$

with $S(\omega)$ denoting the undistorted complex spectrum. Thus, the real part of $\tilde{S}(\omega)$ is influenced by both $X(\omega)$ and $Y(\omega)$:

$$\text{Re}[\tilde{S}(\omega)] = X(\omega)\,\text{Re}[S(\omega)] - Y(\omega)\,\text{Im}[S(\omega)]. \tag{2.123}$$

With equations (2.121) inserted, (2.122) reads:

$$\tilde{S}(\omega) = \omega_1 t_p \left(\frac{\sin(\omega_{\text{eff}}t_p)}{\omega_{\text{eff}}t_p} + i\,\frac{\omega}{\omega_{\text{eff}}}\,\frac{\cos(\omega_{\text{eff}}t_p) - 1}{\omega_{\text{eff}}t_p} \right) S(\omega). \tag{2.124}$$

The effect on the phase of the spectrum can be partly removed by linear phase correction $e^{i\omega t_p}$, since for small $\omega_{\text{eff}}t_p$, the excitation factor in equation (2.124) becomes:

$$\frac{\sin(\omega_{\text{eff}}t_p)}{\omega_{\text{eff}}t_p} + i\,\frac{\omega}{\omega_{\text{eff}}}\,\frac{\cos(\omega_{\text{eff}}t_p) - 1}{\omega_{\text{eff}}t_p} \cong 1 - i\omega t_p \cong e^{-i\omega t_p}$$

Figure 2.23 displays the similarity of the correct equation (2.124) and the result of the FT approach, equation (2.116); the equations become identical for small flip angles and for $\omega_{\text{eff}} \to \omega$, which is valid for large resonance offsets $\omega > \omega_1$.

The simultaneous action of B_1 field and quadrupolar couplings of deuterons with oscillation frequency $\omega = \omega_Q$ can also be calculated exactly. For instance, the 3×3 matrix representation (2.86) can be applied to $\hat{H} = -\gamma B_1 \hat{I}_z + \hat{H}_Q$. The resulting $\mathbf{H}$ matrix is not diagonal, so that the treatment of Section 2.5.4 is not immediately applicable. However, the matrix for $\exp(-i\mathbf{H}t)$ can be obtained by diagonalization as outlined in Appendix D.4: with the matrix $\mathbf{P}$ diagonalizing $\mathbf{H}$, the matrix $\mathbf{U}^D = \mathbf{P}\exp(-i\mathbf{H}t)\mathbf{P}^{-1}$ is a diagonal matrix with $U^D_{jj} = \exp(-i\varepsilon_j t)$, where ε_j is the jth eigenvalue of $\mathbf{H}$. This allows the following transformation of equation (2.73):

$$\boldsymbol{\rho}(t) = \exp(-i\mathbf{H}t)\boldsymbol{\rho}(0)\exp(i\mathbf{H}t) = \mathbf{P}^{-1}\mathbf{U}^D\mathbf{P}\boldsymbol{\rho}(0)\mathbf{P}^{-1}\mathbf{U}^{D*}\mathbf{P}. \tag{2.125}$$

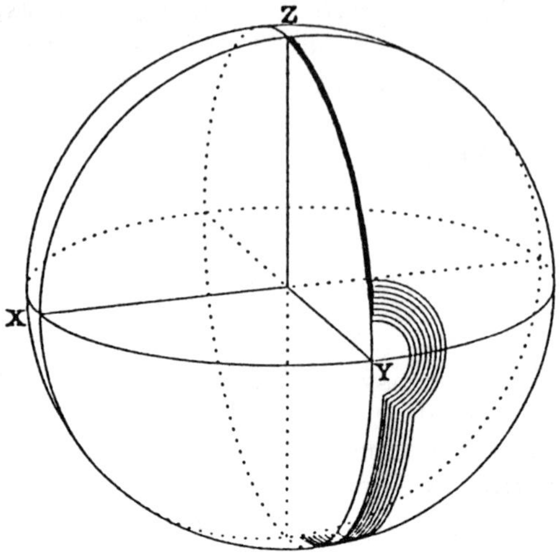

Figure 2.24 Visualization of the pulse-length compensation in the composite 180° *y* pulse consisting of nominal (90° *x*)(180° *y*)(90° *x*) pulses. Adapted from Levitt and Freeman (1979).

The time evolution can thus be calculated in terms of a product of the known matrices $\mathbf{P}$, $\mathbf{U}^{\mathrm{D}}$, and $\boldsymbol{\rho}(0)$. The result of an equivalent technique (Bloom *et al.*, 1980) is:

$$D(\omega) = \omega_1 t_{\mathrm{p}} \frac{\sin(t_{\mathrm{p}}\sqrt{\omega_1^2 + \frac{1}{2}\omega^2}}{t_{\mathrm{p}}\sqrt{\omega_1^2 + \frac{1}{4}\omega^2}}. \tag{2.126}$$

This curve has the same shape as the spin-$\frac{1}{2}$ curve $X(\omega)$ of equation (2.121), but is two times wider, which means that the excitation is twice as good, relatively speaking. The first null point occurs at $\omega = 3.5\omega_1$ for an on-resonance 90° pulse. The FID appears to start in the centre of the pulse (Bloom *et al.*, 1980).

The detailed analysis shows that after the two pulses of the solid echo, a triple product of excitation profiles occurs (namely, that of the second pulse with the square of expression (2.126)). The echo maximum is predicted to occur at $2\tau + t_{\mathrm{p}2} - \frac{1}{2}t_{\mathrm{p}1}$.

2.6.6 *Composite 180° pulses*

So far, only isolated pulses have been considered. By suitably combining several pulses as a block, certain features of the pulse can often be improved. In ^{13}C solution NMR, such composite pulses play an important role, for instance in connection with broadband excitation. In solid-state ^{13}C NMR, the excitation problem is not so relevant due to the higher power levels employed anyway to overcome the dipolar couplings. For quadrupolar nuclei, the performance of traditional composite pulses is not particularly convincing, as is witnessed by the fact that they are not widely used.

The most commonly used composite pulse in solid-state NMR of spin-$\frac{1}{2}$ nuclei is the 180° pulse consisting of (90° *x*)(180° *y*)(90° *x*) pulses (or the corresponding pulses with a common phase shift) (Levitt, 1986). Figure 2.24 indicates how a 180° rotation results from this composite pulse even for imprecise pulse lengths. As can be checked by applying the rotations of the three component pulses to *x*, *y* and *z* magnetization, they effect $M_x \rightarrow -M_x$, $M_y \rightarrow M_y$ and $M_z \rightarrow -M_z$. This shows that the composite pulse is equivalent to a 180° $\pm y$ pulse: the phase of the central 180° pulse counts.

2.7 References

Abragam, A. (1961). *The Principles of Nuclear Magnetism*, Chapter VIII. Oxford University Press, London.

Andrew, E. R., Bradbury, A. and Eades, R. G. (1958). NMR Spectra from a Crystal Rotated at High Speed. *Nature (London)*, **182**, 1659.

Bloembergen, N. and Rowland, T. S. (1955). 'Nuclear Spin Exchange in Solids: Tl^{203} and Tl^{205} NMR. *Phys. Rev.*, **97**, 1679.

Bloom, M., Davis, J. H. and Valic, M. I. (1980). Spectral distortion due to finite pulse widths in 2H NMR. *Can. J. Phys.*, **58**.

Bovey, F. A. (1988). *Nuclear Magnetic Resonance Spectroscopy*. Academic Press, San Diego.

Buckley, P. D., Jolley, K. W. and Pinder, D. N. (1975). Density Matrix in NMR Line-Shape Calculations. *Progr. NMR Spectroscopy*, Pergamon, Oxford.

Davis, J. H., Jeffrey, K. R., Bloom, M., Valic, M. I. and Higgs, T. P. (1976). Quadrupolar echo ^2H NMR in ordered hydrocarbons. *Chem. Phys. Lett.* **42**, 390.

Ernst, R. R. and Anderson, W. A. (1966). FT Spectroscopy in Magnetic Resonance, *Rev. Sci. Instr.*, **37**, 93.

Ernst, R. R., Bodenhausen, G. and Wokaun, A. (1987). *Principles of Nuclear Magnetic Resonance in One and Two Dimensions*. Clarendon, Oxford.

Feynman, R. P., Leighton, R. B. and Sand, M. (1965). *Lectures on Physics*, Vol. III. Addison-Wesley, Reading MA.

Gerstein, B. C. and Dybowski, C. R. (1985). *Transient Techniques in NMR of Solids*. Academic Press, Orlando.

Haeberlen, U. (1976). High Resolution NMR of Solids. *Adv. Magn. Reson.: Supplement 1*. Academic Press, San Diego.

Hahn, E. L. (1950). Spin Echoes, *Phys. Rev.*, **80**, 580.

Jackson, J. D. (1975). *Classical Electrodynamics*, 2nd ed. Wiley, New York, Chapter 6.

Komoroski, R. A. (ed.) (1986). *High Resolution NMR of Synthetic Polymers in Bulk*. VCH Publishers, Deerfield Beach, FL.

Kutzelnigg, W., Fleischer, U. and Schindler, M. (1990). The IGLO Method: Ab-initio Calculation of Chemical Shifts. *NMR Basic Principles and Progress* **23**, 165.

Levitt, M. H. and Freeman, R. (1979). Population Inversion Using a Composite Pulse. *J. Magn. Reson.*, **33**, 473; Levitt, M. H. 'Composite Pulses. *Progr. Nucl. Magn. Reson. Spectry.*, **18**, 61.

Linder, M., Höhener, A. and Ernst, R. R. (1980). Orientation of interactions from 2D NMR powder spectra. *J. Chem. Phys.*, **73**, 4959.

Lowe, I. J. (1979). Free Induction Decays in Rotating Solids. *Phys. Rev. Lett.*, **2**, 285.

Maricq, M. M. and Waugh, J. S. (1979). NMR in rotating solids. *J. Chem. Phys.*, **70**, 3300.

Mehring, M. (1983). *High Resolution NMR of Solids*. Springer, Berlin.

Pake, G. E. (1948). NMR in Hydrated Crystals: Fine Structure of the ^1H Line. *J. Chem. Phys.*, **16**, 327.

Pham, Q.-T., Petiaud, R. and Waton, H. (1983). *Proton and Carbon NMR Spectra of Polymers*. Wiley, New York.

Pines, A., Chang, J. J. and Griffin, R. G. (1974). ^{13}C NMR in solid ammonium tatrate. *J. Chem. Phys.*, **61**, 1021.

Powles, J. G. and Strange, J. H. (1963). Zero Time Resolution Nuclear Magnetic Transients in Solids. *Proc. Phys. Soc.*, **82**, 60.

Sakurai, J. J. (1985). *Modern Quantum Mechanics*. Addison-Wesley, Redwood City.

Schaefer, J. and E. O. Stejskal, (1976). ^{13}C NMR of Polymers Spinning at the Magic Angle. *J. Am. Chem. Soc.*, **98**, 1031.

Slichter, C. P. (1990). *Principles of Magnetic Resonance*. 3rd edn, Springer, Berlin.

Spiess, H. W. (1978). Dynamic NMR Spectroscopy. *NMR Basic Principles and Progress*, **15**, 55.
Terao, T., Miura, H. and Saika, A. (1986). SASS NMR: Separation of Heteronuclear Dipolar Powder Patterns in Solids. *J. Chem. Phys.*, **85**, 3816.
Tonelli, E. (1989). *NMR Spectroscopy and Polymer Microstructure*. VCH publ., New York.
VanderHart, D. L. (1976). Methylene ^{13}C chemical-shift tensor. *J. Chem. Phys.*, **64**, 830.
VanderHart, D. L. (1986). Field-dependent ^{13}C chem. shifts: A 2nd-order dipolar perturbation. *J. Chem. Phys.*, **84**, 1196.
Veeman, W. S. (1984). ^{13}C Chemical Shift Anisotropy. *Progr. NMR Spectr.*, **16**, 193.

CHAPTER THREE

High-Resolution NMR Techniques for Solids

Keywords: *average-Hamiltonian theory, toggling frame, multipulse homonuclear decoupling, WHH-4, MREV-8, Magnus expansion, second averaging, cross polarization, heteronuclear decoupling; homonuclear coupling: two spins, chain of spins, spin diffusion; MAS: time signals, sidebands, mirror-image condition, TOSS, deTOSS; coherent vs. relaxation effects; line-broadening: homogeneous vs. inhomogeneous, coherent vs. incoherent.*

3.1 Introduction

In the preceding chapter, the basics of pulsed NMR were presented, as well as exact calculations of the time evolution and of the spectra for simple one- or two-spin systems in solids.

For most experiments in solid-state NMR, multispin couplings must also be considered, which produce line broadenings that made high-resolution solid-state NMR impossible for two decades. In solution, by the fast isotropic tumbling of molecules, these couplings are averaged to zero; in order to obtain high resolution in solids, special techniques for coherent averaging of the couplings 'in spin space' and 'in real space', by pulse sequences and by sample rotation, respectively, are essential.

A theory that has proved useful for understanding most decoupling and coupling schemes in solid-state NMR experiments is the average-Hamiltonian theory (Haeberlen and Waugh, 1968). We will restrict our description to the points which appear essential for an understanding of this approach, while leaving out further details which have been described comprehensively in the literature (Haeberlen, 1976; Mehring, 1983).

Once dipolar couplings and angular dependences have been 'tamed', one can also exploit them. This holds in particular for multidimensional experiments, while one-dimensional experiments often require suppression of nearly all interactions in order to keep the spectra sufficiently simple.

Dipolar couplings permit the measurement of distances, either via the distance-dependent strength of the coupling of spin pairs or through the diffusive transfer of magnetization ('spin diffusion'). This allows for determining domain sizes in heterogeneous samples. In other experiments, the partial averaging of dipolar couplings by molecular motions is exploited.

A decisive feature of all these pulsed-NMR experiments, and in particular of multidimensional ones, is the switching on and off of interactions as required, for example by applying dipolar coupling and decoupling specifically in different periods of an experiment.

In addition to the dipolar couplings, the anisotropies of the chemical shift and the quadrupolar interaction are of great importance. Magic-angle spinning to reduce or eliminate the anisotropic line broadening was introduced only very briefly in Section 2.4.5; we shall discuss it in more detail here, and analyse pulse sequences for spinning-sideband suppression. Other techniques for obtaining highly resolved spectra in solids, including dynamic-angle spinning, double rotation, and zero-field NMR will be described briefly at the end of the chapter, after a short discussion of relaxation and line-broadening.

3.2 Phenomenological descriptions of pulsed techniques

In trying to explain the effects of pulse sequences on systems of coupled spins, it is important to note that the familiar picture of precessing magnetizations breaks down. The pair of coupled spins, treated in Section 2.5.6, provided a first example for such a more complex dipolar behaviour. New 'archetypes' are required particularly for the description of decoupling techniques. They have to be calculated by means of a valid theory in the first place; this is analogous to the extraction of the idea of precessing magnetization from the solution of the equation of motion for isolated spins. A valid and widely applicable description can be given in terms of the simplest form of average-Hamiltonian theory. This is the main concern of this chapter.

For many experiments, however, in a first approach it is sufficient to be just familiar with the effects of the decoupling and coupling (polarization-transfer) techniques. These are listed in this section (with a restriction to pulse techniques; a relatively simple account of MAS and OMAS has been given in Section 2.4.5):

Spin lock

Magnetization aligned along a strong $\vec{\mathbf{B}}_1$ field is spin-locked: it does not dephase, neither by chemical shifts nor by dipolar couplings. It only relaxes with the $T_{1\rho}$ relaxation time (see Section 3.13.1).

Hartmann–Hahn cross polarization (CP)

Magnetization is transferred from protons to carbons if ^{1}H magnetization has been spin-locked and continuous irradiation is applied simultaneously at the ^{13}C frequency, with the Hartmann–Hahn condition for the B_1 field strengths:

$$\gamma^{^1\mathrm{H}} B_1^{^1\mathrm{H}} = \gamma^{^{13}\mathrm{C}} B_1^{^{13}\mathrm{C}}. \tag{3.1}$$

In a somewhat simplistic picture, this can be explained as energy-level matching $\hbar\omega_1^{^1\mathrm{H}} = \hbar\omega_1^{^{13}\mathrm{C}}$ in the rotating frame (for a more elaborate description see Section 3.4.3). The length of the resulting ^{13}C magnetization component is proportional to that of the protons that have generated it, that is, those close to the carbon site. The gain relative to single-pulse excitation from Boltzmann equilibrium can reach $\gamma^{^1\mathrm{H}}/\gamma^{^{13}\mathrm{C}} \cong 4$. In addition, the recycle delays, and thus the measuring times, are reduced due to the shorter T_1 relaxation times of protons. Those are due to the efficiency of ^{1}H spin diffusion, which transfers the relaxation behaviour of the ^{1}H nuclei with the shortest relaxation times to the rest of the system.

High-power decoupling or dipolar decoupling (DD)
Continuous high-power irradiation on one spin species eliminates the heteronuclear couplings to the other spin species, regardless of the states of the magnetizations.

Dipolar dephasing or gated decoupling
This serves for selectively observing ^{13}C sites with small heteronuclear dipolar couplings. By switching off (i.e. gating) the heteronuclear decoupling for some 30–150 μs, the ^{13}C magnetization in protonated sites is dephased by their dipolar couplings to the protons (strength about 20 kHz ≙ 50 μs). The chemical shift evolution of the unprotonated sites during the gated decoupling is most suitably refocused by a 180° ^{13}C pulse to generate a Hahn spin echo.

Multipulse sequences
These are used mainly for homonuclear decoupling of protons and consist of specifically designed cycles of several pulses, possibly separated by windows without irradiation. They are repeated many times with stroboscopic observation of the magnetization in certain windows.

CRAMPS (combined rotation and multiple-pulse spectroscopy)
Multipulse homonuclear decoupling combined with MAS yields the highest available resolution (0.2 to 2 ppm) of proton spectra in rigid solids.

TOSS (total suppression of sidebands)
Spinning sidebands in MAS spectra are removed by destructive interference if a TOSS pulse sequence of four (or $2n+2$) properly spaced 180° pulses is applied before the start of the acquisition (see Section 3.9.3). The samples must exhibit cylindrical symmetry with respect to the rotor axis; 'powder' samples in particular fulfil this condition. For single isochromats, as observed in many 2D experiments, TOSS does not work properly (see Section 6.5.1). The number of sidebands should not be too large, otherwise small or even negative intensities result after TOSS. The last pulse is separated from the start of the acquisition by more than 100 μs, so no receiver dead-time problems occur in the signal detection.

Chemical-shift-scaling sequences
Sequences of alternating $+x$ and $-x$ pulses with stroboscopic detection between them result in a scaled effective precession frequency. By scaling the anisotropies to less than the MAS rotor frequency, these sequences can be used for sideband suppression.

Spin diffusion
Spatial transfer of z magnetization between dipolar-coupled spins has been termed 'spin diffusion' by Bloembergen (Bloembergen, 1949). Its characteristics and efficiency depend on the dipolar couplings and the average distances between coupled spins. For protons in organic solids, the continuum approach of diffusion provides a suitable description of the experimental data, while for ^{13}C in natural abundance (1.1%) a considerable fraction of isolated few-spin 'clusters' with oscillatory behaviour exists.

Spin diffusion occurs not only for z magnetization but also for spin-locked transverse ^{1}H magnetization, for instance under Hartmann–Hahn CP. The locking field scales the dipolar couplings by $-\frac{1}{2}$ (see Section 3.4.2) so that, in static samples, spin diffusion evolves (apparently) backwards in time. Under sample rotation, the effect of the negative sign is soon cancelled and spin diffusion effectively proceeds with half the coupling strength.

3.3 Averaging of Hamiltonians

The 'average-Hamiltonian theory' developed by Haeberlen and Waugh (Haeberlen and Waugh, 1968) facilitates the analysis of the action of pulse sequences when the B_1 field strength, averaged over the duration of the pulse sequence, exceeds the local fields and the resonance offset. Then, in an interaction representation that transforms away the immediate action of the pulses, the effect of the local fields can be calculated to first order as a simple average. The interaction representation can be visualized as a transition into a frame 'toggling' (flipping) with the pulses. This can be exploited to determine the Hamiltonians averaged in the toggling frame without calculations by means of simple sketches of coordinate systems subject to rotations.

The 'magic zero' of homonuclear dipolar couplings:

$$(3\hat{I}_x^j\hat{I}_x^k - \hat{\mathbf{I}}^j \cdot \hat{\mathbf{I}}^k) + (3\hat{I}_y^j\hat{I}_y^k - \hat{\mathbf{I}}^j \cdot \hat{\mathbf{I}}^k) + (3\hat{I}_z^j\hat{I}_z^k - \hat{\mathbf{I}}^j \cdot \hat{\mathbf{I}}^k) = 0 \tag{3.2}$$

will be of particular importance for the elimination of these interactions.

Sometimes, homonuclear decoupling by multiple-pulse irradiation is termed 'MAS in spin space': the pulses induce rotations on the spin operators. It should be noted that it is not the magnetization that is subject to the decisive 'magic condition'. Rather, the dipolar Hamiltonian is rotated or 'hopped' around a $(\pm 1, \pm 1, \pm 1)$ axis (see also Section 6.4.2), which makes the 'magic angle' of $54.74°$ with the z axis. Subject to this motion in spin space, the average of $\frac{1}{2}(3\hat{I}_z^j\hat{I}_z^k - \hat{\mathbf{I}}^j \cdot \hat{\mathbf{I}}^k)$, which is reminiscent of $\frac{1}{2}(3 \cos\theta \cos\theta - 1 \cdot 1)$, produces the 'magic zero' equation (3.2). (Actually, in 'real' MAS it is also not the magnetization but the Hamiltonian which is suitably manipulated through the modulations of its spatial factors.)

3.3.1 Homonuclear decoupling by multipulse sequences

As an important and simple example for the action of decoupling pulse sequences and averaging of operators in the toggling frame, we treat the 'classical' WHH-4 (WAHUHA, Waugh–Huber–Haeberlen) pulse sequence (Waugh *et al.*, 1968). It consists of many cycles, each made up of two solid-echo pulse pairs: that is, four $90°$ pulses separated by 'windows' of duration τ or 2τ (see Figure 3.1). In short, WHH-4 can be written as:

$$\text{WHH-4: } (\tau, -x, \tau, y, \tau, \tau, -y, \tau, x, \tau)$$

where $-x$, y, $-y$ and x denote the pulse phases. The duration of the pulses is assumed to be negligible. The density operator after a cycle of 6τ is then:

$$\hat{\rho}(t_0 + 6\tau) = \hat{U}(6\tau)\hat{\rho}(t_0)\hat{U}^{-1}(6\tau) \tag{3.3}$$

with the time-evolution operator, or 'propagator', $\hat{U}$ resulting from the sequence of the interactions (from right to left):

$$\hat{U}(6\tau) = \hat{L}_z\hat{P}_x\hat{L}_z\hat{P}_{-y}\hat{L}_z\hat{L}_z\hat{P}_y\hat{L}_z\hat{P}_{-x}\hat{L}_z. \tag{3.4}$$

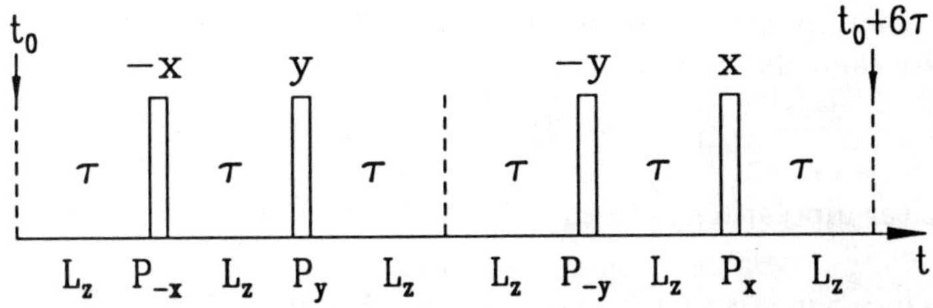

Figure 3.1 Propagators of a WHH-4 pulse sequence with strong pulses $\hat{P}_\beta$ and windows of durations τ and 2τ with free evolution $\hat{L}_z$.

The propagators of the rf pulses are designated by $\hat{P}_\alpha$ and those of the internal Hamiltonians by $\hat{L}_z(\tau) = \exp(-\mathrm{i}\hat{H}_z\tau)$. (The index z of $\hat{L}_z$ indicates that $\hat{I}_z$ in the truncated Hamiltonians appears indeed as $\hat{I}_z$; if $\hat{L}_\beta$ or $\hat{H}_\beta$ occur in the formulas below, then these $\hat{I}_z$ have to be replaced by $\hat{I}_\beta$.)

Pulses $\hat{P}_\alpha$ and couplings $\hat{L}_\beta$ can be separated by inserting $\hat{1} = \hat{P}_\alpha^{-1}\hat{P}_\alpha$ (this corresponds to a transition into an 'interaction representation'; see below). In the calculation, we will use identities like:

$$\hat{P}_x\hat{L}_z\hat{P}_x^{-1} = \hat{P}_x\,\mathrm{e}^{-\mathrm{i}\hat{H}_z\tau}\hat{P}_x^{-1} = \mathrm{e}^{-\mathrm{i}\hat{P}_x\hat{H}_z\hat{P}_x^{-1}\tau} = \mathrm{e}^{-\mathrm{i}\hat{H}_y\tau} = \hat{L}_y$$

(proof as for relation (D.9)). This can be considered as a (left-handed) rotation $\hat{P}_x$ transforming $\hat{L}_z$ into $\hat{L}_y$. First steps in the transformation of $\hat{U}(6\tau)$ are:

$$\begin{aligned}\hat{U}(6\tau) &= \hat{L}_z\hat{P}_x\hat{L}_z(\hat{P}_x^{-1}\hat{P}_x)\hat{P}_{-y}\hat{L}_z\hat{L}_z(\hat{P}_{-y}^{-1}\hat{P}_{-y})\hat{P}_y\hat{L}_z(\hat{P}_y^{-1}\hat{P}_y)\hat{P}_{-x}\hat{L}_z(\hat{P}_{-x}^{-1}\hat{P}_{-x})\\ &= \hat{L}_z(\hat{P}_x\hat{L}_z\hat{P}_x^{-1})\hat{P}_x(\hat{P}_{-y}\hat{L}_z\hat{L}_z\hat{P}_{-y}^{-1})\hat{P}_{-y}(\hat{P}_y\hat{L}_z\hat{P}_y^{-1})\hat{P}_y(\hat{P}_{-x}\hat{L}_z\hat{P}_{-x}^{-1})\hat{P}_{-x}\\ &= \hat{L}_z\quad\hat{L}_y\quad\hat{P}_x\quad\hat{L}_x\hat{L}_x\quad\hat{P}_{-y}\quad\hat{L}_{-x}\quad\hat{P}_y\quad\hat{L}_{-y}\quad\hat{P}_{-x}.\end{aligned}\tag{3.5}$$

Comparison of (3.4) and (3.5) shows that by each $\hat{P}_\alpha^{-1}\hat{P}_\alpha$, the corresponding $\hat{P}_\alpha$ has been shifted past its neighbour $\hat{L}_\beta$, rotating it from $\hat{L}_\beta$ to $\hat{L}_{\tilde{\beta}}$. Repetition of this simple procedure finally yields:

$$\hat{U}(6\tau) = \hat{L}_z\hat{L}_y\hat{L}_x\hat{L}_x\hat{L}_y\hat{L}_z\qquad\hat{P}_x\hat{P}_{-y}\hat{P}_y\hat{P}_{-x}.\tag{3.6}$$

Such a separation of $\hat{L}$ and $\hat{P}$ operators can be performed for any pulse sequence. For WHH-4 in (3.6) the product of the pulses cancels itself on the basis of $\hat{P}_{-\alpha} = \hat{P}_\alpha^{-1}$ (therefore WHH-4 is called a 'cyclic' sequence) (Haeberlen and Waugh, 1968). The resulting $\hat{U}(6\tau) = \hat{L}_z\hat{L}_y\hat{L}_x\hat{L}_x\hat{L}_y\hat{L}_z$ is still an exact expression. To analyse it further, approximations for short τ are used:

$$\begin{aligned}\hat{U}(6\tau) &= \hat{L}_z\hat{L}_y\hat{L}_x\hat{L}_x\hat{L}_y\hat{L}_z\\ &= \mathrm{e}^{-\mathrm{i}\hat{H}_z\tau}\,\mathrm{e}^{-\mathrm{i}\hat{H}_y\tau}\,\mathrm{e}^{-\mathrm{i}\hat{H}_x\tau}\,\mathrm{e}^{-\mathrm{i}\hat{H}_x\tau}\,\mathrm{e}^{-\mathrm{i}\hat{H}_y\tau}\,\mathrm{e}^{-\mathrm{i}\hat{H}_z\tau}\\ &\cong \hat{1} - \mathrm{i}(\hat{H}_z + \hat{H}_y + 2\hat{H}_x + \hat{H}_y + \hat{H}_z)\tau + \mathrm{Order}(\tau^2)\\ &\cong \mathrm{e}^{-\mathrm{i}\bar{\hat{H}}\cdot 6\tau}.\end{aligned}\tag{3.7a}$$

This shows that for sufficiently short pulse-spacings τ, the spins effectively evolve under the average Hamiltonian:

$$\bar{\hat{H}} = (\hat{H}_z + \hat{H}_y + 2\hat{H}_x + \hat{H}_y + \hat{H}_z)/6.\tag{3.7b}$$

This average Hamiltonian can be written explicitly in terms of the chemical shift or resonance offset and the homonuclear dipolar interactions:

$$\bar{\hat{H}} = -\tfrac{1}{3}\sum_j \omega^j(\hat{I}_z^j + \hat{I}_y^j + \hat{I}_x^j) + \tfrac{1}{3}(\hat{H}_z^{II} + \hat{H}_y^{II} + \hat{H}_x^{II}). \tag{3.7c}$$

The homonuclear dipolar terms in equation (3.7c) yield the 'magic zero' (see equation (3.2)):

$$(\hat{H}_z^{II} + \hat{H}_y^{II} + \hat{H}_x^{II}) = \sum_{k<j}\sum \omega_{II}^{jk}(\theta^{jk})(3\hat{I}_z^j\hat{I}_z^k - \hat{\mathbf{I}}^j\cdot\hat{\mathbf{I}}^k)$$

$$+ (3\hat{I}_y^j\hat{I}_y^k - \hat{\mathbf{I}}^j\cdot\hat{\mathbf{I}}^k) + (3\hat{I}_x^j\hat{I}_x^k - \hat{\mathbf{I}}^j\cdot\hat{\mathbf{I}}^k)$$

$$= \sum_{k<j}\sum \omega_{II}^{jk}(\theta^{jk})(3(I_x^jI_x^k + I_y^jI_y^k + I_z^jI_z^k) - 3\vec{\mathbf{I}}^j\cdot\vec{\mathbf{I}}^k) = 0! \tag{3.8}$$

This is valid for completely arbitrary spatial prefactors $\omega_{II} = (-\mu_0\hbar/4\pi)(\gamma^2/r_{jk}^3) \times \frac{1}{2}\{3\cos^2(\theta^{jk}) - 1\}$. Thus, at times $n\,.\,6\tau$ the propagator does not exhibit dipolar couplings up to terms of order τ^2. Stroboscopic detection at these times (Figure 3.2) therefore produces a time signal that is to a first approximation modulated only by terms linear in $\hat{I}_z^j$: the chemical shift, frequency offset, and heteronuclear dipole couplings. The latter have not been treated explicitly here for the sake of simplicity.

Equation (3.7) shows that during the multiple-pulse sequence the precession no longer occurs around the $\vec{\mathbf{B}}_0$ direction; instead, the first-order average Hamiltonian:

$$\bar{\hat{H}}\cdot 6\tau = -2\tau\sum_j \omega^j(\hat{I}_z^j + \hat{I}_y^j + \hat{I}_x^j) \overset{!}{=} -\gamma\sum_j \vec{\mathbf{B}}_{\text{eff}}^j\cdot\hat{\mathbf{I}}^j\,.\,6\tau$$

$$\Rightarrow \gamma\vec{\mathbf{B}}_{\text{eff}}^j = \frac{\omega^j}{\sqrt{3}}\frac{1}{\sqrt{3}}\begin{pmatrix}1\\1\\1\end{pmatrix} \tag{3.9}$$

causes a precession with an effective frequency $\omega_{\text{eff}}^j = \omega_j/\sqrt{3}$ around the effective-field unit vector $(1/\sqrt{3})$ (1,1,1). The ratio $\omega_{\text{eff}}^j/\omega^j$ is generally known as the frequency scaling factor (but be aware that some authors use this term for the inverse ratio (Burum and Rhim, 1979)). Heteronuclear dipolar interactions are linear in $\hat{I}_z^j$ and therefore scale like the chemical shift.

The preconditions and approximations in the derivation already indicate the special requirements for the use of 'multipulse' techniques: pulses with accurate flip angles and relative phases are needed. The pulse lengths and the spacings τ must be small

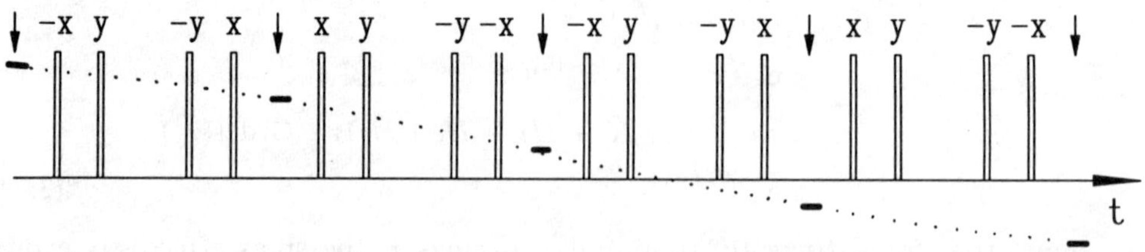

Figure 3.2 Pulses, windows, and stroboscopic data acquisition in multipulse NMR. In certain windows, the magnetization appears as if it had evolved under the average Hamiltonian. The particular pulse sequence shown represents two cycles of MREV-8.

(1–4 μs) in order to keep terms with higher orders of τ small. At the same time, within 2τ the NMR signal must be acquired without being masked by the ring-down of the radio-frequency pulses.

On the basis of WHH-4, pulse sequences have been developed which reduce the higher-order terms and the effects of pulse imperfections (Haeberlen, 1976; Rhim *et al.*, 1973; Burum and Rhim, 1979). Even small error terms in the multiply repeated basic pulse cycle can produce large effects if they are cumulative. Among the various multiple-pulse sequences, the MREV-8 (Mansfield–Rhim–Elleman–Vaughan, 8 pulses per cycle) (Rhim *et al.*, 1973) sequence proves to be particularly robust and versatile. It consists of two phase-cycled WHH-4 sequences:

$$\text{MREV-8: } (\tau, -x, \tau, y, \tau, \tau, -y, \tau, x, \tau, \qquad \tau, x, \tau, y, \tau, \tau, -y, \tau, -x, \tau)$$

as shown in Figure 3.2. By means of the geometric analysis described in the following section, the effective field direction (1, 0, 1) and the scaling factor of $\sqrt{2}/3 \simeq 0.5$ for MREV-8 are easily ascertained:

$$\vec{\mathbf{B}}^{j}_{\text{eff}} = \frac{\sqrt{2}}{3}\frac{\omega^j}{\gamma}\frac{1}{\sqrt{2}}\begin{pmatrix}1\\0\\1\end{pmatrix}. \tag{3.10}$$

The signal can be sampled twice per MREV-8 cycle, in every other large window.

Using an elaborate algorithm for phase cycling the pulses in three partly nested MREV-8 cycles while retaining the favourable properties of the underlying MREV-8 sequences, Burum and Rhim (1979) devised the BR-24 sequence:

$$\begin{aligned}\text{BR-24:}^{*}\ &(\tau, -x, \tau, y, 2\tau, -y, \tau, x, 2\tau, x, \tau, y, 2\tau, -y, \tau, -x, \tau)\\ &(\tau, y, \tau, -x, 2\tau, x, \tau, -y, \tau)^{*}\,(\tau, -y, \tau, -x, \tau)\\ &(\tau, y, \tau, -x, 2\tau, x, \tau, -y, 2\tau, -y, \tau, -x, 2\tau, x, \tau, y, \tau)\ (\tau, x, \tau, y, \tau)\end{aligned}$$

which averages out the homonuclear dipolar interaction up to the third order in the Magnus expansion (as given in Section 3.3.3). The stars indicate the timing of the signal sampling. The spacing of these sampling windows is quite large, which results in a relatively small spectral width. The effective field is along the (1, 1, 1) direction, with a scaling factor of $2/(3\sqrt{3}) = 0.385$. In practice, BR-24 is found to be superior to MREV-8 only on very well tuned spectrometers.

3.3.2 How to determine average Hamiltonians

The rotations of the $\hat{L}_z$ operators that came about as a consequence of the inserted $\hat{P}^{-1}\hat{P}$ terms can be viewed as a transition into a frame 'toggling' with the pulses such that in this frame the pulses are transformed away (Haeberlen, 1976). However, in return the Hamiltonians 'look different'. Formally, this can be written as an interaction representation with the index T marking operators seen from the toggling frame:

$$\hat{H}_T(t) := \hat{P}^{-1}(t)\hat{H}_z\hat{P}(t) = (\hat{H}_{z,\text{rot}})_T \tag{3.11a}$$

$$\hat{\rho}_T(t) := \hat{P}^{-1}(t)\hat{\rho}(t)\hat{P}(t) \tag{3.11b}$$

$$\dot{\hat{\rho}}_T = -\mathrm{i}[\hat{H}_T, \hat{\rho}_T]. \tag{3.11c}$$

$\hat{P}(t)$ denotes the product of the pulse propagators at time t. Due to $\hat{P}(n\,.\,6\tau) = \hat{1}$ for WHH-4 and MREV-8, we have $\hat{\rho}(n\,.\,6\tau) = \hat{\rho}_T(n\,.\,6\tau)$: the magnetization detected in every second 'large window' looks indeed as if it had evolved in the toggling frame under the action of $\hat{H}_T$. $\hat{H}_T$ is the $\hat{H}_z$ of the rotating frame seen from the toggling frame, and the average Hamiltonian is the average of $\hat{H}_T$ (Ernst *et al.*, 1987). This is visualized in Figure 3.3. Indeed, evaluation of equation (3.11a) confirms that $\hat{H}_T$ for WHH-4 takes the states $\hat{H}_z$, $\hat{H}_y$, $\hat{H}_x \hat{H}_x$, $\hat{H}_y$, and $\hat{H}_z$ as obtained above in the propagator product (3.6).

The average Hamiltonian of any pulse sequence can thus be readily determined by flipping the sketched toggling frame with the pulses in the rotating frame (left-handed rotations according to equation (2.109), around the axes of the rotating frame) and noting for every 'window' the axis of the toggling frame on which the z axis of the rotating frame is located. This determines the 'direction' ($\hat{I}_{\pm x}$, $\hat{I}_{\pm y}$ or $\hat{I}_{\pm z}$) that $\hat{I}_{z,\text{rot}}$ has acquired in the toggling frame. Finally, the Hamiltonians $\hat{H}_T$ with these rotated $\hat{I}_{z,\text{rot}}$ states are summed up for the average, taking into account the magic zero $\hat{H}_x^{II} + \hat{H}_y^{II} + \hat{H}_z^{II} = 0$.

The average Hamiltonian under continuous rf irradiation can be determined according to the same principles. The toggling frame in this case rotates continuously. It is equivalent to the doubly rotating frame introduced by Redfield (1955) and used in various circumstances that involve spin-locking fields (Hartmann and Hahn, 1962; Rhim *et al.*, 1971).

As a simple example, consider how an $\hat{I}_{z,\text{rot}}$ operator in the rotating frame is seen from a frame toggling continuously around a $\vec{\mathbf{B}}_1$ field applied along the rotating-frame x axis, with angular velocity $\omega_1 = -\gamma B_1$:

$$(\hat{I}_{z,\text{rot}})_T = \hat{I}_z^j \cos \omega_1 t + \hat{I}_y^j \sin \omega_1 t. \tag{3.12}$$

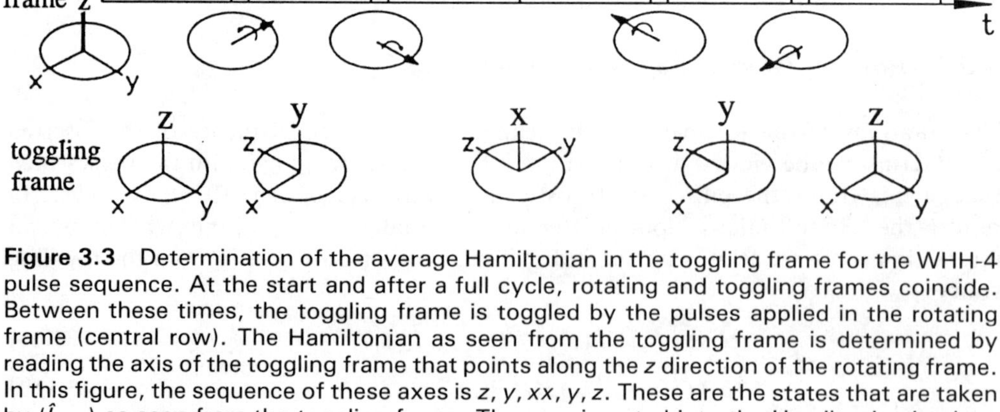

Figure 3.3 Determination of the average Hamiltonian in the toggling frame for the WHH-4 pulse sequence. At the start and after a full cycle, rotating and toggling frames coincide. Between these times, the toggling frame is toggled by the pulses applied in the rotating frame (central row). The Hamiltonian as seen from the toggling frame is determined by reading the axis of the toggling frame that points along the z direction of the rotating frame. In this figure, the sequence of these axes is z, y, xx, y, z. These are the states that are taken by $(\hat{I}_{z,\text{rot}})$ as seen from the toggling frame. They are inserted into the Hamiltonian in place of $\hat{I}_z$, and the Hamiltonians are then summed up to yield the average Hamiltonian.

After the interaction frame has toggled by an angle $\gamma B_1 t_p$, the average Hamiltonian is given by:

$$\overline{(\hat{I}_z^{\mathrm{rot}})_T} = \frac{1}{\gamma B_1 t_p} \int_0^{\gamma B_1 t_p} \mathrm{d}(\omega t)\,[\hat{I}_z^j \cos \omega t + \hat{I}_y^j \sin \omega t]. \qquad (3.13)$$

For full periods $\gamma B_1 t_p = n \,.\, 2\pi$, this average vanishes, and for long times $t_p \gg 2\pi/\gamma B_1$, tends to zero. This is the basis of heteronuclear dipolar decoupling, which is discussed in more detail in Section 3.4.1.

Effects of finite pulse lengths in multiple-pulse sequences can also be treated in this way. In practice, the pulse lengths are often quite comparable to the lengths of the windows in between; in fact, it is not uncommon to work in the limit of vanishingly small windows (contiguous pulses, windowless or semi-windowless sequences). Then, the timings are not the same as in the δ-pulse sequence, but have to be recalculated. With WHH-4 and MREV-8, for instance, in the long window a delay as long as the 90° pulse length must be inserted (Haeberlen, 1976; Mehring, 1983). Otherwise, the Hamiltonian in the toggling frame remains in the $\hat{H}_y$ state for too long a time.

A preparation pulse is often placed in front of the multipulse sequence to flip the magnetization into a direction perpendicular to or along the effective field. In calculations this pulse is for convenience not included in the rotations of the toggling frame but rather applied directly to the initial density operator.

3.3.3 The Magnus expansion

Up to this point, the treatment was concerned only with the simple average Hamiltonian, which applies well for short pulses and short delays between them. This level of theory is sufficient for understanding most of the common NMR techniques and for developing or modifying NMR experiments. However, for finding the multipulse sequences with the best performance, higher order corrections to the simple averaged Hamiltonian have to be taken into account. This is described in detail in standard texts (Haeberlen, 1976; Mehring, 1983; Gerstein and Dybowski, 1985). Here, we only sketch the most important features of this extensive theory.

The objective now is to replace the time evolution in the toggling frame over a certain cycle time t_c, which is the product of many successive propagators:

$$\hat{U}(t_c) \cong \mathrm{e}^{-\mathrm{i}\hat{H}(t^c - \tau)\tau} \ldots \mathrm{e}^{-\mathrm{i}\hat{H}(2\tau)\tau}\, \mathrm{e}^{-\mathrm{i}\hat{H}(\tau)\tau}\, \mathrm{e}^{-\mathrm{i}\hat{H}(0)\tau} \qquad (3.14)$$

by a propagator with an effective Hamiltonian:

$$\hat{U}(t_c) \stackrel{!}{=} \mathrm{e}^{-\mathrm{i}\hat{H}_{\mathrm{eff}} t_c}. \qquad (3.15)$$

To get an idea what this effective Hamiltonian $\hat{H}_{\mathrm{eff}}$ involves, consider the simple case of:

$$\begin{aligned}\mathrm{e}^{-\mathrm{i}\hat{H}_b\tau}\, \mathrm{e}^{-\mathrm{i}\hat{H}_a\tau} &\cong \{\hat{1} - \mathrm{i}\hat{H}_b\tau - \tfrac{1}{2}(\hat{H}_b\tau)^2 - \ldots\}\{\hat{1} - \mathrm{i}\hat{H}_a\tau - \tfrac{1}{2}(\hat{H}_a\tau)^2 - \ldots\} \\ &\cong \{\hat{1} - \mathrm{i}(\hat{H}_b + \hat{H}_a)\tau - \tfrac{1}{2}(\hat{H}_b + \hat{H}_a)^2\tau^2 - \tfrac{1}{2}[\hat{H}_b, \hat{H}_a]\,\tau^2 + \ldots\}.\end{aligned}$$

In addition to the terms of the average Hamiltonian $(\hat{H}_b + \hat{H}_a)$, we find a term proportional to $[\hat{H}_b, \hat{H}_a]$, that is, a term which is due to the noncommutation of the operators. This is the first nontrivial term in the Baker–Campbell–Hausdorff relation:

$$\mathrm{e}^{\hat{B}}\, \mathrm{e}^{\hat{A}} = \mathrm{e}^{\hat{B} + \hat{A} + 1/2[\hat{A},\hat{B}]] + 1/12([\hat{B},[\hat{B},\hat{A}]] + [[\hat{B},\hat{A}],\hat{A}]) + \cdots}. \qquad (3.16)$$

By applying this relation many times to the propagator product (3.14) and taking the limit $\tau \to 0$, the propagator can indeed be written as $\hat{U}(t_c) \stackrel{!}{=} e^{-i\hat{H}_{eff}t_c}$ with an effective Magnus-Hamiltonian ('Magnus expansion') (Gerstein and Dybowski, 1985; Ernst *et al.*, 1987):

$$\hat{H}_{eff} = \bar{\hat{H}} + \hat{H}^{(1)} + \hat{H}^{(2)} + \hat{H}^{(3)} + \dots. \tag{3.17}$$

It consists of the average Hamiltonian:

$$\bar{\hat{H}} = \frac{1}{t_c} \int_0^{t_c} dt_1 \, \hat{H}(t_1) \tag{3.18}$$

and further terms which are effectively proportional to increasing powers of t_c, in particular:

$$\hat{H}^{(1)} = \frac{-i}{2t_c} \int_0^{t_c} dt_1 \int_0^{t_1} dt_2 \, [\hat{H}(t_1), \hat{H}(t_2)] \tag{3.19a}$$

$$\hat{H}^{(2)} = \frac{-1}{6t_c} \int_0^{t_c} dt_1 \int_0^{t_1} dt_2 \int_0^{t_2} dt_3 \, \{[\hat{H}(t_1), [\hat{H}(t_2), \hat{H}(t_3)]] + [\hat{H}(t_3), [\hat{H}(t_2), \hat{H}(t_1)]]\}. \tag{3.19b}$$

(Note that for multipulse sequences the time-dependent $\hat{H}(t)$ is the Hamiltonian $\hat{H}_T(t)$ as seen from the toggling frame; we have omitted the index for the sake of clarity in the formulas.) For pulse sequences with pulse lengths short compared with the pulse spacings, these expressions can be evaluated pulse spacing by pulse spacing (the integrals are effectively replaced by sums over spacings between pulses, during each of which $\hat{H}_T(t)$ is not time dependent).

The performance of multipulse trains cannot be improved simply by repeating a good basic sequence, like WAHUHA or MREV-8, with the phases of all pulses simply shifted by $n \times 90°$. This can completely wipe out the chemical shift, or on a more subtle level, reintroduce cross-terms that had been eliminated in the basic pulse sequence. Burum and Rhim (1979) have developed impressive techniques of combining simple pulse sequences without losing their original favourable properties. This eventually led to the discovery of the BR-24 sequence discussed above.

3.3.4 Second averaging

Second averaging denotes the phenomenon that a weak interaction $\hat{H}_1$ in the presence of a stronger first-order-averaged interaction $\hat{H}_s$ is effectively reduced to the components of $\hat{H}_1$ that commute with $\hat{H}_s$. For instance, second averaging is one of the reasons why off-resonance detection of multiple-pulse spectra is advisable (Haeberlen, 1976): small residual error terms are suppressed through their second-averaging by the off-resonance Hamiltonian. Formally, second averaging is similar to the averaging-out of the local fields' nonsecular components by the Zeeman interaction (see Section 2.4.1).

A visualization of the second-averaging effect is provided by considering the simultaneous action of a strong field $\vec{\mathbf{B}}_s$ and a weak field $\vec{\mathbf{B}}_1$ with components B_{1a}

along $\vec{\mathbf{B}}_s$ and B_{1p} perpendicular to $\vec{\mathbf{B}}_s$. The length of the resultant field vector $\vec{\mathbf{B}}_{tot} = \vec{\mathbf{B}}_s + \vec{\mathbf{B}}_1$ is given by:

$$|\vec{\mathbf{B}}_{tot}| = (B_s + B_{1a})\sqrt{1 + \left(\frac{B_{1p}}{B_s + B_{1a}}\right)^2} \tag{3.20a}$$

(Pythagorean theorem: $(B_s + B_{1a})^2 + B_{1p}^2 = B_{tot}^2$). For $B_{1p}/B_s \ll 1$, this relation (3.20a) reduces to:

$$|\vec{\mathbf{B}}_{tot}| \cong (B_s + B_{1a}). \tag{3.20b}$$

Obviously, the effect of the perpendicular component B_{1p} is eliminated.

More formally, second averaging can be treated in complete analogy to truncation, which is described in Appendix G. The strong Hamiltonian $\hat{H}_s$ is now given by the dominating part of the first-order average, while $\hat{H}_1$ is a weaker, usually undesired, contribution to the Hamiltonian. According to equations (G.4) and (G.7), the second-average Hamiltonian:

$$\hat{H}_1^s = \overline{\exp(i\hat{H}_s t)\hat{H}_1 \exp(-i\hat{H}_s t)} \tag{3.21}$$

is the effective part of $\hat{H}_1$ since under the stated conditions:

$$e^{-i(\hat{H}_s + \hat{H}_1)t} \cong e^{-i\hat{H}_s t}\, e^{-i\hat{H}_1^s t}. \tag{3.22}$$

3.4 Further applications of average-Hamiltonian theory

In the preceding sections, the average-Hamiltonian theory was used for analysing the coherent averaging of homonuclear dipolar couplings. This is the application for which the theory was originally developed. Nevertheless, it can also be applied to many other NMR problems (Pines, 1988). In this section, we will use it in particular in the context of heteronuclear dipolar couplings between I and S spins when they are manipulated by rf pulses resonant for one or both species; this involves heteronuclear decoupling and cross polarization.

3.4.1 Heteronuclear decoupling

Heteronuclear decoupling by continuous irradiation is a standard procedure in the majority of NMR experiments on organic solids. Simplistic misconceptions concerning the explanation of the decoupling effect are widespread. For instance, it is believed that the decoupling is based on 'saturation of the proton resonance'. However, the state of the proton 'resonance' or magnetization (spin system, density operator) is completely irrelevant for the decoupling. What counts are the Hamiltonians of the $\vec{\mathbf{B}}_1$ field and the heteronuclear (to some extent also the homonuclear) dipolar interactions in the common propagator. This propagator can be analysed by means of average-Hamiltonian theory, which shows that the average heteronuclear dipolar Hamiltonian vanishes. That provides a valid and relatively simple explanation of heteronuclear dipolar decoupling.

The spin parts $\hat{I}_z^j \hat{S}_z^k$ of the truncated heteronuclear coupling $\hat{H}_{IS}$, equation (2.20), are considerably simpler to average out than those of homonuclear couplings. The sign of $\hat{I}_z^j \hat{S}_z^k$ seen from the toggling frames of I and S spins can easily be inverted by toggling the I-frame with I-pulses that flip by 180° ($\hat{I}_z^j \rightarrow -\hat{I}_z^j$). (By contrast, the $\hat{I}_\alpha^j \hat{I}_\alpha^k$ terms of the homonuclear couplings are unaffected by an inversion because $(-\hat{I}_\alpha^j)(-\hat{I}_\alpha^k) = \hat{I}_\alpha^j \hat{I}_\alpha^k$.) The simplest and, in some ways, most robust decoupling 'sequence' is continuous irradiation at the I resonance with constant pulse phase, for instance with the $\vec{\mathbf{B}}_1$ field along the x direction in the rotating frame of the I spins. Their toggling frame rotates or precesses continuously around $\vec{\mathbf{B}}_1$ (see Section 3.3.2). Seen from this frame, the operator $\hat{I}_z$ of the rotating frame precesses from the z direction through y, $-z$ and $-y$, back to z, $(\hat{I}_{z,\mathrm{rot}}^j)_T = \hat{I}_z^j \cos \omega t + \hat{I}_y^j \sin \omega t$ (see equation (3.12) and Figure 3.4), so that:

$$\overline{(\hat{H}_{\mathrm{rot}}^{SI})_T} = \sum_{j,k} \omega_{SI}^{jk} \hat{S}_z^k \int_0^{n.2\pi} \frac{\mathrm{d}(\omega t)}{n.2\pi} (\hat{I}_z^j \cos \omega t + \hat{I}_y^j \sin \omega t). \tag{3.23}$$

This means that the average of the dipolar Hamiltonian vanishes over integer precession periods ($n \times 360°$ pulses) and tends to zero for long irradiation times t_p (see the t_p-dependence in equation (3.13)). Only a single pulse phase is required, the exact 90°-pulse lengths play no role for the averaging, and the pulses are not separated by windows. Therefore, this decoupling 'sequence' is very robust and efficient, provided that the duration of a free-precession period is short enough, or, equivalently, that the B_1 field strength is sufficiently large. In fact, it must exceed not only the heteronuclear, but also the homonuclear couplings in order to dominate the Hamiltonian to such an extent that the average-Hamiltonian approach given above is justified.

Carbon–proton decoupling during the observation of protons under MREV-8, desirable in some heteronuclear experiments, can be achieved by means of 180° ^{13}C pulses applied in every fourth long window of the ^{1}H MREV-8 cycles. Thus, in the toggling-frame Hamiltonian, alternating $\hat{S}_z$ and $-\hat{S}_z$ terms are generated with equal $\hat{I}_z$ terms so that $\overline{(\hat{I}_z \hat{S}_{z,\mathrm{rot}})_T} = 0$.

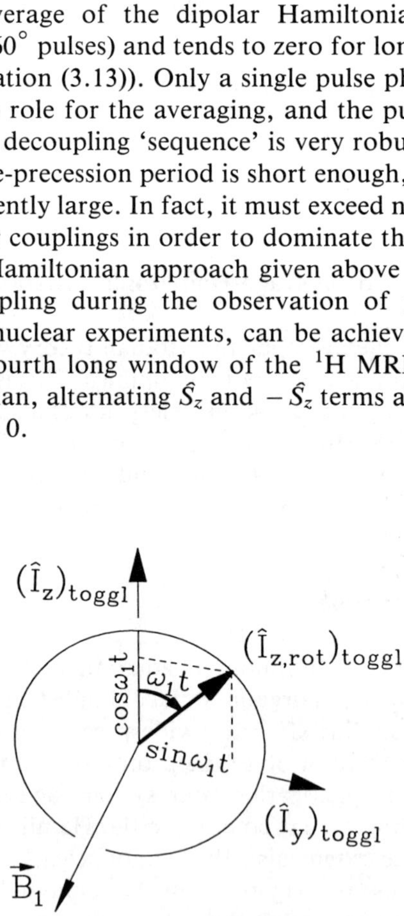

Figure 3.4 Decoupling of the heteronuclear dipolar interactions $\omega_{SI}\,\hat{I}_z\hat{S}_z$ by continuous irradiation (a constant $\vec{\mathbf{B}}_1$ field in the rotating frame) ($\hat{I}_{z,\mathrm{rot}}$) is visualized in a frame 'toggling continuously' around the $\vec{\mathbf{B}}_1$ field. The average over full cycles is zero, and after several completed cycles, the relative contribution of the last incomplete cycle in the average also becomes small. The duration of a full cycle is equal to four 90°-pulse lengths ($4 \times 90° = 360°$).

3.4.2 Spin lock

In many experiments, the magnetization is 'spin locked' by a $\vec{\mathbf{B}}_1$ field exceeding the dipolar couplings (see Figure 3.5): for the magnetization component along the locking field, there is no measurable precession around $\vec{\mathbf{B}}_1$, and also no dipolar dephasing as will be shown here. The magnetization can thus be held in this state for a long time, limited only by the longitudinal relaxation in the rotating frame with a time constant $T_{1\rho}$ (see Section 3.13.1). The dipolar couplings are nevertheless not really averaged out. For instance, spin diffusion (Section 3.5) does occur, which is relevant in Hartmann–Hahn cross-polarization and some proton spin diffusion experiments. In fact, the homonuclear dipolar couplings of the protons under continuous irradiation with the $\vec{\mathbf{B}}_1$ field along the x axis of the rotating frame are averaged to:

$$\overline{(\hat{H}^{II}_{z,\mathrm{rot}})_T} = \sum_{k<j}\sum \omega_{\mathrm{d}}^{jk} \int_0^{n\,.\,2\pi} \frac{\mathrm{d}(\omega t)}{n\,.\,2\pi} \times \tfrac{1}{2}\{3\underbrace{(\hat{I}^j_z \cos\omega t + \hat{I}^j_y \sin\omega t)}_{(\hat{I}_{z,\mathrm{rot}})_T}(\hat{I}^k_z \cos\omega t + \hat{I}^k_y \sin\omega t) - \hat{\vec{\mathbf{I}}}^j \cdot \hat{\vec{\mathbf{I}}}^k\}$$

$$= \sum_{k<j}\sum \omega_{\mathrm{d}}^{jk}\, \tfrac{1}{2}(3\tfrac{1}{2}\hat{I}^j_z\hat{I}^k_z + 3\tfrac{1}{2}\hat{I}^j_y\hat{I}^k_y - \hat{\vec{\mathbf{I}}}^j \cdot \hat{\vec{\mathbf{I}}}^k)$$

$$= \tfrac{1}{2}(\hat{H}^{II}_z + \hat{H}^{II}_y) = -\tfrac{1}{2}\hat{H}^{II}_x \qquad (3.24)$$

where the 'magic zero' (3.2), $\hat{H}^{II}_x + \hat{H}^{II}_y + \hat{H}^{II}_z = 0$, has been used in the last step. A dipolar coupling remains where each spatial prefactor $\omega_{\mathrm{d}}^{jk} = (-\mu_0\hbar/4\pi)\,(\gamma^2/r^3_{jk}) \times \frac{1}{2}\{3\cos^2(\theta^{jk}) - 1\}$ is scaled by $-\frac{1}{2}$. But because $\hat{I}_x$ and $\hat{H}^{II}_x$ commute, an initial density matrix $\hat{\rho}(0) = \hat{I}_x$, corresponding to magnetization along the lock field, remains invariant, $\dot{\hat{\rho}} = [\hat{H}^{II}_x, \hat{I}_x] = 0$. This means that these remaining dipolar couplings do not dephase magnetization along the x ($\vec{\mathbf{B}}_1$ field) direction. $[\hat{H}^{II}_x, \hat{I}_x] = 0$ is directly equivalent to $[\hat{H}^{II}_z, \hat{I}_z] = 0$, which is obvious because the truncation has left only those parts of the dipolar coupling in $\hat{H}^{II}_z$ that commute with $\hat{I}_z$.

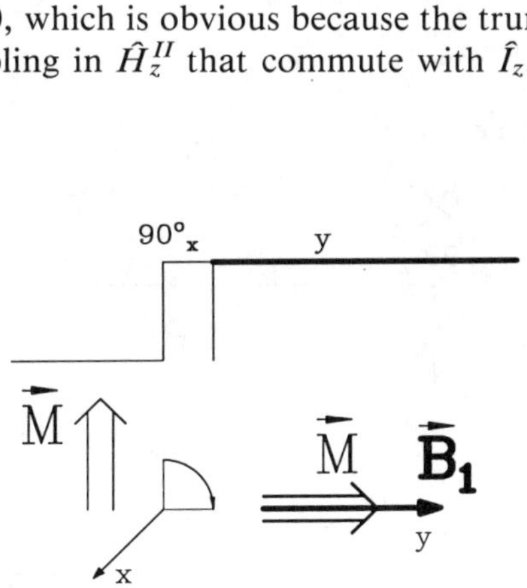

Figure 3.5 Schematic representation of spin locking of magnetization on a $\vec{\mathbf{B}}_1$ locking field in the rotating frame. The magnetization component along $\vec{\mathbf{B}}_1$ remains locked if the locking field is much stronger than local fields (dipolar or quadrupolar couplings, or the chemical shift). In the figure, the spin lock is preceded by a 90° pulse which flips z magnetization onto the locking field.

3.4.3 Cross polarization

Cross polarization (CP) from protons to rare spins takes an important place in standard as well as multidimensional ^{13}C, ^{29}Si, and ^{15}N NMR spectroscopy. It is also an indispensable element of multidimensional heteronuclear (in particular, $^{1}H-^{13}C$) experiments.

In our discussion, we shall concentrate on a description of the decisive step of the CP experiment, namely the creation of a coupling that causes the transfer of magnetization between *S* and *I* spins. We will discuss it in terms of average-Hamiltonian theory, which is equivalent to the original treatment by Hartmann and Hahn. The thermodynamic aspects of cross polarization have been analysed at length by Pines *et al.* (1973), and Mehring (1983).

As indicated in Figure 3.6, Hartmann–Hahn cross polarization (Hartmann and Hahn, 1962) is obtained by continuous simultaneous irradiation at both resonance frequencies with B_1 field strengths fulfilling the Hartmann–Hahn condition:

$$\gamma^I B_1^I \overset{!}{=} \gamma^S B_1^S \Leftrightarrow \omega_1^I \overset{!}{=} \omega_1^S. \tag{3.25}$$

This means that the toggling frames of the *I* and the *S* spins precess synchronously. The Hamiltonians are averaged over full periods of rotations of $\hat{I}_z$ and $\hat{S}_z$ around the resonant $\vec{\mathbf{B}}_1$ fields in their respective rotating frames. With the fields along the *x* axes, in the same way as in the preceding sections, it is found that $\overline{(\hat{H}^I_{\mathrm{cs}})_T}=0$, $\overline{(\hat{H}^S_{\mathrm{cs}})_T}=0$, and $\overline{(\hat{H}^{II}_{z,\mathrm{rot}})_T}=-\frac{1}{2}\hat{H}^{II}_x$ (equation (3.24)). The heteronuclear average Hamiltonian is crucial for the polarization transfer:

$$\overline{(\hat{H}^{IS})_T}=\sum_{j,k}\omega_{IS}^{jk}\,2\int_0^{n.2\pi}\frac{\mathrm{d}(\omega t)}{n.2\pi}\,(\hat{I}_z^j\cos\omega t+\hat{I}_y^j\sin\omega t)(\hat{S}_z^k\cos\omega t+\hat{S}_y^k\sin\omega t)$$

$$=\sum_{j,k}\omega_{IS}^{jk}\,(\hat{I}_z^j\hat{S}_z^k+\hat{I}_y^j\hat{S}_y^k). \tag{3.26}$$

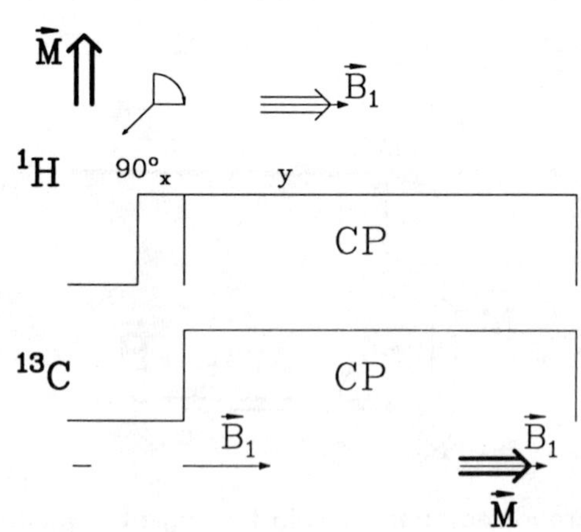

Figure 3.6 Hartmann–Hahn cross polarization pulse sequence with schematic representation of $^{1}H/^{13}C$ $\vec{\mathbf{B}}_1$ fields and magnetization vectors at different points in the sequence.

where $\omega := \omega_1^I = \omega_1^S$. The result (3.26) represents a form of the heteronuclear Hamiltonian (with spatial prefactors $\omega_{IS}^{jk} = (-\mu_0\hbar/4\pi)(\gamma^j\gamma^k/r_{jk}^3)\frac{1}{2}\{3\cos^2(\theta^{jk}) - 1\}$) that leads to polarization transfer between I and S spins due to cross terms between its components $\hat{I}_z^j\hat{S}_z^k$ and $\hat{I}_y^j\hat{S}_y^k$. This is to be exemplified now for a pair of nuclei with $I = \frac{1}{2}$ and $S = \frac{1}{2}$. We assume that initially only x magnetization of the I nucleus exists:

$$\hat{\rho}(0) = \hat{I}_x$$
$$\hat{H} = \omega_{IS}(\hat{I}_z\hat{S}_z + \hat{I}_y\hat{S}_y) \tag{3.27}$$

where $\omega_{IS} = (-\mu_0\hbar/4\pi)\,(\gamma^I\gamma^S/r_{IS}^3)\frac{1}{2}(3\cos^2(\theta^{IS}) - 1)$. For evaluating the time-evolution series (2.75), we need:

$$\hat{r}_1 = [\hat{H}, \hat{\rho}(0)] = \omega_{IS}[\hat{I}_z\hat{S}_z + \hat{I}_y\hat{S}_y, \hat{I}_x] = \mathrm{i}\omega_{IS}(\hat{S}_z\hat{I}_y - \hat{S}_y\hat{I}_z)$$
$$\hat{r}_2 = [\hat{H}, \hat{r}_1] = \omega_{IS}^2\{(\hat{S}_y^2 + \hat{S}_z^2)\hat{I}_x - (\hat{I}_y^2 + \hat{I}_z^2)\hat{S}_x\}. \tag{3.28}$$

In $\hat{r}_2$, a term proportional to $\hat{S}_x$ appears which represents the transfer of polarization $\hat{I}_x \rightarrow \hat{S}_x$. To evaluate the expression for $\hat{r}_2$ we can use the properties of the Pauli matrices since $I = S = \frac{1}{2}$ (see Section 2.5.3):

$$\mathbf{S}_x^2 = \mathbf{S}_y^2 = \mathbf{S}_z^2 = \tfrac{1}{4}\mathbf{1}_S$$
$$\mathbf{I}_x^2 = \mathbf{I}_y^2 = \mathbf{I}_z^2 = \tfrac{1}{4}\mathbf{1}_I \tag{3.29}$$

which simplify the commutators (3.28) to:

$$\mathbf{r}_2 = [\mathbf{H}, \mathbf{r}_1] = \omega_{IS}^2\tfrac{1}{2}(\mathbf{I}_x - \mathbf{S}_x)$$
$$\mathbf{r}_3 = [\mathbf{H}, \mathbf{r}_2] = \omega_{IS}^2\mathbf{r}_1 \tag{3.30}$$
$$\mathbf{r}_4 = [\mathbf{H}, \mathbf{r}_3] = [\mathbf{H}, \omega_{IS}^2\mathbf{r}_1] = \omega_{IS}^2\mathbf{r}_2, \ldots$$

where $\mathbf{1}_I$ and $\mathbf{1}_S$ have been omitted because they have no effect whatsoever. Inserting this into the time-evolution series (2.75) yields:

$$\mathbf{I}_x \xrightarrow{\mathbf{I}_z\mathbf{S}_z + \mathbf{I}_y\mathbf{S}_y} \mathbf{I}_x\tfrac{1}{2}(1 + \cos\omega_{IS}t) + \mathbf{S}_x\tfrac{1}{2}(1 - \cos\omega_{IS}t) + (\mathbf{I}_y\mathbf{S}_z - \mathbf{I}_z\mathbf{S}_y)\sin\omega_{IS}t. \tag{3.31}$$

The polarization transfer is the general effect here; the oscillatory behaviour is rather uncommon, resulting from the restriction to a pair of spins. Nevertheless it has indeed been observed in special cases (Müller *et al.*, 1974). It is usually inconspicuous due to the destructive interference arising from the broad range of coupling strengths ω_{IS} and is smeared out completely by the coupling between the I spins which is not averaged out according to equation (3.24).

The prefactor in the partial Boltzmann density matrix $(\gamma^I B_0/kT)\mathbf{I}_z$ is also transferred to the amplitude of the $\mathbf{S}_x$ term in (3.31), making its maximum larger by γ^I/γ^S than the density matrix $(\gamma^S B_0/kT)\mathbf{S}_x$ that would result from a 90° pulse on equilibrium S magnetization. The γ^I/γ^S enhancement can also be derived on the basis of thermodynamics, which is appropriate considering the large system of coupled I spins in organic solids.

The treatment given here shows that the sign of the ^{13}C magnetization depends on the sign of the ^{1}H magnetization. The analysis tells us that the signs of the magnetization vectors have to be specified only relative to the direction of the corresponding locking field (Figure 3.7). This must be so since the ^{1}H and ^{13}C frames

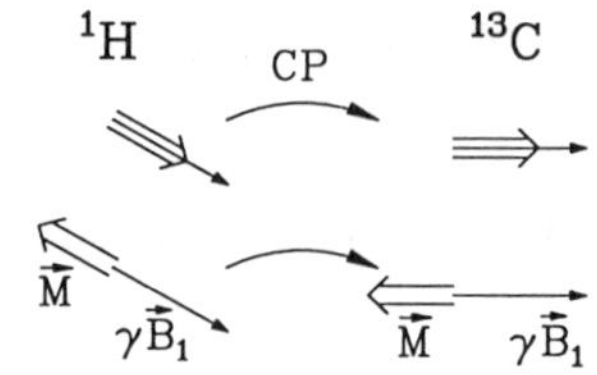

Figure 3.7 Relation between locking fields ($\vec{\mathbf{B}}_1$) and magnetization directions ($\vec{\mathbf{M}}$ under cross polarization): a change of the sign of the relative directions of ^{1}H locking field and magnetization inverts the direction of the ^{13}C magnetization relative to the ^{13}C locking field. Otherwise, no fixed phase relation exists between the ^{1}H and ^{13}C field or magnetization vectors in their respective rotating frames.

rotate relative to each other, so no fixed phase relation exists between them. Generally, if the ^{1}H magnetization is antiparallel to the ^{1}H locking field, the ^{13}C magnetization is also generated antiparallel to the ^{13}C locking field. The direction of the S magnetization depends on the sign of γ. For a negative γ^S, the toggling frame of the S spins moves in the opposite direction. That produces a negative sign in front of the term $\hat{I}^i_y\hat{S}^j_y$ of the average Hamiltonian (3.26) which is transferred to $\hat{S}^j_x$ in the density operator.

Hartmann–Hahn CP, with simultaneous spin-lock of proton and ^{13}C spins, is not the only possible technique for heteronuclear polarization transfer. The necessary transfer Hamiltonians of a structure $\hat{I}_\alpha\hat{S}_\alpha + \hat{S}_\beta\hat{I}_\beta$ can be generated by synchronous multipulse irradiation on both rf channels with equal 90° pulse lengths so that the toggling-frame Hamiltonians rotate in synchrony. In order to prevent destruction of the ^{1}H magnetization, the pulse sequence must also provide homonuclear decoupling of protons. Since the magic zero (3.2) requires that the $\hat{I}_z$ operator as seen from the toggling frame has equal x, y and z components on the average, the heteronuclear Hamiltonian reads:

$$\begin{aligned}\bar{\hat{H}}_{IS} &\sim \omega_{IS}[(\pm\hat{I}_z)(\pm\hat{S}_z) + (\pm\hat{I}_y)(\pm\hat{S}_y) + (\pm\hat{I}_x)(\pm\hat{S}_x)] \\ &= \omega_{IS}[\hat{I}_z\hat{S}_z + \hat{I}_y\hat{S}_y + \hat{I}_x\hat{S}_x]. \end{aligned} \tag{3.32}$$

The chemical shift has to be prevented from dephasing the magnetization, which must therefore be aligned along the effective field, unless the average CS interaction is zero; the windowless isotropic mixing (WIM) sequence (Caravatti *et al.*, 1983), for instance, decouples all homonuclear interactions on the average. The average Hamiltonian (3.32) is symmetric in x, y and z ($\hat{H} \sim \hat{\mathbf{I}} \cdot \hat{\mathbf{S}}$ is 'isotropic'), so when we consider a density-operator component $\hat{I}_x$ the results are equivalently valid for $\hat{I}_y$ and $\hat{I}_z$. The density operator $\hat{\rho}(0) = \hat{I}_x$ commutes with $\hat{I}_x\hat{S}_x$ and $\exp(i\alpha\hat{I}_x\hat{S}_x)$, and $\hat{I}_x\hat{S}_x$ commutes with $\hat{I}_z\hat{S}_z + \hat{I}_y\hat{S}_y$ (for spins $\frac{1}{2}$, see equation (3.38) below). Therefore, the relevant Hamiltonian is essentially the same as given in equation (3.26), which was shown to produce polarization transfer.

Under WIM, due to the absence of an effective field, the phase of the magnetization is also 'transferred'. This may seem to contradict the above statement that the ^{1}H and ^{13}C systems are 'independent'. That impression is not correct: shifting the ^{1}H pulses in the sequence as well as the transverse components of the magnetization all by the same phase (e.g. $+90°$: $x \to y$, $y \to -x$, $-x \to -y$, and $-y \to x$) leaves the ^{13}C magnetization unchanged. Only the orientation of the ^{1}H magnetization relative to the ^{1}H $\vec{\mathbf{B}}_1$ field orientation can play a role and can actually be transferred.

3.4.4 *Chemical-shift scaling*

In the preceding sections, it was shown how dipolar interactions can be averaged, or scaled to zero in the average-Hamiltonian approximation. In certain experiments, it is desirable to scale the (^{13}C) chemical shift. In particular, removal of spinning sidebands without intensity losses can be achieved by scaling the anisotropic broadening below the spinning rate. All multiple-pulse sequences scale down the chemical shift at least to some extent. By means of $(x, \tau, -x, \tau)_n$ sequences, the chemical-shift interaction can be scaled particularly conveniently. The scaling factor, which can be calculated in the average-Hamiltonian approximation, depends on the flip angles of the pulses; it is unity for 0° pulses, and zero for (short) 180° pulses, since these refocus the chemical shift completely. With short 90° pulses, the average $\hat{I}_z$ Hamiltonian is toggled between $\hat{I}_z$ and $\hat{I}_y$ by the x and $-x$ pulses. Therefore, the effective $\vec{\mathbf{B}}$ field is given by:

$$\vec{\mathbf{B}}_{\text{eff}} = \frac{\omega_{\text{cs}}}{\gamma}\frac{1}{2}\left(\begin{bmatrix}0\\0\\1\end{bmatrix} + \begin{bmatrix}0\\1\\0\end{bmatrix}\right) = \frac{\omega_{\text{cs}}}{\gamma}\frac{1}{\sqrt{2}}\frac{1}{\sqrt{2}}\begin{bmatrix}0\\1\\1\end{bmatrix}$$

which results in a scaling factor of $1/\sqrt{2}$. Finite pulse lengths can quite easily be incorporated by integrating over the 'orientations' $\hat{I}_z$ assumes as seen from the toggling frame during the course of a pulse as was demonstrated in Section 3.3.2. Chemical-shift scaling is a multiple-pulse technique. Correspondingly, the typical multipulse problems arise: the sequence requires some care in tuning up the spectrometer (actually the ^{13}C channel, so ^{13}C enriched methanol or a similar tune-up material is nearly indispensable). Also, artifacts from pulse imperfections occur, and there is a dependence of the scaling factor on the off-resonance position. Due to these flaws, the scaling technique is not widely used.

3.5 Homonuclear polarization transfer

In Section 3.4.3, heteronuclear polarization transfer under rf irradiation was described. Under homonuclear coupling, transfer of z magnetization, known as spin exchange or spin diffusion (Bloembergen, 1948), proceeds without rf irradiation. We start out with the simple phenomenological description of spin diffusion that is sufficient for most practical purposes. We shall discuss the discrete and continuous versions of the diffusion equation and the relation of the diffusivity to microscopic parameters. Then, as the simplest example of quantum-mechanical treatment of the transfer of z magnetization, we shall calculate the behaviour of two coupled like spins-$\frac{1}{2}$, and subsequently consider systems consisting of many similar spins. As an example, we shall present the quantum-mechanical evolution of magnetization on a chain of 40 spins-$\frac{1}{2}$ with next-neighbour dipolar couplings. Spin diffusion in the presence of other local fields (chemical shifts, heteronuclear couplings or homonuclear couplings to 'spectator' spins) will be treated later, after an introduction to time-dependent perturbation theory in Section 3.6.

3.5.1 *Phenomenological description of spin diffusion*

In order to describe the experimentally observed diffusive behaviour of magnetization transfer in a simple manner, the following *ansatz* can be made, which leads to a master equation for the z magnetization M (the index z is dropped for convenience here) that is a discrete version of the diffusion equation: per unit time Δt, through dipolar couplings every nuclear magnet gives a certain fraction $f_m = \Delta t \,.\, \Omega$ of its magnetization to its neighbours, and accordingly receives magnetization from them. In the simple case of a linear array of spins this results in:

$$\Delta M_j / \Delta t = -2\Omega M_j + \Omega M_{j+1} + \Omega M_{j-1}. \tag{3.33}$$

This formula can be employed to calculate the diffusion of any initial distribution of z magnetization on a computer (see Section 13.3.3). The discrete equation (3.33) can be transformed into the continuous one-dimensional diffusion equation (Fick's second law):

$$\frac{\partial M(x,t)}{\partial t} = D\,\frac{\partial^2 M(x,t)}{\partial x^2} \tag{3.34}$$

by taking the limits $\Delta t \to 0$ and $a \to 0$:

$$\lim_{\Delta t \to 0} \frac{\Delta M_j}{\Delta t} = \frac{\partial M}{\partial t} \qquad \lim_{a \to 0} \frac{(M_{j+1} - M_j) - (M_j - M_{j-1})}{a^2} = \frac{\partial^2 M}{\partial x^2}. \tag{3.35}$$

From this, a relation between the transition rate Ω, the distance a of neighbours in the array, and the diffusivity D can be determined as:

$$D = \Omega a^2. \tag{3.36}$$

The transition rate Ω should correspond to the rate of the dipolar coupling, which is approximately equivalent to half the linewidth ω_σ of the proton wideline spectrum. As only the couplings of spins in one of six directions (the forward direction) are relevant for Ω, ω_σ will have to be scaled by a factor of about $1/\sqrt{6}$ (the coupling strengths add up roughly statistically). For values typical for polymers ($2\omega_\sigma = 2\pi \times 50$ kHz, $a \cong 0.25$ nm), we therefore expect a diffusivity of about:

$$D \cong 1\ \text{nm}^2/\text{ms} = 10^{-15}\ \text{m}^2/\text{s}.$$

The unit nm^2/ms is particularly suited for characterizing this process, which within a millisecond causes a redistribution of magnetization from a given spin over a length of about one nanometer.

3.5.2 *Polarization transfer between two spins*

In the previous section, it was simply assumed that the dipolar couplings produce transfer of z magnetization between like spins. This effect of the homonuclear interaction can indeed be derived quantum-mechanically on the basis of the time-evolution equations of Section 2.5). In this section, we shall perform the calculation for the simple case of two spins-$\frac{1}{2}$, which we label I and J instead of I^a and I^b to simplify the notation.

The secular Hamiltonian reads (see equation (2.41)):

$$\hat{H}_{\mathrm{D}}^{IJ} = \omega_{\mathrm{D}}(3\hat{I}_z\hat{J}_z - \hat{\mathbf{I}}\cdot\hat{\mathbf{J}}) \tag{3.37}$$

$$\omega_{\mathrm{D}} := -\frac{\mu_0}{4\pi}\,\gamma^2\hbar\tfrac{1}{2}\,\frac{3\cos^2\theta - 1}{r^3}$$

(without any averaging other than high-field truncation). For the calculation, we use commutation relations of the Pauli matrices (prove by writing the commutators explicitly, e.g. $[\mathbf{I}_y\mathbf{J}_y, \mathbf{I}_z\mathbf{J}_z] = \mathbf{I}_y\mathbf{J}_y\mathbf{I}_z\mathbf{J}_z - \mathbf{I}_z\mathbf{J}_z\mathbf{I}_y\mathbf{J}_y = (\mathbf{I}_y\mathbf{I}_z)(\mathbf{J}_y\mathbf{J}_z) - (\mathbf{I}_z\mathbf{I}_y)(\mathbf{J}_z\mathbf{J}_y)$, and multiply the matrices or equivalently exploit the product relations (2.80c) without mixing I and J matrices):

$$\begin{aligned}
&[\mathbf{I}_y\mathbf{J}_y, \mathbf{I}_z\mathbf{J}_z] = [\mathbf{I}_x\mathbf{J}_x, \mathbf{I}_z\mathbf{J}_z] = [\mathbf{I}_y\mathbf{J}_y, \mathbf{I}_x\mathbf{J}_x] = 0 \; (!)\\
&[\mathbf{I}_y\mathbf{J}_y, \mathbf{I}_x\mathbf{J}_z] = 0 \; (!)\\
&[\mathbf{I}_y\mathbf{J}_y, \mathbf{I}_y\mathbf{J}_z] = \mathbf{I}_y^2\,\mathrm{i}\,\mathbf{J}_x = \tfrac{1}{4}\mathrm{i}\,\mathbf{J}_x\\
&[\mathbf{I}_y\mathbf{J}_y, \mathbf{I}_y] = 0\\
&[\mathbf{I}_y\mathbf{J}_y, \mathbf{I}_z] = \mathrm{i}\,\mathbf{I}_x\mathbf{J}_y.
\end{aligned} \tag{3.38}$$

Further relations are obtained, for instance, by interchanging I and J, or, with the necessary change of signs, by interchanging indices y and z. On the basis of the commutation relations, it is straightforward to check that certain triplets of operator combinations can be identified as angular momentum components which fulfil the angular-momentum commutation relations (2.23). Examples are:

$$\begin{aligned}
&\mathbf{I}_x^{(0)} := \mathbf{I}_x\mathbf{J}_x + \mathbf{I}_y\mathbf{J}_y \qquad \mathbf{I}_y^{(0)} := \mathbf{I}_y\mathbf{J}_x - \mathbf{I}_x\mathbf{J}_y \qquad \mathbf{I}_z^{(0)} := \tfrac{1}{2}(\mathbf{I}_z - \mathbf{J}_z)\\
&\mathbf{I}_x^{(1)} := \mathbf{I}_x\mathbf{J}_x - \mathbf{I}_y\mathbf{J}_y \qquad \mathbf{I}_y^{(1)} := \mathbf{I}_y\mathbf{J}_x + \mathbf{I}_x\mathbf{J}_y \qquad \mathbf{I}_z^{(1)} := \tfrac{1}{2}(\mathbf{I}_z + \mathbf{J}_z).
\end{aligned} \tag{3.39}$$

These operators are so-called single-transition operators (Ernst *et al.*, 1987). With a cyclic permutation of the indices x, y, z of $\mathbf{I}$ and $\mathbf{J}$, further such spin operators can be created.

Now the time evolution of the system with only one of the two spins polarized is to be considered, that is, $\boldsymbol{\rho}(0) = \mathbf{I}_z$. Because $\mathbf{I}_x\mathbf{J}_x + \mathbf{I}_y\mathbf{J}_y$ and $\mathbf{I}_z\mathbf{J}_z$ commute (see equation (3.38)), we can write:

$$\begin{aligned}
\exp(-\mathrm{i}\mathbf{H}_{\mathrm{D}}^{IJ}t) &= \exp(-\mathrm{i}\omega_{\mathrm{D}}(2\mathbf{I}_z\mathbf{J}_z - \mathbf{I}_x\mathbf{J}_x - \mathbf{I}_y\mathbf{J}_y)t)\\
&= \exp(-\mathrm{i}\omega_{\mathrm{D}}(-\mathbf{I}_x\mathbf{J}_x - \mathbf{I}_y\mathbf{J}_y)t)\,\exp(-\mathrm{i}\omega_{\mathrm{D}}2\mathbf{I}_z\mathbf{J}_z t).
\end{aligned}$$

The matrix $\exp(-\mathrm{i}\omega_{\mathrm{D}}2\mathbf{I}_z\mathbf{J}_z t)$ commutes with $\mathbf{I}_z$ and is therefore without effect on it. The relevant Hamiltonian thus reads:

$$\mathbf{H}_{\mathrm{R}} = -\omega_{\mathrm{D}}(\mathbf{I}_x\mathbf{J}_x + \mathbf{I}_y\mathbf{J}_y) \tag{3.40}$$

and with a calculation identical to that of Section 3.4.3 except for a cyclical shift of the indices x, y, z, we obtain:

$$\mathbf{I}_z \xrightarrow{\omega_{\mathrm{D}}(3\mathbf{I}_z\mathbf{J}_z - \vec{\mathbf{I}}\cdot\vec{\mathbf{J}},} = \mathbf{I}_z\tfrac{1}{2}(1 + \cos\omega_{\mathrm{D}}t) + \mathbf{J}_z\tfrac{1}{2}(1 - \cos\omega_{\mathrm{D}}t) + (\mathbf{I}_y\mathbf{J}_x - \mathbf{I}_x\mathbf{J}_y)\sin\omega_{\mathrm{D}}t. \tag{3.41}$$

That is, a homonuclear transfer $\mathbf{I}_z \rightarrow \mathbf{J}_z$ is realized. For the same problem expressed in terms of the single-transition operators (3.39), we find that around the 'field' $\mathbf{H}_\mathrm{R} = -\omega_\mathrm{D}\mathbf{I}_x^{(0)}$, a typical 'precession' occurs:

$$\mathbf{I}_z^{(1)} + \mathbf{I}_z^{(0)} \xrightarrow{-\omega_\mathrm{D}\mathbf{I}_x^{(0)}} \mathbf{I}_z^{(1)} + \mathbf{I}_z^{(0)} \cos \omega_\mathrm{D} t + \mathbf{I}_y^{(0)} \sin \omega_\mathrm{D} t. \tag{3.42}$$

This demonstrates that the single-transition operators provide another, and quite elegant, possibility of determining the time evolution for a system of interacting spins-$\frac{1}{2}$.

At this point, we can easily calculate the effect of the homonuclear dipolar Hamiltonian (3.37) on transverse magnetization of the dipolar spin pair and thus prove the prefactor 3/2 in the homonuclear dipolar anisotropy parameter (2.37). A calculation of $\boldsymbol{\rho}(t)$ according to equation (2.77) can be performed quite conveniently for a pair of spins $-\frac{1}{2}$, employing the commutator relations (3.38). The result reads:

$$\mathbf{I}_x + \mathbf{J}_x \xrightarrow{\omega_\mathrm{D}(3\mathbf{I}_z\mathbf{J}_z - \vec{\mathbf{I}}\cdot\vec{\mathbf{J}})} (\mathbf{I}_x + \mathbf{J}_x) \cos (\tfrac{3}{2}\omega_\mathrm{D} t) + 2(\mathbf{I}_y\mathbf{J}_z + \mathbf{I}_z\mathbf{J}_y) \sin (\tfrac{3}{2}\omega_\mathrm{D} t). \tag{3.43}$$

Note that here the oscillation frequency of the transverse magnetization differs by the factor of 3/2 from the homonuclear z-magnetization-transfer frequency in (3.41).

3.5.3 Spin exchange in systems of abundant spins

Spatially inhomogeneous distributions of z magnetization are changed by coherent dipolar couplings (not to be confused with T_1 relaxation). The oscillations in the two-spin system of the preceding section represent the simplest example; for systems of many spins, the complicated network of couplings cancels all oscillations to produce an effectively diffusive behaviour.

That ^{1}H spin diffusion is indeed a coherent, reversible process has been demonstrated experimentally by Zhang *et al.* (1992). To produce a polarization echo, the sign of the average dipolar Hamiltonian was inverted using a spin-lock field, which produces a prefactor $-\frac{1}{2}$ as discussed in Section 3.4.2. To detect the reversibility of the spin diffusion, ^{1}H magnetization was generated selectively in the vicinity of each ^{13}C nucleus by cross polarization from ^{1}H to ^{13}C and back to the protons. How the proton magnetization diffuses away from the ^{13}C nuclei, and returns after application of the spin-lock field, was monitored in the ^{13}C spectrum after a third CP transfer back to the ^{13}C spins.

Spatially homogeneous z magnetization is stationary (except for T_1 relaxation which is disregarded here). However, for inhomogeneous distributions of magnetization, that is, for $\hat{\rho}(0) = \Sigma_n a_n \hat{I}_z^n$ where not all a_n are equal, $\hat{H}_\mathrm{D}^{II} = \Sigma_{j,k}\, c_{jk}(3I_z^j I_z^k - \vec{\mathbf{I}}^j \cdot \vec{\mathbf{I}}^k)$ does not commute with $\hat{\rho}$ (the proof is analogous to relation (2.24)). Thus:

$$\dot{\hat{\rho}} = -\mathrm{i}[\hat{H}_\mathrm{D}^{II}, \hat{\rho}] = -\mathrm{i}\left[\sum_{j,k} A^{jk}(3I_z^j I_z^k - \vec{\mathbf{I}}^j \cdot \vec{\mathbf{I}}^k),\ \sum_n a_n(t)\hat{I}_z^n\right] \neq 0 \tag{3.44}$$

which results in a time evolution. In this process, the total z magnetization $\langle \hat{I}_z \rangle = \langle \Sigma_n \hat{I}_z^n \rangle$ is a conserved quantity because $\hat{H}_\mathrm{D}^{II}$ commutes with $\hat{I}_z$ (remember that the secular Hamiltonian $\hat{H}_\mathrm{D}^{II}$ commutes with the truncating Zeeman interaction $\omega_0 \hat{I}_z$).

There have been several attempts to derive the spin diffusion equation. In most cases, the calculations require assumptions that are not fully justified in the case of proton spin diffusion. Only below shall we discuss the classical approach by Bloembergen (Bloembergen, 1949) which was taken up by Abragam (1961); the foundations of Fermi's golden rule on which this theory relies will be introduced in the following chapter. In any case, this classical approach is based on a relatively 'naive' picture (Bloembergen, 1949), and subsequently several other theories were developed to derive the spin-diffusion equation and to obtain a useful formula for the dependence of the diffusivity D on molecular parameters.

In a density-matrix approach by Lowe and Gade (1967) it is assumed that the density matrix should be essentially diagonal during the spin diffusion process (due to T_2 relaxation). However, this assumption does not appear to be justified in the network of strongly coupled protons where nondiagonal zero-quantum coherences of the type $(\mathbf{I}_y\mathbf{J}_x - \mathbf{I}_x\mathbf{J}_y)$ (see equation (3.41)) are created much faster ($\sim 100\ \mu$s) than true T_2 relaxation can destroy them: irreversible T_2 relaxation times of protons are found to exceed 10 ms in multipulse experiments. Likewise, derivations based on methods of statistical physics, such as the Mori formalism (Mori and Kawasaki, 1962; Cheung, 1981), often use approximations or ad-hoc assumptions (unknown correlation functions are replaced by 'convenient' functions, etc.) such that finally a diffusion equation results.

In order to obtain some insight into the quantum-mechanical nature of the spin diffusion process, the calculation of the coherent dipolar evolution of large spin systems should be useful. The results of such a calculation for chains of up to 40 spin-$\frac{1}{2}$ nuclei (with nearest-neighbour couplings only) are displayed in Figure 3.8. At time $t = 0$, only the first spin carries z magnetization. In Figure 3.8(a), the time-dependence of the overall magnetization on the other spins is displayed. For two spins, the oscillations of equation (3.41) are observed, while for the larger systems the magnetization is seen to leave the first spin for a time that increases with the size of the spin system. In Figure 3.8(b), the spatial and time dependence of z magnetization on a chain of 40 spins is shown. A spin wave is seen to run through the system. This differs from the solutions of the empirical diffusion equation (3.34) which are presented in Figure 3.8(e). Even when fast relaxation is applied to terms in the density operator which do not represent z magnetization (zero-quantum coherence of the type $(\hat{I}_y^i\hat{I}_x^j - \hat{I}_x^i\hat{I}_y^j)$), a spin wave is still observed (Figure 3.8(c)). Only when the coupling strengths vary along the chain does a diffusive behaviour result (Figure 3.8(d)), which can be attributed to multiple scattering of the wave. If the restriction to a linear chain and nearest-neighbour couplings is lifted, an even more diffusive behaviour may be expected.

Some technical points concerning the spin-chain simulation of Figure 3.8 include the following: instead of diagonalizing a $2^{40} \times 2^{40}$ matrix, which is the 'direct product' of 40 2×2 Pauli-type matrices, $\hat{\rho}(t)$ was calculated by means of the time-evolution expansion (2.75) up to the t^{10} term for $t \leqslant t_1$, where $t_1 \cong 1/\omega$ is the inverse coupling strength. $\hat{\rho}(t_1)$ was taken as the initial $\hat{\rho}(0)$ in the next step of the calculation, again using the series (2.75), to proceed up to $2t_1$. The operator $\hat{\rho}(2t_1)$ was used as the new starting point, and so on, without accumulation of errors. The commutators required in (2.75) were evaluated by means of Pauli-matrix relations as given in (3.38), with $\hat{\rho}(t)$ represented as a sum of products of 40 spin operators, one operator for each spin. These products could be treated as 40-digit expressions, each digit containing a 4-bit information ('0' $\triangleq \hat{1}^j$, '1' $\triangleq \hat{I}_x^j$, '2' $\triangleq \hat{I}_y^j$, '3' $\triangleq \hat{I}_z^j$). One digit per spin is always sufficient, as the commutators never produce a product of operators of the same spin j (i.e. no $\hat{I}_\alpha^j\hat{I}_\beta^j$). The book-keeping with choice of appropriate commutators, and combination of equal terms in the sum after each step is easily handled by a computer.

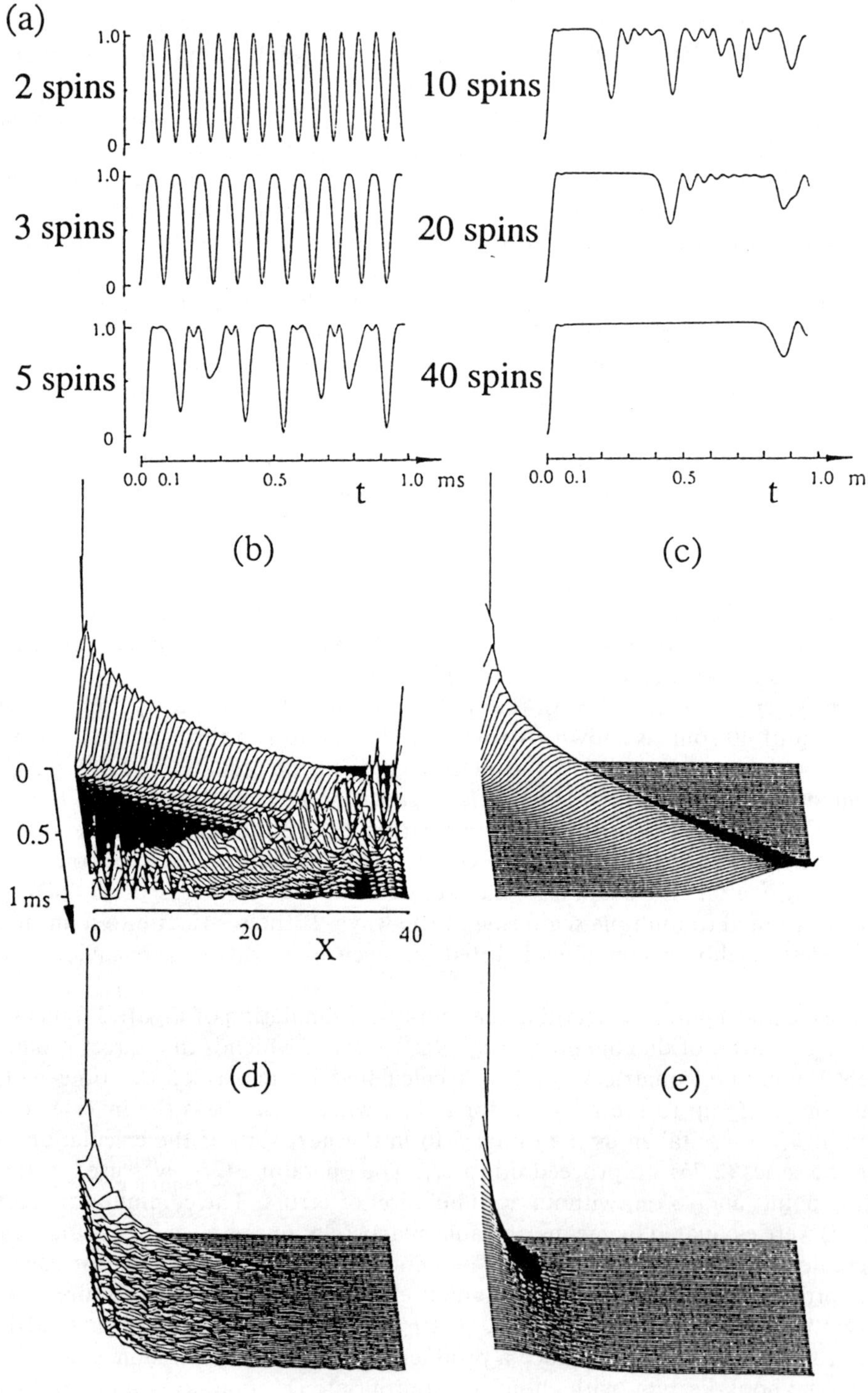
(a)
1.0
2 spins
0
1.0
3 spins
0
1.0
5 spins
0
0.0 0.1
0.5
1.0 ms
t
10 spins
20 spins
40 spins
0.0 0.1
0.5
1.0 ms
t
(b)
(c)
0
0.5
t
1 ms
0
20
40
X
(d)
(e)

3.6 Time-dependent perturbation theory

In the preceding discussions, it was assumed that the dipolar coupling that produces the spin diffusion was the only relevant interaction. In many cases, this assumption is not justified: for ^{13}C in natural abundance, the $^{1}H-^{13}C$ couplings and the chemical-shift differences are much larger than the $^{13}C-^{13}C$ couplings (which are typically 15 Hz). Even in the discussion of ^{1}H spin diffusion, it is customary to split the homonuclear interaction into the interaction between the 'special' spins that exchange magnetization and the dipolar action of the other spins onto the 'special' spins; with this trick, it is intended to make the system suitable for a perturbation treatment. To discuss this approach, in this section the results of time-dependent perturbation theory will be reviewed. We shall not derive the whole framework of perturbation theory since it can be found in all standard texts on quantum mechanics (Sakurai, 1985). Instead, our discussion will be based on the relation of the perturbation-theoretical results to some readily available exact solutions of the time-dependent Schrödinger equation.

For a system subject to a time-independent Hamiltonian $\hat{H}_0$, we want to determine what change is induced by switching on an otherwise time-independent perturbation Hamiltonian $\hat{H}_1$ at the time $t = 0$. In the case of spin diffusion, this time-independent perturbation is the dipolar coupling Hamiltonian that generates the magnetization transfer. In the following, the quantity to be calculated is $P_{\mathrm{if}}(t)$, the transition probability (i.e. the probability of having made a transition) from the initial state i to the final state f at a time t after the perturbation was switched on. In order to simplify the treatment, it is assumed that only state i is initially populated. For a two-level system, the Schrödinger equation can be written as:

$$-\frac{\hbar}{\mathrm{i}}\begin{bmatrix}\dot{c}_{\mathrm{i}}(t)\\ \dot{c}_{\mathrm{f}}(t)\end{bmatrix}=\begin{bmatrix}\frac{1}{2}\Delta\omega_{\mathrm{if}}+\delta\omega_0 & \frac{1}{2}\omega_1^{\mathrm{if}}\\ \frac{1}{2}\omega_1^{\mathrm{fi}} & -\frac{1}{2}\Delta\omega_{\mathrm{if}}+\delta\omega_0\end{bmatrix}\begin{bmatrix}c_{\mathrm{i}}(t)\\ c_{\mathrm{f}}(t)\end{bmatrix} \tag{3.45}$$

where the $c_n(t)$ are the coefficients of the eigenfunctions $|n\rangle$ of $\hat{H}_0$, and $\frac{1}{2}\omega_1^{\mathrm{if}} = \langle \mathrm{i} | \hat{H}_1 | \mathrm{f} \rangle$ is the transition-matrix element (the factor of $\frac{1}{2}$ is introduced here for later convenience). $\Delta\omega_{\mathrm{if}} = \langle \mathrm{i} | \hat{H}_0 | \mathrm{i} \rangle - \langle \mathrm{f} | \hat{H}_0 | \mathrm{f} \rangle$ is the zeroth-order energy difference between states i and f (recall that our Hamiltonians are given in frequency units), and $\delta\omega_0 = \langle \mathrm{i} | \hat{H}_1 | \mathrm{i} \rangle = \langle \mathrm{f} | \hat{H}_1 | \mathrm{f} \rangle$. Under the specified conditions,

$$P_{\mathrm{if}}(t) = |c_{\mathrm{f}}(t)|^2. \tag{3.46}$$

While the exact formulas given below refer to such a two-level system, the perturbation-theory results, which represent approximations to the exact two-level

Figure 3.8 Spin dynamics on chains of N spins-$\frac{1}{2}$ with only nearest-neighbour couplings of 15 kHz. (a) Time evolution for $N = 2, 3, 5, 10, 20, 40$, with the initial condition that at $t = 0$ ms, only the first spin carries z magnetization. The time dependence of the total magnetization of the other spins is plotted. For $N = 2$, the $\frac{1}{2}(1 - \cos \omega_{\mathrm{D}} t)$ oscillations of equation (3.41) are observed. (b) Space–time dependence of the magnetization on the $N = 40$ chain. A spin wave is observed. (c) Like (b), but with fast relaxation of the zero-quantum coherences. A slower spin wave results. (d) Like (b) but with random variation (between 5 and 25 kHz) of the coupling strengths along the chain (average over 128 chains of 40 spins). (e) Same system behaving according to the discrete spin-diffusion equation (3.33), with $a = 0.2$ nm, and $D = 0.8$ nm^2/ms. The spatial profiles at a fixed time are Gaussians, broadening proportional to $\sqrt{t}$.

results for short times, apply for any state f in a multilevel system: the initial behaviour is always equivalent to an i–f two-level system; the couplings between different initially unpopulated levels appear only in higher-order terms of the theory.

I $\Delta\omega_{if} = 0$: *transitions between two states of equal energy (in NMR: of equal Larmor frequency)*

Exact calculation:

$$P_{if}(t) = \sin^2(\tfrac{1}{2}\omega_1^{if}t) = \tfrac{1}{2}\{1 - \cos(\omega_1^{if}t)\}. \qquad (3.47)$$

Perturbation-theory result:

$$P_{if}(t) \cong (\tfrac{1}{2}\omega_1^{if}t)^2. \qquad (3.48)$$

The perturbation-theory result represents the initial behaviour of the exact relation (3.47) $[\sin(\tfrac{1}{2}\omega_1^{if}t) = \tfrac{1}{2}\omega_1^{if}t + \ldots]$. It applies only for $t \ll 1/\omega_1^{if}$.

II Transitions between states of different energy with a separation $\Delta\omega_{if}$ *(caused, for instance, by a difference in chemical shift)*

This is an off-resonance condition since the oscillation frequency of the 'irradiated' energy $\hat{H}_1$ is zero according to the constancy of $\hat{H}_1$. According to a widespread simplistic notion, no transitions are possible for such an off-resonance situation, at least after a time Δt with $\Delta E \,.\, \Delta t \cong \hbar$ and $\Delta E = \hbar\Delta\omega_{if}$. However, transitions do occur for infinite times.

Exact solution, 'Rabi formula' (Atkins (1983) and the derivation indicated below):

$$P_{if}(t) = \tfrac{1}{2}\{1 - \cos[t\sqrt{(\Delta\omega_{if})^2 + (\omega_1^{if})^2}]\}\,\frac{(\omega_1^{if})^2}{(\Delta\omega_{if})^2 + (\omega_1^{if})^2}. \qquad (3.49)$$

Perturbation-theory result:

$$P_{if}(t) \cong \tfrac{1}{2}\{1 - \cos(\Delta\omega_{if}t)\}\left(\frac{\omega_1^{if}}{\Delta\omega_{if}}\right)^2. \qquad (3.50)$$

This approximation (3.50) applies as long as $P_{if}(t) \ll 1$. Consequently, if we have $(\omega_1^{if}/\Delta\omega_{if})^2 \ll 1$, $P_{if}(t)$ of equation (3.50) represents a valid approximation to the exact solution (3.49) for arbitrarily long times. Only when the energy difference $\Delta\omega_{if}$ becomes comparable to or smaller than the coupling ω_1^{if} is the validity of the approximation (3.50) restricted to short times $t \ll 1/\omega_1^{if}$. Note that in the perturbation result (3.50) the oscillatory behaviour involves the 'energy-level-difference' frequency $\Delta\omega_{if}$ rather than the 'Hamiltonian-magnitude' frequency ω_1^{if} as in the exact on-resonance solution (3.47). However, this is not a fundamental difference, but rather reflects limiting cases of the general effective frequency $\sqrt{(\Delta\omega_{if})^2 + (\omega_1^{if})^2}$ in equation (3.49). The 'on-resonance' equations (3.47) and (3.48) are obtained from (3.49) and (3.50) for $\Delta\omega_{if} = 0$, or rather $\Delta\omega_{if} \to 0$, employing $\cos(\Delta\omega_{if}t) = 1 - \tfrac{1}{2}(\Delta\omega_{if}t)^2 + \ldots$. The factor $(\omega_1^{if}/\Delta\omega_{if})^2$ in equation (3.50) shows that the transfer becomes very inefficient (but not impossible) for energy or chemical-shift differences $\Delta\omega_{if}$ that are larger than the perturbing field ω_1^{if}.

The Rabi formula has a visualization that is easily understood and instructive in the NMR context: $P_{if}(t)$ can be expressed as:

$$P_{if}(t) = \frac{1}{2}\frac{M_0 - M_z(t)}{M_0} \tag{3.51}$$

where $M_z(t)$ is the z component of a magnetization vector $\vec{\mathbf{M}}$ precessing around a field:

$$\vec{\mathbf{B}}_{tot} = \Delta\vec{\mathbf{B}}_0 + \vec{\mathbf{B}}_1 \qquad \omega_{eff} = \gamma\,|\,\vec{\mathbf{B}}_{tot}\,| = \gamma\sqrt{(\Delta\vec{\mathbf{B}}_0)^2 + (\vec{\mathbf{B}}_1)^2} \tag{3.52a}$$

where

$$\gamma\Delta\vec{\mathbf{B}}_0 = \Delta\omega_{if}\begin{bmatrix}0\\0\\1\end{bmatrix} \qquad \gamma\vec{\mathbf{B}}_1 = \omega_1^{if}\begin{bmatrix}1\\0\\0\end{bmatrix}. \tag{3.52b}$$

The relation between the Rabi oscillations and the precession around a transverse field (Figure 3.9) is based on the fact that the precessing magnetization describes the quantum-mechanical time evolution of a two-level system of a spin-$\frac{1}{2}$ nucleus under the action of an off-resonance field $\Delta\vec{\mathbf{B}}_0$ and a perturbing field $\vec{\mathbf{B}}_1$. The Hamiltonian matrices for this problem are easily found by employing the Pauli matrices (see equation (2.80a)):

$$\mathbf{H}_0^{\Delta} = \gamma\Delta\vec{\mathbf{B}}_0\cdot\vec{\mathbf{I}} = \gamma\Delta B_0\mathbf{I}_z = \gamma\begin{bmatrix}\frac{1}{2}\Delta B_0 & 0\\ 0 & -\frac{1}{2}\Delta B_0\end{bmatrix} \tag{3.53a}$$

$$\mathbf{H}_1 = \gamma\vec{\mathbf{B}}_1\cdot\vec{\mathbf{I}} = \gamma B_1\mathbf{I}_x = \gamma\begin{bmatrix}0 & \frac{1}{2}B_1\\ \frac{1}{2}B_1 & 0\end{bmatrix}. \tag{3.53b}$$

The solution of the corresponding Schrödinger matrix equation (see equation (3.45)) yields the Rabi formula; in particular, the eigenvalue of $\mathbf{H}_0^{\Delta} + \mathbf{H}_1$ is the precession frequency $\omega_{pr} = \gamma\sqrt{(\Delta B_0)^2 + (B_1)^2}$. This is recognized as the effective precession frequency around $\vec{\mathbf{B}}_{tot}$ as given in equation (3.51), and as the oscillation frequency in the Rabi formula if the substitutions (3.52b) are made. In Figure 3.9, with the angle

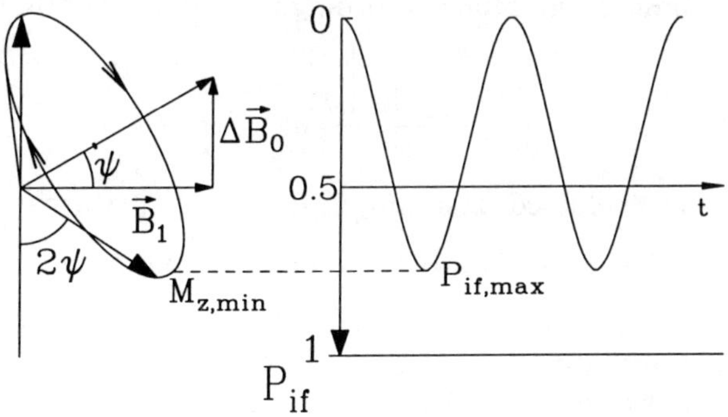

Figure 3.9 Precession of magnetization around the sum of a $\vec{\mathbf{B}}_1$ field and an off-resonance field $\Delta\vec{\mathbf{B}}_0$ and resulting Rabi oscillations of the z magnetization. The corresponding P_{if} values can be read off the vertical scale: when the magnetization is fully along the positive z axis, the system is fully in the initial state, $P_{if} = 0$; when $\vec{\mathbf{M}}$ points along $-z$, the final state is fully populated, $P_{if} = 1$.

ψ between $\vec{\mathbf{B}}_1$ and $\vec{\mathbf{B}}_{tot}$, the z component (or projection) of the magnetization at its lowest point on the cone of precession is found to be:

$$M_{z,\min} = -M_0 \cos 2\psi. \tag{3.54a}$$

On this basis, using equation (3.51) the amplitude factor:

$$P_{if,\max} = \tfrac{1}{2}\frac{M_0 - M_{z,\min}}{M_0} = \tfrac{1}{2}\{1 + \cos(2\psi)\}$$

$$= \cos^2\psi = \left(\frac{B_1}{B_{tot}}\right)^2 = \frac{(\omega_1^{if})^2}{(\Delta\omega_{if})^2 + (\omega_1^{if})^2} \tag{3.54b}$$

of the Rabi oscillation (Figure 3.9) is easily derived. It may be interesting to note that the Rabi formula is essentially the square to the pulse excitation profile as discussed in Section 2.6.5; this is again a consequence of the relation between the Rabi oscillations and the precession under a transverse field off-resonance.

III Transitions into a band of energies

Often, the transitions involve not just two definite states with a single energy difference, but indeed states with a distribution of energy differences $\rho(\Delta\omega_{if})$, as, for instance, in a chemical-shift powder pattern.

Exact solution: The Rabi-formula transition probabilities of equation (3.49), weighted with $\rho(\Delta\omega_{if})$, are integrated over $\Delta\omega_{if}$.

In perturbation theory, the equivalent integration is performed employing $P_{if}(t; \Delta\omega_{if})$ of equation (3.50). With a distribution $\rho(\Delta\omega_{if})$ of width δ_f, for $t \gg 1/\delta_f$ one can write:

$$\int_{-\infty}^{\infty} \rho(\omega) P_{if}(t;\omega)\, d\omega \cong \rho(0) \int_{-\infty}^{\infty} P_{if}(t;\omega)\, d\omega \tag{3.55}$$

since the portions of $P_{if}(t; \Delta\omega_{if})$ very close to $\Delta\omega_{if} = 0$ increase with t^2 (see equation (3.48)) and thus become dominant for sufficiently long times (which must still obey $t \ll 1/\omega_1^{if}$); for $\Delta\omega_{if}$ values further from the zero point, the transition probabilities oscillate only with a limited amplitude according to equation (3.50). Using the definite integral:

$$\int_{-\infty}^{\infty} \frac{1 - \cos \omega t}{\omega^2}\, d\omega = \pi t \tag{3.56}$$

equation (3.55) can be evaluated, resulting in the simple perturbation-theory result:

$$P_{if}(t) \cong \tfrac{1}{2}\pi\, (\omega_1^{if})^2 \rho(0) t \qquad \text{'Fermi's golden rule'.} \tag{3.57}$$

In this formula, only $\rho(0)$ appears: if the system 'has a choice', it prefers the resonant (energy-conserving) transitions with $\Delta\omega_{if} = 0$. $P_{if}(t)$ in equation (3.57) increases only linearly in t although its height $P_{if}(t; \Delta\omega_{if} = 0)$ increases as t^2 (see equation (3.48)); this can be understood by considering that it is not only the exact point $\Delta\omega_{if} = 0$ that counts but also $P_{if}(t; \Delta\omega_{if})$ in its environment, which decreases in its $\Delta\omega_{if}$-width proportional to $1/t$. For very short times, $t < 1/\delta_f$, indeed $P_{if}(t) \sim t^2$ is found. It should be noted that Fermi's golden rule is not necessarily a 'relaxation expression'. The comparison with the exact solution proves that it describes the initial behaviour of a completely reversible time evolution with a time-independent Hamiltonian. In order to generate an irreversible relaxation behaviour, the Hamiltonian must fluctuate stochastically in time (see Section 3.13).

When introducing $\rho(\Delta\omega_{\rm if})$, it is customary to assume that only the final-state energies are distributed. However, there is no reason why the initial-state energies should not be distributed in a similar way. This can be easily incorporated into the theory, since $\rho(\Delta\omega_{\rm if})$ involves only the probability distribution of energy differences. With an initial-state energy distribution $\rho_{\rm i}(\omega_{\rm i})$ and a final-state distribution $\rho_{\rm f}(\omega_{\rm f})$, the probability density $\rho(\Delta\omega_{\rm if})$ for finding a given energy difference $\Delta\omega_{\rm if}$ is the product of the probabilities $\rho_{\rm i}$ and $\rho_{\rm f}$ with the difference of the arguments $\omega_{\rm f}-\omega_{\rm i}$ being equal to $\Delta\omega_{\rm if}$, integrated over the distributions:

$$\begin{aligned}\rho(\Delta\omega_{\rm if}) &= \iint \rho_{\rm i}(\omega_{\rm i})\,\rho_{\rm f}(\omega_{\rm f})\,\delta(\omega_{\rm f}-\omega_{\rm i}-\Delta\omega_{\rm if})\,{\rm d}\omega_{\rm i}\,{\rm d}\omega_{\rm f}\\ &= \int \rho_{\rm i}(\omega_{\rm f}-\Delta\omega_{\rm if})\rho_{\rm f}(\omega_{\rm f})\,{\rm d}\omega_{\rm f}. \qquad (3.58)\end{aligned}$$

If there is only a single initial-state energy $\omega_{\rm i0}$, so that $\rho_{\rm i}(\omega)=\delta(\omega-\omega_{\rm i0})$, equation (3.58) reduces to:

$$\rho(\Delta\omega_{\rm if}) = \int \delta(\omega_{\rm f}-\Delta\omega_{\rm if}-\omega_{\rm i0})\rho_{\rm f}(\omega_{\rm f})\,{\rm d}\omega_{\rm f} = \rho_{\rm f}(\Delta\omega_{\rm if}+\omega_{\rm i0})$$

which represents the customary identification of $\rho(\Delta\omega_{\rm if})$ and the distribution of final states.

3.7 Proton-driven ^{13}C spin diffusion, and ^{1}H spin diffusion

In this section, it will be shown that in organic solids the transition probability for ^{13}C spin-diffusion processes, or more precisely between two nuclei A and B with effective Larmor frequencies ω_A and ω_B, is given by:

$$P_{\rm if}(t) = \tfrac{1}{2}\pi g_0^{AB}(\omega_A-\omega_B)\omega_{\rm D}^2 t. \qquad (3.59)$$

This formula (Abragam, 1961), derived from Fermi's golden rule (3.57), contains all the important factors for 'proton-driven' ^{13}C spin diffusion: the ^{13}C–^{13}C coupling frequency $\omega_{\rm D}$, the chemical-shift difference $\omega_A-\omega_B$ and the 'zero-quantum lineshape' $g_0^{AB}(\omega)$, which is determined by the ^{1}H–^{13}C couplings and has a width corresponding to the average ^{1}H–^{13}C next-neighbour couplings.

This relation will be derived on the basis of the expressions of the previous section, after having related those to the results on dipolar couplings obtained in Section 3.5.

First, we consider two ^{13}C–^{13}C spins with identical chemical shift (i.e. $\omega_{\rm A}-\omega_{\rm B}=\Delta\omega_{\rm if}$) and only the ^{13}C–^{13}C coupling being effective. For such a two-spin system, oscillations corresponding exactly to those of equation (3.47) were found in Section 3.5.1, with $\omega_1^{\rm if}=\omega_{\rm D}$. This indicates that we can use the results of Section 3.6 as a basis for the treatment of more complicated cases of homonuclear dipolar couplings.

When a chemical-shift difference $\Delta\omega_{\rm if}$ is introduced (but with protons still decoupled), case (II) applies. From the factor $\omega_{\rm D}^2/\{(\Delta\omega_{\rm if})^2+\omega_{\rm D}^2\}$ in the transition probability (3.49), it is evident that magnetization transfer becomes inefficient if the ^{13}C–^{13}C dipolar coupling $\omega_{\rm D}$ is smaller than the difference $\Delta\omega_{\rm if}$ of the resonance frequencies. Since ω_D is only 15 Hz for the average nearest-neighbour distance of 0.8 nm, ^{13}C spin diffusion is very inefficient under these circumstances. For a distribution of chemical shifts, as in a powder pattern under proton decoupling, Fermi's golden rule (3.57) tells us that spin exchange is only efficient for sites with $\Delta\omega_{if} = 0$, that is, identical chemical shifts.

By means of special techniques, the effect of the chemical-shift differences can be removed. One approach relies on averaging the chemical-shift Hamiltonian to zero (see Section 3.4.2) by spin-locking the carbons, while the dipolar coupling is only scaled by $-\frac{1}{2}$ ('rf-driven' spin diffusion). It is also possible to compensate the effect of the chemical-shift difference by magic-angle spinning with a spinning speed which is a multiple of the shift difference ('rotor-driven' spin diffusion). These techniques will be discussed in Section 13.5.

In 'natural' spin diffusion, that is, without the ^{1}H decoupling that had been assumed so far in this section, it is the ^{1}H–^{13}C dipolar couplings which can produce a resonance shift that cancels the chemical-shift difference $\omega_A - \omega_B$ and thus makes ^{13}C spin diffusion possible. In this manner, the protons 'drive' the ^{13}C spin diffusion. Note that proton-driven spin-diffusion does not at all involve a transfer of magnetization via the protons; the transfer occurs through space directly from ^{13}C to ^{13}C.

To indicate the quantum-mechanical nature of the compensation of the chemical-shift difference by the dipolar coupling, we consider a pair of ^{13}C nuclei labeled A and B, coupled to a proton. The relevant part of the Hamiltonian reads:

$$\hat{H} = \omega_{S-S}\tfrac{1}{2}(3\hat{S}_Z^A\hat{S}_Z^B - \hat{\mathbf{S}}^A\cdot\hat{\mathbf{S}}^B) + (\omega_A - \omega_B)(\hat{S}_Z^A - \hat{S}_Z^B) + 2\omega_{IS}^A\hat{I}_Z\hat{S}_Z^A + 2\omega_{IS}^B\hat{I}_Z\hat{S}_Z^B. \quad (3.60)$$

The term $(\omega_A + \omega_B)(\hat{S}_Z^A + \hat{S}_Z^B) = (\omega_A + \omega_B)\hat{S}_Z$ has not been kept since it commutes with all the other parts of the Hamiltonian as well as with the z magnetization of the density matrix and is thus without effect (see equation (2.71)).

The $\hat{I}$ part of the Hamiltonian in equation (3.60) can be written in terms of Pauli matrices (see equation (2.80a)); discarding again terms proportional to $\hat{S}_Z^A + \hat{S}_Z^B$, we have:

$$\hat{\mathbf{H}} = \begin{bmatrix} \hat{H}_{S,S} + (\omega_A - \omega_B + \omega_{IS}^A - \omega_{IS}^B)(\hat{S}_Z^A - \hat{S}_Z^B) & 0 \\ 0 & \hat{H}_{S,S} + (\omega_A - \omega_B - \omega_{IS}^A + \omega_{IS}^B)(\hat{S}_Z^A - \hat{S}_Z^B) \end{bmatrix}. \quad (3.61)$$

If $(\omega_A - \omega_B) = -(\omega_{IS}^A - \omega_{IS}^B)$, one of the matrix elements reduces to $\hat{H}_{S,S}$. The exponential of a diagonal matrix is a diagonal matrix of exponentials (see equation (D.8)), so the propagator relevant for the magnetization transfer reads:

$$\exp(-\mathrm{i}\hat{\mathbf{H}}t) = \begin{bmatrix} \exp(-\mathrm{i}\hat{H}_{S,S}t) & 0 \\ 0 & \exp\{-\mathrm{i}\hat{H}_{S,S}t - \mathrm{i}2(\omega_{IS}^B - \omega_{IS}^A)(\hat{S}_Z^A - \hat{S}_Z^B)t\} \end{bmatrix}. \quad (3.62)$$

Obviously, due to the compensation of the chemical-shift difference $(\omega_A - \omega_B)$ by the dipolar couplings $(\omega_{IS}^B - \omega_{IS}^A)$, in the propagator matrix element $\exp(-\mathrm{i}\hat{H}_{S,S}t)$ the ^{13}C–^{13}C coupling can act directly.

The various heteronuclear dipolar couplings in a real solid produce distributions of resonance frequencies $g_{IS}^A(\omega - \omega_A)$ and $g_{IS}^B(\omega - \omega_B)$ for spins A and B; these distributions will be centred on ω_A and ω_B, respectively. According to equation (3.58), the distribution of energy differences is given by:

$$\rho(\Delta\omega_{\text{if}}) = \int g_{IS}^{A}(\omega - \omega_A - \Delta\omega_{\text{if}})\, g_{IS}^{B}(\omega - \omega_B)\, d\omega$$
$$= \int g_{IS}^{A}(\omega' - \{\omega_A - \omega_B + \Delta\omega_{\text{if}}\})\, g_{IS}^{B}(\omega')\, d\omega'. \qquad (3.63)$$

This is related to the zero-quantum line-shape function:

$$g_0^{AB}(\omega) = \iint g_{IS}^{A}(\omega_{\text{i}})g_{IS}^{B}(\omega_{\text{f}})\, \delta(\omega - \{\omega_{\text{i}} - \omega_{\text{f}}\})\, d\omega_{\text{i}}\, d\omega_{\text{f}}$$
$$= \int g_{IS}^{A}(\omega' - \omega)\, g_{IS}^{B}(\omega')\, d\omega' \qquad (3.64)$$

which represents the probability that, disregarding the Larmor frequencies, the energies of spin A and B are different by ω due to the dipolar couplings. Equation (3.64) reveals that $g_0^{AB}(\omega)$ is the (cross-)correlation function of $g_{IS}^{A}(\omega)$ and $g_{IS}^{B}(\omega)$. Since $g_{IS}^{A}(\omega)$ and $g_{IS}^{B}(\omega)$ are usually rather featureless Gaussian-type lines, $g_0^{AB}(\omega)$ will also be approximately Gaussian with a width $2\sigma_{g_0}$ which depends on the widths $2\sigma_A$ and $2\sigma_B$ of $g_{IS}^{A}(\omega)$ and $g_{IS}^{B}(\omega)$, respectively, according to $2\sigma_{g_0} \cong 2\sqrt{\sigma_A^2 + \sigma_B^2}$.

Combining $\rho(\Delta\omega_{\text{if}})$ of equation (3.63) and $g_0^{AB}(\omega)$ of equation (3.64), we find:

$$\rho(\Delta\omega_{\text{if}}) = g_0^{AB}(\Delta\omega_{\text{if}} + \omega_A - \omega_B) \qquad (3.65)$$

so that Fermi's golden rule (3.57), with $\rho(0) = g_0^{AB}(\omega_A - \omega_B)$ inserted, results in:

$$P_{\text{if}}(t) = \tfrac{1}{2}\pi\, g_0^{AB}(\omega_A - \omega_B)\, \omega_{\text{D}}^2 t \qquad (3.66)$$

as given at the beginning of this section. The factor $g_0^{AB}(\omega_A - \omega_B)$ is the bottle-neck of the spin-diffusion process, and cannot become very large since it is subject to conflicting requirements: in order to exhibit appreciable intensity at $(\omega_A - \omega_B)$, the $g_0^{AB}(\omega)$ lineshape must be relatively broad; on the other hand, the area of $g_0^{AB}(\omega)$ is restricted since the underlying $g_{IS}(\omega)$ probability densities are normalized to unity, and as a consequence with increasing breadth, the absolute height of the $g_0^{AB}(\omega)$ curve decreases.

Fermi's golden rule in the form of equation (3.66) has also been applied to ^1H spin diffusion where $\omega_A - \omega_B \cong 0$ due to the small chemical shifts of protons. Here, both ω_{D} and the distribution $g_0^{AB}(\omega)$ are due to proton–proton couplings. While ω_{D} produces spin diffusion, the couplings generating $g_0^{AB}(\omega)$ should not; this is in fact possible if they are the couplings between ^1H spins with similar polarization (i.e. along the spin-diffusion front). While equation (3.66) appears to be applicable for couplings ω_{D} that are weak because the angle between the A–B internuclear vector and $\vec{\mathbf{B}}_0$ is close to 54.7°, the largest couplings ω_{D} are comparable to the typical coupling δ_{f} between all the other spins and the spin pair. This case of $\omega_{\text{D}} \cong \delta_{\text{f}}$ violates the preconditions $1/\omega_{\text{D}} \gg t$ and $t \gg 1/\delta_{\text{f}}$ in the derivation of Fermi's golden rule (Suter and Ernst, 1985). Nevertheless, this does not make the ^1H–^1H magnetization transfer uncertain as such; as shown in Section 3.5.3, it can proceed well without additional fields, and even then, a phenomenologically diffusive behaviour was found if the coupling strengths ω_{D} vary along the spin-diffusion pathway.

According to equation (3.66), the influence of the internuclear distances on the transition rate is different for diffusion among rare spins and among abundant spins.

For the rare spins, the $1/r_{jk}^6$ dependence resulting from the dipolar-frequency factor ω_{jk}^2 remains unchanged. For the abundant spins, however, in the evaluation of equation (3.66) the average transition rate is found to be only proportional to $1/a^3$ where a is the averaged distance:

$$a = \left\langle \frac{1}{r_{jk}^3} \right\rangle^{-\frac{1}{3}} \qquad \frac{1}{a^3} = \left\langle \frac{1}{r_{jk}^3} \right\rangle. \tag{3.67}$$

Due to the fast decay of $1/r^3$ with r, this a is essentially the nearest-neighbour distance. The $1/a^3$ dependence of $\langle \omega_D^2 g_0^{AB}(0) \rangle$ arises via the multiplication of $\omega_D^2 \sim 1/r^6$ by $g_0^{AB}(0) \sim r^3$, which derives from the proportionality of δ_f (the width of $F(\omega)$) to the strength of the homonuclear dipolar couplings, $\delta_f \sim 1/a^3$. This relationship between $g_0^{AB}(0)$ and δ_f is due to the fact that the area under the $g_0^{AB}(\omega)$ curve (which is normalized to unity) is proportional to the width δ_f times the height $g_0^{AB}(0)$:

$$g_0^{AB}(0)\delta_f \sim \int_0^\infty d\omega \; g_0^{AB}(\omega) = 1$$
$$\Rightarrow g_0^{AB}(0) \sim 1/\delta_f \sim r^3. \tag{3.68}$$

With the rate Ω in $D = \Omega a^2$ (equation (3.36)) given by $\Omega = P_{if}(t)/t$, we find:

$$D \sim \frac{1}{a} \tag{3.69}$$

for ^1H spin diffusion in rigid solids; this makes spin diffusion much more efficient than direct through-space dipolar couplings ($\sim 1/r^3$) or through-space cross relaxation ($\sim 1/r^6$).

3.8 Selective excitation by hard pulses

In the application of hard pulses discussed so far, a dependence of the excitation on the frequency offset from the irradiation frequency is undesirable. In several experiments, however, selective excitation is of interest. In particular, one-dimensional experiments with frequency-selection of a single line bear a close conceptual relation to two-dimensional experiments. This will be outlined in some detail in Sections 7.2 and 13.2.4.

According to the discussion given in Section 2.6.5, the simplest form of selective excitation consists in continuous-wave irradiation, or a long soft pulse, which can also be shaped to produce the desired excitation profile. In solid-state NMR, however, in addition to technical difficulties in reducing the rf power to the desired low level, homonuclear decoupling, if required during the precession of the magnetization, interferes with the soft pulse. Under ^1H multiple-pulse decoupling conditions in particular, a soft pulse cannot be implemented. Therefore, pulsed methods of selective excitation are necessary. Two methods based on different principles have been employed in the experiments discussed below.

The DANTE sequence (Bodenhausen *et al.*, 1976; Morris and Freeman, 1978) employs many short pulses of only one phase separated by a delay of free precession. The selectivity is based on the fact that the effects of the pulses add up only for those isochromats that precess by an angle of exactly $n \times 360^\circ$ during the free-precession delay. This is visualized in Figure 3.10, where for clarity only the differences with the 360° precession angles are shown. The effect of a pulse is a small rotation of the magnetization. These rotations appear as short vertical lines in the representation in Figure 3.10.

The principle of the DANTE sequence has also been exploited for selective excitation under multiple-pulse irradiation (Caravatti *et al.*, 1986). In immediate analogy to the normal DANTE sequence, short pulses of a unique phase can be inserted between MREV-8 sequences. However, due to the dipolar coupling during these extra pulses, the average Hamiltonian does not vanish any more. Therefore, Caravatti *et al.* have

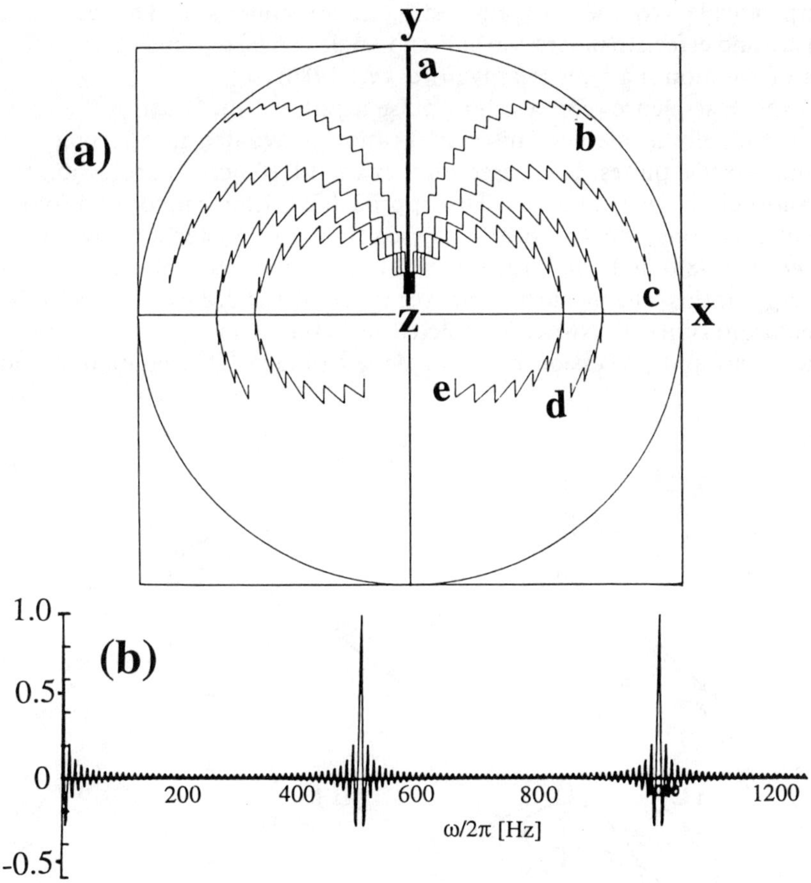

Figure 3.10 Visualization of selective excitation by the DANTE pulse sequence. (a) For twenty*x* pulses of 4.5° flip angle each, spaced by 2 ms, trajectories of the tips of various isochromats are shown projected onto the *xy* plane. The 4.5° pulses produce the small recurring displacements along the *y* direction (upward in the figure). The on-resonance isochromat *a* is rotated fully onto the *y* axis. So are the isochromats for offsets $\Delta\omega/2\pi = n/2$ ms, because they return to their original direction at the end of the 2 ms pulse spacing. The other isochromats (examples (b)–(e)) move away from the *y* axis during the free-precession periods in between the pulses. (b) Frequency profile of the *y* magnetization after DANTE excitation by fifty 1.8° pulses spaced by 2 ms. Adapted from Bodenhausen *et al.* (1976).

introduced a sequence with the last pulses in certain MREV-8 cycles extended as for a short extra pulse. This is compensated by shortening the last pulse in other MREV-8 cycles, which corresponds to a short pulse with the opposite sign of rotation. This is the equivalent of a DANTE sequence with phase-alternated pulses where isochromats precessing by 180° (instead of the 360° above) are flipped most efficiently. In the average-Hamiltonian description most appropriate for multiple-pulse experiments, these short pulses change the axes of precession. Thus, the magnetization is subject to alternating precession periods around two slightly different axes. Initially, by a 45° pulse the magnetization is set between the two precession axes. For $\Delta\omega = 0$, the magnetization remains along its original direction. For precession angles of 180° (i.e. $\Delta\omega t_{pr} = \pi$), the magnetization descends in large steps. The effective precession axis is then perpendicular to the two alternating precession axes. The general relations between $\Delta\omega$ and effective precession axis and angle can be obtained from a formulation in terms of rotation matrices (Caravatti *et al.*, 1986).

The DANTE sequence requires short pulse lengths of the order of 0.2 μs, which are often not routinely available. This complication is avoided in a technique involving only x and $-x$ 90° pulses. It is based directly on differences in precession frequencies and the short effective T_2 in solids. This approach has been employed unsystematically on several occasions and has been analysed and used systematically for ^{13}C NMR (Tekely *et al.*, 1988) and for ^{1}H CRAMPS applications (Schmidt-Rohr *et al.*, 1990). The principle of this 'chemical-shift filter', termed SELDOM (selectivity by destruction of magnetization) in ^{13}C NMR, is depicted in Figure 3.11: after 90° pulse excitation from the z axis and precession by an angle ωt_s (under ^{1}H decoupling), another 90°

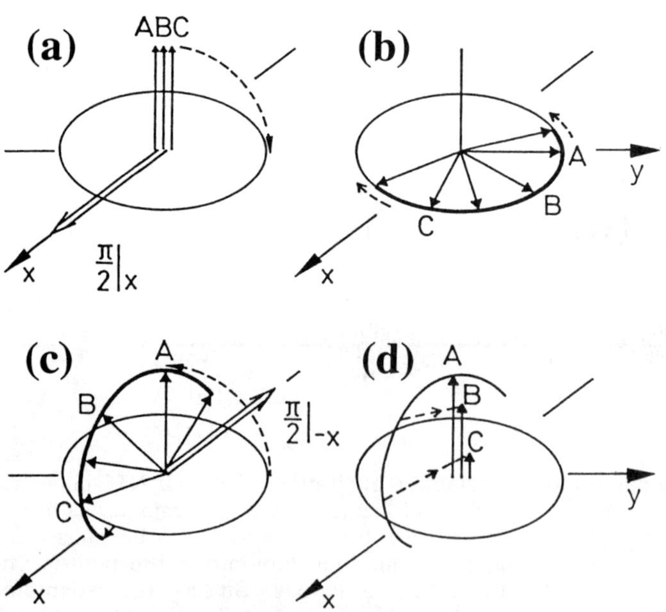

Figure 3.11 Principle of SELDOM (chemical-shift filter). The isochromats are separated on the basis of their different precession frequencies in the transverse plane (b), while the pulses are simple, nonselective 90° pulses. In (d), transverse relaxation reduces each isochromat to its projection $M_0(\omega)\cos(\omega t_s)$.

pulse flips one magnetization component, of length $M_0 \cos(\omega t_s)$, back to the z axis. The component along the $\vec{\mathbf{B}}_1$ field is unchanged and destroyed by effective T_2 relaxation during the subsequent delay t_r which is without ^{1}H decoupling. This results in a modulation of the magnetization of the original spectrum by $\cos(\omega t_s)$. In particular, signals with a phase ωt_s of $\pi/2 + n\pi$ are completely suppressed. If necessary, n-fold repetition of the procedure modulates the signal by $[\cos(\omega t_s)]^n$, which for $n > 5$ becomes an excitation profile with a broad range of vanishing intensity and narrow spikes at $\omega t_s = n\pi$. The use of various t_s values and coadding of signals with n and $n+1$ repetitions can help to make the sequence more efficient.

The same chemical-shift filter principle can be employed under CRAMPS conditions. However, the precession axis here is generally not the (0, 0, 1) direction; for

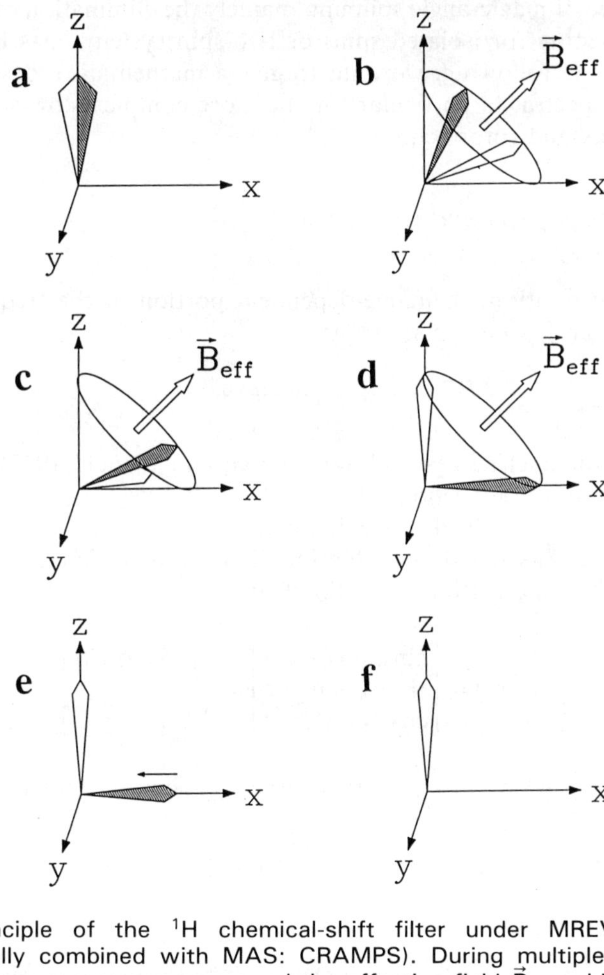

Figure 3.12 Principle of the ^{1}H chemical-shift filter under MREV-8 homonuclear decoupling (usually combined with MAS: CRAMPS). During multiple-pulse irradiation, the isochromats precess on a cone around the effective field $\vec{\mathbf{B}}_{eff}$, which for MREV-8 is along the (1, 0, 1) direction. The irradiation frequency is set such that after a full revolution of the component to be retained, the other is fully transverse in the xy plane. This transverse magnetization component is dephased by dipolar couplings in (e) where no pulses are applied.

MREV-8, the effective field points along (1, 0, 1). The magnetization isochromats start on the z axis. As portrayed in Figure 3.12, after a precession time t_s under MREV-8 decoupling some magnetization components reach the z axis again, while others end up in or near the xy plane and are dephased soon after the multipulse decoupling is switched off. Although a preparation pulse is not required, a small-angle flip away from the precession axis improves the performance of the experiment: two null points occur under these conditions and the region with small z components of the magnetization is much wider.

3.9 Magic-angle spinning (MAS)

The basic principle of magic-angle spinning, namely the elimination of inhomogeneous anisotropy broadening for isolated spins or two-spin systems, has been discussed in Section 2.4.5. In the following, we want to give a mathematical description of MAS time signals and spectra, in particular for the more complex slow-spinning case with spinning sidebands and rotor echoes.

3.9.1 MAS time signals and sidebands

Subject to sample rotation, the angle-dependent portion of the frequency acquires a time dependence $\omega(t)$, so a replacement:

$$e^{i\omega t} \rightarrow e^{i\int_0^t \omega(t')\, dt'} \tag{3.70}$$

must be made in the nuclear signal (2.84) (Maricq and Waugh, 1979) (for a quantum-mechanical derivation, see Section 3.9.5).

The rotation can be described conveniently in the rotor frame (RF) where we choose the rotor axis as the $\vec{\mathbf{Z}}_{RF}$ axis. For evaluating the formula (2.28), $\boldsymbol{\omega} = \vec{\mathbf{b}}_0 \boldsymbol{\sigma} \vec{\mathbf{b}}_0 \gamma B_0$, in the RF, we express the $\vec{\mathbf{B}}_0$ direction $\vec{\mathbf{b}}_0$ in this frame:

$$\vec{\mathbf{b}}_0^{RF}(t) = \begin{bmatrix} \sin\theta_m \cos\omega_r t \\ \sin\theta_m \sin\omega_r t \\ \cos\theta_m \end{bmatrix} = \sqrt{\frac{2}{3}} \begin{bmatrix} \cos\omega_r t \\ \sin\omega_r t \\ \sqrt{\frac{1}{2}} \end{bmatrix} \tag{3.71}$$

with $\theta_m = 54.74^\circ$. Inserting this vector into the frequency formula (2.28) we obtain:

$$\frac{\omega(t)}{\gamma B_0} = \vec{\mathbf{b}}_0^{RF} \boldsymbol{\sigma}^{RF} \vec{\mathbf{b}}_0^{RF} = \sigma_{iso} + \tfrac{2}{3}[\tfrac{1}{2}(\sigma_{11}^{RF} - \sigma_{22}^{RF})\cos(2\omega_r t)$$

$$+ \sigma_{21}^{RF} \sin(2\omega_r t) + \sqrt{2}\sigma_{31}^{RF} \cos\omega_r t + \sqrt{2}\sigma_{32}^{RF} \sin\omega_r t] \tag{3.72}$$

where we have used several trigonometric addition theorems (see Appendix A.1) and $(\sigma_{11}^{RF} + \sigma_{22}^{RF} + \sigma_{33}^{RF})/3 = \sigma_{iso}$; in the following, we shall omit this isotropic part.

As the next step, the $\sigma_{mn}^{\mathrm{RF}}$ elements are expressed in terms of the principal values σ_j of $\boldsymbol{\sigma}$, that is, the $\sigma_{jj}^{\mathrm{PAS}}$ elements, and in terms of the Euler angles (α, β, γ) which characterize the relative orientation of PAS and RF. The transformation (B.17):

$$\boldsymbol{\sigma}^{\mathrm{RF}} = \tilde{\mathbf{R}}(\alpha, \beta, \gamma)\boldsymbol{\sigma}^{\mathrm{PAS}}\tilde{\mathbf{R}}^{-1}(\alpha, \beta, \gamma) \tag{3.73}$$

is performed by means of the Euler rotation matrix $\tilde{\mathbf{R}}$ as given in (B.14). The resulting frequency formula can be written as:

$$\omega(t) = C_1 \cos(\gamma + \omega_r t) + C_2 \cos(2\gamma + 2\omega_r t)$$
$$+ S_1 \sin(\gamma + \omega_r t) + S_2 \sin(2\gamma + 2\omega_r t). \tag{3.74}$$

The C and S coefficients, which depend on δ, η, α and β, are listed in Table 3.1. When $\eta = 0$, as for the dipolar interaction of isolated spin pairs, the coefficients simplify to $S_1 = S_2 = 0$, $C_1(\beta) = -\delta\sqrt{2}\frac{1}{2}\sin 2\beta$ and $C_2(\beta) = \delta\frac{1}{2}\sin^2\beta$. The fact that the γ and $\omega_r t$ dependences in equation (3.74) can always be written in terms of the sum $\gamma + \omega_r t$ is due to the fact that both γ and $\omega_r t$ correspond to rotations around the rotor axis. Note, however, that in contrast to the rotor phase $\omega_r t$, the molecular-orientation angle γ is not fully under the control of the experimenter, taking different values for different molecular orientations. Therefore, we really have to consider γ explicitly in addition to $\omega_r t$. Relation (3.71) and the definition of the Euler rotations establish that for $\omega_r t = 0$ the x axis of the rotor frame is in the plane containing the $\vec{\mathbf{B}}_0$ field and the rotor axis. If $\gamma = 0$, the z principal axis is located in that same plane. This will sometimes be referred to below. In addition to γ, pairs of angles $\pm\alpha$ belonging to the same β value will occasionally be needed in derivations of general features of MAS spectra.

Sometimes it is also convenient to express $\omega(t)$ in terms of the Euler angles $(\alpha_M, \beta_M, \gamma_M)$ which relate the molecular-chain frame MF to the RF. This is achieved

Table 3.1

$$C_1(\alpha, \beta) = -\delta\sqrt{2}\tfrac{1}{2}\sin 2\beta(1 + \tfrac{1}{3}\eta\cos 2\alpha)$$

$$C_2(\alpha, \beta) = \delta\{\tfrac{1}{2}\sin^2\beta - \tfrac{1}{6}\eta(1 + \cos^2\beta)\cos 2\alpha\}$$

$$S_1(\alpha, \beta) = \delta\eta\tfrac{1}{3}\sqrt{2}\sin\beta\sin 2\alpha$$

$$S_2(\alpha, \beta) = \delta\eta\tfrac{1}{3}\cos\beta\sin 2\alpha$$

$$C_k(-\alpha) = C_k(\alpha) \quad S_k(-\alpha) = -S_k(\alpha)$$

$$C_1^{\mathrm{nd}} = \tfrac{2}{3}\sqrt{2}\{\tfrac{1}{4}(\sigma_{11} + \sigma_{22} - 2\sigma_{33})\sin 2\beta_M - \tfrac{1}{4}(\sigma_{22} - \sigma_{11})\sin 2\beta_M\cos 2\alpha_M$$
$$+ \tfrac{1}{2}\sigma_{12}\sin 2\beta_M\sin 2\alpha_M + \sigma_{13}\cos 2\beta_M\cos\alpha_M + \sigma_{23}\cos 2\beta_M\sin\alpha_M\}$$

$$C_2^{\mathrm{nd}} = \tfrac{1}{3}\{\tfrac{1}{2}(-\sigma_{11} - \sigma_{22} + 2\sigma_{33})\sin^2\beta_M - \tfrac{1}{2}(\sigma_{22} - \sigma_{11})(1 + \cos^2\beta_M)\cos 2\alpha_M$$
$$+ \sigma_{12}(1 + \cos^2\beta_M)\sin 2\alpha_M - \sigma_{13}\sin 2\beta_M\cos\alpha_M - \sigma_{23}\sin 2\beta_M\sin\alpha_M\}$$

$$S_1^{\mathrm{nd}} = \tfrac{2}{3}\sqrt{2}\{\tfrac{1}{2}(\sigma_{22} - \sigma_{11})\sin\beta_M\sin 2\alpha_M + \sigma_{12}\sin\beta_M\cos 2\alpha_M - \sigma_{13}\cos\beta_M\sin\alpha_M$$
$$+ \sigma_{23}\cos\beta_M\cos\alpha_M\}$$

$$S_2^{\mathrm{nd}} = \tfrac{2}{3}\{\tfrac{1}{2}(\sigma_{22} - \sigma_{11})\cos\beta_M\sin 2\alpha_M + \sigma_{12}\cos\beta_M\cos 2\alpha_M$$
$$+ \sigma_{13}\sin\beta_M\sin\alpha_M - \sigma_{23}\sin\beta_M\cos\alpha_M\}$$

by expressing the σ_{mn}^{RF} of equation (3.72) in terms of the σ_{k1}^{MF} elements as well as in terms of α_M, β_M and γ_M:

$$\omega(t) = C_1^{nd} \cos(\gamma_M + \omega_r t) + C_2^{nd} \cos(2\gamma_M + 2\omega_r t) + S_1^{nd} \sin(\gamma_M + \omega_r t) + S_2^{nd} \sin(2\gamma_M + 2\omega_r t). \quad (3.75)$$

The C^{nd} and S^{nd} are also given in Table 3.1, taking into account all elements of the chemical-shift tensor, which is usually non-diagonal in this frame (for clarity, the indices MF of the σ_{jk}^{MF} are omitted in Table 3.1). The formal similarity of equations (3.74) and (3.75) is obvious, and therefore in the following relations, expressed in terms of the C and S coefficients and γ, these can always be replaced by the C^{nd} and S^{nd} coefficients and γ_M, respectively.

Alternatively to equation (3.74), one can express the frequency as:

$$\omega(t) = \tilde{C}_1 \cos(\omega_r t) + \tilde{C}_2 \cos(2\omega_r t) + \tilde{S}_1 \sin(\omega_r t) + \tilde{S}_2 \sin(2\omega_r t) \quad (3.76)$$

separating γ and $\omega_r t$ according to the trigonometric addition theorems (A.5) and (A.6). The following simple relationships between the C, S and $\tilde{C}$, $\tilde{S}$ coefficients result:

$$\begin{aligned} \tilde{C}_1 &= C_1 \cos \gamma + S_1 \sin \gamma & \tilde{C}_2 &= C_2 \cos 2\gamma + S_2 \sin 2\gamma \\ \tilde{S}_1 &= S_1 \cos \gamma - C_1 \sin \gamma & \tilde{S}_2 &= S_2 \cos 2\gamma - C_2 \sin 2\gamma. \end{aligned} \quad (3.77)$$

Setting $\omega_r = 0$ in equation (3.76), we find $\omega = \tilde{C}_1 + \tilde{C}_2$ which must represent the frequency in the static ($\omega_r = 0$) sample. This fact has been exploited in a 2D experiment by Tycko *et al.* for obtaining undistorted inhomogeneously broadened patterns, equivalent to static spectra, under conditions of fast MAS (see Section 6.4.4).

For a single segment, or a site in a single crystal(lite), the MAS signal is given by:

$$g(t) = e^{i\int_0^t \omega(t')\,dt'} = e^{i\{\Phi(t) - \Phi(0)\}} = e^{i\Phi(t)} e^{-i\Phi(0)}. \quad (3.78)$$

The phase angle:

$$\Phi(\alpha, \beta, \gamma + \omega_r t) := \text{`}\int^t \omega(\alpha, \beta, \gamma + \omega_r \tilde{t})\, d\tilde{t}\text{'}$$

$$= \frac{\delta}{\omega_r}\left(\frac{C_1}{\delta} \sin(\gamma + \omega_r t) + \frac{C_2}{2\delta} \sin(2\gamma + 2\omega_r t) - \frac{S_1}{\delta} \cos(\gamma + \omega_r t) - \frac{S_2}{2\delta} \cos(2\gamma + 2\omega_r t)\right)$$

$$= \frac{\delta}{\omega_r}\left(\frac{\tilde{C}_1}{\delta} \sin(\omega_r t) + \frac{\tilde{C}_2}{2\delta} \sin(2\omega_r t) - \frac{\tilde{S}_1}{\delta} \cos(\omega_r t) - \frac{\tilde{S}_2}{2\delta} \cos(2\omega_r t)\right) \quad (3.79)$$

is proportional to δ/ω_r, which reduces the anisotropy effects with increasing relative spinning speed ω_r/δ. The full expression for $g(t)$, combining equations (3.78), and (3.79), is:

$$g(t) = e^{(-i/\omega_r) C_1 \sin \gamma + \frac{1}{2} C_2 \sin 2\gamma - S_1 \cos \gamma - \frac{1}{2} S_2 \cos 2\gamma\}} \times e^{(i/\omega_r)\{C_1 \sin(\gamma + \omega_r t) + \frac{1}{2} C_2 \sin(2\gamma + 2\omega_r t) - S_1 \cos(\gamma + \omega_r t) - \frac{1}{2} S_2 \cos(2\gamma + 2\omega_r t)\}}. \quad (3.80)$$

In order to concentrate on the important time dependence in this expression, it is convenient to define an 'f function':

$$f(\alpha, \beta, \gamma + \omega_r t) := e^{i\Phi(\alpha,\beta,\gamma+\omega_r t)} \tag{3.81}$$

so that according to (3.78) we can express the time signal simply as:

$$g(t) = f^*(\gamma) f(\gamma + \omega_r t) \tag{3.82}$$

(for clarity, we sometimes drop arguments of minor importance, α and β in this case). As the signal $g(t)$ is periodic with $t_r = 2\pi/\omega_r$ according to the harmonic time dependences in equation (3.80), intensity can only appear at ω_r and its harmonics $N\omega_r$ (referenced to the isotropic resonance frequency). The intensity of the sideband at $N\omega_r$ is obtained by Fourier transformation of $g(t)$:

$$I_N(\alpha, \beta, \gamma) = f^*(\gamma) \int_0^{t_r} \frac{dt}{t_r} f(\gamma + \omega_r t)\, e^{-i\omega_r N t} \text{ (with } \gamma' = \gamma + \omega_r t)$$

$$= f^*(\gamma)\, e^{i\gamma N} \int_0^{2\pi} \frac{d\gamma'}{2\pi} f(\gamma')\, e^{-i\gamma' N}$$

$$\Rightarrow I_N(\alpha, \beta, \gamma) = f^*(\alpha, \beta, \gamma)\, e^{i\gamma N} F_N(\alpha, \beta) \tag{3.83}$$

with:

$$F_N(\alpha, \beta) := FT_\gamma\{f(\alpha, \beta, \gamma)\} = \int_0^{2\pi} \frac{d\gamma}{2\pi} e^{-i\gamma N} f(\alpha, \beta, \gamma). \tag{3.84}$$

The $I_N(\alpha, \beta, \gamma)$ coefficients are usually complex, and rotor-synchronized spectra of single crystals can therefore contain negative and dispersive sidebands (Maricq and Waugh, 1979). For powder samples, however, experience shows that the sidebands are all absorptive and positive. This can be easily proved, actually for any sample that is transverse isotropic about the rotor axis, and also for the signal from arbitrary samples subject to random, unsynchronized signal averaging. In any of these cases, all γ values are equally probable so that we can simply integrate equation (3.83) over γ. This integral effectively produces a second FT over γ:

$$\mathcal{I}_N(\alpha, \beta) = \int_0^{2\pi} \frac{d\gamma}{2\pi} I_N(\alpha, \beta, \gamma) \overset{(3.84)}{=} F_N^*(\alpha, \beta) F_N(\alpha, \beta). \tag{3.85}$$

$F_N^*(\alpha, \beta) F_N(\alpha, \beta) = |F_N(\alpha, \beta)|^2$ is always real and positive; this property is also preserved under possible α and β integrations, for instance in a powder sample. Averages over all segmental orientations in the sample (also including the integral over α and d cos β) will often be indicated by a bar over the corresponding quantity (e.g. $\bar{g}(t)$ or $\bar{I}_N$).

The sideband intensities contain information on the anisotropic chemical shift. For isotropic samples ('powders'), the spectra can be characterized by three parameters: the isotropic chemical shift ω_{iso}, which is well known from the position of the central band, the anisotropy parameter δ, and the asymmetry parameter η (see Section 2.4.3). A graphical method for determining η and δ from the relative sideband intensities $\bar{\mathcal{I}}_N/\bar{\mathcal{I}}_0$ was developed by Herzfeld and Berger (1980). An example for this procedure is given in Figure 3.13. The problem of determining η and δ can also be solved on a computer in terms of an iterative fitting procedure (de Groot *et al.*, 1991).

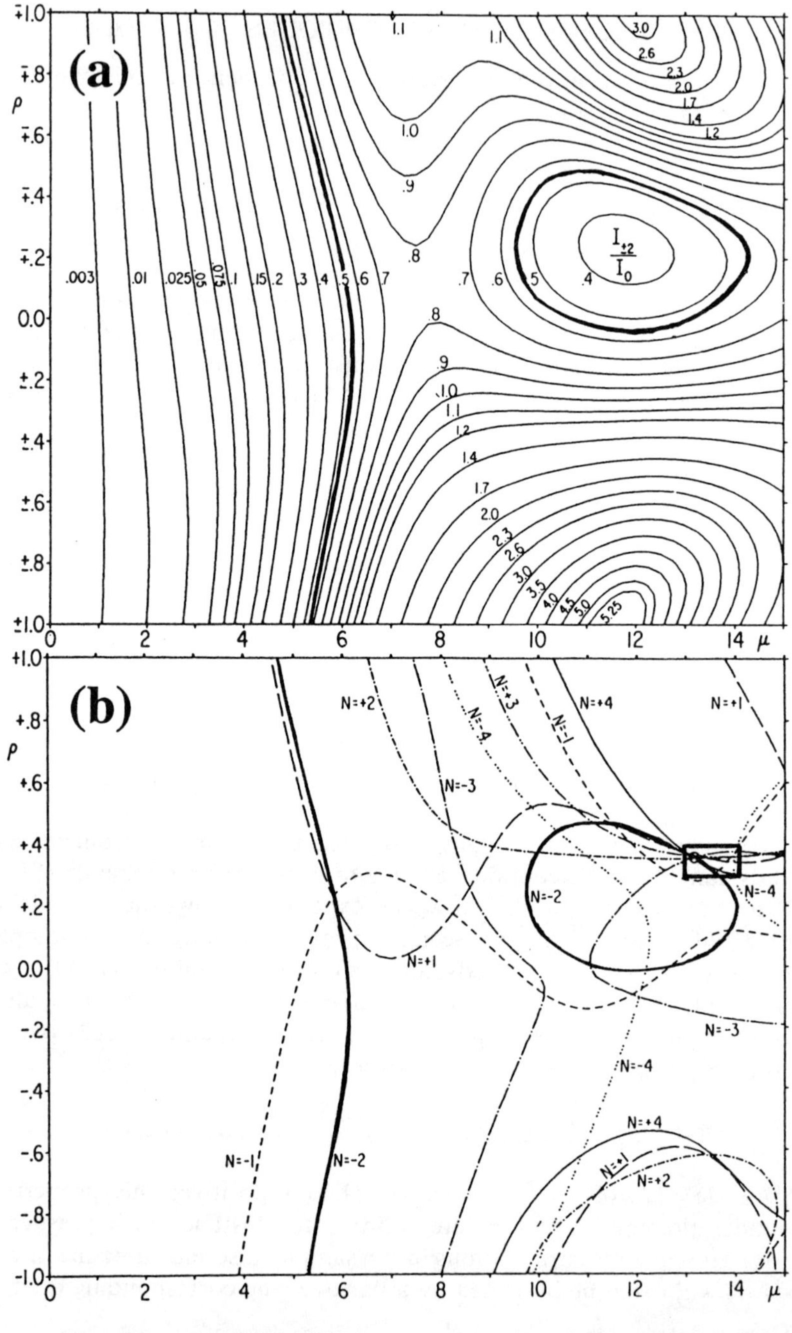

Figure 3.13 Principle of the Herzfeld–Berger graphical determination of the chemical-shift principal values from MAS sideband patterns of powders. Theoretically expected sideband-to-centreband intensity ratios are plotted in a map as a function of a spinning-speed-dependent width parameter $\mu = \gamma B_0(\sigma_{33} - \sigma_{11})/\omega_r$ and a shape parameter $\rho = (\sigma_{11} + \sigma_{33} - 2\sigma_{22})/(\sigma_{33} - \sigma_{11})$. For each sideband order, a separate map exists. In (a), the map for the second sideband order is shown. (b) The lines corresponding to the measured ratios of sideband-to-centreband intensities as obtained from the various maps are plotted in one graph and intersect in one point (μ_0, ρ_0). From μ_0 and ρ_0, η and δ can be determined. The higher the density of contour levels in the map (a), the higher the precision in that region of the map. Adapted from Herzfeld and Berger (1980).

From the definition (3.81) of the f function, it is obvious that:

$$f(\alpha, \beta, \gamma) f^*(\alpha, \beta, \gamma) = 1. \qquad (3.86)$$

This relation can be exploited in 1D and 2D experiments for cancelling all the anisotropy effects described by $f(\alpha, \beta, \gamma)$ (see Section 3.9.4).

3.9.2 The mirror-image condition for absorptive sidebands

In several 1D and 2D MAS experiments, it is of interest to prove that the real parts of the spectra are absorptive, or equivalently that the Fourier transform $\bar{I}_N$ of the periodic anisotropy signal $\bar{g}(t)$ is real ($\bar{I}_N = \bar{I}_N^*$). According to basic Fourier theory, equation (4.16), for the Fourier pair $\bar{I}_N$ and $g(t)$ one has the relation:

$$\bar{I}_N = \bar{I}_N^* \Leftrightarrow \bar{g}(t) = \bar{g}^*(-t). \qquad (3.87)$$

In the theory of MAS experiments, the time signal $g(t)$ is more readily accessible than the sideband intensities I_N. Therefore, it is easier to check on the time signal according to (3.87) whether the sidebands in the spectrum are absorptive.

For that purpose, the relation:

$$f^*(\alpha, \beta, \gamma) = f(-\alpha, \beta, -\gamma) \qquad (3.88)$$

is very useful (note that (α, β, γ) describe the relative orientation of PAS and RF; for Euler angles $(\alpha_M, \beta_M, \gamma_M)$ relating MF and RF, relation (3.88) is usually not valid). For a proof of (3.88), consider that (3.81) results in:

$$f(\alpha, \beta, \gamma) = e^{i\Phi(\alpha,\beta,\gamma)} \Rightarrow f^*(\alpha, \beta, \gamma) = e^{-i\Phi(\alpha,\beta,\gamma)}$$

so we only have to prove that:

$$\Phi(\alpha, \beta, \gamma) = -\Phi(-\alpha, \beta, -\gamma). \qquad (3.89)$$

Using $C_k(-\alpha) = C_k(\alpha)$, $S_k(-\alpha) = -S_k(\alpha)$ (see Table 3.1) for Φ as given in equation (3.79):

$$\Phi(-\alpha, \beta, -\gamma) = \frac{1}{\omega_r} \{C_1(-\alpha)\sin(-\gamma) + \tfrac{1}{2}C_2(-\alpha)\sin(-2\gamma)$$
$$- S_1(-\alpha)\cos(-\gamma) - \tfrac{1}{2}S_2(-\alpha)\cos(-2\gamma)\}$$

we find that the signs of the C_1 and C_2 terms are inverted relative to those in $\Phi(\alpha, \beta, \gamma)$ due to $\sin(-k\gamma) = -\sin(k\gamma)$, and the signs of the S_1 and S_2 terms due to $S_k(-\alpha) = -S_k(\alpha)$.

The PAS orientations specified by the Euler angles (α, β, γ) and $(-\alpha, \beta, -\gamma)$ relative to the rotor frame have a definite geometric relationship that is important for the analysis of 1D and 2D MAS spectra. In fact, segment orientations that are mirror images of one another, with the xz plane of the rotor frame as the mirror plane, provide a pair of (α, β, γ) and $(-\alpha, \beta, -\gamma)$ principal axes systems. This can be derived from the graphical analysis as in Figure 3.14, where in particular the invariance of β under the reflection is easily recognized. For an analytical proof, we express the three coordinate axes $\vec{\mathbf{X}}_P^{\pm}$, $\vec{\mathbf{Y}}_P^{\pm}$, and $\vec{\mathbf{Z}}_P^{\pm}$ (superscript '+' for (α, β, γ), '−' for $(-\alpha, \beta, -\gamma)$) of either PAS in the RF. They are obtained by rotating the RF

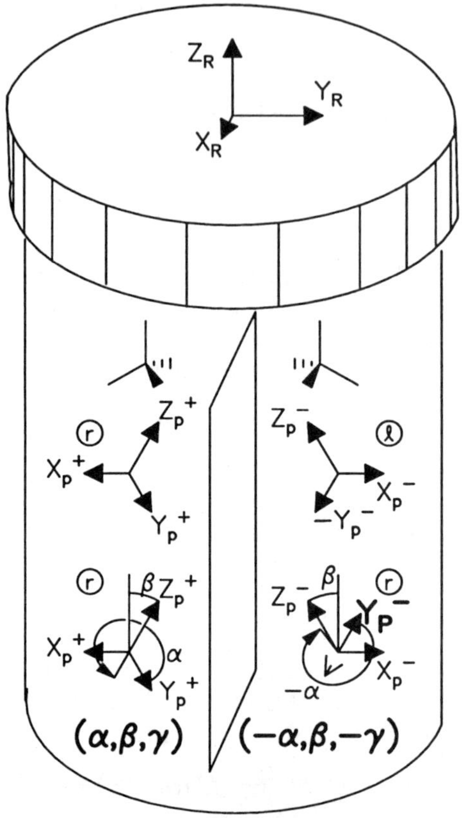

Figure 3.14 Mirror-image principal axes orientations in a sample, as required for absorptive MAS sidebands. The orientations (α, β, γ) and $(-\alpha, \beta, -\gamma)$ of principal-axes systems relative to the rotor frame correspond to two right-handed (r) PASs; these derive from two mirror-symmetric PASs, one right- (r), the other left-handed (l), after inversion of the $\vec{\mathbf{Y}}_P$ axis of the left-handed system.

coordinates axes (1, 0, 0), (0, 1, 0) and (0, 0, 1) through the inverse Euler rotations (see Appendix B.2):

$$\vec{\mathbf{X}}_P^{\pm} = \tilde{R}^{-1}(\pm\alpha, \beta, \pm\gamma)\begin{bmatrix}1\\0\\0\end{bmatrix} = \begin{bmatrix}\cos\alpha\cos\beta\cos\gamma - \sin\alpha\sin\gamma\\ \pm(\sin\alpha\cos\beta\cos\gamma + \cos\alpha\sin\gamma)\\ -\sin\beta\cos\gamma\end{bmatrix}$$

$$\vec{\mathbf{Y}}_P^{\pm} = \tilde{R}^{-1}(\pm\alpha, \beta, \pm\gamma)\begin{bmatrix}0\\1\\0\end{bmatrix} = \begin{bmatrix}\pm(-\cos\alpha\cos\beta\sin\gamma - \sin\alpha\cos\gamma)\\ -\sin\alpha\cos\beta\sin\gamma + \cos\alpha\cos\gamma\\ \pm\sin\beta\sin\gamma\end{bmatrix} \tag{3.90}$$

$$\vec{\mathbf{Z}}_P^{\pm} = \tilde{R}^{-1}(\pm\alpha, \beta, \pm\gamma)\begin{bmatrix}0\\0\\1\end{bmatrix} = \begin{bmatrix}\cos\alpha\sin\beta\\ \pm\sin\alpha\sin\beta\\ \cos\beta\end{bmatrix}.$$

These vectors are simply the columns of $\tilde{R}^{-1}(\pm\alpha, \beta, \pm\gamma) = \tilde{R}^{T}(\pm\alpha, \beta, \pm\gamma)$, that is, the rows of $\tilde{R}(\pm\alpha, \beta, \pm\gamma)$ given in (B.14). Generally, a reflection at the xz plane inverts the y component of a vector, while the x and z components are invariant. Thus, the axes vectors $\vec{\mathbf{Z}}_P^+$ and $\vec{\mathbf{Z}}_P^-$ as well as $\vec{\mathbf{X}}_P^+$ and $\vec{\mathbf{X}}_P^-$, having different signs only for their y components, are indeed such mirror images of each other, and so are $\vec{\mathbf{Y}}_P^+$ and $(-\vec{\mathbf{Y}}_P^-)$. The minus sign for the $\vec{\mathbf{Y}}_P^\pm$ axis must arise here because the coordinate systems generated from a right-handed system by rotations are right-handed, while the mirror image of a right-handed system is left-handed, which means that an odd number of axes is inverted compared with a right-handed system. The NMR frequencies are invariant under the inversion of a principal axis. Therefore, the right-handed principal-axes systems of two mirror-image-pair segments are indeed specified by (α, β, γ) and $(-\alpha, \beta, -\gamma)$ (see Figure 3.14).

On the basis of $f(\alpha, \beta, \gamma + \omega_r t) = f^*(-\alpha, \beta, -\gamma - \omega_r t)$ (equation (3.88)), the signal $g_P(t)$ from a pair of segments that are mirror images of one another generally fulfils $g_P(t) = g_P^*(-t)$. As an example, consider the single-segment signal $g(t) = f^*(\gamma) f(\gamma + \omega_r t)$. The mirror-image-pair signal is:

$$g_P(t) = f^*(\alpha, \beta, \gamma) f(\alpha, \beta, \gamma + \omega_r t) + f^*(-\alpha, \beta, -\gamma) f(-\alpha, \beta, -\gamma + \omega_r t). \quad (3.91)$$

Taking its complex conjugate (using $(f^*)^* = f$) and replacing t by $(-t)$ produces:

$$\begin{aligned} g_P^*(-t) &= f(\alpha, \beta, \gamma) f^*(\alpha, \beta, \gamma - \omega_r t) + f(-\alpha, \beta, -\gamma) f^*(-\alpha, \beta, -\gamma - \omega_r t) \\ &\overset{\text{(using (3.88))}}{=} f^*(-\alpha, \beta, -\gamma) f(-\alpha, \beta, -\gamma + \omega_r t) + f^*(\alpha, \beta, \gamma) f(\alpha, \beta, \gamma + \omega_r t). \quad (3.92) \end{aligned}$$

Comparison of (3.91) and (3.92) shows that indeed $g_P(t) = g_P^*(-t)$.

The signal of any sample exhibiting reflection symmetry with respect to the xz-plane of the rotor frame can be decomposed into mirror-image-pair signals $g_P(t)$. As these fulfil $g_P(t) = g_P^*(-t)$, the total signal $\bar{g}(t)$ generated by a superposition of the pair signals also has the property $\bar{g}(t) = \bar{g}^*(-t)$, and this leads to absorptive sidebands.

3.9.3 Total suppression of spinning sidebands (TOSS)

If the time signal was just:

$$g(t) = f(\gamma + \omega_r t) \quad (3.93)$$

instead of $f^*(\gamma) f(\gamma + \omega_r t)$, all spinning sidebands would be removed: instead of equation (3.83) we would obtain:

$$I_N(\alpha, \beta, \gamma) = e^{i\gamma N} F_N(\alpha, \beta) \quad (3.94)$$

and for unsynchronized data acquisition, equation (3.85) would be replaced by:

$$\begin{aligned} \mathscr{I}_N(\alpha, \beta) &= F_N(\alpha, \beta) \frac{1}{2\pi} \int_0^{2\pi} d\gamma \, e^{i\gamma N} \\ &= F_N(\alpha, \beta) \, \delta_{N,0}. \quad (3.95) \end{aligned}$$

The appearance of the Kronecker $\delta_{N,0}$ tells that all sidebands ($N \neq 0$) would indeed be eliminated, in particular for powder samples!

Dixon (1982) has introduced sequences of four (or generally $2m + 2$) properly spaced 180° pulses (Figure 3.15) that do essentially suppress the factor $f^*(\gamma)$ in the time signal and thus achieve total suppression of sidebands (TOSS, Dixon *et al.*, 1982). With every

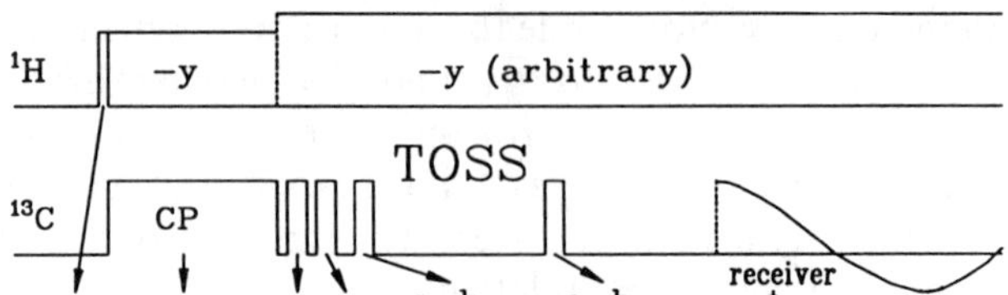

−x	y	y −y	y −y	−x x	x
x	y	y −y	y −y	−x x	−x
−x	−x	−x x	−x x	y −y	y
x	−x	−x x	−x x	y −y	−y
−x	−y	−y y	−y y	x −x	−x
x	−y	−y y	−y y	x −x	x
−x	x	x −x	x −x	−y y	−y
x	x	x −x	x −x	−y y	y

Figure 3.15 Four-pulse sequence for TOSS (total suppression of sidebands). For all times after the last 180° pulse, the sum of all isochromats in a powder sample precesses with the isotropic chemical shift, so that no spinning sidebands appear in the spectrum. Below the pulse sequence, a full phase cycle for compensation of experimental imperfections is indicated.

180° pulse, applied at a time $t^{(n)}$, the sign of the precession frequency of each isochromat is inverted (view from the frame toggling with the pulses). The effective phase is thus:

$$\Phi_{\text{eff}}(t) = \int_0^{t^{(1)}} \omega(\tilde{t})\, d\tilde{t} - \int_{t^{(1)}}^{t^{(2)}} \omega(\tilde{t})\, d\tilde{t} + \int_{t^{(2)}}^{t^{(3)}} \omega(\tilde{t})\, d\tilde{t} - \int_{t^{(3)}}^{t^{(4)}} \omega(\tilde{t})\, d\tilde{t} + \int_{t^{(4)}}^{t^{(5)}+t} \omega(\tilde{t})\, d\tilde{t}$$

$$\Phi_{\text{eff}}(t) = \Phi(t + t^{(5)}) - 2\Phi(t^{(4)}) + 2\Phi(t^{(3)}) - 2\Phi(t^{(2)}) + 2\Phi(t^{(1)}) - \Phi(0). \tag{3.96}$$

We demand $\Phi_{\text{eff}}(t) \stackrel{!}{=} \Phi(t + t^{(5)})$ in order to obtain:

$$g(t) = e^{i\Phi_{\text{eff}}(t)} \stackrel{!}{=} f(\tilde{\gamma} + \omega_r t)$$

with the 'initial' phase $\tilde{\gamma}$ at the time $t^{(5)}$, $\tilde{\gamma} = \gamma + \omega_r t^{(5)}$. To this end, we must fulfil the condition:

$$-2\Phi(t^{(4)}) + 2\Phi(t^{(3)}) - 2\Phi(t^{(2)}) + 2\Phi(t^{(1)}) - \Phi(0) \stackrel{!}{=} 0. \tag{3.97}$$

Using the expression (3.79) for Φ, equation (3.97) can be expressed more explicitly as:

$$\begin{aligned} & -\tilde{C}_1(\alpha, \beta, \gamma)\, 2 \sum_n (-1)^n \sin(\omega_r t^{(n)}) \\ & + \tilde{S}_1(\alpha, \beta, \gamma) \{1 + 2 \sum_n (-1)^n \cos(\omega_r t^{(n)})\} \\ & - \tfrac{1}{2}\tilde{C}_2(\alpha, \beta, \gamma)\, 2 \sum_n (-1)^n \sin(2\omega_r t^{(n)}) \\ & + \tfrac{1}{2}\tilde{S}_2(\alpha, \beta, \gamma) \{1 + 2 \sum_n (-1)^n \cos(2\omega_r t^{(n)})\} \stackrel{!}{=} 0. \end{aligned} \tag{3.98}$$

Each of the four sums over the pulses in this equation (3.98) must vanish *individually* since the action of the pulse sequence is to be independent of the orientations (α, β, γ) and therefore of the $\tilde{C}_n(\alpha, \beta, \gamma)$ and $\tilde{S}_n(\alpha, \beta, \gamma)$. A visualization of the resulting transcendental equations for the pulse timings has been indicated by Dixon. Unit vectors $\vec{\mathbf{v}}_n := (\cos \omega_r t^{(n)}, \sin \omega_r t^{(n)})$ and $\vec{\mathbf{w}}_n = (\cos 2\omega_r t^{(n)}, \sin 2\omega_r t^{(n)})$ are considered;

their directions depend on the pulse timings $t^{(n)}$. The individual terms in equation (3.98) vanish if the sum vectors:

$$\begin{pmatrix}1\\0\end{pmatrix} + \sum_n (-1)^n 2\begin{pmatrix}\cos(\omega_r t^{(n)})\\ \sin(\omega_r t^{(n)})\end{pmatrix} \stackrel{!}{=} \begin{pmatrix}0\\0\end{pmatrix} \tag{3.99a}$$

$$\begin{pmatrix}1\\0\end{pmatrix} + \sum_n (-1)^n 2\begin{pmatrix}\cos(2\omega_r t^{(n)})\\ \sin(2\omega_r t^{(n)})\end{pmatrix} \stackrel{!}{=} \begin{pmatrix}0\\0\end{pmatrix} \tag{3.99b}$$

are simultaneously zero.

In addition to the condition for the phase Φ generated by the anisotropic interaction, the isotropic chemical shift must also be refocused by the pulses. This is equivalent to demanding a Hahn echo at $t^{(5)}$, or $\Phi_{\text{iso}} \stackrel{!}{=} 0$ with:

$$\Phi_{\text{iso}} = \omega_{\text{iso}}(t^{(5)} - t^{(4)}) - \omega_{\text{iso}}(t^{(4)} - t^{(3)}) + \omega_{\text{iso}}(t^{(3)} - t^{(2)}) - \omega_{\text{iso}}(t^{(2)} - t^{(1)}) + \omega_{\text{iso}} t^{(1)}.$$

This is immediately found to require:

$$t^{(5)} \stackrel{!}{=} +2t^{(4)} - 2t^{(3)} + 2t^{(2)} - t^{(1)}. \tag{3.100}$$

For $t^{(1)}$, $t^{(2)}$, $t^{(3)}$ and $t^{(4)}$ determined by the relation (3.98), the Hahn spin-echo condition can be fulfilled simply by an appropriate choice of $t^{(5)}$, which marks the start of the signal detection. TOSS generates an echo at $t^{(5)}$, many μs after the last pulse; therefore, signal detection free of pulse dead-time problems is conveniently achieved.

A few sets of pulse timings for TOSS are listed in Table 3.2. In order to compensate experimental pulse imperfections (Raleigh *et al.*, 1990), pulse-phase cycling (see Figure 3.15) and proper timings of the pulse centres are relevant. Composite 180° pulses (see Section 2.6.6) can also help to reduce the residual sidebands at low spinning speeds, where little rotation occurs during the composite pulse.

The intensity of the spectral line resulting after TOSS is $\bar{F}_0$ according to (3.95). This is in general unequal to the MAS centre line intensity $\overline{|F_0|^2}$. Only if the anisotropy is 'spun out' (i.e. $\delta/\omega_r \rightarrow 0$), do both become equal because in this limit we find $F_0(\alpha, \beta) \rightarrow 1$ and thus also $|F_0(\alpha, \beta)|^2 \rightarrow 1$.

The application of TOSS to spectra with very large anisotropies using insufficient spinning speeds is limited because $\bar{F}_0$ becomes small and even negative. In principle, $\bar{F}_0(\alpha, \beta)$ can also contain an imaginary part, which produces a dispersive component in the lineshape. However, for uniaxial or isotropic samples, the dispersive components cancel out in the integral over the different orientations, as can be shown according to Section 3.9.2.

Table 3.2 *Timing of widely used TOSS sequences. The delay* Δt_1 *is the timing of the first pulse after the start of free precession,* Δt_j*, j = 2, 3, 4 are spacings between the centres of the pulses, and* Δt_5 *is the delay between the fourth pulse and the start of data acquisition* (i.e. $\Delta t_j = t^{(j)} - t^{(j-1)}$). *The times are given in units of the rotation period* t_r.

	$\Delta t_1/t_r$	$\Delta t_2/t_r$	$\Delta t_3/t_r$	$\Delta t_4/t_r$	$\Delta t_5/t_r$	$\sum_j \Delta t_j/t_r$
I	0.1226	0.0773	0.2236	1.0433	0.7744	2.2412
II	0.1885	0.0412	0.5818	0.9588	0.2297	2

3.9.4 Time-reversed TOSS

In several 2D experiments, all anisotropy effects in an evolution time t_1 are removed by exploiting relation (3.86):

$$f(\gamma + \omega_r t_1) f^*(\gamma + \omega_r t_1) = 1. \tag{3.101}$$

The evolution under the isotropic shifts, given by $\exp(i\omega_{iso} t_1)$, is to remain unaffected. The term $f(\gamma + \omega_r t_1)$ in (3.101) can be generated by TOSS and an evolution during a time t_1. The second factor $f^*(\gamma + \omega_r t_1)$ is produced by applying a 'time-reversed TOSS' sequence (deTOSS) (Kolbert and Griffin, 1990) after the t_1 evolution time. Here, $\gamma + \omega_r t_1$ is just the rotor phase γ_1 at the start of the deTOSS pulse sequence. So the pulses are to produce:

$$g(t^{(5)}) = e^{i\Phi_{eff}(0)} \stackrel{!}{=} f^*(\gamma_1) = e^{-i\Phi(0)}$$

$$\Leftrightarrow \Phi_{eff}(0) \stackrel{!}{=} -\Phi(0)$$

with the Φ_{eff} of the sequence of four π pulses, relation (3.96):

$$\Phi_{eff}(0) = \underbrace{\Phi(0 + t^{(5)}) - 2\Phi(t^{(4)}) + 2\Phi(t^{(3)}) - 2\Phi(t^{(2)}) + 2\Phi(t^{(1)}) - \Phi(0)}_{\stackrel{!}{=} 0}$$

$$\Rightarrow \Phi(t^{(5)}) - 2\Phi(t^{(4)}) + 2\Phi(t^{(3)}) - 2\Phi(t^{(2)}) + 2\Phi(t^{(1)}) \stackrel{!}{=} 0.$$

If we relabel ('reverse') the timings by $\tilde{t}^{(n)} = t^{(5)} - t^{(5-n)}$, the problem acquires a form identical to the TOSS problem (equation 3.97):

$$\Phi(0) - 2\Phi(\tilde{t}^{(1)}) + 2\Phi(\tilde{t}^{(2)}) - 2\Phi(\tilde{t}^{(3)}) + 2\Phi(\tilde{t}^{(4)}) = 0$$

where all the functions $\Phi(t) = \Phi(\alpha, \beta, \gamma + \omega_r t)$ were subject to an irrelevant shift of γ to $\gamma + \omega_r t_5$ and to an inversion of the sign of ω_r. As indicated in Figure 3.16, the pulse spacings in the deTOSS sequence are reversed relative to those in TOSS.

The TOSS–t_1 – deTOSS sequence (Figure 3.16) can be used to generate spectra equivalent to fast-spinning spectra; that is, without sidebands and with relative centreband areas that can be evaluated quantitatively (unless altered by CP efficiencies). However, the signal cannot be measured in a single scan. To measure the

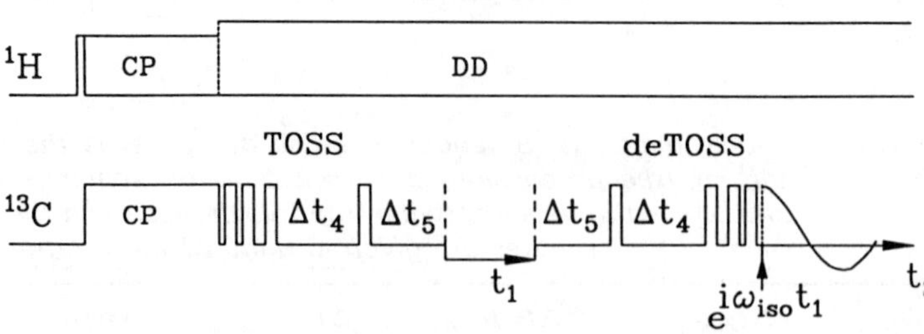

Figure 3.16 Pulse sequence for completely isotropic evolution, employing the TOSS and deTOSS sequences. At the end of the TOSS–t_1–deTOSS sequence, each isochromat has acquired only a phase $\omega_{iso} t_1$, without any effects of the angle-dependent part of the NMR frequency.

full time signal, points from about 15 scans must be combined in a simple way: if the time signal after TOSS–t_1–deTOSS:

$$s(t_1 + t_2) = e^{i\omega_{iso}t_1} f(\gamma')f^*(\gamma' + \omega_r t_2)\, e^{i\omega_{iso}t_2}\, e^{-(t_1+t_2)/T_2} \tag{3.103}$$

is sampled at the rotor echoes $t_2 = nt_r$, when $f(\gamma')f^*(\gamma' + \omega_r t_2) = 1$, with $t := t_1 + t_2$ we obtain the desired isotropic signal:

$$s(t) = e^{i\omega_{iso}t}\, e^{-t/T_2}.$$

Thus, several points of $s(t)$ can be sampled at once; for each t_1, the points for $t = t_1$, $t_r + t_1$, $2t_r + t_1, \ldots$ are filled into the 1D time-domain data array. If the spectral width is $15\omega_r$, 15 such partial time signals have to be combined. This accordingly increases the measuring time, equivalent to a reduction of the S/N by a factor of $\sqrt{15}$. Still, the S/N can be better than for the normal MAS centreband in a spectrum with 5 or more strong sidebands, since the signal per band is reduced by about a factor of 5 in such a conventional MAS spectrum.

3.9.5 Hamiltonians under MAS

Under conditions of sample rotation, the Hamiltonian acquires a time dependence which can seriously affect the spectrum, as was shown in the preceding sections. If the rotation frequency is much larger than the width of the anisotropic broadening produced by the overall Hamiltonian, only the isotropic effects remain. Under conditions of spinning slow compared with the overall linewidth, spinning sidebands occur if the Hamiltonian commutes with itself at different orientations. To understand this, consider the propagator after a time $t = N\Delta t$, $\hat{U}(N\Delta t)$, with the time increment Δt so much shorter than the rotor period, $\Delta t \ll t_r$, that the change of the Hamiltonian $\hat{H}(t)$ during Δt is negligible. For each period Δt we can thus apply the time-evolution expression (2.74):

$$\hat{U}(N\Delta t) \cong e^{-i\hat{H}(N\Delta t)\Delta t} \ldots e^{-i\hat{H}(2\Delta t)\,\Delta t}\, e^{-i\hat{H}(\Delta t)\Delta t}. \tag{3.104}$$

A Hamiltonian that commutes with itself at all different rotor orientations $\omega_r\, n\Delta t$ (i.e. that fulfils $[\hat{H}(t), \hat{H}(t')] = 0$ for all t and t') has been termed inhomogeneous by Maricq and Waugh (1979) (see also Section 3.13.4). For such a Hamiltonian, we can combine the exponents of equation (3.104):

$$\hat{U}(t) \cong \exp\left[-i \sum_{n=1}^{N} \hat{H}(n\Delta t)\Delta t\right] \xrightarrow{\Delta t \to 0} \exp\left[-i \int_0^t \hat{H}(t')\, dt'\right]. \tag{3.105}$$

This formula is applicable in particular to the chemical shift, which indeed commutes with itself at different times:

$$\left[\sum_j \omega_{cs}^j(t)\hat{I}_z^j, \sum_k \omega_{cs}^k(t')\hat{I}_z^k\right] = 0.$$

This is the actual basis of the fundamental equation (3.70) for the chemical-shift MAS signal. The propagator is:

$$\hat{U}_{cs}(t) = \exp\left[-i \int_0^t \hat{H}_{cs}(t')\, dt'\right] = \exp\left[-i\hat{I}_z \int_0^t \omega_{cs}(t')\, dt'\right] \tag{3.106}$$

which causes a rotation of the magnetization around the z axis with a time-dependent precession frequency, resulting in the signal of equation (3.70), which is periodic with ω_r or in its anisotropic part and thus produces rotor echoes and spinning sidebands. In the same fashion, for all single-particle and dipole-pair interactions acting without significant disturbances by other Hamiltonians, the occurrence of rotor echoes and sidebands can be derived. This includes in particular the quadrupolar interaction of deuterons: MAS breaks up the broad quadrupolar pattern into sidebands with an envelope similar to the static (Pake) pattern.

Equations (3.104) and (3.105) also explain how isotropic averaging is achieved under fast spinning, where the Hamiltonian $\hat{H}(t)$ varies just as much as under slow-spinning conditions. The crucial point is that the exponents $\hat{H}(t)\Delta t$ in equation (3.104) are small since by definition of the fast-spinning regime, the frequency shift $\omega_{\hat{H}}$ produced by $\hat{H}$ fulfils $\omega_{\hat{H}} t_r \ll 1$. For small $\hat{H}(t)\Delta t$, $\exp(-i\hat{H}(t)\Delta t) \cong 1 - i\hat{H}(t)\Delta t$, such that the propagators of equation (3.104) commute to first order and can be combined according to equation (3.105).

So far, only cases of commuting propagators have been treated. As an example of a simple Hamiltonian that does not commute with itself (homogeneous Hamiltonian), consider a system of two spins with dipolar coupling and moderately different chemical shifts for the two spins, $\hat{H}^a_{cs} + \hat{H}^b_{cs} + \hat{H}^{ab}_{SS}$. The sample rotation only modulates the orientation-dependent prefactors of $\hat{H}^a_{cs}$, $\hat{H}^b_{cs}$ and $\hat{H}^{ab}_{SS}$, so we can write quite concisely:

$$\hat{H}_{tot}(t) = a(t)\hat{H}^a_{cs}(0) + b(t)\hat{H}^b_{cs}(0) + c(t)\hat{H}^{ab}_{SS}(0). \tag{3.107}$$

This Hamiltonian would commute with itself at any time if $\hat{H}^a_{cs}(0)$, $\hat{H}^b_{cs}(0)$ and $\hat{H}^{ab}_{SS}(0)$ commuted with one another. Actually, however, the dipole-coupling Hamiltonian does not commute with the sum of the non-identical chemical-shift Hamiltonians. Nevertheless, $\hat{H}_{tot}(t)$ would commute with itself at different times if $a(t)$, $b(t)$ and $c(t)$ were mutually proportional, $a(t) = a_0 d(t)$, $b(t) = b_0 d(t)$ and $c(t) = c_0 d(t)$, so that:

$$[\hat{H}_{tot}(t), \hat{H}_{tot}(t')] = d(t)d(t')[\hat{H}', \hat{H}'] = 0$$

where $\hat{H}' = a_0\hat{H}^a_{cs}(0) + b_0\hat{H}^b_{cs}(0) + c_0\hat{H}^{ab}_{SS}(0)$. If the orientation-dependent prefactors $a(t)$, $b(t)$ and $c(t)$ vary in different ways, the commutator does not vanish. Writing it down for our special case in terms of the coefficients $a(t)$, $b(t)$, $c(t)$ and the commutators of dipolar coupling and chemical shifts is in principle straightforward but lengthy and does not provide much further insight. It should be noted that although the Hamiltonian (3.107) is homogeneous, in practice 'second averaging', see Section 3.3.4, of the dipolar interaction by a large difference in the isotropic chemical shifts removes the dipolar effects so that the Hamiltonian appears to be inhomogeneous. The second averaging is only prevented if the shift difference is an integer multiple of the rotation frequency; this rotational-resonance case is discussed in Section 13.5.6. At this point, it is perhaps worth noting that even if made up of noncommuting constituent Hamiltonians, the static total Hamiltonian always commutes with itself: for an arbitrary time-independent operator $\hat{H}$, we always have $[\hat{H}, \hat{H}] = \hat{H}\hat{H} - \hat{H}\hat{H} = 0$.

The same principles as outlined for the dipolar coupling and the chemical shift lead to the non-commutation of the homonuclear coupling of abundant spins. Already for three interacting spins, labelled a, b and c, the Hamiltonian consists of three terms, $\hat{H}^{ab}_{II}(t)$, $\hat{H}^{ac}_{II}(t)$, and $\hat{H}^{bc}_{II}(t)$, which do not commute with one another: for instance,

$\hat{H}_{II}^{ab}(t)$ does not commute with $3\hat{I}_z^b\hat{I}_z^c$ of $\hat{H}_{II}^{bc}(t)$:

$$[3\hat{I}_z^a\hat{I}_z^b - \hat{\mathbf{I}}^a \cdot \hat{\mathbf{I}}^b, 3\hat{I}_z^b\hat{I}_z^c] = \mathrm{i}(-\hat{I}_x^a\hat{I}_y^b + \hat{I}_y^a\hat{I}_x^b)3\hat{I}_z^c.$$

Usually, the factors $(3\cos^2\{\theta_{ab}(\gamma_{ab} + \omega_r t)\} - 1)$, $(3\cos^2\{\theta_{bc}(\gamma_{bc} + \omega_r t)\} - 1)$ and $(3\cos^2\theta_{ac}(\gamma_{ac} + \omega_r t) - 1)$ in $\hat{H}_{II}^{ab}(t)$, $\hat{H}_{II}^{ac}(t)$, and $\hat{H}_{II}^{bc}(t)$, respectively, vary in different fashions with the rotor orientation, so that the total Hamiltonian does not commute with itself at a different orientation and no pronounced rotor echoes of sharp spinning sidebands occur. There is one notable exception: if all the coupled spins are on a straight line, the angular dependences of the dipolar couplings are equal, since the relevant angles θ_{jk} of the internuclear vectors relative to the $\vec{\mathbf{B}}_0$ field are the same, as the internuclear vectors are all directed along the same line. In this case, the interaction is refocused by MAS and the dipolar spectrum is split into sidebands.

If the broadening is much larger than the rotation frequency and the Hamiltonian is strong and does not at all commute with itself at different orientations due to its complicated spatial structure, the spectrum remains essentially unaffected by the sample rotation. This is valid for ^{1}H wideline spectra of rigid organic polymers. When rotation frequency and coupling strength of an homogeneous Hamiltonian are comparable, the system is difficult to describe. Experimentally, it is found that ^{1}H wideline spectra start to develop broad spinning sidebands when the rotor frequency exceeds about one quarter of the full width at half maximum.

Under MAS conditions, the averaging behaviour of a weak inhomogeneous Hamiltonian $\hat{H}_1$ can be changed by a strong Hamiltonian $\hat{H}_s$ that commutes neither with itself at different orientations nor with $\hat{H}_1$. Since the total Hamiltonian $\hat{H}_{tot} = \hat{H}_1 + \hat{H}_s$ does not commute with itself at different times, $\hat{H}_1$ is not averaged out by MAS: $\hat{H}_s$ 'protects' $\hat{H}_1$ against MAS averaging. In spectroscopy, this effect is not so relevant since the strong Hamiltonian also dominates the spectrum which makes the effect of MAS on the weak Hamiltonian hardly detectable. In ^{13}C spin diffusion experiments, however, the effective strength of the ^{13}C$-^{13}$C dipolar coupling can be monitored even in the presence of the strong ^{13}C$-^1$H and ^{1}H$-^1$H couplings. These actually 'protect' ^{13}C$-^{13}$C couplings of only a few Hertz from being spun out; this subject is taken up again in Section 13.5.

3.10 Averaging of second-order quadrupolar couplings

Internuclear dipolar couplings and chemical-shift interactions are small compared with the Zeeman interaction in the high-fields typically used today, and their Hamiltonians can therefore be truncated or, equivalently, the energy shifts treated in first-order perturbation theory. In contrast, quadrupolar couplings can become comparable to or even by far exceed the Zeeman interaction in many isotopes (of spin $I > 1$). Higher order perturbation terms have to be taken into account, or the Zeeman-perturbation approach even breaks down completely. Deuteron quadrupolar coupling as discussed above actually represents a rather untypical case favourable for NMR experiments: the ^{2}H quadruple moment is one of the smallest found in nature (for $I > \frac{1}{2}$).

For half-integer spins, the so-called central transition dominates the spectra. It is of central importance since it is independent of the first-order quadrupolar coupling. Therefore only the relatively small second-order anisotropies have to be removed in order to detect the isotropic line positions. To understand the occurrence of this ω_Q-independent signal, we calculate the density-matrix time-evolution under the first-order (truncated) Hamiltonian (2.22) for the example of $I = \frac{3}{2}$. We use the following matrix representation of the spin operators obtained from equation (2.81) (for clarity, only the non-vanishing off-diagonal elements are given):

$$\mathbf{I}_z = \begin{bmatrix} \frac{3}{2} & & & \\ & \frac{1}{2} & & \\ & & -\frac{1}{2} & \\ & & & -\frac{3}{2} \end{bmatrix} \quad \mathbf{I}_x = \begin{bmatrix} 0 & \sqrt{\frac{3}{4}} & & \\ \sqrt{\frac{3}{4}} & 0 & 1 & \\ & 1 & 0 & \sqrt{\frac{3}{4}} \\ & & \sqrt{\frac{3}{4}} & 0 \end{bmatrix} \quad \mathbf{I}_y = \mathrm{i}\begin{bmatrix} 0 & -\sqrt{\frac{3}{4}} & & \\ \sqrt{\frac{3}{4}} & 0 & -1 & \\ & 1 & 0 & -\sqrt{\frac{3}{4}} \\ & & \sqrt{\frac{3}{4}} & 0 \end{bmatrix}$$

$$\mathbf{H}_Q = \tfrac{1}{6}\omega_Q\{3\mathbf{I}_z^2 - I(I+1)\mathbf{1}\} = \tfrac{1}{6}\omega_Q \begin{bmatrix} 3 & & & \\ & -3 & & \\ & & -3 & \\ & & & 3 \end{bmatrix}. \tag{3.108a}$$

Comparing this expression to the secular Hamiltonian of equation (2.43), ω_Q for $I = \frac{3}{2}$ is found as:

$$\tfrac{1}{6}\omega_Q = \frac{eqeQ}{3.2\hbar}\,\frac{1}{2}\,\tfrac{1}{2}(3\cos^2\theta - 1 - \eta\sin^2\theta\cos 2\phi). \tag{3.108b}$$

In the Hamiltonian matrix (3.108a), the differences of neighbouring $\mathbf{H}_Q$ diagonal elements (energy levels) are ω_Q, 0 and $-\omega_Q$. These will be the oscillation frequencies of the magnetization according to equation (2.83):

$$\mathbf{I}_x \xrightarrow{\mathbf{H}_Q} \boldsymbol{\rho}(t) = \begin{bmatrix} 0 & \sqrt{\frac{3}{4}}\exp(-\mathrm{i}\omega_Q t) & & \\ \sqrt{\frac{3}{4}}\exp(\mathrm{i}\omega_Q t) & 0 & 1 & \\ & 1 & 0 & \sqrt{\frac{3}{4}}\exp(\mathrm{i}\omega_Q t) \\ & & \sqrt{\frac{3}{4}}\exp(-\mathrm{i}\omega_Q t) & 0 \end{bmatrix}. \tag{3.109}$$

From this, we obtain the x and y NMR signals:

$$\begin{aligned} &\mathrm{tr}\{\boldsymbol{\rho}(t)\mathbf{I}_x\} = 3\cos(\omega_Q t) + 2\cos(0.t) \\ &\mathrm{tr}\{\boldsymbol{\rho}(t)\mathbf{I}_y\} = 0. \end{aligned} \tag{3.110}$$

In the x signal, we have substituted 1 by $\cos(0.t)$ to stress that the oscillation frequency of the second term is indeed zero. The corresponding orientation-independent and hence narrow central line at $\omega = 0$ is the spectral feature observed in most of the experiments, since the intensities at the satellites $\pm\omega_Q$, from the $\cos(\omega_Q t)$ term, are spread over a wide range by the angular dependence of ω_Q.

If ω_Q is not very small compared with the Larmor frequency, the truncated Hamiltonian has to be supplemented by higher-order terms (3.19) in the Magnus expansion (3.17) of $\hat{H}_Q$ in the $\hat{I}_z$ interaction representation (see Section 3.3.3 and

Appendix G). With a quadrupolar frequency $\Omega_Q = eqeQ/[2I(2I-1)\hbar]$, the strength of the second-order coupling is characterized by the second-order quadrupolar frequency:

$$\Omega_Q^{(2)} = \frac{\Omega_Q}{\omega_0}\,\Omega_Q(I-\tfrac{1}{2})(I+\tfrac{3}{2}) \tag{3.111}$$

which is in particular inversely proportional to the Larmor frequency ω_0. With the even orders vanishing due to the symmetry of the field-gradient tensor **V**, the second-order term consists of zeroth-, second-, and fourth-rank tensors, whose transformational behaviour under fast uniaxial rotation is determined by the zeroth, second and fourth Legendre polynomials, respectively. While the P_0 term produces an isotropic shift of the resonance frequency depending on ω_0 via equation (3.111), the P_2 and P_4 terms result in anisotropic linebroadening. Two sample-rotation approaches (Figure 3.17) have been taken to eliminate this second-order anisotropy of the central transition of half-integer-spin nuclei, leaving only the isotropic chemical shift and the isotropic second-order quadrupolar shift.

Double rotation (DOR) (Samoson *et al.*, 1988; Llor and Virlet, 1988) is a 1D technique with simultaneous averaging of the second- and fourth-rank tensors by simultaneous sample spinning at the magic angles (null points) of both $P_2(\cos\theta)$ and $P_4(\cos\theta)$. This requires a double rotor, with an inner sample rotor spinning fast inside a larger rotor (Figure 3.17(a)). Such a complex rotor system has indeed been built, with the outer rotor spinning at about 1 kHz and the inner rotor at about 5 kHz. The line narrowing achieved by DOR is more efficient than in the dynamic-angle spinning approach (see next paragraph) because dipolar couplings are also eliminated by the magic-angle spinning of the outer rotor. Moreover, the technique is applicable to materials with short T_1 relaxation times. Drawbacks of the DOR technique are spinning sidebands due to the relatively low rotation frequency of the outer rotor (Figure 3.18, right column) and a relatively small filling factor.

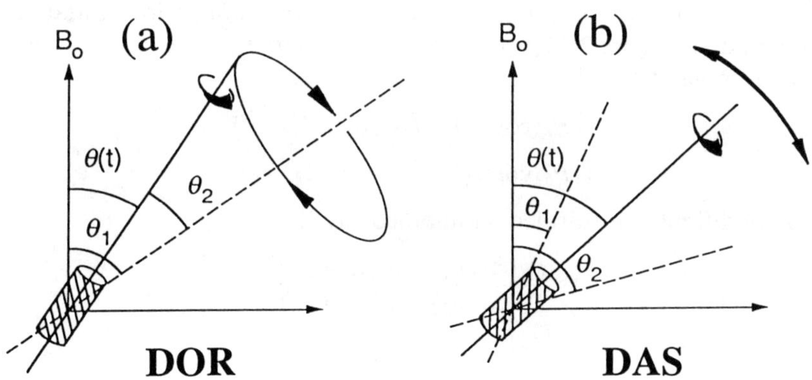

Figure 3.17 Schematic representation of (a) double rotation (DOR) and (b) dynamic-angle spinning (DAS) sample motions for averaging second-order quadrupolar interactions (adapted from Mueller *et al.* 1991). In DOR, a small rotor spins inside a large rotor. The axis of the outer rotor is at 54.74°, the magic angle of $P_2(\cos\theta)$, with respect to the $\vec{\mathbf{B}}_0$ field, while the inner rotor spins at 30.12°, a 'magic angle of $P_4(\cos\theta)$', relative to the axis of the outer rotor. In DAS, evolution and detection occur at two different but fixed orientations of the rotation axis, such that the signs of both the P_2 and the P_4 terms in the second-order anisotropy are inverted.

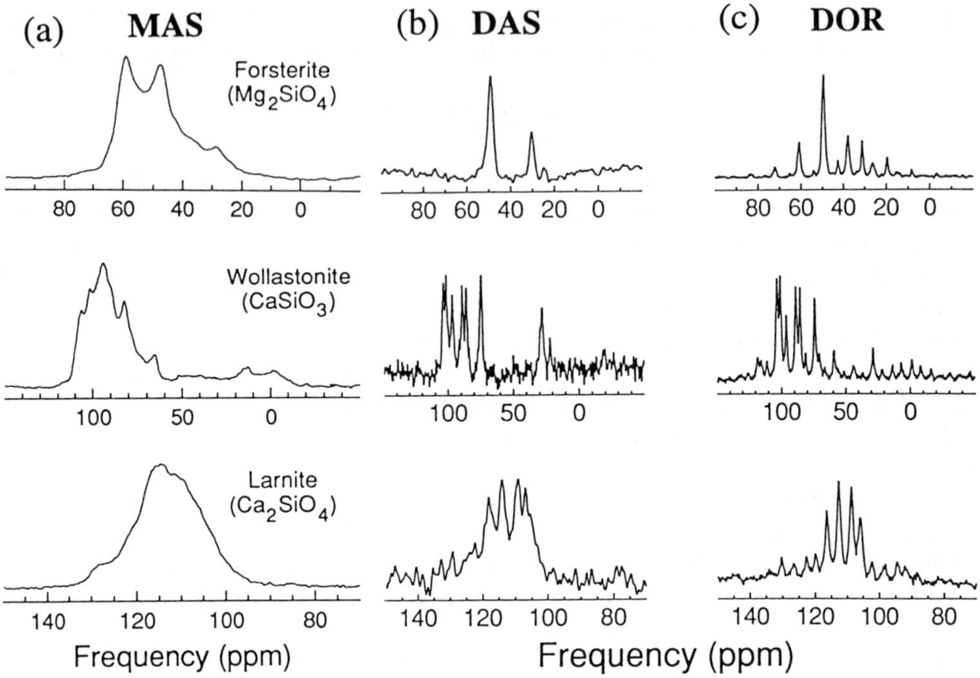

Figure 3.18 ^{17}O spectra of various silicate minerals. (a) MAS spectra. (b) DAS spectra. (c) DOR spectra (adapted from Mueller *et al.*, 1991). While the DOR spectra are usually more highly resolved, the DAS spectra have fewer sidebands.

Both the second-order-quadrupolar and the dipolar couplings can be averaged out by the dynamic-angle spinning (DAS) approach (Mueller *et al.*, 1990, 1991) where the sample is spun at the angle θ_1 relative to $\vec{\mathbf{B}}_0$ in the evolution period $\tilde{t}_1$ and at θ_2 in the detection period t_2 (Figure 3.17(b)), so that the P_2 and P_4 terms vanish (individually) in the time average at the time $\tilde{t}_2 = t_2$:

$$\tilde{t}_1 P_2(\cos\theta_1) + \tilde{t}_2 P_2(\cos\theta_2) \stackrel{!}{=} 0 \tag{3.112a}$$

$$\tilde{t}_1 P_4(\cos\theta_1) + \tilde{t}_2 P_4(\cos\theta_2) \stackrel{!}{=} 0. \tag{3.112b}$$

These two conditions are fulfilled simultaneously if:

$$\frac{P_2(\cos\theta_1)}{P_2(\cos\theta_2)} = \frac{P_4(\cos\theta_1)}{P_4(\cos\theta_2)}.$$

Then, at the time:

$$t_2 = \tilde{t}_2 = \tilde{t}_1\,\frac{P_2(\cos\theta_1)}{P_2(\cos\theta_2)} = \tilde{t}_1\,\frac{P_4(\cos\theta_1)}{P_4(\cos\theta_2)}$$

an echo occurs which is modulated by the isotropic interactions only. Figure 3.18 shows MAS, DAS, and DOR ^{17}O spectra in comparison. The second, anisotropic dimension in the DAS spectrum can be exploited for further structural analysis. For instance, information on the bond-angle distribution in amorphous silicate glasses has been obtained (Farnan *et al.*, 1992).

Since half-integer quadrupolar nuclei (except, perhaps, for ^{17}O, which must, however, be introduced by isotopic labelling) play much less of a role in polymers than in many inorganic materials, the treatment of the DOR and DAS techniques given here is relatively terse and incomplete. For a more extensive description, the reader is referred to the original literature (Mueller *et al.*, 1990, 1991; Sun *et al.*, 1992).

For the sake of completeness, other methods for obtaining structural information from quadrupolar nuclei are to be mentioned here too. The SATRAS (satellite transition spectroscopy) technique works at high spinning rates (about 15 kHz) where the intensities of the spinning sidebands of the satellite transitions become appreciable. These sidebands are approximately three times less broadened by the quadrupolar interaction than the centreband (Samoson, 1985). The true chemical shift can be extracted in a single experiment without measuring at different B_0 field strengths and the sideband intensities contain information on the quadrupolar coupling constant (Samoson, 1985; Skibsted *et al.*, 1991; Jäger, 1992).

Nutation spectroscopy (Yannoni and Kendrick, 1981; Samoson and Lipmaa, 1983; Kentgens *et al.*, 1987) is a two-dimensional technique that probes the quadrupolar evolution under the influence of an rf field. The resulting nutation spectra are free of chemical-shift effects. Information on the quadrupolar coupling constant and the asymmetry parameter η can be extracted by comparison with simulations of the relatively characteristic nutation lineshapes. However, the spectra become more complex rather than being simplified; therefore, the application seems to be restricted to systems with few sites. In addition, for a nutation spectrum to consist of more than just a single featureless line, the quadrupolar coupling Ω_Q should not be too small or too large compared with the strength of the rf field:

$$\omega_{rf} \overset{!}{\cong} \Omega_Q = \frac{eqeQ}{2I(2I-1)\hbar}.$$

It should be noted here that ω_{rf} need not be comparable to the quadrupole coupling constant $eqeQ/\hbar$ itself, because the effect of the factor $[2I(2I-1)]^{-1}$ becomes quite appreciable for large I: it is 0.083 for $I=\frac{3}{2}$, 0.025 for $I=\frac{5}{2}$, and as small as 0.012 for $I=\frac{7}{2}$. For instance, for an $I=\frac{5}{2}$ nucleus $\omega_{rf}=\Omega_Q=2\pi\times 40$ kHz corresponds to $eqeQ/\hbar=2\pi\times 1.6$ MHz.

3.11 Zero-field NMR

The usefulness of anisotropies is stressed duly above and below. For the important aim of distinguishing structurally different sites, however, the broadening introduced by the anisotropies has to be removed. As discussed in some detail above, one solution is to make the interactions appear isotropic on average by rotation at the magic angle in high field. As an alternative, the magnetic environment of the sample can be made isotropic by removing the field. While the chemical shift vanishes with the field, the dipolar and quadrupolar couplings are still present in zero field to generate energy differences; due to the isotropic environment, no angular dependences can arise and sharp lines without anisotropic broadening result even in a powder sample.

A strong $\vec{\mathbf{B}}_0$ field is, however, still required to generate a great magnetization vector according to the Boltzmann factor in equation (2.66) and to detect the signal with high sensitivity at a high Larmor frequency according to equation (2.10). Therefore, zero-field NMR experiments consist of at least three periods: preparation in high field, evolution in zero field, and detection in high field (Figure 3.19). Thus, the zero-field experiment is inherently two-dimensional.

The principal phenomenology of zero-field evolution (Zax *et al.*, 1985) can be understood quite easily as follows: subject to an axially symmetric interaction (e.g. the static dipole pair coupling), the magnetization 'precesses' around the local field with the single dipolar frequency δ_D. The unique axis of the interaction, e.g. an internuclear vector, is the axis of precession. If the magnetization is parallel to this axis, the magnetization precesses 'around itself', that is, it is invariant. Magnetization perpendicular to this axis precesses with its full amplitude while, for angles in between, both static and precessing components are present.

This intuitively appealing picture can be confirmed by considering the full (nontruncated) dipolar Hamiltonian (2.14) for two spins a and b in its principal-axes system ($D_{zz}^{\mathrm{PAS}} = (\vec{\mathbf{r}}_{ab})^{\mathrm{T}}\mathbf{D}\vec{\mathbf{r}}_{ab} = 2\omega_{\mathrm{d}}$ according to (2.14) and (2.91a)):

$$\hat{H}_{\mathrm{D}}^{\mathrm{PAS}} = (\hat{\vec{\mathbf{I}}}^a)^{\mathrm{T}} \begin{bmatrix} -\omega_{\mathrm{d}} & & \\ & -\omega_{\mathrm{d}} & \\ & & 2\omega_{\mathrm{d}} \end{bmatrix}_{\mathrm{D}} \hat{\vec{\mathbf{I}}}^b = \omega_{\mathrm{d}}(3\hat{I}_z^a\hat{I}_z^b - \hat{\vec{\mathbf{I}}}^a \cdot \hat{\vec{\mathbf{I}}}^b). \tag{3.113}$$

In complete formal analogy to the calculation for the high-field dipolar interaction (see equation (3.43)), for a pair of spins $\frac{1}{2}$ this results in an oscillation of the transverse magnetization component with $\cos(\frac{3}{2}\omega_{\mathrm{d}}t)$. These frequencies $\pm\frac{3}{2}\omega_{\mathrm{d}}$ correspond to the locations $\omega = \pm\,\delta_{\mathrm{D}}^{II}$ of the limiting edges of the high-field Pake pattern, since $\frac{3}{2}\omega_{\mathrm{d}} = \delta_{\mathrm{D}}^{II}$ according to equations (2.37b) and (3.43). The component of the magnetization directed along the z axis, which is the z axis of the PAS here, is time-invariant. The overall time dependence of the magnetization therefore corresponds to the projection of a vector precessing around $\vec{\mathbf{Z}}_{\mathrm{P}}$, projected into the plane spanned by $\vec{\mathbf{Z}}_{\mathrm{P}}$ and $\vec{\mathbf{M}}_0$. In a more general description including $\eta \neq 0$, for spin pairs or quadrupolar nuclei (where $a \triangleq b$), the Hamiltonian in the PAS reads:

$$\hat{H}_{\mathrm{D}}^{\mathrm{PAS}} = (\hat{\vec{\mathbf{I}}}^a)^{\mathrm{T}} \begin{bmatrix} -\frac{1}{2}\delta - \frac{1}{2}\delta\eta & & \\ & -\frac{1}{2}\delta + \frac{1}{2}\delta\eta & \\ & & \delta \end{bmatrix}_{\mathrm{D}} \hat{\vec{\mathbf{I}}}^b = \tfrac{1}{2}\delta(3\hat{I}_z^a\hat{I}_z^b - \hat{\vec{\mathbf{I}}}^a \cdot \hat{\vec{\mathbf{I}}}^b) + \tfrac{1}{2}\delta\eta(\hat{I}_y^a\hat{I}_y^b - \hat{I}_x^a\hat{I}_x^b)$$

and the density matrix evolves in the local field with the frequencies:

$$\omega^+ = \delta + \tfrac{1}{3}\eta\delta \quad \omega^- = \delta - \tfrac{1}{3}\eta\delta \quad \omega_0 = \tfrac{2}{3}\eta\delta = \omega^+ - \omega^- \tag{3.114}$$

which correspond to the energy differences between the three energy levels in zero field.

The actual form of the density matrix depends on the way in which the field is removed from the sample. If the drop of the field is sudden enough, the density matrix remains high-field magnetization. On the other hand, the field can be removed adiabatically, that is, so slowly that the coefficients of the high-field eigenstates follow the state as it changes into a low field eigenstate (Kreis *et al.*, 1985). This is discussed in more detail in the literature (compiled, for instance, in the review by Thayer and Pines (1987)).

Since zero-field NMR can detect the η and δ values of dipolar and quadrupolar interactions with high accuracy, it is also suited to detect motional averaging through the corresponding $\bar{\delta}$ and $\bar{\eta}$. At first sight, however, it may appear as surprising that pure reorientations can be detected through an orientation-independent frequency. The solution of this 'paradox' is the fact that the amplitudes of the oscillating and of

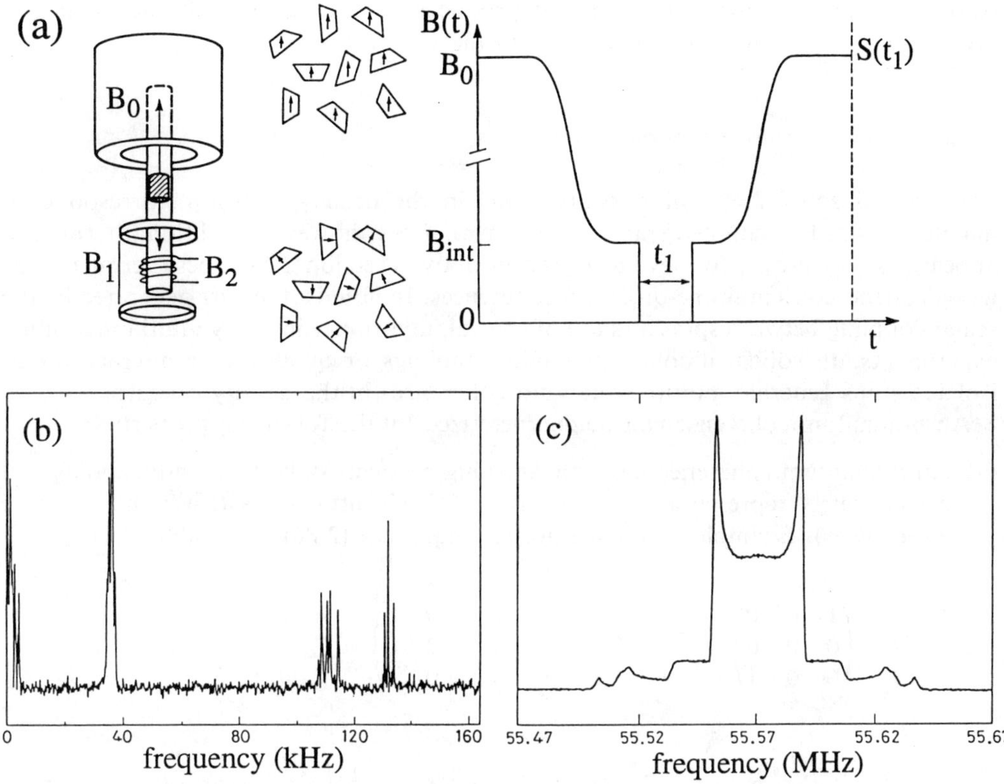

Figure 3.19 Zero-field NMR. (a) Basic field cycle. (b) Zero-field ^{2}H spectrum of polycrystalline perdeuterated diethylterephthalate. Four groups of lines are observed, corresponding to phenylene, methylene, and methyl sites and ω_0 lines (from right to left). Quadrupolar coupling constants and small η values of five sites have been established from the spectrum. (c) High-field powder pattern of the same material for comparison. Adapted from Thayer and Pines, 1987.

the static signal components are orientation dependent: in the time domain, the exchange process can be viewed as an exchange of the 'precession' axis between different directions, one for each site in the exchange process. For each orientation of the 'precession' axis, the angle χ with the magnetization is different, therefore the amplitudes of the static and of the oscillating part are subject to sudden changes. In the fast-exchange limit, the averaged principal values and PAS determine the time evolution.

Dipolar zero-field spectra become rather featureless for large spin systems since the number of couplings grows exponentially with the number of coupled spins. This makes the application to organic polymers with their large abundance of protons unattractive. In addition, the special instrumentation required for zero-field NMR is available only in a few laboratories. For these reasons, we touch on zero-field NMR only briefly in this text.

Recently, an elaborate technique for obtaining zero-field spectra at high field under MAS conditions has been developed by Tycko (1990). It is based on a rotor-synchronized pulse sequence that produces an average Hamiltonian which is proportional to the zero-field Hamiltonian (3.113).

3.12 Multiple-quantum coherences

For isolated spins $I = \frac{1}{2}$, all possible terms in the density operator correspond to magnetization. For coupled spins or spins with $I > \frac{1}{2}$, however, additional terms can appear, as was shown for several instances above (Section 2.5). These terms can be globally denoted as multiple-quantum coherences. In liquids, they are generated by the scalar coupling between spins and are of central importance in many multidimensional experiments. In solids, dipolar spin–spin couplings or quadrupolar interactions of isolated spins generate multiple-quantum coherences in the density operator.

An n-quantum coherence can be characterized by the following properties:

(a) An n-quantum coherence has nonvanishing elements only in the nth subdiagonal of its matrix representation in a basis of $\hat{I}_z$ eigenfunctions (for nondegenerate eigenvalues). Examples for $I = 1$ are (see equation (2.86) and Table C.1):

$$n = 0: \quad \begin{pmatrix} 1 & 0 & 0 \\ 0 & 0 & 0 \\ 0 & 0 & 1 \end{pmatrix} = \hat{T}_{10} = \hat{I}_z \qquad \frac{1}{\sqrt{6}} \begin{pmatrix} 1 & 0 & 0 \\ 0 & -2 & 0 \\ 0 & 0 & 1 \end{pmatrix} = \hat{T}_{20}$$

$$n = 1: \quad \frac{1}{\sqrt{2}} \begin{pmatrix} 0 & 1 & 0 \\ 1 & 0 & 1 \\ 0 & 1 & 0 \end{pmatrix} = \frac{-1}{\sqrt{2}} (\hat{T}_{11} - \hat{T}_{1-1}) = \hat{I}_x \qquad \frac{1}{\sqrt{2}} \begin{pmatrix} 0 & -1 & 0 \\ -1 & 0 & 1 \\ 0 & 1 & 0 \end{pmatrix} = \hat{T}_{21} - \hat{T}_{2-1}$$

$$n = 2: \quad \begin{pmatrix} 0 & 0 & 1 \\ 0 & 0 & 0 \\ 1 & 0 & 0 \end{pmatrix} = \hat{T}_{22} + \hat{T}_{2-2}. \tag{3.115}$$

(b) An n-quantum coherence oscillates with a frequency $n\Delta\omega$ under the action of the Hamiltonian $\Delta\omega\hat{I}_z$ in the rotating frame of the pulses. By this criterion, $\hat{I}_z$ is a zero-quantum coherence and $\hat{I}_x$ is a one-quantum coherence. Note, however, that according to the preceding list of matrices there are zero- and one-quantum coherences that do *not* correspond to magnetization. The relation of this condition (b) to the condition (a) is provided by equation (2.83).

(c) An n-quantum coherence corresponds to a $\hat{T}_{m\pm n}$ spherical tensor operator. Examples are provided in (3.115).

In an energy-level picture, n-quantum coherences correspond to (forbidden) transitions between energy levels separated by n times the single-spin Zeeman energy $\hbar\omega_0$. Due to the correspondence of such transition frequencies with the oscillation frequencies in the density matrix (see equation (2.83)), we expect that n-quantum coherences oscillate with $n\omega_0$. This is in fact true: in the laboratory frame, where $\hat{H}_0 = \omega_0\hat{I}_z$ is not transformed away, the n-quantum elements of the lab-frame density matrix oscillate with $n\omega_0$ according to equation (2.83). Naively, one could expect that in the frame rotating with ω_0, these elements should oscillate with $n\omega_0 - \omega_0$. This is, however, incorrect. Upon transition into the rotating frame, the oscillation with the n-fold Larmor frequency is transformed away, simply because the action of the Zeeman propagator $\exp(-i\omega_0\hat{I}_z)$ is cancelled by $\exp(i\omega_s\hat{I}_z)$ in the interaction representation (3.20). Only for the residual Hamiltonian $\Delta\omega\hat{I}_z$ with the frequency offset $\Delta\omega = \omega_0 - \omega_s$, the n-fold precession frequency $n\Delta\omega$ remains.

Acting on a multiquantum-coherence term $\hat{T}_{mn}$, free precession under $\Delta\omega\hat{I}_z$ changes neither m nor $|n|$, only the prefactors of $\hat{T}_{mn}$ and $\hat{T}_{m,-n}$ vary harmonically with the frequency $n\Delta\omega$. Radiofrequency pulses can change n by -1, 0 or $+1$, leaving m invariant. Evolution under $\omega\hat{T}_{20}$ changes m but leaves $|n|$ unchanged. Therefore, starting with z magnetization, at least two pulses with intermittent evolution are required to generate multiquantum coherences with $|n| > 1$, and a third pulse is required to transform these coherences back into observable magnetization with $|n| = 1$.

For a fast understanding of multiple-quantum (MQ) NMR of solids, it may be helpful to make a distinction between MQ spectroscopy and MQ spin counting. Multiple-quantum spectroscopy can be applied in solution NMR or for small molecules dissolved and oriented in deuterated liquid-crystalline matrices, where the (scalar or dipolar) couplings produce narrow lines or at least heterogeneous line-broadening. In most organic systems, including polymers, in the solid state the number of couplings between protons (or fluorines) is usually so large that homogeneously broadened featureless multiple-quantum spectral lines result. Nevertheless, in such systems 'spin counting' can be achieved by observing the time evolution of the integral intensity of the different multiple-quantum orders (Pines, 1988; Baum *et al.*, 1985; Baum and Pines, 1986). This is somewhat similar to spin diffusion experiments. In particular, limited spin-cluster sizes can be detected through a 'saturation behaviour' in the time dependence of the multiple-quantum coherences.

Both MQ dipolar spectroscopy and spin counting are based on the same principal pulse sequence and can be performed simultaneously in principle. The relevance of the different periods of the experiments, however, is somewhat different. The basic pulse sequence consists of four periods, to be labelled (A) to (D) here (Fig. 3.20). Period (A) lasts for a time τ during which from z magnetization under a suitable average Hamiltonian (e.g. $\frac{1}{3}(\hat{H}_{yy} - \hat{H}_{xx})$ as generated by a sequence of x and $-x$ 90° pulses),

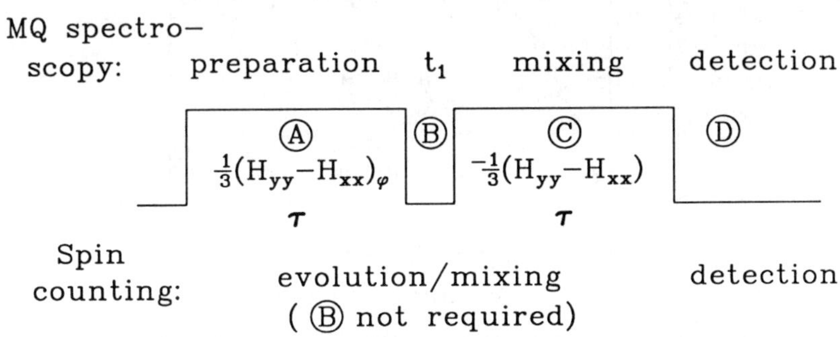

Figure 3.20 Schematic of the pulse sequence for multiple-quantum spectroscopy and spin counting. In the periods (A) and (C), multiple-pulse sequences are applied which yield the average Hamiltonians given in the figure. The index ϕ indicates that the phases of all pulses in period (A) are incremented, making angles ϕ with respect to those in period (C). Adapted from Shykind *et al.*, 1988.

multiple-quantum coherences are generated or gradually arise, depending whether MQ spectroscopy or spin counting is considered. The coherences do not all appear at once; high orders can become appreciable only when τ is long (compared with the inverse of the typical dipolar couplings). The contributions of the different coherence orders are distinguished by their different behaviour under phase shifts $\Delta\phi$ of the pulses in period (A). This phase shift corresponds to a rotation by $\Delta\phi$ around the z axis, $\exp(-i\,\Delta\phi\hat{I}_z)$, which is felt as a phase shift $n\Delta\phi$ by an n-quantum coherences (Shykind *et al.*, 1988). The next period, (B), with free time-evolution of the coherence is required only for MQ spectroscopy, where it represents the evolution time, which is incremented by Δt_1.

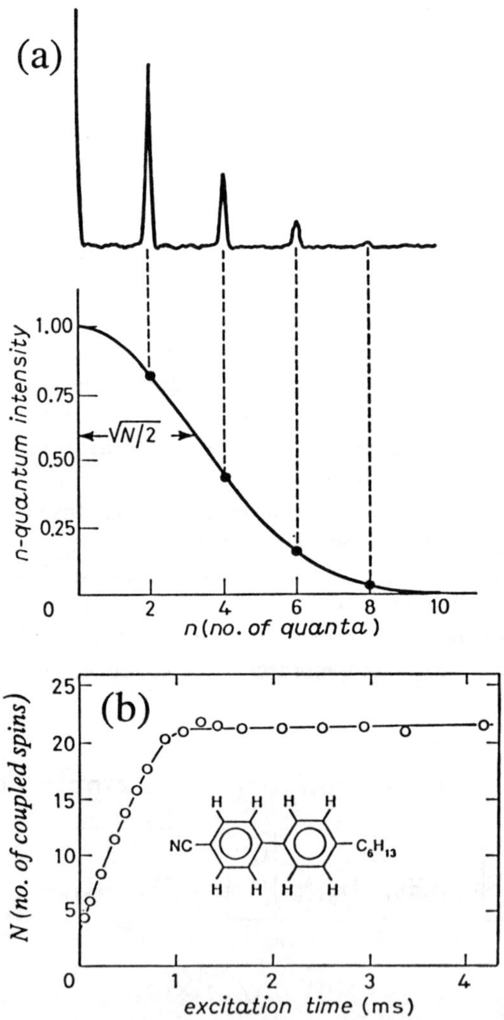

Figure 3.21 Multiple-quantum spin counting for the liquid crystal 5CB, containing 21 protons per molecule. (a) Measured multiple-quantum intensity distribution. (b) Width of the Gaussian distribution as a function of the multiple-quantum mixing time. The plateau at 21 indicates that clusters of 21 spins are coupled. This is consistent with intermolecular couplings being eliminated by the translational motion of the molecules in the liquid-crystalline phase. Adapted from Baum and Pines (1986).

For spin counting, this period can be omitted since the phase increment $\Delta\phi$ in period (A) corresponds to the t_1 increment of the evolution period. Period (C) provides the reconversion of the multiple-quantum coherence into z magnetization which is detected after a 90° pulse.

The labelling preparation/evolution/mixing/detection for the periods (A) to (D) which applies for MQ spectroscopy does not suit the spin-counting experiment, in particular if a spin diffusion 2D experiment is viewed in comparison. As mentioned above, the 'evolution' period (B) can be discarded completely for spin counting; the modulation corresponding to the evolution time in a 2D experiment is provided by the pulse-phase increment $\Delta\phi$ (Shykind *et al.*, 1988). In addition, the most immediate analogue to the mixing time t_m in a spin diffusion experiment is the period (A), during which 'intensity diffuses into higher coherence orders', while in the period (C) the coherence-orders are reduced.

Figure 3.21 shows how the number of protons in liquid-crystalline molecules has been counted. It is exploited that each molecule represents an independent dipolar

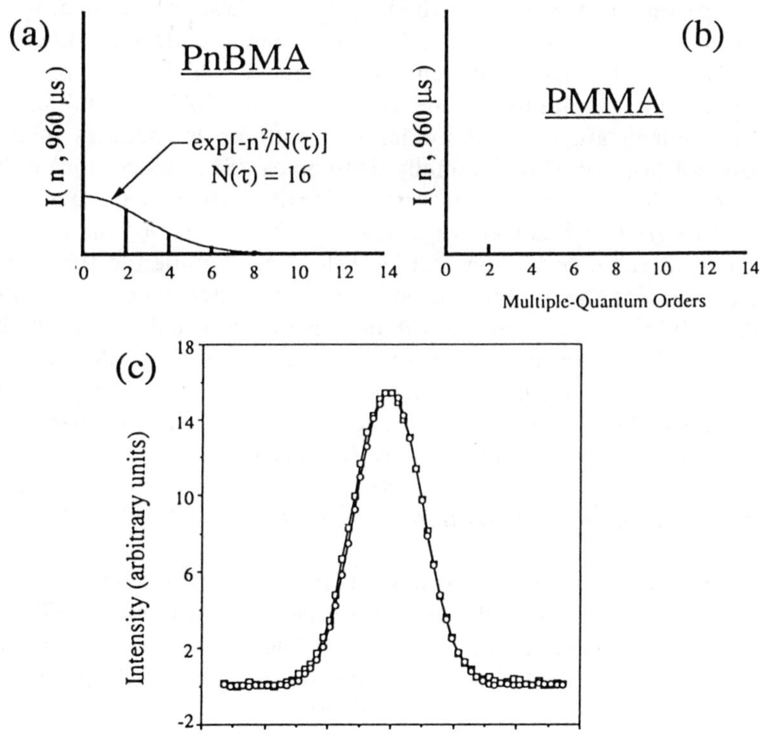

Figure 3.22 ^{19}F multiple-quantum intensity distributions of 10% fluorine-containing photosensitive salt molecules in (a) PnBMA and (b) PMMA. The higher-order coherences in PnBMA indicate that the initiator molecules must be segregated in clusters. The absence of four- or higher quantum coherences in the spectrum for PMMA shows that most of the salt molecules are dispersed. (c) The normal ^{19}F wideline spectra of the two materials are virtually identical. Adapted from Scruggs and Gleason (1992).

cluster due to the fast relative motion of the molecules. An application of spin counting in polymeric materials is provided by the investigation of the clustering behaviour of fluorine-containing photosensitive salt molecules in various polymer matrices (Scruggs and Gleason, 1992). Figure 3.22 compares the ^{19}F multiple-quantum spectra of 10% salt in PnBMA and in PMMA. The absence of four- or higher quantum coherences in the spectrum for PMMA shows that most of the salt molecules are dispersed in PMMA. On the other hand, the higher-order coherences of salt in PnBMA indicate that in that polymer the initiator molecules must be segregated. Their multiple-quantum time evolution is found to be identical to that of bulk salt (Scruggs and Gleason, 1992), which indicates that the structure of the segregated clusters must be similar to that of salt crystals.

An appealing and detailed treatment of multiple-quantum coherences with emphasis on solid-state NMR has been given by Munowitz (1988).

3.13 Incoherent effects in solid-state NMR

So far, only coherent processes described by a well-defined propagator have been treated; these are indeed spectroscopically most relevant. Several other important effects in NMR are due to incoherent processes; in particular, these relaxation phenomena represent the limits to the coherent evolution. Certain coherent and incoherent phenomena are even so similar that they are sometimes mixed up: for instance, cross polarization is occasionally denoted as 'cross relaxation' although CP is a coherent, completely reversible process (as is witnessed by its coherent averaging by high-speed MAS). This incorrect or at least 'diffuse' usage of the term relaxation can produce considerable confusion and is misleading, hampering the acceptance of the valid description of the coherent process. On the other hand, erroneous notions also beset the folklore on relaxation: for instance, it is widely believed that at the temperature of the T_1 minimum, the local fields oscillate, or the molecules rotate, with a frequency close to the Larmor frequency; the fact that molecular motion is a stochastic process to be characterized by a rate (inverse of the correlation time) and not by a motional frequency is often not rationalized.

3.13.1 Phenomenology of relaxation in solid-state NMR

Relaxation times have played an important role in nuclear magnetic resonance of bulk matter ever since its discovery (and even before: Gorter's experiments (Slichter, 1990) probably failed due to excessive relaxation times). The study of NMR relaxation is long established as a field of research, overlapping often with NMR spectroscopy. The relaxation times are sensitive to the spectral density of molecular dynamics with rates in the range of the characteristic frequency $\omega_r = \gamma B_r$ (Abragam, 1961; Slichter, 1990). B_r denotes the strength of the relevant $\vec{\mathbf{B}}$ field. For the longitudinal relaxation time in the laboratory frame (T_1), this is the $\vec{\mathbf{B}}_0$ field; for the longitudinal relaxation time in the rotating frame ($T_{1\rho}$), B_r is the $\vec{\mathbf{B}}_1$ field of the coil that 'locks' the magnetization (see Section 3.4.2). T_1 thus probes the spectral density at the Larmor frequency, at tens to hundreds of MHz, while $T_{1\rho}$ is sensitive in the region of tens of kHz. When the rate of the motion is approximately equal to the characteristic frequency γB_r, the corresponding relaxation rate ($1/T_1$) has a maximum. In a very simplistic picture this can be seen as the result of a 'resonant' coupling of the local fluctuating fields with the spin system that speeds up the return to the Boltzmann equilibrium.

The correct description considers the correlation function $C(t)$ (see Chapter 9) of the fluctuating local fields B_{loc} and its Fourier transform, the spectral density $J(\omega)$; the Bloembergen–Purcell–Pound (BPP) theory shows that the relaxation rate is essentially proportional to the spectral density at the Larmor frequency, $1/T_1 \cong (\gamma B_{loc})^2 J(\omega_L)$ (Bloembergen *et al.* 1948). For $C(t) = e^{-t/\tau}$ with the correlation time τ, the Fourier transform $J(\omega)$ is a Lorentzian $J(\omega, \tau) = \tau/(\omega^2 + \tau^{-2})$. Plotting $-\ln\{(\gamma B_{loc})^2 J(\omega, \tau)\} \cong \ln(T_1)$ as a function of $\ln\tau$, we obtain a 'typical T_1 curve' with the minimum at $\tau = 1/\omega_L$ (Figure 3.23). The T_1 curve as a function of the temperature T, or rather $1/T$, is immediately obtained through the Arrhenius relationship $\tau = \tau_0 \exp(-E_a/kT)$, or $\ln(\tau/\tau_0) = (E_a/k)(1/T)$; due to this proportionality of $\ln(\tau/\tau_0)$ and $1/T$, the T_1 curves

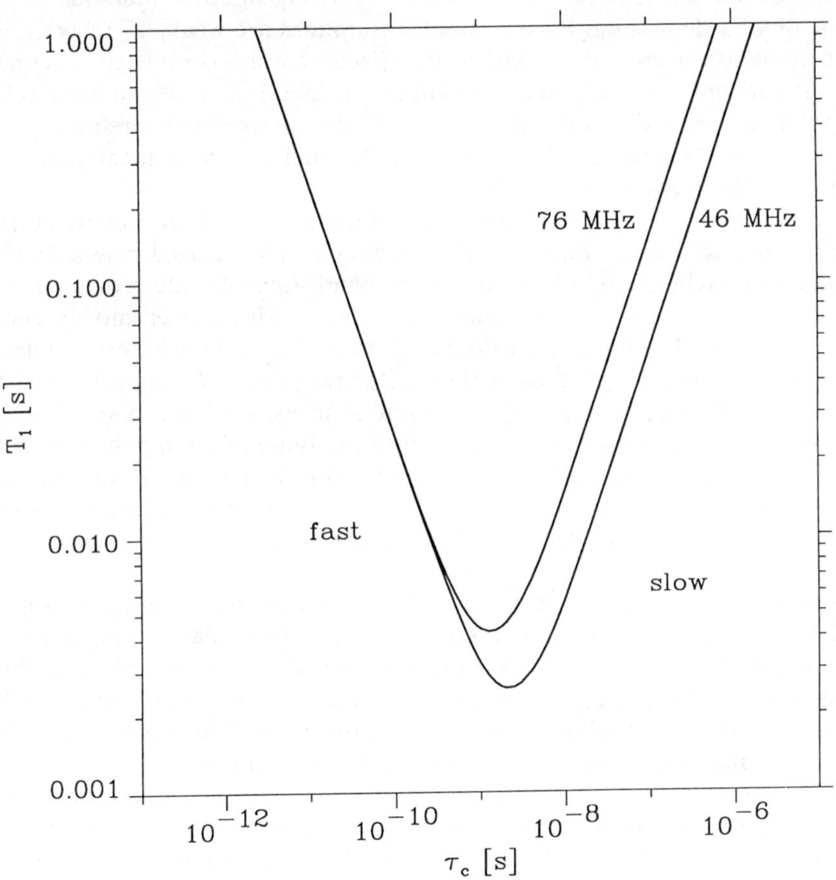

Figure 3.23 Typical correlation-time dependence of T_1 relaxation times. The T_1 minimum occurs at a motional rate (inverse correlation time) which is approximately equal to the Larmor frequency. The curves shown were calculated for quadrupolar relaxation resulting from isotropic reorientation, at Larmor frequencies of 46 and 76 MHz. The experimentally measurable dependence of T_1 on temperature in an Arrhenius representation ($\ln(T_1)$ vs. $1/T$) has the same shape if the correlation times obey an Arrhenius law. At the temperature of the T_1 minimum, the motional rate is approximately equal to the Larmor frequency.

as a function of $\ln \tau$ and $1/T$ are completely analogous. When specific motional models are considered, the theory becomes considerably more complicated (Torchia and Szabo, 1982).

As already indicated in the Bloch equations (2.8), of the three magnetization components, only $\langle \hat{I}_z \rangle$ relaxes with T_1. This longitudinal relaxation is coupled to changes in energy (while T_2 relaxation only affects the phase of the magnetization). The elements of the density matrix (in the $\hat{I}_z$ eigenbasis) that react to energy changes are located on the diagonal: these elements correspond to populations of energy levels. Therefore, only diagonal elements are affected by T_1 relaxation. In addition to $\hat{I}_z$ itself, the operators $3\hat{I}_z^i \hat{I}_z^j - \hat{\mathbf{I}}^i \cdot \hat{\mathbf{I}}^j$ and $\hat{T}_{20}$ are diagonal in this basis; their relaxation times are denoted as T_1^{D} and T_1^{Q}, respectively.

The off-diagonal elements in the density matrix relax with the 'transverse' relaxation time T_2, which is usually shorter than T_1. Real T_2 relaxation is exploited less often in solids: the irreversible dipolar T_2 relaxation is usually unobservable due to the fast reversible dipolar dephasing. Even if dipolar couplings are weak, T_2 relaxation in the sense of the BPP theory can be masked by effects that are completely dominated by secular interactions. In particular, incomplete refocusing of Hahn spin echoes or quadrupolar echoes will result from motion-induced stochastic frequency changes during the delay 2τ between the first 90° pulse and the echo maximum. The 'T_2 minimum' of ^2H NMR relies on this effect.

In NMR spectroscopy, short T_1 times of less than 1 second are usually desirable in order to be able to average many scans with almost fully relaxed magnetization in a short time. For exchange NMR, on the other hand, long T_1 values of many seconds can be desirable to facilitate exchange experiments with longer mixing times (see Chapters 7 and 13). For multidimensional ^{13}C exchange NMR with cross polarization from the protons to the ^{13}C spins, a short effective proton T_1 and a long carbon T_1 is optimum: the former permits fast repetition of scans, the latter makes long mixing times possible. Such a combination is indeed found quite often in polymers, since due to the efficient ^1H spin diffusion, the ^1H relaxation behaviour is governed by fast-relaxing fractions in the sample, such as rotating methyl groups or mobile amorphous regions. Long $T_{1\rho}$ times are beneficial for the magnetization not to decay under the spin-lock required during cross polarization.

Cross relaxation can lead to nuclear Overhauser enhancements (NOEs) (Neuhaus and Williamson, 1989) or magnetization transfers. Those play an important role in solution NMR, in particular in the two-dimensional NOESY experiment, which allows the measurement of proton–proton distances. This homonuclear proton NOE plays hardly any role in proper solids, as it is masked by ^1H spin diffusion which is a much more efficient magnetization-transfer process among protons.

The heteronuclear NOE (Neuhaus and Williamson, 1989) between protons and S spins in certain solids, however, can be appreciable if there are motions with rates approaching or surpassing the Larmor frequencies. The enhancement factors found here can even exceed the maximum isotropic NOE factor of $1 + (\gamma_{^1H}/\gamma_S)$ if the motion is appropriately anisotropic. The observed heteronuclear NOE effect consists in an enhancement of the S-spin polarization if during S-spin T_1 relaxation, the proton spin system is saturated or inverted, either by continuous irradiation or by pulses spaced by about a proton T_1 or less. The NOE relies on dipolar cross relaxation and is reduced when other relaxation pathways (chemical shift, quadrupolar) are efficient. As a relaxation effect, the NOE is definitely different from cross polarization, which is a coherent effect. This will be discussed in the following section.

3.13.2 Coherent vs. incoherent processes

Coherent evolution of spin systems in the NMR of organic solids relies on the secular parts of spin Hamiltonians, as outlined in these chapters, and it can occur in essentially rigid molecules. The secular Hamiltonians relevant here are usually time-independent, or effectively so (averaging of interactions), or exhibit a definite, slow time dependence (MAS). The time evolution is reversible; it can be inverted by pulse sequences or rotations to produce an echo (though in certain cases, irreversible relaxation of certain components of the density matrix can affect the reversibility of a process that is in principle coherent, e.g. spin diffusion).

Incoherent processes, also termed relaxation, are determined by both secular and nonsecular Hamiltonians. They rely on stochastic modulations of local fields by molecular motions, and in particular on their spectral density at the Larmor frequency of interest. Due to the stochastic time dependence of the interactions, the dephasing under relaxation cannot be refocused.

Relaxation usually results in time dependences that are made up of one or a few exponentials. Coherent processes in small systems with uniform couplings usually produce oscillatory behaviour. However, this difference is washed out in more complex systems by the superposition of many coherent signals with different periods (destructive interference), or because of multiple couplings. This can result in different time-dependences with characteristic initial regimes: t^2 for Gaussian, t for pseudo-exponential, or $\sqrt{t}$ for diffusive behaviour.

In addition to the reversibility, the dependence of a change of the density operator on the rate of molecular motions is a primary distinction between coherent and incoherent processes: to be effective, relaxation requires fluctuations, while coherent processes can occur in essentially rigid solids and are usually slowed down by motions.

In many cases, the classification of a process as coherent, or as a relaxation phenomenon, is obvious. There are, however, some exceptions. Indeed, pairs of processes exist that exhibit a similar phenomenology, but one is a relaxation process while the other is a complicated coherent evolution. These include the following:

(a) cross polarization (coherent) vs. heteronuclear NOE (cross relaxation)
(b) dipolar dephasing (coherent) vs. irreversible T_2 (relaxation)
(c) dipolar spin diffusion (coherent in principle) vs. spin diffusion by multistep homonuclear NOE (cross relaxation).

The coherent processes are usually governed by propagators $\exp(-i\hat{H}t)$ (see Section 2.5.2). They are in principle, and to a good extent in practice, reversible by inverting the sign of the Hamiltonian (i.e. by creating an appropriate average Hamiltonian) (Rhim *et al.*, 1971). Along the same lines, the coherent processes can be averaged out by very fast sample spinning while the relaxation will be hardly affected.

The distinction of coherent and relaxation processes is a basic aspect in the understanding of these effects. Mixing them up can at least result in confusion. For instance, the predictions of very fast spin–spin relaxation in rigid solids made by extrapolation of the Bloembergen–Purcel–Pound (BPP) theory are not correct: under high-power decoupling, crystalline materials can exhibit narrow lines, corresponding to T_2 times of tens of milliseconds (longer than the corresponding minimum T_1). This misconception concerning transverse relaxation times in solids is caused by confusing reversible, coherent, static dipolar couplings with incoherent, irreversible effects of fluctuating dipolar couplings.

A mixture of coherent and incoherent processes is involved in the generation of the 'T_2 minimum' of ^{2}H NMR. Stochastic, incoherent changes of the purely secular, coherent Hamiltonians are sufficient to produce this effect. The stochastic nature of the fluctuations makes this process usually indistinguishable from 'genuine' T_2 relaxation.

3.13.3 Homogeneous, inhomogeneous, coherent line-broadening

Line-broadening can arise from different origins, which are also related to the differences between coherent and incoherent processes. Usually, one distinguishes inhomogeneously and homogeneously broadened lines (Portis, 1953). Inhomogeneously broadened lineshapes consist of a continuum of independent lines that can be manipulated individually, for instance through spectral hole burning by selective excitation (Figure 3.24). Homogeneously broadened lines react to all pulses as an indivisible entity. Thirdly, if irradiation at one frequency affects the intensity at several other distinct frequency positions, the broadening can be termed heterogeneous. However, this term is rarely employed in practice: if the coupled lines are few, the situation can usually be treated similarly to inhomogeneous broadening; if there are very many coupled lines spread over the whole spectrum, this represents the transition to the homogeneous broadening.

Note that from the simple time signal, the nature of the broadening cannot be determined. If the spectra are identical, their (inverse) Fourier transforms in the time domain are also indistinguishable. For instance, in the case of inhomogeneous broadening, the destructive interference of the large number of closely spaced lines leads to a fast decay of the time signal, indistinguishable from the signal corresponding to a homogeneously broadened line of the same shape.

The distinction between T_2 relaxation and dipolar dephasing discussed above corresponds to a distinction between coherent and incoherent homogeneous broadening (inhomogenous broadening is always due to many parallel coherent processes). Incoherent broadening, being due to incoherent effects of stochastically fluctuating local fields, cannot be reduced by pulses or the corresponding decay of the magnetization reversed. By contrast, coherent homogeneous broadening generated by multibody interactions, such as dipolar couplings among protons, can be eliminated

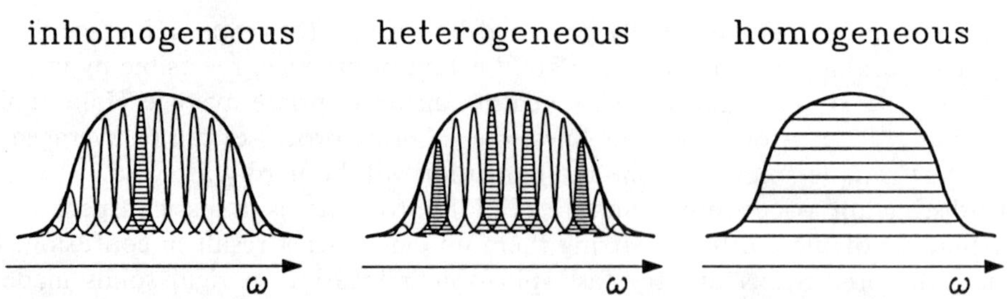

Figure 3.24 Visualization of inhomogeneous, heterogeneous, and homogeneous line-broadening. Some coupled lines that behave as one entity are indicated by hatching. This means, for instance, that irradiation at any frequency located in the hatched range will affect all of the hatched regions identically. On the basis of that definition, in the homogeneous case the whole spectrum is hatched.

by appropriate multiple-pulse sequences. These either average the Hamiltonian to zero, or they refocus the time decay into an echo by a change of the sign in the average Hamiltonian; this latter effect can also be viewed as a 'time reversal' in the propagator on the basis of $\exp\{-i(-\hat{H})t\} = \exp\{-i\hat{H}(-t)\}$ (Rhim *et al.*, 1971). Examples of inhomogeneous, coherent homogeneous, and incoherent homogeneous line-broadenings in NMR are displayed in Figure 3.25.

An instructive, tractable example of heterogeneous broadening and the transition to coherent homogeneous broadening is provided by the dipolar coupling of a single S spin to I spins without homonuclear couplings, as treated in Section 2.5.6. If there is only one I spin, a pair of sharp lines results. In a powder, the continuum of such lines generates essentially inhomogeneous line-broadening. If N I spins are coupled to the S spin, 2^N pairs of lines are observed. Being all due to one term in the density operator (see equation (2.98)) these lines cannot easily be individually inverted and do not evolve independently, as can be detected, for instance, in 2D experiments. The broadening is

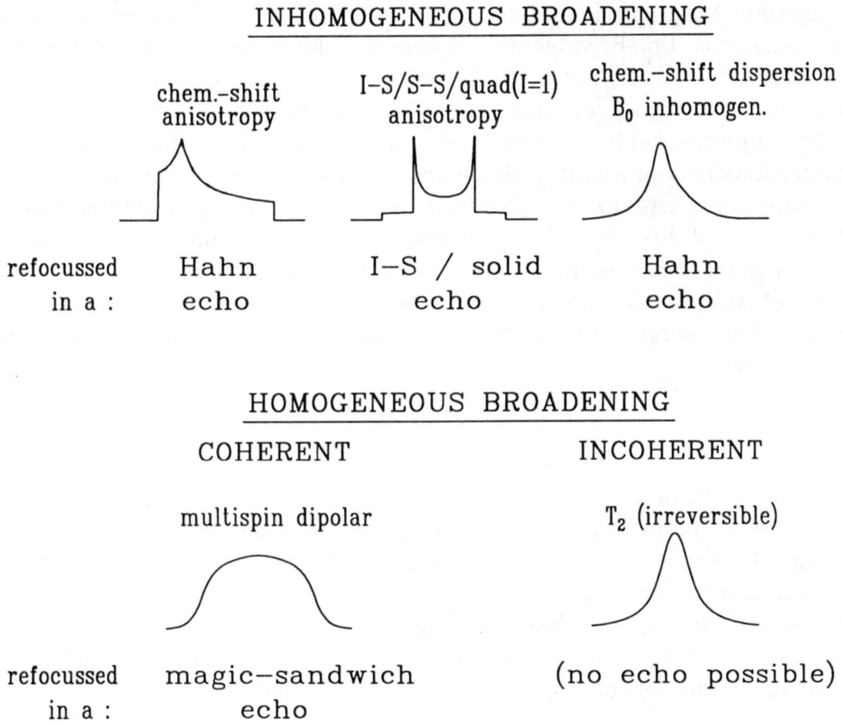

Figure 3.25 Typical examples of inhomogeneous, homogeneous coherent, and homogeneous incoherent line-broadenings. Chemical-shift, two-spin-dipolar, spin-1 quadrupolar anisotropies, as well as chemical-shift dispersion (a distribution of isotropic chemical shifts) and $\vec{\mathbf{B}}_0$ inhomogeneity result in inhomogeneous broadening. Multispin dipolar couplings produce homogeneous and coherent broadening, while the broadening caused by (irreversible) T_2 relaxation is homogeneous and incoherent. The spectral broadening is directly linked to a faster decay of the signal in the time domain. For coherent interactions, it can be refocused into a spin echo, by use of pulsed techniques indicated below the spectra. The magic-sandwich echo (Rhim *et al.*, 1971) refocuses multispin dipolar couplings.

thus heterogeneous. If the multitude of these lines overlap strongly, the broadening has become essentially homogeneous. It is coherent since it can be inverted by an 180° pulse applied either to the I or to the S spins.

3.13.4 (In)homogeneous: line-broadening vs. Hamiltonians

The terminology for inhomogeneous and homogeneous line-broadening as discussed in the previous section has been widely used in spectroscopy. Maricq and Waugh (1979) started a usage of the terms 'homogeneous' and 'inhomogeneous' in an apparently analogous way when they applied them to characterize NMR Hamiltonians under sample rotations. In their nomenclature, if $\hat{H}(\omega_r t)$ commutes with itself at all different orientations (times) then it is inhomogeneous, otherwise it is homogeneous (see Section 3.9.5). In the former case, spinning sidebands are observed so that the system seems to behave inhomogeneously; in the second case, the line is hardly affected by MAS. Indeed, the Hamiltonians that fulfil the commutation condition are usually simple one- or two-particle interactions, which also produce inhomogeneous line-broadening in the static spectrum. Multispin interactions generally produce homogeneous Hamiltonians and homogeneous line-broadening. However, closer inspection reveals that this equivalence is only a rule, not a fundamental law. As was recognized by Maricq and Waugh themselves (below equation (26) of Maricq and Waugh (1979)), for an isolated chain of dipolar-coupled spins the MAS spectrum splits into sidebands (inhomogeneous Hamiltonian) while no holeburning is possible for the static lineshape (homogeneous broadening). This can be demonstrated by comparing the behaviour of an isolated chain of dipolar-coupled spins with that of the same spins arranged in some three-dimensional array. As mentioned in Section 3.9.5, the dipolar Hamiltonian for the chain of spins is inhomogeneous, since all couplings have the same angular dependence. On the other hand, the static dipolar Hamiltonians are of similar form for both geometries:

$$\hat{H} = \sum_{n<m} \sum c_{mn}(3\hat{I}_z^n \hat{I}_z^m - \hat{\mathbf{I}}^n \cdot \hat{\mathbf{I}}^m).$$

Nothing is special about the coefficients c_{mn} for the chain of spins here, since these spatial factors in the Hamiltonian depend not only on the orientation of internuclear vectors but also on the internuclear distances. This tells us that the multiple dipolar interactions lead to typical homogeneously broadened lines in both geometries.

In summary, this means that in using the terms 'homogeneous' and 'inhomogeneous', one should specify whether one talks about homogeneous/inhomogeneous Hamiltonians or homogeneous/inhomogeneous line-broadening.

3.14 References

Abragam, A. (1961) *The Principles of Nuclear Magnetism*, Chapter V. Oxford University Press, London.

Atkins, P. W. (1983). *Molecular Quantum Mechanics, 2nd ed.*, Chapter 8. Oxford University Press, Oxford.

Baum, J., Munowitz, M., Garroway, A. N. and Pines, A. (1985). Multiple-quantum dynamics in solid state NMR. *J. Chem. Phys.*, **83**, 2015.

Baum, J. and Pines, A. (1986). NMR Studies of Clustering in Solids. *J. Am. Chem. Soc.*, **108**, 7447.

Bloembergen, N. (1949) On the Interaction of Nuclear Spins. *Physica*, **XV**, 386.

Bloembergen, N., Purcell, E. M. and Pound, R. V. (1948). Relaxation Effects in NMR Absorption. *Phys. Rev.*, **73**, 679.

Bodenhausen, G., Freeman, R. and Morris, G. A. (1976). Simple Selective Excitation in FT NMR. *J. Magn. Reson.*, **23**, 171.

Burum, D. P. and Rhim, W. K. (1979). Analysis of multiple pulse NMR in solids. III. *J. Chem. Phys.*, **71**, 944.

Caravatti, P., Braunschweiler, L. and Ernst, R. R. (1983). I-S Correlation Spectroscopy in Rotating Solids. *Chem. Phys. Lett.*, **100**, 305.

Caravatti, P., Levitt, M. H. and Ernst, R. R. (1986). Selective Excitation in the Presence of Multiple-Pulse Line Narrowing. *J. Magn. Reson.*, **68**, 323.

Cheung, T. T. P. (1981). Diffusion in NMR in Solids. *Phys. Rev. B: Condens. Matter*, **23**, 1404.

de Groot, H. J. M., Smith, S. O. Kolbert, A. C., Courtin, J. M. L, Winkel, C., Lugtenburg, J., Herzfeld, J. and Griffin, R. G. (1991). Iterative Fitting of MAS NMR Spectra. *J. Magn. Reson.*, **91**, 30.

Dixon, W. T. (1982). NMR Spectra in Spinning Samples (TOSS). *J. Chem. Phys.*, **77**, 1800.

Dixon, W. T., Schaefer, J., Sefcik, M. D., Stejskal, E. O. and McKay, R. A. (1982). Total Suppression of Sidebands in CPMAS C-13 NMR. *J. Magn. Reson.*, **49**, 341.

Ernst, R. R., Bodenhausen, G. and Wokaun, A. (1987). *Principles of Nuclear Magnetic Resonance in One and Two Dimensions*. Clarendon Press, Oxford (1987).

Farnan, I., Grandinetti, P. J., Baltisberger, J. H., Stebbins, J. F., Werner, U., Eastman, M. A. and Pines, A. (1992). Quantification of Disorder in Silicate Glasses. *Nature*, **358**, 31.

Gerstein, B. C. and Dybowski, C. R. (1985). *Transient Techniques in NMR of Solids*. Academic Press, Orlando.

Haeberlen, U. (1976). High Resolution NMR of Solids. *Adv. Magn. Reson.: Supplement 1*, Academic Press, San Diego.

Haeberlen, U. and Waugh, J. S. (1968). Coherent Averaging Effects in Magnetic Resonance. *Phys. Rev.*, **175**, 453.

Hartmann, S. R. and Hahn, E. L. (1962). Nuclear Double Resonance in the Rotating Frame. *Phys. Rev.*, **128**, 2042.

Herzfeld, J. and Berger, A. E. (1980). Sideband Intensities in NMR Spectra. *J. Chem. Phys.*, **73**, 6021.

Jäger, C. (1992). ^{27}A1 MAS NMR by High-Speed SATRAS. *J. Magn. Reson.*, **99**, 353.

Kentgens, A. P. M., Lemmens, J. J. M., Geurts, F. M. M. and Veeman, W. S. (1987). 2D Nutation NMR of Half-Integer Quadrupolar Nuclei. *J. Magn. Reson.*, **71**, 62.

Kolbert, A. C. and Griffin, R. G. (1990). 2D Resolution of Isotropic and Anisotropic Chemical Shifts in MAS NMR. *Chem. Phys. Lett.*, **166**, 87.

Kreis, R., Suter, D. and Ernst, R. R. (1985). Time-Domain Zero-Field NMR with Field Pulses. *Chem. Phys. Lett.*, **118**, 120.

Llor, A. and Virlet, J. (1988). Sample Spinning with Time-Dependent Spinner Axis Angle. *Chem. Phys. Lett.*, **152**, 248.

Lowe, I. J. and Gade, S. (1967). Density-Matrix Derivation of Spin Diffusion. *Phys. Rev.*, **156**, 817.

Maricq, M. M. and Waugh, J. S. (1979). NMR in Rotating Solids. *J. Chem. Phys.*, **70**, 3300.

Mehring, M. (1983). *High Resolution NMR of Solids*. Springer-Verlag, Berlin.

Mori, H. and Kawasaki, K. (1962). Theory of Dynamical Behaviors of Ferromagnetic Spins. *Progr. Theor. Phys.*, **27**, 529.

Morris, G. A. and Freeman, R. (1978). Selective Excitation in FT NMR. *J. Magn. Reson.*, **29**, 433.

Müller, L., Kumar, A., Baumann, T. and Ernst, R. R. (1974). Transient Oscillations in CP in Solids. *Phys. Rev. Lett.*, **32**, 1402.

Mueller, K. T., Sun, B. Q., Chingas, G. C., Zwanziger, J. W., Terao, T. and Pines, A. (1990). DAS of Quadrupolar Nuclei. *J. Magn. Reson.*, **86**, 470.

Mueller, K. T., Wu, Y., Chmelka, B. F., Stebbins, J. and Pines, A. (1991). ^{17}O NMR of Solid Silicates. *J. Am. Chem. Soc.*, **113**, 32.

Munowitz, M. (1988). *Coherence and NMR*. Wiley, New York.
Neuhaus, D. and Williamson, M. P. (1989). *The NOE and Conformational Analysis*. VCH Publishers, New York.
Pines, A., Gibby, M. G. and Waugh, J. S. (1973). Proton-Enhanced NMR of Dilute Spins in Solids. *J. Chem. Phys.*, **59**, 569.
Pines, A. (1988). Lectures on Pulsed NMR. *Proc. 100th Fermi School*. North Holland, Amsterdam.
Portis, A. M. (1953). Homogeneous and Inhomogeneous Line-Broadening. *Phys. Rev.*, **91**, 1971.
Raleigh, D. P., Kolbert, A. C. and Griffin R. G. (1990). Effect of Imperfections on TOSS Spectra. *J. Magn. Reson.*, **89**, 1.
Redfield, A. G. (1955). NMR Saturation and Rotary Saturation in Solids. *Phys. Rev.*, **98**, 1787.
Rhim, W.-K., Elleman, D. D. and Vaughan R. W. (1973). Analysis of Multiple-Pulse NMR in Solids. *J. Chem. Phys.*, **59**, 3740.
Rhim, W.-K., Pines, A. and Waugh, J. S. (1971). Time-Reversal Experiments in Dipolar-Coupled Spin Systems. *Phys. Rev. B*, **3**, 684.
Sakurai, J. J. (1985). *Modern Quantum Mechanics*. Addison-Wesley, Redwood City.
Samoson, A. (1985). Satellite Transition NMR of Quadrupolar Nuclei in Powders. *Chem. Phys. Lett.*, **119**, 29.
Samoson, A. and Lipmaa, E. (1983). Central Transition NMR Excitation Spectra of Quadrupole Nuclei. *Chem. Phys. Lett.*, **100**, 205.
Samoson, A., Lippmaa, E. and Pines A. (1988). Averaging of Second-Order Effects by Means of a Double-Rotor. *Mol. Phys.*, **65**, 1013.
Schmidt-Rohr, K., Clauss, J., Blümich, B. and Spiess, H. W. (1990). Miscibility of Polymers: ^{1}H Spin Diffusion and ^{13}C Detection *Magn. Reson. Chem.*, **28**, 3.
Scruggs, B. E. and Gleason, K. K. (1992). Photosensitive Salt Distribution in Polymers: MQ NMR. *Macromolecules*, **25**, 1864.
Shykind, D. N., Baum, J., Liu, S.-B., Pines, A. and Garroway, A. N. (1988). Phase-Incremented Multiple-Quantum NMR. *J. Magn. Reson.*, **76**, 149.
Skibsted, J., Nielsen, N. C., Bildsoe, H. and Jakobsen, H. J. (1991). Satellite Transitions in MAS of Quadrupolar Nuclei. *J. Magn. Reson.*, **95**, 88.
Slichter, C. P. (1990). *Principles of Nuclear Magnetic Resonance, 3rd ed*. Springer, Berlin.
Sun, B. Q., Baltisberger, J. H., Wu, Y., Samoson, A. and Pines, A. (1992). Sidebands in DAS and DOR. *Solid State Nucl. Magn. Reson.*, **1**, 267.
Suter, D. and Ernst, R. R. (1985). Spin Diffusion in Solid-State NMR Spectra. *Phys. Rev. B*, **32**, 5608.
Tekely, P., Brondeau, J., Elbayed, K., Retournard, A. and Canet, D. (1988). Simple Pulse Train for Selective NMR Excitation in Solids. *J. Magn. Reson.*, **80**, 509.
Thayer, A. M. and Pines, A. (1987). Zero-Field NMR. *Acc. Chem. Research*, **20**, 47.
Torchia, D. A. and Szabo, A. (1982). Spin-Lattice Relaxation in Solids. *J. Magn. Reson.*, **49**, 107.
Tycko, R. (1990). Zero-Field NMR at High Field. *J. Chem. Phys.*, **92**, 5776.
Yannoni, C. S. and Kendrick, R. D. (1981), Nutation Spectroscopy for Determining Molecular Geometry. *J. Chem. Phys.*, **74**, 747.
Waugh, J. S., Huber, L. M. and Haeberlen, U. (1968). Approach to High-Resolution NMR in Solids. *Phys. Rev. Lett.*, **20**, 180.
Zax, D. B., Bielecki, A., Zilm, K. W., Pines, A. and Weitekamp, D. P. (1985). Zero-field NMR and NQR. *J. Chem. Phys.*, **83**, 4877.
Zhang, S., Meier, B. H. and Ernst, R. R. (1992). Polarization Echoes in NMR. *Phys. Rev. Lett.*, **69**, 2149.

CHAPTER FOUR

Fourier Theory for 1D and 2D NMR

Keywords: *FT, convolution, symmetries, apodization, Fourier pairs; quadrature detection, TPPI; phasing, baseline problems; 2D phase-problem solutions, phase problems in 2D MAS; digitization, 'delta comb' III(t), DFT, Nyquist frequency, aliasing, first-point effect; S/N, dwell time and filter effects.*

4.1 Motivation

Fourier transformations play a key role in pulsed NMR. They serve to obtain the NMR frequency spectrum from the measured nuclear-induction time signal. In this approach, all the signals of the NMR system, with various frequencies, are measured simultaneously in the time domain. The Fourier transformation then sorts the signals out according to their frequencies, with the spectral intensity reflecting the weight of the frequency components in the total signal. This procedure provides a multiplex advantage that reduces the measuring time drastically (Ernst and Anderson, 1966) compared with continuous-wave spectroscopy where the responses for the various frequencies are measured sequentially.

For obtaining highly resolved, artifact-free spectra with a low noise level, the time signal has to be acquired with sufficient care, taking special effects of the Fourier transformation into account. This applies in particular for multidimensional spectroscopy, where the proper application of Fourier theory is crucial for obtaining highly resolved spectra free of artifacts. In this chapter, after an introduction of some fundamental relations between time and frequency signals, these practical aspects of Fourier theory will be stressed. In that context, the relation between the continuous nuclear induction signal and the digitized signal as actually acquired by the spectrometer will be treated explicitly. Finally, the dependence of the signal-to-noise-ratio on certain experimental parameters will be considered on the basis of Fourier theory.

The theory of Fourier transformations (Fourier, 1822) with all its different aspects is a quite extensive subject that requires a whole book to cover thoroughly. In this chapter, we shall restrict the discussion to certain topics relevant for NMR. For the mathematical relations given, we will only indicate proofs if these are very simple or not well documented in the literature. Books on general Fourier theory are available with different levels of difficulty and thoroughness of argumentation. The descriptive book by Ramirez (1985) could serve as a relatively simple introduction. A more complete and formal, but still quite useful, treatment has been given by Bracewell (1986). For derivations we will sometimes refer to that work. The reader not so familiar with complex numbers and complex exponentials may profit from refreshing the memory by reading through Appendix A.2 before tackling this chapter.

4.2 Basic definitions

For different types of data, different definitions of the Fourier transformation are required.

First of all, for continuous and sufficiently steady complex functions with finite integral of $|f(t)|$ over the full range from $-\infty$ to $+\infty$, there is the Fourier-integral relation:

$$f(t) \xleftrightarrow{\mathrm{FT}} F(\omega) = \int_{-\infty}^{\infty} \mathrm{d}t\, f(t)\, \mathrm{e}^{-\mathrm{i}\omega t} \tag{4.1}$$

and its inversion:

$$f(t) = \frac{1}{2\pi} \int_{-\infty}^{\infty} \mathrm{d}\omega\; F(\omega)\; \mathrm{e}^{\mathrm{i}\omega t}. \tag{4.2}$$

The choice of the prefactors (unity in (4.1) and $1/2\pi$ in the inversion (4.2)) is only one of several common conventions (Bracewell, 1986, p. 7). In the verbal formulation of FT theorems, these factors will usually be disregarded.

Periodic, continuous, complex functions (defined by $f(t) = f(t + nt_{\mathrm{p}})$ with the period t_{p}) can be expressed by Fourier series, which have discrete spectral components only at multiples of the basic frequency $\omega_{\mathrm{f}} = 2\pi/t_{\mathrm{p}}$:

$$f(t) \xleftrightarrow{\mathrm{FT}} F^{\ddagger}(\omega_{\mathrm{f}}k) := F_k = \frac{1}{t_{\mathrm{p}}} \int_{0}^{t_{\mathrm{p}}} \mathrm{d}t\, f(t)\; \mathrm{e}^{-\mathrm{i}\omega_{\mathrm{f}}kt} \tag{4.3}$$

inverse transformation:

$$f(t) = \sum_{k=0}^{\infty} F_k\; \mathrm{e}^{\mathrm{i}\omega_{\mathrm{f}}kt} \tag{4.4}$$

A third type of data is encountered in the practice of pulsed NMR. Here one usually has noncontinuously sampled (discrete, digitized) data, taken point by point with a fixed increment t_{dw} ('dwell time') over a limited total time t_{aq} ('acquisition time'). Out of N complex time data points, by means of discrete Fourier transformation (DFT), a discrete spectrum, traced out by N complex points, is generated. With $\omega_{\mathrm{f}} = 2\pi/t_{\mathrm{aq}} = (1/N)2\pi/t_{\mathrm{dw}}$ (due to $t_{\mathrm{aq}} = Nt_{\mathrm{dw}}$) the DFT relations read (Bracewell, 1986, p. 362):

$$f_j := f(jt_{\mathrm{dw}}) \quad F_k := F^{\ddagger}(\omega_{\mathrm{f}}k) \quad j = 0, \ldots, N-1, k = 0, \ldots, N-1$$

$$f_j \xleftrightarrow{\mathrm{DFT}} F_k = \frac{1}{N} \sum_{j=0}^{N-1} f_j\; \mathrm{e}^{-\mathrm{i}\omega_{\mathrm{f}}kjt_{\mathrm{dw}}} \tag{4.5}$$

inversion:

$$f_j = \sum_{k=0}^{N-1} F_k\; \mathrm{e}^{\mathrm{i}\omega_{\mathrm{f}}kjt_{\mathrm{dw}}}. \tag{4.6}$$

The data in both domains can be assumed to repeat periodically, that is, $f_{j+mN} = f_j$ and $F_{j+mN} = F_j$.

Of course, the three types of Fourier transformations are related, and most theorems of Fourier theory apply to all three, requiring only different notations; of these, the most transparent ones will be picked in the following discussions.

In principle, it is less obvious what relation exists between the spectrum $F(\omega)$ and the DFT spectrum $F^{\ddagger}(\omega)$ (i.e. the Fourier transforms of a continuous time function $f(t)$ and of a set of points $f_k := f(kt_{\mathrm{dw}})$ sampled from it). Simply from experience with sampled time signals and spectra in NMR, it is clear that $F^{\ddagger}(\omega)$ must be a good approximation to the spectrum $F(\omega)$. Nevertheless, there are some more or less well known effects of the digitization which will be discussed below after a mathematical description of digitization has been outlined.

To conclude this discussion on Fourier transformation procedures, we want to mention the so-called 'fast Fourier transformation', FFT (Cooley and Tuckey, 1965). This refers to particularly fast algorithms for DFT in which the calculation time rises only as $N\log_2(N)$ instead of increasing with N^2 as in a brute-force DFT approach. While the general DFT is formulated for arbitrary numbers of points N, the simplest and most common FFT ('butterfly') algorithms (Cooley and Tuckey, 1965; Higgins, 1976) require $N = 2^n$.

4.3 Useful FT relations

Here we compile various relations between time and frequency signals needed in other sections below. Time signals and their corresponding spectra are denoted by the same letter in upper case and lower case, respectively, for instance, by $f(t)$ and $F(\omega)$. Because of the similarity of FT (equation (4.1)) and inverse FT (equation (4.2)), most of the FT relations apply also with time and frequency domains interchanged; this is explicitly stated in most cases.

Linearity
FT is a linear operation: $f(t) + g(t) \xleftrightarrow{\mathrm{FT}} F(\omega) + G(\omega)$:

$$(z_1 + \mathrm{i}z_2)f(t) \xleftrightarrow{\mathrm{FT}} (z_1 + \mathrm{i}z_2)F(\omega). \tag{4.7}$$

Integral relation
The value of a function at zero equals the integral over its Fourier transform:

$$f(0) = \frac{1}{2\pi}\int_{-\infty}^{\infty} \mathrm{d}\omega\, F(\omega) \qquad F(0) = \int_{-\infty}^{\infty} \mathrm{d}t\, f(t) \tag{4.8}$$

$$f_0 = \sum_{k=0}^{N-1} F_k \qquad F_0 = \frac{1}{N}\sum_{j=0}^{N-1} f_j. \tag{4.9}$$

This follows directly from the basic definitions (4.1) or (4.2) for $t = 0$ or $\omega = 0$. This relation is needed quite frequently in the analysis of 2D data sets and their projections (integrals over one frequency dimension).

Convolution theorem

The Fourier transform of a product of two functions is the convolution of their Fourier transforms (Bracewell, 1986, p. 25):

$$f(t)g(t) \longleftrightarrow \frac{1}{2\pi}(F*G)(\omega) := \frac{1}{2\pi}\int_{-\infty}^{\infty} d\omega' F(\omega')G(\omega-\omega') \tag{4.10}$$

$$(f*g)(t) \longleftrightarrow F(\omega)G(\omega). \tag{4.11}$$

This is used, for instance, to perform a fast convolution of F and G. By an inverse FT, f and g are obtained, and the FT of the product $f(t)g(t)$ yields the desired $(F*G)(\omega)$. Often, instead of the awkward $(F*G)(\omega)$, we write the less precise, but more suggestive, $F(\omega)*G(\omega)$. For a more thorough discussion and a visualization of the convolution procedure the reader is referred to Bracewell's book (p. 24). Only one simple case, reflecting an important feature of convolution, is to be addressed here: assume $G(\omega')$ is a 'smoothing function' which exhibits an appreciable amplitude only close to $\omega' = 0$, such as a relatively narrow Gaussian or Lorentzian with a width $\Delta\omega$. Then, the convolution provides approximately the average, taken over an environment of ω with a width of $\Delta\omega$, of the spectrum $F(\omega)$. According to the convolution theorem, such a smoothing of $F(\omega)$ in the frequency domain is simply achieved by multiplying the time signal $f(t)$ with the inverse transform $g(t)$ of the smoothing function $G(\omega)$. This is the basis of the spectral smoothing effect of the 'Gaussian or exponential multiplication' of the time signal that is commonly applied in NMR.

An important convolution relation involves the delta function $\delta(\omega-\tilde{\omega})$:

$$F(\omega)*\delta(\omega-\tilde{\omega}) = \int_{-\infty}^{\infty} d\omega' F(\omega')\,\delta(\omega-\tilde{\omega}-\omega') = F(\omega-\tilde{\omega}). \tag{4.12}$$

This relation can actually serve as a definition of the δ function. For a simplistic visualization, it could be said that the delta function is an infinitely narrow Gaussian with an area of unity which is zero where its argument is nonzero, i.e. $\delta(\omega-\tilde{\omega}-\omega')$ vanishes for $\omega' \neq \omega-\tilde{\omega}$. In the integral (4.12), $\delta(\omega-\tilde{\omega}-\omega')$ therefore picks out the value of $F(\omega')$ at $\omega' = \omega-\tilde{\omega}$.

Power theorem

The integral 'power' is equal in both domains:

Rayleigh's theorem:
$$\int_{-\infty}^{\infty} dt\,|f(t)|^2 = \frac{1}{2\pi}\int_{-\infty}^{\infty} d\omega\,|F(\omega)|^2 \tag{4.13}$$

Parseval's theorem:
$$\frac{1}{N}\sum_{l=0}^{N-1}|f_l|^2 = \sum_{k=0}^{N-1}|F_k|^2. \tag{4.14}$$

Shift theorem

A shift of $f(t)$ by t_0 results in a modulation of the spectrum by $\exp(-i\omega t_0)$:

$$f(t-t_0) \xleftrightarrow{\text{FT}} e^{-i\omega t_0}F(\omega) \tag{4.15}$$

as can be proven quite easily by inserting $f(t-t_0)$ into equation (4.1). For DFT, the periodic repetition of the array of data points must be taken into account here. In other words, points left-shifted to $t = -mt_{dw}$ appear at $t = (N-m)t_{dw}$ near the right end of the data array.

Symmetry

For a real time signal, the spectrum is symmetric (Figure 4.1(b)). More precisely, the real part of the spectrum is symmetric, while the imaginary part is antisymmetric:

$$\begin{array}{ccc} f(t) = f^*(t) & \Longleftrightarrow & \text{Re}\, f(t) = f(t), \text{Im}\, f(t) = 0 \\ \Updownarrow & & \\ F(\omega) = F^*(-\omega) & \Longleftrightarrow & \text{Re}\ F(\omega) = \text{Re}\ F(-\omega), \text{Im}\ F(\omega) = -\text{Im}\ F(-\omega). \end{array} \tag{4.16}$$

For instance, this means that by setting the imaginary part of a time-domain signal to zero, the (real part of the) spectrum is symmetrized. The same is of course true with t and ω interchanged.

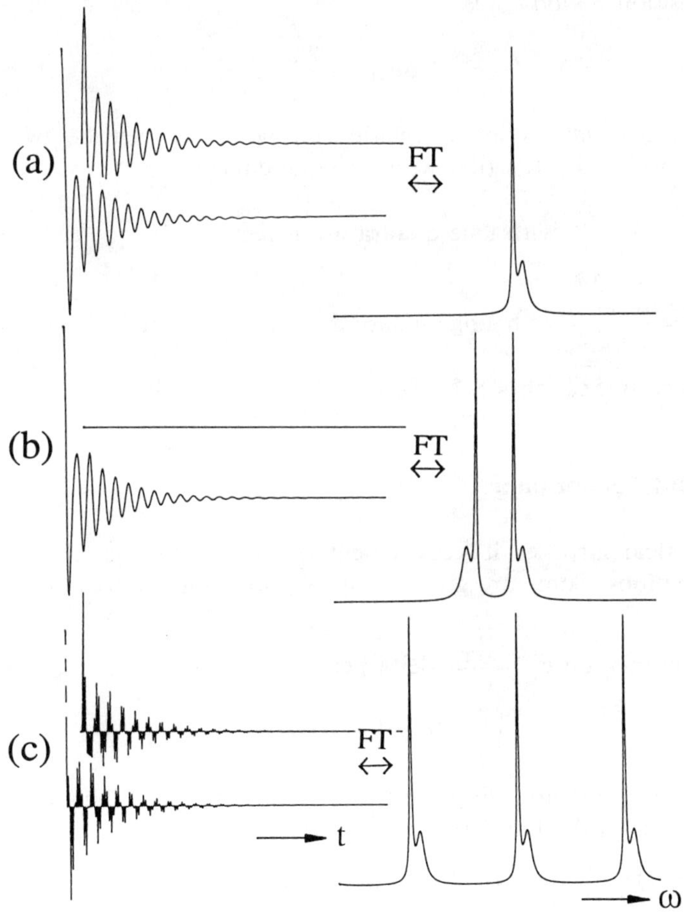

Figure 4.1 Representation of some Fourier relations between the complex time signal and the real part of the spectrum. (a) Original situation. (b) Setting the imaginary part of the time signal to zero results in a symmetrization of the spectrum relative to $\omega = 0$. (c) Sampling of the data results in a periodic repetition of the spectrum.

Inverse-widths relation

In Fourier theory one often encounters relations between characteristic widths or periodicities of a function and its Fourier transform which have the following form:

$$\Delta\omega \cong \frac{2\pi}{\Delta t}. \tag{4.17}$$

The simplest example is the relation between the resonance frequency and the period of the time signal.

In NMR this principle also appears in the relation between the decay constant of a time signal and the linewidth(s) in the spectrum. This can be roughly summarized:

$$\begin{aligned} &\text{Narrow (fast decaying) time signal} \leftrightarrow \text{broad spectral lines} \\ &\text{broad (slowly decaying) time signal} \leftrightarrow \text{narrow spectral lines.} \end{aligned} \tag{4.18}$$

Similarly, the optimum spectral resolution obtained from a time-domain signal taken over an acquisition period t_{aq} is:

$$\Delta\omega_{opt} \cong \frac{2\pi}{t_{aq}}. \tag{4.19}$$

For the sampling of data, a similar relationship holds between the dwell time t_{dw} and the useful spectral width ω_{sw} (i.e. without redundance):

$$\frac{\omega_{sw}}{2\pi} = \frac{1}{t_{dw}} \quad \text{with true quadrature detection}$$

$$\frac{\omega_{sw}}{2\pi} = \frac{1}{2}\frac{1}{t_{dw}} \quad \text{with single-phase detection, or TPPI (Section 4.5.1).}$$

This is discussed in Section 4.8.1 below.

4.4 Important Fourier pairs

For many practical purposes, it is convenient to know the Fourier transforms of certain important functions. Some are given in this section, with short hints regarding their derivations.

- harmonic oscillation $e^{i\tilde{\omega}t} \xleftrightarrow{\text{FT}}$ delta peak at $\tilde{\omega}$:

$$\int_{-\infty}^{\infty} dt\, e^{i\tilde{\omega}t}\, e^{-i\omega t} = 2\pi\, \delta(\omega - \tilde{\omega}). \tag{4.20}$$

  This fundamental relationship can formally be derived on the basis of the concept of 'transforms in the limit' (Bracewell, 1986).

- cosine $\cos(\tilde{\omega}t) \xleftrightarrow{\text{FT}}$ pair of delta peaks $\pm\,\tilde{\omega}$:

$$\int_{-\infty}^{\infty} dt\, \cos(\tilde{\omega}t)\, e^{-i\omega t} = 2\pi\tfrac{1}{2}\{\delta(\omega - \tilde{\omega}) + \delta(\omega + \tilde{\omega})\} \tag{4.21}$$

  because of $\cos(\tilde{\omega}t) = \frac{1}{2}(e^{i\tilde{\omega}t} + e^{-i\tilde{\omega}t})$ and (4.20).

– i $\sin(\tilde{\omega}t) \xleftrightarrow{\text{FT}}$ pair of delta peaks with opposite signs:

$$\int_{-\infty}^{\infty} \mathrm{d}t \; \mathrm{i} \sin(\tilde{\omega}t) \; \mathrm{e}^{-\mathrm{i}\omega t} = 2\pi\tfrac{1}{2}\{\delta(\omega - \tilde{\omega}) - \delta(\omega + \tilde{\omega})\} \tag{4.22}$$

because of i $\sin(\tilde{\omega}t) = \frac{1}{2}(\mathrm{e}^{\mathrm{i}\tilde{\omega}t} - \mathrm{e}^{-\mathrm{i}\tilde{\omega}t})$ and (4.20).

– Gaussian $\xleftrightarrow{\text{FT}}$ Gaussian:

$$\int_{-\infty}^{\infty} \mathrm{d}t \; \exp\left(-\frac{t^2}{2\sigma_t^2}\right) \exp(-\mathrm{i}\omega t) = \sqrt{2\pi} \; \sigma_t \; \exp\left(\frac{-\omega^2\sigma_t^2}{2}\right). \tag{4.23}$$

For a derivation use

$$\left(\frac{t^2}{2\sigma_t^2} + \mathrm{i}\omega t\right) = \left(\frac{1}{\sqrt{2}}\frac{t}{\sigma_t} + \frac{\mathrm{i}\omega}{2}\sqrt{2}\sigma_t\right)^2 - \left(\frac{\mathrm{i}\omega}{2}\sqrt{2}\sigma_t\right)^2.$$

$\omega_{\mathrm{fwhm}} = 2\sqrt{2\ \ln 2}/\sigma_t = 2.355/\sigma_t = 2\pi \times 0.3748/\sigma_t$.

– half Gaussian $\xleftrightarrow{\text{FT}}$ Gaussian with dispersive part:

$$\int_{0}^{\infty} \mathrm{d}t \; \exp\left(\frac{-t^2}{2\sigma_t^2}\right) \mathrm{e}^{-\mathrm{i}\omega t} = \sqrt{2\pi} \; \sigma_t \tfrac{1}{2} \exp\left(-\frac{\omega^2\sigma_t^2}{2}\right) \{1 + \mathrm{erf}(\mathrm{i}\omega\sigma_t/\sqrt{2})\}. \tag{4.24}$$

– exponential $\xleftrightarrow{\text{FT}}$ Lorentzian:

$$\int_{0}^{\infty} \mathrm{d}t \; \mathrm{e}^{-t/\tau} \; \mathrm{e}^{-\mathrm{i}\omega t} = \frac{1}{\mathrm{i}\omega + 1/\tau} = \frac{1/\tau}{\omega^2 + 1/\tau^2} - \mathrm{i}\,\frac{\omega}{\omega^2 + 1/\tau^2} \tag{4.25}$$

$$\omega_{\mathrm{fwhm}} = 2\,\frac{1}{\tau} = 2\pi\,\frac{1}{\pi\tau} = 2\pi \times 0.3183\,\frac{1}{\tau}.$$

– box $\xleftrightarrow{\text{FT}}$ sinc

Symmetric box $[-t_0 \dots t_0]$:

$$\int_{-t_0}^{t_0} \mathrm{d}t \; \mathrm{e}^{-\mathrm{i}\omega t} = \frac{1}{-\mathrm{i}\omega}\,(\mathrm{e}^{-\mathrm{i}\omega t_0} - \mathrm{e}^{\mathrm{i}\omega t_0}) = 2\,\frac{\sin(\omega t_0)}{\omega} \tag{4.26}$$

$$\omega_{\mathrm{fwhm}} = 1.8955\,\frac{1}{t_0} = 2\pi \times 0.3017\,\frac{1}{t_0}.$$

Box $[0 \dots t_0]$:

$$\int_{0}^{t_0} \mathrm{d}t \; \mathrm{e}^{-\mathrm{i}\omega t} = \frac{\mathrm{e}^{-\mathrm{i}\omega t_0} - 1}{-\mathrm{i}\omega} = \frac{\sin(\omega t_0)}{\omega} + \mathrm{i}\,\frac{\cos(\omega t_0) - 1}{\omega}. \tag{4.27}$$

ω_{fwhm} denotes the full width at half maximum of the spectral functions $F(\omega)$. In the cases where $f(t = 0) = 1$ and the time integral extends from $-\infty$ to ∞, according to relation (4.8) the integral over the spectral function $F(\omega)$ amounts to 2π. If the function has a step at $t = 0$, or equivalently if the lower integration limit is $t = 0$, the integral is π. This means that at the step, the inverse Fourier transform of the spectral function $F(\omega)$ takes the intermediate value $\frac{1}{2}(f(t > 0) + f(t < 0) = \frac{1}{2}$.

4.5 Specific applications of Fourier theory in NMR

In principle, in pulsed NMR the spectrum is obtained by applying a short resonant radio-frequency pulse to the nuclear spins in the sample and Fourier transforming the nuclear-induction response, the free induction decay (FID). In practice, a more complicated procedure is required, as shown schematically in Figure 4.2: the radio-frequency signal (with a frequency ω_L of 50 to 500 MHz) is amplified, then the spectrometer frequency ω_{sp} is subtracted in a phase-sensitive detector (double-balance mixer, DBM) followed by a low-pass filter, and the resulting audio signal (with a frequency ω_r of about 1 to 100 KHz) is digitized by an analog-to-digital converter (ADC). The spectrum is obtained from these digitized time data by a discrete Fourier transformation (DFT). Special precautions must be taken that after these various steps a highly resolved spectrum without artifacts is obtained. These points will be discussed in the following subsections. The quality of the experimental data which can thus be achieved is even more important in multidimensional spectra, where insufficiencies are much harder to correct than in 1D spectra.

4.5.1 Quadrature detection and TPPI

In the simplest case, the harmonic oscillation $\cos(\omega_L t)$ of a single isochromat is picked up in the rf coil. After the signal processing mentioned above, the spectrometer frequency ω_{sp} has been subtracted from ω_L, so that the magnetization is essentially observed to precess with $\omega_r = \omega_L - \omega_{sp}$ in the frame that rotates with ω_{sp}.

In the simplest set-up, only the cosine component $\cos(\omega_r t)$ in the rotating frame is detected. Then, due to $\cos(\omega_r t) = \cos(-\omega_r t)$ lines at ω_r and $-\omega_r$ are not distinguishable.

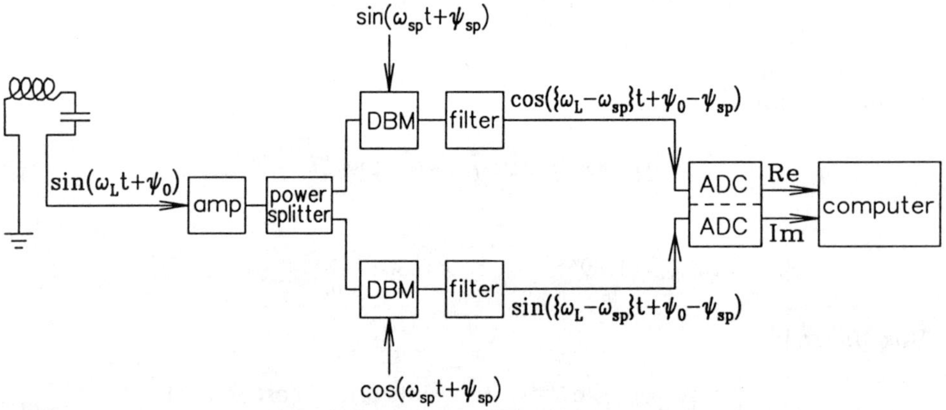

Figure 4.2 Scheme of the signal acquisition system in FT NMR. The NMR signal with its radio frequency ω_L is amplified and mixed down in phase-sensitive rectifiers (double-balance mixers, DBMs) with spectrometer reference signals of frequency ω_{sp} which differ by 90° in phase (sine and cosine). The audio frequency remaining after a low-pass filter is the difference $\omega_L - \omega_{sp}$, and the phase information is preserved. The signals are digitized (sampled) by two analog-to-digital converters (ADCs) in the case of full quadrature detection, and alternately by a single ADC in the case of TPPI detection. The digitized data are passed on to the computer for further processing, in particular the discrete Fourier transformation that yields the NMR spectrum.

This problem can be remedied by taking care that ω_r has the same sign for all signals (this is denoted as measuring 'off resonance'). However, with signals appearing over a broad spectral range, this aggravates pulse-excitation problems (Section 2.6.5). In addition, noise components at negative and positive frequencies add up, with the consequence of a deterioration of the signal-to-noise ratio ('S/N') by $\sqrt{2}$.

These disadvantages of single-phase detection can be avoided by utilizing the second double-balance mixer in Figure 4.2 with the phase of the spectrometer reference signal shifted by 90°. This phase shift is also imparted to the resulting signal, which is thus $\sin(\omega_r t)$, so that one can form $\exp(i\omega_r t)$ to distinguish between positive and negative ω_r. Quadrature (phase) detection is the technical term for such a measurement of both components of $\exp(i\omega_r t)$. The improvement of the S/N by $\sqrt{2}$ resulting from the distinction between noise components at positive and negative frequencies can also be explained by noting that two signals, $\cos(\omega_r t)$ and $\sin(\omega_r t)$, with independent, partially compensating noise components are simultaneously measured and added.

If the spectrometer has two mixers, but only one ADC, sine and cosine can be measured alternately to retain the quadrature information. Usually, time-proportional receiver phase incrementation (TPPI) is applied (Redfield and Kunz, 1975), by rotating the receiver phase 90° between the sampling of successive data points in the time domain. Thus the receiver effectively performs single-phase detection in a system rotating with an angular frequency of $\omega_{\text{tppi}} = 2\pi/(4t_{\text{dw}})$ (2π after four increments) relative to the conventional rotating coordinate system. This shifts a signal with $+\Delta\omega$ to an apparent frequency $\omega_{\text{tppi}} + \Delta\omega$, and one with $-\Delta\omega$ to $\omega_{\text{tppi}} - \Delta\omega$, thereby making them distinguishable. The S/N is the same as with real quadrature detection, even though the useful spectral width is only $\omega_{\text{sw}}^{\text{eff}}/2\pi = \frac{1}{2}(1/t_{\text{dw}})$. The low-pass filtering of the analog signal cancels the noise components that would appear at negative frequencies when seen from the TPPI frame. Quadrature and TPPI in the evolution period of 2D spectra will be discussed in Section 4.6.

4.5.2 *Absorptive and dispersive signals, and constant phase correction*

Generating fully absorptive lineshapes is important for obtaining highly resolved spectra, in particular in two and more dimensions. The absorptive part of a spectrum can be defined as the superposition (sum) of the symmetric portions $A(\omega - \omega_0)$ of the lineshapes of the individual, just homogeneously broadened signals. The symmetry is considered with respect to the centre-of-gravity ω_0 of each individual signal. The dispersive part $D(\omega - \omega_0)$ of the spectrum is the sum of the corresponding antisymmetric components. The terminology goes back to continuous-wave spectroscopy: the absorptive signal is equivalent to the absorption component of the complex susceptibility measured in a continuous-wave NMR experiment, and the dispersive signal to the dispersion component, as can be shown by linear-response theory.

The lineshape function around the centre-of-gravity ω_0 of the individual line is the Fourier transform $H(\omega)$ of the time-domain damping function $h(t)$ (e.g. e^{-t/T_2} or $e^{-t^2/(2\sigma^2)}$) that multiplies the time signal; this relation follows directly from the convolution theorem (4.10):

$$\exp(i\tilde{\omega}_0 t)h(t) \xleftrightarrow{\text{FT}} 2\pi\,\delta(\omega - \omega_0) * H(\omega)\,\frac{1}{2\pi} \overset{(4.12)}{=} H(\omega - \omega_0). \tag{4.28}$$

The spectral signal appears around the resonance frequency ω_0 with the lineshape $H(\omega)$. As long as the damping function (e.g. exponential e^{-t/T_2} or half-Gaussian $e^{-t^2/(2\sigma^2)}$) has no imaginary component, the real part of its Fourier transform $H(\omega)$ is symmetric and the imaginary part antisymmetric (see relation (4.16)). This provides an identification of the real part of the spectral line $\mathrm{Re}\{H(\omega-\omega_0)\}$ with the ω_0-symmetric, absorptive portion of the signal, $A(\omega-\omega_0)=\mathrm{Re}\{H(\omega-\omega_0)\}$.

It should be noted, however, that a general identification of the real part of the spectrum and the absorptive lineshape is not correct. A simple phase shift ψ in the time-domain signal $\exp(\mathrm{i}\omega_0 t+\mathrm{i}\psi)=\exp(\mathrm{i}\psi)\exp(\mathrm{i}\omega_0 t)$ produces a multiplication of the spectrum by a phase factor $\exp(\mathrm{i}\psi)=\cos\psi+\mathrm{i}\sin\psi$ (since the constant prefactor $\exp(\mathrm{i}\psi)$ remains unaffected by the FT). The spectrum thus acquires dispersive components in the real part:

$$\begin{aligned}\mathrm{Re}\ F(\omega) &= \mathrm{Re}\{(\cos\psi+\mathrm{i}\sin\psi)(A+\mathrm{i}D)\}\\ &= \cos\psi\ A(\omega-\omega_0)-\sin\psi\ D(\omega-\omega_0)\end{aligned} \tag{4.29}$$

while in turn absorptive components appear in the imaginary part. This effect can be cancelled through multiplying the spectrum by $e^{-\mathrm{i}\psi}$. This procedure is called a constant phase correction (frequency-independent, or zeroth-order, phase correction).

Usually, the correct value of ψ is not known beforehand. For spectra consisting of separated individual lines (i.e. without inhomogeneous broadening), the spectrum is multiplied by $e^{-\mathrm{i}\psi_c}$ with assumed phase values ψ_c which are varied until the real part of the spectrum consists of symmetric lines and is therefore the absorptive part of the spectrum. For inhomogeneously broadened spectra, this symmetry criterion cannot be applied since the absorptive part of the overall spectrum, consisting by definition of a continuum of lines, can be asymmetric (see, for instance, the powder spectra of Figure 2.8). Here, the phasing procedure exploits the fact that the real parts of the two most important lineshape functions, the Lorentzian and the Gaussian, are positive throughout. The phasing by $e^{-\mathrm{i}\psi_c}$ is now aimed at producing a fully positive signal in the real part of the spectrum. With general line-shape functions, maximizing the integral of the real part can serve as a criterion for automated constant phase correction.

Written in components, the constant phase correction $\tilde{F}=\exp(-\mathrm{i}\psi_c)F$ involves:

$$\begin{aligned}\mathrm{Re}(\tilde{F}) &= \cos\psi_c\ \mathrm{Re}(F)+\sin\psi_c\ \mathrm{Im}(F)\\ \mathrm{Im}(\tilde{F}) &= \cos\psi_c\ \mathrm{Im}(F)-\sin\psi_c\ \mathrm{Re}(F).\end{aligned} \tag{4.30}$$

Ideally, the constant phase correction can be applied either in the time domain (by actual multiplication or by adjustment of the receiver phase) or in the frequency domain. In practice, correct adjustment of the receiver phase is preferable, as will be pointed out in Section 4.5.3.

Why do we insist on removing dispersive lineshapes from the spectrum? The reasons are related to the characteristics of the dispersive lineshape: it is more complicated than the absorptive lineshape, since the dispersive signal, being antisymmetric, always exhibits both positive and negative intensity which superimposes in a complicated way in spectra containing several lines. Again due to the antisymmetry, the integral over the dispersive lineshape vanishes, which represents the loss of important information. Figure 4.3 shows that the dispersive signal also has broader wings than the absorption component, resulting in worse resolution. In one-dimensional spectroscopy, the dispersive component is rarely of interest (while the imaginary part of the spectrum is required in the case of $\psi\neq 0$; then, it is partially absorptive and must be available in

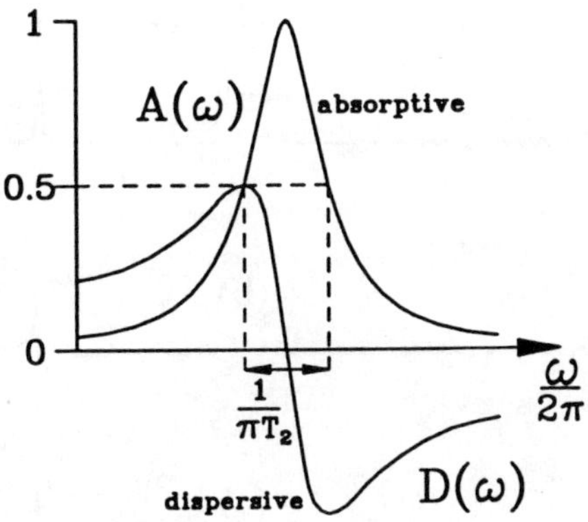

Figure 4.3 Absorption and dispersion lineshapes: real and imaginary parts of a single Lorentzian. The dispersion signal exhibits a larger linewidth than the absorption signal.

the phase correction (4.30) to produce a completely absorptive real part). In multidimensional spectroscopy, the dispersive part is even troublesome; special care must be taken to remove it from the spectra. This will be discussed in Section 4.6.

The appearance of the dispersive part is coupled to the fact that the time signal vanishes for $t < 0$: When the time signal $\Sigma_j \exp(i\tilde{\omega}_j t) h_j(t)$ is extended smoothly to negative times, by continuing the real (cosine) part symmetrically and the imaginary (sine) part antisymmetrically, according to relation (4.16) the imaginary part of the spectrum, which is the dispersion signal here, is cancelled.

If for a truly periodic signal the acquisition time is a multiple of the period, as is the case in the evolution time of certain rotor-synchronized MAS experiments, no dispersion signal appears because the periodicity implicit in DFT continues the signal periodically for all times.

4.5.3 Dead time, linear phase correction, and baseline distortions

In pulsed NMR, acquisition of the undistorted signal can start only some time after the last rf pulse, as the dissipation (ring-down) of the pulse power requires a finite time. In this 'dead time' t_0, the beginning of the FID is lost. Sometimes it is claimed that the dead-time effect could be eliminated by a linear phase correction $\exp(-i\omega t_0)$. We will analyse now to what extent this is true.

With the acquisition starting at a time t_0 after the pulse, the signal $\tilde{f}(t) = f(t + t_0)$ is acquired for $t \geqslant 0$, instead of $f(t)$, as indicated in Figure 4.4(b). This results in a spectrum:

$$\tilde{F}(\omega) = \int_0^\infty dt\, \tilde{f}(t)\, e^{-i\omega t} \overset{\substack{\text{with}\\ t'=t+t_0}}{=} \int_{t_0}^\infty dt'\, \tilde{f}(t'-t_0)\, e^{-i\omega(t'-t_0)}$$

$$= \int_0^\infty dt'\, f(t')\, e^{-i\omega(t'-t_0)} - \int_{0(-t_0)}^{t_0} dt'\, f(t')\, e^{-i\omega(t'-t_0)}$$

$$= e^{i\omega t_0}\left\{F(\omega) - F(\omega) * \frac{\sin(\omega t_0)}{\omega}\frac{1}{\pi}\right\}. \tag{4.31}$$

A narrow line $F(\omega) = A(\omega - \tilde{\omega})$ centred at $\tilde{\omega}$ with an area (integral) I and a small

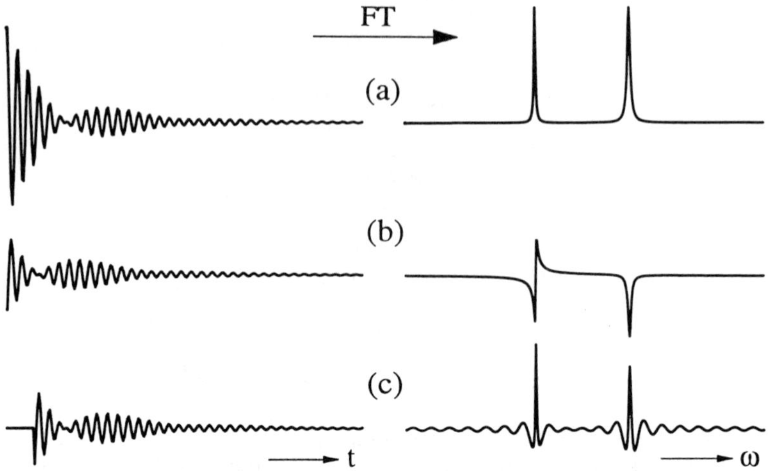

Figure 4.4 Spectral effects of a deadtime at the beginning of the time signal. (a) Without deadtime (reference spectrum). (b) With signal detection starting after a deadtime t_0. Major phase errors are visible in the spectrum. (c) With deadtime and signal detection starting at $t = 0$. The peaks themselves are absorptive here, but strong sinc-shape baseline distortions are observed. The spectrum in (c) is obtained from that of (b) by a linear phase correction $e^{-i\omega t_0}$.

width $\ll 2\pi/t_0$, can be approximated by $A(\omega - \tilde{\omega}) \cong I\delta(\omega - \tilde{\omega})$ so that we can evaluate the last term in (4.31):

$$A(\omega - \tilde{\omega}) * \frac{\sin(\omega t_0)}{\omega} \cong I\{\sin(\omega - \tilde{\omega})t_0\}/(\omega - \tilde{\omega}). \tag{4.32}$$

This represents a sinc baseline distortion with a width of $\sim 2\pi/t_0$ (Figure 4.4(c)). The narrower the line, the smaller the area I for a given signal height, so that the relative height of the distortion decreases since it is proportional to I according to equation (4.32). Therefore, for narrow peaks in a wide ω_{sw} range, the dominating effect of the deadtime t_0 is the prefactor $\exp(i\omega t_0)$ in equation (4.31) which can be easily compensated by a phase-correction factor $\exp(-i\omega t_0)$. Since the phase $-t_0\omega$ varies linearly with the frequency ω, this compensation is referred to as a linear phase correction (frequency dependent, first-order phase correction).

Note, however, that the discussion of the second term in equation (4.31) has disregarded the prefactor $e^{i\omega t_0}$, or, in the time domain, assumed that the time signal is zero between $t = 0$ and $t = t_0$ (like the time signal in Figure 4.4(c)). Figure 4.4(b) shows that with detection starting at t_0 as assumed above in the definition of $\tilde{f}(t)$, the baseline distortion is much less pronounced, and certainly not a sinc function. However, in order to phase all the peaks truly absorptive, the linear phase correction must be applied, and brings back the sinc wiggles.

With broad lines, the extended baseline distortion by the second term in (4.31) becomes dominant. There are no simple analytical formulas for its compensation. Various techniques have been developed for numerical baseline correction (Heuer and Haeberlen, 1989). Most of these require input of additional information, such as defining ranges in the spectrum where only baseline is observed. In general, these techniques work better for isolated lines than for the broad powder patterns often observed in solid-state NMR. Therefore, in many cases it is more convenient to circumvent the dead-time problem by generating a spin echo separated by more than t_0 from the refocusing pulse; this is the standard procedure in ^{2}H NMR (quadrupole echo), and also useful with chemical shift anisotropies (Hahn spin echo).

In practice, the largest peak ω_m in the spectrum is often taken as the reference point for a linear phase correction $\exp(-it_0(\omega-\omega_m))$, instead of the $\omega=0$ point with $\exp(-it_0\omega)$ as indicated in the preceding discussion. Both factors are made equal by a constant phase correction $\exp(it_0\omega_m)$.

As exemplified by the foregoing discussion, baseline and phasing problems are often related, resulting usually from the same insufficiencies of a practical measurement. In 2D spectra, they are particularly serious since they can occur in both dimensions and are more difficult to correct if they involve complicated distortions of the baseplane.

Baseline and phasing problems can be due to many effects:

- *Deadtime problem*. As shown above, curved baselines result if the beginning of the time signal cannot be properly detected, and in many important cases these effects cannot be compensated by linear phase correction.
- *Filter-bandwidth problem*. Near the limits of the spectral region, the hardware low-pass filter introduces phase errors. Usually, the filter can be set to a larger width, but that is costly in terms of noise aliased into the spectrum (see Section 4.8.1).

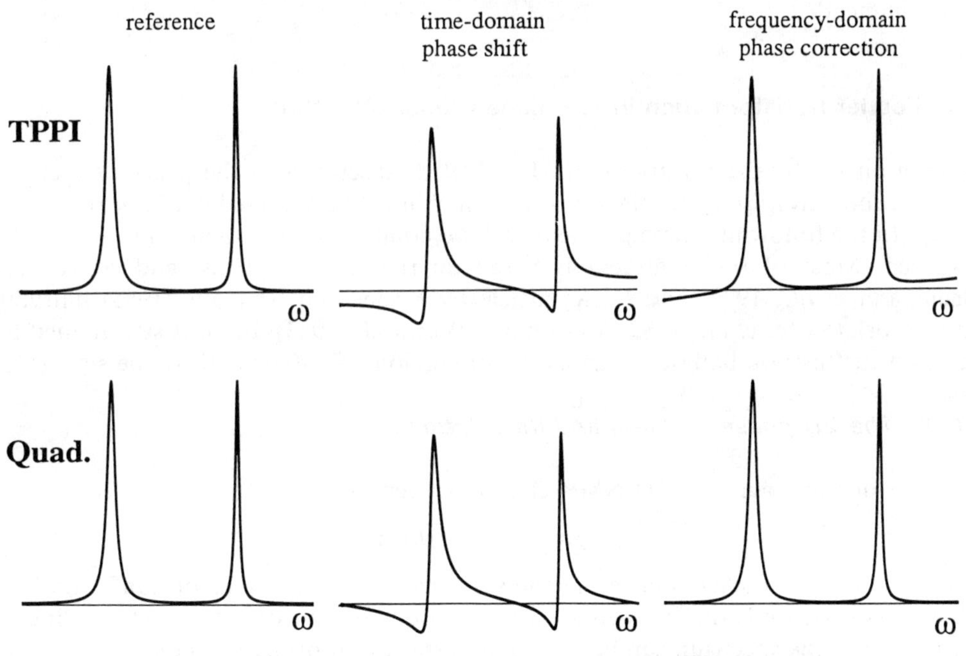

Figure 4.5 Baseline slope due to aliasing of the dispersive wings of the spectrum with TPPI detection. (a) TPPI detection. (b) Full quadrature for comparison. (Left) Time signal in phase. (Middle) Time signal out-of-phase by 45°. From these spectra, those on the right were obtained after a constant phase correction $e^{-i\pi/4}$. In the TPPI case, a baseline slope remains (upper right).

- *Filter rise-time problem*. The finite rise-time of the filter function in the time domain often severely suppresses the intensity of the first two time data points (Schaefer and Stejskal, 1974), resulting in corresponding distortions in the spectrum. (Filter bandwidth and rise-time problems are related but not identical: the latter can be completely obviated by generating a spin echo, starting acquisition before the echo maximum, and removing the first two points by left-shifts; in contrast the spectral filter-bandwidth effects will remain unaffected by that procedure.)
- *Probe bandwidth problem*. The resonance curve of the resonance circuit has a finite width $\Delta\omega$ (in order to attain a sufficiently high quality factor $\omega_0/\Delta\omega$ which is important for the amplification of the resonant signal). The phase between driving NMR voltage and detected signal, passing through null on resonance, varies across the resonance curve, resulting directly in a frequency-dependent phase error in the spectrum.
- *Excitation problem*. Excitation with pulses of finite duration leads to linear and higher-order phase errors, since the initial direction of the magnetization component in the *xy* plane, and thus its phase, depends on the frequency offset (see Section 2.6.5).
- *Aliasing of dispersive wings*. Time domain and frequency domain zero-order phase corrections can be inequivalent under TPPI detection. This is caused by aliasing (Section 4.8.1) of dispersive wings of the signals, leading to a baseline slope which cannot be eliminated by phase correction (Stejskal and Schaefer, 1974), see Figure 4.5. Therefore, with broad signals, in-phase acquisition is to be strongly recommended.

4.6 Fourier transformation in two-dimensional (2D) NMR

In addition to the complications of 1D NMR as discussed in the previous section, specific problems arise in 2D NMR spectra which must be analysed in terms of Fourier theory. In the following, these problems will be pointed out and their solutions will be discussed. Most of the treatment is based on the work of Ernst and co-workers (Bachmann *et al.*, 1977) and on an article by Keeler and Neuhaus (1985) unifying earlier work (States *et al.*, 1982, Marion and Wüthrich, 1983), but it is streamlined by omission of 'historic ballast', such as the discussion of *p*-type and *n*-type signals.

4.6.1 The 2D phase problem and its solutions

The time-domain signal in 1D NMR can be written as:

$$f(t) := \langle \exp(i\tilde{\omega}t)h(t) \rangle \tag{4.33}$$

with the angular brackets indicating the sum over all orientations or conformations present, as $\tilde{\omega}$ depends on the segmental orientations or conformations. According to Section 4.5.2, the spectrum can be written in terms of absorptive and dispersive signal components:

$$S(\omega) = \mathrm{FT}(f(t)) = \langle A(\omega - \tilde{\omega}) + iD(\omega - \tilde{\omega}) \rangle \tag{4.34}$$

where the FT, as a linear operation, has been drawn into the brackets. While both

$A(\omega - \tilde{\omega})$ and $D(\omega - \tilde{\omega})$ will have to be considered explicitly in the following equations, the damping functions like $h(t)$ will be omitted to keep the treatment as clear as possible.

In two-dimensional NMR spectroscopy, the time signals are functions of the general form:

$$f_p(t_1, t_2) := \langle e^{i\tilde{\omega}_1 t_1} e^{i\tilde{\omega}_2 t_2} \rangle. \tag{4.35}$$

A typical 2D time signal is shown in Figure 4.6. From the 2D time signal, the 2D spectrum can be generated by two successive Fourier transformations, over t_2 ('FT_2') and t_1 ('FT_1'). The real part of the 2D spectrum is to be purely absorptive $A(\omega_1 - \tilde{\omega})A(\omega_2 - \tilde{\omega})$, in short A_1A_2, in order to have optimum resolution and no spectral distortions.

However, in the simplest approach of two successive FTs directly applied one finds that the spectrum:

$$\begin{aligned} F_p(\omega_1, \omega_2) &= FT_1(FT_2\langle e^{i\tilde{\omega}_1 t_1} e^{i\tilde{\omega}_2 t_2} \rangle) = \langle (A_1 + iD_1)(A_2 + iD_2) \rangle \\ &= \langle (A_1A_2 - D_1D_2) + i(D_1A_2 + A_1D_2) \rangle \end{aligned} \tag{4.36}$$

contains a mixture of absorptive and dispersive contributions in both the real and imaginary parts; this results in 'phase twist' of spectral lines (Bax, 1982) as displayed in Figure 4.7 for isolated lines and for a powder diagonal spectrum. Such distortions of 2D spectra are usually intolerable (Schmidt *et al.*, 1988).

To solve this phase problem, the data processing must be modified. At first, it is preferable to disregard the phase-modulated signal of (4.35) (referred to as such because it can be written as $\langle e^{i(\tilde{\omega}_1 t_1 + \tilde{\omega}_2 t_2)} \rangle$, so that the modulation in t_1 and t_2 concerns

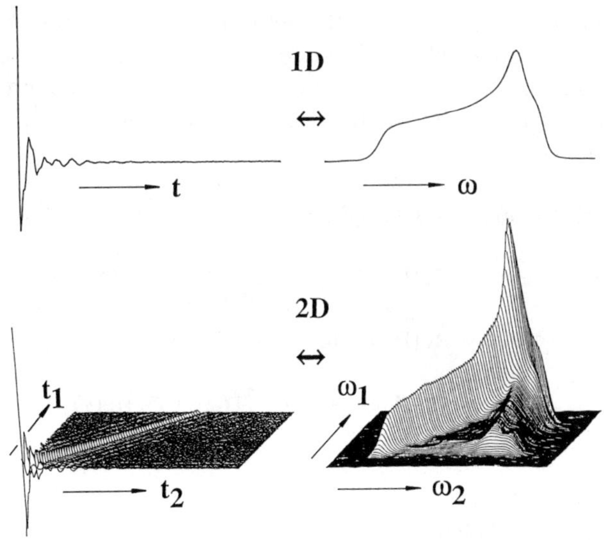

Figure 4.6 Typical 1D and 2D time signals and corresponding (absorptive) powder spectra. Only the real parts of the signals are shown. The relationship between the train of stimulated echoes in the 2D time signal and the intensity close to the diagonal of the frequency spectrum is discussed in Section 10.2.3.

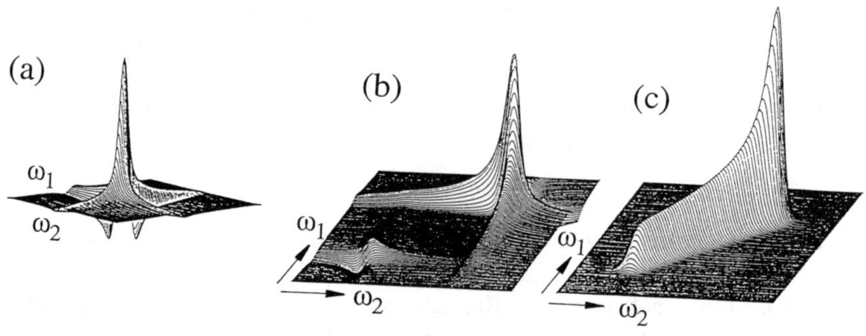

Figure 4.7 Phase-twist in 2D spectra. (a) Single Lorentzian line; note the complicated peak structure with absorptive and mixed-phase cross-sections; the latter generate broad dispersive wings. (b) 'Diagonal' 2D powder pattern (continuous distribution of Gaussian lines with centres on the diagonal; no frequency changes occur during the mixing time, i.e. $\tilde{\omega}_1 = \tilde{\omega}_2$). The off-diagonal intensity is due purely to the dispersive components. (c) Corresponding absorptive 2D diagonal powder pattern for comparison. In (b), the features of the spectrum are completely distorted due to cancellation of positive and negative lobes of the individual phase-twisted Gaussians which make up the powder spectrum.

only the phase of the complex exponential). Instead, we want to treat amplitude modulation:

$$f_c(t_1, t_2) = \langle \cos(\tilde{\omega}_1 t_1)\ e^{i\tilde{\omega}_2 t_2} \rangle \tag{4.37}$$

$$f_s(t_1, t_2) = \langle i\ \sin(\tilde{\omega}_1 t_1)\ e^{i\tilde{\omega}_2 t_2} \rangle \tag{4.38}$$

which is actually encountered for most 2D techniques in solid-state NMR. Both types of modulation are of course closely related, $f_p = f_c + f_s$. In the processing of the amplitude-modulated signal (4.37), it is essential to set the imaginary part to zero before the second Fourier transformation in order to obtain a signal free of phase twist:

$$\begin{aligned} F_c(\omega_1, \omega_2) &= \mathrm{Re}(\mathrm{FT}_1(\mathrm{Re}(\mathrm{FT}_2\langle \cos(\tilde{\omega}_1 t_1)\ e^{i\tilde{\omega}_2 t_2} \rangle))) \\ &= \mathrm{Re}(\mathrm{FT}_1(\mathrm{Re}\langle \cos(\tilde{\omega}_1 t_1)(A_2 + iD_2) \rangle)) \\ &= \mathrm{Re}(\mathrm{FT}_1 \langle \tfrac{1}{2}(e^{i\tilde{\omega}_1 t_1} + e^{-i\tilde{\omega}_1 t_1}) A_2 \rangle) \\ &= \tfrac{1}{2}\langle (A(\omega_1 - \tilde{\omega}_1) + A(\omega_1 + \tilde{\omega}_1)) A_2 \rangle \end{aligned} \tag{4.39}$$

which reduces to:

$$F_c(\omega_1, \omega_2) = \tfrac{1}{2}\langle A_1 A_2 \rangle \text{ for } \omega_1 \geqslant 0 \tag{4.40}$$

if $A(\omega_1 + \tilde{\omega}_1) = 0$ for all $\omega_1 \geqslant 0$. This is equivalent to demanding that $A(\omega_1 - \tilde{\omega}_1) = 0$ for all $\omega_1 \leqslant 0$, which can be fulfilled by setting the detection frequency ω_R slightly outside the range of spectral intensity ('off-resonance technique').

By measuring both f_c and f_s, which together represent a 'hypercomplex' dataset, the off-resonance condition can in principle be avoided, because the application of the above procedures to f_s yields:

$$F_s(\omega_1, \omega_2) = \tfrac{1}{2}\langle (A(\omega_1 - \tilde{\omega}_1) - A(\omega_1 + \tilde{\omega}_1))A_2 \rangle \qquad (4.41)$$

and therefore the absorptive spectrum:

$$F_c(\omega_1, \omega_2) + F_s(\omega_1, \omega_2) = \langle A_1 A_2 \rangle \qquad (4.42)$$

can be generated, providing sign discrimination of frequencies in both dimensions.

In practice, it often proves to be sufficient or advantageous to work off-resonance with a single pulse sequence producing $f_c(t_1, t_2)$. For instance, the difference of the large and, in practice, usually non-identical terms $A_c(\omega_1 + \omega_a) - A_s(\omega_1 + \omega_a)$ occurring in the combination of the datasets in (4.42) often leaves artifacts, most pronounced along the subdiagonal. Quite often, at $\omega_1 = 0$ artifacts appear from constant background signals; these can be discarded in the off-resonance measurement where no signal appears at $\omega_1 = 0$. Also, it is more convenient to work with a single pulse program and data set. Of course, one can measure the t_1 data points of f_c and f_s interleaved and store them as one data set, or produce TPPI in t_1, by changing the phase of the second pulse in the pulse programs by 90°.

In certain cases, only the time-signals:

$$\langle \cos(\tilde{\omega}_1 t_1)\cos(\tilde{\omega}_2 t_2) \rangle \text{ and } \langle \sin(\tilde{\omega}_1 t_1)\sin(\tilde{\omega}_2 t_2) \rangle \qquad (4.43)$$

can be generated, for instance in ^{2}H NMR. Here, it turns out that working on-resonance is virtually inevitable (see Section 10.2.4). Although it may not be evident at first sight, the signals (4.43) represent TPPI or hypercomplex data:

$$\begin{aligned} \cos(\tilde{\omega}_1 t_1)\cos(\tilde{\omega}_2 t_2) &= \tfrac{1}{2}\cos(\tilde{\omega}_1 t_1)e^{i\tilde{\omega}_2 t_2} + \tfrac{1}{2}\cos(-\tilde{\omega}_1 t_1)e^{i(-\tilde{\omega}_2)t_2} \\ -\sin(\tilde{\omega}_1 t_1)\sin(\tilde{\omega}_2 t_2) &= \tfrac{1}{2}i\sin(\tilde{\omega}_1 t_1)e^{i\tilde{\omega}_2 t_2} + \tfrac{1}{2}i\sin(-\tilde{\omega}_1 t_1)e^{i(-\tilde{\omega}_2)t_2}. \end{aligned} \qquad (4.44)$$

This is identical to the hypercomplex f_c and f_s signals (4.37) and (4.38) with the contributions for $(\tilde{\omega}_1, \tilde{\omega}_2)$ and $(-\tilde{\omega}_1, -\tilde{\omega}_2)$ added up. This pair of signals at $(\pm\tilde{\omega}_1, \pm\tilde{\omega}_2)$ represents the two inseparable 'transitions' of the quadrupolar interaction of the deuteron (see equation (2.90)).

In 3D spectroscopy, the practical advantage of the off-resonance measurement is particularly pronounced. Here only:

$$\cos_1 \cos_2 \text{expi}_3 := \cos(\tilde{\omega}_1 t_1)\cos(\tilde{\omega}_2 t_2)\exp(i\tilde{\omega}_3 t_3) \qquad (4.45)$$

is required, while the on-resonance approach requires four components, written in short as $\cos_1 \cos_2 \text{expi}_3$, $\sin_1 \cos_2 \text{expi}_3$, $\cos_1 \sin_2 \text{expi}_3$ and $\sin_1 \sin_2 \text{expi}_3$. That off- and on-resonance measurements are equivalent in terms of S/N will be shown in Section 4.9.5 below.

The solution to the phase problem in the case of phase modulation can be traced back to that found for amplitude modulation. From:

$$f_{\rm p}(t_1, t_2) = \langle {\rm e}^{{\rm i}\tilde{\omega}_1 t_1}\, {\rm e}^{{\rm i}\tilde{\omega}_2 t_2} \rangle$$

and

$$f_{\rm m}(t_1, t_2) = \langle {\rm e}^{-{\rm i}\tilde{\omega}_1 t_1}\, {\rm e}^{{\rm i}\tilde{\omega}_2 t_2} \rangle \qquad (4.46)$$

created through 'time reversal' in t_1 (by a 180° pulse in the simplest case, Section 2.6.3), one can generate $f_{\rm c} = \frac{1}{2}(f_{\rm p} + f_{\rm m})$ and $f_{\rm s} = \frac{1}{2}(f_{\rm p} - f_{\rm m})$. These can be transformed into an absorptive 2D spectrum with quadrature in both dimensions as described above.

For theoretical considerations in particular, it is useful to note that straightforward Fourier transformation of $\langle {\rm e}^{{\rm i}\tilde{\omega}_1 t_1}\, {\rm e}^{{\rm i}\tilde{\omega}_2 t_2} \rangle$ can provide an absorptive spectrum with sign discrimination of both frequencies, if only t_1 covers positive as well as negative time values symmetrically. The dispersive part of the lineshape vanishes, $D_1 = 0$, since the data for negative t_1 make $h(t_1)$ symmetrical around $t_1 = 0$, so that according to equation (4.16) with t and ω interchanged, the imaginary part of the Fourier transform of $h(t_1)$ vanishes, $D_1 = 0$. Fourier transformations of $\langle {\rm e}^{{\rm i}\tilde{\omega}_1 t_1}\, {\rm e}^{{\rm i}\tilde{\omega}_2 t_2} \rangle$ over t_1 and t_2 (without setting anything to zero) therefore yield a spectrum $A_1(A_2 + {\rm i}D_2)$ with a fully absorptive real part A_1A_2. In practice, the data for negative t_1 are obtained by taking the time-reversed signal with ${\rm e}^{-{\rm i}\tilde{\omega}_1 t_1'} = {\rm e}^{{\rm i}\tilde{\omega}_1(-t_1')}$ as the signal ${\rm e}^{{\rm i}\tilde{\omega}_1 t_1}$ for $-t_1' = t_1 < 0$.

4.6.2 The phase problems in multidimensional MAS sideband spectra

The treatment of the phase properties of multidimensional MAS spectra with sidebands in two or more dimensions requires special care. For simplicity, we restrict the discussion to two dimensions here; the extension to more dimensions is straightforward.

On the one hand, in time signals the coefficients $I_{M,N}$ of ${\rm e}^{{\rm i}\tilde{\omega}_1 t_1}\, {\rm e}^{{\rm i}\tilde{\omega}_2 t_2}$ (with $\tilde{\omega}_1 = M\omega_{\rm r}$ and $\tilde{\omega}_2 = N\omega_{\rm r}$) can be complex numbers. This is a complication compared with the spectra for static samples, where the corresponding $S(\tilde{\omega}_1, \tilde{\omega}_2)$ is a probability (of finding a certain segment with frequencies $\tilde{\omega}_1$ and $\tilde{\omega}_2$) and therefore non-negative and real. With complex $I_{M,N}$, the scheme of discarding D_2 by cancelling the imaginary part after the first FT fails: A_2 and D_2 are inextricably mixed in both real and imaginary parts by multiplication with the complex coefficients $I_{M,N}$. This is discussed in more detail below.

On the other hand, strictly periodic signals can occur in the evolution periods of 2D MAS experiments, if only the rotor phase is varied, without evolution of transverse magnetization; an example is given in Section 12.4. The Fourier transforms of these periodic signals have no dispersive parts, being simply $\delta(\omega_1 - M\omega_{\rm r})$ functions. This avoids phase twist for products of complex exponentials in either dimension (${\rm e}^{{\rm i}\tilde{\omega}_1 t_1}$ ${\rm e}^{{\rm i}\tilde{\omega}_2 t_2}$), in particular if the coefficients $I_{M,N}$ are real numbers. In this case, phase-sensitive sign-discriminating 2D spectra can be obtained with a single pulse sequence, with no t_1-time-reversal or sine/cosine datasets required.

Under sample rotation, the angle-dependent part of the signal is strictly periodic with $\omega_{\rm r}$ and can be represented by a Fourier series $\Sigma_N I_N {\rm e}^{{\rm i}N\omega_{\rm r}t}$. Including these anisotropic interactions, the isotropic chemical shift, and T_2 relaxation, the total 2D time signal can be written as:

$$s(t_1, t_2) = \sum_{M,N} I_{M,N}\, \mathrm{e}^{\mathrm{i}M\omega_r t_1}\, \mathrm{e}^{\mathrm{i}N\omega_r t_2}\, \mathrm{e}^{\mathrm{i}\omega_{\mathrm{iso}} t_1}\, \mathrm{e}^{\mathrm{i}\omega_{\mathrm{iso}} t_2}\, \mathrm{e}^{-t_1/T_2}\, \mathrm{e}^{-t_2/T_2}. \tag{4.47}$$

The coefficients $I_{M,N}$ are the Fourier coefficients of the periodic functions $g(t_1, t_2)$ describing the action of the anisotropic interaction alone. These $g(t_1, t_2)$ functions are treated at length in Sections 6.5, 10.4 and 12.4, and their one-dimensional prototypes in Section 3.9. Here, we should only be aware of the Fourier-series relations between the $g(t_1, t_2)$ functions and the $I_{M,N}$ coefficients:

$$\begin{aligned} g(t_1, t_2) &= \sum_{M,N} I_{M,N}\, \mathrm{e}^{\mathrm{i}M\omega_r t_1}\, \mathrm{e}^{\mathrm{i}N\omega_r t_2} \\ I_{M,N} &= \int_0^{t_r} \frac{\mathrm{d}t_1}{t_r} \int_0^{t_r} \frac{\mathrm{d}t_2}{t_r}\, g(t_1, t_2)\, \mathrm{e}^{-\mathrm{i}M\omega_r t_1}\, \mathrm{e}^{-\mathrm{i}N\omega_r t_2}. \end{aligned} \tag{4.48}$$

For the following discussion, we distinguish two cases:

I Fully periodic signal during the evolution t_1:

$$s(t_1, t_2) = \sum_{M,N} I_{M,N}\, \mathrm{e}^{\mathrm{i}M\omega_r t_1}\, \mathrm{e}^{\mathrm{i}N\omega_r t_2}\, \mathrm{e}^{\mathrm{i}\omega_{\mathrm{iso}} t_2}\, \mathrm{e}^{-t_2/T_2}. \tag{4.49}$$

The spectrum results after double FT:

$$\begin{aligned} S(\omega_1, \omega_2) &= \mathrm{FT}_1\{\mathrm{FT}_2\{s(t_1, t_2)\}\} \\ &= \sum_{M,N} I_{M,N}\, \delta(\omega_1 - M\omega_r)\{A(\omega_2 - \omega_{\mathrm{iso}} - \omega_r t_2) \\ &\qquad + \mathrm{i}D(\omega_2 - \omega_{\mathrm{iso}} - \omega_r t_2)\}. \end{aligned} \tag{4.50}$$

The sequence of the Fourier transformations is arbitrary. In between, no parts of the data are set to zero or swopped between real and imaginary parts. Only after the double FT is phasing applied to the data. The real parts of the peak intensities are then:

$$\mathrm{Re}(I_{M,N})A_2 - \mathrm{Im}(I_{M,N})D_2. \tag{4.51}$$

Thus, they are fully absorptive if $\mathrm{Im}(I_{M,N}) = 0$. In some cases, the $I_{M,N}$ are not real as such, but rather we have $I_{M,N} = \mathrm{e}^{\mathrm{i}M\gamma_0}\tilde{I}_{M,N}$ with real $\tilde{I}_{M,N}$. After phasing (multiplication) of the data by $\mathrm{e}^{-\mathrm{i}M\gamma_0}$, the spectrum is also fully absorptive in that case.

Here, it should be noted that a spectrum can be fully 'absorptive' but still have an imaginary part: $\mathrm{e}^{\mathrm{i}\phi}A(\omega - \omega_0)$ is the simplest general example (see also the discussion in Section 4.5.2).

II Non-periodic signals in both t_1 *and* t_2:

$$s(t_1, t_2) = \sum_{M,N} I_{M,N}\, \mathrm{e}^{\mathrm{i}M\omega_r t_1}\, \mathrm{e}^{\mathrm{i}N\omega_r t_2}\, \mathrm{e}^{\mathrm{i}\omega_{\mathrm{iso}}(t_1+t_2)}\, \mathrm{e}^{-(t_1+t_2)/T_2}. \tag{4.52}$$

Here, the phase-twist problem arises again as in the previous section. If the $I_{M,N}$ are real, the procedure for the solution of the problem is similar as outlined there: a 't_1-time-reversed' data set:

$$s^{\mathrm{tr}}(t_1, t_2) = \sum_{M,N} I_{M,N}\, \mathrm{e}^{\mathrm{i}M\omega_r(-t_1)}\, \mathrm{e}^{\mathrm{i}N\omega_r t_2}\, \mathrm{e}^{\mathrm{i}\omega_{\mathrm{iso}}(-t_1+t_2)}\, \mathrm{e}^{-(t_1+t_2)/T_2} \tag{4.53}$$

must be created by appropriate pulse sequences (application of a 180° pulse often does not achieve this under MAS conditions since it does not reverse the sample rotation). Signals with $\cos(M\omega_r t_1)$ and $\sin(M\omega_r t_1)$ modulation are then generated by adding and subtracting $s(t_1, t_2)$ and $s^{\mathrm{tr}}(t_1, t_2)$. The further procedure goes as in the previous section: FT_2 of both signals is performed separately, ω_2-phasing with values determined on the $\cos(t_1 = 0)$ slice, followed by combining the real parts of the cosine and sine signals as the real and imaginary parts, respectively, of the data then subject to FT_1. Of these, after ω_1 phasing the real part is taken as the final absorptive spectrum with quadrature in both dimensions.

If the $I_{M,N}$ are not real (or cannot be phased to be real in the ω_2 or N dimension), again the phase problem cannot be solved because both the real part:

$$\mathrm{Re}[s_{\mathrm{s}}^{\mathrm{c}}(t_1, \omega_2)] = \sum_{M,N} [\mathrm{Re}(I_{M,N})A_2 - \mathrm{Im}(I_{M,N})D_2] \begin{matrix}\cos(\{M\omega_r + \omega_{\mathrm{iso}}\}t_1)\\ \sin(\{M\omega_r + \omega_{\mathrm{iso}}\}t_1)\end{matrix}\, \mathrm{e}^{-t_1/T_2} \tag{4.54}$$

and the imaginary part:

$$\mathrm{Im}[s_{\mathrm{s}}^{\mathrm{c}}(t_1, \omega_2)] = \sum_{M,N} [\mathrm{Im}(I_{M,N})A_2 + \mathrm{Re}(I_{M,N})D_2] \begin{matrix}\cos(\{M\omega_r + \omega_{\mathrm{iso}}\}t_1)\\ \sin(\{M\omega_r + \omega_{\mathrm{iso}}\}t_1)\end{matrix}\, \mathrm{e}^{-t_1/T_2} \tag{4.55}$$

contain dispersive components D_2 that cannot be cancelled.

In discussing MAS signals in the sections below, it will be helpful to express Fourier transformation of time signals not just as FT_2 or FT_1 but more explicitly along the lines of the basic definitions given at the beginning of this chapter. The time data actually sampled are always discrete, therefore discrete Fourier transformation (see equation (4.5)):

$$I_N = \sum_m g(mt_{\mathrm{dw}})\, \mathrm{e}^{-\mathrm{i}N(mt_{\mathrm{dw}})}$$

is applied in practice. However, Fourier transformation with a sum over time domain data points is not such a common notion. Therefore, we will instead write the Fourier transformation in terms of the more widely known Fourier integrals:

$$I_N = \frac{1}{t_r} \int_0^{t_r} \mathrm{d}t\, g(t)\, \mathrm{e}^{-\mathrm{i}Nt}. \tag{4.56}$$

That this is not quite correct may become particularly apparent in the context of 2D data sets where the evolution time is sampled only at eight or 16 t_1 values; however, we feel that it is more important to have the FT easily recognized and therefore choose the integral notation (4.56).

4.6.3 The projection/cross-section theorem

In 2D NMR, the two-dimensional Fourier transformation is usually performed by two successive one-dimensional FTs, over t_2 and t_1, since this is the simplest approach and since the data have to be manipulated between the two transformations. Sometimes, however, it can also be useful to consider certain relations directly between 2D time-domain and spectral data. Often, this involves the projection/cross-section theorem (Nagayama *et al.*, 1978), which is the most important theorem in multidimensional Fourier transformation without a meaningful analog in 1D FT. It states that the intensity distribution along a one-dimensional cross-section line passing through the origin in one domain is the Fourier transform of the projection onto the same direction in the other domain. The projection involves the integral over the intensity along the directions perpendicular to the selected line. This is related to the notion of a plane wave in one domain and the corresponding $\vec{\mathbf{k}}$ vector in the other domain, where $\vec{\mathbf{k}}$ is the normal of the wave fronts.

An example for the validity of the theorem is provided by the projection of a 2D spectrum on the ω_2 axis that can be characterized as the line $(0, x)$ in the frequency domain. Such a projection is known to be the ω_2 spectrum for $t_1 = 0$, which is the Fourier transform of the time-signal $f(t_1 = 0, t_2)$. The corresponding points $(0, t_2)$ define the $(0, x)$ cross-section in the time domain. Thus, the projection on the $(0, x)$ axis in the frequency plane is indeed the Fourier transform of the time signal along the $(0, x)$ cross-section in the time domain.

4.6.4 *Shearing of 2D spectra*

Various isotropic–anisotropic separation 2D experiments (see Section 6.4) are based on refocusing the anisotropic interaction in the evolution period. In the conceptually simplest version of these experiments, acquisition is started at the point where the refocusing is completed. It is advantageous in most of these experiments to start acquisition directly after the last pulse, observing how an echo forms at the point $t_2 = ct_1$ where the anisotropic effects are fully refocused. The constant c depends on the experiment; in many cases, $c = -1$. If the acquisition is started before the echo maximum, the signal in the left half of the echo is also exploited. The shift theorem (4.15) tells us that the t_1-proportional shift of the time origin in t_2 can be compensated by a t_1-proportional phase correction $e^{-i\omega_2 ct_1}$ linear in ω_2:

$$
\begin{aligned}
s'(t_1, t_2) &= s(t_1, t_2 + ct_1) \\
&\big\updownarrow \mathrm{FT}_2 \\
S'(t_1, \omega_2) &= e^{i\omega_2 ct_1} S(t_1, \omega_2) \\
&\big\updownarrow \mathrm{FT}_1 \\
S'(\omega_1, \omega_2) &= S(\omega_1 + c\omega_2, \omega_2)
\end{aligned}
\tag{4.57}
$$

The last expression describes shearing of the 2D spectrum parallel to the ω_1 axis: the ω_2 projection remains unchanged, while points are shifted in ω_1 linear with the corresponding ω_2 values. For instance, if we have a spectrum at $\omega_1 = 0$ parallel to the ω_2 axis (i.e. perpendicular to the direction of the shearing):

$$S(\omega_1, \omega_2) = S(\omega_2)\ \delta(\omega_1)$$

with $c = -1$ in equation (4.57) it is sheared to become a diagonal spectrum:

$$S'(\omega_1, \omega_2) = S(\omega_1 - \omega_2, \omega_2) = S(\omega_2)\ \delta(\omega_1 - \omega_2).$$

Equation (4.57) is the basis of a convenient technique for shearing a given 2D spectrum $S'(\omega_1, \omega_2) = S(\omega_1 + c\omega_2, \omega_2)$: the spectrum $S'(t_1, \omega_2)$, obtained directly from the time data $s'(t_1, t_2)$, or by inverse FT over t_1 from $S'(\omega_1, \omega_2)$, can be automatically phased with the exactly known factors $e^{-i\omega_2 c t_1}$, and FT over t_1 produces the sheared 2D spectrum $S(\omega_1, \omega_2)$ in a straightforward way.

4.7 Mathematical formulation of data sampling

Nuclear magnetic resonance signals are in principle continuous, but for storage and evaluation on a computer they have to be sampled (made discrete). For a 2D signal, the t_1 dimension can in fact only be measured point by point. How well the Fourier transform of the digitized data represents the true spectrum of the original continuous function will be discussed now. *A priori* this is by no means obvious, and indeed there can be major effects of the digitization on the spectrum. Before discussing these, a mathematical description of the sampling will be given.

Consider a simple periodic array of delta functions (Bracewell, 1986, p. 77) with a spacing t_{dw} ('delta array'):

$$\mathrm{III}(t/t_{dw}) := \sum_{n=-\infty}^{\infty} \delta(t - nt_{dw}). \tag{4.58}$$

Multiplied on a function $f(t)$, this delta array picks out points at $t = nt_{dw}$, that is, it samples $f(t)$ with a spacing of t_{dw}. The sampling can therefore be written as:

$$f^{\ddagger}(t) = \mathrm{III}(t/t_{dw}) f(t). \tag{4.59}$$

To determine the effect of this time-domain multiplication on the spectrum, we have to know the Fourier transform of the delta array. It is again a delta array (Bracewell, 1986, p. 192 and p. 214):

$$\mathrm{III}(t/t_{dw}) \xleftrightarrow{\mathrm{FT}} 2\pi\mathrm{III}(\omega/(2\pi/t_{dw})) \tag{4.60}$$

and the periods are inversely proportional. This is like an infinite 1D lattice with lattice spacing t_{dw}, which has a reciprocal lattice with spacing $2\pi/t_{dw}$ as its Fourier transform. The perfect periodicity in one domain is reflected in the infinitely narrow peaks in the other domain.

According to the convolution theorem (4.10) we can find the Fourier transform of the sampled data (4.59):

$$\mathrm{III}(t/t_{dw}) f(t) \xleftrightarrow{\mathrm{FT}} \mathrm{III}(\omega/(2\pi/t_{dw})) * F(\omega). \tag{4.61}$$

The right-hand side represents a convolution of $F(\omega)$ with a delta array with period $\omega_{sw} = 2\pi/t_{dw}$. This results in:

$$(\mathrm{III} * F)(\omega) = \int_{-\infty}^{\infty} d\omega' F(\omega') \sum_{n=-\infty}^{\infty} \delta(\omega - n\omega_{sw} - \omega')$$

$$\overset{(4.12)}{=} \sum_{n=-\infty}^{\infty} F(\omega - n\omega_{sw}) \tag{4.62}$$

which represents a periodic repetition of $F(\omega_{sw})$ with a repetition period of ω_{sw}. This proves the sampling theorem: spectra can be reproduced correctly from sampled time data only if the spectral intensity is restricted to a zone $\omega_{sw} = 2\pi/t_{dw}$ wide, since only one such zone is reproduced without overlapping on itself. Out of the spectral function made periodic by the sampling, usually the 'Nyquist zone' between the positive and negative Nyquist frequencies $\pm\omega_{ny} = \pm\frac{1}{2}\omega_{sw}$ is chosen to represent the spectrum completely. It corresponds to the first Brillouin zone of the one-dimensional reciprocal lattice. 'Aliasing' or 'cyclic return' of a signal from outside the Nyquist zone will be discussed in the following section. With single-phase detection, only one half of the Nyquist zone represents the useful spectral width since positive and negative frequencies cannot be distinguished.

The sampling as described so far has a relationship to the theory of Fourier series: in one domain we have a periodic function, while in the other the intensity is discrete, restricted to multiples of a fundamental period. However, time and frequency domains are interchanged for the data sampling compared to the customary treatment of Fourier series, where the time-signal is periodic and continuous, and the spectrum is discrete.

Up to this point, it has been assumed that the data are acquired point by point up to infinite times so that a periodically repeated but continuous Fourier transform is obtained. In practice, however, one can only measure up to a finite time t_{aq}. Assuming a periodic continuation of the time signal, delta array considerations again show that the spectrum is sampled (digitized) with a spacing $\Delta\omega = 2\pi/t_{aq}$, but otherwise not altered. This is exactly the spectrum that results from the discrete Fourier transformation (equation (4.63)), since DFT indeed assumes periodic time data.

4.8 Consequences of data sampling

Sampling of data over a finite time can lead to aliasing, cut-off effects, or baseline lifting.

4.8.1 Aliasing

Aliasing denotes the appearance, inside the Nyquist zone, of a signal that has actually a high frequency outside the Nyquist zone. As indicated in Figure 4.8(a), the periodic repetition of the spectrum by time-domain sampling causes a line outside the Nyquist zone at $\omega = \omega_{ny} + \delta\omega$ to reappear inside the Nyquist zone at $\omega = -\omega_{ny} + \delta\omega$.

Quite commonly, it is believed that the line is always 'folded back', appearing at $\omega = \omega_{ny} - \delta\omega$. This is an effect of single-phase or TPPI detection, where positive and negative frequencies are not distinguished such that $\omega = -\omega_{ny} + \delta\omega$ and $\omega = \omega_{ny} - \delta\omega$ are equivalent as indicated in Figure 4.8(b) (with TPPI this refers to frequencies measured in the frame rotating with $\omega_{sw}/4$; see Section 4.5.1). However, as stated above, under complex Fourier transformation a signal that is slightly beyond the positive Nyquist frequency reappears at the negative Nyquist limit.

4.8.2 Truncation effects and apodization

'Cut-off', or truncation, effects can occur for narrow lines in the spectrum as a consequence of the acquisition time t_{aq} being finite. The cut-off effects manifest themselves in terms of negative and positive side-lobes of the narrow lines. For a

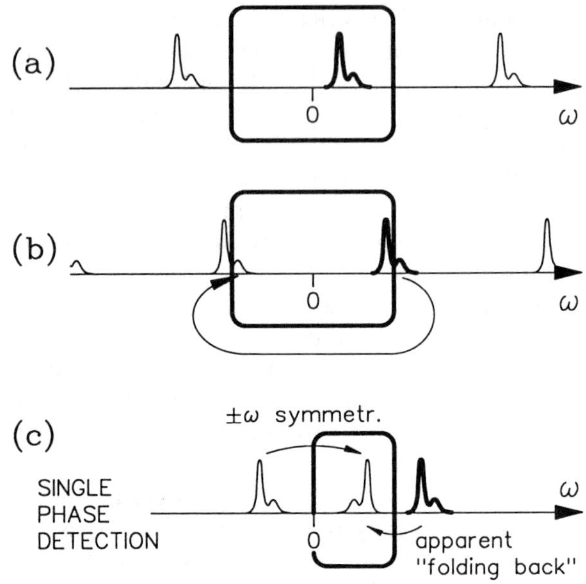

Figure 4.8 Spectral effect of aliasing. (a) Periodic spectrum of a digitized time signal shown for reference. The width of the black frame delimits the Nyquist zone. (b) Shift of the spectrum; the signal, moving out of the Nyquist zone to the right, appears to 'return cyclically' to the left from outside the Nyquist zone. (c) Apparent 'folding-back' due to single-phase detection: signals at ω and $-\omega$ are not distinguished, so that the cyclically returning signal at the left is reflected over to the right limit of the Nyquist zone. Therefore, the signal moving to the right out of the Nyquist zone seems to be folded back in.

mathematical description, the actual acquired signal can be viewed as the product of the 'true', infinite time signal and the box function of extension t_{aq} (at $t > 0$). Since $f(t)$ vanishes for $t < 0$, the extension of the box function into negative times is arbitrary; for convenience, we choose the box symmetric between $-t_{aq}$ and t_{aq}. Then, according to the box–sinc Fourier relationship (4.26) and the convolution theorem, the Fourier transform of the product of $f(t)$ and the box function is:

$$\tilde{F}(\omega) = F(\omega) * \frac{2\sin(\omega t_{aq})}{\omega}\frac{1}{2\pi} = \frac{1}{\pi}\int_{-\infty}^{\infty} d\omega' F(\omega')\,\frac{\sin(t_{aq}(\omega-\omega'))}{\omega-\omega'}. \quad (4.63)$$

The sinc function has a full width at half maximum of about $2/t_{aq}$ (this is a relatively small width). For lines much broader than $2/t_{aq}$, over the region close to $\omega' = \omega$ where the sinc function is nonnegligible, $F(\omega')$ is essentially constant, equal to $F(\omega)$, and can be drawn out of the integral:

$$\tilde{F}(\omega) \cong F(\omega)\int_{-\infty}^{\infty} d\omega' \,\frac{1}{\pi}\,\frac{\sin(t_{aq}(\omega-\omega'))}{\omega-\omega'} = F(\omega) \quad (4.64)$$

so that the original spectrum results. Seen from the time domain, this is obvious since the time signals corresponding to the broad lines have decayed to very small values before the termination of the acquisition at t_{aq} so that their spectrum cannot be changed appreciably. For signals with decay constants exceeding t_{aq}, cut-off effects must occur (except for the case of fully periodic signals, see below). In the limit of very narrow spectral lines, when $F(\omega) \cong I_0\,\delta(\omega - \omega_0)$, we find:

$$\tilde{F}(\omega) \cong \int_{-\infty}^{\infty} \mathrm{d}\omega'\, I_0 \delta(\omega' - \omega_0)\, \frac{1}{\pi}\, \frac{\sin(t_{\mathrm{aq}}(\omega - \omega'))}{\omega - \omega'} = I_0\, \frac{1}{\pi}\, \frac{\sin(t_{\mathrm{aq}}(\omega - \omega_0))}{\omega - \omega_0}. \qquad (4.65)$$

Thus, the cut-off results in replacing each sharp positive line by a sinc function of width $\cong 2/t_{\mathrm{aq}}$ and negative sidelobes which attain -20% of the maximum height of the line (Figure 4.9).

In order to avoid such disturbing artificial features, the time-signal can be 'apodized' by multiplication with an apodization function that decreases so fast at long times that the cut-off effects at t_{aq} become negligible. Common well-behaved apodization or damping functions are half-Gaussians $\exp(-t^2/2\sigma_t^2)$ and decaying exponentials. The effect of the apodization on the spectrum can be described as a convolution or smoothing of the original spectrum with the Fourier transform of the apodization function (according to the theorem (4.10)). Since the Lorentzian, the Fourier transform of the exponential, has broader 'wings' than the Gaussian of equal height and halfwidth, Gaussian multiplication is preferable for retaining good resolution without negative sidelobes. This is true in particular for 2D spectra, where the Lorentzian convolution produces star-shaped (or rather cross-shaped) contours with long prongs parallel to either axis, while the Gaussian gives rise to narrow circular profiles. A simple technique of resolution enhancement can be employed for removing the Lorentzian lineshape and replacing it by a Gaussian: the decay by $\exp(-t/T_2)$ that gives rise to the Lorentzian is cancelled or reduced by multiplication with $\exp(+t/\tau)$ in the time domain, where $\tau \geqslant T_2$. Cut-off effects are prevented by multiplication with a suitable Gaussian which will determine the lineshape.

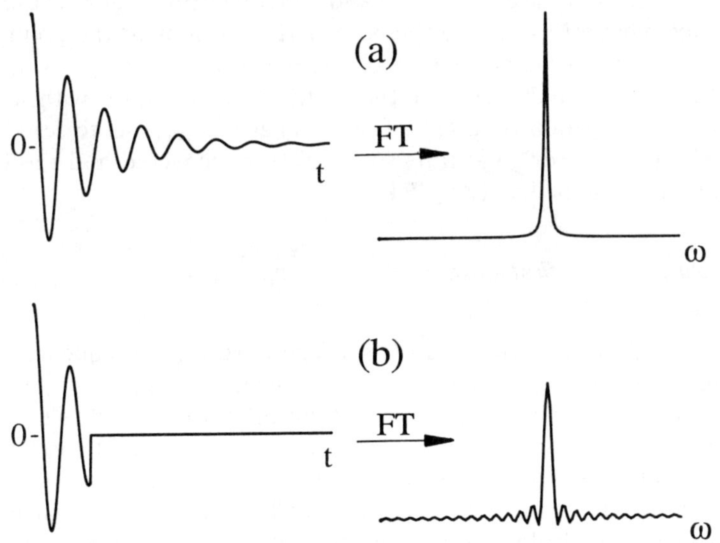

Figure 4.9 Time-domain cut-off (truncation) effects on a narrow spectral signal. The signal is convoluted with a sinc-function. It has negative lobes that can cancel or distort small signals close to a large resonance. In addition, the spectral line is broadened.

If the signal is truly periodic, with t_{aq} being an exact multiple of its period t_p ($t_{aq} = nt_p$), the cut-off effects must vanish; they do because the array of spectral points around each line, at $n \times 2\pi/t_{aq}$, only probe the null points of the sinc function.

4.8.3 *Zero-filling*

Zero-filling is the term for appending zeros to the end of the time-domain signal before the DFT. It serves mainly for producing smoother spectra, since it achieves a trigonometric interpolation of the spectral data points in a truly simple manner. Zero-filling $f(t) = 0$ for $t > t_{aq}$ is a valid extrapolation of the time signal when the time signal has decayed or been apodized to 'zero' (in practice, to less than 5% of the maximum of the signal) before t_{aq}. In multidimensional spectra, zero-filling provides flexibility in the number of t_1 increments that must be measured: time-domain data arrays of any size can be zero-filled up to the $n_{td} = 2^m$ points required for standard FFT algorithms.

As was shown by Ernst (Ernst, 1969; Bartholdi and Ernst, 1973), doubling the data size to $2t_{aq}$ by zero-filling can even increase the signal-to-noise ratio. Zero-filling ensures that the information of the imaginary part of the spectrum is introduced into the real part, so that the S/N is improved by $\sqrt{2}$ if the signal does not decay appreciably during t_{aq}. This can be rationalized by proving that after zero-filling the information contained in the real and imaginary parts of the Fourier transform is identical. This is reminiscent of the Kramers–Kronig relations (Abragam, 1961); for data sampled over a limited time interval, it is safer to base the proof on a simple Fourier algorithm which allows us to generate the imaginary part of the spectrum from its real part: inverse FT of the pure real part produces a time signal $f_s(t)$ with symmetric real part and antisymmetric imaginary part (see relation (4.16)), which are the symmetric and antisymmetric portions of the original real and imaginary time signals, respectively (the symmetry or antisymmetry refers to the point t_{aq}, as well as to the point $t = 0$ which is identical to $2t_{aq}$ due to the periodicity of the data). In the range $t < t_{aq}$, the signal $f_s(t)$ is equal to the original $f(t)$ since the (anti) symmetrization with the zero-filled data points reflects only nulls onto $f(t)$. The signal at $t > t_{aq}$ is set to zero to reproduce the zero filling; thus, the original time signal $f(t)$ is recovered and the full complex spectrum is obtained immediately by FT.

4.8.4 *Sampling of the first datapoint*

The digitization of typical NMR signals has a further subtle consequence. The original continuous time signal is zero (or virtually zero) at negative times, and jumps to a finite value at $t = 0$. At this discontinuity, special attention must be paid to the sampling. It will be shown that for an optimum reproduction of the original spectrum, the first point must be multiplied by $\frac{1}{2}$; otherwise, the baseline in the spectrum is lifted. In conventional 1D spectroscopy this effect is usually inconspicuous while it produces disturbing artifacts in multidimensional spectra (Otting *et al.*, 1986). For sampling the first point, consider the delta function at $t = 0$ as the limit of Gaussians $g(t; \Delta t)$ or similar functions whose width Δt is sent to zero while the area is held normalized to

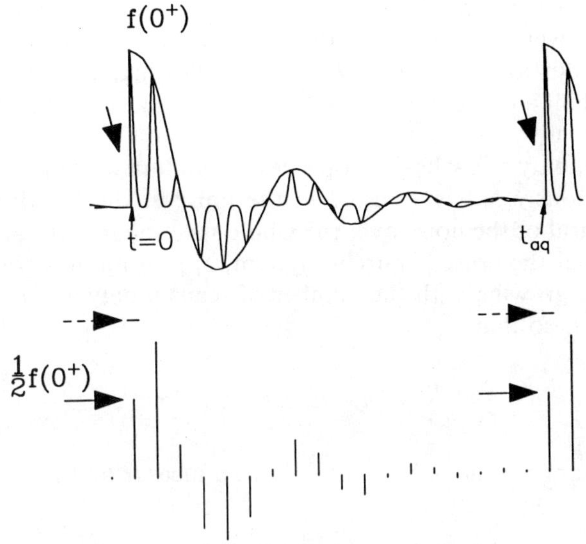

Figure 4.10 When the NMR time signal is digitized (sampled) at $t = 0$ the signal has to be set to half the amplitude of $f(t \geqslant 0)$ to avoid a lifted baseline or spurious ridges in a 2D spectrum. In (a), the product of the time signal with narrow Gaussian functions is shown. In (b) the limit to infinitely narrow Gaussians (δ-functions) has been taken, which is the basis of the mathematical description of digitization.

unity. As indicated in Figure 4.10, at the discontinuity the sampling peak picks up intensity from $f(t)$ only on its right half ($t > 0$):

$$f^{\ddagger}(0) = \lim_{\Delta t \to 0} \int_{-\infty}^{\infty} \mathrm{d}t\, f(t) g(t; \Delta t) = \lim_{\Delta t \to 0} \int_{0}^{\infty} \mathrm{d}t\, f(t) g(t; \Delta t)$$

$$= f(0^+) \lim_{\Delta t \to 0} \int_{0}^{\infty} \mathrm{d}t\, g(t; \Delta t) = f(0^+) \cdot \tfrac{1}{2} \qquad (4.66)$$

using $f(t) = 0$ for $t < 0$. Correct sampling of the point at $t = 0$ is thus achieved by taking only half of its height at $t \geqslant 0$. Only with correct sampling of the first point is the spectrum of the sampled time domain function identical with the original one (apart from the periodization). Subjecting the time data with the first point at its full height to the DFT lifts the baseline in the spectrum (according to relation (4.8)) so far that the sum over all points in the spectrum is exactly doubled. Shifting the discontinuity of the time signal to $t < 0$ does not eliminate the effect; aliasing of the high-frequency spectral components of the discontinuity produces the same lifting of the baseline.

4.9 Signal and noise

In spite of numerous important improvements, high-resolution solid-state NMR has not reached a particularly high sensitivity yet. The signal-to-noise ratio, S/N, is therefore an important factor for the success of an NMR experiment, in particular for multidimensional spectroscopy. It is directly coupled to the duration of an experiment because for almost every solid-state NMR measurement, at least a few scans have to be averaged to attain a sufficient S/N. In the following, by making a few simple comparisons, we shall analyse how the S/N changes with experimental parameters, or whether it is independent of them.

For this purpose, only a few basic properties of random noise will be required. First of all, we are obviously only interested in the noise at the location of the spectral intensity. The integral of the noise over the whole spectral area is not relevant as such. We shall assume that the noise is purely random, appearing uncorrelated in different scans and therefore growing with the number of scans n only as $\sqrt{n}$. The useful signal grows linear with n, so that:

$$\frac{S}{N} \sim \frac{n}{\sqrt{n}} = \sqrt{n} \sim (\text{measuring time})^{\frac{1}{2}}. \tag{4.67}$$

Consequently, to improve the S/N by a factor k, a measuring time longer by k^2 times is required.

4.9.1 The 'dwell-time theorem'

'The signal-to-noise ratio in a NMR spectrum is usually independent of the dwell time'; also in multidimensional spectroscopy (Aue *et al.*, 1978; Levitt *et al.*, 1984; Bax and Mareci, 1983). (4.68)

This statement may appear trivial at first sight. It does, however, prove helpful and it is not so self-evident in certain cases (Turner, 1982, 1984). In the time-domain, for instance, it does not apply: there, the signal-to-noise ratio per point generally deteriorates with decreasing dwell time (see below). This is relevant, for example, in the measurement of echo heights. The S/N in the spectrum becomes worse if every second point is removed from a time signal so that the dwell time is increased artificially (e.g. to decrease the amount of data). There have also been approaches for measuring 2D spectra with a larger t_1 increment, and therefore smaller spectral width in ω_1, so that the spectrum is aliased several times in the 2D plane (Turner, 1982, 1984). According to the dwell-time theorem (4.68), this does not make sense in terms of S/N (Bax and Mareci, 1983). Only if the number of scans is determined by the minimum phase cycle, or if the amount of data becomes too large otherwise (as with 3D spectroscopy in liquids) does such an approach appear reasonable. Considerations of such secondary effects which are irrelevant in NMR of solids also tend to make it difficult to extract the essential dwell-time theorem from the original discussions of sensitivity in 2D solution NMR (Aue *et al.*, 1978; Levitt *et al.*, 1984).

This should serve as a sufficient motivation for discussing the dwell-time theorem. For its proof, direct and indirect detection (i.e. t_2 and t_1 dimensions in a 2D experiment) must be treated separately, and the assumptions must be given specifically. In t_2 or ω_2 (i.e. also in 1D spectra) the choice of an appropriate width of the low-pass filter (after the signal demodulation) is relevant to prevent noise from being aliased from outside the Nyquist zone. With the spectral NMR intensity located sufficiently close to the centre of the spectral range, it is obvious that an increase of the spectral width, or a decrease of the dwell time, does not change the S/N at the spectral NMR intensity. The high-frequency noise components that appear additionally deteriorate the S/N only in the time signal.

With indirect detection (t_1 or ω_1 dimensions in 2D), the time signal is probed point by point at macroscopically separated times, so that there is no correlation between the points. The filter width plays no separate role for this dimension; it can even be chosen smaller than the spectral width in ω_1; it must only be properly selected for the ω_2/t_2 dimension. In fact, in wideline-separation 2D spectroscopy (see Section 6.6.2) the filter width is often considerably smaller than the spectral width in ω_1.

Concerning the indirect-detection dimension, we can state that the overall number of scans that can be measured in a given total measuring time is independent of t_{dw}, and therefore the S/N is also unaffected by t_{1dw}, which only determines how the scans are distributed to the various t_1 values. This is particularly obvious for a signal on resonance, whose Fourier-transform intensity (height):

$$S(\omega = 0) = \sum s(t_j)\, e^{i0\, t_j} = \sum s(t_j) \tag{4.69}$$

is just the sum over all time domain points (for a general off-resonance signal see the corresponding equation (4.74)). Since the noise in each scan is independent, it is also independent of the mode of addition produced by the various possible dwell-time arrangements.

It is interesting to note that in multiple-pulse NMR with detection in short windows between pulses, due to violations of its preconditions the dwell-time theorem does not apply: the S/N depends on the dwell time t_{dw} which is the spacing between these detection windows. In fact, for detecting the signal during the short windows of length t_w the filter width ω_{fw} must be chosen so large that its inverse, the response time of the detection system, is sufficiently short ($2\pi/\omega_{fw} < t_w$). The Nyquist range ω_{sw} is thus much smaller than the filter width ω_{fw} here:

$$\omega_{sw} = 2\pi/t_{dw} < 2\pi/t_w < \omega_{fw}. \tag{4.70}$$

This means that the preconditions for the dwell-time theorem are violated. Viewed in the frequency range, the noise from outside the Nyquist range but inside the filter width is aliased into the spectrum and deteriorates the S/N by a factor of approximately $t_{dw}\omega_{fw}/2\pi$.

4.9.2 Signal increase by line narrowing

Assuming a very long T_2 relaxation time (very small homogeneous linewidth), the height of an inhomogeneously broadened spectrum rises by the factor $1/w$ when the width is scaled down by w ($w \leqslant 1$) because the integral intensity is constant. Naively, from this $1/w$ increase of the signal height one might expect an enhancement of the S/N also by $1/w$. However,

$$\text{the actual improvement is only } \sqrt{1/w}. \tag{4.71}$$

It has to be taken into account that the time signal must be measured longer by a factor $1/w$ for obtaining the same relative resolution. With the spectral width unchanged, this leads to an increase of the noise per frequency interval by $\sqrt{1/w}$. Together, $(1/w)/\sqrt{1/w}$ yield the factor of $\sqrt{1/w}$. Although not attaining the full enhancement by $1/w$ expected naively, this factor enters significantly into the comparison of the S/N

ratios expected for spectra with large differences in widths, such as for lines anisotropically broadened by ^{13}C chemical shift or ^{2}H quadrupolar interaction.

If the desired resolution approaches the homogeneous linewidth, the scaling arguments cannot be applied any more and eventually the narrowed spectrum becomes useless because of insufficient resolution. (Alternative derivation of the $\sqrt{1/w}$ gain in S/N: in order to obtain spectra appearing identical, the time signal corresponding to the narrowed spectrum must be an expanded reproduction of the original one. In particular, the dwell time is increased by $1/w$. To match it, the filter width can be reduced by the factor w. Consequently, the noise is reduced by $\sqrt{1/w}$ in the expanded time signal and in the corresponding spectrum.) The factor $\sqrt{1/w}$ is also relevant for the $\vec{\mathbf{B}}_0$-field dependence of the S/N of static spectra broadened by the $\vec{\mathbf{B}}_0$-dependent chemical-shift anisotropy.

Practical examples of line narrowing which reduces, but does not eliminate, the inhomogeneous anisotropic broadening are provided by molecular motional averaging, for instance, due to methyl group rotation, which scales the ^{2}H quadrupolar pattern by a factor of $-\frac{1}{3}$, or by off-magic-angle spinning (see Section 2.4.5).

4.9.3 Damping, resolution, and measuring time

The relationship of the signal-to-noise ratio (or the measuring time) and the maximum acquisition time t_{aq} is of some practical interest. The acquisition time is inversely proportional to the resolution that can be achieved in the spectrum, and often a certain spectral resolution is to be attained. For this potential resolution, a certain price must be paid in terms of the signal-to-noise ratio. Similar to this effect of the maximum acquisition time is the effect of time-domain damping functions (Gaussian or exponential multiplication) on the S/N.

We consider a single harmonic signal:

$$s(t) = a(t)\, e^{i\tilde{\omega} t} \tag{4.72}$$

with some slowly decaying function $a(t)$ which is generated by T_2 relaxation or by multiplication with a damping function. According to basic FT theory, the spectrum is:

$$S(\omega) = \int_0^{t_{aq}} s(t)\, e^{-i\omega t}\, dt \tag{4.73}$$

and consequently the signal height at the resonance position $\omega = \tilde{\omega}$ is:

$$S(\tilde{\omega}) = \int_0^{t_{aq}} a(t)\, e^{i\tilde{\omega} t}\, e^{-i\tilde{\omega} t}\, dt = \int_0^{t_{aq}} a(t)\, dt \tag{4.74}$$

the integral over the weighting function $a(t)$. In the discrete FT, the exact frequency $\tilde{\omega}$ may fall between frequency points but this represents no problems when the linewidths are larger than the spacing of frequency points.

With the simple formula (4.74) for the signal height, we can consider the effect of natural damping of the signal by T_2 relaxation (or residual dipolar couplings), with a weighting function $a(t)$ which is assumed to be exponential with the time constant T_2:

$$S(\tilde{\omega}) \sim \int_0^{t_{aq}} e^{-t/T_2}\, dt = T_2(1 - e^{-t_{aq}/T_2}) \xrightarrow{t_{aq} \gg T_2} T_2. \tag{4.75}$$

Considering now the noise, it is clear that it increases when more time-domain data are acquired due to an increase of t_{aq}. With the spectral width unchanged, the noise per frequency interval grows like:

$$N(\omega) \sim \sqrt{t_{aq}}. \tag{4.76}$$

(Note that according to the dwell-time theorem, more datapoints due to a smaller dwell-time increase both the noise and the signal, so that this effect can be disregarded here.) For the signal-to-noise ratio at the spectral peak, we obtain:

$$\frac{S(\tilde{\omega})}{N(\tilde{\omega})} \sim \frac{T_2(1 - e^{-t_{aq}/T_2})}{\sqrt{t_{aq}}} \xrightarrow{t_{aq} \gg T_2} \frac{T_2}{\sqrt{t_{aq}}}. \tag{4.77}$$

This means that for excessive acquisition times, the S/N becomes worse $t_{aq}^{-\frac{1}{2}}$.

The effect of a damping function $g(t)$ multiplied onto the detected signal is different from the natural damping just discussed; such an apodization function also decreases the noise at large t values, according to:

$$\text{Noise} \sim \left\{ \int_0^{t_{aq}} \{g(t)\}^2 \, dt \right\}^{\frac{1}{2}}. \tag{4.78}$$

We consider only $g(t)$ functions with a decay constant τ_g which fulfils $\tau_g < T_2$, that is, the $g(t)$ damping dominates the signal decay. For a Gaussian $g(t) = e^{-t^2/(2\tau_g^2)}$ for $t_{aq} \gg \tau_g$ we obtain:

$$\frac{S(\tilde{\omega})}{N(\tilde{\omega})} \sim \sqrt{\tau_g}. \tag{4.79}$$

The noise-reducing effect of damping functions expresses itself in the fact that for long acquisition times $t_{aq} \gg T_2 > \tau_g$ this S/N (4.79) is more favourable than the expression (4.77) for the signal without apodization: for $t_{aq} \gg \tau_g$ the S/N is nearly independent of t_{aq} (while the S/N in equation (4.77) becomes worse with increasing t_{max} when $t_{max} \gg T_2$). This independence of t_{aq} in equation (4.79) is easily rationalized: at large t values, both signal and noise have been damped to zero so that an extension of the t range changes neither signal nor noise. Thus, for large t_{aq} the damping improves the signal-to-noise ratio. It should also be noted that contrary to common preconceptions, 'strong' damping (i.e. with short τ_g), makes the signal-to-noise ratio worse in the limit $\tau_g \ll T_2$ considered here: the S/N in equation (4.79) improves proportional to $\sqrt{\tau_g}$. This is due to the fact that the damping function suppresses the signal at $t > \tau_g$. Only when the signal decreases intrinsically such that $T_2 \leqslant \tau_g$ does the damping improve the S/N (for a fixed t_{aq}).

The improvement of the S/N by damping the points with weak signals is related to the optimization in the signal averaging of scans that have signals of equal shapes but different amplitudes, and equal noise per scan. This case is met, for instance, in multiple-contact cross polarization, where the signal amplitude decreases with increasing number of previous contacts, or in the averaging of trains of echo-refocused time signals with decaying echo amplitudes. Standard averaging by simple addition of these signals can actually make the S/N worse: consider adding the scan $s_1(t)$ with a large nuclear signal amplitude $\tilde{s}_1(t)$ and a scan $s_2(t)$ with a low nuclear signal $\tilde{s}_2(t) \ll \tilde{s}_1(t)$. In the sum $s_1(t) + s_2(t)$, the nuclear signal $\tilde{s}_1(t) + \tilde{s}_2(t)$ is not much larger than $\tilde{s}_1(t)$, while the noise is increased by $\sqrt{2}$.

By weighting each scan $s_j(t)$ in the sum with its maximum intensity $s_j(0)$ (assumed to be dominated by the nuclear signal and to occur at $t = 0$ for simplicity) according to:

$$s(t) = \sum_j s_j(t) s_j(0) \tag{4.80}$$

the optimum total S/N is obtained: large signals contribute much, weak ones bring in only little noise; since each scan $s_j(t)$ is weighted as a whole, the signal $s(t)$ is by no means distorted. This weighting prescription can be derived by searching the a_j that maximizes the signal-to-noise ratio of the weighted sum:

$$\frac{S}{N} \sim \frac{\Sigma_j a_j \tilde{s}_j(0)}{\sqrt{\Sigma_j (a_j)^2}} \tag{4.81}$$

where we have considered $s(t) = \Sigma_j a_j \tilde{s}_j(t)$ at $t = 0$ without restricting the generality of the treatment. The maximum is found via taking the derivative:

$$\frac{\partial (S/N)}{\partial a_j} \stackrel{!}{=} 0 \quad \Rightarrow \quad \sum_j (a_j)^2 \stackrel{!}{=} \sum_j a_j \tilde{s}_j(0).$$

The last formula is fulfilled by $a_j = \tilde{s}_j(0)$ (for all j). Since $\tilde{s}_j(0)$ represents the nuclear signal at its maximum, it is approximately equal to the signal including the noise, $\tilde{s}_j(0) \cong s_j(0)$ (this can always be ensured by sufficient signal averaging for each 'scan' $s_j(t)$). In summary, we have thus proved that for the optimum weighting of unequal scans $s_j(t)$, weighting factors $a_j \cong s_j(0)$ should be chosen.

It is interesting to compare the S/N of a time signal taken in one scan with the S/N obtained for indirect detection of the 1D time signal, or equivalently to determine the increase in the measuring time required. To sample the signal point by point, t_{max}/t_{dw} points must be sampled; this takes $(t_{max}/t_{dw})^{-1}$ times longer than the single-scan detection (i.e. the dwell-time theorem does not apply here). The minimum measuring-time factor is thus given by:

$$\left(\frac{S}{N_{scan}} : \frac{S}{N_{pointwise}}\right)^{-\frac{1}{2}} = (t_{max}/t_{dw}^{max})^{-1} = 1/(2\omega_{sw}^{min} t_{max}) \tag{4.82}$$

with the maximum allowable dwell time $t_{dw}^{max} = 1/(2\omega_{sw}^{min})$, where ω_{sw}^{min} is the minimum spectral width that contains all relevant signals. (Actually, if the filter width is kept at ω_{sw}^{min}, for any $t_{dw} < t_{dw}^{max}$ within a given measuring time the same S/N is attained so that the dwell-time theorem again applies.) The dependence (4.82) on ω_{sw}^{min} does not apply for indirect detection in the t_1 dimension of a 2D experiment where the dwell-time theorem was seen to hold (see Section 4.9.1). Why the signal in a 2D experiment can be much larger than for 1D indirect detection, and even larger than for 1D direct detection, will be outlined in the following section.

4.9.4 Measuring times: 1D vs. 2D spectra

It is commonly believed that the signal-to-noise ratio in a 2D spectrum is necessarily much worse than in the corresponding 1D spectrum. However, as was shown early in the history of 2D spectroscopy, this is not generally true (Aue *et al.*, 1978). As a very general argument, one should consider that in many 2D experiments essentially the signal of the whole sample is detected in each scan, so that the overall spectrum contains the signal equivalent to all scans in the 2D experiment. (Projections on one of the spectral axes (integrals over the other dimension), on the other hand, correspond to the Fourier transforms of the single $t_1 = 0$ or $t_2 = 0$ time-domain slices and therefore have a lower signal-to-noise ratio.) Under special circumstances, in 2D spectroscopy of solids the S/N in the 2D experiment can even be better than in the comparable standard 1D experiment. To show this, we shall compare measuring times, or equivalently signal-to-noise ratios, in 1D and corresponding 2D experiments. Separation 2D experiments and diagonal 2D exchange spectra will be considered in particular. In passing, we note that selective 1D experiments which yield spectra equivalent to slices out of 2D spectra are usually less favourable than 2D experiments since most of the signal is actually discarded in the signal selection, while in the 2D experiment, as already mentioned, every scan excites essentially the whole spectrum. This relation, but also certain practical advantages of the selective 1D experiments, are discussed in Section 13.2.4.

As a simple example, we consider a separation experiment where the magnetization evolves only under the isotropic shift during one period (t_1 or t_2) and under the anisotropic chemical shift in the other period. For simplicity, we consider only a single site. In one dimension, we thus obtain a sharp isotropic line; in the other, we obtain a powder pattern. The question is now whether obtaining the powder pattern with a certain S/N in this 2D experiment takes longer than in the corresponding 1D experiment. For simplicity, we assume phase modulation; as discussed in Section 4.9.5, amplitude modulation deteriorates the S/N by a factor of $\sqrt{2}$.

With MAS evolution and $t_{1\,\max} \ll T_2$, every t_2 scan for any of the t_1 values produces the same signal as would be observed in the 1D experiment. This is particularly obvious for $\tilde{\omega}_1 = \omega_{\text{iso}} = 0$ where the magnetization does not precess during t_1 so that, for all t_1 values, the '1D signal' is detected in t_1. For general $\tilde{\omega}_1$, the precession during t_1 produces only a t_1 dependent phase factor $\exp(i\tilde{\omega}_1 t_1)$ which multiplies the detected t_2 signal; at $\omega_1 = \tilde{\omega}_1$, this factor is cancelled by the FT over t_1 as shown in equation (4.74), so that at $\omega_1 = \tilde{\omega}_1$ the ω_2 powder spectrum is detected with the same S/N as in the 1D experiment.

The situation is more complex when the anisotropic evolution modulates the detected isotropic-chemical-shift signal. Considering for simplicity an $\tilde{\omega}_2 = 0$ on-resonance signal with an extremely long T_2, it is obvious that the line at $\omega_2 = \tilde{\omega}_2$ will become very intense. The powder spectrum along ω_1 results from this line's modulation, and therefore the signal-to-noise ratio of our powder spectrum rises with that of the isotropic line. To relate the S/N in this $\omega_2 = \tilde{\omega}_2$-slice powder pattern to that of the conventional 1D spectrum, we refer to the indirect-detection results of Section 4.9.3. The long T_2 during the signal detection corresponds to very narrow spectral line, so that $\omega_{\text{sw}}^{\min}$ in equation (4.82) can be extremely reduced, or equivalently $t_{\text{dw}}^{\max}$ extremely increased. There is no physical reason limiting this $t_{\text{dw}}^{\max}$ to less than the maximum acquisition time $t_{\max}$ in the 1D detection of the powder spectrum. Consequently, the signal in the signal-bearing slice of the 2D experiment can even be considerably stronger than in the conventional 1D experiment! (Note that in the 1D experiment, a similar signal gain from long T_2 times would be obtained by multi-spin-echo experiments, where the time-signal is detected many times by refocusing the magnetization by 180° pulses at $t_{\max}$ and at the time $t_{\max}$ after each echo maximum.)

The spectral lineshape along the diagonal of a 2D exchange spectrum without exchange is equivalent to the corresponding 1D spectrum. In addition, the S/N of a purely diagonal spectrum is similar to the S/N of a 1D spectrum taken with the same resolution and the same total number of scans, assuming that $T_2 \gg t_{aq}$. As a simple argument, it can be said that the stimulated echo (see Section 10.2.3) in every scan for the 2D data field is nearly as intense as the 1D time signal. T_2 relaxation (including coherent dipolar effects) broadens the diagonal proportional to $1/T_2$, thus decreasing its signal height proportional to T_2 (if the broadening is larger than the distance of spectral points $2\pi/t_{aq}$); equivalently, in the time domain T_2 relaxation is seen to reduce the signal intensity (of the stimulated echoes) for large t_1 values. Exchange peaks in a 2D spectrum are always smaller than the diagonal intensity at the same ω_1 in the corresponding spectrum without exchange. The S/N of the exchange intensity can be estimated in many cases from that of the diagonal spectrum by taking into account the number of exchanging spectral sites and the relaxation effects.

4.9.5 Signal-to-noise ratio and 2D data acquisition

In this section we shall present some considerations concerning the S/N for comparable modes of 2D data acquisition.

Transformation of phase modulation into amplitude modulation is often possible by flipping the magnetization to the z axis for a certain period of time. This leads to a loss of half of the useful signal, as one keeps only a projection $\cos(\tilde{\omega}_1 t_1)$ of the complex magnetization $\exp(i\tilde{\omega}_1 t_1)$ while the orthogonal component $\sin(\tilde{\omega}_1 t_1)$ is destroyed (Bachmann *et al.*, 1977); the intensity of the Fourier transform of $\exp(i\tilde{\omega}_1 t_1)$ at $\tilde{\omega}_1$ is twice as large as that of $\cos(\tilde{\omega}_1 t_1) = \frac{1}{2}\exp(i\tilde{\omega}_1 t_1) + \frac{1}{2}\exp(-i\omega_1 t_1)$. The noise is reduced by $1/\sqrt{2}$, so overall there is a $1/\sqrt{2}$ deterioration of the S/N.

There is a widespread belief that the 'off-resonance' solution of the phase problem with just one 2D data set (i), compared to the method with quadrature in both dimensions through measurement of cosine and sine data sets (ii), wastes a factor of $\sqrt{2}$ in the S/N. In fact, however, both approaches are equivalent in terms of S/N or measuring time. As a short argumentation it can be stated that in case (ii) there is no quadrature advantage for the ω_1 dimension because the $\cos(\tilde{\omega}_1 t_1)$ and $\sin(\tilde{\omega}_1 t_1)$ components cannot be measured simultaneously. For the explicit proof, a series of equivalent spectral situations will be considered as sketched in Figure 4.11. According to the dwell-time theorem (4.68), the dwell time and its inverse, the spectral width, can be changed without affecting the S/N at the spectral NMR intensity. Shifts of the NMR spectrum in the 2D plane also leave S/N unchanged. Thus, without negative consequences, in the experiment (ii) the detection ('carrier') frequency and the dwell time can be set to values required for version (i) where all spectral intensity must appear at $\omega_2 > 0$. Now the $\cos(\tilde{\omega}_1 t_1)$ component of (ii) is identical to the data set (i). The sine component of (ii) also produces the same spectrum in the region $\omega_1 > 0$ (marked by the thick square frames in Figure 4.11). In the time required for measuring both components of (ii), experiment (i) can be performed twice. Thus, both cases have the same S/N in the same total measuring time.

Setting the imaginary part to zero before the second Fourier transformation to solve the phase problem may appear like wasting half of the signal, as in the first paragraph of this section, and suggests that more adept data processing could gain a factor $\sqrt{2}$ in the S/N. Unfortunately, this is not true; the real and imaginary parts of the Fourier transforms of a signal zero-filled to at least twice the original number of datapoints are not independent, and the noise is not partially averaged if their signals are coadded; this fact is proved in Section 4.8.3.

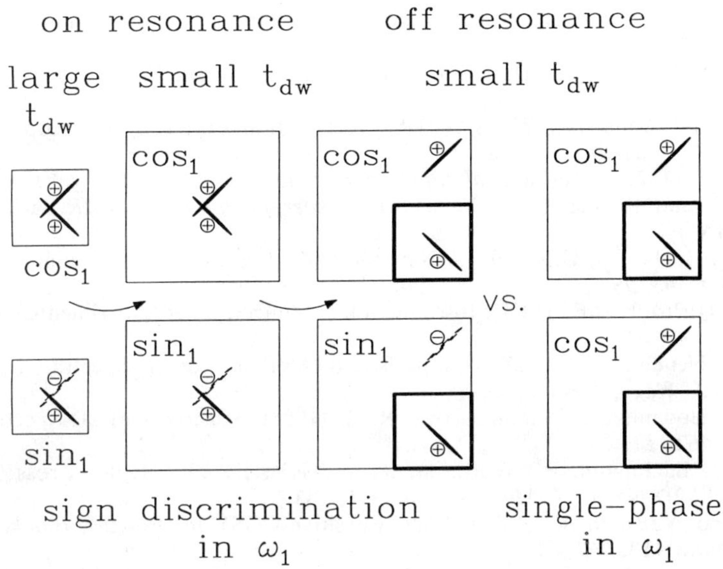

Figure 4.11 Scheme for proving the equivalence of on-resonance and off-resonance 2D measurements in terms of the measuring time (i.e. the signal-to-noise ratio). All the spectra shown have the same *S/N* ratio if they are taken with the same overall measuring time each. This fact is based on the dwell-time theorem: the spectral width is irrelevant for the *S/N*.

4.10 Concluding remarks

Since 2D spectroscopy generally involves 2D Fourier transformation, for a profound understanding of multidimensional NMR, as well as for actually setting up the experiments and the data evaluation, a good working knowledge of Fourier theory is of importance. Fourier theorems will in fact be referred to repeatedly in the following chapters. In some respect, this chapter will be continued by the discussion of practical aspects of 2D data acquisition and evaluation in Section 10.5.

4.11 References

Abragam, A. (1961). *The Principles of Nuclear Magnetism*. Oxford University Press, London.

Aue, W. P., Bachmann, P. and Ernst, R. R. (1978). Sensitivity in 2D NMR Spectroscopy. *J. Magn. Reson.*, **29**, 523.

Bachmann, P., Aue, W. P., Müller, L. and Ernst, R. R. (1977). Phase Separation in 2D Spectroscopy. *J. Magn. Reson.*, **28**, 29.

Bartholdi, E. and Ernst, R. R. (1973). Fourier Spectroscopy and the Causality Principle. *J. Magn. Reson.*, **11**, 9.

Bax, A. (1982). *Two-Dimensional Nuclear Magnetic Resonance in Liquids*. Delft University Press, Delft.

Bax, A. and Mareci, T. H. (1983). Practical Aspects of ^{13}C Double-Quantum NMR. *J. Magn. Reson.*, **53**, 360.

Bracewell, R. N. (1986). *The Fourier Transform and Its Applications, 2nd rev. ed.* McGraw-Hill, New York.

Cooley, J. W. and Tuckey, J. W. (1965). Machine Calculation of Complex Fourier Series. *Math. Comput.*, **19**, 297.

Ernst, R. R. (1969). Numerical Hilbert Transform and Automatic Phase Correction. *J. Magn. Reson.*, **1**, 7.

Ernst, R. R. and Anderson, W. A. (1966). Application of FT Spectroscopy to Magnetic Resonance. *Rev. Sci. Instrum.*, **37**, 93.

Fourier, J. B. J. (1822). *Theorie analytique de la chaleur*. Firmin Didot, père et fils, Paris.

Fukushima, E. and Roeder, S. B. W. (1981). *Experimental Pulse NMR*. Addison-Wesley, Reading, MA.

Heuer, A. and Haeberlen, U. (1989). A New Method for Suppressing Baseline Distortions. *J. Magn. Reson.*, **85**, 79.

Higgins, R. J. (1976). Fast FT (FFT): Introduction & Minicomputer Experiments. *Am. J. Phys.*, **44**, 766.

Keeler, J. and Neuhaus, D. (1985). Comparison of Methods for 2D Spectra with Absorption Lineshapes. *J. Magn, Reson.*, **63**, 454.

Levitt, M. H., Bodenhausen, G. and Ernst, R. R. (1984). Sensitivity of 2D Spectra. *J. Magn. Reson.*, **58**, 462.

Nagayama, K., Bachmann, P., Wüthrich, K. and Ernst, R. R. (1978). Cross Sections and Projections in 2D NMR. *J. Magn. Reson.*, **31**, 133.

Marion, D. and Wüthrich, K. (1983). Phase-Sensitive COSY in Proteins. *Biochem. Biophys. Res. Commun.*, **113**, 967.

Otting, G., Widmer, H., Wagner, G. and Wüthrich, K. (1986). t_1 and t_2 Ridges in 2D Spectra and Procedures for Suppression. *J. Magn. Reson.*, **66**, 187.

Ramirez, R. W. (1985). *The FFT Fundamentals and Concepts*. Prentice-Hall, Englewood Cliffs, NJ.

Redfield, A. G. and Kunz, S. D. (1975). Quadrature FT NMR: Multiplex for Dual Detection. *J. Magn. Reson.*, **19**, 250.

Schaefer, J. and Stejskal, E. O. (1974). Baseline Artifacts in FT-NMR Spectra. *J. Magn. Reson.*, **15**, 173.

Schmidt, C., Blümich, B. and Spiess, H. W. (1988). ^{2}H 2D NMR in Solids. *J. Magn. Reson.*, **79**, 269.

States, D. J., Haberkorn, R. A. and Ruben, D. J. (1982). 2D NOE Experiments with Pure Absorption in 4 Quadrants. *J. Magn. Reson.*, **48**, 286.

Stejskal, E. O. and Schaefer, J. (1974). Comparison of Quadrature and Single-Phase FT NMR. *J. Magn. Reson.*, **14**, 160.

Turner, D. L. (1982). ^{13}C NMR Using Double-Quantum Coherence. *J. Magn. Reson.*, **49**, 175.

Turner, D. L. (1984). Sensitivity of 2D NMR Spectroscopy. *J. Magn. Reson.*, **58**, 500.

CHAPTER FIVE

Short Review of Structure and Dynamics of Solid Polymers

5.1 Motivation

The intention of this short review of polymer structure and dynamics is to provide the background required for understanding the applications of multidimensional NMR to polymers, in particular those of Chapters 7, 12 and 13. It is addressed particularly to readers who are not polymer experts. The treatment is by no means complete and the selection of topics is certainly biased. More detailed introductions to polymers are found, for instance, in the books by Mandelkern (1983), and Young (1983). The chapter is concluded by a defining list of polymer acronyms used in this book together with very short notes on the properties of the materials and some typical applications.

5.2 Chain and phase structure of semicrystalline polymers

A large number of polymers are semicrystalline with comparable amounts of crystalline and amorphous regions. The crystallites yield relatively sharp reflections in X-ray scattering (which are nonetheless much broader than those observed for low-molar mass crystals); the amorphous regions are largely disordered, producing only a broad halo in the scattering pattern. The crystallites usually form lamellae which are only 4–40 nm thick along the c axes of the lattice but often extend over hundreds of nanometers in the other directions; this is derived mainly from electron micrographs. A structure to be considered typical is displayed in Figure 5.1.

As shown in Figure 5.2, the chain structures in the crystallites are often helical (Natta, 1960; Takahashi and Tadokoro, 1973). They are commonly classified as m_n helices with m denoting the number of structural repeat units ('monomer units') per repeat structure ('period') of the helix, and n the corresponding number of turns around the helix axis. In the 9_5 helix of POM, for instance, the kth and $(k+9)$th OCH_2 units have the same orientation but are separated by five turns of the helix, that is, a displacement by 5 helix pitches along the helix axis. There are 'typical' or 'simple' helices whose projections on a plane perpendicular to the chain axis have m-fold rotation symmetry around their centres. In these simple helices, a translation of the chain by one repeat unit, combined with a rotation of the chain by $360^\circ\ n/m$, reproduces the structure (if we disregard the ends of the chain). Other m_n helices have a structure not immediately recognized as helical. For instance, even the all-trans chain of polyethylene can be classified as a 2_1 helix (note, however, that not every 2_1 helix has an all-trans structure). Beyond that, structures repeating every m monomer units

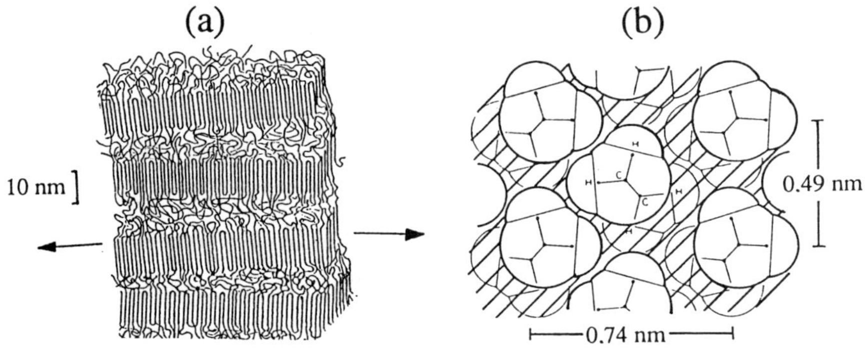

Figure 5.1 (a) Typical lamellar structures of a semicrystalline polymer. Thin crystalline lamellae alternate with amorphous regions. Note that the chain axes are approximately parallel to the short dimension of the lamellae. (b) View of polyethylene chains in a crystallite along the *c* (chain) axis. Note the considerable amount of space outside the van-der-Waals radii (hatched at the height of the upper CH_2 groups), which allows for large-angle molecular motions at temperatures where shorter *n*-alkanes rotate fast or are already molten.

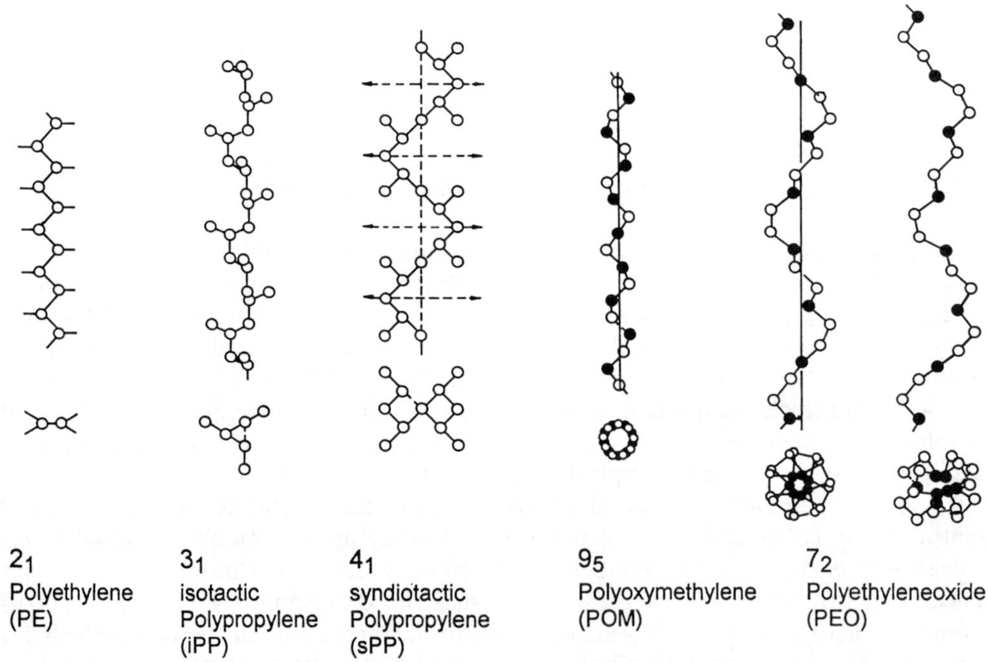

Figure 5.2 Helical chain structures of simple polymers (in their crystallites), as determined by X-ray diffraction.

and somehow winding n times around the centre of gravity of the projection also fulfil the basic requirements of an m_n helix.

The c axis in the lattice is usually chosen along the chain axis. This axis is often approximately perpendicular to the lamellar surfaces. Thus, consecutively linked sections of a chain are located in the amorphous and crystalline regions (see Figure 5.1(a)).

A single chain typically extends over several lamellae. There are indications that the chains have an increased probability of folding back in the amorphous regions with a re-entry into the crystallite relatively close to the exit point. This represents some compromise between models of regular, 'adjacent re-entry', possibly 'tight re-entry', and of 'random re-entry' (Peterlin, 1980). The latter is closely related to the 'switch-board model' (the chains are assumed to be entangled like cables on a switch-board) where 'tie molecules' between different lamellae are also considered. There are certainly differences in the re-entry behaviour of solvent-cast materials, where the chains can order more easily, and of melt-cast specimens, where the solidification more or less 'on the spot' can be expected to result in a quite random behaviour. Beyond the two-phase model presented here, in PE and other polymers there are indications of a substantial interfacial layer or 'interphase' with noncrystalline character but reduced mobility. In fact, such a transition layer must result in order to avoid an increase of the density by close packing of chain folds at the interface between amorphous and crystalline regions (Flory, 1953).

In melt-crystallized samples, the lamellae are part of nearly spherical (polyhedral) superstructures called 'spherulites', which have diameters ranging from micrometres to millimetres (Mandelkern, 1983). Being larger than the wavelength of visible light, the spherulites are responsible for much of the light scattering and the white appearance of most semicrystalline polymers (amorphous homopolymers, which lack such superstructures, are usually transparent). The centres of the spherulites are the nucleation centres of the material. The lamellae grow radially away from the centre, and the chains are oriented tangentially. Certain structural details in this mesoscopic range are, however, not yet fully understood. For instance, in the periphery of the spherulite there must be many more lamellae than can possibly grow outwards in the centre. How these additional crystalline lamellae are induced to fit into the radial structure of the spherulite is not completely clear.

Upon heating, 'premelting' occurs; that is, the crystallinity gradually decreases, in polyethylene polypropylene and other simple polymers. At the same time, however, X-ray data show nearly constant (Tanabe *et al.*, 1986) or even increasing (Strobl *et al.*, 1980) average crystallite thicknesses (of course, the amorphous regions grow more strongly with temperature). Explanations of these observations, which are relevant in the context of chain-diffusion NMR measurements in Section 7.14.2, range from selective melting of small crystallites to various grain-boundary melting mechanisms.

Scattering techniques, in particular X-ray methods, have focused the physical description of solid polymers on the crystalline parts. The amorphous parts were often treated only as filling material, tacitly assumed as rigid and without structure, possibly apart from mobile chain ends. The notion of a dynamic amorphous phase in many polymers, with effects on macroscopic material properties, is often not appreciated. As discussed in more detail below, NMR even shows that in some polymers material exchange occurs between the mobile amorphous and the crystalline regions; this means that the material is not rigidly phase separated.

Even though via spin diffusion, NMR can provide information on structures measuring up to 200 nm (and is thus comparable to small-angle X-ray scattering), the focus of NMR spectroscopy is usually on individual types of segments. The unique power of NMR in differentiating between chemical structures often leads to a less generalistic approach to polymer structure. The diversity between classes of structural elements is thus more promptly realized, as is the importance of certain chemical

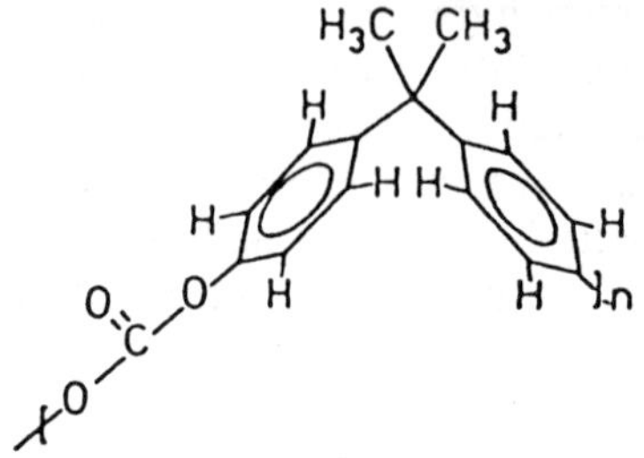

Figure 5.3 Structure of the repeat unit bisphenol-A polycarbonate. The kink that it introduces into a polymer chain and the chain dynamics related to the phenylene-ring flips appear to be related to the favourable mechanical properties of bisphenol-A polycarbonate.

building blocks such as structurally rigid kinks that prevent the formation of linear or planar structures or even of flexible coils. A particular example for such a building block is the bisphenol-A unit (Figure 5.3), which in addition to its structural rigidity exhibits internal dynamics with its flipping phenylene rings. As another example, alkyl sidechains are found by NMR to exhibit considerable dynamics, though with various amplitudes and rates. Conformational statistics, packing, and dynamics in fully amorphous polymers are further fields where NMR plays an important role in clarifying conceptions.

5.3 Dynamical processes in polymers

Molecular dynamics represents an important feature in the microscopic behaviour of solid polymers, expanding the traditional, static picture of their microstructures: a typical methyl group at room temperature jumps more than 10^9 times per second around its three-fold axis; phenylene rings can flip by 180° in glassy polymers (more than 10^6 times per second at room temperature in polycarbonate, cf. Sections 6.6.1 and 7.13). Even such a 'local' motion has to be related to the polymer structure: it must somehow be made possible by the arrangements of chains around the phenylene ring. Even in the relatively well-organized crystalline lamellae of various polymers, large-amplitude motions can occur (Boyd, 1985). For example, long chain segments have been shown to undergo helical jumps involving rotations in steps of more than 100° (Section 7.12). The passage of the defect that causes such a rotation also generates a translation of the chain by at least one monomer unit. Many such displacements result in chain diffusion or chain transport (Sections 7.12 and 7.14.2). This dynamic process again entails a microstructural feature: the chain translation through the crystallites must be accommodated by the chain portions in the amorphous regions, so the lengths of the chain loops fluctuate in time and space. Quite generally, it should be kept in mind that above their glass-transition temperature, the chains in amorphous parts or in soft segments of composite polymers do not occupy fixed positions in space but rather can exhibit considerable displacement amplitudes.

Molecular dynamics also influences macroscopic material properties like brittleness or impact strength, mechanical moduli (Ward, 1983), creep, solid-state extrudability (Aharoni and Sibilia, 1979), absorption of energy from alternating electric fields, etc.

This is known from dynamic-mechanical and dielectric experiments (McCrum *et al.*, 1967), where a change in the dynamic moduli or compliances is related to molecular motion with rates in the range of the frequency of the alternating mechanical or electric fields. In many polymeric materials, motional processes determine the temperature range of use: on cooling down, brittleness sets in when energy-dissipating local molecular motions 'freeze'; at higher temperatures, large-amplitude dynamics, usually involving chain translation, can lead to a drastic decrease of the strength of the material. The requirements for a hard and tough polymeric material can be summarized as follows: *restricted motion is good, too much motion is bad.*

Dielectric and dynamic-mechanical relaxation studies have established that typical polymers exhibit a limited number of relaxation processes of appreciable amplitude with correlation times exceeding the picosecond range. These exhibit relatively well-defined temperature dependencies of the correlation times. Those relaxation processes are traditionally labelled α, β, γ, etc., with α appearing highest in temperature and lowest in frequency. This notation does not take into account the molecular nature of those processes. Actually, before the advent of NMR spectroscopic techniques, the molecular basis of many of these relaxations was rather speculative (McCrum *et al.*, 1967). The α process in amorphous polymers is directly related to the glass transition (e.g. in PMMA), while in semicrystalline polymers it was usually assigned to motions in the crystallites or in the chain folds of the interface. Exchange NMR has proven that in many polymers (e.g. in POM or iPP, see Section 7.12) it is actually due to helical jumps of the chain stems in the crystallites. The β relaxation can be strong and due to the glass transition in the amorphous regions (as in iPP), or weak and not yet fully analysed (as in POM and PE), or related to side-group motion (as in PMMA). Even the γ relaxation is sometimes based on the glass transition (e.g. in POM and PE) as was demonstrated by NMR after considerable controversy in the literature.

The coupling or merging of motional mechanisms is a question not yet investigated in molecular detail. However, these issues are starting to be addressed. For example, if the α-process dynamics based on helical jumps in the crystallites (cf. Sections 7.9 and 7.12) are to occur, the amorphous regions must be above their T_g (β or γ relaxation) in order to permit the necessary translations and rearrangements of chain segments.

5.4 Solid biopolymers

The term 'biopolymers' can encompass quite an array of different types of macromolecules. Examples include materials such as proteins, polysaccharides, nucleic acids, and other macromolecules produced by living organisms. Many of these biopolymers are of great scientific and practical interest. Some of these materials, such as the biodegradable poly(hydroxy butyrate) produced by certain bacteria, are quite similar to the synthetic polymeric materials discussed above. They are semicrystalline or amorphous and can be treated like synthetic semicrystalline nonaromatic polymers. Cellulose (e.g. in cotton), a polysaccharide, also exhibits a relatively simple secondary structure and is semicrystalline in the bulk. Other important fibrous biopolymers with solid-state properties similar to those of synthetic polymeric materials are silk and wool, both made up of fibrillar proteins with special superstructures. On the other hand, polymeric nucleic acids like DNA are relevant for biological information storage and replication while their solid-state materials properties are not so important.

Due to their unique biological importance, the investigation of structural details of their double-helix structure is of great relevance. For many proteins, in particular those with globular tertiary structures, the behaviour of single chains, possibly in an aqueous surrounding, is of interest. Some of these can be oriented in magnetic fields and thus acquire anisotropic properties, which can be best studied by solid-state NMR. More detailed accounts on the structure and properties of biopolymers can be found in the literature (Mark *et al.*, 1985).

5.5 Heterogeneous polymer systems

Many important polymeric materials are actually not homogeneous, but rather heterogeneous in terms of their chemical and/or physical structure. This applies for polymer blends or alloys, block copolymers, semicrystalline polymers, and polymers mixed with filler materials. The combination of components can result in a combination of properties that are nearly mutually exclusive in homopolymers. For instance, hardness and toughness are combined in the ABS (acrylonitrile–butadiene–styrene) systems, where the butadiene contributes the molecular mobility required for impact-energy dissipation, and the styrene-acrylonitrile (SAN) matrix provides the hardness. Similarly, thermoplastic synthetic 'rubbers' are formed by using thermoplastic hard-segment aggregates as physical cross links for rubbery soft segments, for instance in polyurethanes or styrene-butadiene-styrene (SBS) block copolymers. It is of great interest to elucidate the morphology of those multicomponent materials in terms of domain sizes, the degree of mixing in homogeneous blend systems, and the structure at the interface or extended interphase between component phases.

Polymer blends are often heterogeneous since polymers are usually immiscible: the entropy of mixing decreases as the number of molecules in a given volume becomes smaller, or equivalently, as the molecular weight increases. The Gibbs free energy of mixing per volume in the Flory–Huggins lattice theory (Flory, 1953) is given as

$$\Delta G_m/V = RT\{n_1 \ln(v_1) + n_2 \ln(v_2) + \chi_{12} v_1 v_2\} \tag{5.1}$$

where n_j represents the number of moles of macromolecules j per volume, and v_j is the volume fraction of species j. The last two terms result from the entropy of mixing, while the last term with the χ_{12} parameter characterizes the energetic effect. The beauty of the theory consists in the fact that result (5.1) does not contain any lattice variables, although the derivation is based on a liquid-lattice theory. Within this relatively simple framework, thermodynamic miscibility of blends indicates a negative χ_{12} or equivalently the fact that the interactions in the mixed system are more favourable (or less unfavourable!) than those within the pure components themselves. Although this condition should be expected to be fulfilled only rarely, a considerable number of compatible systems have been found, including PS/PXE, PS/PVME, PMMA/PVF_2, PMMA/PC, PEI/polyetherketones, and PVP/PBI (to decode the acronyms, see Table 5.1). This miscibility is often attributed to favourable 'specific' interactions between the components. However, simple geometric effects, such as unfavourable packing in one of the pure components that causes those molecules to prefer the packing in the mixture, also have to be considered. The elucidation of these factors, which are often not directly apparent from the chemical structures, is still at an early stage. Modern solid-state NMR methods promise to play a major role in this field.

Two polymers can be thermodynamically miscible at one temperature and composition, and immiscible at another. Figure 5.4 shows a schematic phase diagram, with both an upper and a lower critical solution temperature (UCST and LCST, respectively) (Flory, 1953). (Note that the *lower* critical solution temperature corresponds to an *upper* gap in the miscibility, and vice versa.) The spinodal, which is the limit between the instable and the metastable two-phase region, is indicated in the figure. In polymers, generally a LCST is found. An UCST (lower gap of miscibility) often cannot be observed because phase separation cannot occur for a homogeneous blend below the T_g of the system, since the structure becomes rigid at that temperature. Generally, the study of the phase diagram must either occur in the melt or on samples quenched at the temperature of interest. By temperature jumps in the opposite direction, the kinetics of phase separation, or spinodal decomposition, can be investigated, being of general interest in the study of mixtures.

Block copolymers of incompatible or partly compatible amorphous components are interesting self-organizing systems. They form microphase-separated structures that minimize the Gibbs free energy by demixing, but under the constraints due to the bonding between the components; this can be rationalized at least partly in terms of a minimization of the surface tensions at the internal surfaces. Finite domains result which are typically lamellar or cylindrical (Woodward, 1989), depending on the ratio of the block lengths as indicated in Figure 5.5(a). The arrangement of these domains is often regular. Even cubic or ordered bicontinuous double-diamond (OBDD) (Herman *et al.*, 1987) superstructures (Figure 5.5(c)) have been observed for noncrystalline polymer components. For the scaling of the domain sizes with the molecular weight, exponents between 0.5 and 0.8 are found experimentally and theoretically.

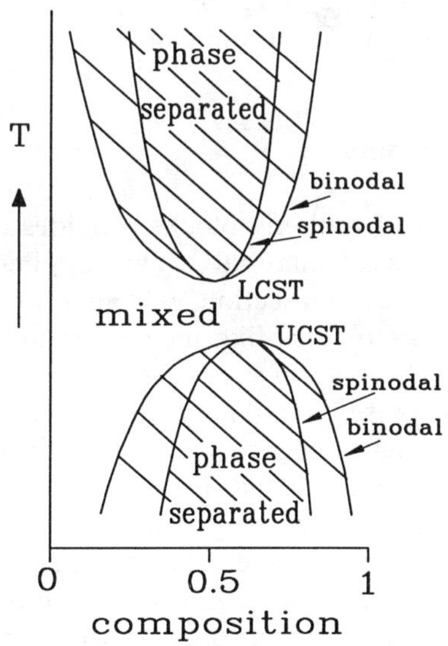

Figure 5.4 Schematic phase diagram of a miscible polymer blend with upper (UCST) and lower (LCST) critical solution temperatures.

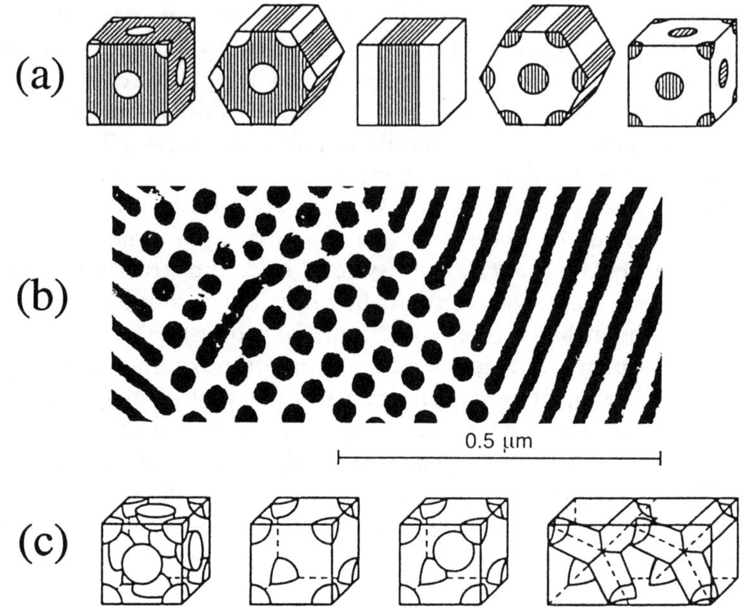

Figure 5.5 (a) Typical phase structures in diblock copolymers as a function of the relative composition. (b) Electron micrograph of an annealed PS-polybutadiene block copolymer ($M_w = 150\,000$ g/mol) (adapted from Hashimoto *et al.*, 1983). Hexagonally packed cylindrical domains are seen. (c) Schematic representation of cubic and ordered bicontinuous double-diamond (OBDD) superstructures.

In practice, it should also be kept in mind that kinetic effects can result in incomplete separation of slightly incompatible systems. Even more trivially, incomplete mixing of miscible blends can arise by insufficient annealing.

5.6 Liquid-crystalline polymers

Supramolecular organization of building blocks provides an important route for the development of novel polymeric materials. This applies not only for the block copolymers discussed in the previous section, but also for liquid-crystalline polymers (LCPs). They contain rod-like or disk-like mesogenic units which are able to form liquid-crystalline phases. The mesogens can be linked together directly (rigid-rod systems) or via flexible spacers (semiflexible systems). Moreover, the mesogens can be incorporated in the main chains or be attached as side groups (Fig. 5.6). The interest in these materials comes from the fact that in the liquid-crystalline phase they can be macroscopically oriented by electric or magnetic fields or by mechanical forces. The order thus generated can be frozen in the solid state. Applications range from high-tensile-strength fibres to materials with nonlinear optical properties to be exploited in optoelectronics. In order to relate their properties to molecular structure, the degree of alignment as well as the mobility of the different structural units has to be determined. This calls for highly selective analytical methods like NMR. In fact, solid-state

Table 5.1 *Polymer acronyms used in the text*

PC	bisphenol-A 'polycarbonate' $[-O-(CO)-O-C_6H_4-C(CH_3)_2-C_6H_4-]_n$	glassy; tough, material for compact discs
PDMS	poly(dimethyl siloxane) $[-O-Si(CH_3)_2-]_n$	viscous
PE	poly(ethylene) $[-CH_2-]_n$	semicrystalline; widespread: packaging, housewares, electric insulations
LDPE	low density PE (high-pressure polym.)	
HDPE	high density PE (low-pressure polym.)	
UHMWPE	ultra-high molecular weight PE	films, fibres, artificial joints
PEO (PEOx)	poly(ethylene oxide) $[-O-CH_2-CH_2-]_n$	semicrystalline, waxy
PET	poly(ethylene terephthalate) $[-(OC)-C_6H_4-(CO)-O-CH_2-CH_2-O-]_n$	semicrystalline, fibres, films, substrate of magnetic tapes, etc.
PI	poly(isoprene) $[-CH{=}C(CH_3)-CH_2-]_n$	rubbery
PIB	poly(isobutylene) $[-CH_2-C(CH_3)_2-]_n$	rubbery; sealing
PMMA	poly(methyl methacrylate) $[-CH_2-C(CH_3)(CO)OCH_3-]_n$	glassy
PnBMA	poly(*n*-butyl methacrylate) $[-CH_2-C(CH_3)(CO)(OCH_2CH_2CH_2CH_3)-]_n$	amorphous
POM	poly(oxymethylene) $[-O-CH_2]_n$	semicrystalline; hard mechanical parts
PP	poly(propylene) $[-CH_2-CH(CH_3)-]_n$	
aPP (atactic)		rubbery/viscous
iPP (isotactic)		semicrystalline; food and medical containers
sPP (syndiotactic)		semicrystalline
PPTA	poly(paraphenyleneterephthalamide) $[-C_6H_4-NH-(CO)-]_n$	frozen-in liquid-crystalline structure; strong, heat-resistant fibres
PS	poly(styrene) $[-CH_2-CH(C_6H_5)-]_n$	glassy; widespread: containers, in many composites
PTFE	poly(tetrafluoroethylene) $[-CF_2-]_n$	semicrystalline; relatively soft, sealing
PVAc	poly(vinyl acetate) $[-CH_2-CH(O(OC)CH_3)-]_n$	glassy; glue, in latex paints
PVF_2	poly(vinylidene fluoride) $[-CH_2-CF_2-]_n$	semicrystalline; piezoelectric
PVME	poly(vinyl methylether) $[-CH_2-CH(OCH_3)-]_n$	rubbery
PXE (PPE, PPO)	poly(phenylene oxide) $[-O-C_6H_2(CH_3)_2-]_n$	glassy; homogeneous blend with PS

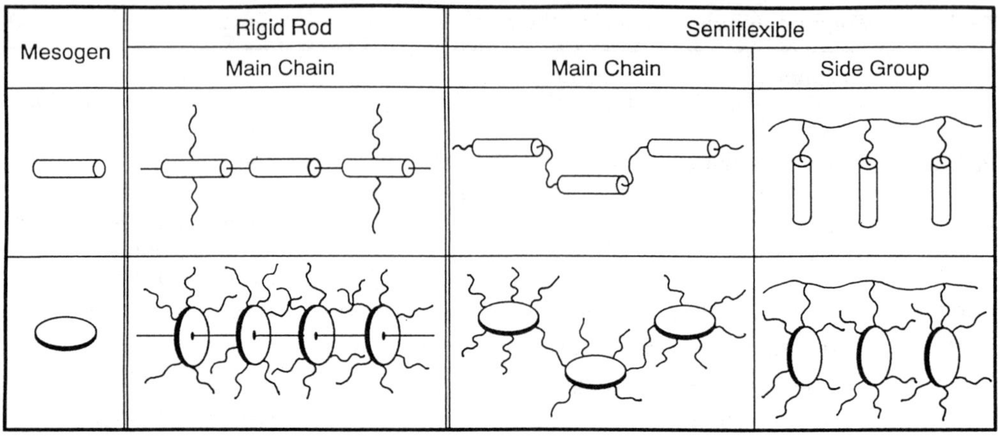

Figure 5.6 Structures of liquid-crystalline polymers (schematic) with calamitic (rod-shaped) or discotic (disc-shaped) mesogens, forming either the main chain or the sidegroups of the polymer. If the mesogenic units are separated by sufficiently long spacers, the structures will become semiflexible.

NMR, including two-dimensional techniques, has already provided a wealth of information about structures and dynamics of LCPs.

5.7 List of polymer acronyms

In Table 5.1, the acronyms of polymers referred to in this text are compiled, together with a very short characterization of their structure and applications (for a more detailed overview, see for instance, Carraher and Seymour (1988)). Concerning the nomenclature, it is interesting to note that several polymers are actually more widely known in terms of certain trademarks; for instance, PTFE is usually referred to as 'Teflon', PMMA is generally known as 'Plexiglass', and PPTA fibres as 'Kevlar'.

5.8 References

Aharoni, S. M. and Sibilia, J. B. (1979). Solid-State Extrudability of Crystalline Polymers. *Polymer Engineering and Science*, **19**, 450.

Boyd, R. H. (1985). Relaxations in crystalline polymers: Molecular interpretation. *Polymer*, **26**, 1123.

Carraher, C. E. and Seymour, R. B. (1988). Polymer Structure—Organic Aspects (Definitions). *J. Chem. Ed.* **65**, 314.

Flory, P. J. (1953). *Principles of Polymer Chemistry*, Chapter XII. Cornell University Press, Ithaca.

Hashimoto, T., Tsukahara, Y., Tachi, K. and Kawai, H. (1983). Structure of Tapered Block Polymers. 4. *Macromolecules*, **16**, 648.

Herman, D. S., Kinning, D. J., Thomas, E. L. and Fetters, L. J. (1987). Morphology of 18-Armed P(S-I) Star Block Copolymers. *Macromolecules*, **20**, 2940.

Mandelkern, L. (1983). *An Introduction to Macromolecules*, 2nd ed. Springer, New York.

Mark, H. F., Bikales, N. M., Overberger, C. G. and Menges, G. (1985). *Encyclopedia of Polymer Science and Engineering*, 2nd ed. Wiley, New York.

McCrum, N. G., Read, B. E. and Williams, G. (1967). *Anelastic and Dielectric Effects in Solids*. Wiley, New York.

Natta, G. (1960). Progress in Stereospecific Polymerization. *Makromolekulare Chem.*, **35**, 94.

Peterlin, A. (1980). Review on Chain Folding in Lamellar Crystals. *Macromolecules*, **13**, 777.

Strobl, G. R., Schneider, M. J. and Voigt-Martin, I. G. (1980). Partial Crystallization and Melting: SAXS and Electron Microscopy on LDPE. *J. Polym. Sci., Phys. Ed.*, **18**, 1361.

Takahashi, Y. and Tadokoro, H. (1973). Structural Studies of Polyethers. X. PEO. *Macromolecules*, **6**, 672.

Tanabe, Y., Strobl, G. R. and Fischer, E. W. (1986). Surface melting in melt-crystallized linear polyethylene. *Polymer*, **27**, 1147.

Ward, I. M. (1983). *Mechanical Properties of Solid Polymers*. Wiley, Chichester.

Woodward, A. E. (1989). *Atlas of Polymer Morphology*. Hanser Verlag, München.

Young, R. J. (1983). *Introduction to Polymers*. Chapman and Hall, New York.

CHAPTER SIX

Multidimensional Separation and Correlation

Keywords: *preparation/evolution/mixing/detection; separation vs. correlation; 'iso–aniso' separation of powder and MAS patterns, PASS, VACSY; SLF, 'dip-shift' separation; bond lengths, phenylene-ring flips; WISE, heteronuclear, homonuclear correlation; 'dip-shift' correlation, 2D ridge patterns, tensor orientations, SASS.*

6.1 Introduction

In this chapter, a variety of multidimensional techniques for separation and correlation of spectral patterns in solid-state NMR are presented. Based on the concepts and techniques introduced in Chapters 2 and 3, the multidimensional experiments can be described and analysed quite efficiently. Actually, they enhance the possibilities of coupling and decoupling described in those previous chapters even further: for instance, the anisotropies in one dimension can be removed while they are retained (or reintroduced) in the other. Thus, we can 'eat the cake and have it, too'!

Certain 2D experiments, such as the heteronuclear correlation of chemical shifts, are rather similar to those used in solution NMR in terms of the type of the resulting spectra (though not with respect to the resolution attained and the pulse sequences, due to the dipolar decoupling required in solids). Other multidimensional experiments are specific for solid-state NMR, in particular those which involve inhomogeneous anisotropic line-broadenings. Several features typical of these solid-state experiments will be discussed in more detail, such as the appearance of characteristic exchange patterns, or the problems encountered in transferring 1D MAS techniques into higher dimensions.

The chapter starts with a short introduction to the principles of 2D spectroscopy and the fundamental notions concerning two-dimensional NMR time signals and their transformation into 2D spectra. After a short definition of separation and correlation spectra in solid-state NMR, various techniques for 2D and 3D separation and correlation in solids are described and illustrated by applications, if available, to polymers. The chapter is concluded by a comparison of the definitions of separation/correlation spectra in solid-state versus solution NMR.

6.2 General aspects of 2D spectra and time signals

In most solid-state NMR experiments, a 2D spectrum with its pairs of frequency coordinates (ω_1, ω_2) correlates the frequencies ω_1 and ω_2 for individual isochromats or molecular segments. The spectral intensity $S(\omega_1, \omega_2)$ is the probability (more precisely, the joint probability density, see Section 9.3) of finding a frequency ω_1 in a first, and a frequency ω_2 in a later, period of the 2D experiment. Equivalently, a one-dimensional slice (cross-section) through the 2D datafield at a fixed $\omega_1 = \tilde{\omega}_1 = \text{const.}$ corresponds to the 1D spectrum $S(\omega_2; \tilde{\omega}_1)$ that is obtained in a one-dimensional experiment where only the magnetization with $\omega_1 = \tilde{\omega}_1$ is selected by means of some selective excitation technique. More specific interpretations of the spectral intensity will be given for each multidimensional experiment described in the following. Due to the great variety and diversity of 2D experiments developed already, even the general correlation description given here fails for certain 2D techniques, such as the VACSY, the wideline-separation, or the sync-MAS 2D experiments.

The actual measurement of 2D data is performed in the time domain; the 2D spectrum is only afterwards generated by double Fourier transformation and a few other procedures as described in Sections 4.6.1 and 10.5. Following Ernst and co-workers (Müller *et al.*, 1975; Aue *et al.*, 1976; Ernst *et al.*, 1987), the course of the experiment in the time-domain can be partitioned into the periods of:

preparation, evolution (t_1), mixing (t_m) and detection (t_2)

(Figure 6.1). Basically, the changes of the magnetization (or more generally, the density operator) during the evolution period t_1, which is incremented in successive sub-experiments, are detected indirectly as modulations of the nuclear signal that is observed during the detection period t_2. By means of the variety of solid-state NMR techniques listed in Table 6.1, the interactions determining the time evolution of the spin system can be manipulated during these different periods.

The preparation period usually consists of a recycle delay to allow z magnetization to build up by T_1 relaxation, followed by direct single-pulse excitation, or cross polarization from abundant to rare nuclei, to generate transverse magnetization.

The evolution time t_1 is incremented with a dwell time t_{dw1}, and at each t_1 value, several t_2 scans are taken for signal averaging and phase cycling. The evolution time serves to modulate the signal that is directly observed in the detection period. We have to distinguish amplitude modulation, which occurs in the majority of multidimensional solid-state NMR experiments, and phase modulation. These modulations of the detected signal can be most easily visualized for the precessing magnetization of isolated spin-$\frac{1}{2}$ nuclei (Figures 6.2 and 6.3). We consider the xy components of one isochromat of length M_0, that is, one constituent of the total magnetization vector. It precesses during the evolution time as $M_0 e^{i\tilde{\omega}_1 t_1}$

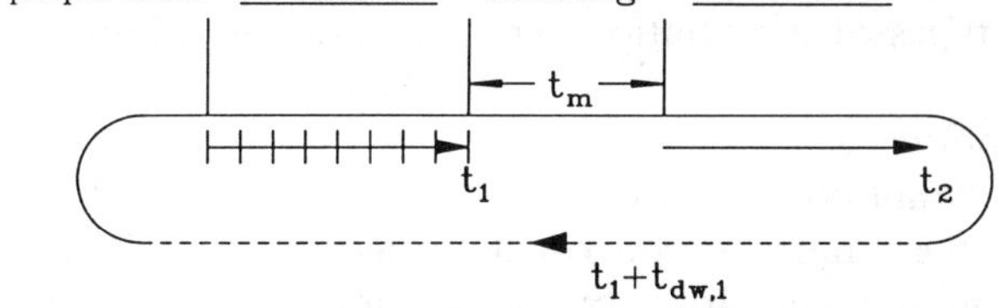

Figure 6.1 Principle of 2D experiments with preparation, evolution time t_1, mixing time t_m and the detection period t_2. As indicated in the figure, the duration of t_1 is incremented. Usually, for each t_1 value several scans are taken.

Table 6.1

(i) 'Processes' during evolution or detection
^{1}H Bloch decay
^{1}H Bloch decay and MAS
^{1}H multiple-pulse irradiation
^{1}H CRAMPS
^{13}C anisotropic chemical shift with ^{1}H dipolar decoupling (DD)
^{13}C fast MAS and DD
^{13}C sideband MAS and DD
^{2}H quadrupolar coupling
(ii) Processes predominantly employed in the evolution period
^{13}C subject to static heteronuclear dipolar couplings
^{13}C subject to heteronuclear dipolar couplings under MAS
^{13}C under fast MAS, rotor-synchronized 180° pulses, and DD
^{13}C under fast MAS, rotor-synchronized 90° pulses, and DD
^{13}C scaled chemical shift
^{13}C purely isotropic chemical shift (TOSS–deTOSS)
^{13}C OMAS (fast or slow)
(iii) Only accessible in the evolution period
Rotor phase (synchronized start of acquisition)
^{1}H or ^{2}H couplings in zero field
(iv) Only useful in the detection period
^{13}C after TOSS
(v) Mixing 'processes'
Stop spinning
Start spinning
Sample flip
Flip of rotation axis
Hartmann–Hahn cross-polarization
Multiple-pulse cross-polarization
Molecular reorientation
Conformational changes
^{1}H spin diffusion
^{13}C spin diffusion (proton driven)
^{13}C rf-driven spin diffusion
^{13}C rotor-driven spin diffusion, rotational resonance

with a certain frequency $\tilde{\omega}_1$. This magnetization vector at the end of the evolution period t_1 modulates the signal detected during t_2 which usually has a frequency $\tilde{\omega}_2 \neq \tilde{\omega}_1$. The two types of modulations are generated as follows:

(i) Amplitude modulation

In this case, the t_1-evolution may have been governed solely by dipolar or quadrupolar couplings, so that the magnetization oscillates as $\cos \tilde{\omega}_1 t_1$ along its initial direction (see Figures 2.17 and 6.2(a)). More commonly, amplitude modulation is generated if only one component of a precessing magnetization vector, either $M_0 \cos \tilde{\omega}_1 t_1$ or $M_0 \sin \tilde{\omega}_1 t_1$, i.e. the real or imaginary part of $M_0 e^{i\tilde{\omega}_1 t_1}$, is stored along the z axis during the mixing period (see Figure 6.2(b)). If the initial flip into the xy plane at the

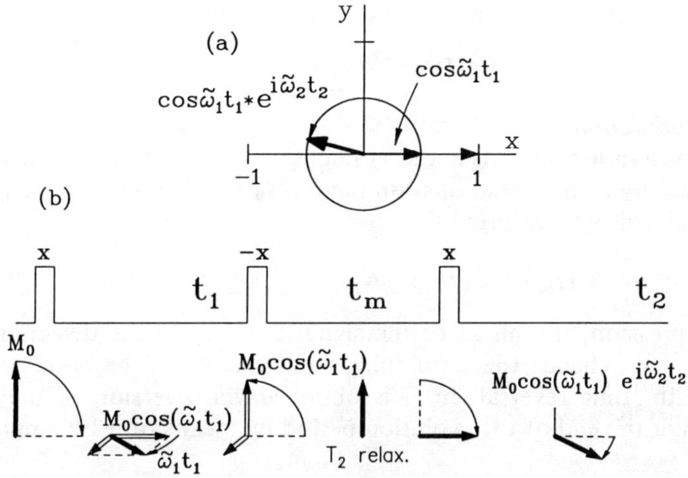

Figure 6.2 Amplitude modulations $f(t_1)$ of the detected signal $e^{i\tilde{\omega}_2 t_2}$. By several possible causes, the magnetization can be modulated only in terms of its amplitude, but not its phase, at the start of the detection period. (a) Under dipolar or quadrupolar couplings in the evolution time, the magnetization evolves as $\cos\tilde{\omega}_1 t_1$ along its initial direction, so only its amplitude changes. (b) For precessing magnetization, one of the two orthogonal transverse magnetization components can be selected by storing it along the z direction at the end of evolution, and flipping it back after the other component has been dephased.

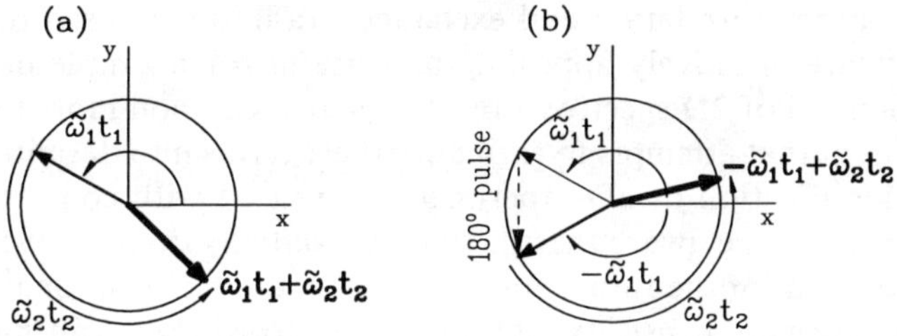

Figure 6.3 Phase modulation. (a) Transverse magnetization vectors are shown with the phases $\tilde{\omega}_1 t_1$ and $\tilde{\omega}_1 t_1 + \tilde{\omega}_2 t_2$ which they have acquired in evolution and detection periods. (b) 'Time reversal' of t_1 by a 180° pulse at the end of the evolution period inverts the phase $\tilde{\omega}_1 t_1$ to $-\tilde{\omega}_1 t_1$, so the phase at the detection time t_2 is $-\tilde{\omega}_1 t_1 + \tilde{\omega}_2 t_2$, yielding a magnetization vector described by $\exp(-i\tilde{\omega}_1 t_1 + i\tilde{\omega}_2 t_2)$.

start of evolution is effected by an x pulse and the storage after t_1 by a $-x$ pulse, the projection onto the y axis, $M_y = M_0 \cos\tilde{\omega}_1 t_1$, is stored along z during t_m. The remaining transverse x component is usually dephased quickly by dipolar couplings, so that the length of the magnetization vector at the end of t_m is $M_0 \cos\tilde{\omega}_1 t_1$. After having been flipped from the z axis back into the xy plane, the magnetization is described by:

$$s^c(t_1, t_2) = \underbrace{(M_0 \cos\tilde{\omega}_1 t_1)}_{\text{amplitude}}\ e^{i\tilde{\omega}_2 t_2}. \tag{6.1a}$$

If the second pulse, at the end of the evolution period, is 90° out of phase with the first, the component $M_0 \sin\tilde{\omega}_1 t_1$ is stored and modulates the amplitude of the signal detected during t_2. As a function of t_1 and t_2, we obtain the signal of the 'sine dataset':

$$s^s(t_1, t_2) \sim (M_0 \sin \tilde{\omega}_1 t_1)\, e^{i\tilde{\omega}_2 t_2}. \tag{6.1b}$$

(ii) Phase modulation

Here, the magnetization remains in the *xy* plane, so that the phase of the magnetization vector and the fully complex modulation factor $M_0\, e^{i\tilde{\omega}_1 t_1}$ is retained (see Figure 6.3(a)). The detected signal is therefore:

$$s^p(t_1, t_2) \sim (M_0\, e^{i\tilde{\omega}_1 t_1})\, e^{i\tilde{\omega}_2 t_2} = M_0\, e^{i(\tilde{\omega}_1 t_1 + \tilde{\omega}_2 t_2)}. \tag{6.2}$$

In the last expression, the phase of the signal detected in the detection period t_2 is modulated by $\tilde{\omega}_1 t_1$, hence the term 'phase modulation'. The corresponding signal $e^{-i\tilde{\omega}_1 t_1}\, e^{i\tilde{\omega}_2 t_2}$ with 'time reversal' in t_1 is obtained via inversion of the phase of the magnetization at the end of the evolution period by means of a 180° pulse (see Figure 6.3(b)).

Table 6.1 gives an overview of building blocks of 2D experiments on solids. For clarity, we specify the nuclei under consideration explicitly as ^1H, ^2H and ^{13}C. In most practical cases, ^{29}Si, ^{15}N or ^{31}P can be treated like ^{13}C, and ^{19}F is similar to ^1H except for its larger chemical shifts.

6.3 Separation versus correlation – Part I

The terms separation, correlation and exchange, traditionally used for characterizing 2D experiments, are intuitively appealing and useful for a simple description of the purpose and potential of 2D spectroscopy. Nevertheless, upon more thorough enquiry it becomes apparent that attempts to formalize these terms into definitions for a precise and complete classification of 2D experiments are faced with considerable problems. First of all, one should be aware that these terms actually do not cover all existing 2D NMR techniques, and for the future further novel and unconventional offspring of the original ideas of Jeener, Ernst, Waugh and their co-workers can be expected. This applies in particular to solid-state NMR with its wealth of possibilities of manipulating various interactions.

Formalized definitions have the tendency to produce unsuitable labelling for experiments which were not considered when the definitions were conceived. For instance, the lack of inhomogeneous line-broadening in solution NMR appears to have biased definitions given with special regard to 2D solution NMR. In addition, it should be noted that the terms separation, correlation and 2D exchange are not mutually exclusive: as will be discussed below, exchange NMR experiments, for instance, can be regarded as a subclass of correlation methods in many respects. And while for some 2D spectra either separation or correlation can be recognized as the main feature, others exhibit aspects of both. In fact, a correlation of spectral intensities at $\tilde{\omega}_1$ and $\tilde{\omega}_2$ is established in nearly all 2D spectra. At the same time, for several solid-state 2D techniques it is intuitively obvious and more instructive to describe the 2D spectrum in terms of a separation of broad signals which overlap in the 1D NMR spectrum. For these reasons, rather than using the definitions for separation, correlation and exchange as introduced by Ernst and co-workers (1987) in view of 2D solution NMR we employ a nomenclature that we find more suitable for solid-state 2D NMR spectra.

In our terminology, a typical 2D separation spectrum exhibits isolated single lines essentially without overlap in one dimension, usually at the isotropic chemical shift positions. For each of these lines, in the other dimension a more complicated spectrum is portrayed, usually representing an isolated 1D spectrum of the site picked in the single-line dimension. Typical examples are shown in Figure 6.4. It should be noted that this separation approach is much more relevant in solid-state than in solution NMR: the inhomogeneous line-broadenings in solids, caused by anisotropic chemical-shift, dipolar or quadrupolar interactions, produce continuous frequency distributions for every site that often overlap heavily in the conventional 1D spectra. The information provided by the 2D separation spectrum is partially contained in that 1D spectrum but removal of the overlap always increases the information content of a spectrum. On the other hand, it is obvious that for a single site the separation spectrum does not provide any new information.

2D correlation spectroscopy according to our terminology establishes relations between signals of two one-dimensional spectra each made up of lines with small homogeneous linewidths. As indicated in Figure 6.5, these may be sharp lines in both dimensions, or they may be merged to produce continuous patterns with large

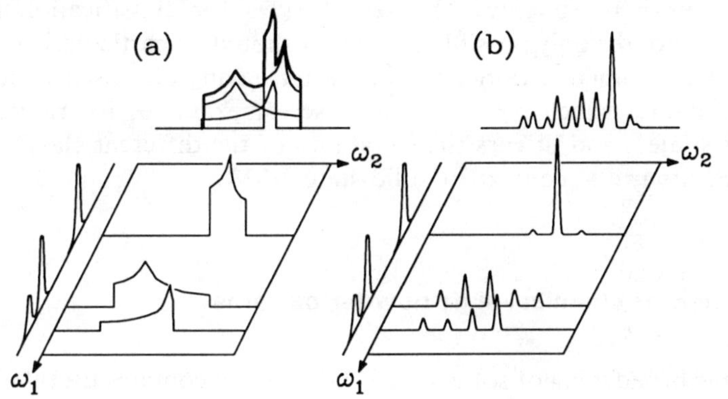

Figure 6.4 Typical 2D separation spectra. A spectrum of individual peaks in one dimension separates the broad patterns in the other. (a) Continuous powder patterns. (b) Spinning-sideband patterns.

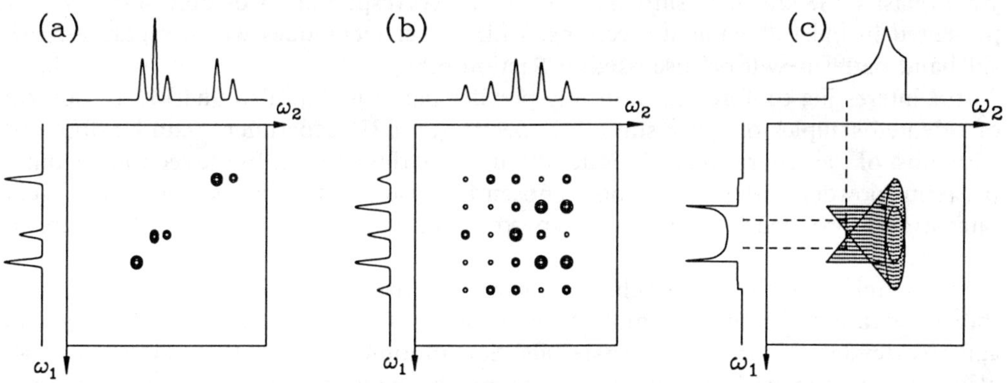

Figure 6.5 Typical 2D correlation spectra. (a) Correlation of individual peaks. (b) Correlation of sideband-patterns. (c) Correlation of inhomogeneously broadened lines.

inhomogeneous linewidths in both dimensions. In all these cases, no particularly relevant separation aspect is observed. It is interesting to note that even for a single site, the correlation of two inhomogeneously broadened patterns resulting from different interactions usually produces information not contained in the corresponding 1D spectra. On the other hand, if each dimension exhibits only a single line without inhomogeneous broadening, the correlation is naturally trivial.

2D exchange NMR is a special type of correlation spectroscopy (in our nomenclature) where the same 1D spectrum is correlated with itself at two different times for a system subject to dynamics of some kind. Since the interactions in evolution and detection are equal, for an exchange spectrum the projections on either axis (obtained by integration over the other dimension) are equal or at least of similar nature. It is a necessary condition for exchange spectra that in the absence of relevant dynamical processes, the intensity is restricted to a straight line, usually the diagonal of the spectral plane.

As discussed above, in several cases the distinction between separation and correlation is not so clear. Actually, for solving practical problems it is usually of little relevance; nevertheless, such considerations on the nature of a technique may help to understand it more thoroughly. The set of rules for classification that we have described here is not the only possible one, but it should be particularly suited for solid-state 2D NMR experiments. In Section 6.9, after having discussed various solid-state NMR 2D techniques which provide a large set of examples for reference, we shall return to this subject and discuss the suitability of the different classifications of 2D NMR experiments in the context of solid-state NMR.

6.4 Separation of chemical-shift powder patterns

The anisotropic broadening of solid-state NMR spectra contains useful information on structure, dynamics and orientation of molecular segments. However, for all but the simplest materials, overlap of patterns from different groups prevents this information from being extracted out of one-dimensional spectra. For eliminating this overlap, a variety of two-dimensional separation techniques have been proposed. Those which yield quasi-static chemical-shift patterns at the corresponding isotropic shifts will be presented in the following subsections, while other techniques which separate MAS sideband patterns will be discussed in Section 6.5.

It is interesting to note that not only powder patterns, but also quasi-static spectra of oriented samples or quasi-static 2D spectra (in a 3D experiment) can be obtained with most of the separation techniques discussed in this section. In the section heading, the term 'powder patterns' was only chosen because it seems best suited to represent 'anisotropy broadened quasi-static spectra', in distinction from MAS sideband patterns.

By obtaining the anisotropic chemical-shift pattern for each isotropic chemical shift, the investigation of molecular motion and order by analysis of static 1D NMR spectra can be extended to complicated materials. Separation techniques that achieve this can also serve as the basis for an extension of the powerful exchange or correlation experiments available for static samples (see Chapter 7): after introducing a second CSA dimension, one could obtain a 3D spectrum consisting of separate 2D exchange or correlation spectra for the various molecular sites, at the corresponding isotropic chemical shifts in the third dimension. However, it should be noted that the practical limitation of the 2D exchange or correlation experiments often lies not so much in the resolution, but rather in the *S/N* of the exchange intensity. Among other things, the signal per site decreases with the number of sites over which the signal is distributed. On ^{13}C or ^{29}Si in natural abundance, with the current technology an increase of the measuring times by more than a factor of ten is hardly tolerable for multidimensional experiments. A suitable separation technique must therefore produce little decrease of the signal intensity; in principle, this is possible since the signal of each site is concentrated in a narrow line along the isotropic-shift dimension, so that loss of signal height by spectral spreading is avoided (see Section 4.9).

A considerable number of CSA-separation experiments have been proposed, involving a variety of clever ideas; several of these were introduced by Maciel, Bax, and co-workers. Some techniques rely on static or quasi-static detection and produce the isotropic evolution by sample motion: stop-and-go, magic-angle hopping during evolution, axis-flipping from MAS to OMAS, variable-angle correlation spectroscopy, or ultraslow MAS with three-period evolution. The reliability and good resolution of the quasi-static spectrum that can be obtained with those techniques is important for extending the technique to a second quasi-static dimension in a 3D experiment as mentioned earlier in this section. However, it must be noted that except for the recently introduced ultra slow MAS approach, all the techniques cited cannot be performed on standard MAS equipment.

This severe limitation is avoided in other approaches that are based on reintroducing the anisotropic interaction by 180° pulses synchronized with the magic-angle spinning. These pulses usually involve the evolution period of the experiments, while the detection period samples the normal MAS signal. In general, due to the multiple 180° pulses applied during the evolution, these techniques are very sensitive to experimental imperfections, which can severely distort the powder patterns; the resolution is also reduced compared with the methods with static or quasi-static detection.

Several techniques of both categories rely on the fast-spinning regime (negligible sideband intensities). Often, they were originally performed at lower B_0 fields where this condition is more easily fulfilled because the chemical-shift anisotropy is smaller (as measured in Hertz). In higher fields, for carbon sites with larger anisotropies, such as aromatics or carbonyls, these particular techniques are not directly applicable.

6.4.1 Stop-and-go

A conceptionally simple but mechanically demanding 'stop-and-go' technique has been implemented by Maciel and co-workers (Zeigler *et al.*, 1988). After an evolution period for a static sample, the spinner is turned on and the signal under MAS detected. In principle, the experiment could also be performed with MAS evolution and stopping of the rotor to detect the signal from the static sample in t_2. The experiment yields reliable static spectra in the anisotropic dimension (Figure 6.6). Clearly, there are limits on either the spinning speed or the ease and swiftness of the change between stopping and spinning. In the literature, rotation frequencies of about 2 kHz and turn-on times of 3 seconds have been reported (Zeigler *et al.*, 1988). Due to the frictional forces involved in the rapid speed changes, wear of the spinner assembly appears to be unavoidable. The spinning sidebands that appear with slow rotation at high B_0 field

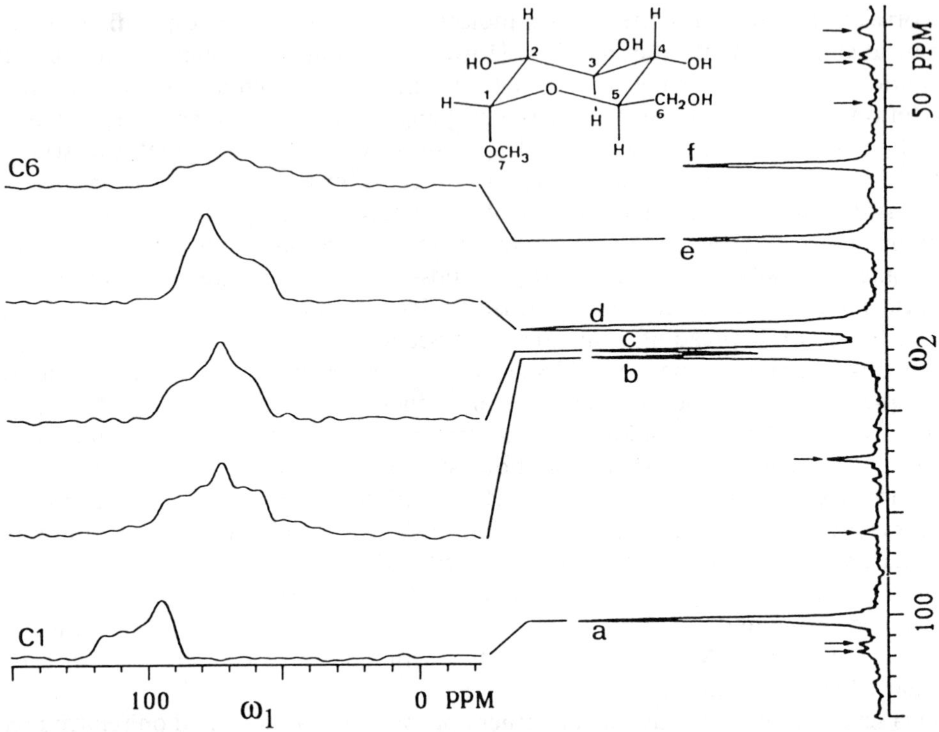

Figure 6.6 Stop-and-go spectrum (adapted from Zeigler *et al.*, 1988) of methyl-α-D-glucopyranoside taken at a Larmor frequency of 50 MHz and a spinning speed of 1.7 kHz; arrows mark spinning sidebands. The turn-on time of the rotor was three to five seconds.

complicate the data analysis considerably and decrease the signal-to-noise ratio. Therefore, it might be useful to remove the sidebands and concentrate all of the intensity in the centrebands by application of TOSS–t_1–deTOSS (Kolbert and Griffin, 1990) as described in Section 6.5.2, in an MAS evolution period.

6.4.2 Magic-angle hopping and ultraslow MAS

Magic-angle hopping (Bax *et al.*, 1983b) is a conceptually interesting approach that consists in averaging the chemical-shift anisotropy to zero by letting the magnetization evolve at three suitable orientations of the sample relative to the $\vec{\mathbf{B}}_0$ field (Figures 6.7(a) and 6.8(a)). After such an effectively isotropic evolution, the static anisotropic pattern is observed in the detection period. Figure 6.7 and the pulse sequence in Figure 6.8(a) indicate the three pieces of dipolar-decoupled evolution and the intermittent sample-flipping periods during which one component of the magnetization is aligned along the z direction. The rotation axis is at the magic angle with respect to the $\vec{\mathbf{B}}_0$ field, which points to the intrinsic relationship with magic-angle spinning: for a second-rank tensor, the averaging by rotational jumps between N sites in steps of $360°/N$ around a fixed axis for $N \geqslant 3$ produces the same averaged tensor as continuous rotation; this fact was already mentioned in Section 2.4.4 in the context of averaging by molecular rotation.

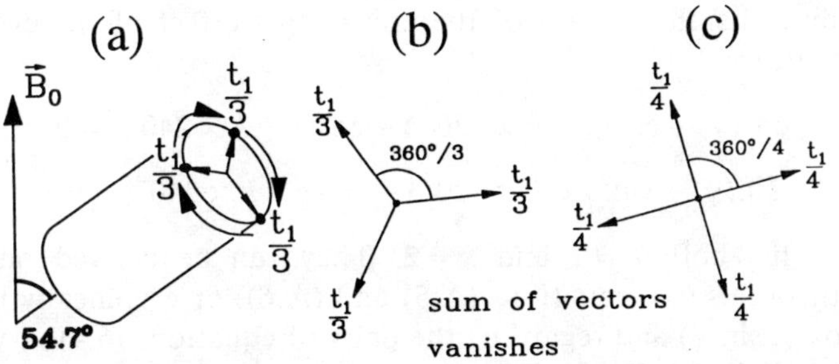

Figure 6.7 (a) Schematic representation of magic-angle hopping evolution. (b) and (c): Geometric visualization of the vanishing vector sums of equation (6.6) for three- and four-period evolution.

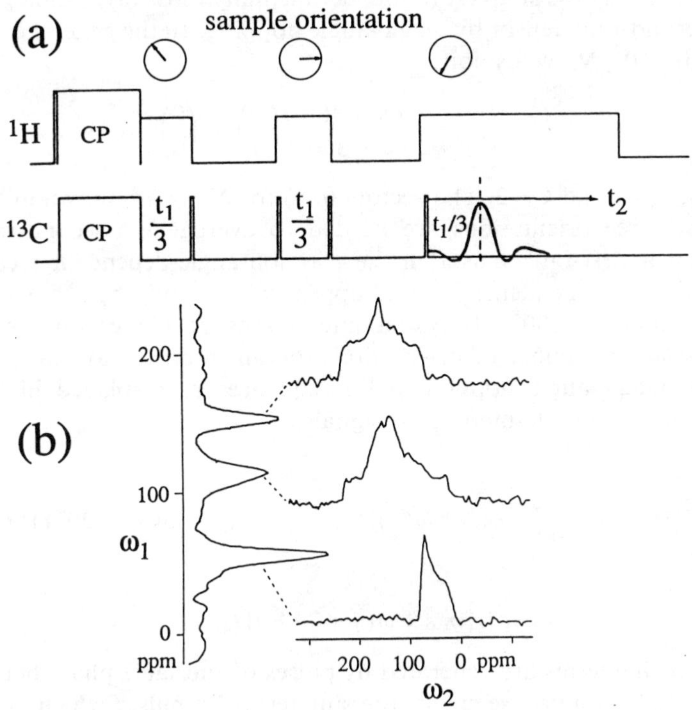

Figure 6.8 Magic-angle hopping (adapted from Bax *et al.*, 1983b). (a) Pulse sequence. (b) Slices along the anisotropic-shift dimension, and absolute-mode projection onto the isotropic-shift axis, of a 2D spectrum of 1,4-dimethoxybenzene taken at a ^{13}C Larmor frequency of 25 MHz.

For a proof, we consider the frequency of precession due to the anisotropic chemical shift as a function of the rotational-position angle γ, which is given by the MAS frequency formula (see equation 3.74)):

$$\omega(\gamma) = C_1 \cos \gamma + C_2 \cos 2\gamma + S_1 \sin \gamma + S_2 \sin 2\gamma. \tag{6.3}$$

For understanding the elimination of the anisotropic effect of the evolution, use is made of the identities:

$$\cos(\gamma) + \cos(\gamma + k \times 120^\circ) + \cos(\gamma + k \times 240^\circ) = 0 \qquad (6.4a)$$

$$\sin(\gamma) + \sin(\gamma + k \times 120^\circ) + \sin(\gamma + k \times 240^\circ) = 0 \qquad (6.4b)$$

which are valid for both $k = 1$ and $k = 2$. They can be derived on the basis of trigonometric theorems (see equations (A.5) and (A.6)) or graphically by considering unit vectors ($\cos\chi$, $\sin\chi$) and regarding the pairs of equations (6.4) as vector sums as displayed in Figure 6.7(b). With $\omega(\gamma)$ of equation (6.3):

$$\omega(\gamma) + \omega(\gamma + 120^\circ) + \omega(\gamma + 240^\circ) = 0 \qquad (6.5)$$

is immediately derived from the trigonometric-sum equations (6.4). The effects of the chemical-shift anisotropy thus vanish. As indicated above, this three-fold hopping in steps of $360^\circ/3$ is only the simplest (but experimentally most favourable) possibility of averaging a second-rank tensor by magic-angle hopping. In the general case, with hops by an angle of $360^\circ/N$, we exploit:

$$\sum_{n=1}^{N} \begin{pmatrix} \cos(\gamma + kn \times 360^\circ/N) \\ \sin(\gamma + kn \times 360^\circ/N) \end{pmatrix} = \begin{pmatrix} 0 \\ 0 \end{pmatrix} \qquad (6.6)$$

again valid for $k = 1$ and $k = 2$. The vectors (6.6) for $N = 4$ are shown in Figure 6.7(c). While at least three orientations are needed to average the second-rank tensorial interaction, for a first-rank tensor in the rotation-angle-dependent frequency $\omega(\gamma)$, only $\cos\gamma$ and $\sin\gamma$ dependencies would appear which could be averaged out by two orientations γ and $\gamma + 180^\circ$. Tensorial interactions of higher rank would require evolution at a larger number of orientations to vanish on the average.

In practice, magic-angle hopping pulse sequences as displayed in Figure 6.8(a) generate twofold amplitude-modulated signals:

$$g^{\begin{array}{c}\mathrm{cc}\\ \mathrm{cs}\\ \mathrm{sc}\\ \mathrm{ss}\end{array}}(t_1, t_2) = \begin{array}{c}\cos\\ \sin\end{array}[\{\omega_{\mathrm{iso}} + \omega(\gamma)\}\tfrac{1}{3}t_1] \begin{array}{c}\cos\\ \sin\\ \cos\\ \sin\end{array}[\{\omega_{\mathrm{iso}} + \omega(\gamma + 120^\circ)\}\tfrac{1}{3}t_1]$$

$$\times \exp[\mathrm{i}\{\omega_{\mathrm{iso}} + \omega(\gamma + 240^\circ)\}t_2]. \qquad (6.7)$$

As usual, the cosine terms are generated by pulses of the same phase before and after the respective $t_1/3$-evolution segment, the sine terms by pulses which are 90° out-of-phase. On the basis of the addition theorems (A.5) and (A.6), and using equation (6.5) as $\omega(\gamma) + \omega(\gamma + 120^\circ) = -\omega(\gamma + 240^\circ)$, sums and differences of the signals (6.7) produce data sets:

$$g_{\mathrm{s}}^{\mathrm{c}}(t_1, t_2) = \begin{array}{c}\cos\\ \sin\end{array}[\omega_{\mathrm{iso}}t_1 - \omega(\gamma + 240^\circ)\tfrac{1}{3}t_1]\ \mathrm{e}^{\mathrm{i}\{\omega_{\mathrm{iso}} + \omega(\gamma + 240^\circ)\}t_2} \qquad (6.8)$$

from which absorption-mode spectra can be generated according to Section 4.6.1. The spectral patterns are sharp ridges exhibiting the intensity distribution of the static 1D powder pattern with their centres of gravity at ω_{iso} and a slope of $-\frac{1}{3}$. The spectrum can be sheared, to look like a conventional separation pattern without slope, by application of the well-defined phase correction $\exp(i\frac{1}{3}t_1\omega_2)$ before the FT over t_1 (see Section 4.6.4).

The magic-angle hopping technique requires a special flip apparatus and has therefore found little application. Recently, it was realized by Gan (1992) that the principle of three-period evolution at rotor orientations γ, $\gamma + 120^\circ$, and $\gamma + 240^\circ$ can also be applied under ultra-slow MAS conditions, actually so slow that during the detection of the NMR signal the rotor appears quasi-static. For instance, with a rotation frequency of 50 Hz, a rotor period is 20 ms long, which results in a rotation by only 18° during 1 ms. The higher the $\vec{\mathbf{B}}_0$ field, the better the quasi-static approximation is valid, since the chemical-shift anisotropies increase proportional to B_0 and the decay of the time signal becomes correspondingly faster. The other crucial point in the technique is the realization that the average anisotropic frequency at the end of the three $t_1/3$-pieces of evolution vanishes exactly even under continuous sample spinning. The acquired phase is zero:

$$\bar{\omega}_{\text{aniso}} t_1 = \int_0^{\omega_r \frac{1}{3} t_1} \frac{d\gamma'}{2\pi} \left[\omega(\gamma + \gamma') + \omega(\gamma + \gamma' + 120^\circ) + \omega(\gamma + \gamma' + 240^\circ)\right] = 0 \quad (6.9)$$

because $\omega(\tilde{\gamma}) + \omega(\tilde{\gamma} + 120^\circ) + \omega(\tilde{\gamma} + 240^\circ) = 0$ according to equation (6.5), for arbitrary $\tilde{\gamma} = \gamma + \gamma'$ in equation (6.9). The acquisition and processing of the data, including shearing, is the same as outlined in the previous section for magic-angle hopping.

The ultraslow magic-angle rotation technique has potential for a wide application, since it yields well-resolved anisotropic patterns, as shown in Figure 6.9, in standard MAS equipment without rotor synchronization. Disadvantages are the loss of signal

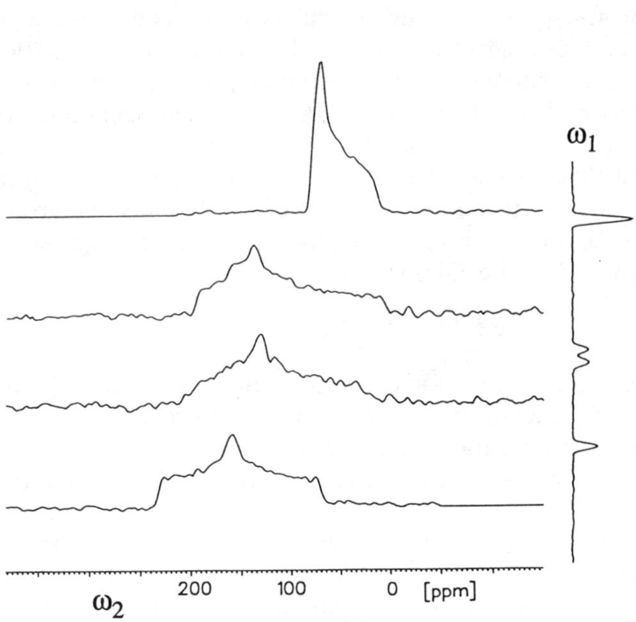

Figure 6.9 Ultraslow-MAS separation experiment (adapted from Gan, 1992). The pulse sequence is the same as in Figure 6.8(a), but the sample rotates continuously. Shown are slices and the isotropic-shift projection of a 2D spectrum of 1,4-dimethoxybenzene taken with a spinning speed of 100 Hz, at a ^{13}C Larmor frequency of 75 MHz.

intensity by the relaxation of the transverse magnetization components in the two storage periods of the evolution, and the fact that the rotor still moves too fast for certain desirable extensions of the technique. For instance, to measure isotropic-shift-separated 2D exchange patterns (for exchange spectroscopy, see Chapter 7) in a 3D experiment with ultraslow-spinning evolution and detection, the sample would have to be quasistatic on the order of 2 ms to 10 ms for an undistorted diagonal in the exchange spectrum to be obtained.

6.4.3 Variable-angle spinning 2D experiments

Quasi-static chemical-shift-anisotropy spectra can be obtained in rotating samples if the sample is spun off the magic angle (OMAS, see Section 2.4.5). This suggests a switching-angle sample spinning (SASS) 2D experiment with OMAS evolution, flip of the sample axis to the magic angle, and MAS detection (Bax *et al.*, 1983c; Terao *et al.*, 1984; Maciel *et al.*, 1985). The magnetization is stored along the z axis during the sample flip, producing a $\cos(\tilde{\omega}_1 t_1)$ or $\sin(\tilde{\omega}_1 t_1)$ amplitude modulation of the magnetization detected during t_2. With an angle θ_r between the rotation axis and $\vec{\mathbf{B}}_0$, the OMAS spectrum corresponds to the static spectrum scaled by $\frac{1}{2}(3\cos^2\theta_r - 1)$ if the signal is sampled at multiples of the rotor period (i.e. $t_{dw} = t_r$). Note that this applies even if the spinning rate is much smaller than the anisotropy parameter! The anisotropic spectrum can always be fitted into the restricted spectral range of $1/t_r$ by choosing a sufficiently small OMAS scaling factor ($\theta_r = \theta_m + 4^\circ$ is typically used). This is actually convenient in practice, since it involves only small flip angles. If the isotropic shift is outside the $1/t_r$ range, the spectrum appears aliased, but the information is not lost (see Section 4.8.1). The 'cyclic return' is easily compensated by repeating or rearranging the ω_1 data cyclically to obtain the standard representation. Well-defined powder patterns are obtained with this technique, as is exemplified in Figure 1.2(a).

Recently, it was realized (Frydman *et al.*, 1992) that separation of isotropic and anisotropic interactions could also be achieved in an efficient variable-spinning approach, termed VACSY (variable-angle correlation spectroscopy), though it has the character of a separation technique. When the sample spins fast around an axis at an angle θ_r with the $\vec{\mathbf{B}}_0$ field, the chemical-shift signal after a simple excite-and-detect pulse sequence (with CP and DD) is:

$$e^{i\omega_{iso}t + i\omega_{ani}P_2(\cos\theta_r)t} = e^{i\omega_{iso}t_1}\, e^{i\omega_{ani}t_2}. \tag{6.10}$$

By incrementing θ_r, the time factors $t_1 = t$ and $t_2 = P_2(\cos\theta_r)t$ of isotropic and anisotropic time dependencies can be varied independently to produce a two-dimensional time-domain datafield. For a given $t_1 = t$, a part of the t_2 range is not accessible: Due to the $[-\frac{1}{2}, \ldots, 1]$ range covered by $P_2(\cos\theta_r)$, t_2 is restricted to $-\frac{1}{2}t_1 < t_2 < t_1$. In practice, however, it is found that the accessible t_2 range is sufficient for obtaining meaningful 2D spectra with a limited artifact level (Figure 6.10). The variable $t_2 = P_2(\cos\theta_r)t$ is not sampled in regular increments and must therefore be appropriately interpolated. An important advantage of the experiment is the simplicity of the pulse sequence, consisting just of CP and acquisition under proton decoupling. This provides nearly the full sensitivity of a 1D experiment; the signal reduction by $\sin\theta_r$ due to the tilt of the coil is approximately compensated by the scaling of the inhomogeneous linewidths (see Section 4.9.2).

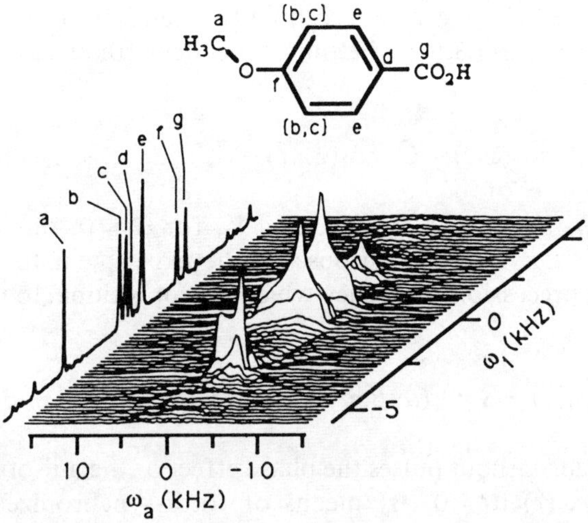

Figure 6.10 VACSY 2D spectrum (adapted from Frydman *et al.*, 1992) of p-anisic acid taken with a spinning speed of about 5 kHz at a ^{13}C Larmor frequency of 75 MHz.

6.4.4 MAS with chemical-shift patterns recovered by 180° pulses

Even under conditions of fast magic-angle spinning, the complete anisotropic information can be reintroduced by one or several pairs of rotor-synchronized 180° pulses per rotation period. This approach was first taken by Lippmaa and co-workers (Alla *et al.*, 1978). Subsequently, several techniques were introduced that gradually decreased the spectral distortions and the sensitivity to pulse imperfections. A technique that results in reasonably reliable spectra was introduced by Bax *et al.* (1983a) on the basis of work by Waugh and co-workers (Yarim-Agaev, 1982). At the times $t_0 + nt_r$, evolution of the isochromats is inverted by a pair of 180° pulses for a time τ. This delay τ should be short enough to allow to approximate the rotor as being static during τ, that is:

$$\tau \ll t_r. \tag{6.11}$$

Sampled at multiples of the rotor period t_r, the phase generated by the anisotropic interaction, instead of vanishing as in undisturbed MAS, has a value of $-2\tau\tilde{\omega}$, where $\tilde{\omega} = \omega(\alpha, \beta, \gamma + \omega_r t_0)$. The factor of 2 arises because during τ an isochromat both covers the phase angle $-\tau\tilde{\omega}$ and loses the phase $\tau\tilde{\omega}$ that it would have acquired had the pulses not been applied. The scaling factor compared with the static spectrum is:

$$\chi = \frac{2\tau}{t_r}. \tag{6.12}$$

Due to the condition $\tau \ll t_r$, χ is much smaller than unity, which reduces the spectral resolution. On the other hand, if the condition (6.11) is relaxed, distortions in the powder spectra arise.

Such distortions can be avoided in a technique recently introduced by Tycko *et al.* (1989). According to Section 3.9.1, the time-dependent frequency under MAS can be written as:

$$\omega(t) = \tilde{C}_1 \cos(\omega_r t) + \tilde{C}_2 \cos(2\omega_r t) + \tilde{S}_1 \sin(\omega_r t) + \tilde{S}_2 \sin(2\omega_r t) \tag{6.13}$$

(see equation (3.76)). In a static sample, that is, for $\omega_r = 0$, this frequency formula reduces to $\omega_{st} = \tilde{C}_1 + \tilde{C}_2$. It is the purpose of the experiment to use 180° pulses to produce an effective precession frequency which is proportional to the static frequency $\omega_{st}(\alpha, \beta, \gamma)$:

$$\bar{\omega}(\alpha, \beta, \gamma) \stackrel{!}{=} \chi[\tilde{C}_1(\alpha, \beta, \gamma) + \tilde{C}_2(\alpha, \beta, \gamma)] = \chi\omega_{st}(\alpha, \beta, \gamma). \tag{6.14}$$

In a rotating sample, without pulses the phase effect of the anisotropy over one rotor period vanishes, $\int_0^{t_r} \omega(t)\, dt = 0$. By means of rotor-synchronized 180° pulses, the anisotropy effects of $\tilde{C}_1$ and $\tilde{C}_2$ are now to be reintroduced with equal prefactors, while those of $\tilde{S}_1$ and $\tilde{S}_2$ are still to remain equal to zero in the average over a rotor period. The effect of a 180° pulse can be treated as an inversion of the sign of the precession frequency. In order to account for this, following Tycko *et al.* (1989), it is convenient to introduce a function $p(t)$ which takes values +1 and −1 characterizing the actual precession frequency as $p(t)\omega(t)$ (Figure 6.11). The overall phase after a full rotor period is then given by:

$$\Phi(t_r) - \Phi(0) = \int_0^{t_r} p(t)\omega(t)\, dt \tag{6.15}$$

which no longer vanishes. The integral over the sine terms of (6.13) vanishes as desired if $p(t)$ is an even function with respect to $t = \frac{1}{2} t_r$, since $\sin(\omega_r t)$ and $\sin(2\omega_r t)$ are odd with respect to $t = \frac{1}{2} t_r$. The prefactors for $\tilde{C}_1$ and $\tilde{C}_2$ are equal in the precession phase after one rotor period if:

$$\int_0^{\frac{1}{2}t_r} p(t)\cos[\omega_r t]\, dt \stackrel{!}{=} \int_0^{\frac{1}{2}t_r} p(t)\cos[2\omega_r t]\, dt = \chi \tfrac{1}{2} t_r. \tag{6.16}$$

The restriction of the integral relies on $p(t)$ being even. The integrals for $2N$ 180° pulses can be rewritten according to:

$$\int_0^{\frac{1}{2}t_r} p(t)\cos(k\omega_r t)\, dt = \int_0^{t_a} \cos(k\omega_r t)\, dt + \int_{t_a}^{t_b} (-1)\cos(k\omega_r t)\, dt + \cdots$$

$$+ \int_{t_N}^{\frac{1}{2}t_r} (-1)^N \cos(k\omega_r t)\, dt = \frac{2}{k\omega_r} \sum_{n=1}^{N} (-1)^n \sin(k\omega_r t_n). \tag{6.17}$$

Thus, the condition (6.14) for equal prefactors in front of $\tilde{C}_1$ and $\tilde{C}_2$ becomes:

$$\sum_{n=1}^{N} (-1)^n [\sin(\omega_r t_n) - \tfrac{1}{2} \sin(2\omega_r t_n)] \stackrel{!}{=} 0. \tag{6.18}$$

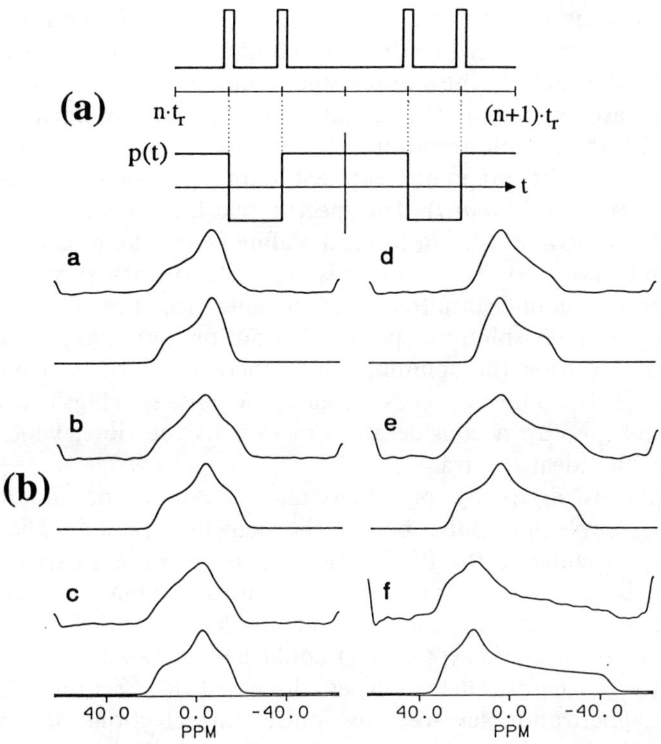

Figure 6.11 Recovery of the chemical-shift anisotropy under fast MAS by rotor-synchronized 180° pulses (adapted from Tycko *et al.*, 1989). (a) Pulse sequence and the corresponding function $p(t)$ which describes the effect of the 180° pulses on the phase of the magnetization. (b) Slices out of a 2D spectrum of methyl-α-D-glucopyranoside (for the chemical structure and the MAS spectrum see Figure 6.6) taken with a spinning speed of 5 kHz at a ^{13}C Larmor frequency of 100 MHz. Simulated powder patterns are shown for comparison.

The minimum number of 180° pulses required is four. If these pulse-timing conditions are met, the precession frequency averaged over full rotor periods is:

$$\frac{\Delta\Phi}{t_r} = \xi\omega_{\rm iso} + \chi(\tilde{C}_1 + \tilde{C}_2) \tag{6.19}$$

where we have included the isotropic shift that was neglected so far. The scaling factors for the isotropic and anisotropic shifts:

$$\begin{aligned} \xi &= \frac{1}{t_r}\int_0^{t_r} p(t)\,\mathrm{d}t \\ \chi &= \frac{1}{t_r}\int_0^{t_r} p(t)\cos(\omega_r t)\,\mathrm{d}t \end{aligned} \tag{6.20}$$

depend on the particular solutions of equations (6.18). There is a continuum of solutions for sequences of four 180° pulses with χ between 0 and 0.412. Of these, a single one remains if $\xi = 0$ is demanded, and its scaling factor χ is 0.393.

Note that the scaling relation $\bar{\omega}(\alpha, \beta, \gamma) = \chi\omega_{st}(\alpha, \beta, \gamma)$, equation (6.14), applies for each individual segment. Consequently, the technique is not only useful for measuring static powder lineshapes but is also applicable to measuring the lineshapes of oriented samples. In that case, equal γ must be ensured in every scan by starting the acquisition in synchrony with the sample rotation.

As is apparent from the foregoing treatment, data points must only be sampled after full rotor periods (i.e. at $t_1 = mt_r$). The spectral width is thus $\omega_{sw} = 2\pi/t_{dw1} = 2\pi/t_r = \omega_r$. The larger the spectral width, the larger a scaling factor can be chosen and the better a resolution can be attained. Therefore, it is desirable to work at high spinning speeds ω_r. This is somewhat counterintuitive since the anisotropic interactions are spun out more completely at high spinning speeds; this notion, however, is valid only if no pulses are applied during the spinning. Still, there is a practical limitation to the spinning speeds: as the spinning speeds increase, the pulse spacings become smaller and the pulse lengths take up a considerable fraction of the time, which increases the deviations from the ideal spectra.

The 2D signal is originally phase-modulated. Amplitude modulation can be generated by a simple 90° pulse before the detection period. The magnetization component perpendicular to the $\vec{\mathbf{B}}_1$ field is flipped onto the z axis, leaving only the projection onto $\vec{\mathbf{B}}_1$, $M_0 \cos \tilde{\omega}_1 t_1$ or $M_0 \sin \tilde{\omega}_1 t_1$, in the xy plane to contribute further to the signal. As an alternative yielding a S/N better by a factor of $\sqrt{2}$, phase-modulated signals $M_0 \exp(\mathrm{i}\tilde{\omega}_1 t_1)$ and $M_0 \exp(-\mathrm{i}\tilde{\omega}_1 t_1)$ could be generated by single 0° and 180° pulses, respectively, before detection as discussed in Section 6.2. With these, longitudinal magnetization created by pulse imperfections during t_1 remains longitudinal, while the 90° pulses at the end of the evolution period rotate it into the transverse plane so that it must be phase-cycled away (Tycko *et al.*, 1989).

Generally, the experiment with its multiple 180° pulses is quite susceptible to experimental pulse imperfections. Careful tune-up of the ^{13}C channel is required, usually performed on a ^{13}C-labelled simple organic liquid, such as methanol. As usual, positioning the centres of the finite 180° pulses on the time of the ideal δ pulse produces relatively good results. Unavoidable B_1 inhomogeneity and pulse-length errors distort the spectra and can produce artifacts. These are shifted to the edges of the spectral range by choosing the phase sequence of the 180° pulses as $(-x, -x, x, x, x, x, -x, -x)$.

Powder patterns measured with this technique are shown in Figure 6.11(b). The resolution is only slightly reduced compared with the spectrum of the same material with truly static evolution as shown in Figure 6.6. The multiple-180°-pulse MAS method has the important practical advantage that it can be performed on standard MAS equipment, and that it can be applied to samples with short ^{13}C T_1 relaxation times.

6.5 Separation of MAS chemical-shift sideband patterns

For characterizing the chemical-shift anisotropy in terms of its principal values, it is not necessary to measure static spectra. Rather, it suffices to achieve 2D separation of 1D MAS sideband patterns (see Section 3.9.1) according to the isotropic shifts. At first sight, this may appear a simple task; one could think of combining TOSS (Section 3.9.2) in one dimension with simple MAS in the other. However, simple TOSS does not suppress sidebands in 2D spectra as desired, and serious 2D phase problems are also encountered. Only recently were convenient solutions to this problem presented.

One very simple but not widely applicable approach consists in sampling t_1 only at the rotor echoes to eliminate the sidebands in ω_1 (Aue *et al.*, 1981). As was mentioned at the end of Section 2.4.5 and can be easily checked by inserting $\omega_r t = m \times 2\pi$ into equation (3.80), the effects of the angle-dependence vanish at times $t = m \times 2\pi/\omega_r$. However, this rotor-echo sampling can be performed only at the cost of a very small spectral width, which spans, in fact, only one sideband spacing. Therefore, more involved methods are usually necessary to separate isotropic and anisotropic shifts.

6.5.1 Why simple combinations of TOSS and MAS fail

For 1D spectra, TOSS with unsynchronized signal averaging will remove the sidebands for all samples, and the centrebands are absorptive and positive in most relevant cases. In 2D separation spectroscopy, by means of such a sideband suppression one would obtain just a single peak for each site in one dimension, thus separating complex 1D spectra. In particular, it seems straightforward at first sight to retain the MAS sidebands in one dimension and suppress them in the other by means of TOSS. However, as we shall show now, simple application of TOSS before evolution or detection does not result in sideband suppression; TOSS and MAS require independent integrals over the rotor phase to produce the desired patterns.

Although they turn out to be of little practical utility, it is instructive to analyse the MAS–TOSS and TOSS–MAS 2D experiments to point out some problems that must be avoided in useful multidimensional MAS techniques. For reference, we start with a description of the desired isotropic-shift-separated sideband patterns. Such a 2D spectrum would have to exhibit only isotropic peaks (centrebands $c_0\delta_{M,0}$ of intensity c_0) in one dimension and the 1D MAS intensity of equation (3.83) in the other. Altogether, it would thus read $c_0\delta_{M,0}f^*(\gamma)\,\mathrm{e}^{\mathrm{i}M\gamma}F_N(\alpha,\beta)$, which becomes:

$$I_{M,N} = c_0\delta_{M,0}\,|\,F_N(\alpha,\beta)\,|^2 \tag{6.21}$$

after integration over γ (see equation (3.85)).

The time signal of an experiment with MAS evolution and TOSS detection, as displayed in Figure 6.12(a), is given by

$$g(t_1,t_2) = \overbrace{f^*(\gamma)f(\gamma+\omega_r t_1)}^{\text{MAS}}\ \overbrace{f([\gamma+\omega_r t_1]+\omega_r t_2)}^{\text{TOSS}}\ \mathrm{e}^{\mathrm{i}\omega_{\mathrm{iso}}(t_1+t_2)} \tag{6.22}$$

(see Section 3.9). In the TOSS signal, we have taken into account that the start phase is $[\gamma+\omega_r t_1]$. The sideband intensities $I_{M,N}$ relative to $(\omega_{\mathrm{iso}}, \omega_{\mathrm{iso}})$ are obtained by double Fourier transformation of $g(t_1,t_2)$:

$$I_{M,N} = \frac{1}{t_r}\int_0^{t_r}\mathrm{d}t_1\,\mathrm{e}^{-\mathrm{i}M\omega_r t_1}\,\frac{1}{t_r}\int_0^{t_r}\mathrm{d}t_2\,\mathrm{e}^{-\mathrm{i}N\omega_r t_2}g(t_1,t_2). \tag{6.23a}$$

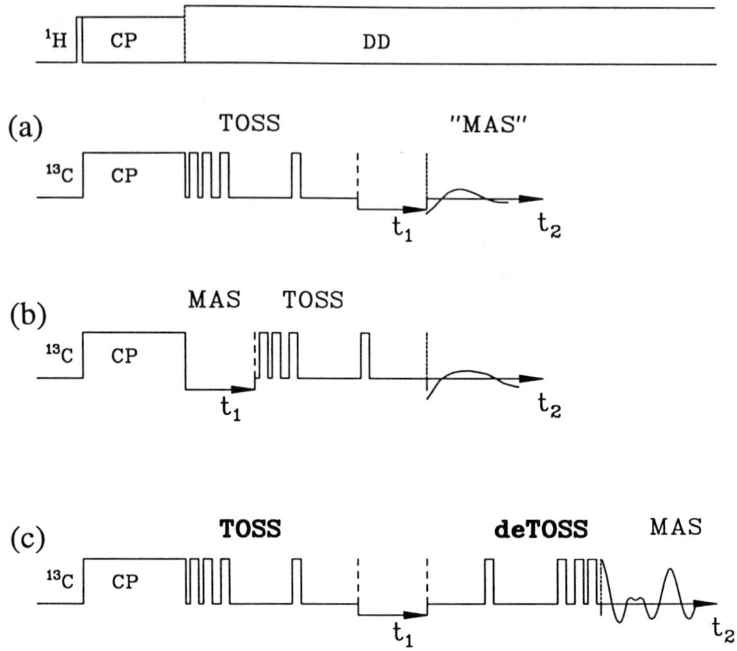

Figure 6.12 Pulse sequences for 2D experiments with TOSS and MAS. Note that the versions (a) and (b) are not useful in practice. Only the TOSS–t_1–deTOSS sequence under (c) provides a separation of MAS sideband patterns by the isotropic chemical shifts.

With the integration variables $\gamma_1 := \gamma + \omega_r t_1$, $\gamma_2 := \gamma + \omega_r t_1 + \omega_r t_2 = \gamma_1 + \omega_r t_2$, and omitting $e^{i\omega_{iso}(t_1+t_2)}$ for clarity, this can be rewritten as:

$$I_{M,N} = f^*(\gamma)\, e^{iM\gamma} \overbrace{\underbrace{\int \frac{d\gamma_1}{2\pi} f(\gamma_1)\, e^{-iM\gamma_1}\, e^{iN\gamma_1}}_{F_{M-N}}}^{FT_1} \overbrace{\underbrace{\int \frac{d\gamma_2}{2\pi} e^{-iN\gamma_2} f(\gamma_2)}_{F_N(\alpha,\beta)}}^{FT_2}$$

$$= f^*(\gamma)\, e^{iM\gamma}\, F_{M-N}(\alpha, \beta)\, F_N(\alpha, \beta). \tag{6.23b}$$

Intensity occurs for all N and M: no sideband suppression is achieved (see also equation (6.21)).

The time signal in the simplest TOSS–MAS 2D experiment (Figure 6.12(b)) would be:

$$g(t_1, t_2) = \overbrace{f(\gamma + \omega_r t_1)}^{TOSS}\, \overbrace{f^*(\gamma + \omega_r t_1)\, f(\gamma + \omega_r t_1 + \omega_r t_2)}^{MAS}\, e^{i\omega_{iso}(t_1+t_2)}$$

$$= f(\gamma + \omega_r t_1 + \omega_r t_2)\, e^{i\omega_{iso}(t_1+t_2)}. \tag{6.24}$$

As in the previous example, the sideband intensities $I_{M,N}$ are determined from $g(t_1,t_2)$ by double FT with the same integration variables γ_1 and γ_2 as above:

$$I_{M,N} = e^{iM\gamma} \underbrace{\overbrace{\int \frac{d\gamma_1}{2\pi}\, e^{-iM\gamma_1}\, e^{iN\gamma_1}}^{FT_1}}_{\delta_{M,N}} \underbrace{\overbrace{\int \frac{d\gamma_2}{2\pi}\, e^{-iN\gamma_2} f(\gamma_2)}^{FT_2}}_{F_N(\alpha,\beta)}$$

$$= e^{iM\gamma}\, \delta_{M,N}\, F_N(\alpha,\beta). \tag{6.25}$$

This looks better than before, but with an average over the initial rotor phase γ, $\int d\gamma\, e^{iM\gamma} = \delta_{M,0}$, we have:

$$\bar{I}_{M,N} = \delta_{M,0}\, \delta_{M,N} \bar{F}_N = \delta_{M,0}\, \delta_{N,0} \bar{F}_0. \tag{6.26}$$

Here we have lost the sidebands in both dimensions. The comparison of the results of these MAS–TOSS and TOSS–MAS experiments shows that the interchange of TOSS and MAS between evolution and detection does not simply result in an interchange of t_1 and t_2, or M and N; this is a typical feature of multidimensional sideband-MAS experiments.

As a potential alternative method of obtaining TOSS–MAS 2D spectra, we could average over all values of the rotor phase before TOSS detection so that the sidebands in that dimension compensate. In Section 3.9.2, this average, which corresponds to an integral over γ, has been shown to eliminate the sidebands for any sample. In practice, such averaging can be achieved by storing the magnetization along z in a short mixing time after MAS evolution and incrementing this mixing time through one rotor period. However, the intensities in the resulting sideband pattern $\int d\alpha \int d\cos\beta\, |F_N(\alpha,\beta)|^2 F_0(\alpha,\beta)$ are not equal to those of the MAS 1D spectrum $\int d\alpha \int d\cos\beta\, |F_N(\alpha,\beta)|^2$. Note that it is not just the absolute values that differ, but rather the relative sideband intensities. The product of the MAS spectrum and the TOSS intensity would actually be $\int d\alpha \int d\cos\beta\, |F_N(\alpha,\beta)|^2 . \int d\alpha' \int d\cos\beta' F_0(\alpha',\beta')\, \delta_{M,0}$.

6.5.2 *Isotropic-shift separation via TOSS and deTOSS*

Instead of trying to compensate the sidebands by means of simple TOSS, it is more advantageous to remove all anisotropy effects in the evolution time by combination of TOSS and time-reversed TOSS (Kolbert and Griffin, 1990; see Section 3.9.4). The corresponding pulse sequence is shown in Figure 6.12(c). The evolution time t_1 is set between TOSS and deTOSS sequences. This results in a fully isotropic t_1-modulation of the MAS signals detected during t_2:

$$g(t_1,t_2) = e^{i\omega_{iso}t_1}[g(t_2)\, e^{i\omega_{iso}t_2}]. \tag{6.27}$$

The spectrum in Figure 1.2(b), taken from Kolbert and Griffin (1990), demonstrates the first application of this technique. It is not absorptive, as it was generated from a single phase-modulated dataset corresponding directly to the signal (6.27) (the problems arising from phase twist are hidden by choosing high contour levels). Since the evolution in equation (6.27) appears as fully isotropic with a simple phase modulation, the phase problem can be solved by measuring a second dataset with a

180° pulse right before the start of t_2 (see Sections 6.2 and 4.6.1). This additional pulse inverts the phase of the signal before the pulse, which appears like time reversal of t_1:

$$g^{\mathrm{tr}}(t_1, t_2) = \mathrm{e}^{-\mathrm{i}\omega_{\mathrm{iso}}t_1}[g(t_2)\ \mathrm{e}^{\mathrm{i}\omega_{\mathrm{iso}}t_2}]. \tag{6.28}$$

Adding and subtracting g and g^{tr} produces $\cos(\tilde{\omega}t_1)$ and $\sin(\tilde{\omega}t_1)$ modulation so that the phase problem can be solved as outlined in Section 4.6.1.

Since the effects of the anisotropy during t_1 are actually refocused by the TOSS–t_1–deTOSS sequence and not just delicately compensated as in the case of simple TOSS, the full signal strength is retained in the ω_1 dimension (apart from relaxation by e^{-4t_r/T_2} during the TOSS–t_1–deTOSS sequences). Therefore, this TOSS–t_1–deTOSS technique is superior to a hypothetical 2D experiment with separation of MAS sideband patterns by normal TOSS. The projection on the ω_1 axis exhibits the full intensity of each site at its isotropic shift, without signal reductions for large anisotropies as encountered for TOSS.

The TOSS–t_1–deTOSS evolution is a superior alternative to the previously introduced technique of ^{13}C chemical-shift scaling (see Section 3.4.4) during t_1 (Harbison *et al.*, 1984). The scaling approach actually involves restrictions in the number of sidebands in ω_2 and in the optimum resolution in ω_1. It relies on scaling the ^{13}C evolution during t_1; with a scaling factor of $1/n$, the experiment is restricted to observing only n or fewer sidebands in ω_2 if all the sidebands in ω_1 are to be eliminated. Scaling factors of 0.3 are attainable but too small a scaling factor must not be used in order to preserve the spectral resolution in ω_1. For applications and a simple extension of this experiment for probing isotropic chemical-shift exchange see Section 7.14.1.

6.5.3 Separation of sidebands by their order, and PASS

A slightly different approach to the separation of isotropic and anisotropic shifts exploits the fact that in MAS sideband spectra the signal intensity appears only at the discrete sideband locations $\omega_{\mathrm{iso}} + N\omega_r$. These can be fully displayed in 2D spectra with typically just 8 or 16 sideband slices in the ω_1 dimension. The reduced data size is particularly advantageous for extending the experiment into a third dimension as discussed in Section 12.5.

The actual experiment for such a separation by sideband order can be based on a time signal:

$$s(t_1, t_2) = f^*(\gamma)f(\gamma + \omega_r t_1 + \omega_r t_2)\ \mathrm{e}^{\mathrm{i}\omega_{\mathrm{iso}}t_2}\ \mathrm{e}^{-t_2/T_2}. \tag{6.29}$$

In the t_1 dimension, this is fully periodic with ω_r (due to the absence of factors like $\mathrm{e}^{\mathrm{i}\omega_{\mathrm{iso}}t_1}$ and e^{-t_1/T_2}). Because of this periodicity, it is sufficient to sample t_1 within one rotor period. The spectral intensities $I_{M,N}$ of the time signal (6.29), with frequencies referenced to ω_{iso} in ω_2, are obtained after double Fourier transformation ($\int \mathrm{d}t_1\ \mathrm{e}^{-\mathrm{i}M\omega_r t_1} \int \mathrm{d}t_2\ \mathrm{e}^{-\mathrm{i}N\omega_r t_2}$) of $s(t_1, t_2)$ without the T_2 relaxation term. Substituting $\gamma_1 := \gamma + \omega_r t_1$ and $\gamma_2 := \gamma + \omega_r t_1 + \omega_r t_2 = \gamma_1 + \omega_r t_2$, we have:

$$I_{M,N} = f^*(\gamma)\ \mathrm{e}^{\mathrm{i}M\gamma} \overbrace{\underbrace{\int \frac{\mathrm{d}\gamma_1}{2\pi}\ \mathrm{e}^{-\mathrm{i}M\gamma_1}\ \mathrm{e}^{\mathrm{i}N\gamma_1}}_{\delta_{M,N}}}^{\mathrm{FT}_1} \overbrace{\underbrace{\int \frac{\mathrm{d}\gamma_2}{2\pi}\ \mathrm{e}^{-\mathrm{i}N\gamma_2} f(\gamma_2)}_{F_N(\alpha,\beta)}}^{\mathrm{FT}_2}$$

$$= f^*(\gamma)\ \mathrm{e}^{\mathrm{i}M\gamma}\ \delta_{M,N}F_N(\alpha, \beta) \overset{(3.83)}{=} \delta_{M,N}I_N(\alpha, \beta, \gamma). \tag{6.30}$$

The Mth ω_1 slice contains the sideband of order M, as displayed in a spectrum of polycarbonate in Figure 6.13. The sideband intensities $I_N(\alpha, \beta, \gamma)$ are the same as in the conventional 1D MAS spectrum.

Such a separation of sidebands by their order was first achieved by Dixon (1982) in the PASS (phase-altered separation of sidebands) experiment. In his approach, the time signals of equation (6.29) are generated by means of four π pulses spaced by appropriate delays, which vary with t_1 in a moderately nonlinear fashion. In the original treatment, these signals are characterized in terms of their rotor-echo shifts ('pitches') which correspond to $\omega_r t_1$ in our notation. In fact, the signal (6.29):

$$f^*(\gamma) f(\gamma + \omega_r t_1 + \omega_r t_2)\ e^{i\omega_{iso} t_2}\ e^{-t_2/T_2}$$

is nearly identical to the normal MAS signal:

$$f^*(\gamma) f(\gamma + \omega_r t_2)\ e^{i\omega_{iso} t_2}\ e^{-t_2/T_2}$$

only with the rotor echoes shifted by t_1 (recall that these echoes occur due to $f^*(\gamma) f(\gamma + n \times 2\pi) = 1$ when $\omega_r t_1 + \omega_r t_2 = 2\pi$). In the data processing of the PASS data as originally described by Dixon, the ω_2 spectra (Fourier transforms of equation (6.29) with respect to t_2) for different $\omega_r t_1$ values are added up after 'phasing' by $e^{-iM\omega_r t_1}$:

$$I_M(\omega_2) = \sum_{n=0}^{n_{max}} g(\omega_2, n t_{dw1})\ e^{-iM\omega_r n t_{dw1}}. \tag{6.31}$$

This is nothing less than the FT (precisely speaking the discrete FT, see Section 4.2) against t_1 in equation (6.30) done by hand. As this procedure was quite time-consuming and appeared awkward, the PASS technique was never really widely used. Because of the unsystematic variation of the delays, automation of the data acquisition

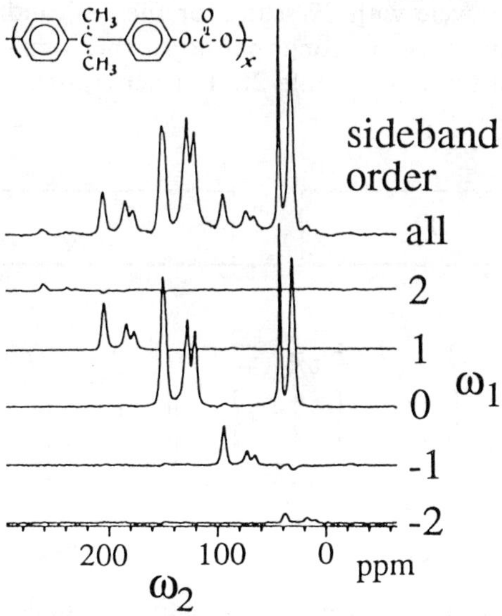

Figure 6.13 PASS spectrum of polycarbonate. The various sideband orders of the 1D spectrum appear in different ω_1 slices. Adapted from Dixon (1982).

is difficult, too. In addition, due to differences in the delays, particularly in the overall duration, the t_1-periodic signal (6.29) is generated with reduced accuracy for the time signals corresponding to different $\omega_r t_1$ values.

These problems have been eliminated in a more recent, fully two-dimensional implementation of the sideband-order separation experiment (Féaux de Lacroix *et al.*, 1992). It employs two TOSS pulse sequences separated by the evolution time t_1, as shown in Figure 6.14. During this t_1, the magnetization is fixed along the z axis, but the rotor phase is evolving; the anisotropic interaction determining the precession of the magnetization during the second TOSS sequence thus changes as a function of t_1. For this pulse sequence, the time signal is obtained according to the principles outlined in Section 3.9:

$$s_{\substack{\mathrm{Re}\\ \mathrm{Im}}}(t_1, t_2) = \substack{\mathrm{Re}\\ \mathrm{Im}} \overbrace{\{f(\gamma)\}}^{\text{after TOSS}_1} \overbrace{f([\gamma + \omega_r t_1] + \omega_r t_2)\, \mathrm{e}^{\mathrm{i}\omega_{\mathrm{iso}} t_2}\, \mathrm{e}^{-t_2/T_2}}^{\text{after TOSS}_2}. \tag{6.32}$$

Calculating $s_{\mathrm{Re}} - \mathrm{i}s_{\mathrm{Im}}$, the signal $f^*(\gamma)\ldots$ and thus the desired form (6.29) is generated. The experiment is more easily implemented than the original PASS experiment, requiring only regular increments of t_1. However, the signal is reduced by a factor $2\mathrm{e}^{+2t_r/T_2}$ compared with the PASS experiment: one component of the magnetization is discarded during t_1, and to obtain one complex s signal, two scans (for s_{Re} and s_{Im}) are required; this produces $\sqrt{2}$ times more noise and takes twice as long. In addition, the evolution with its two TOSS sequences takes $2t_r$ longer than in PASS, a time during which relaxation reduces the magnetization by e^{-2t_r/T_2} (for the sake of simplicity, these factors of relaxation during TOSS have not been included in the time signal of equation (6.32)).

The small number of t_1 increments required in (both versions of) this experiment permits the measurement of the 2D spectrum with a total of about 1024 scans (16 t_1 increments and a phase cycle with 32 scans for the real and the imaginary parts of equation (6.32)) if the signal is so strong that no further signal averaging is necessary. With data processing in terms of simple 2D Fourier transformation and with phase-

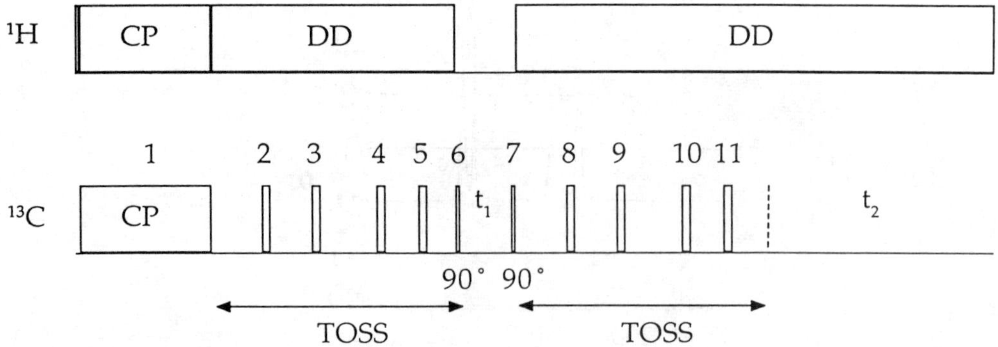

Figure 6.14 Pulse sequence for separating sidebands by their order with systematically incremented t_1. Note that during t_1, the only relevant magnetization component is z magnetization. Still, the rotor evolves, and the desired MAS signal results, as shown in the text. Adapted from Féaux de Lacroix *et al.* (1992).

cycled and possibly composite 180° pulses, the PASS technique might find applications in experiments on nuclei with large CS anisotropies $\delta > 5\omega_r$ where TOSS produces small or even negative intensities, or in 3D MAS experiments as described in Section 12.5.

The centreline ($M = 0$) spectrum, containing only isotropic peaks, is the sum over spectra for all t_1 values according to $I_{0,N} \sim \int dt_1\, e^{i0\omega_r t_1}\, g(t_1, \omega_2) = \int dt_1\, g(t_1, \omega_2)$. It can thus be generated simply by coadding the signals for all different $\omega_r t_1$ values directly in the FID, as a 1D experiment. Compared to TOSS, for spectra with many sidebands this PASS 1D spectrum of isotropic lines has the advantage of always being equal to the MAS centreband intensity, which is $\int d\alpha\, d\cos\beta\, |F(\alpha, \beta)|^2$ and therefore always positive, while the TOSS intensity $\int d\alpha\, d\cos\beta F(\alpha, \beta)$ can become negative or even vanish.

As for the signal-to-noise ratio, the PASS and the TOSS$-t_1-$deTOSS experiments for separating sidebands are comparable. Both contain TOSS-like sequences of comparable duration and consequently similar signal reduction by T_2 relaxation (effective T_2 including residual dipolar couplings). The single-t_r method for separation of sidebands by their order loses a factor of $\sqrt{2}$ by discarding one magnetization component during t_1 (see Section 4.9.5). On the other hand, no signal reduction by T_2 relaxation occurs during t_1 in this experiment. For the TOSS$-t_1-$deTOSS experiment, measuring up to a longer maximum t_1 evolution time t_{1aq} increases the signal, the noise, and the measuring time. The resulting signal-to-noise for the same total measuring time is approximately proportional to T_2/t_{1aq}, according to Section 4.9.3. Typically, $t_{1aq} = 2T_2$ is chosen to have the sidebands well represented. Then, the two factors of two cancel so that the overall signal-to-noise ratios in the experiments are similar.

6.6 Separation of dipolar patterns by isotropic shifts

In the preceding sections, we mainly considered 2D separation of chemical-shift anisotropy patterns. There are also other relevant anisotropies that have to be separated by the isotropic chemical shift in order to be related to the individual sites. In unlabelled polymers, the dipolar couplings of ^{13}C nuclei and the protons close by are of particular interest. This applies for pure heteronuclear $^{13}C-^{1}H$ couplings, which are well defined but require multiple-pulse techniques, but also for homonuclear (and heteronuclear) ^{1}H couplings close to a ^{13}C nucleus, which are particularly easy to detect. These experiments represent typical examples of the concept of 'separated local field' (SLF) spectroscopy, introduced by Waugh (1976) and first demonstrated by a separation of $^{13}C-^{1}H$ dipolar spectra by chemical-shift peaks in single crystals (Hester *et al.*, 1975, 1976; Rybaczewski *et al.*, 1977).

6.6.1 Separation of heteronuclear dipolar sideband patterns

The heteronuclear coupling between isolated ^{13}C spins and nearby protons, with decoupling of the $^{1}H-^{1}H$ interactions, can be related directly to the molecular structure: the heteronuclear dipole coupling is highly sensitive to the distances of the interacting nuclei as well as to the number and arrangement of protons coupled to the carbon considered. It can also serve to detect molecular motions since fast rotations average the dipolar interaction tensors. For the analysis, it is advantageous that the orientations of the dipolar interaction tensors (see Sections 2.4.2 and 2.5.6) with respect to the molecular structure are well known.

As indicated in Figure 6.15, ^{13}C magnetization that was generated by cross polarization evolves during t_1 under the action of the heteronuclear $^{1}H-^{13}C$ dipolar couplings, with homonuclear decoupling of protons effected by a multiple-pulse sequence such as MREV-8. By means of the 180° pulse, the ^{13}C chemical shift during evolution is refocused at the beginning of the detection period t_2. Under MAS conditions, the pulse must be applied only at a time nt_r, i.e., on a rotor echo, for a Hahn spin echo to form together with the next rotor echo. Therefore, between the end of the heteronuclear dipolar evolution time t_1 and the rotor echo with the 180° pulse, heteronuclear ^{1}H decoupling is applied. The carbon magnetization is observed during t_2 under standard ^{1}H heteronuclear decoupling. Thus, under MAS conditions heteronuclear dipolar sideband patterns for each distinguishable ^{13}C site are obtained along ω_1 in a 2D spectrum with the isotropic ^{13}C chemical shift along ω_2 ('dip-shift' separation). The envelope of the heteronuclear sideband intensities can be used to analyse the dipolar coupling for the purpose of detecting motional averaging. It can also be used for measuring $^{13}C-H$ or $^{15}N-H$ bond lengths (Roberts *et al.*, 1987), which is particularly interesting in systems with hydrogen bonds. An example concerning the base-pairing hydrogens in a model DNA fragment is shown in Figure 6.16 (DiVerdi and Opella, 1982). The spectrum proves directly that the H-bond is asymmetric. Comparison with simulations shows that the average distance of the proton to the nitrogen in the hydrogen-donor N–H moiety of the guanine base is 1.13 ± 0.05 Å, and that the proton is at least 1.5 Å away from the hydrogen-acceptor nitrogen on the cytosine in the base pair. This means that the hydrogen is not delocalized between donor and acceptor sites; a shift of the proton to the acceptor nitrogen is completely ruled out (DiVerdi and Opella, 1982).

Suppression of the chemical shift in the evolution period is achieved by simultaneous 180° pulses on both the *I* and the *S* spins in the middle of the evolution period. They

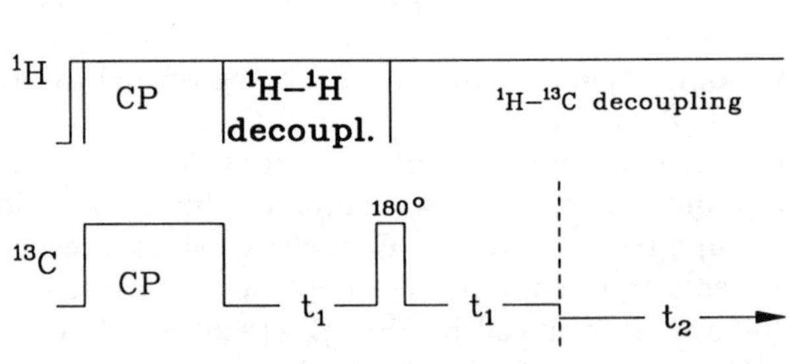

Figure 6.15 Pulse sequence for separated-local-field spectroscopy, with evolution of ^{13}C magnetization subject only to heteronuclear dipolar couplings, and detection of the chemical-shift spectrum. For homonuclear decoupling, various multiple-pulse sequences (MREV-8 etc.) can be used. Under fast MAS, the experiment can be used for separating heteronuclear dipolar sideband patterns by isotropic chemical shifts; here, the 180° pulse in the evolution period and the start of detection must be synchronized with rotational echoes; during the additional delays before and after the 180° pulse, $^{1}H-^{13}C$ decoupling is applied. Under conditions of slow MAS, the same sequence provides a correlation of heteronuclear-dipolar and chemical-shift sideband patterns. On static samples, or under OMAS, the sequence as shown in the figure serves to correlate (quasi)static powder patterns.

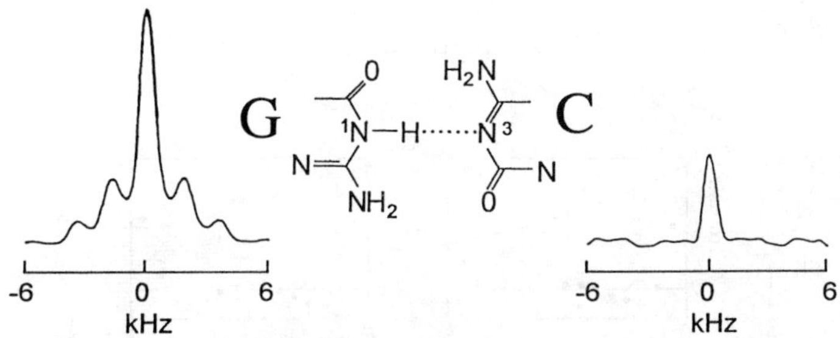

Figure 6.16 $^{15}N-^{1}H$ dipolar 2D separation spectroscopy of ^{15}N-labelled B-DNA (adapted from DiVerdi and Opella, 1982). The slices shown are taken at the ^{15}N isotropic chemical shifts of guanine (left) and cytosine (right) residues which interact by hydrogen bonds. The larger number of dipolar sidebands for the guanine is direct proof of the small distance between the guanine nitrogen and its bonded proton. Simulations of the sideband intensities yield a N—H bond length of 1.13 Å. A shift of the proton further towards the acceptor cytosine site can be directly excluded.

refocus the chemical shift, inverting $\hat{S}_z \rightarrow -\hat{S}_z$ in the interaction frames that toggle with the 180° pulses, but leave the dipolar couplings unaffected, $\hat{I}_z\hat{S}_z \rightarrow (-\hat{I}_z)(-\hat{S}_z)$. Since the observable time signal of the dipolar coupling is simply a cosine (Section 2.5.6), measuring the $\langle \cos \tilde{\omega}_1 t_1 \exp(i\tilde{\omega}_2 t_2) \rangle$ component of the signal is sufficient.

The signal for a single segment (crystallite) is:

$$
\begin{aligned}
g(t_1,t_2) &= \cos\left[\int_0^{t_1} \omega_d(\tilde{t})\, d\tilde{t}\right] \exp\left\{i \int_{2t_r}^{2t_r+t_2} \omega_c(t')\, dt'\right\} e^{i\omega_{iso}t_2} \\
&= \mathrm{Re}\left\{\exp\left[i \int_0^{t_1} \omega_d(\tilde{t})\, d\tilde{t}\right]\right\} f_c^*(\gamma + \omega_r 2t_r) f_c(\gamma + \omega_r 2t_r + \omega_r t_2)\, e^{i\omega_{iso}t_2} \\
&= \mathrm{Re}\{f_d^*(\gamma_d) f_d(\gamma_d + \omega_r t_1)\} f_c^*(\gamma) f_c(\gamma + \omega_r t_2)\, e^{i\omega_{iso}t_2} \qquad (6.33)
\end{aligned}
$$

where we have used the periodicity of the f functions, $f(\gamma + \omega_r 2t_r) = f(\gamma + 4\pi) = f(\gamma)$. In the separation experiment, the sidebands of the chemical-shift spectrum should be spun out, that is, $f_c^*(\gamma) f_c(\gamma + \omega_r t_2) = 1$. Then, Fourier transformation with respect to t_2 produces a spectrum of isotropic lines in which the individual lines are weighted with the t_1-dependent prefactor. As this factor is a real number, the phase-problem can be easily solved by setting the imaginary part to zero before the second FT. This eliminates the dispersive components in the chemical-shift dimension. The FT against t_1 produces the dipolar sideband pattern for each chemical-shift line, a typical separation spectrum.

The spectral information in ω_1 is contained in the t_1 signal sampled over a single rotor period. This was exploited by Schaefer, Stejskal and co-workers (1983, 1984b, 1985) to implement a shortened version of the experiment. Due to the 180° pulse between evolution and detection, the isotropic chemical shifts do not modulate the magnetization at t_1. Therefore, under ideal conditions the evolution is exactly periodic with t_r. In reality it is not, due to relaxation with $\exp(-t_1/T_2^*)$. Nevertheless, it can be made to appear periodic after compensation of the signal decay by multiplication with $\exp(+t_1/T_2^*)$, such that the intensities at $t = 0$ and $t = t_r$ are equal (Figure 6.17). Thus, t_1 must be sampled only within one rotor period. With typically 16 t_1 points sampled, 16 'delta-peak' sidebands spaced by ω_r appear as the sideband spectrum after the

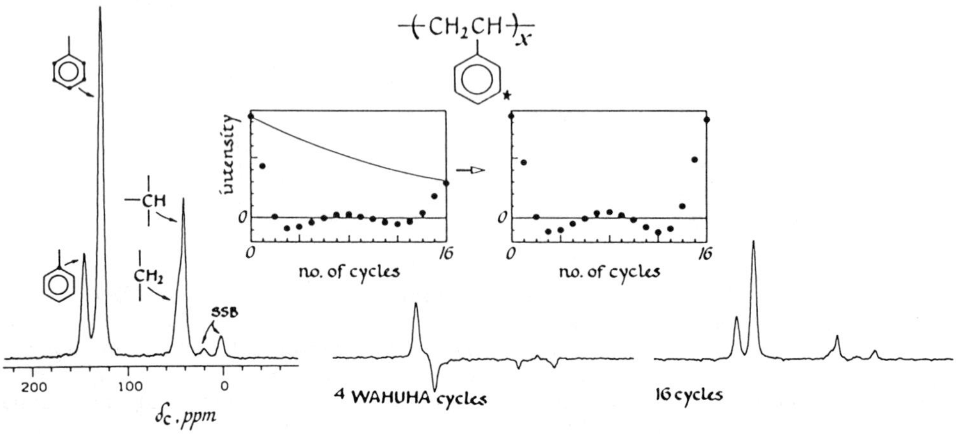

Figure 6.17 Dipolar rotational spin-echoes, resolved for different isotropic-chemical shift resonances, demonstrated for polystyrene (adapted from Schaefer *et al.*, 1983). The pulse sequence is essentially the same as the one shown and described in Figure 6.15. Note the different behaviour of the protonated and unprotonated aromatic sites. As shown in the inserts, before Fourier transformation over the evolution time to yield the spectrum in Figure 6.18(b), the measured points are made periodic with a full rotor period by multiplication with $\exp(t_1/T_2)$.

discrete FT, which automatically assumes a periodic continuation of the data points (see Section 4.2). Such stick spectra of various phenol groups are shown in Figure 6.18.

The procedure of making the signal periodic is based on the assumption of equal linewidths for all dipolar sidebands. This assumption cannot be expected to be generally valid, even though the sample rotation in general varies the dipolar coupling so that a given segment contributes to several sidebands. The dipolar centreband, for instance, is dominated by the signal from the C—H bonds close to the rotor axis which are always near the magic angle with the $\vec{\mathbf{B}}_0$ field, and therefore experience small heteronuclear dipolar couplings. Compared with this line, a significant increase in line-broadening is expected for the outer sidebands because these correspond to stronger heteronuclear dipolar couplings for which first-order average-Hamiltonian theory starts to break down.

The measurement of C—H bond lengths in synthetic polymers by means of the dip-shift MAS experiment is usually of relatively little interest. This is in part due to the error margins of the sideband patterns with their restricted resolution and the problems of the multipulse scaling factor, which amount to about 5% uncertainties for the bond lengths. Therefore, the applications of dip-shift separation to these materials have mainly involved the analysis of fast molecular motion of phenylene rings. As indicated in Section 2.5.6, Pake spectra with their characteristic 'horns' result only for isolated $^{1}H-^{13}C$ pairs. According to Section 2.5.6, CH_2 groups yield spectra with $\eta = 1$ which lack large splittings characteristic of the dipolar coupling. Heteronuclear dipolar spectra of rotating CH_3 groups consist of two Pake spectra with different splittings, which can hardly be exploited with the limited resolution of the anisotropy under MAS conditions. Only for methines and protonated phenylene carbons is the dipolar coupling to a single bonded proton dominating (still, the interaction of the ^{13}C phenylene spin to the neighbouring ortho-proton amounts to more than 3 kHz or 15%

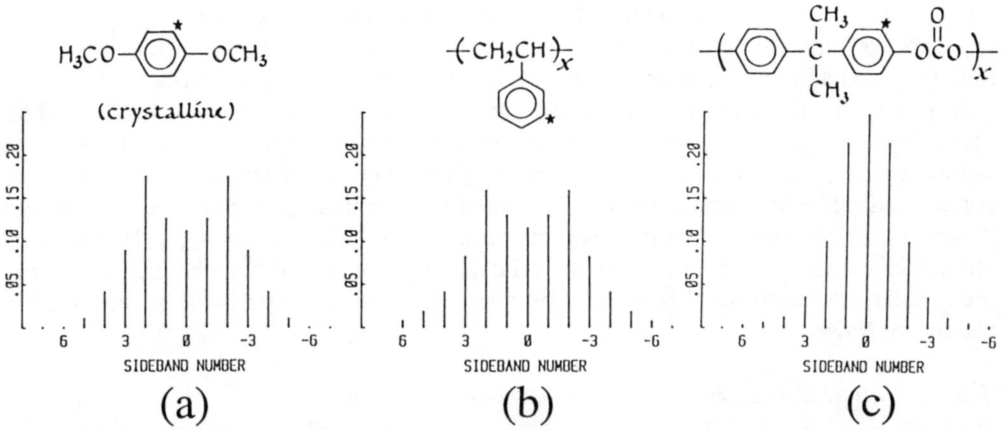

Figure 6.18 Dipolar $^{13}C-^{1}H$ MAS stick spectra of protonated aromatic carbons in phenylene rings, at ambient temperature (adapted from Schaefer *et al.*, 1984b). The spectra were obtained with 16 t_1 points, corrected as shown in Figure 6.17. The 'sticks', spaced by the rotor frequency, represent the spinning centreband and sidebands of the dipolar $^{13}C-^{1}H$ spectrum. (a) Spectrum of a rigid crystalline model compound. Note that the stickend-to-stickend envelope of the spectra, though resembling a Pake pattern with two horns, shows significant deviations from the Pake lineshape; this is due to the sample rotation. (b) Spectrum of polystyrene, showing only minor lineshape changes compared with the spectrum for rigid segments; this is in agreement with ^{2}H NMR spectra. (c) Spectrum of bisphenol-A polycarbonate. The observed significant lineshape change compared to the model compound can be attributed to phenylene-ring flips.

of the coupling of the bonded $^{1}H-^{13}C$ pair which is not independently spun out (see the sum and difference frequencies of Section 2.5.6), and leads to a distortion of the Pake pattern). Schaefer *et al.* (1984a, 1984b, 1985) have investigated the dynamics of the phenylene groups in various polymers by dip-shift separation. They have detected large-angle dynamics which must be attributed to phenylene ring flips (Figure 6.18). In addition, indications of small-angle motions are seen. By referencing the spectra to those of a crystalline model compound (Figure 6.18(a)), effects of possible experimental imperfections are considerably mitigated.

In terms of their relevance for characterizing molecular motions, the heteronuclear dipolar spectra have to compete with ^{2}H NMR spectra, which provide higher spectral resolution, yielding about 100 distinguishable lines in the spectrum (200 kHz/ 2 kHz = inhomogeneous/homogeneous linewidths) instead of the 12 lines typically available in a $^{13}C-^{1}H$ MAS experiment. More importantly, ^{2}H NMR can be employed over a wide range of motional rates (Spiess and Sillescu, 1981; Spiess, 1985) (see Sections 2.4.4 and 2.4.7), whereas the MAS experiment only shows whether the lineshape is motionally averaged by large-amplitude motion occurring with rates exceeding the 20 kHz of the dipolar couplings. The MAS experiment has the considerable advantage of not requiring isotopic enrichment. On the other hand, ^{2}H NMR is not restricted to phenylene rings or methine carbons and is relatively simple in terms of the pulse sequence and data evaluation. The widespread application of ^{2}H NMR even in industrial laboratories shows that the potential and simplicity of the ^{2}H experiments outweigh the enrichment aspect. The ^{13}C experiment has recently been applied to study the phenylene dynamics at the interface of heterogeneous blends of

^{13}C-enriched PC and ^{13}C-depleted PS which was also doped with a free-radical complex for dynamic nuclear polarization (Afeworki and Schaefer, 1992). It was found that PC chains at the interface have less motion than PC chains in the bulk.

If sidebands are tolerated in the chemical-shift dimension of the dip-shift MAS experiment, the potential for spectral separation is reduced. In that case, in order to obtain the dipolar powder pattern for a given site, all chemical-shift sidebands corresponding to the same isotropic shift must be summed up, which may be difficult if overlap of the chemical-shift sidebands occurs. On the other hand, the 2D sideband intensities contain information on the relative orientations of the chemical-shift and the dipolar interactions. This correlation aspect of the experiment will be discussed in Section 6.7.5.

6.6.2 Separation of heteronuclear powder patterns

Terao and co-workers (1986) applied the switching-angle sample spinning (SASS, see Section 6.4.3) technique to separate $^{1}H-^{13}C$ dipolar powder patterns (see Section 2.5.6) by isotropic ^{13}C chemical shifts. In the experiment, carbon magnetization evolves in a sample spinning off the magic angle under the action of heteronuclear $^{1}H-^{13}C$ dipolar couplings, while the $^{1}H-^{1}H$ dipolar couplings are suppressed by a suitable multiple-pulse sequence (e.g. MREV-8). After an integer number of rotation periods, a 90° pulse stores the magnetization along the z direction. Subsequently, the rotation axis is flipped to the magic angle. Following another 90° pulse, the ^{13}C magnetization is detected under MAS and ^{1}H dipolar decoupling. The evolution time is incremented by full rotation periods in order to avoid spinning sidebands in the ω_1 dimension. To prevent aliasing of the dipolar patterns, a sufficiently small OMAS scaling factor $\frac{1}{2}(3\cos^2\theta_r - 1)$ is chosen, that is, a rotation angle θ_r close to the magic angle. This also keeps flipping times, flipping angles and the concomitant probe-detuning problems small.

6.6.3 $^{1}H-^{13}C$ Pake patterns in unlabelled solids

For measuring internuclear distances, dipolar Pake patterns are particularly suitable since the 'splitting', that is, the distance of the 'horns' of the Pake spectrum, can be determined quite reliably even in the presence of considerable line-broadening. At first thought, one would expect to obtain these in the SLF experiment with separation of the dipolar sideband patterns by the isotropic ^{13}C shifts as described in the previous section. However, for ^{13}C subject to the dipolar coupling to the protons, Pake patterns are only observed in the case of relatively isolated CH groups. For CH_2 groups, the dipolar ^{13}C pattern consists of a superposition of two $\eta = 1$ patterns (see Section 2.5.6), which with some line-broadening becomes a featureless broad line. For CH_3 groups, the analysis is also difficult since the lineshape is a superposition of Pake patterns with different splittings.

Those complex non-Pake lineshapes for CH_2 groups arise from the simultaneous coupling of several protons to the ^{13}C spin under consideration (see Section 2.5.6). In contrast to this, under $^{1}H-^{1}H$ decoupling each proton close to the ^{13}C nucleus is subject only to the heteronuclear dipolar coupling of the single ^{13}C spin. The spectrum of the magnetization of such protons will thus be a Pake pattern (Nakai and Terao, 1992). Being by definition close to the ^{13}C nucleus, these protons are the first to cross-

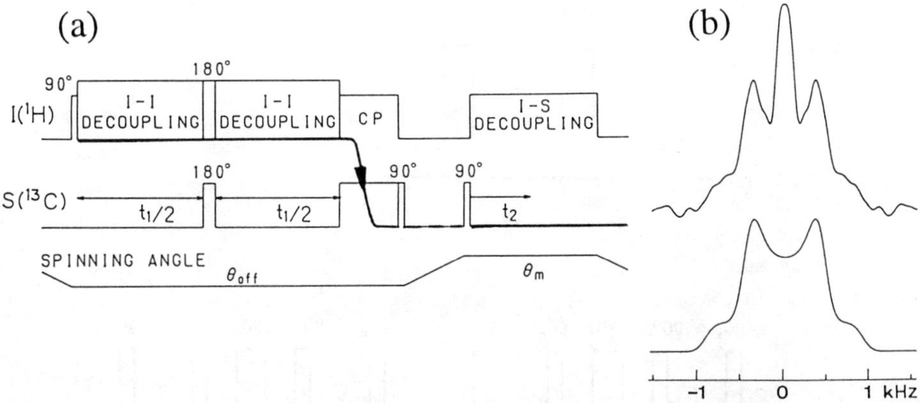

Figure 6.19 $^1H-^{13}C$ heteronuclear dipolar Pake spectrum from CH_2 groups. (a) Pulse sequence with switching-angle sample spinning. Note that in contrast to the SLF techniques discussed above, ^1H magnetization is evolving under the dipolar couplings during t_1. (b) Spectrum of POM. The dominating Pake powder spectrum yields a C—H bond length of 1.14 Å. The narrow peak in the centre of the spectrum is due to weaker $^{13}C-^1H$ couplings (adapted from Nakai and Terao, 1992).

polarize. Therefore, after a short CP contact time of the order of the inverse heteronuclear coupling, possibly with $^1H-^1H$ decoupling (see Section 3.4.1), their magnetization can be selectively recorded in terms of the amplitude modulation of the ^{13}C magnetization. The resulting Pake spectrum for POM (Nakai and Terao, 1992) is shown in Figure 6.19(b). The central peak in the experimental spectrum is due to the more distant protons subject to not so strong heteronuclear dipolar coupling. As indicated in Figure 6.19(a), the experiment by Nakai and co-workers (1992) was performed with SASS, i.e. ^1H heteronuclear dipolar evolution and $^1H-^{13}C$ cross polarization under off-magic-angle spinning and ^{13}C isotropic-shift detection under MAS, so that quasistatic dipolar patterns for individual ^{13}C sites are separated.

6.6.4 $^{13}C-^{13}C$ dipolar recovery at the magic angle (DRAMA)

Tycko and Dabbagh (1990) have introduced a technique for recovering full continuous (in the best case quasi-static) powder patterns for the homonuclear dipolar interaction of isolated spin pairs (in particular $^{13}C-^{13}C$ pairs) under fast MAS and heteronuclear decoupling. Rotation-synchronized 90° pulses are applied during the evolution period, changing the homonuclear dipolar Hamiltonian in the toggling frame in such a way that MAS is prevented from averaging out this dipolar interaction. As shown in Figure 6.20(a), per rotor cycle a pair of 90° pulses, of phases x and $-x$, is applied, spaced by a delay τ and centred around a time $\tilde{t}$. They toggle the toggling-frame Hamiltonian between $\hat{H}_{zz}(t)$ and $\hat{H}_{yy}(t)$, where the t-dependence is caused by the sample rotation. In the limit of $\tau \ll t_r$, the evolution is quasistatic during τ, under a toggling-frame propagator $e^{-i\tau\hat{H}_{yy}(\tilde{t})}$. At the end of the rotor period, the 'gap' in the $\hat{H}_{zz}$ evolution amounts to $e^{i\tau\hat{H}_{zz}(\tilde{t})}$. Consequently, the average-Hamiltonian propagator per rotor cycle is $e^{-i\tau\{\hat{H}_{yy}(\tilde{t})-(\hat{H}_{zz}(\tilde{t})\}}$, which results in a quasi-static dipolar pattern. In order to average out chemical-shift and resonance-offset effects, 180° ^{13}C pulses are applied at

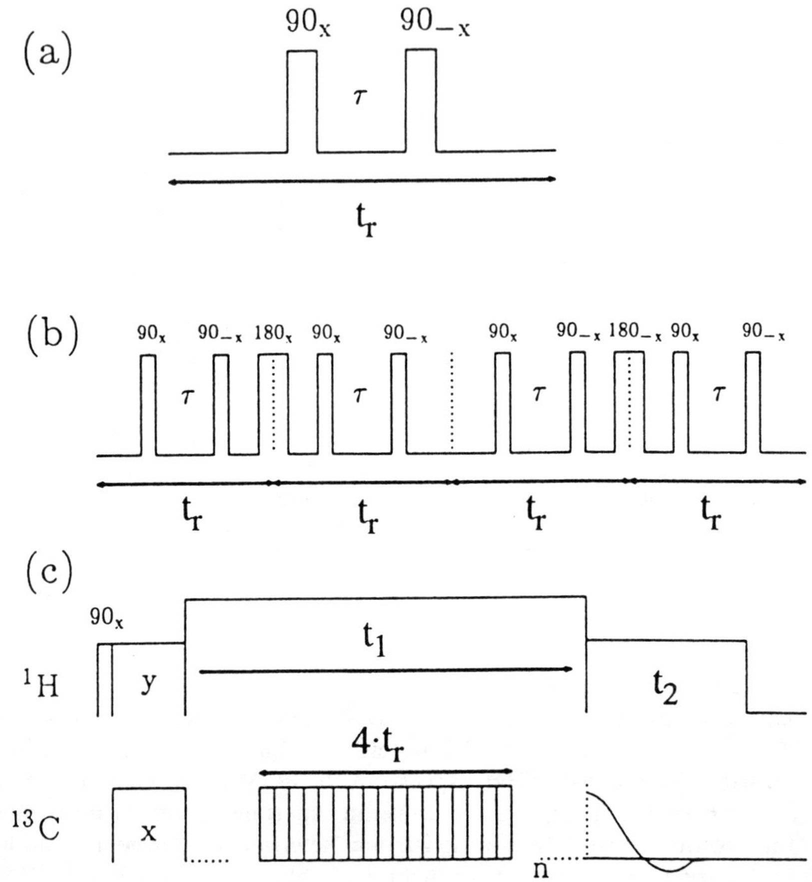

Figure 6.20 Pulse sequence for the DRAMA experiment (adapted from Tycko and Dabbagh, 1990). (a) Basic cycle with two 90° pulses within one rotation period. (b) Four rotation periods with the cycle of (a) and two 180° pulses for refocusing the chemical shift. (c) 2D pulse sequence with the sequence of (b) applied during t_1.

times $n \times 2t_r$ (Figure 6.20(b)). If τ is made longer, the quasistatic approximation breaks down, but the effective dipolar evolution becomes longer, which is advantageous for measuring small dipolar couplings. The internuclear distance is then determined by comparison of experimental and simulated spectra. Figure 6.20(c) shows the DRAMA sequence included in a 2D experiment, with MAS detection providing separation of the dipolar DRAMA patterns. In practice, the DRAMA approach will generally require doubly labelled materials in order to have a sufficient number of isolated homonuclear spin pairs.

6.6.5 Rotational-echo double resonance (REDOR)

While the experiments described so far aim at determining $^1H-^{13}C$, $^1H-^1H$ or $^{13}C-^{13}C$ dipolar couplings, rotational-echo double resonance (REDOR) experiments (Gullion and Schaefer, 1989) are designed to measure weak heteronuclear dipolar couplings, for example, of ^{15}N and ^{13}C (Gullion and Schaefer, 1989), or ^{2}H and ^{13}C (Schmidt *et al.*, 1992), in doubly labelled materials. The principle of the REDOR experiment is to change the heteronuclear dipolar interaction under MAS conditions through 180° pulses on the ^{15}N spins in $^{15}N-^{13}C$ spin pairs, thereby measurably decreasing the height of the following ^{13}C rotational echoes since the pulses prevent MAS from completely refocusing the weak heteronuclear $^{13}C-^{15}N$ dipolar interactions. The stronger this interaction, the more pronounced the echo decay will be. The reference echo height is measured without pulsing on the ^{15}N channel. In practice, a problem arises from the background signal from the uncoupled ^{13}C spins, which distorts the ratio of the echo intensities.

The REDOR technique is demanding both in terms of the hardware and sample preparation. Though the acronym contains three letters for double resonance, REDOR in practice requires triple-resonance equipment since protons must be decoupled while transverse magnetization of the observed spin species evolves. Selective double labelling of a certain percentage of the sites of interest is required to obtain a sufficiently strong change in the rotational-echo intensity.

6.6.6 *^{1}H-Wideline separation (WISE)*

The widths of proton lines, as caused by the homonuclear dipolar couplings, have long been used to obtain qualitative information on molecular dynamics in polymers and other organic materials: as indicated in Section 2.5.7, rigid materials exhibit broad ^{1}H lines due to the static dipolar couplings, while molecular mobility averages these couplings and thus reduces the linewidth. In the limit of extremely fast isotropic motion, narrow lines result as known from solution NMR. In heterogeneous systems, however, the dynamics of the individual components, though of considerable interest, cannot be derived from the 1D proton spectrum without *a priori* assumptions. In particular, immobilized 'soft segments' cannot be distinguished from the intrinsically rigid 'hard segments'. This problem can be overcome by wideline separation (WISE) (Schmidt-Rohr *et al.*, 1992) via ^{13}C isotropic chemical shifts. The simplest version of this convenient experiment derives from the standard CP–MAS(–TOSS) experiment simply by placing an incremented delay t_1 after the initial ^{1}H 90° pulse. This allows the detection of the decay of the proton magnetization through the amplitude modulation of the ^{13}C signal generated by cross polarization, possibly with sideband suppression by TOSS. Fourier transformations over t_1 and t_2 provide ^{1}H wideline spectra separated by the corresponding ^{13}C isotropic shifts. Figure 6.21 displays the application to a diblock copolymer of PS and PDMS, where the PS is found to be rigid and the PDMS highly mobile in spite of the chemical connectivity and 20-nm-scale spatial proximity of the two components. WISE spectra of the aliphatic $C_{16}H_{35}$ sidechains of various stiff macromolecules, Figure 6.22 (Clauss *et al.*, 1992), show a surprising variety of ^{1}H wideline shapes, revealing differences in dynamics and structure of the short aliphatic chains, as indicated schematically in the figure.

It may be interesting to note that it is not only the $^{1}H-^{1}H$, but also the $^{1}H-^{13}C$ coupling that affects the magnetization decay of those protons that are close enough to the ^{13}C nuclei to cross-polarize them. In fact, a ^{13}C nucleus bonded to a proton structurally isolated from the other protons can exhibit sharp spinning sidebands with a Pake-like envelope, since the $^{13}C-^{1}H$ pair interaction dominates the evolution of the single proton cross-polarizing the carbon. The information on the dynamical state of

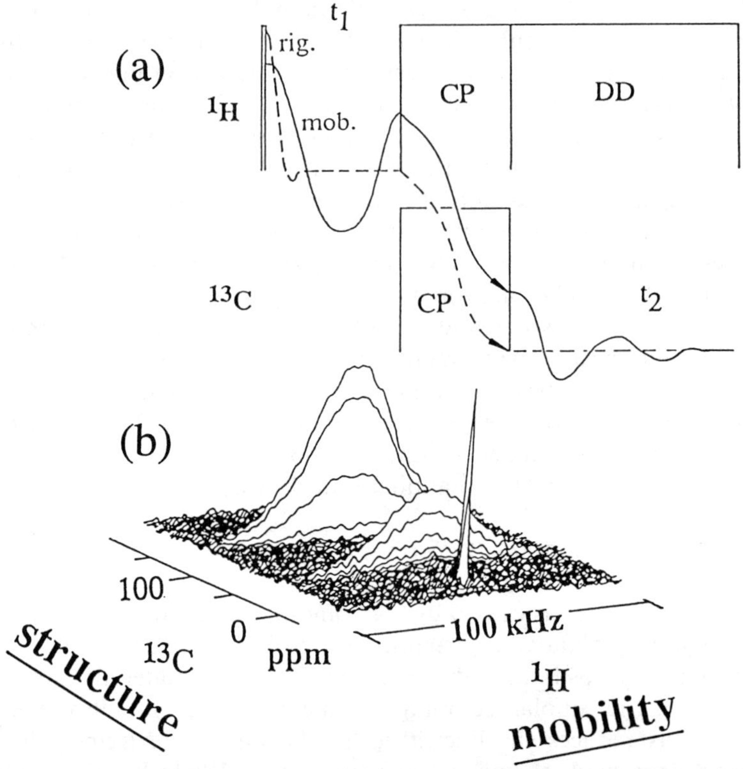

Figure 6.21 (a) Simplest pulse sequence for WISE 2D NMR. Note that it differs from the standard CP-MAS sequence only by the incremented t_1 delay before the cross polarization. To suppress spinning sidebands in the ^{13}C dimension, the TOSS sequence can be appended after CP. (b) WISE spectrum of a symmetric PS-b-PDMS block copolymer (total $M_w = 60$ kg/mol). The polystyrene is shown to be completely rigid, and the poly(dimethyl siloxane), with its single peak close to 0 ppm, highly mobile, in spite of their chemical connection and close spatial proximity on a 20 nm length scale also proven by spin diffusion; see Section 13.2.5 (adapted from Schmidt-Rohr *et al.*, 1992).

the segments is not impaired by these $^1H-^{13}C$ couplings, since $^1H-^1H$ and $^1H-^{13}C$ dipolar interactions are affected by motional averaging in a similar way. ^1H spin diffusion in the WISE experiment, both undesirable during CP and desirable during a proper mixing time, is discussed in Section 13.2.5.

The experimental technique useful for wideline separation can also be employed in a somewhat different manner, namely for determining whether the protons detected in narrow proton lines are in close proximity to the ^{13}C or ^{29}Si nuclei detected in a ^{13}C or ^{29}Si spectrum. If this is the case, the 2D spectrum shows a ^1H lineshape identical to the 1D ^1H wideline spectrum itself. From this perspective, the experiment was

Figure 6.22 WISE spectra of aliphatic side chains of various stiff macromolecules, showing characteristic lineshape differences (adapted from Clauss *et al.*, 1992). Structural models obtained from WISE as well as from ^{13}C chemical-shift and ^1H spin diffusion data are shown below the spectra.

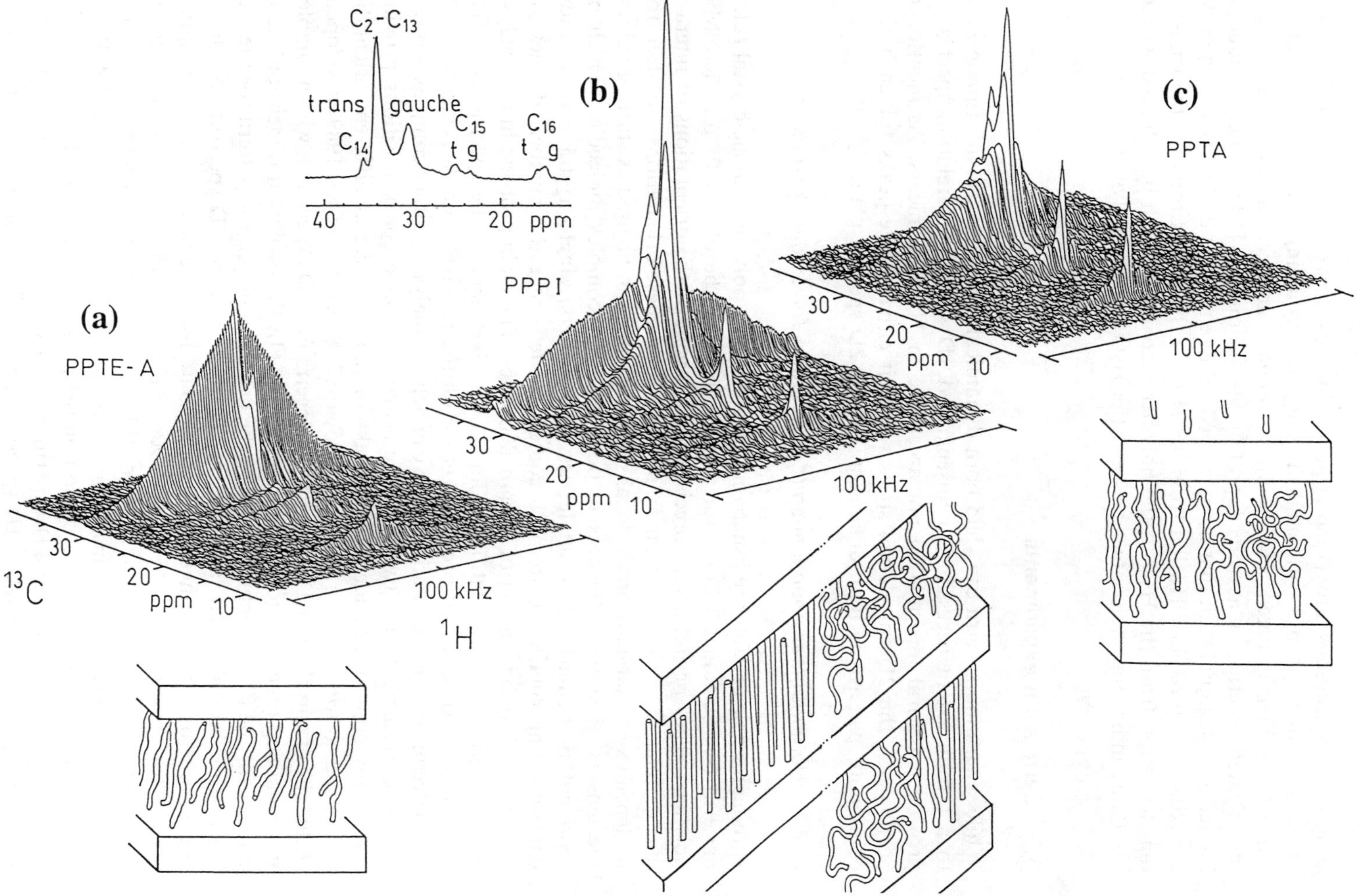

(a)
PPTE- A
(b)
PPPI
(c)
PPTA
13C
C2-C13
trans
gauche
C14
C15
C16
t g
t g
40
30
20
ppm
10
100 kHz
1H

conceived by Zumbulyadis (1986) who could thus prove the close proximity of all protons, including those with small homonuclear couplings, to the silicon nuclei in amorphous hydrogenated silicon. In a similar manner, in polymers it can be investigated whether a narrowed proton line is due to H_2O, which produces no immediate CP signal such that no narrowed ^1H line appears in the 2D spectrum. With a mixing time for spin diffusion from the H_2O to the matrix protons, the transfer of magnetization from the H_2O to the polymer matrix can often be achieved within 10 to 50 ms, indicating the contact between the H_2O and the polymer.

6.7 Correlation experiments

While separation spectra can yield both dynamical and structural information that is to a certain extent accessible in unseparated 1D spectra, correlation spectra often provide structural information not accessible in 1D experiments. As mentioned in Section 6.3, the 2D correlation spectra exhibit either single peaks for each site, or continuous patterns in both dimensions of the 2D spectral plane.

6.7.1 Homonuclear chemical-shift correlation for single crystals

Although single crystals are usually not available for polymers, single-crystal rotation patterns (see Section 2.4.2) of other compounds can be of benefit for the NMR of polymers. They permit a very precise measurement of the orientations of interaction tensors relative to the molecular frame, and since these orientations are often found to depend systematically on the local structure, they can be used as starting values for the tensors of similar segments in polymers. For complex molecules, which often represent particularly close analogues to polymers, single-crystal spectra and the corresponding rotation patterns become crowded since the number of lines grows proportional to the number of inequivalent sites. The main problem in the analysis of such spectra is the correlation of lines of a given site in the spectra for different orientations. Although fit programs based on the relatively simple dependence (2.44) of the frequency on the rotation angle can help in analysing rather complex sets of line positions (Heuer, 1990), two-dimensional correlation of peak positions provides a convincing approach to identifying peaks of the same site at different orientations. Grant and co-workers (Carter *et al.*, 1985, 1987; Sherwood *et al.*, 1989) developed the technique of correlating 1D single-crystal spectra in a 2D experiment with a sample flip in the mixing time, as shown in Figure 6.23(a). In the evolution dimension the peak pattern for the first orientation, and in the detection period the pattern for the second orientation, are obtained, and they are correlated in the 2D spectrum as shown in Figure 6.23(b). The flip angle most appropriate for the analysis is 45°: arranging 2D spectra for flips $0° \rightarrow 45°$, $45° \rightarrow 90°$, $90° \rightarrow 135°$, and $135° \rightarrow 180°$ in a square array such that like projections are adjacent (Figure 6.24), the four peaks of each site are located at the corners of a square (the proof is on the basis of equation 2.44)), which provides a redundancy of information on the peak positions that allows it to tolerate overlap of peaks. The four 2D experiments provide the correlations of the six flip experiments that would relate 1D spectra taken at 0° (= 180°), 45°, 90°, and 135°. This information is sufficient for determining the full information available from 1D spectra for rotation around one axis, providing three of the six elements of $\boldsymbol{\sigma}$ in the

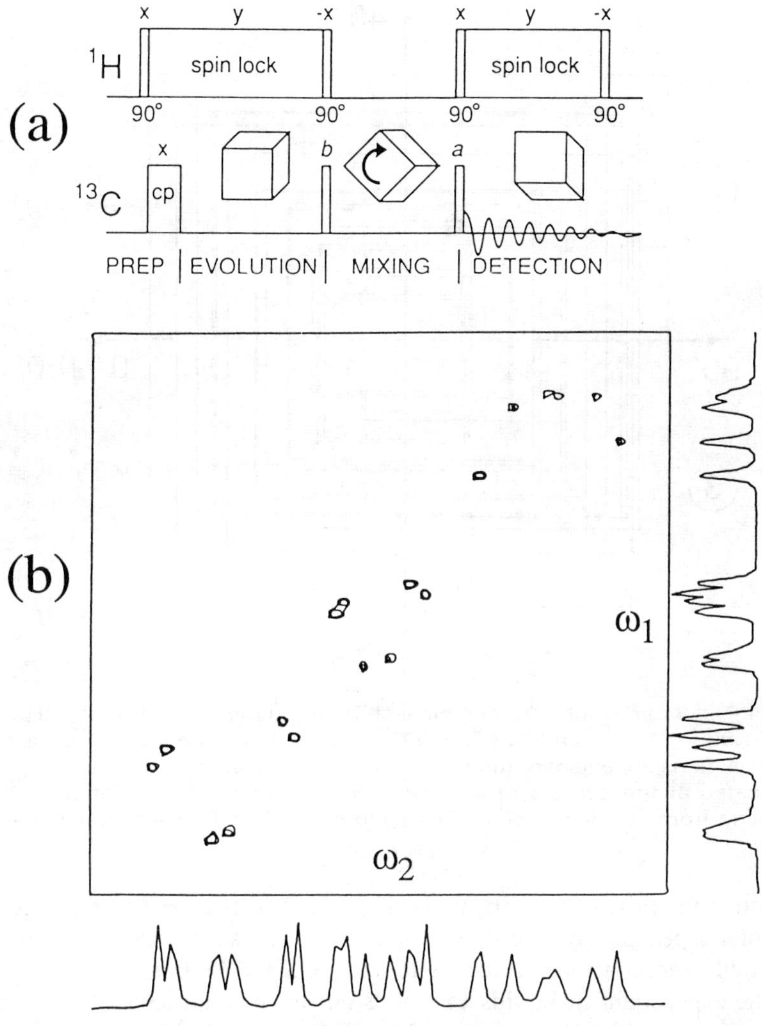

Figure 6.23 Homonuclear ^{13}C chemical-shift correlation experiment on single crystals (adapted from Carter *et al.*, 1985). (a) Pulse sequence with sample flip in the mixing time. (b) Typical chemical-shift correlation 2D spectrum. The single-crystal 1D spectra at two different orientations (shown on the axes of the 2D spectrum) are correlated by the peak positions in the 2D plane.

crystal-fixed frame. Rotations around further two orthogonal axes yield the other elements, so that after diagonalization, the tensor in the principal-axes system is obtained. Under optimum conditions, an estimated number of 100 carbon sites can be analysed by this technique.

6.7.2 Heteronuclear correlation of isotropic chemical shifts

In liquids, two-dimensional correlation of ^{1}H and ^{13}C 1D spectra, which are both highly resolved, is of considerable relevance since it can provide additional structural

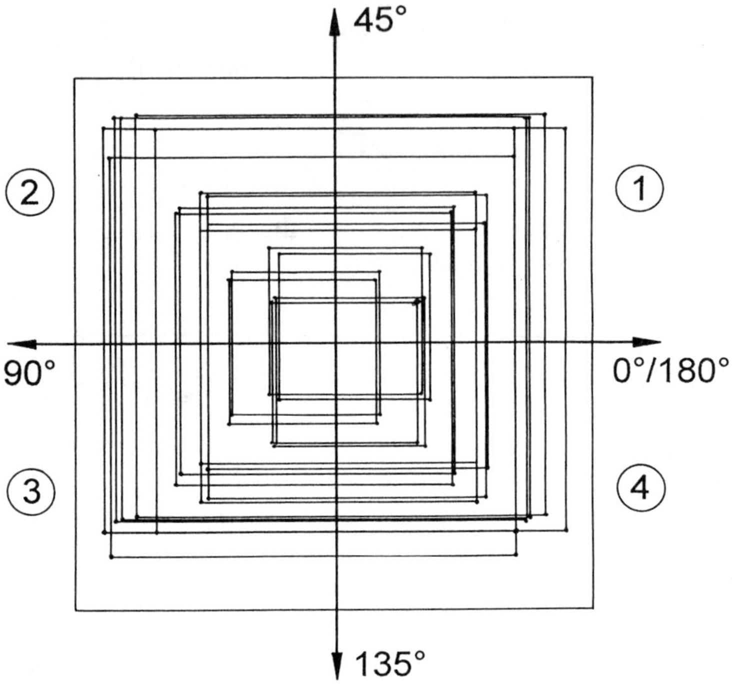

Figure 6.24 Four different ^{13}C chemical-shift correlation spectra, for flips $0° \rightarrow 45°$, $45° \rightarrow 90°$, $90° \rightarrow 135°$, and $135° \rightarrow 180°$, are shown arranged in a square array such that like projections are adjacent (adapted from Carter *et al.*, 1987). The four peaks of each site are located at the corners of a square, which reduces peak identification problems arising from accidental peak overlap in one of the four individual spectra.

information. In solids, even with the best CRAMPS, the ^1H spectrum contains only a few resolved peaks. Therefore, this heteronuclear shift-correlation experiment in solids partially serves to separate overlapping CRAMPS lines. A further difference between the experiment in liquids vs. solids lies in the character of the heteronuclear interaction that provides the transfer of magnetization. The heteronuclear J coupling in liquids is a through-bond interaction, while the dipolar interaction in solids acts through space. The transfer in solids thus can be of a less local nature, which may complicate the analysis of the spectrum. On the other hand, $^1H-^{13}C$ correlation peaks between unprotonated carbons and (nonbonded) protons, indicating the spatial proximity of those nuclei, can provide valuable information on the conformation of a molecule. In addition, the experiment can be complemented by a mixing time for the ^1H magnetization before the CP. This allows ^1H spin diffusion to be followed with the high resolution of the ^{13}C spectrum in the detection period (see Section 13.2.4).

The experiment as schematically shown in Figure 6.25 requires proton evolution under dipolar decoupling, homonuclear (e.g. by means of MREV-8; Roberts *et al.*, 1984) as well as heteronuclear (e.g., by ^{13}C 180° pulses as mentioned at the end of Section 3.4.1). $^{13}C-^1H$ decoupling is required because protons detected via the ^{13}C magnetization are necessarily close to the ^{13}C nuclei and therefore subject to the $^1H-^{13}C$ coupling which can reach up to 25 kHz (scaled to about 12 kHz by MREV-8).

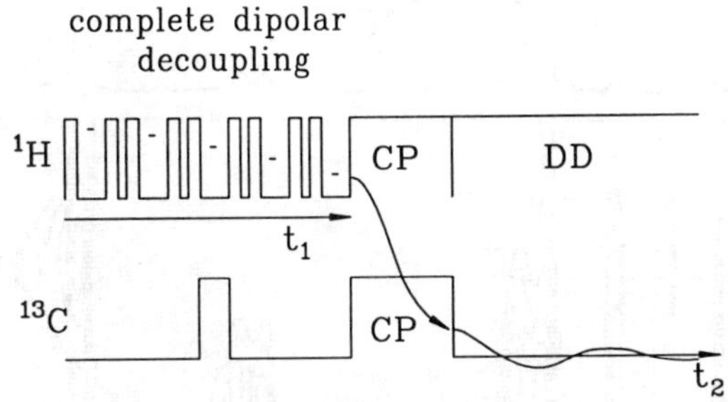

Figure 6.25 Pulse sequence for ^{1}H–^{13}C heteronuclear correlation of chemical shifts. Various multiple-pulse decoupling sequences can be employed in the evolution period; that includes windowless ones because the ^{1}H magnetization is not directly detected. During the cross-polarization, no ^{1}H spin diffusion should occur. This can be achieved using either cross-polarization sequences with homonuclear decoupling (see Section 3.4.3) or a short ($\sim 50\ \mu$s) Hartmann–Hahn cross-polarization time. Detection of ^{13}C magnetization is standard with ^{1}H–^{13}C decoupling.

In the original experiments by Caravatti *et al.* (1982, 1983), their own BLEW-12 multiple-pulse sequence was applied for homonuclear decoupling, and the WALTZ sequence was employed for heteronuclear decoupling in the presence of large ^{13}C chemical shifts. The detection is conventional ^{13}C MAS with heteronuclear dipolar decoupling of protons and possibly TOSS for sideband suppression.

The main challenge is the mixing process which should consist of cross polarization from ^{1}H to ^{13}C spins without redistribution of proton magnetization by proton spin diffusion as caused by long Hartmann–Hahn CP (Section 3.4.3). One can try cross-polarization sequences with homonuclear decoupling of protons by means of synchronous multipulse irradiation on both ^{13}C and ^{1}H as discussed at the end of Section 3.4.3. The WIM sequence (Caravatti *et al.*, 1983), which averages all Hamiltonians to zero, except for the ^{13}C–^{1}H coupling, provides only very inefficient polarization of unprotonated carbons because MAS refocuses ^{13}C–^{1}H pair couplings. They are usually much stronger under synchronous ^{1}H and ^{13}C MREV-8 pulsing after 'locking' the ^{1}H magnetization along the effective field of MREV-8; MAS cannot average out the heteronuclear Hamiltonian that interferes with the offset Hamiltonian not eliminated under MREV-8. This interference, on the other hand, results in a troublesome dependence of ^{13}C intensities on the irradiation frequencies. A cross-polarization procedure producing essentially the same results as isotropic mixing, but in a convenient standard experiment, consists simply of a short mixing time with the ^{1}H magnetization along z (for a duration of one or two 360° pulse lengths) followed by Hartmann–Hahn cross-polarization of twice that duration: the dipolar coupling in the z period compensates the dipolar coupling scaled by $-\frac{1}{2}$ during the Hartmann-Hahn CP, because during that period of less than 50 μs the rotor can be treated as approximately static.

The heteronuclear correlation experiment has been applied to study polymer conformations and polymer blends (see also section 13.2.4). Figure 6.26 portrays a spectrum of poly(phenylene oxide) PXE (Bielecki *et al.*, 1991). The correlation peaks of unprotonated carbon sites in particular can provide useful structural information since they establish the proximity of those sites to certain protons in the structure.

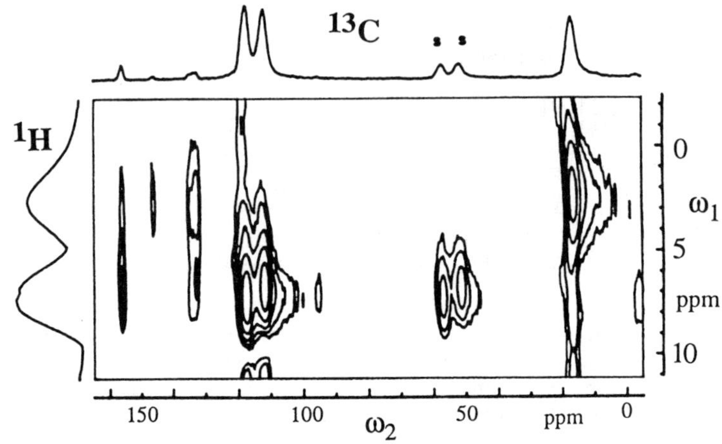

Figure 6.26 2D ^{1}H–^{13}C isotropic-shift correlation spectrum of PXE (adapted from Bielecki *et al.*, 1991). The bands in the ^{1}H dimension are quite broad, so the ^{13}C dimension also provides separation. Different ^{1}H peak positions for two sets of aromatic protons can be read off the 2D spectrum.

6.7.3 Correlation of static ^{13}C chemical-shift and ^{13}C–^{1}H-dipolar patterns

The orientation of chemical-shift tensors can be obtained, even in powders, by correlating the chemical-shift anisotropy with the dipolar ^{13}C–^{1}H interaction whose tensor orientation is precisely known. By contrast, tensor-orientation analysis from rotation patterns as well as the related 2D chemical-shift correlation experiment (Section 6.7.1) require single crystals of macroscopic size which are rarely obtainable from synthetic polymers. While dip-shift correlation data had previously been analysed in the t_1–ω_2 domain only (Stoll *et al.*, 1976), absorptive two-dimensional dip-shift correlation spectroscopy as introduced by Linder *et al.* (1980) was the first truly two-dimensional experiment performed on static powder samples. These authors have also provided an analysis of the ridge patterns of the two-dimensional spectral intensity which can be applied to other 2D powder spectra as well.

The ridge patterns in these two-dimensional powder spectra characterize the relative orientation of the interaction tensors probed in the experiment. Similar patterns, often with elliptical ridges, are found in anisotropy exchange experiments (Chapter 7). There, the orientations of the same tensor at different times are correlated. Consequently, those 2D patterns contain information on the geometry of the molecular reorientation (see Section 7.4) or on molecular orientation if the change in orientation is induced externally by flipping the sample (Sections 6.7.1 and 12.7). In particularly simple cases, for instance, when both tensors are axially symmetric, a simple explanation for the elliptical ridges can be given without calculations (Section 7.3). Generally, the calculation and analysis of two-dimensional lineshapes rely on similar principles as those invoked for the 1D powder patterns discussed in Section 2.4.3. The spectral intensity $S(\omega_1, \omega_2)$ is first of all proportional to the number of those molecules for which $\omega_1 = \tilde{\omega}_1(\Theta, \Phi)$ and $\omega_2 = \tilde{\omega}_2(\Theta, \Phi)$, where (Θ, Φ) characterize the $\vec{\mathbf{B}}_0$ field orientation in some frame that is fixed in a unique way relative to the respective crystallite or chain unit; this frame could be the PAS of the interaction relevant during

t_1 such that $\Theta = \theta_1$ and $\Phi = \phi_1$. However, it is a second factor that leads to the ridge pattern: namely, the transformation from the angular surface element $d\Omega(\Theta, \Phi, d\omega_1, d\omega_2)$ to the corresponding spectral area $d\omega_1\, d\omega_2$. Conservation of integral intensity in corresponding angular and spectral surface elements requires:

$$S(\tilde{\omega}_1(\Theta, \Phi),\ \tilde{\omega}_2(\Theta, \Phi))\ d\omega_1\ d\omega_2 = P(\Theta, \Phi)\ d\Omega(\Theta, \Phi,\ d\omega_1, d\omega_2). \tag{6.34}$$

The transformation of surface elements between different but related sets of coordinates, $(\tilde{\omega}_1, \tilde{\omega}_2)$ and (Θ, Φ) in our case, is well known from multidimensional integrations and is described in Appendix E. The distortion of the surface elements is given by the Jacobian (determinant):

$$\partial(\tilde{\omega}_1, \tilde{\omega}_2)/\partial(\Theta, \Phi) := \begin{vmatrix} \dfrac{\partial\tilde{\omega}_1}{\partial\Theta} & \dfrac{\partial\tilde{\omega}_2}{\partial\Theta} \\ \dfrac{\partial\tilde{\omega}_1}{\partial\Phi} & \dfrac{\partial\tilde{\omega}_2}{\partial\Phi} \end{vmatrix} \equiv |\,\mathrm{grad}_{\Theta,\Phi}(\tilde{\omega}_1) \times \mathrm{grad}_{\Theta,\Phi}(\tilde{\omega}_2)\,|$$

$$d\omega_1\ d\omega_2 = \partial(\tilde{\omega}_1, \tilde{\omega}_2)/\partial(\Theta, \Phi)\ d\Omega(\Theta, \Phi, d\omega_1, d\omega_2) \tag{6.35}$$

(see again Appendix E). With this relation inserted, the conservation of integral intensity, equation (6.34), yields:

$$S(\tilde{\omega}_1(\Theta, \Phi), \tilde{\omega}_2(\Theta, \Phi)) = P(\Theta, \Phi)/\,[\,\partial(\tilde{\omega}_1, \tilde{\omega}_2)/\partial(\Theta, \Phi\,]\,. \tag{6.36}$$

The ridges in the 2D spectrum are the (integrable) singularities that occur where $\partial(\tilde{\omega}_1, \tilde{\omega}_2)/\partial(\Theta, \Phi) = 0$. A geometric interpretation of their appearance at these points is obtained by writing, according to equation (6.35), the Jacobian as the absolute value of the vector product of the gradients; the singularities appear where:

$$|\,\mathrm{grad}_{\Theta,\Phi}(\tilde{\omega}_1) \times \mathrm{grad}_{\Theta,\Phi}(\tilde{\omega}_2)\,| \equiv \partial(\tilde{\omega}_1, \tilde{\omega}_2)/\partial(\Theta, \Phi) = 0. \tag{6.37}$$

This victor product vanishes in two cases:

(a) where one of the gradients is zero, that is, at the extreme or saddle points of the angle-dependent frequencies
(b) where the gradients are parallel, or equivalently, where the equifrequency lines of the interaction tensors run locally parallel (to understand the equivalence, remember that gradients and equifrequency lines are always perpendicular to one another). It is intuitively clear that for a given $d\omega_1$ and $d\omega_2$ the angular surface element corresponding to the 2D interval $[\tilde{\omega}_1 \ldots \tilde{\omega}_1 + d\omega_1]$. $[\tilde{\omega}_2 \ldots \tilde{\omega}_2 + d\omega_2]$ exhibits a large extension along the equifrequency lines if these are locally parallel. Therefore, the surface area and the spectral intensity assumes an extremum in this case, in fact even a singularity. Figure 6.27(a) shows the equifrequency lines of a pair of chemical-shift and $^{13}C{-}^{1}H$ dipolar interactions on the unit sphere. The thick line marks the locations of all the points where the equifrequency lines of the two interactions run locally parallel. The corresponding ridge pattern is shown in Figure 6.27(b). Superimposed is the pattern for the second transition of the dipolar interaction, to yield an overall pattern symmetric with respect to $\omega_1 = 0$.

Condition (b) yields those ridges which are most sensitive to the relative orientations of the correlated tensors. As seen in the examples of Figure 6.28, these ridges often trace out elliptical patterns. In contrast to the ellipses in 2D exchange powder spectra (see Sections 7.3, 7.7 and 12.7.2), for the ellipses in heteronuclear spectra the exact

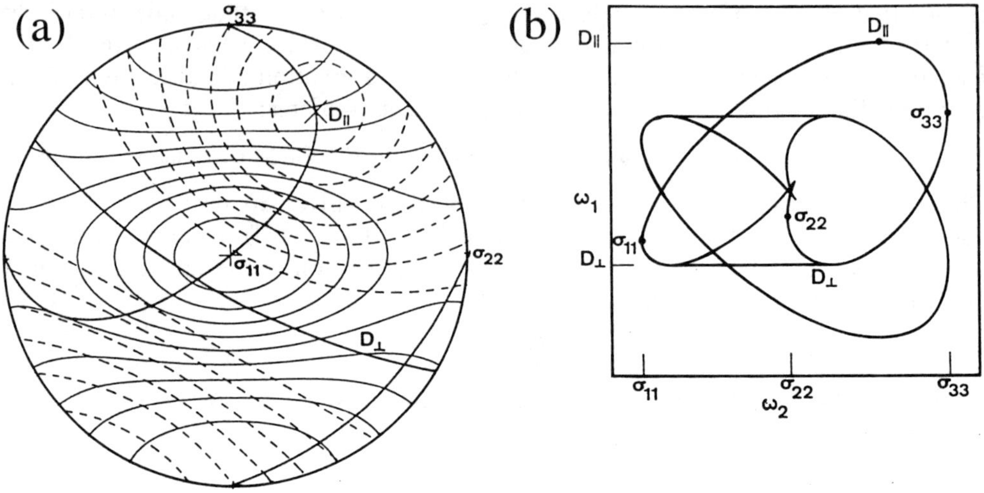

Figure 6.27 (a) Equifrequency lines of the chemical-shift interaction (thinner continuous lines) and of the uniaxial dipolar interaction (dashed lines) shown in a stereographic projection of the unit sphere (adapted from Linder *et al.*, 1980). The heavy lines mark the orientations where the chemical-shift and dipolar equifrequency lines run locally parallel. The corresponding frequency pairs (ω_{dip}, ω_{cs}) define the location of the singularities in the 2D spectrum. These produce characteristic ridge patterns as shown in (b). Since the dipolar interaction always produces doublets of lines symmetric with respect to $\omega = 0$, a second ridge pattern, mirror-symmetric to the first, is also observed.

shapes can be altered by changing the scaling factor of one of the axes in the plot of the 2D spectrum. The axes correspond to different interactions and therefore their relative scales are arbitrary.

To keep the derivation as simple as possible, we have so far disregarded the fact that several (Θ, Φ) values contribute at the same $(\tilde{\omega}_1, \tilde{\omega}_2)$. This is taken into account simply by adding those contributions in (6.34) (see Appendix E).

In the experiment, for isolated C–H pairs there are three unknowns to be determined: the polar angles β and α of the dipolar coupling axis (the $^{13}C-^{1}H$ bond) in the chemical-shift PAS, and the strength δ_D of the dipolar coupling. These parameters also influence the shape of 1D spectra but much less so than the 2D patterns. As was shown in Section 2.5.6, 1D spectra of ^{13}C subject to C–H dipolar coupling and chemical shift consist simply of two powder patterns, for $(\boldsymbol{\sigma} + \frac{1}{2}\mathbf{D})_{zz}$ and $(\boldsymbol{\sigma} - \frac{1}{2}\mathbf{D})_{zz}$, with equal isotropic values $\boldsymbol{\sigma}_{iso}$. Thus, the 1D powder spectrum of C–H groups under homonuclear ^{1}H decouping can yield four parameters: the anisotropy parameters δ^{s} and δ^{d} for $(\boldsymbol{\sigma} + \frac{1}{2}\mathbf{D})_{zz}$ and $(\boldsymbol{\sigma} - \frac{1}{2}\mathbf{D})_{zz}$, respectively, and the two corresponding asymmetry parameters η^{s} and η^{d}. Therefore, in principle the three desired parameters can be extracted from 1D spectra. However, as demonstrated by the examples shown in Figure 2.19, the 1D spectra all look more or less alike so that the accuracy obtained from 1D experiments is very limited. With line-broadening, secondary dipolar interactions, or overlap of other signals, the desired angles can hardly be determined. In the 2D spectrum, the redundancy of the information is so great that these adverse factors can be tolerated. Nonetheless, it must be mentioned that the parameters that can be extracted from the spectrum do not determine the

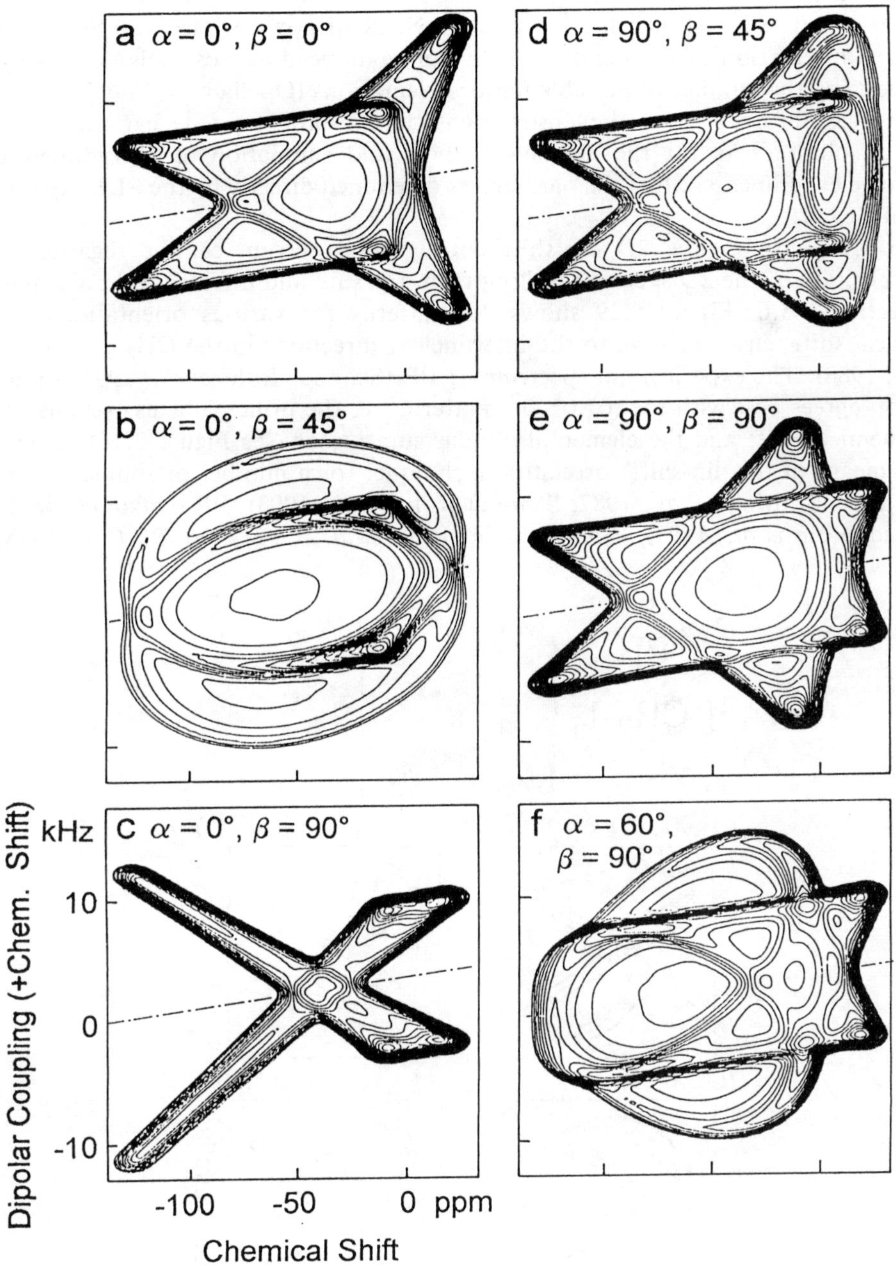

Figure 6.28 Examples of $^{1}H-^{13}C$-pair dip-shift patterns for various relative orientations of chemical-shift and dipolar principal-axes systems, given in terms of the Euler angles α and β, which are the polar coordinates of the C–H internuclear vector in the chemical-shift PAS (adapted from Linder *et al.*, 1980). The third Euler angle γ is irrelevant due to the uniaxiality of the dipolar-pair interaction. The spectra are slanting because the ω_1 frequency was calculated as the sum of dipolar and chemical-shift frequencies, for comparison with experiments without the refocusing 180° pulse in the evolution period.

relative tensor orientations completely. This is due to degeneracies of certain angles, such as α and $180^\circ - \alpha$, which are typical of NMR of powders (see also Section 7.5). Uniaxial orientation of the sample, which can be induced in most polymers, is useful for reducing the number of possible tensor orientations (Opella and Waugh, 1977). In many cases, information on the tensor orientation in similar systems can single out the correct solution. As for the detailed analysis and simulation of multidimensional exchange experiments, often the parameters detemined directly by the SLF experiment are sufficient.

For CH_2 groups, the 2D dip-shift correlation spectrum can be regarded as a superposition of the 2D spectra resulting from the sum and difference dipolar tensors of Section 2.5.6. Figure 6.29 shows 2D patterns for various orientations of the chemical-shift tensor relative to the internuclear directions in the CH_2 group (Nakai *et al.*, 1988). The experimental spectrum of PE (Schmidt-Rohr *et al.*, 1993) shown at the top agrees best with the $(0^\circ, 0^\circ, 0^\circ)$ pattern, i.e. the principal-axes systems of the PE chemical shift and the elemental dipolar sum tensor (see Figure 2.18) coincide.

By means of the dip-shift correlation technique, for a number of simple polymers, such as POM (Maas *et al.*, 1987; Schmidt-Rohr *et al.*, 1993), PE (Nakai *et al.*, 1988; Schmidt-Rohr *et al.*, 1993), PEO (Schmidt-Rohr *et al.*, 1993), and iPP (under OMAS

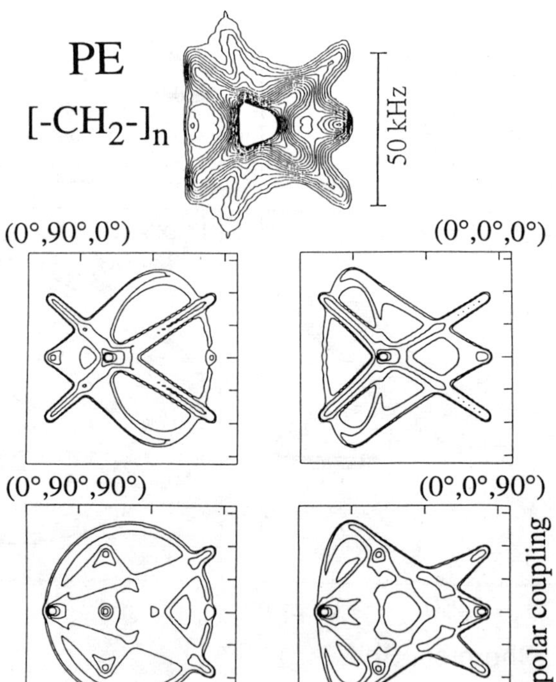

Figure 6.29 Examples of $^1H-^{13}C$ dip-shift patterns in CH_2 groups for various relative orientations of chemical-shift and dipolar principal-axes systems, given in terms of the Euler angles α, β and γ between the PAS of the CH_2 dipolar sum-tensor (see Figure 2.18) and the PAS of the chemical shift (adapted from Nakai *et al.*, 1988). For the experimental PE spectrum shown at the top, α, β and γ must be close to zero.

conditions, see Figure 1.3(a)) (Nakai *et al.*, 1989), the ^{13}C chemical-shift tensor orientations have been measured to be used in the detailed analysis of molecular dynamics in those polymers by multidimensional exchange ^{13}C NMR (Section 7.12).

6.7.4 Correlation of static ^{13}C chemical-shift and $^{13}C-^{1}H$ dipolar patterns under SASS conditions

Terao and co-workers (Nakai *et al.*, 1988, 1989) have introduced several practical improvements to the dip-shift correlation experiment by combining it with off-magic-angle sample spinning. This nearly eliminates the frequency-dependence of the cross polarization (which is due to the angular dependence of the C–H dipolar coupling) and introduces a $\frac{1}{2}(3\cos^2\theta_s - 1)$ scaling of the anisotropies (see equation (2.60)) which reduces the overlap of different powder patterns. These improvements allowed the dip-shift correlation 2D experiment to be tested on PE and applied to iPP (Nakai *et al.*, 1989) (see Figure 1.3(a)). For cases where spectral overlap is a problem in the two-dimensional experiment, a 3D switching-angle sample spinning (SASS) experiment was introduced (Nakai *et al.*, 1988) that achieves separation of isotropic chemical shifts in a third dimension. To this end, the sample must spin at the magic angle during the t_3 period. The corresponding pulse and axis-flip sequence is shown in Figure 6.30. The experiment represents a combination of the SASS separation 2D experiment of Section 6.4.3 with the quasi-static (OMAS) dip-shift correlation experiment of Section 6.7.3. During the two evolution periods, the sample is spun off-the-magic-angle so that the anisotropies are not scaled to zero. The scaled dipolar spectrum is obtained by sampling t_1 at the rotor echoes with a sufficiently small OMAS scaling factor (an angle $\theta_m + 4^\circ$ is typically used). Thus, the spectrum is obtained undistorted even if the spinning rate is much smaller than the static anisotropy (see Section 2.4.5). The same is possible for the OMAS chemical shift evolution t_2 in the 3D experiment. In the ω_3 MAS dimension, spinning sidebands can occur. TOSS is not applicable because the t_1

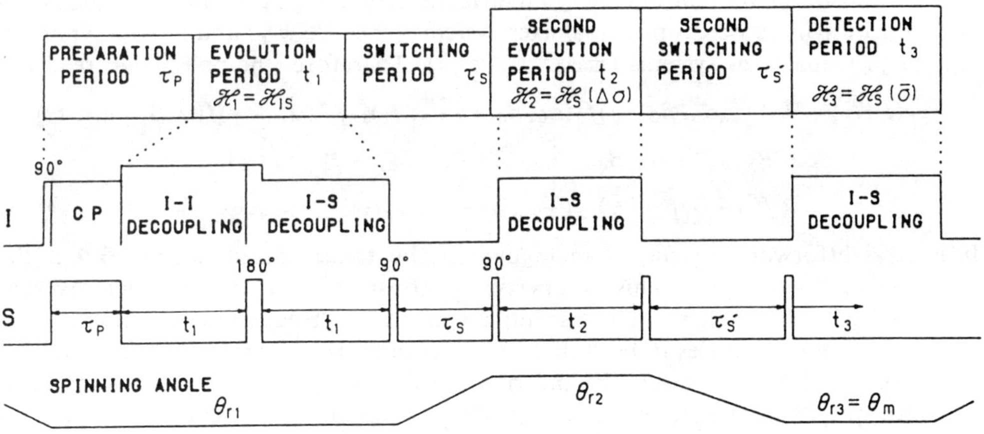

Figure 6.30 3D-SASS pulse sequence (adapted from Nakai *et al.*, 1988) for obtaining isotropic-shift-separated dip-shift correlation 2D patterns, with OMAS and heteronuclear dipolar evolution during t_1, OMAS and chemical-shift evolution during t_2, and MAS detection during t_3. The evolution times t_1 and t_2 must be incremented in multiples of the rotation period.

and t_2 amplitude modulation destroys the symmetry required for the delicate compensation of the sidebands (see Section 3.9.2). To obtain the dip-shift spectrum of a particular carbon site, the corresponding ω_3 sideband signals can simply be summed up. In contrast, spinning faster than the static chemical shift anisotropies is required in the 2D OMAS version of the experiment without MAS detection. In the experiments by Terao and co-workers, data formats were kept small by measuring on resonance only 8 or 10 t_1 increments. This is enough to resolve the major features of the 2D patterns; in fact, due to the large (inhomogeneous) linewidths along ω_1, the t_1-time-domain signal interferes out very fast. Still, for resolving some of the spectral details helpful in determining precise values of orientation and coupling parameters, longer evolution times might be useful. Due to the symmetry of the dipolar patterns, quadrature detection is not needed in t_1.

6.7.5 Correlation of ^{13}C chemical-shift and $^{13}C-^{1}H$ dipolar interactions under slow MAS

The dip-shift correlation experiment has also been performed under pure MAS conditions on standard MAS equipment, yielding sideband-correlation patterns for distinguishable sites (Munowitz *et al.*, 1981, 1982; Munowitz and Griffin, 1982; Kolbert *et al.*, 1989). Sufficiently intense spinning sidebands are required in both dimensions to have anisotropic information for both interactions; therefore, the experiments are performed under slow-MAS conditions. For obtaining high spectral resolution, pure-absorption-mode spectra have to be generated. In order to show that the chemical-shift-dipolar sidebands can be made absorptive (i.e., that the sideband intensities $I_{M,N}$ are real, see Section 4.6.2), we exploit the equivalence:

$$\bar{I}_{M,N} \equiv \bar{I}^*_{M,N} \Leftrightarrow \bar{g}(t_1, t_2) \equiv \bar{g}^*(-t_1, -t_2) \tag{6.38}$$

(see Section 4.6.2). The powder time signal $\bar{g}(t_1, t_2)$ is decomposed into mirror-image-pair signals along the lines of Section 3.9.2. In the case of the dip-shift spectrum, it is important to realize that mirror-symmetric molecular segments are accompanied by both mirror-symmetric CS principal axes systems ($\pm\alpha$, β, $\pm\gamma$) and mirror-symmetric dipolar principal axes systems ($\pm\alpha_d$, β_d, $\pm\gamma_d$). Therefore, the pair signal reads:

$$\begin{aligned} g_p(t_1, t_2) = {} & \mathrm{Re}\{f_d^*(\alpha_d, \beta_d, \gamma_d) f_d(\alpha_d, \beta_d, \gamma_d + \omega_r t_1)\}\; f^*(\alpha, \beta, \gamma) f(\alpha, \beta, \gamma + \omega_r t_2) \\ & + \mathrm{Re}\{f_d^*(-\alpha_d, \beta_d, -\gamma_d)\; f_d(-\alpha_d, \beta_d, -\gamma_d + \omega_r t_1)\} \\ & \times f^*(-\alpha, \beta, -\gamma)\; f(-\alpha, \beta, -\gamma + \omega_r t_2). \end{aligned} \tag{6.39}$$

It is straightforward to show, analogous to the treatment in Section 3.9.2, that $g_p(t_1, t_2) = g_p^*(-t_1, -t_2)$. This property is conserved in the powder average, $\bar{g}(t_1, t_2) = \bar{g}^*(-t_1, -t_2)$, so that absorptive spectra result according to (6.38).

The sideband intensities in both dimensions contain the information on the relative tensor orientations. However, the information is not as directly accessible as in the static (or quasi-static) experiment, where one has characteristic 2D ridge patterns due to definite relations between the orientation and the frequency, so that quite often conclusions on the tensor orientations can be drawn simply upon visual inspection. In addition, MAS reduces the angular resolution. For a C–H pair, the angular range accessible in the experiment is $90^\circ \times 90^\circ$ in the Euler angles α and β. With 6×12 or fewer sidebands, the angular resolution can only amount to about 10° on average, while the static 2D dip-shift spectrum consists of typically 20×80 independent lines (estimate based on the ratios of homogeneous to inhomogeneous linewidths) to yield an average resolution of about 2.5°.

6.7.5 Zero-field correlation experiments

In Section 3.11, it became clear that both zero-field and multiple-quantum experiments are inherently two-dimensional. Therefore, separation of anisotropically broadened high-field lineshapes by zero-field lines is straightforward. However, usually no additional information is obtained since the zero-field line positions are determined by the same interaction (dipolar or quadrupolar) as the maxima and edges of the high-field powder patterns.

For the purpose of relating the ω^+ and ω^- lines observed for each site in zero field, a 2D correlation spectrum has been introduced. It involves two zero-field evolution periods separated by a short 'static' (dc) field pulse that transfers coherence between the energy levels of the spin system. Such a 2D NQR correlation spectrum of diethylterephthalate-d_4 (Thayer *et al.*, 1986) is shown in Figure 6.31.

6.8 2D Relaxation NMR

Anisotropic relaxation times, T_1 (Torchia and Szabo, 1982) as well as T_2, contain information on reorientational motions. In ^{2}H NMR in particular, the 'T_2 minimum', occurring for motions with rates comparable to the quadrupole coupling constant, usually involves relaxation times that vary across the anisotropically (inhomogeneously) broadened spectrum. This anisotropy of T_2 is traditionally measured by means of a quadrupole-echo pulse sequence (see Section 2.6.4) yielding spectra as a function of the quadrupole-echo pulse spacing. On this basis, by systematic incrementation of this delay as t_1 and Fourier transform over this variable, a 2D relaxation ^{2}H NMR spectrum can be generated (Müller and Chan, 1983; Schleicher *et al.*, 1990). To display the frequency dependence of $1/T_2$ in a contour plot, each slice of constant ω_2 is individually scaled such that its maximum is equal to unity. This requires division of the measured spectrum $S(\omega_1, \omega_2)$ by the central slice $S(0, \omega_2)$. The technique is analogous to experiments developed in pulsed EPR (Millhauser and Freed, 1984).

Spectra and relaxation behaviour can be calculated by a brute-force solution of the Liouville–v.-Neumann equation (Millhauser and Freed, 1984; Schleicher *et al.*, 1990):

$$\frac{\partial \hat{\rho}}{\partial t} = -\mathrm{i}[\hat{H}, \hat{\rho}] - \hat{\Pi}(\hat{\rho} - \hat{\rho}_{\mathrm{eq}}). \tag{6.40}$$

Nonsecular components of the Hamiltonian, relevant for the relaxation, are included in this calculation. The stationary Markov operator $\hat{\Pi}$ can be approximated in the grid-point approach by the exchange matrix $\mathbf{\Pi}$ (see Section 11.4.3).

However, the essential features of the time-dependence of the nuclear signal will result already from a simpler calculation that takes into account the secular terms only.

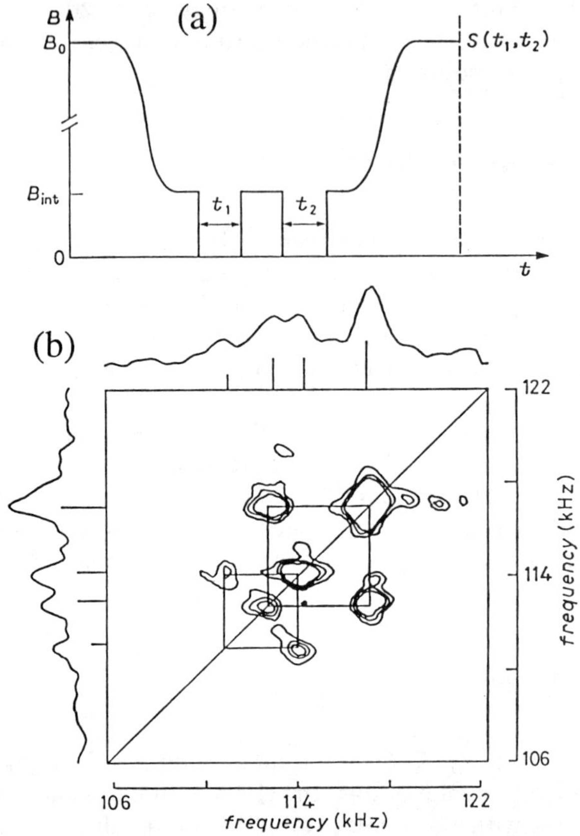

Figure 6.31 2D zero-field correlation spectroscopy. (a) Field cycle. (b) Spectrum with cross-peaks between ω^+ and ω^- transitions, facilitating the interpretation of the 1D spectrum. Adapted from Thayer (1987).

According to Section 2.4.7, this will yield:

$$f(t_1, t_2) = \vec{\mathbf{1}}^{\mathrm{T}} \underbrace{\mathrm{e}^{(\mathrm{i}\omega + \Pi)^{\frac{1}{2}} t_1} \mathrm{e}^{(-\mathrm{i}\omega + \Pi)^{\frac{1}{2}} t_1}}_{\text{evolution}} \underbrace{\mathrm{e}^{(-\mathrm{i}\omega + \Pi) t_2}}_{\text{detection}} \mathbf{P}\vec{\mathbf{1}}\ \mathrm{e}^{-(t_1 + t_2)/T_{2\mathrm{eff}}} \tag{6.41}$$

which can be evaluated as outlined in Section 10.3. Here, $T_{2\mathrm{eff}}$ is an effective T_2 interpolated between the rigid-limit and the fast-limit values. This may even be more realistic than a solution of equation (6.40), since the rigid-limit 'T_2' for deuterons is determined by dipolar couplings, which are totally neglected in the quadrupolar single-particle equation (6.40).

The most distinctive features in these 2D relaxation ^{2}H spectra are the contractions in the ω_1 direction, which correspond to orientations that do not change their frequencies during the motional process. The frequency difference for orientations involving different angles θ_a and θ_b between $\vec{\mathbf{B}}_0$ and a C—^{2}H bond is given by:

$$\Delta\omega = \omega(\theta_a) - \omega(\theta_b) = \frac{3}{8}\frac{eqeQ}{\hbar}(3\cos^2\theta_a - 3\cos^2\theta_b). \tag{6.42}$$

A signal component producing a pair of contractions at frequencies $\pm\omega_{co}$ must fulfil:

$$\Delta\omega \stackrel{!}{=} 0 \tag{6.43}$$

for all θ_a and θ_b occurring during the motional process: if the frequency does not change, the signal is refocused by the quadrupolar echo sequence. According to equation (6.42), $\Delta\omega \stackrel{!}{=} 0$ means that we demand:

$$|\cos\theta_a| \stackrel{!}{=} |\cos\theta_b|. \tag{6.44}$$

(Note that at ω_{co} there are usually also noncontraction signal contributions: from the 'second transition' of the signal at $-\tilde{\omega}_2$, as well as from segments that do change their frequency. These are spread out along the ω_1 axis, and are therefore lower than the signal component contracted at $\omega_1 = 0$.)

With the major restriction to motions of the C$-^2$H bonds on a cone, one can apply relation (A.14) between the rotation angle $\Delta\psi$, the apex angle $2\theta_k$ of the cone, and the angle Θ relating the cone (molecular) axis and the $\vec{\mathbf{B}}_0$ field:

$$\cos\theta_a = \cos\Theta\cos\theta_k + \sin\Theta\sin\theta_k\cos(\psi_0) \tag{6.45a}$$

$$\cos\theta_b = \cos\Theta\cos\theta_k + \sin\Theta\sin\theta_k\cos(\psi_0 + \Delta\psi). \tag{6.45b}$$

One solution is obtained when $\Theta = 0$, that is, when the cone axis is along the $\vec{\mathbf{B}}_0$ field, such that $\theta \equiv \theta_k$ and therefore the frequencies $\omega(\theta)$ do not change. With $\Theta = 0$, we have:

$$|\cos\theta_k| = |\cos\theta(\omega_{co})| = \sqrt{\tfrac{2}{3}}\sqrt{\omega_{co}/\delta + \tfrac{1}{2}}. \tag{6.46}$$

This relation can help to determine the aperture angle θ_k of the cone from the contraction frequency ω_{co}. However, choosing the pair of contractions corresponding to this solution among several contractions possibly found in an experimental spectrum is by no means straightforward. Even if this uncertainty is absent because there is only one pair of contractions, the solution of equation (6.46) for ω_{co} is not single-valued since there are usually four values for θ_k that would fulfil relation (6.44) due to the fourfold degeneracy of $\theta(\omega)$ in ^{2}H spectra, arising from $\cos\theta_k = \cos(-\theta_k)$ and the overlap of the two transitions. The correct θ_k must be picked on the basis of the molecular structure. Figure 6.32 displays 2D relaxation patterns of the methyl-group rotation in methyl-deuterated L-alanine. In the intermediate motional regime, three contractions are observed.

The 2D quadrupole-echo relaxation experiment hardly provides more information than a series of standard 1D quadrupole-echo spectra with various echo delays. However, it allows for a compact representation of the data; the important signal components that possess a long T_2 produce well-distinguished sharp signals, rather than long decays as in the time-domain data set. Nevertheless, the contraction frequency points, as the most distinctive features in the 2D spectrum, can easily be determined in the t_1 time domain as the frequencies that retain signal intensity after long echo delays. The strengths of the 2D experiment are basically those of 1D ^{2}H NMR experiments in the 'intermediate motional regime' (see Section 2.4.7 and Figure 2.16(b)) which are known to produce distinctive spectral features for different well-defined jump motions (Spiess and Sillescu, 1981).

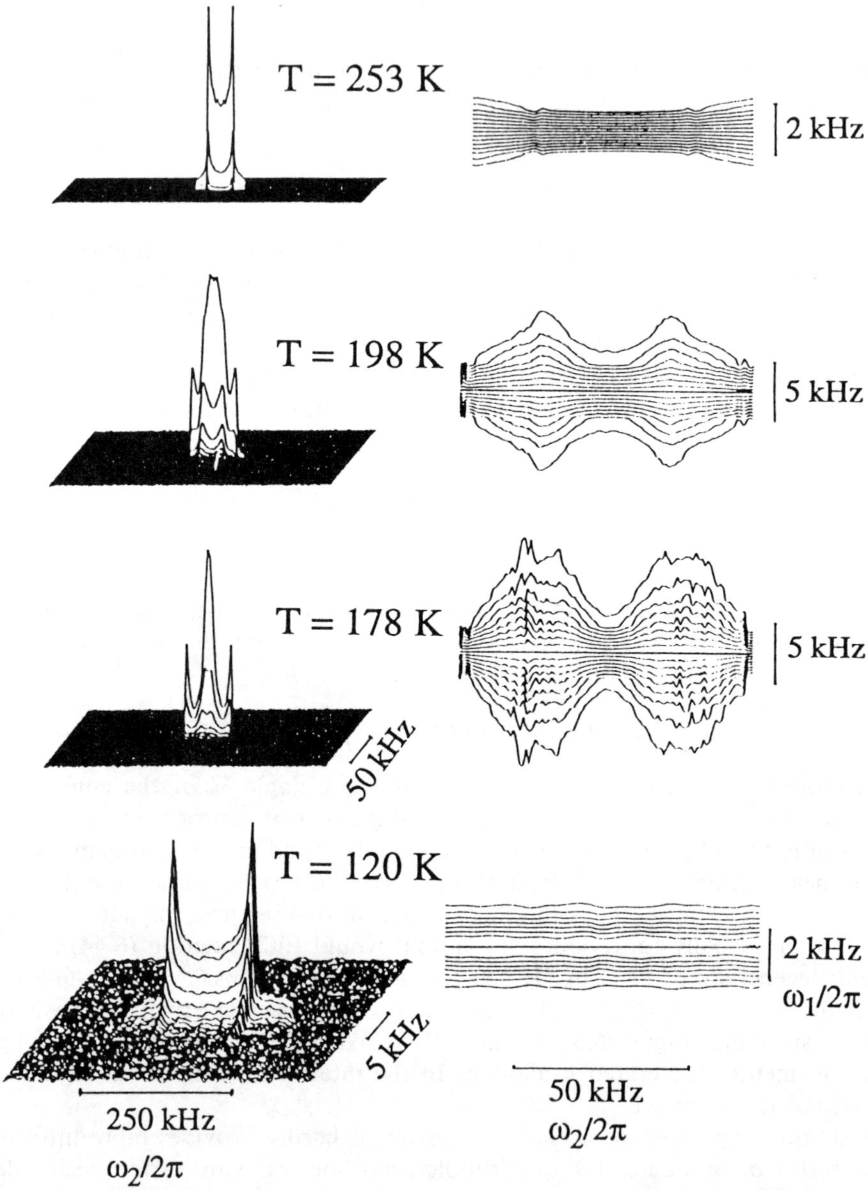

Figure 6.32 2D relaxation patterns for L-alanine-d_3 (deuterated methyl group) showing the spectral effects of methyl-group rotation as a function of temperature (adapted from Schleicher *et al.*, 1990). Left: stacked plots. Right: contour plots after normalizing the maximum of each slice at a fixed ω_2 to be unity. The spectrum at 120 K is in the slow-motion limit, that at 253 K in the fast-motion limit, where the quadrupolar coupling is averaged to $-\frac{1}{3}$ of its rigid-lattice value. In the intermediate dynamical regime, three broad contractions are observed in the patterns.

6.9 Separation versus correlation — Part II

In the preceding sections, many different 2D experiments have been discussed. With this background, we can re-evaluate in more detail the classification of 2D experiments in terms of separation and correlation that was established in Section 6.3. We want to stress again that the point is quite academic, but since the classifications are commonly made, an analysis of definitions and usages may eliminate some confusion.

In solution NMR, definitions of separation, correlation and exchange are commonly used (Ernst *et al.*, 1987) from which the rules introduced in Section 6.3 and similarly applied by other solid-state NMR groups differ to some extent. To compare the definitions, we cite the full wording of the definitions given by Ernst, Bodenhausen and Wokaun (EBW) (Ernst *et al.*, 1987, p. 286):

EBW separation experiments
'Experiments designed to *separate* different interactions (e.g. shifts and couplings) in orthogonal frequency dimensions, with the purpose of resolving 1D spectra by spreading overlapping resonances in a second dimension. These experiments require a means of modifying the effective Hamiltonian, in such a way that the spectra in the evolution and detection periods (determined by $\hat{H}^{(e)}$ and $\hat{H}^{(d)}$, respectively) contain different information.'

EBW correlation experiments
'Experiments designed to *correlate* transitions of **coupled spins**, by transferring transverse magnetization or multiple-quantum coherence from one transition to another in the course of a suitably designed mixing process.'

EBW exchange experiments
'A third class of 2D time-domain experiments is concerned with studying dynamic processes such as chemical exchange, cross relaxation, transient Overhauser effects, and spin-diffusion in solids.'

These definitions focus on the nature of the Hamiltonian and of transfer processes between coupled spins, while our rules presented in Section 6.3 rely mainly on the spectral features where the coherent (usually inhomogeneous) line-broadenings occurring in solid-state NMR play a major role. These coherent broadenings are not taken specifically into account in the EBW definitions. Related to this, it is not clear in those definitions whether the changes of an anisotropic Hamiltonian generated by sample reorientation make it a 'different interaction': after a sample rotation, a chemical-shift Hamiltonian is still a chemical-shift Hamiltonian, but the orientation-dependent prefactors for the various units in the sample are altered.

According to the EBW usage (and ours as well), there are heteronuclear experiments that are correlation experiments (e.g. $^{1}H-^{13}C$ chemical-shift correlation experiments, Section 6.7.2); consequently, cross-polarization should be considered as a 'suitably designed mixing process for transverse magnetization' which makes it a separation experiment according to the above definition. Thus, every heteronuclear experiment would be a correlation experiment, since heteronuclear experiments require cross-polarization in the mixing period. This is somewhat counterintuitive since certain heteronuclear spectra are more appropriately classified as separation spectra: for instance the WISE experiment (Section 6.6.6) obviously separates ^{1}H wideline patterns, or an experiment could separate ^{13}C MAS sideband patterns by the corresponding ^{1}H isotropic shifts. In addition, there appears to be an internal problem of uniqueness in the EBW definitions: heteronuclear experiments usually involve different Hamiltonians for the proton evolution and for the *S*-spin (^{13}C) detection which makes them candidates for EBW separation experiments, while they were just found to represent EBW correlation experiments.

In 2D solid-state NMR, further problems with the EBW definitions become apparent in several cases. For instance, the 'chemical-shift correlation' experiment introduced by Grant and co-workers (Carter *et al.*, 1985) (see Section 6.7.1) does not produce a correlation spectrum according to the EBW definition: the technique simply does not involve coupled spins. Neither is this 'chemical-shift-anisotropy correlation' experiment an EBW exchange experiment, since it aims at studying the CS tensor orientations rather than the dynamic process of the sample flip. The EBW definition of a separation experiment applies only marginally, since the separation of overlapping resonances is only of subordinate importance in the experiment. In summary, if anything, the 'chemical-shift correlation' experiment is an EBW separation experiment, which is obviously at variance with the notions of its inventors. While there are certain 2D techniques that cannot be classified easily by any separation/correlation/exchange nomenclature (see below), the chemical-shift correlation experiment is a relatively typical 2D experiment with correlation of well-defined precessional NMR frequencies, and exhibits the character of a correlation or exchange experiment.

It can also occur that two experiments yielding the same spectra have to be classified as an EBW-separation in one case and as an EBW-correlation experiment in the other. Consider a system made up of two isolated $^{1}H-^{13}C$ spin-pair sites with different ^{13}C isotropic chemical shifts. By proper combination of MAS and OMAS, we have the isotropic ^{13}C chemical shift along ω_2 and the scaled dipolar (e.g. Pake) pattern along ω_1. In the first experiment, both the evolution and the detection period are to involve ^{13}C magnetization. In the second experiment, ^{1}H magnetization evolves during t_1 and is later transferred onto the ^{13}C nuclei for detection. The dipolar patterns and the whole spectra will be essentially identical. Yet the former experiment would have to be designated as an EBW separation experiment, and the latter as an EBW correlation experiment since it involves magnetization transfer. By our definition, both experiments are naturally separation experiments: the dipolar patterns are separated in terms of the ^{13}C chemical shifts.

Our rules laid out in Section 6.3 sometimes do not provide a unique classification of certain experiments which contain aspects of both correlation and separation. For instance, consider the 2D experiment with a CRAMPS ^{1}H spectrum in one dimension and isotropic ^{13}C MAS lines in the other (Section 6.7.2). The CRAMPS spectra of polymers usually consist of broad overlapping bands. Therefore, the 2D spectrum both establishes a correlation of frequencies by the centres of the signals in the 2D plane as well as a separation of the originally overlapping CRAMPS bands. In other cases, it should be noted that spectral separation can involve a correlation of properties. For instance, separation of anisotropically broadened patterns in terms of ^{13}C isotropic chemical shifts can be seen to correlate structure (isotropic shifts) with dynamics (reflected in the motional averaging of anisotropic couplings).

No classification within the above framework appears to apply for the sync-MAS 2D experiment to be described in Section 12.4. It is quite atypical, since there is actually no evolution of the density matrix during t_1; only the phase of the rotor is altered, modulating the spatial prefactor in the chemical-shift Hamiltonian. This is just another example of the great potential and variability of the concepts of multidimensional spectroscopy introduced by Jeener (1971), Ernst (Müller *et al.*, 1975) and Waugh (Hester *et al.*, 1975).

6.10 References

Afeworki, M. and Schaefer, J. (1992). Dynamics of PC at the Interface of PS/PS Blends. *Macromolecules*, **25**, 4097.

Alla, M. A., Kundla, E. I. and Lippmaa, E. T. (1978). Selective Determination of Anisotropic Interactions from Powders. *JETP Lett.*, **27**, 194.

Aue, W. P., Bartholdi, E. and Ernst, R. R. (1976). 2D Spectroscopy. Application to NMR. *J. Chem. Phys.*, **64**, 2229.

Aue, W. P., Ruben, D. J. and Griffin, R. G. (1981). 2D Rotational-Echo NMR: Isotropic vs. Anisotropic Chemical Shifts. *J. Magn. Reson.*, **43**, 472.

Bax, A., Szeverenyi, N. M. and Maciel, G. E. (1983a). CS Anisotropy in Powders Studied by 2D CP/MAS NMR. *J. Magn. Reson.*, **51**, 400.

Bax, A., Szeverenyi, N. M. and Maciel, G. E. (1983b). Isotropic Shifts and CSAs: 2D Magic-Angle Hopping NMR. *J. Magn. Reson.*, **52**, 147.

Bax, A., Szeverenyi, N. M. and Maciel, G. E. (1983c). CS Anisotropy of Powders: 2D NMR with Flip of Spinning Axis. *J. Magn. Reson.*, **55**, 494.

Bielecki, A., Burum, D. P., Rice, D. M. and Karasz, F. E. (1991). Solid-State 2D ^{13}C–^{1}H NMR of PPX, *Macromolecules*, **24**, 4820.

Caravatti, P., Bodenhausen, G. and Ernst, R. R. (1982). *I–S* Solid-State Correlation Spectroscopy. *Chem. Phys. Lett.*, **89**, 363.

Caravatti, P., Braunschweiler, L. and Ernst, R. R. (1983). *I–S* Correlation Spectroscopy in Rotating Solids. *Chem. Phys. Lett.*, **100**, 305.

Carter, C. M., Alderman, D. W. and Grant, D. M. (1985). 2D Chem.-Shift Anisotropy Correlation, *J. Magn. Reson.*, **65**, 183.

Carter, C. M., Alderman, D. W. and Grant, D. M. (1987). 2D CSA Correlation for Complex Single Crystals. *J. Magn. Reson.*, **73**, 114.

Clauss, J., Schmidt-Rohr, K., Adam, A., Boeffel, C. and Spiess, H. W. (1992). Aliphatic Side-Chain Mobility and Conformation from 2D NMR. *Macromolecules*, **25**, 5208.

Dixon, W. T. (1982). Spinning-Sideband Free and Spinning-Sideband Only NMR Spectra (PASS). *J. Chem. Phys.*, **77**, 1800.

DiVerdi, J. A. and Opella, S. (1982). N–H Bond Lengths in DNA. *J. Am. Chem. Soc.*, **104**, 1761.

Ernst, R. R., Bodenhausen, G. and Wokaun, A. (1987). *Principles of NMR in One and Two Dimensions*. Clarendon Press, Oxford.

Féaux de Lacroix, S., Titman, J. J. and Spiess, H. W. (1992). 2D Separation of Sidebands by Order. *J. Magn. Reson.*, **97**, 435.

Frydman, L., Chingas, G. C., Lee, Y. K., Grandinetti, P. J., Eastman, M. A., Barrall, G. A. and Pines, A. (1992). Variable-angle correlation spectroscopy in solid-state NMR. *J. Chem. Phys.*, **97**, 4800.

Gan, Z. (1992). High-Resolution-Shift and CSA Correlation by Slow MAS. *J. Am. Chem. Soc.*, **114**, 8307.

Gullion, T. and Schaefer, J. (1989). Detection of Weak Heteronuclear Couplings by REDOR. *Adv. Magn. Reson.*, **13**, 57 (1989).

Harbison, G., Raleigh, D. P., Herzfeld, J. and Griffin, R. G. (1984). High-Field 2D Exchange NMR in Rotating Solids. *J. Magn. Reson.*, **64**, 284 (1984).

Hester, R. K., Ackerman, J. L., Cross, V. R. and Waugh, J. S. (1975). Resolved Dipolar Spectra of Dilute Spins. *Phys. Rev. Lett.*, **34**, 993.

Hester, R. K., Ackermann, J. L., Neff, B. L. and Waugh, J. S. (1976). Separated-Local-Field Spectra in NMR, *Phys. Rev. Lett.*, **36**, 1081.

Heuer, A. (1990). Automatic Analysis of Rotation Patterns. *J. Magn. Reson.*, **89**, 287.

Jeener, J. (1971). *Proc. Ampere International Summer School II*, Basko Polje, Yugoslavia.

Kolbert, A. C., Raleigh, D. P., Levitt, M. H. and Griffin, R. G. (1989). 2D spin-echo NMR in rotating solids, *J. Chem. Phys.*, **90**, 679.

Kolbert, A. C., and Griffin, R. G. (1990). 2D Resolution of Isotropic and Anisotropic Chemical Shifts in MAS NMR. *Chem. Phys. Lett.*, **166**, 87.

Linder, M., Höhener, A. and Ernst, R. R. (1980). Orientation of Tensorial Interactions from 2D NMR Powder Spectra. *J. Chem. Phys.*, **73**, 4959.

Maas, W. E. J. R., Kentgens, A. P. M. and Veeman, W. S. (1987). ^{13}C Chemical-Shift Tensor in POM. *J. Chem. Phys.*, **87**, 6854.

Maciel, G. E., Szeverenyi, N. M. and Sardashti, M. (1985). CSA Powder Patterns: 2D Angle Flipping. Effects of Crystallite Packing. *J. Magn. Reson.*, **64**, 365.

Millhauser, G. L. and Freed, J. H. (1984). 2D Electron Spin Echo Spectroscopy. *J. Chem. Phys.*, **81**, 37.

Müller, L., Kumar, A. and Ernst, R. R. (1975). 2D ^{13}C NMR. *J Chem. Phys.*, **63**, 5490.

Müller, L. and Chan, S. I. (1983). *J. Chem. Phys.*, **78**, 4341.

Munowitz, M. G. and Griffin, R. G. (1982). 2D NMR in Rotating Solids: CS-Dipolar Spectra. *J. Chem. Phys.*, **76**, 2848.

Munowitz, M. G., Griffin, R. G., Bodenhausen, G. and Huang, T. H. (1981). 2D Rotational NMR: Correlation of Chemical Shift and Dipolar Interactions. *J. Am. Chem. Soc.*, **103**, 2529.

Munowitz, M. G., Aue, W. P. and Griffin, R. G. (1982). 2D separation of dipolar and scaled CS interactions in MAS NMR. *J. Chem. Phys.*, **77**, 1686.

Nakai, T., Ashida, J. and Terao, T. (1988). 2D NMR Powder Patterns in Rotating Solids (3D SASS). *J. Chem. Phys.*, **88**, 6049.

Nakai, T., Ashida, J. and Terao, T. (1989). Determination of ^{13}C CS Tensors in iPP via 2D Powder Pattern. *Magn. Reson. Chem.*, **27**, 666.

Nakai, T. and Terao, T. (1992). *I–S* Dipolar Patterns Due Only to Directly Bonded Couplings. *Magn. Reson. Chem.*, **30**, 42.

Opella, S. J. and Waugh, J. S. (1977). 2D ^{13}C NMR of Oriented PE. *J. Chem. Phys.*, **73**, 4919.

Roberts, J. E., Vega, S. and Griffin, R. G. (1984). 2D *I–S* CS Correlation Spectroscopy in Rotating Solids. *J. Am. Chem. Soc.*, **106**, 2506.

Roberts, J. E., Harbison, G. S., Munowitz, M. G., Herzfeld, J. and Griffin, R. G. (1987). Heteronuclear Bond Distances in Polycrystalline Solids. *J. Am. Chem. Soc.*, **109**, 4163.

Rybaczewski, E. F., Neff, B. L., Waugh, J. S. and Sherfinski, J. S. (1977). ^{13}C Local Fields of CH, CH_2, and CH_3, *J. Chem. Phys.*, **67**, 1231.

Schaefer, J., Stejskal, E. O., McKay, R. A. and Dixon, W. T. (1983). Dipolar Rotational-Echo ^{13}C-NMR of Polymers. *J. Magn. Reson.*, **52**, 123.

Schaefer, J., Sefcik, M. D., Stejskal, E. O., McKay, R. A., Dixon, W. T. and Cais, R. E. (1984a). Motion in Glassy Polystyrenes. *Macromolecules*, **17**, 1107.

Schaefer, J., Stejskal, E. O., McKay, R. A. and Dixon, W. T. (1984b). Motion in Polycarbonate by Dipolar Rotational-Echo ^{13}C NMR. *Macromolecules*, **17**, 1479.

Schaefer, J., Stejskal, E. O., Perchak, D., Skolnick, J. and Yaris, R. (1985). Mechanism of Ring-Flip in PC. *Macromolecules*, **18**, 368.

Schleicher, A., Müller, K. and Kothe, G. (1990). 2D NMR relaxation spectroscopy of solids. *J. Chem. Phys.*, **92**, 6432.

Schmidt, A., McKay, R. A. and Schaefer, J. (1992). Distance Measurement between ^{2}H and a Spin-$\frac{1}{2}$ Nucleus. *J. Magn. Reson.*, **96**, 644.

Schmidt-Rohr, K., Clauss, J. and Spiess, H. W. (1992). Correlation of Structure, Mobility, and Morphology by 2D WISE NMR. *Macromolecules*, **25**, 3273.

Schmidt-Rohr, K., Wilhelm, M., Johansson, A. and Spiess, H. W. (1993). Shift Tensors in Methylene Groups. *Magn. Reson. in Chem.*, **31**, 352.

Sherwood, M. H., Alderman, D. W. and Grant, D. M. (1989). 2D CSA Correlation. Multiple-Axis Reorientation Mechanism. *J. Magn. Reson.*, **84**, 466.

Spiess, H. W. and Sillescu, H. (1981). Solid Echoes in the Slow-Motion Region, *J. Magn. Reson.*, **42**, 381.

Spiess, H. W. (1985). ^{2}H NMR for Studying Mobility and Orientation in Polymers. *Adv. Polym. Sci.*, **66**, 23.

Stoll, M. E., Vega, A. J. and Vaughan, R. W. (1976). Heteronuclear-Dipolar Modulated Chemical-Shift Spectra in Polycrystalline Solids. *J. Chem. Phys.*, **65**, 4093.

Terao, T., Fujii, T., Onodera, T. and Saika, A. (1984). Switching-Angle Spinning NMR: Powder-Pattern-Resolved 2D Spectra. *Chem. Phys. Lett.*, **107**, 145.

Terao, T., Miura, H. and Saika, A. (1986). 2D Separation of 1D *I–S* Dipolar Patterns in Rotating Solids (SASS). *J. Chem. Phys.*, **85**, 3816.

Thayer, A. M., Millar, J. M. and Pines, A. (1986). 2D Zero-Field NMR and NQR. *Chem. Phys. Lett.*, **129**, 55.

Torchia, D. A. and Szabo, A. (1982). Spin-Lattice Relaxation in Solids. *J. Magn. Reson.*, **49**, 107.

Tycko, R., Dabbagh, G. and Mirau, P. A. (1989). CS-Anisotropy Lineshapes in 2D MAS. *J. Magn. Reson.*, **85**, 265.

Tycko, R. and Dabbagh, G. (1990). Measurement of *S–S* Dipole Couplings in MAS NMR (DRAMA). *Chem. Phys. Lett.*, **173**, 461.

Waugh, J. S. (1976). Uncoupled Local-Field Spectra in NMR: Determination of Atomic Positions. *Proc. Natl Acad. Sci. USA*, **73**, 1394.

Yarim-Agaev, Y., Tutunjian, P. N. and Waugh, J. S. (1982). MAS with Rotation-Synchronized rf Pulses. *J. Magn. Reson.*, **47**, 51.

Zeigler, R. C., Wind, R. A. and Maciel, G. E. (1988). Stop-and-Go Spinning in MAS. *J. Magn. Reson.*, **79**, 299.

Zumbulyadis, N. (1986). ^{1}H–Si Correlations in *a*-Si : H : A 2D NMR Study', *Phys. Rev. B*, **33**, 6495.

CHAPTER SEVEN

Polymer Dynamics: Multidimensional Exchange Experiments

Keywords: *2D and 1D exchange NMR, cross peaks, correlation times, (elliptical) 2D ridge patterns, reorientation (or 'jump') angles, distribution* R*(β), model-independent results, 2D patterns for η ≠ 0, advantage of oriented samples; 3D exchange NMR, sequence of jumps, non-Markovian processes; helical jumps, α relaxations, chain diffusion; phenylene flips, β relaxations.*

7.1 Introduction

Molecular dynamics represents an important aspect of solid polymers which influences many macroscopic materials properties, as discussed in Section 5.3.

Multidimensional exchange NMR can be applied to characterize molecular details of the dynamical α, β, γ, ... processes when the correlation times range between 10^{-5} and 10^{2} seconds, most conveniently between 1 ms and 1 s. The basic principle of exchange NMR (Jeener *et al.*, 1979) is the measurement of the NMR frequency of one and the same molecular segment at two different times and the detection of slow dynamics through a change in NMR frequency. This is schematically indicated in Figure 7.1.

In the general context of exchange NMR spectroscopy, dynamics can sometimes be understood in a broader sense. Apart from the cases of actual reorientation of molecules or molecular segments in space, and of chemical exchange, it can refer to 'spin dynamics', namely spin diffusion (see Section 3.5 and Chapter 13), or to dynamics of the sample as a whole, which can involve continuous rotation or flipping (see Sections 6.7.1 and 12.7). The 2D exchange spectra arising from these different processes often bear out similar features, and due cross-references between different sections of the text will be made. In order to keep the interpretation clear and simple, in this chapter we shall restrict the discussion to molecular dynamics. First, exchange within anisotropies will be treated, and later, in Section 7.14, isotropic-shift exchange, chemical exchange and related effects will be covered.

Without any slow dynamics in the mixing time, the frequencies remain unchanged, $\tilde{\omega}_2 = \tilde{\omega}_1$, for all signals such that the intensity is confined to the line $\omega_2 = \omega_1$, which is the diagonal of the (ω_1, ω_2) plane. With frequency changes due to reorientations in the mixing time, off-diagonal signals with $\tilde{\omega}_2 \neq \tilde{\omega}_1$ appear. Figure 7.2(b) displays the archetypical case of two sites, A and B, with equal populations. Cross peaks at (ω_A, ω_B) and (ω_B, ω_A) indicate the exchange process. As they rise, the diagonal peaks lose intensity.

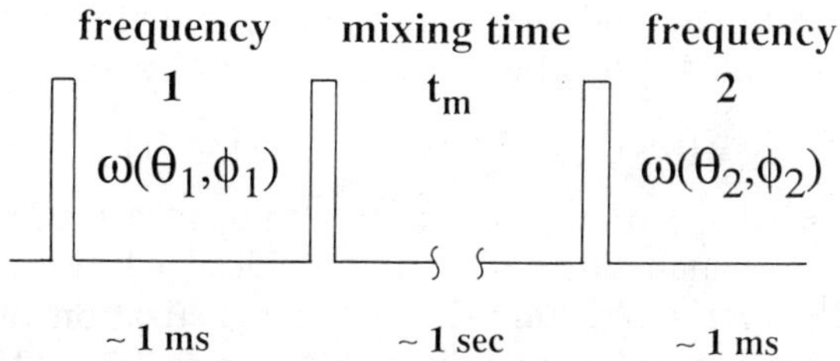

Figure 7.1 Principle of exchange experiments for detecting reorientational motions. The angle-dependent frequency is measured at two points in time, which are separated by the mixing time t_m. A reorientation (a change of the angles (θ, ϕ)) is detected through the concomitant change of the angle-dependent frequency from $\omega(\theta_1, \phi_1)$ to $\omega(\theta_2, \phi_2)$. The frequency measurement in the time domain typically requires evolution and detection times of the order of a millisecond, while the mixing time can be up to tens of seconds long.

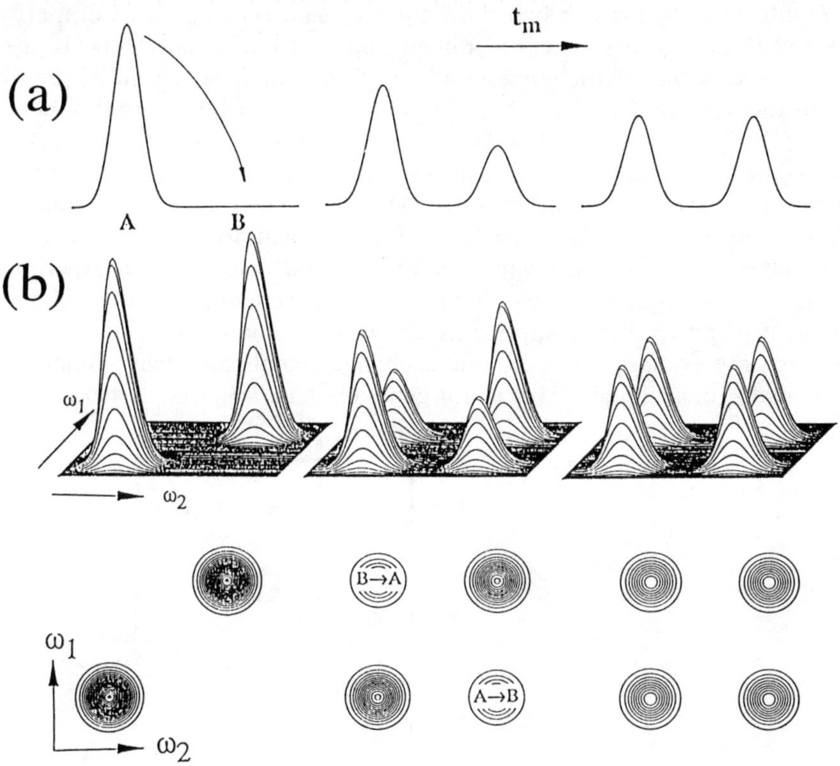

Figure 7.2 Schematic comparison of the detection of slow exchange between two peaks in 1D and 2D exchange experiments. (a) 1D exchange experiment with selective excitation of line A, followed by exchange to line B at longer mixing times t_m. (b) Stacked and contour plots of the corresponding 2D exchange spectra, with cross peaks generated by the exchange process. The spectrum in (a) corresponds to a cross section at $\omega_1 = \omega_A$.

The two-dimensional spectrum $S(\omega_1, \omega_2; t_m)$ represents the probability (density) of finding a segment with a frequency ω_1 and, by a time interval t_m later, with the frequency ω_2 (Wefing and Spiess, 1988). This joint probability (see Section 9.3), which depends only parametrically on t_m, is the product of the *a priori* probability $P(\omega_1)$ that the segment is found at ω_1, and the conditional probability $S(\omega_2; t_m \mid \omega_1)$ that the segment gets to ω_2 provided it has been at ω_1 a time t_m ago:

$$S(\omega_1, \omega_2; t_m) = S(\omega_2; t_m \mid \omega_1)P(\omega_1). \tag{7.1}$$

The conditional probability $S(\omega_2; t_m \mid \omega_1)$ can also be measured in a one-dimensional exchange experiment, which in addition provides helpful hints for a simple interpretation of 2D spectra. Assume that the magnetization of one line has been selected, the rest being suppressed, as indicated in Figure 7.2(a). If exchange occurs in the mixing time after the selection sequence, the other signal(s) will reappear, while the selected line loses intensity correspondingly (Figure 7.2(a)). From the integrated intensities of the peaks, the relative populations of the sites can be determined, while the mixing-time dependence yields the correlation time of the motional process that produces the exchange. Those intensities are proportional to the probabilities $S(\omega_2; t_m \mid \omega_1)$ of finding a segment with ω_2, under the condition imposed by the selection pulse sequence that it had the frequency ω_1 a time t_m ago. Comparison with equation (7.1) shows that the one-dimensional exchange spectrum is essentially identical to a slice out of the corresponding 2D exchange spectrum at the given ω_1, parallel to the ω_2 axis. For the 1D spectrum, the *a priori* term $P(\omega_1)$ appears in the intensity of the line at ω_1 selected in the first place. Thus, a slice out of the 2D spectrum at a given ω_1 can be read like a 1D exchange spectrum, which is often helpful in the interpretation of intensities in 2D spectra. 1D exchange experiments as indicated here, indeed, can be and have been performed experimentally. In Section 13.2.4, the practical advantages and disadvantages of 2D and selective 1D experiments are discussed. For the spectra in this chapter, 2D experiments with their simultaneous measurement of all spectral components are generally preferable.

In the two-site exchange process, the exchange signals are always smaller than or equal to the diagonal signals. This is not generally true. Figure 7.3 demonstrates that

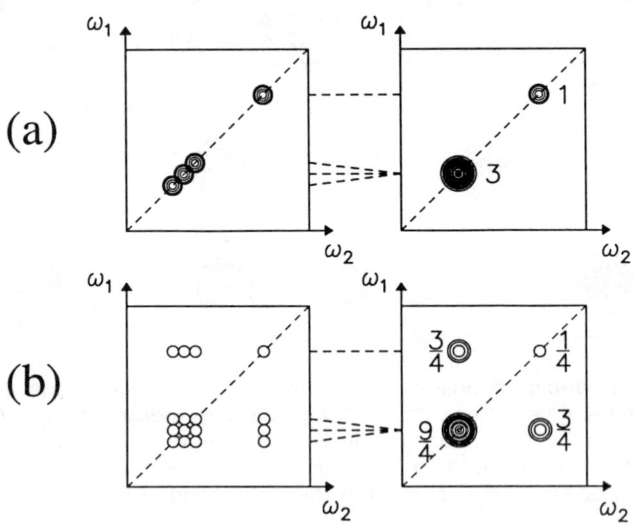

Figure 7.3 The diagonal in an exchange spectrum can be smaller than the exchange intensity if a signal in the spectrum corresponds to a larger number of sites than others. This schematic demonstrates that fact for a system of four equally populated sites by merging the peaks of three sites into one. (a) Diagonal spectra. (b) 'Full exchange', i.e. complete equilibration between all four sites.

the exchange signal can be larger than some of the diagonal signals if certain peaks correspond to several sites. This is also the case in powder spectra, due to the sin θ-dependence of the size of the surface elements and because of the nonlinear relation between the frequency ω and the orientation θ.

The range of correlation times accessible in the exchange experiment covers approximately 500 μs (5×10^{-4} s) to 20 s. With mixing times exceeding 20 s, exchange from spin diffusion usually dominates over exchange from reorientations, T_1 relaxation reduces the magnetization (see also Section 10.2.5), and the measuring times become very long. The lower limit for the correlation times is based on the 'slow-motion' requirement, which demands that orientations, and thus frequencies, should not change while they are being measured in the evolution and detection periods. Sufficiently precise frequency measurements require times of 0.1 ms to 10 ms for typical NMR interactions (corresponding to spectral resolution of 3 KHz to 30 Hz). If the slow-motion condition is not met, the simple, far-reaching interpretation of the 2D exchange spectrum is not applicable any more. Nevertheless, even then, by comparison of ^{2}H 2D spectra with full simulations, information on the nature and timescale of the motional process can be extracted, thus extending the 2D exchange method into the realm of 1D lineshape analysis. Examples are presented in Section 8.2 and the approach for simulating these lineshapes will be discussed in Section 10.3.

7.2 Exchange by segmental reorientations

Due to the dependence of the NMR frequency on the orientation of a given molecular segment relative to $\vec{\mathbf{B}}_0$ (see Section 2.4.2), a change in molecular orientation, in general, causes a frequency change that can be detected in an exchange experiment.

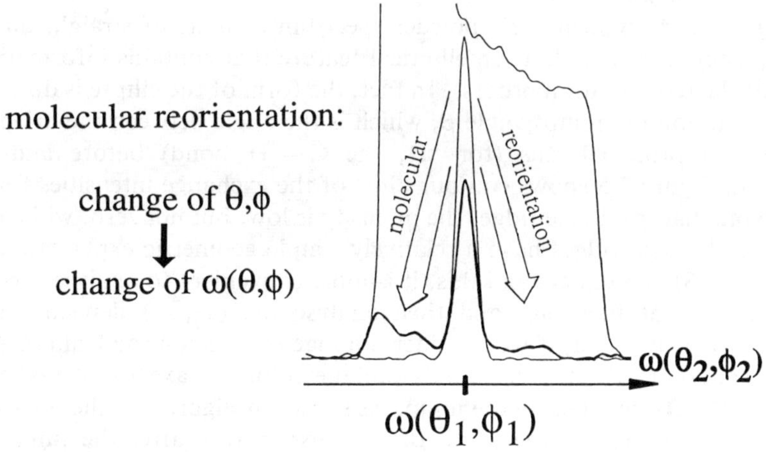

Figure 7.4 Detection of exchange within a powder pattern in a 1D exchange experiment with selective excitation. After the mixing time, the changes of the angle-dependent frequencies are detected in terms of spectral intensity outside of the initially selected band of frequencies.

Figure 7.4 demonstrates this in terms of the 1D exchange experiment. The labelling of the frequency $\omega(\theta_1, \phi_1)$ is achieved by selection of a narrow band of frequencies from the inhomogeneously broadened powder spectrum. Reorientations during t_m involve changes of (θ, ϕ), which are detected as changes of the frequency $\omega(\theta, \phi)$: intensity appears at frequencies $\omega(\theta_2, \phi_2)$ outside the selected band of frequencies. The exchange intensity obviously covers a broad range of frequencies; in fact, discrete cross peaks are not to be expected because the 1D powder spectrum itself is continuous rather than made up of discrete peaks. Still, the exchange intensity exhibits some characteristic features. They become apparent in the full 2D spectrum as characteristic ridge patterns (such as the elliptical features in Figure 7.5).

The frequencies $\tilde{\omega}_1$ and $\tilde{\omega}_2$ in an exchange spectrum depend, in principle, on the Euler angles (ϕ_1, θ_1) and (ϕ_2, θ_2) specifying the orientations of the corresponding PAS_1 and PAS_2 relative to the $\vec{\mathbf{B}}_0$ field according to $\omega(\theta, \phi)$ of equation (2.35). In 2D spectra of isotropic samples, instead of these laboratory-frame angles, which are anyway of little immediate interest for the characterization of the motional process in a molecular frame, the redirection of orientation of the PAS, and thus of the molecular segment, exclusively determines the intensity distribution in 2D exchange powder spectra. This means that such a powder spectrum can depend only on the relative orientation of PAS_1 and PAS_2. The reason is that in an isotropic sample the overall spectrum does not change with the orientation of the sample, and therefore does not depend on the sample orientation. This simplifies the analysis of these 2D spectra in terms of molecular reorientations.

7.3 2D exchange powder spectra for vanishing asymmetry parameter ($\eta = 0$)

The simplest 2D exchange patterns result for vanishing asymmetry parameter, $\eta = 0$ (to be discussed exclusively in this section). This case of axially symmetric coupling is also of great practical importance, as it is met for ^{2}H NMR of aliphatic C—^{2}H bonds (Schmidt *et al.*, 1986, 1988). In addition, up to $\eta = 0.3$, the exchange patterns often resemble those for $\eta = 0$.

The ridge pattern in such a 2D powder spectrum consists of straight lines parallel to the frequency axes as well as an elliptical feature that contains information on the geometry of the reorientation process. In fact, the form of the ellipse is directly related to the reorientation or 'jump' angle β, which is the angle between the orientations of the unique (z) principal axis (for ^{2}H, the C—^{2}H bond) before and after the reorientation. Figure 7.5 shows contour plots of the exchange intensities for different β values. Note that inside the ridges the intensity is low, but not zero, while it vanishes outside. The elliptical ridges have a relatively simple geometric explanation (Schmidt *et al.*, 1986, 1988). To understand this, it is most convenient to analyse slices through the 2D spectrum at fixed ω_1, and then analyse the (ω_1, ω_2) dependence of their singularities (their maxima, if we consider homogeneous line-broadening). A given ω_1 value corresponds to the C—^{2}H bonds (unique principal axes) that make a certain angle θ_1 with the $\vec{\mathbf{B}}_0$ field before the jump. As shown in Figure 7.6, these C—^{2}H bonds start at θ_1 and reorient by the angle β. The distribution after the jump is conical (a circle on the unit sphere) centred on the start orientation (Figure 7.6). The spectrum in ω_2 is determined by the distribution of θ_2 angles made by the C—^{2}H bonds with the $\vec{\mathbf{B}}_0$ field after the jump. The θ_2-distribution ('θ_2-projection') resulting from the conical distribution after the jump is indicated on the top right in Figure 7.6.

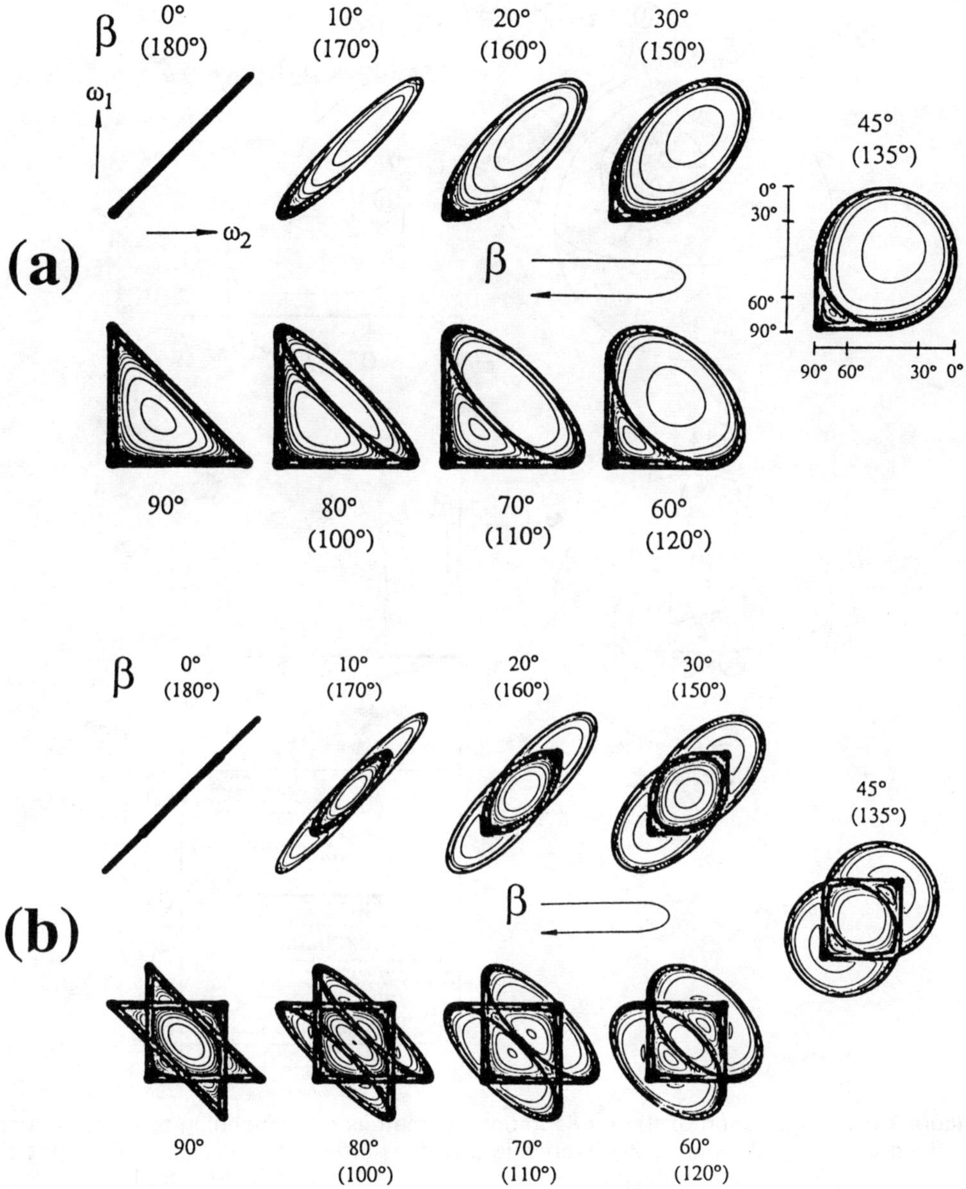

Figure 7.5 2D exchange powder patterns for $\eta = 0$ as a function of the reorientation angle β. (a) Patterns for $I = \frac{1}{2}$. (b) Spectra for ^2H ($I = 1$, two 'transitions'), arising when the patterns in (a) are combined with their reflections about ($\omega_1 = 0, \omega_2 = 0$). The elliptical ridges delimiting most of the spectral-intensity range are prominent features in the spectra and depend characteristically on the reorientation angle. Note, however, that also inside the elliptical ridge appreciable spectral intensity is found, forming a shallow basin. Along parts of the lines ($\omega(90^\circ), \omega_2$) and ($\omega_1, \omega(90^\circ)$), straight '$90^\circ$ ridges' are observed.

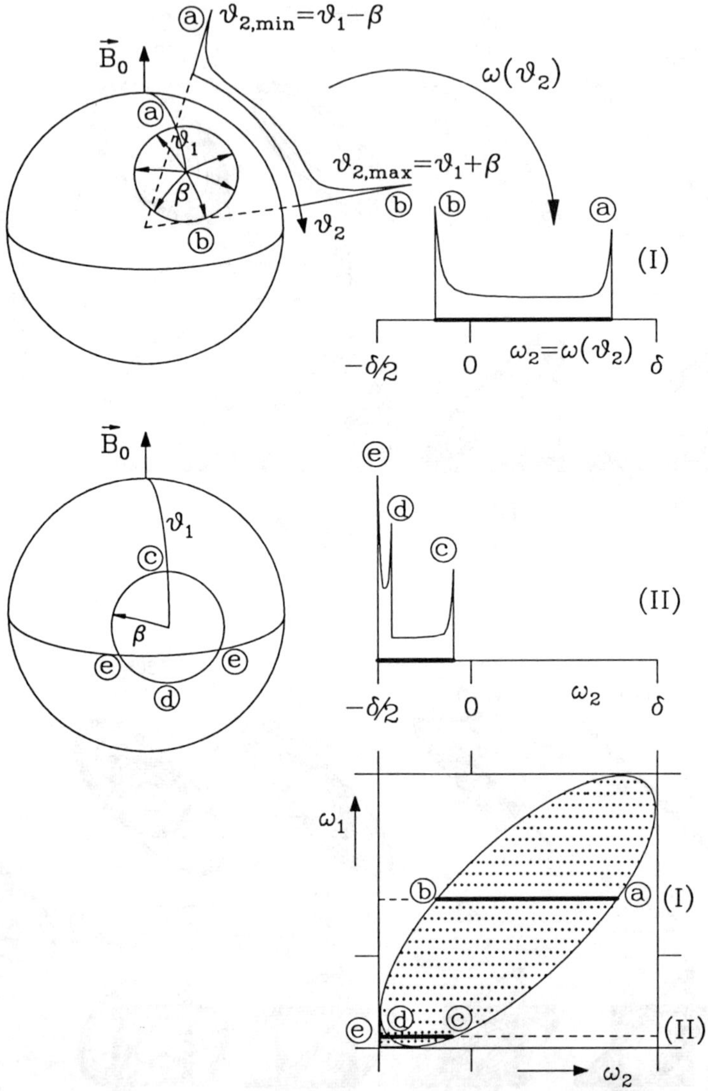

Figure 7.6 Visualization of the reorientation geometries corresponding to cross-sections at fixed $\omega_1 = \omega(\theta_1)$ through a 2D exchange powder pattern shown schematically at the bottom, for $\eta = 0$. The $\vec{\mathbf{Z}}_P$ principal axes located at θ_1 relative to $\vec{\mathbf{B}}_0$ reorient by an angle β, ending up on a circle (cone) centred on the initial (ϕ_1, θ_1) orientation. For the β jump, all directions of rotation are equally probable in a powder. Thus, after the jump a range of angles θ_2 results. (I) At its limits $\theta_2 = \theta_1 \pm \beta$, the θ_2-distribution shows singularities (labelled *a* and *b*). They arise from the parts of the circle running tangential to the $(\theta_2 = \theta_1 \pm \beta)$ lines. These singularities in angle space generate the singularities at $\omega_2 = \omega(\theta_1 \pm \beta)$ which form the elliptical ridges in the 2D spectra (see the 2D spectrum at the bottom). (II) When the circle intersects the line $\theta_2 = 90^\circ$, at $\omega_2 = -\frac{1}{2}\delta$ a third singularity arises (labelled *e*), contributing to the '90° ridge' in the 2D spectrum.

It has singularities (sharp maxima) at its limits, $\theta_2 = \theta_1 - \beta$ and $\theta_2 = \theta_1 + \beta$, as can be shown rigorously on the basis of the principle of differential conservation of the integral (see Section 11.3). These angular singularities transform into singularities (maxima) in the spectral intensity along the $\omega(\theta_1)$ slice, which are consequently located at the frequency points:

$$\begin{pmatrix}\omega_1\\ \omega_2\end{pmatrix} = \tfrac{1}{2}\delta\begin{pmatrix}3\cos^2\theta_1 - 1\\ 3\cos^2(\theta_1 \pm \beta) - 1\end{pmatrix} = \begin{pmatrix}\tfrac{1}{4}\delta\\ \tfrac{1}{4}\delta\end{pmatrix} + \tfrac{3}{4}\delta\begin{pmatrix}\cos(2\theta_1)\\ \cos(2\theta_1 \pm 2\beta)\end{pmatrix}. \tag{7.2}$$

In this equation, $\tfrac{3}{4}\delta(\cos\{2\theta_1\}, \cos\{2\theta_1 \pm 2\beta\})$ can be recognized as the Lissajou parameter representation of an ellipse (see Appendix F) with semi-axes:

$$a = |\tfrac{3}{4}\delta\sqrt{2}\cos\beta| \qquad b = |\tfrac{3}{4}\delta\sqrt{2}\sin\beta| \qquad |\tan\beta| = \frac{b}{a}. \tag{7.3}$$

The ellipse for the tetrahedral angle $\beta = 109^\circ$ (equivalent to 71°, since identical spectra result for β and $180^\circ - \beta$) is found particularly often in deuteron 2D exchange spectra. In the case of phenylene-ring flips, $\beta = 120^\circ$ (equivalent to $\beta = 180^\circ - 120^\circ = 60^\circ$) is observed. There, it should be noted that the rotation angle ψ of 180° around the phenylene 1–4 para-axis is not equal to the jump angle β of the C—^{2}H bonds. This difference between ψ and β is indicated for a general case in Figure 7.7. The angles ψ and β are related through the angle χ between the C—^{2}H bond and the rotation axis according to:

$$\cos\beta = \cos^2\chi + \sin^2\chi\cos\psi. \tag{7.4}$$

This is an application of formula (A.14), with the rotation axis chosen as the z axis and the two C—^{2}H bond orientations having polar coordinates (χ, ψ_0) and $(\chi, \psi_0 + \psi)$.

The straight '90° ridges' in the 2D spectra of Figure 7.5, which extend between the ellipse and the singularity at $(\omega(90^\circ), \omega(90^\circ))$ on the diagonal, arise in a different way from the elliptical ridges. The 90° ridges are due to the vanishing slope of $\omega(\theta)$ at $\theta = 90^\circ$, similar to the singularity in the 1D powder spectrum. In a slice at $\omega_1 = \omega(\theta_1)$, such a singularity at $\omega_2 = \omega(90^\circ) = -\tfrac{1}{2}\delta$ can obviously occur only if the θ_2 distribution extends beyond 90°, that is, if $\theta_1 + \beta = \theta_2 \geqslant 90^\circ$ (see Figure 7.6). Therefore, the 90°

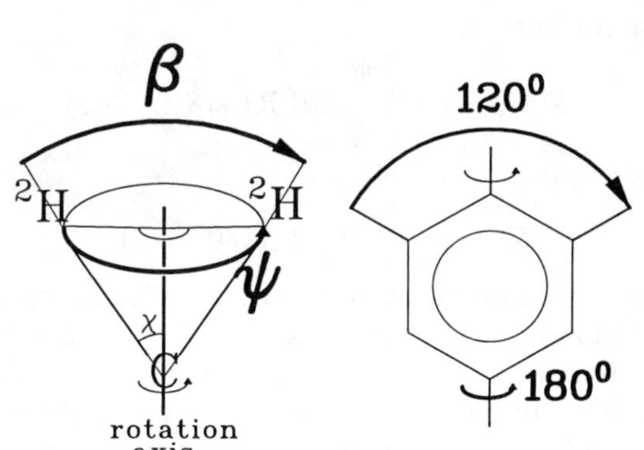

Figure 7.7 Visualization of the distinction to be made between the rotation angle ψ and the NMR-relevant reorientation angle ('jump angle') β. While β is well-defined by the orientations of the $\vec{Z}_P$ axis before and after the jump, the angle ψ also depends on the orientation of the rotation axis.

ridge extends only up to the ω_1 value where $\theta_1 + \beta = 90^\circ$, i.e. where $\theta_1 = 90^\circ - \beta$, which is:

$$\omega_{1,\max} = \omega(90^\circ - \beta) = \tfrac{1}{2}\delta(3\cos^2(90^\circ - \beta) - 1). \tag{7.5}$$

Thus, the length of the 90° ridge depends directly on the jump angle. This can be exploited in the case of ill-defined reorientation processes to estimate the typical reorientation angle. While the elliptical ridges for somewhat varying jump angles do not add up, the 90° ridges do, and therefore they are well detectable. The 90° ridge parallel to the ω_2 axis can be deduced by interchanging ω_1 and ω_2 in the argument just given.

So far, ($\eta = 0$) 2D spectra for the chemical shift interaction, that is, with a single 'transition', have been considered. From those, quadrupolar ^2H and dipolar $^{13}C-^1H$ or $^{15}N-^1H$ spectra can be easily generated: the second 'transition' produces a 2D pattern which is a mirror image of the first with respect to the spectral centre ($\omega_{iso}, \omega_{iso}$). The total quadrupolar or dipolar spectrum can thus be produced simply by combining (i.e. adding) the mirror image and the original single-transition spectrum. Figure 7.5(b) shows the series of simulated 2D exchange ^2H spectra as a function of β, which are to be compared with the corresponding ^{13}C ('single-transition') spectra for $\eta = 0$ displayed in Figure 7.5(a).

7.4 The reorientation-angle distribution

As indicated in the previous section, in the case of $\eta = 0$, only the angle β between the z principal axes ('C—^{2}H bonds') is relevant for the spectrum. Changes of the other Euler angles, α and γ, which correspond to rotations around the C—^{2}H bonds, do not result in changes of frequencies (due to the ϕ-independence of $\omega(\theta, \phi)$ for $\eta = 0$) and therefore have no effect on the spectrum.

This means that every 2D exchange powder spectrum for $\eta = 0$ can be generated by adding subspectra for different reorientation angles β as shown in Figure 7.5, weighted by factors $R_f(\beta)$ with $0 \leqslant \beta \leqslant 180^\circ$. With spectra for β and $180^\circ - \beta$ being identical, the reorientation-angle distribution (RAD) $R(\beta)$ that uniquely determines such a 2D exchange spectrum according to:

$$S(\omega_1, \omega_2) = \int_{0^\circ}^{90^\circ} d\beta\ \mathrm{R}(\beta) S_\beta(\omega_1, \omega_2) \tag{7.6}$$

must be restricted to $0 \leqslant \beta \leqslant 90^\circ$:

$$R(\beta) = R_f(\beta) + R_f(180^\circ - \beta). \tag{7.7}$$

This one-dimensional distribution is encoded in every slice at constant ω_1 (or ω_2) of the two-dimensional spectrum for $\eta = 0$, since the slices $S_\beta(\omega_1 = \text{const.}, \omega_2)$ for different β are found to be linearly independent (Hagemeyer *et al.*, 1990). Thus, the information on the reorientation-angle distribution is contained many times in the spectrum: in other words the information in the 2D spectrum is highly redundant. For determining $R(\beta)$ (Hagemeyer *et al.*, 1990) from an experimental spectrum with finite homogeneous line broadening, it is, however, useful to take various slices into account. The highest resolution for small reorientation angles β is available from the slices in the centre of the spectrum. This is apparent from Figure 7.5, where the ellipses for

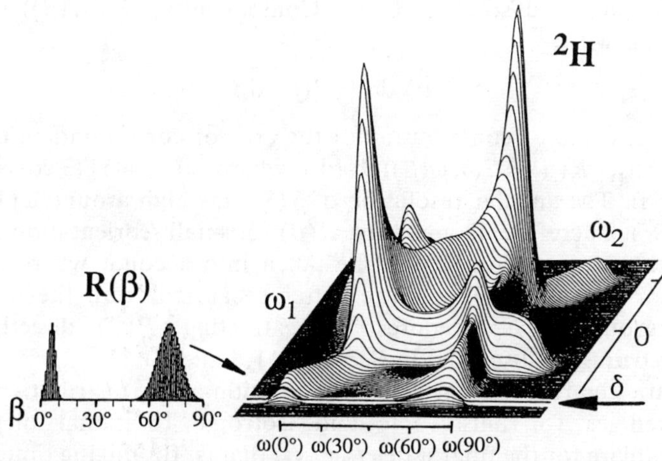

Figure 7.8 Example of the direct relation between the 'edge slice' of a 2D pattern and the reorientation-angle distribution. Both curves show a bimodal shape, with maxima at 0° and 70°, and $\omega(0^\circ)$ and $\omega(70^\circ)$, respectively.

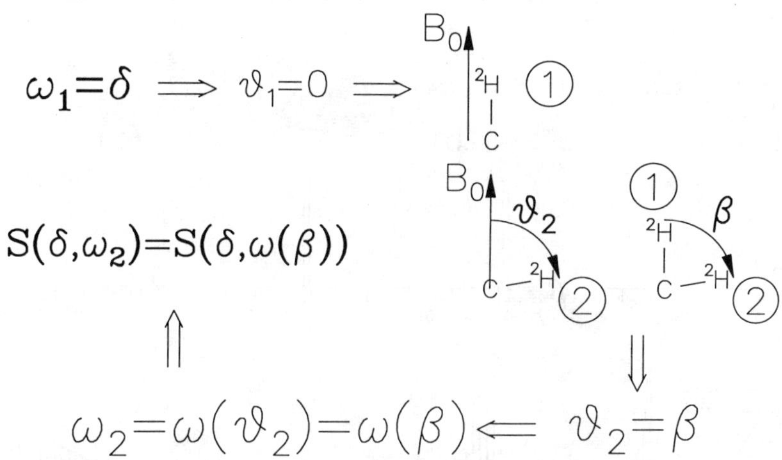

Figure 7.9 This diagram demonstrates the geometric relationships that result in the equality of 'edge slice' $S(\delta, \omega_2)$ with $S(\delta, \omega(\beta))$. This then leads to $R(\beta) = S(\delta, \omega(\beta))\ d\omega/d\beta$, which relates the $R(\beta)$ distribution directly to the edge slice.

small β are most clearly separated from the diagonal of the spectrum, which corresponds to $\beta = 0$. On the other hand, the most direct relation to $R(\beta)$ is exhibited by the 'edge slice' $S(\delta, \omega_2)$ at the low edge of the spectrum. As indicated in Figure 7.8, the edge slice is a direct image of $R(\beta)$ that is just somewhat compressed near $\beta = 0^\circ$ and $\beta = 90^\circ$:

$$R(\beta) = S(\delta, \omega(\beta))\ \sin(2\beta)\tfrac{3}{2}\delta. \tag{7.8a}$$

The derivation of this relation is based on the geometric relations summarized in Figure 7.9: for $\omega_1 = \delta$, we have $\theta_1 = 0$ so that the C—^{2}H bond is initially parallel to the $\vec{\mathbf{B}}_0$ field. After the reorientation, the bond makes an angle β with its initial orientation and thus also with the $\vec{\mathbf{B}}_0$ field so that $\beta = \theta_2$. Consequently, $S(\delta, \omega(\beta))$ is essentially equivalent to $R(\beta)$, with:

$$S(\delta, \omega)\ \mathrm{d}\omega = R(\beta)\ \mathrm{d}\beta \tag{7.8b}$$

according to the principle of conservation of the area of corresponding intervals (see Appendix E). Thus, $R(\beta) = S(\delta, \omega(\beta))\ \mathrm{d}\omega/\mathrm{d}\beta$ where $\omega(\beta) = \delta\frac{1}{2}(3\cos^2\beta - 1)$. This yields relation (7.8). The angular resolution of $S(\delta, \omega)$ is high around $\omega(45°)$ and low at $\omega(0°)$ and $\omega(90°)$. Therefore, in analysing $R(\beta)$ for small reorientation angles β, the centre slices of the 2D spectrum should be taken into account where according to Figure 7.5 the corresponding ellipses are better separated from the diagonal. The theoretical background of procedures for extracting $R(\beta)$ directly from an experimental spectrum are discussed in Sections 11.6 and 11.7.

For illustration, Figure 7.10 portrays reorientation-angle distributions for jump processes, as well as for anisotropic and isotropic rotational diffusion. The distributions shown are for the final states $t_m \to \infty$, that is, for mixing times that exceed the time constant of the dynamic process by far. The full dynamic evolution of $R(\beta, t_m)$ for these processes as a function of t_m is discussed in Chapter 11.

The information on the geometry of the dynamics contained in $R(\beta, t_m)$ makes 2D exchange NMR unique among experimental techniques for studying dynamics in

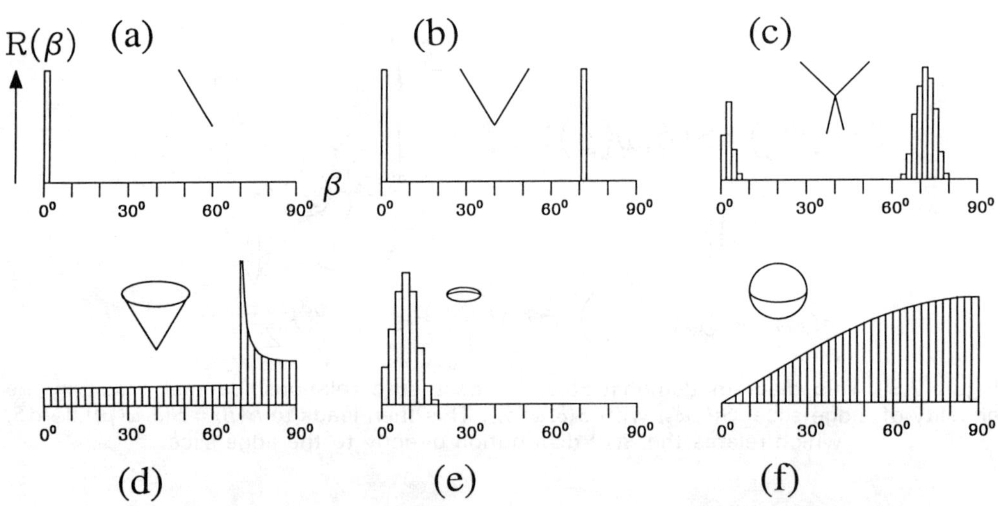

Figure 7.10 Reorientation-angle distributions for various final motional states. (a) No motion. (b) Two-site jump motion. (c) Tetrahedral four-site jumps with some scatter of the jump angles. (d) Uniaxial diffusion (diffusion on a cone). While the final-state distribution of the rotation angle ψ is a constant, the reorientation-angle distribution $R(\beta)$ shows a singularity at the upper limit of the distribution, which results from the differential conservation of the integral ($\mathrm{d}\beta/\mathrm{d}\psi$ in general vanishes at $\psi = 180°$, see Section 11.4.2). (e) Restricted diffusion. (f) Isotropic motion; the sin β curve just reflects the size of the surface element at β. The corresponding C—^{2}H density $R'(\alpha, \beta)$ on the unit sphere is a constant $1/4\pi$.

solids. Relaxation techniques can only measure correlation functions that correspond to weighted integrals over $R(\beta, t_m)$, so that most of the geometric information is lost. These relations are discussed at length in Chapter 9.

7.5 2D exchange powder spectra for $\eta \neq 0$

For $\eta \neq 0$, there are three reorientation angles to be considered which describe the orientation of the principal-axes system (PAS) after the jump relative to its orientation before. If it is known that only a single well-defined jump process occurs, the spectrum for $\eta \neq 0$ usually contains more information about the geometry of the reorientation than does a corresponding $\eta = 0$ spectrum.

For the analysis of 2D spectra in the case of $\eta \neq 0$, it is advantageous to view the situation from the 'unified PAS', made up of all principal-axes systems of the sample superimposed as indicated in Figure 7.11. The original distribution of PAS orientations is now reflected in a distribution of $\vec{\mathbf{B}}_0$ field orientations. This is convenient because $\vec{\mathbf{B}}_0$ vectors are more easily visualized and described than PAS coordinate-system orientations in the lab frame. The NMR frequency corresponding to a certain PAS and $\vec{\mathbf{B}}_0$ field is determined directly in terms of the equifrequency line on which the $\vec{\mathbf{B}}_0$ is located.

For the analysis of 2D spectra in the case of a well-defined jump process, the unified-PAS viewpoint is also particularly useful. Then, the orientation of the PAS after the jump is fixed relative to its orientation before the jump. Thus, both unified PAS_1 and PAS_2 can be displayed in the same figure, while the distribution of segmental

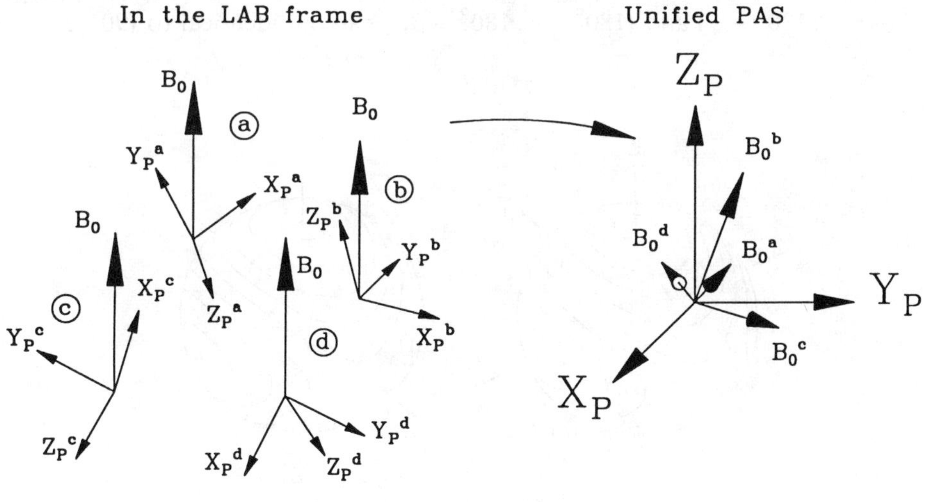

Figure 7.11 Visualization of the transition from the laboratory frame (LF) into the unified PAS. In the LF, the $\vec{\mathbf{B}}_0$ field always points up, while the principal-axes systems are oriented randomly together with the molecular segments to which they are fixed. In the transition to the unified PAS, all principal-axes systems are rotated to coincide, with each $\vec{\mathbf{B}}_0$ field orientation remaining unchanged relative to a given PAS. This results in a distribution of $\vec{\mathbf{B}}_0$ orientations in the unified PAS, which reflects the orientation distribution. For instance, in a powder the distribution is isotropic.

orientations is contained in the distribution of $\vec{\mathbf{B}}_0$ fields (by comparison, a treatment of the problem in the lab frame would involve a complicated analysis of the orientations of many different pairs of principal-axes systems). The frequencies before and after the jump can be determined directly from the orientations of a given $\vec{\mathbf{B}}_0$ field in PAS_1 and PAS_2. The calculation of the spectral intensity, in particular the ridge patterns, can then be performed exactly analogous to the procedure outlined in Section 6.7.3 for the dip-shift correlation experiment. Procedures for spectral simulations are described in Chapter 11.

The 2D exchange patterns are characteristic of the jump angles. Certain sets of reorientations, however, produce identical 2D spectra due to the symmetries of the NMR interactions. Figure 7.12 shows two different PAS_1 that have identical equifrequency lines and are therefore completely equivalent for the 2D spectrum. The coordinate systems shown are two out of four unequal but spectrally equivalent systems, which are related by the inversion of two principal axes. There are also four equivalent PAS_2 orientations, yielding altogether $4 \times 4 = 16$ spectrally equivalent relative orientations of PAS_1 and PAS_2. In terms of Euler angles, they are given as:

$$(n \times 180^\circ + \alpha, \beta, m \times 180^\circ + \gamma) \qquad (n \times 180^\circ - \alpha, \beta, m \times 180^\circ - \gamma)$$

$$(n \times 180^\circ - \alpha, 180^\circ - \beta, m \times 180^\circ + \gamma) \qquad (n \times 180^\circ + \alpha, 180^\circ - \beta, m \times 180^\circ - \gamma) \tag{7.9}$$

$n = 0_1 \pm 1_1 \pm 2$, $m = 0_1 \pm 1_1 \pm 2$. In general, these 16 sets of Euler angles that yield the same spectra describe physically different relative orientations. For certain values of α, β and γ, however, several of the triplets become identical or equivalent. For instance, for $\alpha = 90^\circ$, $\beta = 90^\circ$ and $\gamma = 90^\circ$, the triplets (α, β, γ), $(180^\circ - \alpha, \beta, 180^\circ - \gamma)$ and $(180^\circ - \alpha, 180^\circ - \beta, \gamma)$ are all identical to $(90^\circ, 90^\circ, 90^\circ)$.

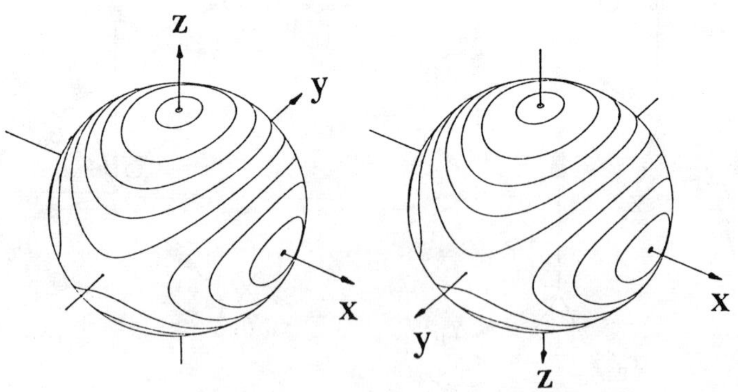

Figure 7.12 Symmetry-equivalent principal-axes systems PAS_1 and PAS_1' which are spectrally equivalent since they correspond to identical equifrequency-lines patterns. Reorientation from PAS_1 to some PAS_2, and from PAS_1' to the same PAS_2, thus result in identical 2D spectra, while the Euler angles of PAS_1 and PAS_1' relative to PAS_2 are different. PAS_1 and PAS_2' are related by an inversion of two principal axes, which leaves the frequency invariant due to the second-rank nature of the interaction. Two axes must be inverted in order to retain right-handed principal-axes systems.

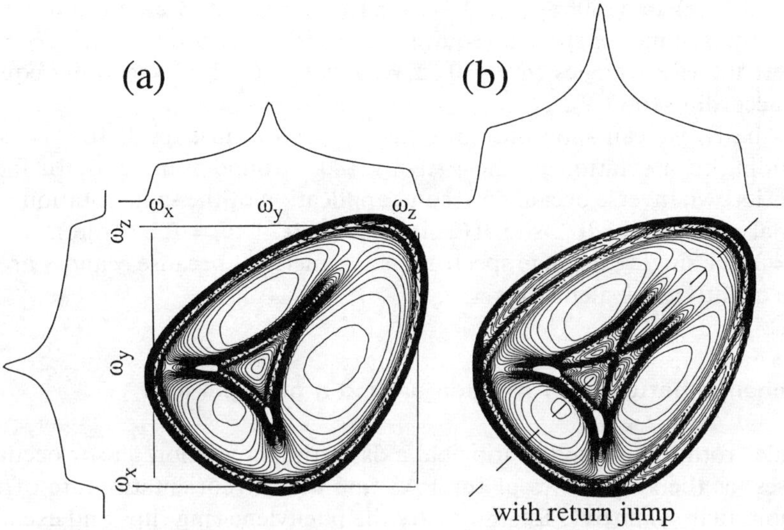

Figure 7.13 Typical 2D exchange powder patterns, calculated for $\eta = 0.85$. (a) Forward jump only. (b) Superposition of forward and backward jump symmetrizes the pattern with respect to the diagonal. The shape of the pattern, and the cross-sections at the ω_x and ω_z edges of the spectrum, allow for an estimate of the reorientation angles: the z axis does not change frequency very much (the ω_z end of the spectrum is close to the diagonal), corresponding to $\beta \cong 25^\circ$ (or equivalent to 25° according to equation (7.9)). The frequency ω_x exchanges far away from the diagonal, corresponding to a reorientation of the x axis by 40° to 70°; for small β, this angle corresponds approximately to $\alpha + \gamma$. The slight asymmetry of the forward jump spectrum shows that α is not equal, but similar to, γ. The values actually input into the simulation were $\alpha = 45^\circ$, $\beta = 30^\circ$, $\gamma = 10^\circ$.

Qualitative information about the jump angles can often be derived directly from a 2D exchange pattern; in particular from the locations of the turning points at the limits ω_x and ω_z of the spectrum as outlined in Figure 7.13.

7.6 The symmetry of the exchange patterns

The theoretical exchange pattern shown in Figure 7.13(a) is not symmetrical with respect to the diagonal. The observed or realistically simulated spectra, however, exhibit such a symmetry. This is a consequence of the jumps forth and back occurring with equal probability in thermodynamic equilibrium, that is, the jump process exhibits detailed balance. This results in equal exchange intensity at (ω_A, ω_B) for $A \to B$, and (ω_B, ω_A) for $B \to A$.

A single exchange pattern is itself symmetric if the reorientation Euler angles fulfil the condition:

$$\alpha = \gamma \tag{7.10}$$

or equivalently, according to (7.9), $\pm\alpha = m \times 180^\circ \pm (-1)^k \gamma$. For a proof, consider that the symmetric pattern is identical to its mirror-image, which is the pattern for the inverse jump. This means that the jump (α, β, γ) and its inverse jump, described by $(-\gamma, -\beta, -\alpha)$ or $(180^\circ - \gamma, \beta, 180^\circ - \alpha)$ (see Section B.2), must be equivalent. Consequently, symmetric spectra require $\pm\alpha = 180^\circ - \gamma$, $\pm\alpha = \gamma$, or $\pm\alpha = 180^\circ + \gamma$, if we insert the Euler angles $(n \times 180^\circ \pm \alpha, \beta, m \times 180^\circ \pm \gamma)$ spectrally equivalent to (α, β, γ) according to (7.9).

On this basis, we can show that symmetric spectra must result for two-site jumps corresponding to a rotation of the PAS by 180° around some axis. In fact, such a rotation is its own inverse because a second application of the same rotation reproduces the original orientation: $\mathbf{R}(\text{2-site})\mathbf{R}(\text{2-site}) = \mathbf{1} \Rightarrow \mathbf{R}(\text{2-site}) = \mathbf{R}^{-1}(\text{2-site})$.

For $\eta = 0$, the single exchange spectrum is symmetrical because α and γ are arbitrary and consequently equivalent to $\alpha = \gamma$.

7.7 Exchange patterns for rotation around a principal axis

A molecular rotation around a principal axis of the interaction tensor occurs in quite a few cases, as the symmetries of the PAS and of the reorientation are often related via the molecular symmetry. Examples are the phenylene ring flips and exchange from interchain ^{13}C spin diffusion (Edzes and Bernards, 1984) in polyethylene (see Figure 13.12). Such a motion around a principal axis is characterized by a single angle ψ, namely the rotation angle around that designated axis. As can be seen in Figure 7.14, characteristic ellipses with $|\tan\psi|$ as the ratio of their half axes (e.g. a circle for $\psi = 45^\circ$) appear in the 2D spectra, similar to the ridge patterns found for $\eta = 0$ (see Figure 7.5). According to Section 6.7.3, the spectral ridges correspond to regions on the unit sphere where equifrequency lines of the PAS before and after the rotation converge to run locally parallel, touching each other without intersecting. For a rotation around the PAS z axis, for instance, this occurs on the equator, where before and after the rotation all the equifrequency lines run perpendicular to the equator line (see Figure 2.5). The frequency on the equator is $\omega(90^\circ, \phi) = \frac{1}{2}\delta\{3 \times 0 - 1 - \eta\ 1\cos(2\phi)\}$. With a rotation by ψ, we find the spectral image of the equator:

$$\begin{pmatrix} \omega_1 \\ \omega_2 \end{pmatrix} = \begin{pmatrix} -\frac{1}{2}\delta \\ -\frac{1}{2}\delta \end{pmatrix} - \frac{1}{2}\delta\eta \begin{pmatrix} \cos(2\phi) \\ \cos(2\phi + 2\psi) \end{pmatrix}. \tag{7.11}$$

This is another Lissajou formula for an ellipse (see Appendix F). Equivalently, the $\phi = 0$ and $\phi = 90^\circ$ meridians produce the elliptical ridges for rotations around the y and the x axis, respectively.

For a molecular orientation with the $\vec{\mathbf{B}}_0$ field aligned along the rotation axis, the frequency remains invariant under the rotation, producing a distinct peak on the diagonal. Therefore, in the case of a rotation around the x axis, the ridges meet at (ω_x, ω_x), for a z-axis rotation at (ω_z, ω_z), and rotations around the y axis produce a sharp and high singularity at (ω_y, ω_y) (in spite of its dominance in the spectrum, this singularity is not indicated in the $\eta = 1$ spectra of Figure 6 of Wefing and Spiess (1988)).

7.8 2D exchange spectra of oriented samples

Highly oriented samples are particularly useful in exchange spectroscopy because they yield significantly higher exchange signals than powders, with better resolution of jump angles and additional restrictions on the jump geometry. This is due to the character

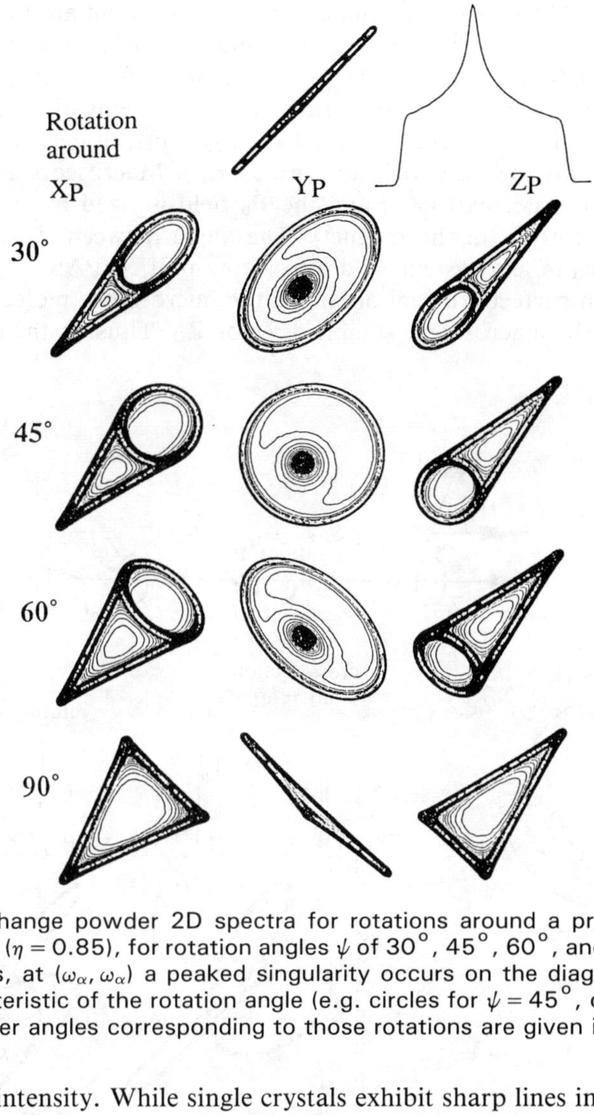

Figure 7.14 Exchange powder 2D spectra for rotations around a principal axis of the interaction tensor ($\eta = 0.85$), for rotation angles ψ of 30°, 45°, 60°, and 90°. For rotation around the α axis, at $(\omega_\alpha, \omega_\alpha)$ a peaked singularity occurs on the diagonal. The elliptical ridges are characteristic of the rotation angle (e.g. circles for $\psi = 45°$, or straight lines for $\psi = 90°$). The Euler angles corresponding to those rotations are given in equation (B.15).

of the exchange intensity. While single crystals exhibit sharp lines in 1D as well as 2D exchange spectra, powders yield inhomogeneously broadened 1D and 2D spectra. The exchange spectra of highly oriented uniaxial samples show inhomogeneously broadened 1D spectra, but 2D spectra with few sharp exchange signals in slices parallel to one of the frequency axes, thus leading to high intensity and resolution of the exchange patterns. This 'mixed' behaviour reflects on the one hand the still continuous distribution of orientations around the sample director. On the other hand, each frequency corresponds to only eight or fewer orientations of principal-axes systems relative to the $\vec{\mathbf{B}}_0$ field. For a well-defined jump process in a sample with perfect uniaxial order, in a slice at constant ω_1 eight or fewer sharp exchange peaks are found instead of the continuous exchange patterns encountered with powders (see Section 7.3).

The geometry of orientation and reorientation is again most conveniently analysed in the unified PAS. For samples exhibiting transverse isotropy (see section 12.3.1), by definition an axis in the molecular structure $\vec{\mathbf{Z}}_M$ can be found around which the PAS systems are uniformly distributed, for example the helix axis in simple helical structures. As visualized in Figure 7.15(a), the angles between this $\vec{\mathbf{Z}}_M$ axis and the principal axes are the same for all the principal axes systems distributed around $\vec{\mathbf{Z}}_M$ (Opella and Waugh, 1977). Therefore, in all PAS systems, in particular in the unified PAS, the $\vec{\mathbf{Z}}_M$ axis has the same orientation (β_P, α_P). Macroscopically, the angle β_0 between the sample order axis ($\vec{\mathbf{Z}}_D$) and the $\vec{\mathbf{B}}_0$ field is fixed by the position of the sample in the laboratory (in the magnet). The angle between $\vec{\mathbf{Z}}_D$ and the $\vec{\mathbf{B}}_0$ field remains equal to β_0 in any system, and also in the unified PAS.

For samples with perfect uniaxial order, all the microscopic molecular unique axes $\vec{\mathbf{Z}}_M$ are parallel to the macroscopic sample director $\vec{\mathbf{Z}}_D$. Thus, in the unified PAS, the

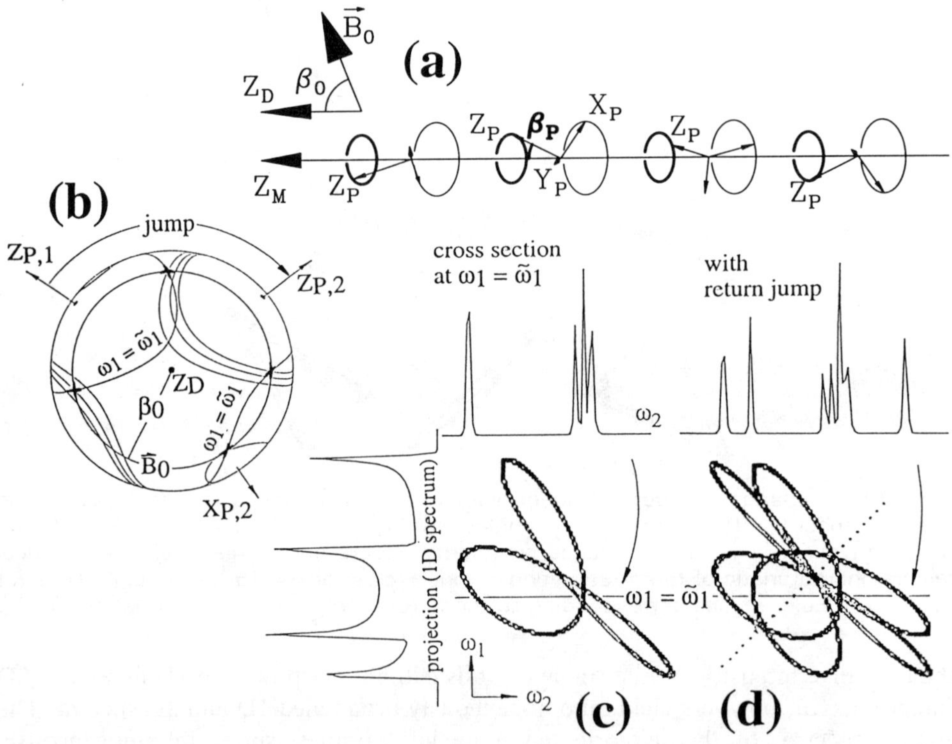

Figure 7.15 Exchange spectrum for a sample with perfect uniaxial orientation (i.e. the molecular and sample directors coincide for all chains, $\vec{\mathbf{Z}}_M = \vec{\mathbf{Z}}_D$). (a) The principal axes for equivalent moieties (e.g. CH_2 groups in POM) are distributed around the directors $\vec{\mathbf{Z}}_M$ on cones such that $\vec{\mathbf{Z}}_D = \vec{\mathbf{Z}}_M$ has the same orientation (α_P, β_P) in every PAS. Thus, in the unified PAS before and after the molecular reorientation, as shown in (b), the sample director has a fixed direction, and all $\vec{\mathbf{B}}_0$ fields are located on a cone (circle on the unit sphere). (c) 2D spectrum, of a forward jump only, for the situation indicated in (b). The spectral pattern is a single closed curve, representing a distorted replica of the circle of $\vec{\mathbf{B}}_0$ field orientations in (a). The cross-section at the frequency $\omega_1 = \tilde{\omega}_1$ marked in (b) consists of four sharp peaks, corresponding to the ω_2 frequencies of the points where the circle of $\vec{\mathbf{B}}_0$ field orientations intersects the $\omega_1 = \tilde{\omega}_1$ line. To the left of the 2D spectrum, its projection is shown, which is identical to the 1D spectrum. (d) Forward and return jump produce a 2D spectrum which is symmetric to the diagonal.

$\vec{\mathbf{B}}_0$ fields are uniformly distributed on a (double) cone at an angle β_0 relative to the axis $\vec{\mathbf{Z}}_M = \vec{\mathbf{Z}}_D$. Spectral intensity at a frequency $\tilde{\omega}$ arises only where the corresponding circle on the unit sphere intersects one of the $\tilde{\omega}$-equifrequency lines (Figure 7.15(b)). At maximum, eight such intersections can occur, corresponding to four distinguishable pairs of points related by the unavoidable inversion symmetry. In a slice at a given ω_1 through the 2D spectrum for a well-defined jump, each of these four distinguishable orientations can give rise to one sharp exchange line (Figure 7.15(c)). Many such ω_1-slices stacked combine these signals into one closed curve in the frequency plane as shown in Figure 7.15(c), thus creating a distorted image of the circle on which the $\mathbf{B}_0$ fields are located. In practice, the return jump cannot be avoided, so that the pattern reflected at the diagonal is superimposed on the original one (see Section 7.6 and Figure 7.15(d)). Often, the exchange pattern is simplified by symmetries, such as the $\vec{\mathbf{Z}}_M$ axis being located in a symmetry plane of the PAS, or cases of equivalent jump and return jump motions (see Section 7.6). Usually, the exchange features are also less complicated if a sufficiently small angle β_0 is chosen. Figure 7.16 shows a spectrum simulated for highly oriented POM where the helix axis $\vec{\mathbf{Z}}_M$ is located nearly in a symmetry plane of the PAS of the (—$O^{13}CH_2$—) units. The complexity of the 2D spectrum for a well-defined jump can usually be predicted on the basis of the 1D spectrum, which is equal to a projection of the 2D spectrum on one of the axes. If more than three singularities appear in the 1D spectrum, as in the projection shown in Figure 7.15(c), eight exchange peaks necessarily occur in certain ω_1 slices through the 2D spectrum. This can be concluded because usually a singularity in the projection corresponds to a turning point of the curve in the 2D plane (see Figure 7.15).

The equivalence of triplets of reorientation angles derived in Section 7.5 for powder samples is lifted if the orientation distribution breaks the underlying symmetry of the equifrequency-line pattern. With oriented samples, more information about the reorientation is available, as the restricted set of orientations also limits the possibilities of reorientations. For instance, in highly ordered samples with transverse isotropy, the only motions which do not destroy the orientational order are rotations around the

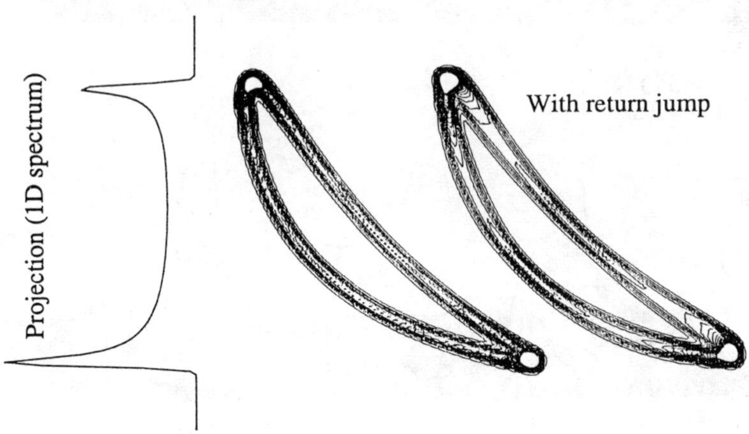

Figure 7.16 Exchange pattern for a perfectly oriented uniaxial sample. The spectrum is much simpler than that in Figure 7.15 because the rotation axis is located nearly in a symmetry plane of the interaction tensor.

molecular director, which coincides with the sample director for these perfectly oriented uniaxial systems. Thus, the motion can be fully analysed in terms of a single angle of rotation instead of the three Euler angles required for the description of reorientations in powders.

Advantages of oriented samples as discussed here for exchange spectroscopy can also be exploited in the dip-shift correlation experiments described in Section 6.7.3.

7.9 Model-independent information from 2D exchange spectra

The preceding sections have shown that two-dimensional exchange NMR spectra contain a wealth of information not only on the time scales, but also about many details of the geometry of slow reorientational motions. A series of 2D exchange ^{13}C spectra of commercial poly(oxymethylene) (Hagemeyer *et al.*, 1989) shown in Figure 7.17 are to demonstrate that many properties of the motional process can be determined in a model-independent manner just by inspection of the spectra.

The intensity in the POM spectrum taken at 252 K, Figure 7.17(a), is confined to the diagonal ($\tilde{\omega}_1 = \tilde{\omega}_2$). This means that in the mixing time of 1 second neither slow motion

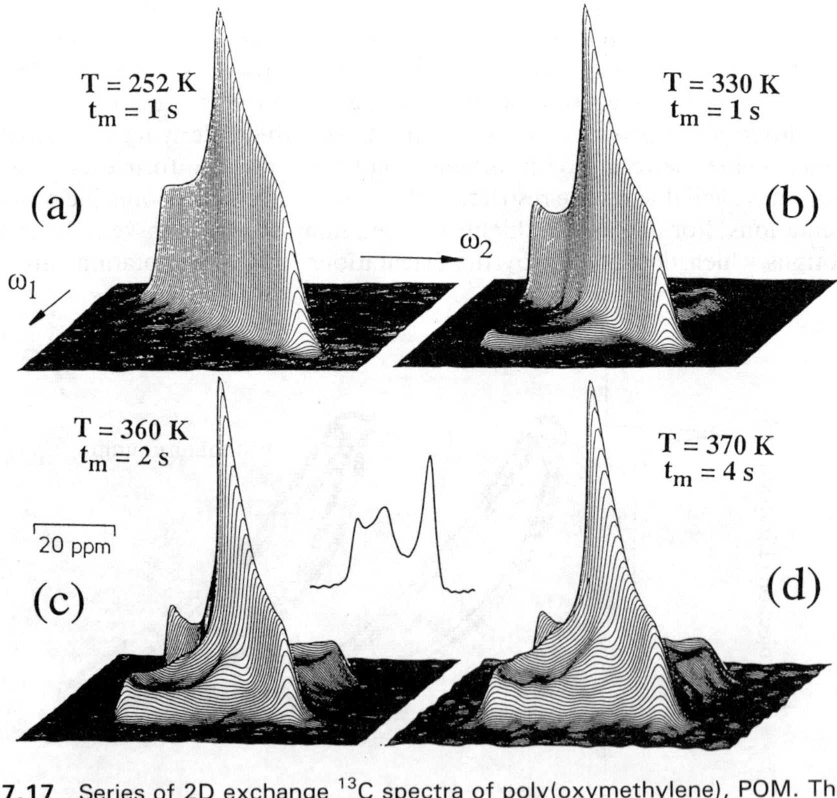

Figure 7.17 Series of 2D exchange ^{13}C spectra of poly(oxymethylene), POM. The signal from the crystallite is observed selectively due to fast ^{13}C T_1 relaxation of the signal from the amorphous fractions of the sample. With increasing temperatures and mixing times, strong and structured exchange intensity appears. Adapted from Hagemeyer *et al.* (1989).

nor spin diffusion is detected. The signal in the spectrum is due only to the crystalline fraction of the polymer. The highly mobile amorphous fraction ($T_g = 200$ K) has such a short T_1 relaxation time that its signal contributions are cancelled by relaxation during t_m and the pulse phase cycle described in Section 10.2.5. Thus, the dynamics generating the exchange intensity detected at 330 K, Figure 7.17(b), must occur in the crystallites. The exchange intensity in this spectrum extends far away from the diagonal. This means that large frequency changes take place, which necessarily imply large-angle reorientations due to the simple relation between frequency and orientation. The exchange intensities in Figures 7.17(b)–(d) exhibit quite well-defined ellipse-like ridge structures, which is a clear indication of relatively well-defined jumps. From the shape of the pattern, the jump angle can be estimated to exceed 50°. The exchange intensity strongly increases with temperature. To estimate its integral intensity relative to the diagonal, a 1D slice at fixed ω_1, as discussed in Section 7.2, is plotted next to the spectrum in Figure 7.17(c). It was taken near one end of the spectrum where exchange and diagonal intensity are best separated. The integration shows that the contribution of the diagonal is only about $\frac{1}{4}$ of the total. Thus, the exchange must occur between at least 4 sites. This strong reduction of the diagonal intensity also shows that nearly all crystalline stems must participate in the process.

Thus, from the series of spectra it can be concluded immediately, without any reference to a model, that:

- in the crystallites of POM
- a thermally activated dynamical process occurs
- with a correlation time of $\simeq 1$ s at 330 K,
- involving large-amplitude (>50°) reorientations
- of essentially all chain stems in the crystallites
- which undergo relatively well-defined jumps
- between at least 4 sites.

These far-reaching conclusions, together with the crystallite structure and highly resolved 2D exchange spectra of oriented POM shown in Section 7.12, conclusively establish helical jumps as the correspondent dynamical process. This mechanism, for which first experimental evidence was seen in POM by Veeman and co-workers (Kentgens *et al.*, 1987) in MAS 2D exchange spectra (see Section 7.11), will be discussed in more detail in Section 7.12.

7.10 3D exchange NMR

7.10.1 The relevance of 3D exchange NMR in complex solid materials

In the 2D exchange experiment, the geometry of a molecular reorientational process is determined by observing segment orientations encoded in the NMR frequencies at two different times. Further information about the motional process, in particular about the sequence of reorientations, is obtained by triple observation of the molecular orientation in a 3D exchange experiment. Inequivalent orientational sites then yield resolved signals, such that accidental or symmetry-based overlap is reduced. In particular, return jumps to the start position can be observed separately. They can be well detected even when so many different large-angle reorientations occur that the corresponding 2D spectrum suggests a rather diffusive process due to the superposition of the large number of exchange signals. In favourable cases, one can determine whether the probability of a return jump is enhanced relative to a further jump forward and thus establish experimentally if the motion is a simple non-Markovian process (for a definition of Markov processes see Section 9.3.1). These possibilities often justify the relatively long measuring times of typically one or two days for a 3D spectrum and the complications which arise because of the large 3D data fields (which are nevertheless small compared with those required in solution 3D NMR: $64 \times 64 \times 128$ complex data points were the maximum required for any of the 3D spectra shown below).

The 3D pulse sequence is obtained from that of the 2D exchange experiment by prefixing an additional evolution time and a second mixing time as indicated in Figure 7.18.

In recent years, a large number of 3D solution-NMR experiments (Griesinger *et al.*, 1989) have been developed for elucidating complicated molecular structures. There, the focus is on the increase of resolution by separation of overlapping lines. Triple correlation of the same interaction is generally of less interest there because completely resolved lines yield no new correlations compared with the 2D spectrum. In the solid state, however, the 3D exchange experiment actually provides new information in two respects. While in solution NMR different exchange sites are characterized by distinct lines, in anisotropy-exchange NMR of solids all positions are first of all equivalent in the inhomogeneously broadened spectrum. In the case of the helical jumps, for instance, there is no frequency position on the diagonal of the spectrum to be attributed exclusively to the first-neighbour position. The overlap of the signals of all positions is complete. Only in the evolution period of the 2D experiment is the start position labelled. In the 2D spectrum, the reorientation relative to the start positions can be read, but not the connectivity between the exchange signals. Whether a jump from the start position occurs directly or in two steps via another position can only be

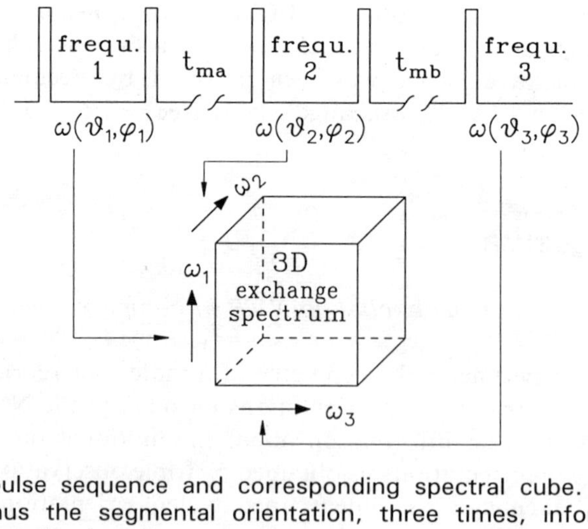

Figure 7.18 3D pulse sequence and corresponding spectral cube. By measuring the frequency, and thus the segmental orientation, three times, information about the sequence of reorientational motions can be obtained. In particular, return jumps result in a return of ω_3 to the initial frequency ω_1, i.e. $\omega_3 = \omega_1$, and are therefore detected separately in the diagonal plane with $\omega_3 = \omega_1$.

concluded indirectly from the t_m dependence of the exchange intensities in a full series of 2D spectra with increasing t_m. In the 3D experiment, this connectivity can be made visible directly, because in the second evolution time the exchange sites are also labelled.

A second aspect making 3D exchange NMR in solids more relevant than in liquids is related to the nature of the exchange processes. In solution, the NOE (nuclear Overhauser effect) and chemical exchange of small molecular segments are most important: to a good approximation both are simple Markov processes between well-defined positions. Their behaviour can therefore be predicted completely on the basis of the exchange rates determined in 2D experiments. In amorphous solids and semi-crystalline polymers, however, the circumstances can be much more complicated. For the dynamics above T_g, for instance, it is *a priori* not clear between what types of sites the exchange occurs: whether a distorted diamond lattice is present because rotationally isomeric states (trans, gauche) are favoured energetically, whether diffusion with very small steps occurs, or whether both processes are coupled. Here the question concerning the occurrence of return jumps back to the start position cannot be answered on the basis of the information available from the 2D spectrum.

In semicrystalline polymers, due to the long-range coupling of different segments and morphological regions along a polymer chain, memory effects in the dynamics can be imagined; in fact, they are quite inevitable. When a chain segment in the crystallite jumps, it also changes the lengths of two loops in the amorphous regions. As a consequence of the tightening of one of these loops, a further jump in the same direction can become less probable, while the return jump in general carries the chain back into an energetically acceptable state. This increases the probability of the return jump with respect to the second jump away from the start position, so the dynamics of a given molecular segment is not strictly Markovian any more (for a more detailed discussion, see Section 9.3.1).

3D exchange NMR is also helpful to analyse molecular dynamics with a superposition of fast and slow processes. For instance, it can clarify whether in a 2D spectrum a diagonal signal that remains strong even after long mixing times is due to the great statistical weight of the start position, or to segments which are trapped so that they cannot participate in the dynamical process. One could think, for example, of small crystallites in a predominantly amorphous polymer, where exchange due to the dynamics above T_g is observed. Similarly, at the glass transition of fully amorphous systems 3D NMR can distinguish to some extent whether slow and fast processes occur in parallel or in series. This is discussed in Sections 8.6 and 9.3.2.

7.10.2 Two-dimensional cross-sections of 3D spectra

The intensity distribution in 3D exchange spectra will obviously depend on the ratio of t_{ma}, t_{mb}, and the correlation time τ of the exchange process. If no exchange occurs during t_{ma} and t_{mb}, the frequencies do not change in the course of the experiment so that the signal is confined to the space diagonal $\omega_1 = \omega_2 = \omega_3$ of the spectral cube (Figure 7.19(a)). If t_{mb} is much shorter than τ, $\omega_2 = \omega_3$ is required and the signal is confined to the diagonal plane thus specified (Figure 7.19(b)). In an analogous manner, for short t_{ma}, signal occurs only in the $\omega_1 = \omega_2$ diagonal plane (Figure 7.19(c)). Return jumps to the start position are obviously characterized by $\omega_1 = \omega_3$, thus defining the third diagonal plane that contains the space diagonal (Figure 7.19(d)).

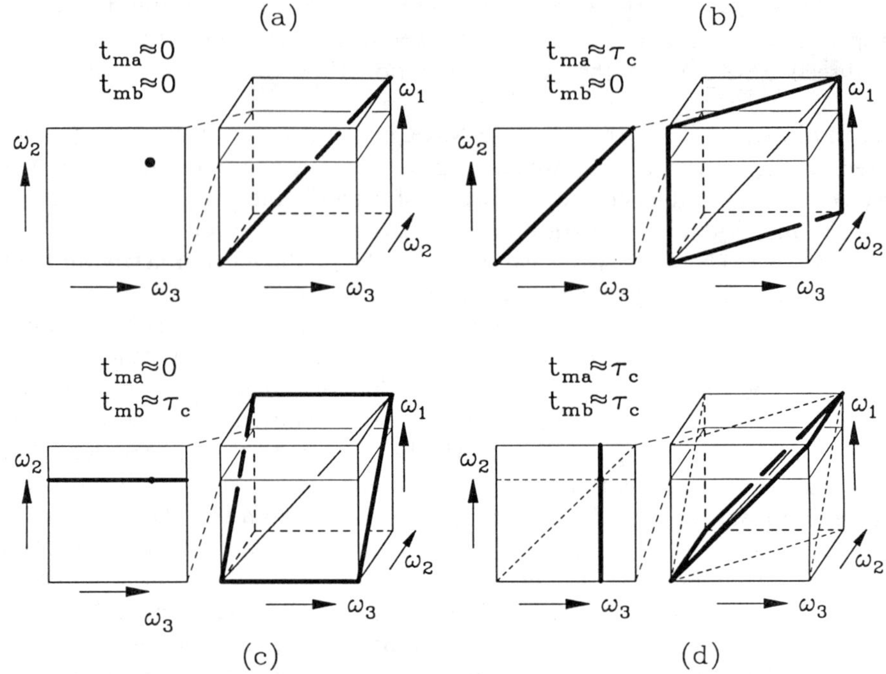

Figure 7.19 3D intensity distribution for various combinations of mixing times in 2D slices at fixed ω_1 and in the 3D cube. Lines or diagonal planes in which spectral intensity can occur are thicker. While (a)–(c) are valid for arbitrary exchange mechanisms, in (d) a two-site jump was assumed.

In the detailed analysis of 3D spectra it is often useful to look at two-dimensional slices at a fixed ω_1 value first. The full 3D intensity distribution can finally be generated by stacking these 2D slices. This is analogous to the relation between 2D and ω_1-selective 1D exchange spectra (see Section 7.2). As shown schematically in Figure 7.19, without exchange, the spectral intensity is confined to the point ($\omega_2 = \tilde{\omega}_1$, $\omega_3 = \tilde{\omega}_1$); exchange only during t_{ma} restricts the intensity to the plane-diagonal line (ω_2, $\omega_3 = \omega_2$); exchange only during t_{mb} produces signal along the line ($\omega_2 = \tilde{\omega}_1$, ω_3) parallel to the ω_3 axis.

Interesting features appear if exchange occurs during both mixing times. Here, as in the following, $t_{ma} = t_{mb}$ will always be chosen for simplicity. In the simplest of exchange cases, namely between two positions, in addition to the features at ($\omega_2 = \tilde{\omega}_1$, ω_3) and (ω_2, $\omega_3 = \omega_2$), signal appears on the line (ω_2, $\omega_3 = \tilde{\omega}_1$) which directly detects the return jumps.

The 3D exchange experiment is practicable within nearly the same measuring time as the corresponding 2D experiment. This is so because the intensity is confined to certain regions in the 3D spectral cube, typically to the three diagonal planes, if the jump angles are reasonably well defined. This splitting up of the signal and the second amplitude modulation reduce the exchange signal only by a factor of four, which can be partially compensated by measuring with lower resolution in the evolution periods: halving the maximum acquisition time in both t_1 and t_2 produces an effective signal gain by up to a factor of four (see Section 4.9). If the intensity was distributed quite uniformly throughout the spectral cube, the 3D experiment would not be feasible on ^{13}C in natural abundance.

7.11 The MAS sideband-exchange 2D experiment

Magic-angle spinning is usually employed to obtain the isotropic-chemical-shift spectrum of a material. If the sample is spun sufficiently slowly, however, spinning sidebands are observed which represent the remainder of the chemical shift anisotropy (see Section 3.9). As such, these sidebands can be used to probe dynamics of the moieties concerned. This slow-MAS exchange experiment represents a compromise between spectral resolution and angular resolution, as well as between angular resolution and the signal-to-noise ratio (SIN). It permits the exchange to be attributed to certain molecular moieties. Due to spectral overlap, this may sometimes be difficult with static samples although techniques for separating different signal contributions, such as short CP, dipolar dephasing (see Section 3.2), or single pulse excitation often help to reduce the overlap. The optimum combination of MAS spectral and static angular resolution would be achieved in a 3D experiment with 2D exchange patterns separated by the isotropic shifts. As yet, however, a technique for performing such experiments in a commercial MAS probe with sufficient resolution and signal-to-noise ratio is not available. The sideband-exchange 2D experiment provides quite good S/N and is relatively robust, but the information on the reorientation process is not as directly accessible as in spectra taken on static samples. For instance, diffusive and jump processes are not immediately distinguishable, and jump angles cannot be read off the spectrum. The information from static and sideband exchange experiments are therefore complementary in several respects.

The most important requirement to make the simplest sideband exchange experiment work is the rotor-synchronized mixing time. If at the start of the detection, the orientation of the rotor is not the same as it was at the end of the evolution period, or in other words if the duration of the mixing time is not chosen as a multiple of the rotation period t_r:

$$t_m = nt_r \tag{7.12}$$

(de Jong *et al.*, 1984) the frequency is changed by the sample rotation which induces off-diagonal sidebands even without a molecular exchange process. The rotor phase also complicates the generation of absorptive sidebands in the spectrum which are required to obtain sufficient spectral resolution (see Section 10.4). The magnetization and rotor phase problems can be solved, as is described in some detail in Section 10.4, by combining two data sets, of which one involves 't_1-time reversal': the relative rotor phase at the start of the detection is chosen as the negative of the relative rotor phase at the end of the evolution period. Like the TOSS pulse sequence or the appearance of positive and absorptive MAS sidebands, the experiment requires isotropic samples, or at least symmetry of the sample orientation distribution with respect to the rotor axis. For uniaxial samples with the order axis at an arbitrary direction, this can be achieved by unsynchronized averaging or regular incrementation of initial rotor orientations.

7.12 Helical jumps in polymer crystallites detected by 2D and 3D NMR

Dielectric and dynamic-mechanical relaxation studies of the α relaxations in semicrystalline polymers have been performed for more than 35 years; however, few definite conclusions on the underlying molecular processes could be drawn. It appears that no technique other than solid-state NMR is available to analyse molecular details of these slow motions; before reaching the rates required for analysis by inelastic neutron or diffuse X-ray scattering, the samples melt or degrade.

For poly(oxymethylene) (POM). in the temperature-frequency region of the α relaxation the 2D exchange spectra discussed in Section 7.9 have proven an activated

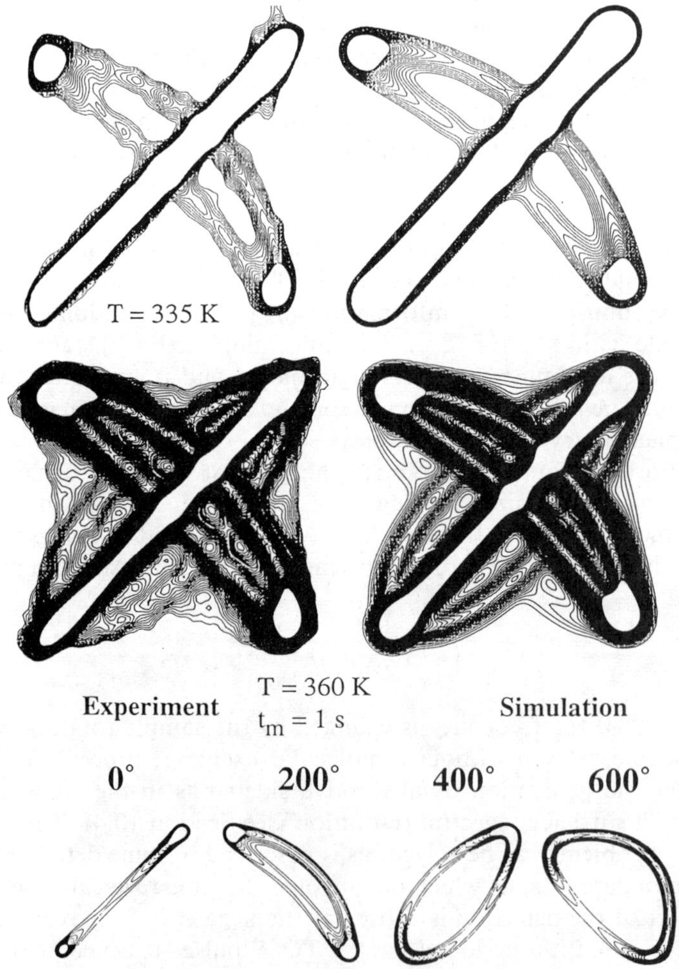

Figure 7.20 Experimental 2D exchange ^{13}C spectra (t_m = 1 s) of highly oriented POM and corresponding simulations for helical jumps. At T = 335 K (top), the bow-shaped exchange pattern corresponding to 200° rotations to next-neighbour positions is observed only. At T = 360 K, further patterns become clearly visible, corresponding to $n \times 200°$ rotation angles. At the bottom, the subspectra for the $n \times 200°$ rotations are shown.

jump process in the crystallites with large, relatively well-defined jump angles involving at least four distinguishable sites. These can indeed be resolved well in 2D ^{13}C spectra on oriented POM as shown in Figure 7.20. They indicate consecutive jumps, with the bow-shaped exchange pattern appearing sooner than the ship-shaped pattern. This sequence of jumps is indeed proven in the 3D spectrum shown below.

Helical jumps as indicated in Figure 7.21(a) represent the simplest motion fulfilling all these requirements, as well as producing no orientational disorder of the stems in the crystallites. A given chain moves such that it is in the energetically most favourable

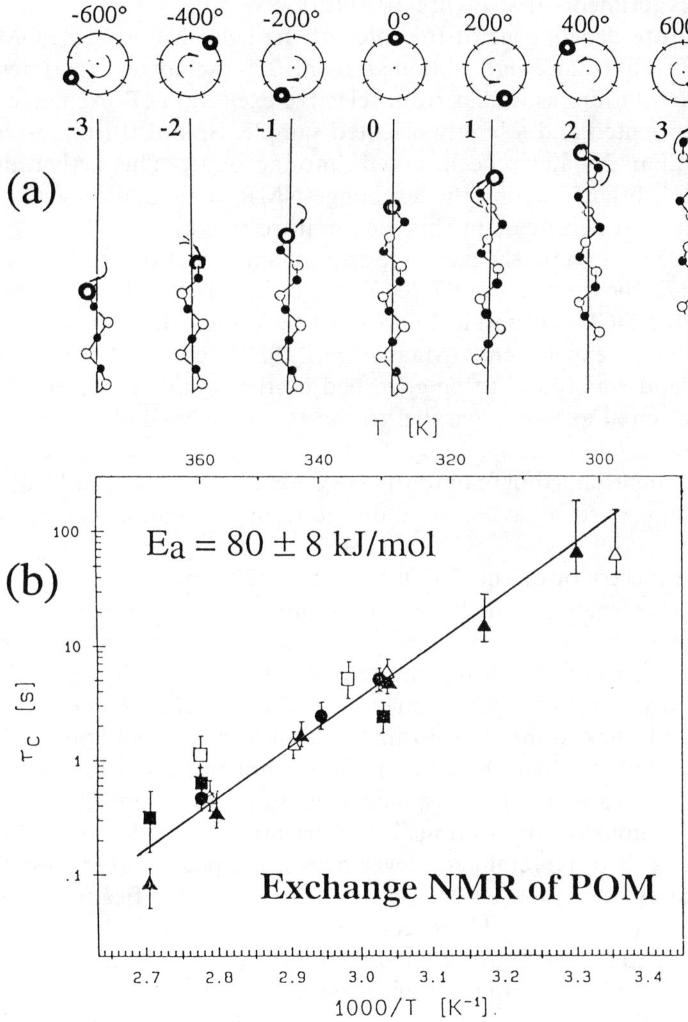

Figure 7.21 (a) Helical jumps of POM chains (g_5 helices) in the crystallites, involving in each step a 200° rotation and a translation by one OCH_2 unit. Note that the chain moves only between energetically equivalent states, which are actually indistinguishable unless a position along the chain is labelled. (b) Arrhenius diagram of the correlation times of the helical jump process in POM as obtained from various exchange ^{13}C spectra. The activation energy of about 85 kJ/mol is in agreement with dynamic-mechanical data for the α relaxation.

position before and after the 'jump'. These positions are determined by the 'corrugations' in the channel formed by the surrounding chains, which are essentially rigid during the time required for the motion of a given chain (less than a nanosecond according to simulations by Mansfield and Boyd (1978)) unless there is strong cooperativity. The simplest process is a shift by one repeat unit and a corresponding rotation by $360^\circ \times n/m$ for an m_n helix. This moves the chain between equivalent positions in the crystallite. Such a motion, which had been suggested to explain dynamic-mechanical and dielectric relaxations in semicrystalline polymers (Hoffmann *et al.*, 1966; Boyd, 1985), was first verified for POM by Kentgens *et al.* (1987) in MAS 2D exchange experiments (see Figure 10.10(a)).

The temperature dependence of the rates of the helical jumps in POM is shown in Figure 7.21(b). The data were obtained from 2D exchange experiments with and without sample rotation, as well as from selective-excitation 1D exchange experiments, on both an unoriented and a highly oriented sample. Spin-diffusion effects at mixing times longer than 2 s have been taken into account. The activation energy of 80 ± 8 kJ/mole obtained from the exchange-NMR data agrees with data for the dynamic-mechanical α process in this temperature range.

For static POM samples, 2D exchange spectra simulated on the basis of this helical jump process are shown in Figure 7.20. For the simulation, the chemical-shift tensor orientation of the OCH_2 carbon in POM must be known. It has been measured in 2D dip-shift correlation experiments (Maas *et al.*, 1987; Schmidt-Rohr *et al.*, 1993) (see Section 6.6.2) and was found to be governed by the symmetry of the CH_2 group. In essence, the principal axes are: parallel to the H–H vector and thus perpendicular to the O–C–O plane (ω_{11}), along the H–C–H bisector (ω_{22}), and parallel to the O–O vector or, equivalently, orthogonal to the H–C–H plane (ω_{33}). The fit of the exchange spectrum was improved by assuming a distribution of rotation angles approximately 8° wide.

3D exchange spectra of oriented POM, Figure 7.22, confirm the sequence of jumps which is a crucial element of the helical jump model as presented here. They establish relations not accessible in the 2D spectra between certain parts of the first and second exchange patterns, and the helical-jump model predicts these correctly. Figure 7.22(c) displays a 2D slice out of the 3D spectrum in Figure 7.22(a), which shows that jumps from position ± 1 back to the start position 0 are not much more probable than jumps to position ± 2. This confirms that the motion is not just a localized ('trapped') 'flip-flop' process, but really involves displacements in terms of angles seen directly here, which must be coupled to translations if the orientational order around the crystallite *c*-axis (chain axis) is to be retained. Nevertheless, the peaks in the spectra indicate on a level of about 10% of their heights, at the verge of insignificance, that jumps back could be preferred. This would be consistent with the PE chain diffusion results discussed in Section 7.14.2 where strong indications of some restrictions against variations of the loop lengths in the amorphous regions are found. For a stem to be able to shift in the crystallite, the loop at one of its ends must not resist being shortened, and the loop at the other end must easily become longer. The jump back leads again into a favourable state taken on previously, while jumping on to the next site can result in a strained loop with a probability of 5 to 10% (according to the PE chain diffusion results, see Section 7.14.2).

In isotactic polypropylene (iPP), ^{13}C (Hagemeyer *et al.*, 1989) and ^{2}H exchange NMR (Schaefer *et al.*, 1990) have identified the motional α process in the crystallites

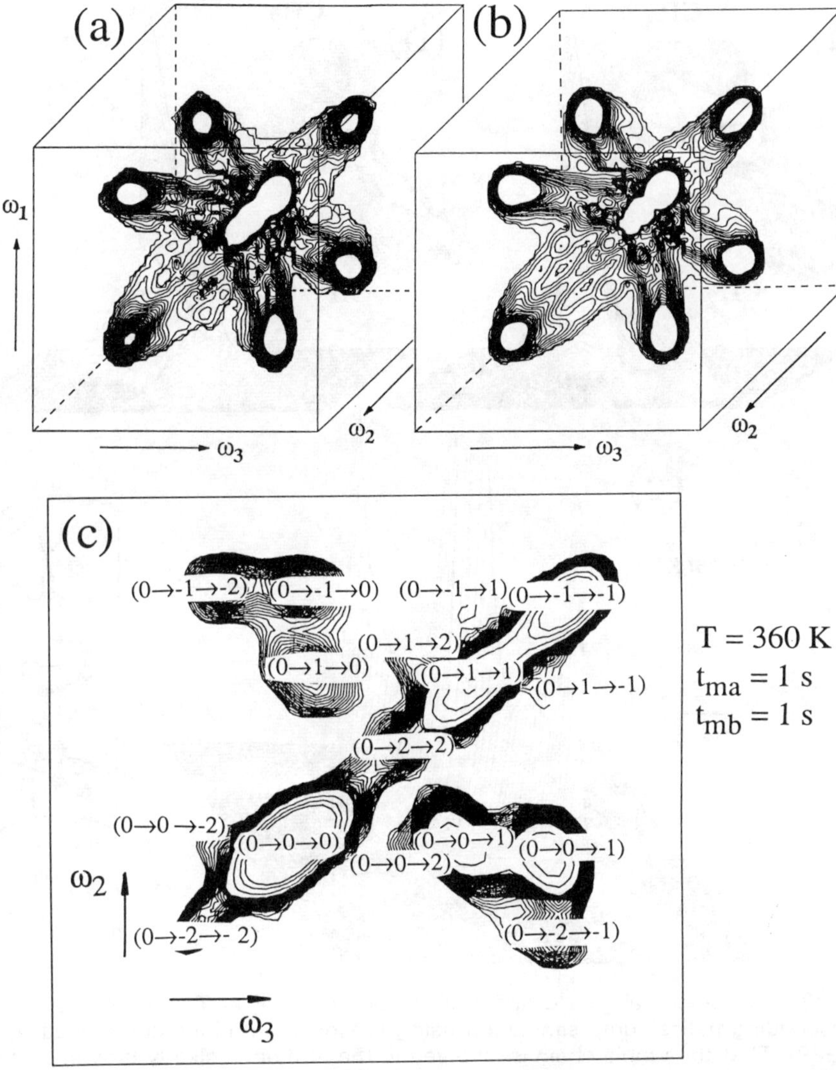

Figure 7.22 (a) 3D spectrum of POM at 360 K and $t_{ma} = t_{mb} = 1$ s. (b) Helical-jump simulation. (c) Two-dimensional slice out of the spectrum shown in (a). The peaks are labelled according to the corresponding positions along the helix (by definition, position 0 is the position in the first evolution time). These connectivities can be determined directly from the spectrum. The $(0 \rightarrow \pm 1 \rightarrow 0)$ and $(0 \rightarrow \pm 1 \rightarrow \pm 2)$ peaks are of particular interest, representing return jumps and jumps to the next position on the helix, respectively.

as helical jumps, in a temperature range between 340 K and 380 K. The ^{13}C spectrum of Figure 7.23(a), for drawn iPP, shows that indeed all carbon sites are involved in the motion with large amplitudes. The experimental spectrum is reproduced properly by a simulation on the basis of the helical jumps as displayed in Figure 7.23(b). For the ^{13}C chemical-shift tensors required in this simulation, values determined by Nakai *et al.* (1989) from 2D dip-shift correlation with OMAS have been used with slight modifications. From the ^{2}H 2D spectrum (Schaefer *et al.*, 1990) shown in Figure

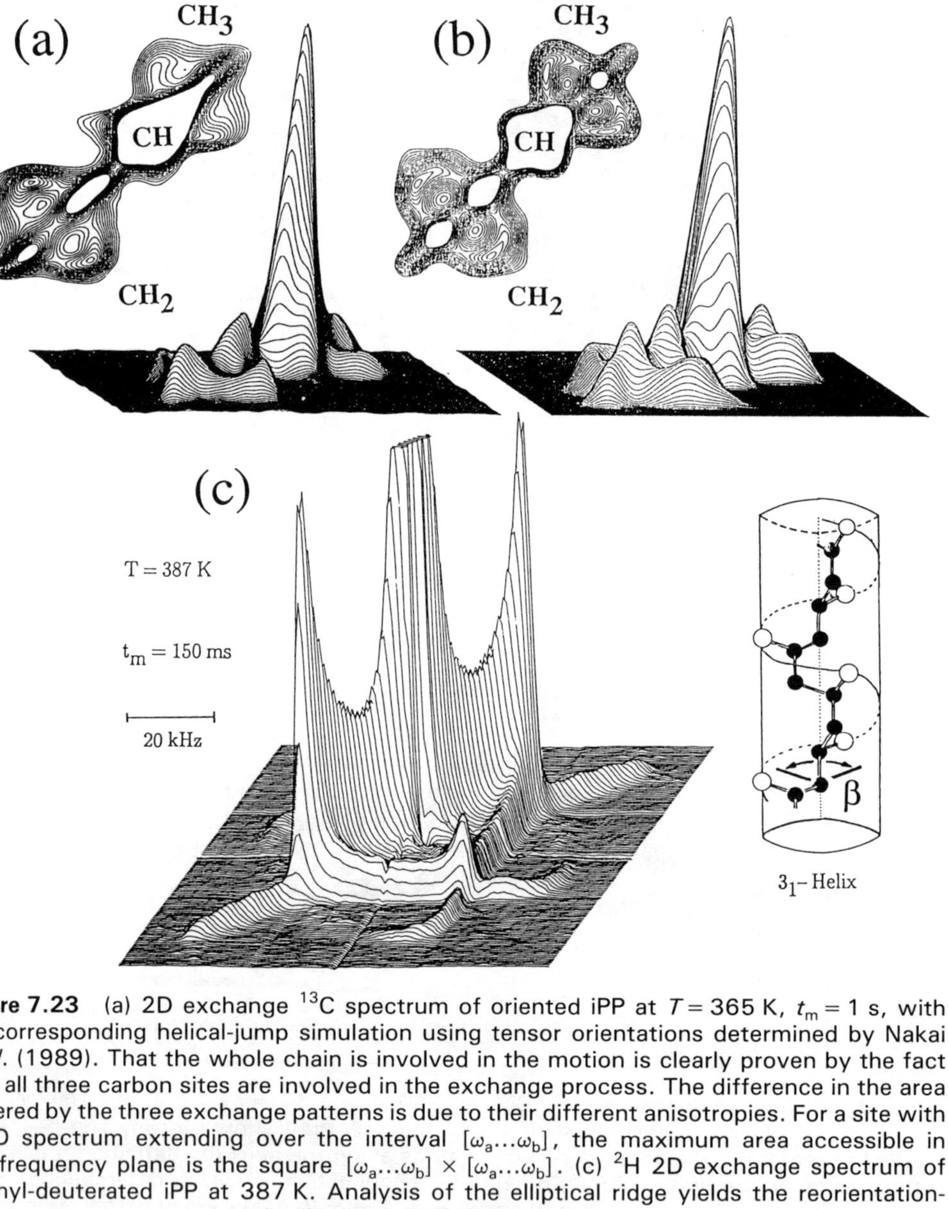

Figure 7.23 (a) 2D exchange ^{13}C spectrum of oriented iPP at $T = 365$ K, $t_m = 1$ s, with (b) corresponding helical-jump simulation using tensor orientations determined by Nakai *et al.* (1989). That the whole chain is involved in the motion is clearly proven by the fact that all three carbon sites are involved in the exchange process. The difference in the area covered by the three exchange patterns is due to their different anisotropies. For a site with a 1D spectrum extending over the interval $[\omega_a \ldots \omega_b]$, the maximum area accessible in the frequency plane is the square $[\omega_a \ldots \omega_b] \times [\omega_a \ldots \omega_b]$. (c) ^{2}H 2D exchange spectrum of methyl-deuterated iPP at 387 K. Analysis of the elliptical ridge yields the reorientation-angle distribution of the methyl groups.

7.23(c), the reorientation-angle distribution has been estimated to be 4° wide (fwhm), which corresponds to a rotation-angle distribution approximately 5° wide (see equation (7.4)). In syndiotactic polypropylene (sPP), with its more complicated 4_1 helix (see Section 5.2), no exchange could be detected, in accordance with the absence of a relaxation of this material in the corresponding rate and temperature regions.

In polyethylene oxide (PEO), the dynamics in the crystallites have not been identified by relaxation techniques or even ^{2}H exchange NMR experiments, since the glass transition dynamics of the amorphous domains occur at similar temperatures. ^{1}H and ^{13}C NMR first established the presence of dynamics in both crystalline and amorphous regions of PEO (VanderHart *et al.*, 1981) with rates in the 50 kHz range near room temperature. With these one-dimensional techniques, however, the separate characterization of the dynamics in the two phases remained a problem.

2D and 3D exchange ^{13}C spectra of PEO as shown in Figure 7.24 exhibit strong, moderately structured exchange patterns which prove large-angle jumps between several sites to occur, with a rate of approximately 3 jumps per second at 225 K. The smearing-out of the 2D exchange patterns shows that some scatter of the jump angles is present, but the 3D spectrum shows that the jumps back to the start position are relatively accurate, proving a definitely nondiffusive rotational process. The intensity in these spectra derives mainly from the crystallites since the signal of the amorphous regions is suppressed by $T_{1\rho}$ relaxation during CP and by inefficient decoupling, which are both due to the glass-transition motions with rates in the 10 to 100 kHz range.

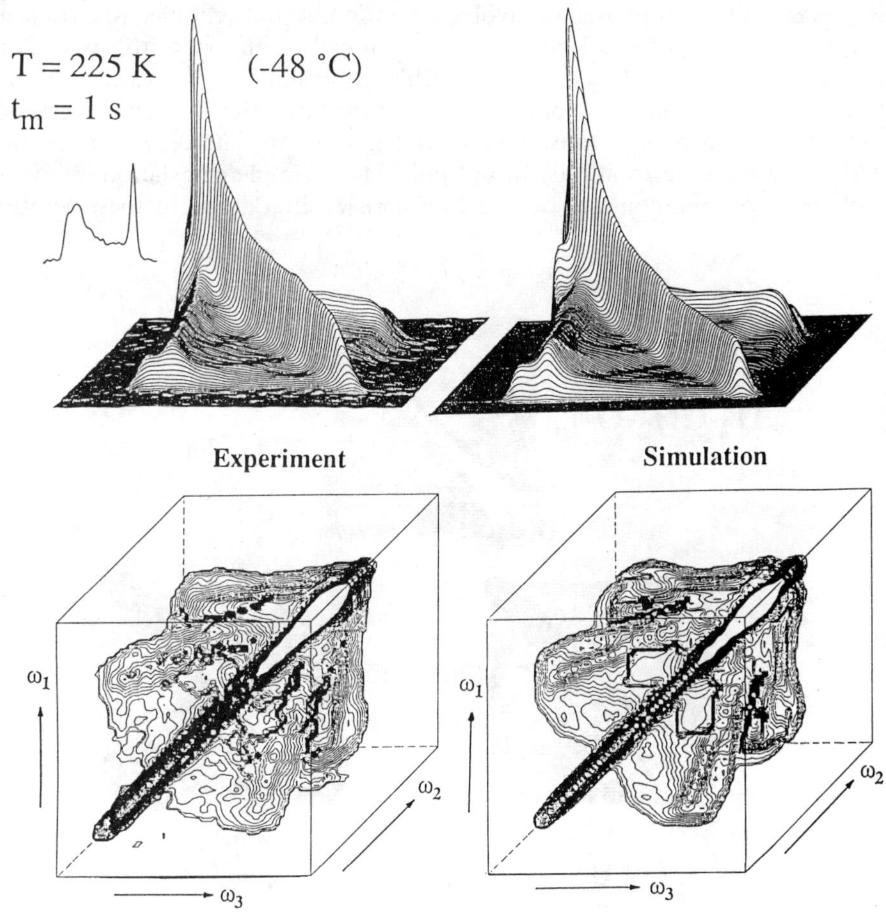

Figure 7.24 2D and 3D exchange ^{13}C spectra of PEO, $T = 225$ K, $t_m = t_{ma} = t_{mb} = 1$ s, with helical-jump simulations, which reproduce the essential features of the experimental spectra.

2D simulations for two different 7_2 helix structures of PEO are compared in Figure 7.25. Here, in addition to establishing the nature of the dynamics in the crystallites, 2D exchange ^{13}C NMR on PEO can indicate the degree of disorder of the helical structure. The 2D spectra confirm the presence of disorder which was first deduced from X-ray diffraction patterns (Takahashi and Tadokoro, 1973). In detail, however, some differences are found. The 'smearing out' of the exchange patterns in experimental spectra at 240 K indicates that long-range order is less pronounced than implicitly assumed in the proposed X-ray structures. On the other hand, in Figure 7.25, the pattern for the jumps to the nearest neighbours is clearer in the NMR 2D experiment than in the 2D spectrum simulated on the basis of the disordered X-ray structure. This indicates that in the X-ray analysis the local deviations from the average positions were overestimated, while the long-range disorder was underestimated.

The helical-jump motions in POM, iPP, and PEO are similar to one of the motional models proposed originally as an explanation of the α relaxation in polyethylene. This motional process involves a rotation of the all-trans PE chain by 180° and a translation by one monomer unit. One of the dynamical processes found in *n*-alkanes is similarly supposed to involve a rotation by 180° coupled to a translation (Ewen *et al.*, 1974; Ungar and Keller, 1979). As already mentioned, this is one of the few types of large-amplitude dynamics possible in crystallites that exhibit no major disorder of chain orientation or conformation. To comply with the ordered crystalline structure, all those processes must involve translations by one or several monomer units with corresponding rotations. In polyethylene, a simple translation by two CH_2 units without a rotation could also occur. Of course, in addition to these rotational-

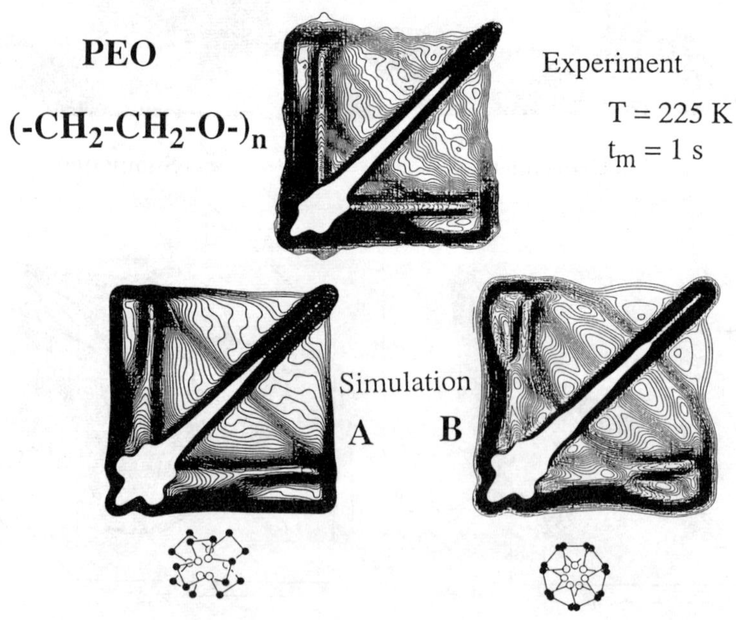

Figure 7.25 Contour plot of experimental 2D exchange ^{13}C spectrum of PEO (T = 225 K), compared with helical-jump simulations for an ideal 7_2 helix and the disordered 7_2 helix proposed on the basis of X-ray data.

jump-and-translation processes, small-amplitude rotations can also occur and have in fact been detected in polyethylene by ^2H NMR (Hentschel *et al.*, 1984).

The 2D and 3D spectra have definitely proven the occurrence of helical jumps in several semicrystalline polymers, as well as characterized the jump angles and sequences of these jumps. The mechanism of the dynamical process in terms of a definite defect, passing through the chain within less than a nanosecond (Mansfield and Boyd, 1978), cannot be determined in the exchange experiment. Only through the jump rates and the activation energy is information on the energetics of the process available. Nevertheless, by a comparison of the behaviour of different polymers, defect models can be checked. The results of the calculations by Mansfield and co-workers (Mansfield and Boyd, 1978; Syi and Mansfield, 1987) in particular are in good agreement with experimental data.

Due to the connection between adjacent crystalline and amorphous regions provided by the polymer chains themselves, the large-amplitude motions in the crystallites proven here also contain some information on the chain loops in the crystallites. With the helical jumps occurring, it is clear that the conformation and length of the loops must not be imagined as static or quasi-static (recall that the amorphous regions are

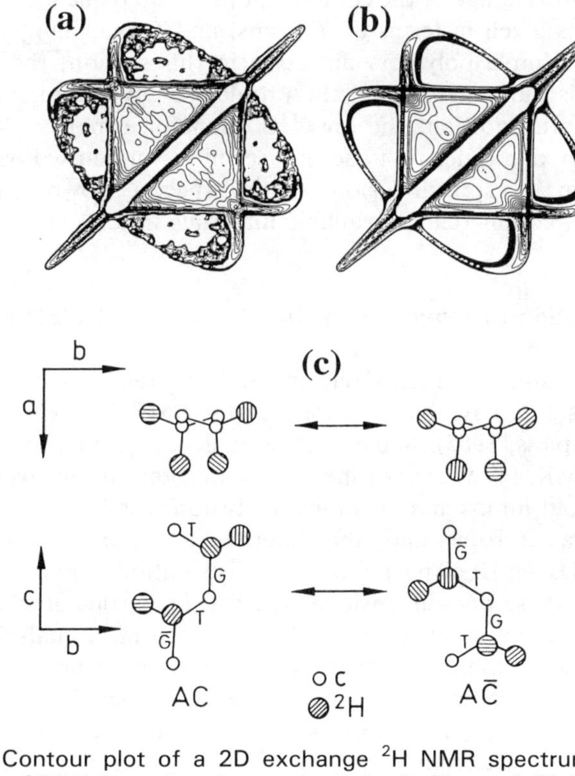

Figure 7.26 (a) Contour plot of a 2D exchange ^2H NMR spectrum of poly(vinylidene fluoride), PVF_2, at 370 K (see also the stacked plot in Figure 1.5(a)). (b) Simulation. (c) Geometry of the conformational jumps in the crystallites of the α modification of PVF_2 from the shape of the elliptical ridges in (a) and from dielectric measurements. *T*, *G*, and $\bar{G}$ refer to trans and gauche conformations of the respective bonds, while *A*, *C*, and $\bar{C}$ describe the direction of the dipole moment with respect to the *a* and *c* axes of the crystal lattice. Adapted from Hirschinger *et al.* (1991).

above their T_g in all cases where helical jumps are observed). In particular, tight re-entry can be ruled out because it would prevent chain rotations, let alone translations in the crystallites (unless an extremely improbable simultaneous translation of all chains connected by tight loops occurs, with alternatingly opposite directions). Independent motions of chain stems through the crystallites will extend nearly every loop sooner or later.

Large-angle reorientations also occur in the (electrically inactive) α modification of poly(vinylidene fluoride) (PVF_2). In contrast to the preceding examples of helical jumps, however, in PVF_2 the chain positions before and after the reorientation are found to be not exactly equivalent. Indications of the corresponding chain disorder around the c axis have indeed been seen in X-ray investigations (Takahashi *et al.*, 1983). The 2D exchange ^{2}H spectrum of deuterated PVF_2 at 370 K shown in Figure 1.5(a) and Figure 7.26(a) (Hirschinger *et al.*, 1991) displays the elliptical exchange pattern for a jump angle of 113° and the intensity ratio characteristic of a two-site jump. On the basis of this jump angle, one proposed crystal structure could be ruled out. In combination with dielectric data from oriented specimens, the conformational change underlying the reorientation could be identified as '$AC \rightarrow A\bar{C}$, $TGT\bar{G} \rightarrow \bar{G}TGT$' (Figure 7.26(b)): it involves an inversion of the dipole moment component along the c axis ('$C \rightarrow \bar{C}$') and no change of the component in the ab plane ('$A \rightarrow A$'); the change of the conformations given in terms of T (trans) and G (gauche).

In the chemically simple polymers discussed in this section, the signal of ^{13}C in natural abundance is quite strong and the problem of overlapping powder patterns absent or tolerable. In addition, the low ^{13}C abundance makes ^{13}C spin diffusion inefficient, so that mixing times up to several seconds can be used without disturbing exchange peaks from ^{13}C spin diffusion. All ^{13}C spectra shown in this section were taken on unlabelled, commercially available materials in a B_0 field of 7 Tesla.

7.13 Characterization of dynamics in the glassy state by 2D exchange NMR

Both the rotational jumps of methyl groups and the 180° flips of para-substituted phenylene rings in solids have been established by NMR, namely 1D lineshape and relaxation studies (Spiess, 1985). Some interesting details remain to be clarified by 2D and 3D exchange NMR: in the case of methyl group rotation, the averaged tensors are identical for three-fold jumps and rotational diffusion (for ^{2}H, $\bar{\delta} = \frac{1}{3}\delta$, $\bar{\eta} = 0$). Only in the T_1-relaxation anisotropy and the intermediate motional regime are they distinguishable by 1D NMR. The jump-angle distribution cannot be extracted with high accuracy from these one-dimensional spectra, in particular if a distribution of correlation times of unknown width complicates the spectral analysis.

In several polymeric materials containing phenylene rings, the strength and frequency location of mechanical (typically β) relaxations are found to be coupled to the number of mobile phenylene rings and the rates of the flips, respectively. The motionally averaged ^{2}H spectrum for mobile phenylene rings is actually not unique for the 180° flip: it could in principle result from restricted oscillations with a root-mean-square amplitude of about 50°. 2D exchange ^{2}H spectra on bisphenol-A polycarbonate (Hansen *et al.*, 1992, 1993) as shown in Figure 7.27 have produced the final proof for the occurrence of the 180° phenylene flip in polymers. Note that the (most probable) rotation angle ψ is 180° while the (most probable) reorientation angle β of the C—^{2}H

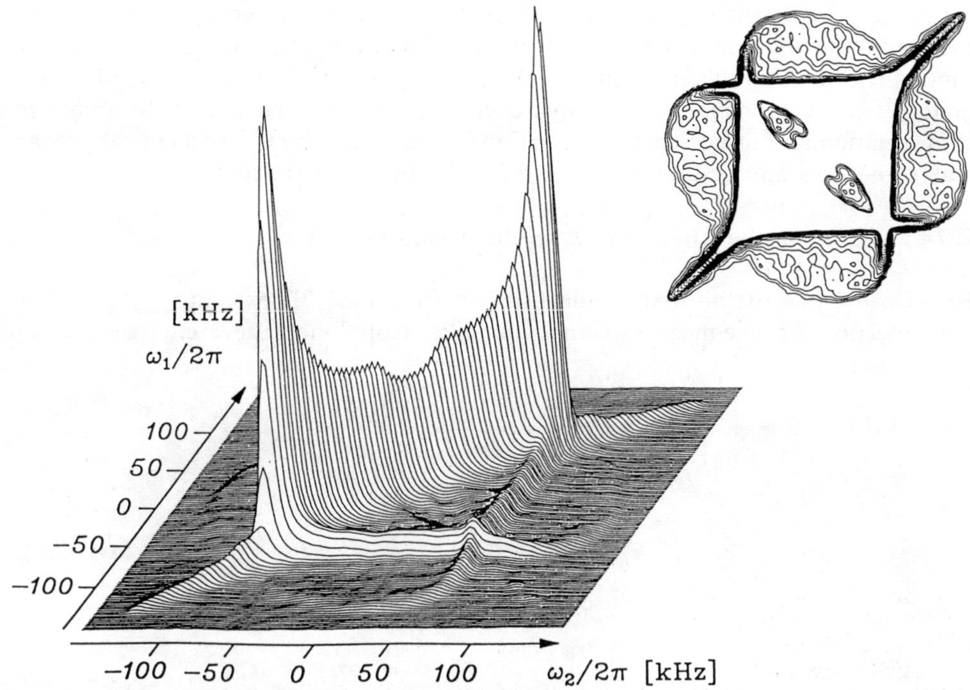

Figure 7.27 2D exchange ^2H spectrum of polycarbonate, PC, at $T = 233$ K, $t_m = 500$ ms. Slightly broadened elliptical exchange ridges are observed, which prove 180° flips of the phenylene rings with slight deviations from the precise 180° rotation angle. Adapted from Hansen *et al.* (1992).

bonds is 120°. The elliptical ridge and the corresponding reorientation-angle distribution $R(\beta)$ for the simulation in Figure 7.27 are somewhat broadened. This means that the flips are not necessarily exact 180° flips. This secondary effect could actually be the source of the mechanical relaxation: only jumps between sites of unequal energy can dissipate mechanical energy put in the material. Exact 180° flips in a completely elastic environment would occur between two identical energy minima. In principle, 3D exchange NMR could make it possible to elucidate whether a return jump actually fails to bring the ring back exactly into its original orientation. In PC under constant stress, the phenylene-ring flips are slightly accelerated (Hansen *et al.*, 1992). In addition, indications for a broadening of $R(\beta)$ are found which corresponds to an increase of the relaxation strength detected by dynamic-mechanical measurements under stress.

7.14 Dynamical exchange of isotropic shifts

So far, exchange has been considered that is caused by changes of the anisotropic part of NMR frequencies. Often, isotropic shifts are also sensitive to the molecular configuration and conformation and can thus change upon dynamical changes of the molecular structure. In solution NMR, processes of this type are termed ‘chemical exchange'. There, the molecular dynamics usually involves chemical changes, that is, changes in the bonding structure. In fact, changes of the isotropic shift always indicate some alterations of the electron distribution in the molecule. In the solid however, quite often the exchange of isotropic chemical shifts is related only to a change of conformation and not of chemical configuration. Thus, the term 'chemical exchange' may sometimes appear inappropriate and will then be avoided.

7.14.1 Chemical exchange in low-molar-mass solids

Historically, one of the first applications of solid-state 2D exchange NMR involved the detection of chemical exchange in solid tropolone (Szeverenyi *et al.*, 1983)

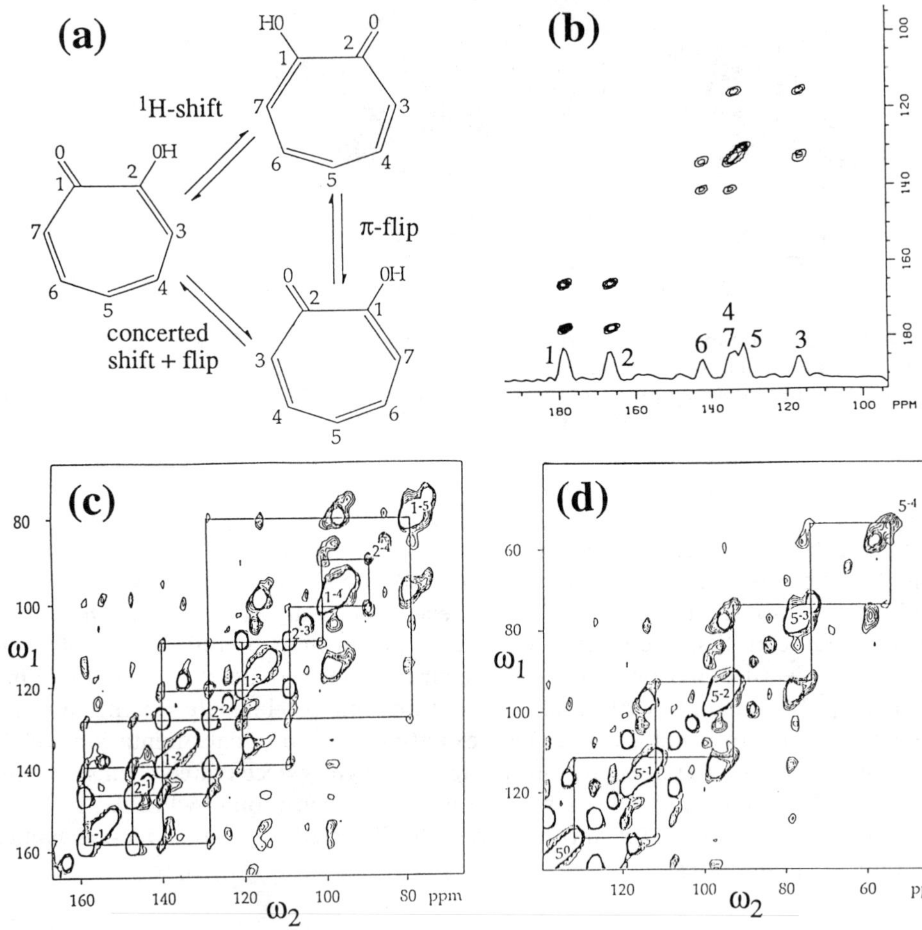

Figure 7.28 Chemical exchange in tropolone. (a) Schematic diagram showing hydrogen shift and/or π flip in solid tropolone. (b) 2D exchange spectrum of isotropic shifts only (fast MAS), which proves that a hydrogen-shift chemical-exchange process is occurring. (c) Exchange sideband pattern (slow MAS) of carbons 1 and 2 which shows that pure molecular reorientation occurs also. (d) Exchange sidebands for carbon 5, which indicate that processes other than those displayed in (a) also occur in solid tropolone. (a), (c) and (d) adapted from Titman *et al.* (1992); (b) adapted from Szeverenyi *et al.* (1983).

(Figures 7.28(a), (b)). Recently, the analysis of this dynamical process has been extended by application of slow-MAS exchange spectroscopy (Figures 7.28(c), (d)), where changes of both the isotropic and anisotropic shifts can be observed. Indications of translational diffusion are detected (Titman *et al.*, 1992). Chemical exchange due to tautomerism in phthalocyanines could also be observed clearly in 2D exchange spectra (Meier *et al.*, 1986). Recently, by application of TOSS–t_1–deTOSS (see Section 3.9.4) in the evolution period and of TOSS during detection, elimination of spinning sidebands in both dimensions has been achieved (Geen and Bodenhausen, 1992). This facilitates the investigation of pure chemical-exchange processes for moieties with large anisotropies at high fields.

^{13}C CP/MAS 2D exchange spectroscopy has also been employed to detect slow flips of phenylene rings with asymmetric substituents that make the two ortho (or the two meta) positions inequivalent in terms of their chemical shift (Harbison *et al.*, 1985).

7.14.2 Isotropic-shift exchange by chain diffusion in polyethylene

The γ-gauche effect on isotropic shifts of aliphatic carbon sites also provides interesting applications for the 2D exchange ^{13}C MAS experiment; detection of trans-gauche isomerizations above T_g will be discussed in the following chapter. Here, an example closely related to the helical jumps discussed above will be treated, namely, exchange of isotropic chemical shifts in polyethylene by chain diffusion (Schmidt-Rohr and Spiess, 1991).

In previous sections of this chapter, for a number of semicrystalline polymers the dynamical α process has been shown to be due to helical jumps on the basis of the clearly detected reorientational jumps and taking into account the crystalline structure. In POM and PEO, displacements by ± 3 positions were clearly detected. In fact, for PEO the exchange intensity at 240 K even indicates displacements over seven and more positions.

In polyethylene, a strong dynamical α process of similar nature as in POM is also detected (McCrum *et al.*, 1967). Though it clearly shows up in NMR $T_{1\rho}$ relaxation data, corresponding off-diagonal intensity is neither found nor expected in exchange ^{13}C or ^{2}H spectra: due to the orientations and symmetries of the interaction tensors in polyethylene (see Sections 2.4.6 and 6.7.3), a helical-jump rotation of the chain by 180° exactly inverts the tensor orientations, which does not involve any frequency changes. Off-diagonal intensity corresponding to a reorientation by 90° around the crystallite c axis is found in ^{13}C spectra only for long mixing times (Edzes and Bernards, 1984; VanderHart, 1987) and is mainly due to ^{13}C spin diffusion between chains.

According to relaxation measurements (McCrum *et al.*, 1967), the rate of the dynamical α process in PE exceeds 10 000 per second at temperatures between 340 K and 380 K depending on the crystallite thickness. If the effect of the helical jumps (Figure 7.29) is assumed to be an unrestricted discrete random walk of each chain along the crystallite c axis, this jump rate would amount to a root-mean-square displacement within 10 seconds by as much as $(2 \times 10 \times 10\,000)^{\frac{1}{2}} \simeq 450$ repeat units, or more than one crystallite thickness. Similar dynamics and diffusion have been observed in n-alkanes (Ewen *et al.*, 1974; Ungar and Keller, 1979).

Under MAS conditions, the γ-gauche effect (Section 2.4.6) separates the ^{13}C signals from amorphous and crystalline fractions by about 2 ppm because of their different

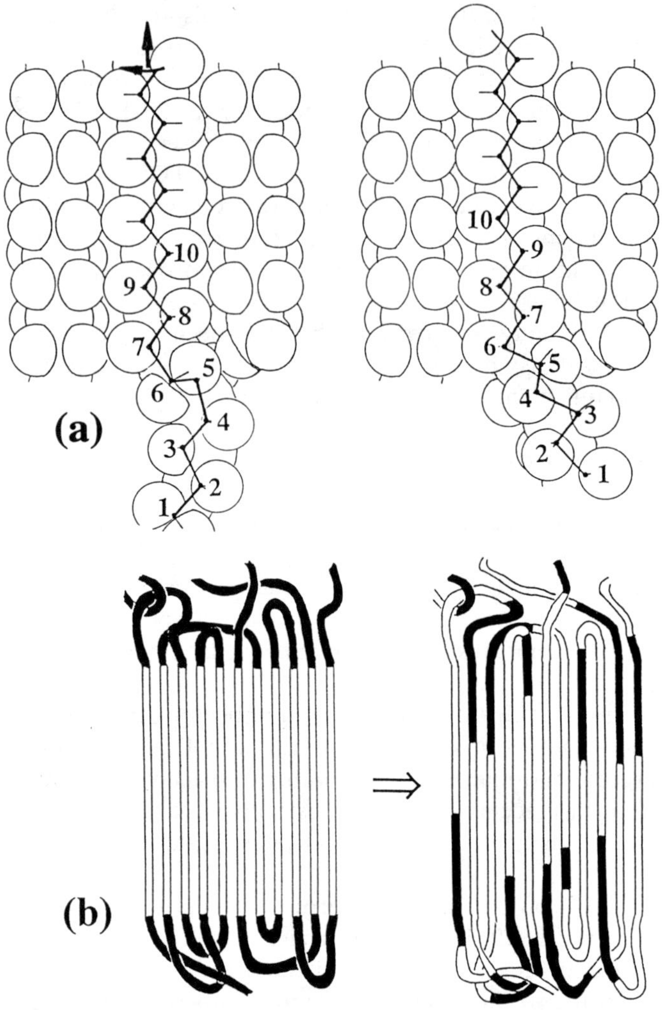

Figure 7.29 Relation of (a) helical jumps and (b) chain diffusion in polyethylene. Translations by one CH_2 unit, which accompany the 180° rotations in the helical jump as shown in (a), occur many times. They displace the chain segments diffusively between crystalline and amorphous regions as indicated in (b), where the segments residing in the amorphous regions at a certain point in time have been labelled. Adapted from Schmidt-Rohr and Spiess (1991).

conformations, namely all-trans in the crystallite and partially gauche in the mobile amorphous material. Cross peaks between those signals are indeed observed after $t_m = 1$ s, though only with about 5% of the total intensity (Schmidt-Rohr and Spiess, 1991). The short ^{13}C T_1 relaxation time of the amorphous component makes 2D experiments with longer mixing times useless. Single pulse excitation after recycle delays t_0 of less than 5 s results in selective observation of components with short T_1: under these conditions, the cross peaks reach 20% of the intensity of the crystalline peak (Figure 7.30) (Schmidt-Rohr and Spiess, 1991). This also indicates that the T_1

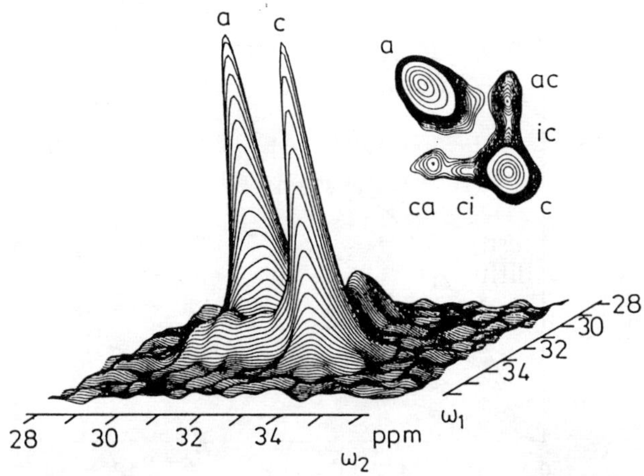

Figure 7.30 Exchange between crystalline and amorphous regions in PE detected in a MAS 2D exchange spectrum with single-pulse excitation. In addition to the crystalline-to-amorphous ('ca') and amorphous-to-crystalline ('ac') cross peaks, indications of signals from an interfacial ('i' in 'ci' and 'ic') region are detected between the signals of crystalline and amorphous regions. Adapted from Schmidt-Rohr and Spiess (1991).

relaxation of the crystalline signal is in fact related to the diffusive exchange process. Indeed, the (apparent) T_1 relaxation of the crystallite signal, taken as a function of the time t_0 after saturation, is strongly nonexponential with a linear initial slope in a plot in terms of $\sqrt{t_0}$ (Figure 7.31(a)), which is typical of a diffusive process. All this supports the conclusion that magnetization from the fast relaxing amorphous regions is transported by a diffusive process into the crystalline lamellae. The strong dependence of the apparent ^{13}C T_1 on the crystallite thickness observed by Axelson *et al.* (1983) is immediately understood on the basis of that process. The exact equality of the NOE (Neuhaus and Williamson, 1989) for both crystalline and amorphous signals in MAS experiments (enhancement by a factor of two), otherwise puzzling considering the completely different T_1 relaxation behaviour, is also easily explained by the chain diffusion: in both peaks, magnetization originating from the amorphous component is detected. The anisotropy of the NOE for both crystalline and noncrystalline material observed on static samples can be understood in terms of the anisotropic mobility of the interphase through which all magnetization from the amorphous domain must pass before reaching the crystallites. ^{13}C spin diffusion as the diffusive exchange process can be ruled out: the temperature dependence of the process is too strong and its efficiency at elevated temperatures too high.

The data points in Figure 7.31(a) were fitted on the basis of a random walk with constraints of a model chain in a lamellar crystallite with 100 crystalline stems and the corresponding amorphous loops (Schmidt-Rohr and Spiess, 1991). Displacements of the chain stems in steps of one repeat unit, with the corresponding extension and shortening of the adjacent loops, were produced in a Monte-Carlo simulation. Restrictions of the fluctuation of the loop lengths had to be included to fit the data. For several types of polyethylene, an average fluctuating loop length of about 20 repeat units was determined. The activation energy of 105 ± 10 kJ/mol (Figure 7.31(b)) as

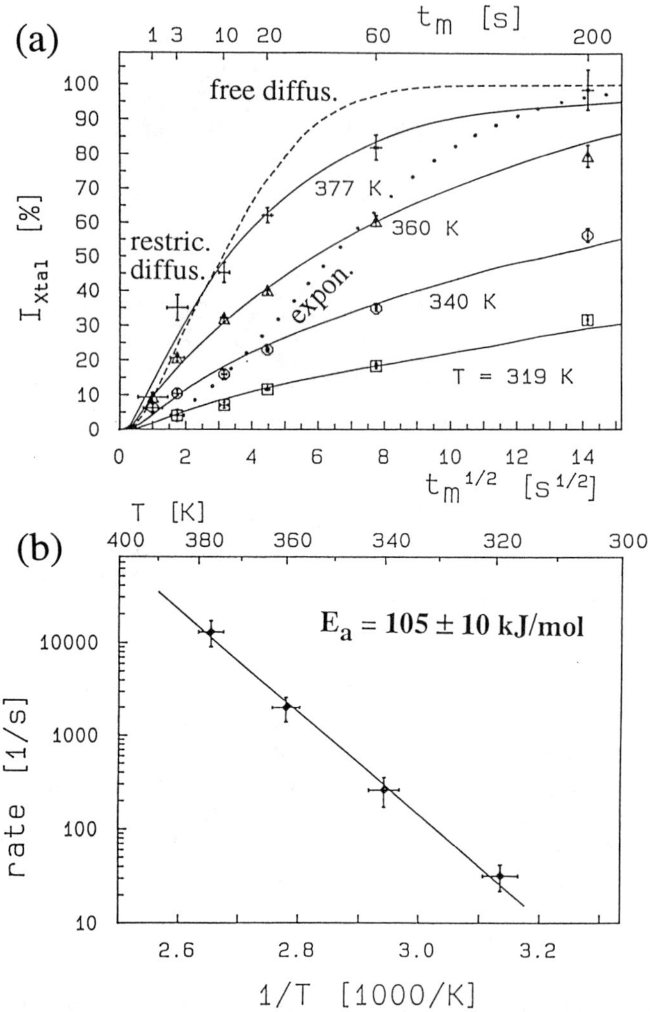

Figure 7.31 (a) PE crystalline-peak intensity in single-pulse ^{13}C MAS experiments, plotted over the square root of the recycle delay (which has the character of a mixing time here). The data indicate a diffusive exchange process, which is strongly temperature dependent. The solid curves were generated by Monte-Carlo simulations of a model chain moving with a certain jump rate. It can move randomly in both directions, except for restrictions imposed on the contractions of the chain loops in the amorphous regions. These lead to the observed slowing-down compared to the free-diffusion behaviour indicated by the dashed curve. (b) Arrhenius plot of the jump rates obtained from the fits in (a). Adapted from Schmidt-Rohr and Spiess (1991).

well as the absolute rates were found to be in good agreement with dielectric data (McCrum *et al.*, 1967).

7.15 References

Axelson, D. E., Mandelkern, L., Popli, R. and Mathieu, P. (1983). ^{13}C NMR of PE's: Correlation of Crystalline-T_1 with Structure. *J. Polym. Sci. Phys. Ed.*, **21**, 2319.

Boyd, R. H. (1985). Relaxations in crystalline polymers: Molecular interpretation. *Polymer*, **26**, 1123.

de Jong, A. F., Kentgens, A. P. M. and Veeman, W. S. (1984). 2D Exchange NMR in Rotating Solids: Study of Slow Molecular Reorientations. *Chem. Phys. Lett.*, **109**, 337.

Edzes, H. T. and Bernards, J. P. C. (1984). 2D NMR in Static Powder: Interchain ^{13}C Spin Exchange in PE. *J. Am. Chem. Soc.*, **106**, 1515.

Ewen, B., Fischer, E. W., Piesczek, W. and Strobl, G. (1974). Molecular motion in *n*-$C_{33}H_{68}$: IR and wide-line NMR. *J. Chem. Phys.*, **61**, 5265.

Geen, H. and Bodenhausen, G. (1992). Chemical exchange NMR with sideband suppression. *J. Chem. Phys.*, **97**, 2928.

Griesinger, C., Sørensen, O. W. and Ernst, R. R. (1987). Practical Approach to 3D NMR. *J. Magn. Reson.*, **73**, 574; 3D High-Resolution NMR. *J. Magn. Reson.*, **84**, 14 (1989).

Hagemeyer, A., Schmidt-Rohr, K. and Spiess, H. W. (1989). 2D NMR for Studying Molecular Order and Dynamics. *Adv. Magn. Reson.*, **13**, 85.

Hagemeyer, A., Brombacher, L., Schmidt-Rohr, K. and Spiess, H. W. (1990). Reconstruction of Angular Distributions from 2D NMR Spectra. *Chem. Phys. Lett.*, **167**, 583.

Hansen, M. T., Blümich, B., Boeffel, C., Spiess, H. W., Morbitzer, L. and Zembrod, A. (1992). Influence of Tensile Stress on Phenylene Flips in PC Studied by 2D NMR. *Macromolecules*, **25**, 5542.

Hansen, M. T., Boeffel, C. and Spiess, H. W. (1993). Phenylene Motion in PC: Tensile Stress, Chemical Modification. *Colloid and Polymer Science*, **271**, 446.

Harbison, G. S., Raleigh, D. P., Herzfeld, J. and Griffin, R. G. (1985). 2D Exchange NMR in Rotating Solids. *J. Magn. Reson.*, **64**, 284.

Hentschel, D., Sillescu, H. and Spiess, H. W. (1984). Chain Motion in PE. *Polymer*, **25**, 1078.

Hirschinger, J., Schaefer, D., Spiess, H. W. and Lovinger, A. J. (1991). Dynamics in α Phase of PVF_2 by 2D ^{2}H NMR. *Macromolecules*, **24**, 2428.

Hoffmann, J. D., Williams, G. and Passaglia, E. J. (1966). Analysis of α, β, γ Relaxations in PCTFE and PE: Dielectric and Mechanical. *J. Polym. Sci.*, **C14**, 173.

Jeener, J., Meier, B. H., Bachmann, T. and Ernst, R. R. (1979). Investigation of Exchange Processes by 2D NMR. *J. Chem. Phys.*, **71**, 4546.

Kentgens, A. P. M., de Boer, E. and Veeman, W. S. (1987). Ultraslow motions in crystalline POM. Complete Elucidation by 2D NMR. *J. Chem. Phys.*, **87**, 6859.

Maas, W. E. J. R., Kentgens, A. P. M. and Veeman, W. S. (1987). ^{13}C Chemical-Shift Tensor in POM. *J. Chem. Phys.*, **87**, 6854.

Mansfield, M. and Boyd, R. H. (1978). Molecular Motions, α Relaxation, Chain Transport in PE Crystals. *J. Polym. Science: Phys. Ed.*, **16**, 1227.

McCrum, N. G., Read, B. E. and Williams, G. (1967). *Anelastic and Dielectric Effects in Polymer Solids.* Wiley, New York.

Meier, B. H., Storm, C. B. and Earl, W. L. (1986). 2D Chemical Exchange NMR: ^{1}H Dynamics in Phthalocyanine. *J. Am. Chem. Soc.*, **108**, 6072.

Nakai, T., Ashida, J. and Terao, T. (1989). ^{13}C Shift Tensors in iPP. *Magn. Reson. Chem.*, **27**, 666.

Neuhaus, D. and Williamson, M. P. (1989). *The NOE and Conformational Analysis.* VCH Publishers, New York.

Opella, S. J. and Waugh, J. S. (1977). 2D ^{13}C NMR of Highly Oriented PE. *J. Chem. Phys.*, **66**, 4919.

Schaefer, D., Spiess, H. W., Suter, U. W. and Fleming, W. W. (1990). 2D NMR Studies of Ultraslow Motion: aPP vs. iPP. *Macromolecules*, **23**, 3431.

Schmidt, C., Wefing, S., Blümich, B. and Spiess, H. W. (1986). Determination of Rotational Angles from 2D NMR of Powders. *Chem. Phys. Lett.*, **130**, 84.

Schmidt, C., Blümich, B. and Spiess, H. W. (1988). ^{2}H 2D Exchange NMR in Solids. *J. Magn. Reson.*, **79**, 269.

Schmidt-Rohr, K., Wilhelm, M., Johansson, A. and Spiess, H. W. (1993). Chemical-shift tensors in CH_2 groups. *Magn. Reson. in Chem.*, **31**, 352.

Schmidt-Rohr, K. and Spiess, H. W. (1991). Chain Diffusion in PE Detected by ^{13}C NMR. *Macromolecules*, **24**, 5288.

Spiess, H. W. (1985). ^{2}H NMR for Studying Mobility and Orientation in Polymers. *Adv. Polym. Sci.*, **66**, 23.

Syi, J.-L. and Mansfield, M. L. (1987). Soliton Model of Crystalline α Relaxation. *Polymer*, **29**, 987.

Szeverenyi, N. M., Bax, A. and Maciel, G. E. (1983). Proton-Exchange in Tropolone Measured by ^{13}C CP/MAS NMR. *J. Am. Chem. Soc.*, **105**, 2579.

Takahashi, Y. and Tadokoro, H. (1973). Structural Studies of Polyethers. X. PEO. *Macromolecules*, **6**, 672.

Takahashi, Y., Matsubara, Y. and Tadokoro, H. (1983). Crystal Structure of Form II of PVF_2. *Macromolecules*, **16**, 1588.

Titman, J. J., Luz, Z. and Spiess, H. W. (1992). ^{13}C MAS 2D Exchange NMR: Self-Diffusion and H-Shift in Tropolone. *J. Am. Chem. Soc.*, **114**, 3757.

Ungar, G. and Keller, A. (1979). Long-range intermixing of paraffin molecules in the crystalline state. *Colloid Polym. Sci.*, **257**, 90.

VanderHart, D. L., Earl, W. E. and Garroway, A. N. (1981). Resolution of ^{13}C NMR of Solids Using ^{1}H Decoupling and MAS. *J. Magn. Reson.*, **44**, 361.

VanderHart, D. L. (1987). Spin Exchange for Natural-Abundance ^{13}C. *J. Magn. Reson.*, **72**, 13.

Wefing, S. and Spiess, H. W. (1988). 2D exchange NMR of powders I. Two-time distribution functions. *J. Chem. Phys.*, **89**, 1219.

CHAPTER EIGHT

Multidimensional Exchange NMR Above the Glass Transition

Keywords: T_g, *WLF/VFT equation, KWW fit; 2D exchange NMR, diffusive reorientation, distribution of correlation times, nonexponential loss of correlation, reduced 4D experiment, nonergodicity; trans–gauche isomerizations; 3D exchange, orientational memory.*

8.1 Dynamics above the glass transition

The α relaxation in most amorphous polymers is related to the glass–rubber transition. This softening process of glassy materials upon heating, or solidification of supercooled melts upon cooling, is of interest far beyond the realm of amorphous polymers. Its most widely known feature is a step in the temperature dependence of the specific heat, at the 'glass transition temperature', T_g. A similar step is observed in the thermal expansion coefficient. Due to the dependence of these features on the cooling rate, the assumption of a simple phase transition as the primary process appears questionable. This rate dependence is related to the strong dynamic α process which has been observed by many methods. Its most striking features are:

(i) the non-Arrhenius $\tau(T)$ temperature dependence of the correlation times, with
(ii) high apparent activation energies ($\geqslant 250$ kJ/mol) close to T_g; and
(iii) the nonexponential loss of correlation, often with
(iv) temperature-dependent nonexponentiality (violation of the time–temperature superposition rule).

Several theories are available that reproduce feature (i). Simple relations (McCrum *et al.*, 1967; Ferry, 1980) are: the Vogel–Fulcher–Tammann(–Hesse) (VFT) equation:

$$\tau(T) = \tau_0 \exp\left(\frac{C_0}{T - T_{\mathrm{VFT}}}\right) \tag{8.1}$$

which was originally proposed to fit the temperature dependence of the viscosity; the Williams–Landel–Ferry (WLF) equation:

$$\tau(T) = \tau(T_g) \exp\left(\frac{-\hat{C}_1(T - T_g)}{T - (T_g - C_2)}\right) \tag{8.2}$$

(Williams *et al.*, 1955) which is equivalent to the VFT equation with:

$$T_{\mathrm{VFT}} = T_g - C_2, \quad C_0 = \hat{C}_1 C_2, \quad \tau_0 = \tau(T_g)\exp(-\hat{C}_1). \tag{8.3}$$

These relations describe many sets of $\tau(T)$ or $\eta(T)$ data over many orders of magnitude. The constants $C_1 := \hat{C}_1/2.303$ (used conventionally instead of $\hat{C}_1$) and C_2 are expected to be universal for polymers. Empirically, mean values of $C_1 = 17.4$ and $C_2 = 52$ K, with some scatter around them, have been determined. The WLF formula can be derived on the basis of the free-volume theory. In a more general view of the glass transition, with their strong curvature of the $\ln[\tau(1/T)]$ curve, polymers usually belong to the class of relatively 'fragile' glass-forming materials (Angell, 1991). In contrast, 'strong' glass formers, such as silicates, show only small deviations from Arrhenius behaviour.

Feature (ii), the large slope of the $\tau(T)$ curve near T_g, is borne out by the WLF/VFT equation. It is usually related to the 'cooperative nature' of the glass transition process. These correlation times τ are usually determined on the basis of the measurement of correlation functions or their Fourier transforms. The theoretical basis of these correlation functions and their relation to 2D exchange spectra will be discussed in Chapter 9.

A strongly nonexponential form of correlation functions, feature (iii), has been observed for many systems above T_g. It indicates that several processes with different correlation times are relevant in the glass-transition dynamics. In most cases, the correlation functions cannot be fitted well by a superposition of two or three exponentials, but rather by a stretched exponential, the Kohlrausch–Williams–Watts (KWW) function (Williams and Watts, 1970):

$$C(t) = \exp\left(-\left[\frac{t}{\tau_{\text{ww}}}\right]^{\beta}\right) \tag{8.4}$$

where $0 \leqslant \beta \leqslant 1$. It has not been possible to clarify the precise nature of this nonexponentiality by means of simple experiments. The most commonly proposed explanations (Davidson and Cole, 1951; Williams *et al.*, 1972; Lindsey and Patterson, 1980; Kaplan and Garroway, 1982; Donth, 1992) are either:

(I) a spatially heterogeneous distribution of correlation times (slow and fast processes in parallel), or

(II) an intrinsically nonexponential loss of correlation in a homogeneous system (slow and fast processes in series).

The simple approach (I) assumes a superposition of exponential relaxations with a broad distribution of correlation times. This situation would be nonergodic. Experimentally, eventual ergodicity of the system, ruling out a time-independent heterogeneous distribution of time-independent correlation times, is indicated by (mono)exponential ^{2}H T_1 relaxation found more than 30 K above T_g (Schnauss *et al.*, 1990). Note that this exponential behaviour concerns the T_1 relaxation of magnetization and not the loss of orientation correlation, which is indeed nonexponential and much faster. This feature of T_1 relaxation shows that on the time scale of about $T_1/10$ ($\gtrsim 10^{-4}$ s), the segments are all subject to the same correlation times. Consequently, during $T_1/10$ they all must have probed the full breadth of the distribution of correlation times, which has a logarithmic average τ_0 more than four orders of magnitude shorter than T_1. Nonetheless, on those shorter time scales, a heterogeneous distribution of correlation times may be a valid characterization of the system with changes of the correlation times occurring only after complete loss of correlation.

The second approach (II) assumes a series of processes with different correlation times but similar temperature dependencies. In order to lead to an overall nonexponentiality, the fastest processes in the hierarchy, taken separately, must not possess an isotropic final state. Rather, they have to be subject to restrictions resulting in a considerable final $C_L(t \to \infty)$ correlation value: one could think of restricted diffusion, or of jumps between few sites, such as trans and gauche conformational states. The complete loss of correlation would occur through slower, finally isotropic processes. An intrinsically nonexponential process is also predicted by mode-coupling theory (Bengtzelius *et al.*, 1984). However, in its current form, the theory does not focus on the time-temperature region close to T_g, and is therefore of little relevance for exchange NMR close to T_g.

The case of nonexponential loss of correlation due to fast diffusion of restricted amplitude and slow isotropic diffusion can serve as an example to show that a series of 2D spectra indeed contains more information than a simple correlation function. This process would first induce a significant, but restricted, broadening of the intensity along the diagonal due to the restricted diffusion. Only later would the slow isotropic diffusion distribute intensity into all accessible regions of the spectrum. The experimental 2D spectra shown below exclude this process close to T_g, exhibiting a definitely different behaviour: intensity distributes over the full spectral range even while the diagonal remains narrow. The correlation function extracted from the 2D spectra is definitely nonexponential, but this nonexponentiality is found to be dominated by a heterogeneous distribution of correlation times.

The nonexponentiality, often expressed in terms of the β parameter of the KWW function (8.4), usually decreases with increasing temperature (feature (iv)), equivalent to an increase of the β values. In fact, in the limit of the low-viscosity liquid, exponential Debye behaviour, or $\beta = 1$, is expected.

8.2 2D Exchange experiments

The first application of 2D exchange ^{2}H NMR above the glass transition involved deuterated polystyrene (PS) (Wefing *et al.*, 1988). In the spectra shown in Figure 8.1, despite marked excitation problems (cf. Section 2.6.5), the simultaneous presence of essentially isotropic reorientations (fully extended 90° ridges) and diffusive small-angle motions (slightly broadened diagonal component) is clearly detected. The spectra cannot be explained in terms of a process with a single correlation time (the corresponding orientational correlation functions, cf. Section 9.1.5, would be definitely nonexponential). Instead, they have been interpreted in terms of isotropic rotational diffusion with a broad distribution of correlation times (3.5 decades half width at 10 K above T_g). The plot of $\log(\tau_0)$ vs. $1/T$ for PS shown in Figure 8.2 combines the results of different NMR techniques (Pschorn *et al.*, 1991), covering 12 orders of magnitude for the correlation times. The data exhibit typical 'WLF behaviour', with a strong increase of the slope (apparent activation energy) as T_g is approached. 2D exchange ^{2}H spectra of deuterated atactic polypropylene (aPP) display similar diffusive dynamics with WLF character (Schaefer *et al.*, 1990).

More complete information on the reorientation angle distribution can be extracted from spectra of methyl-deuterated poly(isoprene) as shown in Figure 8.3 (Schaefer and Spiess, 1993). The narrowing of the spectra by the fast methyl-group rotation, and

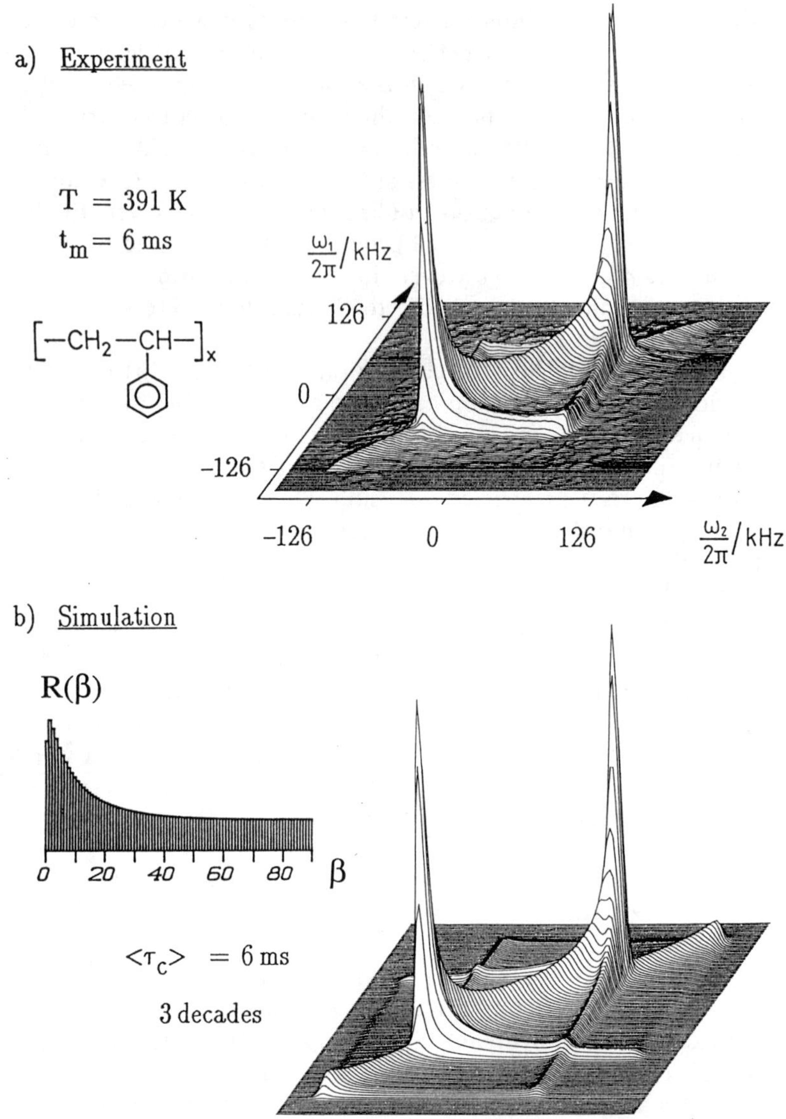

Figure 8.1 (a) 2D exchange ^2H spectra of chain-deuterated polystyrene above T_g. The spectra exhibit no elliptical exchange patterns that would indicate strongly preferred reorientation angles. The gradual broadening of the diagonal indicates a predominantly diffusive nature of the reorientation. (b) Simulation by isotropic rotational diffusion with a distribution of correlation times 3 decades wide and a logarithmic-average correlation time of 6 ms. Adapted from Kaufmann *et al.* (1990).

application of short (2.5 μs) 90° pulses strongly reduces the excitation problem, so that a reliable analysis of the 'edge slices' (outer parts of the spectra) can be performed (see section 7.4). The intensity in these slices is seen to spread over the full accessible range. This directly means that the reorientation-angle distribution is continuous, without special preference for certain reorientation angles. Description of the motion in terms

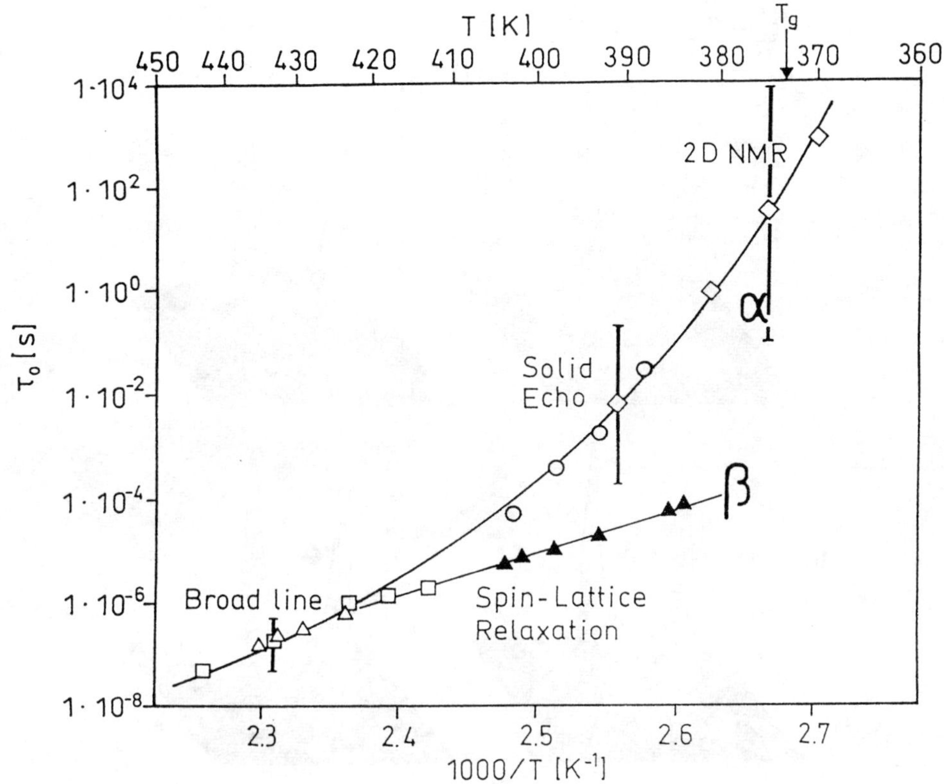

Figure 8.2 Arrhenius plot of the main-chain dynamics of chain-deuterated polystyrene obtained by various ^2H NMR techniques, including 2D exchange spectroscopy, 1D lineshape analysis, and T_1 relaxation measurements (adapted from Pschorn *et al.*, 1992). The full curve is a WLF fit of the data. The widths of the distributions of correlation times required in the simulations are indicated by vertical bars.

of dominating trans-gauche conformational changes (jumps on a diamond lattice) can thus be immediately excluded.

The motion in PI was followed in the 2D exchange experiment over 7 decades of the correlation times (cf. Figure 8.4) (Schaefer and Spiess, 1992), from the slow-motion limit through the intermediate dynamic regime up to a temperature where only a narrow isotropic line remains. When the slow-motion limit is left, the beautifully simple interpretation of the 2D spectra in terms of the frequencies before and after the mixing time is lost. Nevertheless, some information on the character of the motion, on the correlation times and the nature of their distribution, is retained, and even some additional insight on complex motions may be gained. The description of the 2D signal in the intermediate regime is discussed in Section 10.3.

Figure 8.5 shows spectra of the deuterated methyl groups of aPP and PI in comparison (Schaefer and Spiess, 1992). In the aPP spectrum 30 K above T_g, a narrow 'isotropic' peak in the centre of the spectrum is clearly observed in addition to the dominant slow-to-intermediate exchange features. Consequently, there must also be segments that are subject to fast isotropic reorientation with correlation times shorter than 1 μs both before and after the mixing time. This proves a heterogeneity of the

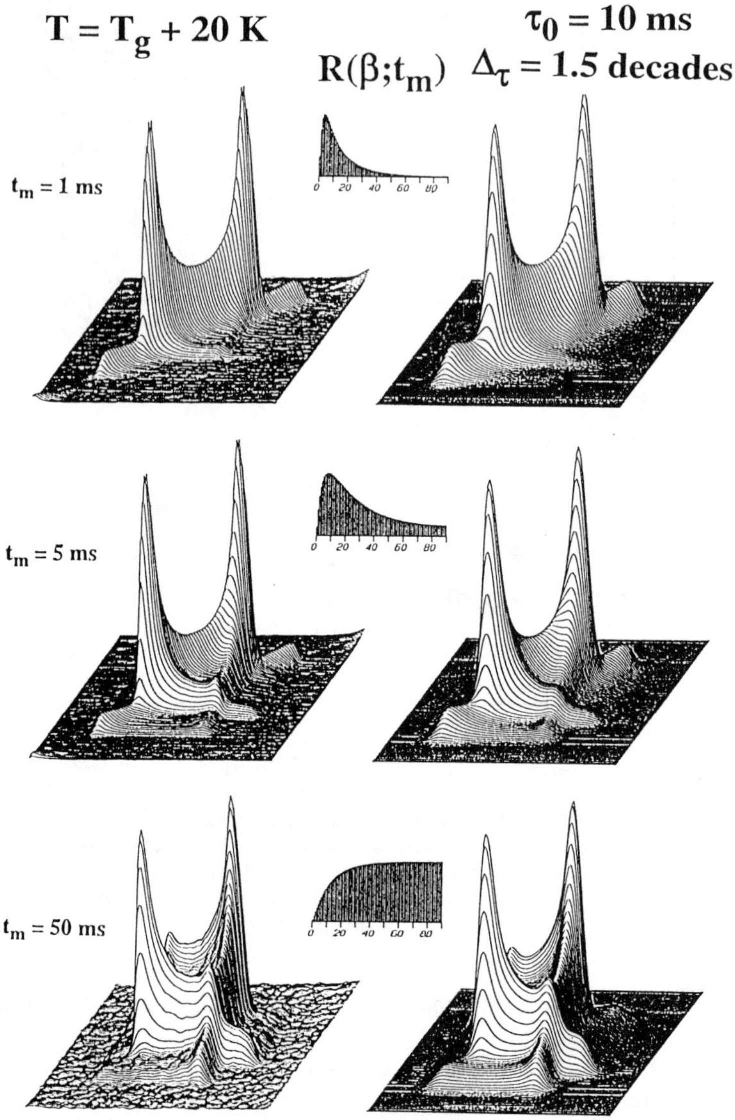

Figure 8.3 Series of 2D exchange ^2H spectra of the C^2H_3 groups of polyisoprene (PI) 20 K above T_g with corresponding simulations (isotropic rotational diffusion with distribution of correlation times 1.5 decades wide).

distribution of correlation times on the time scale of t_m. A distribution of correlation times two decades wide can fit this aPP spectrum. With a broader distribution, either the narrow peak would be higher (if the distribution extends more to the short correlation times) or the signal on the diagonal would be stronger (if the distribution covers more of the long correlation times. By contrast, in PI also at 30 K above T_g, no isotropic peak is observed (see Figure 8.5), which sets an even smaller upper limit to the width of the distribution of correlation times. The study of the lineshape and the reduction factor as a function of the echo delay can produce relatively precise

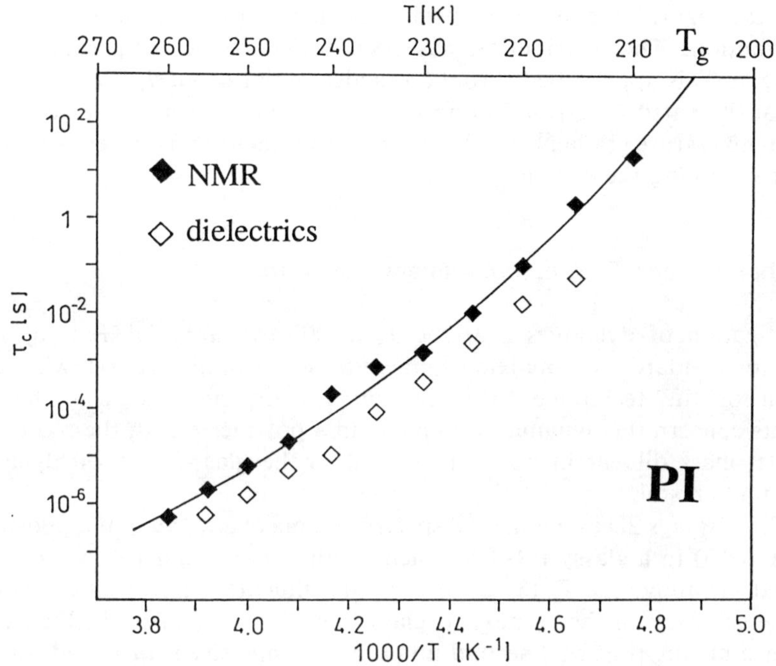

Figure 8.4 Arrhenius plot of the correlation times from 2D exchange ^2H spectra of PI. The NMR data cover 7 orders of magnitude in the correlation time. Solid line; WLF fit. Also included are correlation times from dielectric relaxation. Note that the differences of the NMR and dielectric data correspond to a relative temperature shift of only 5 K. Adapted from Schaefer and Spiess (1992).

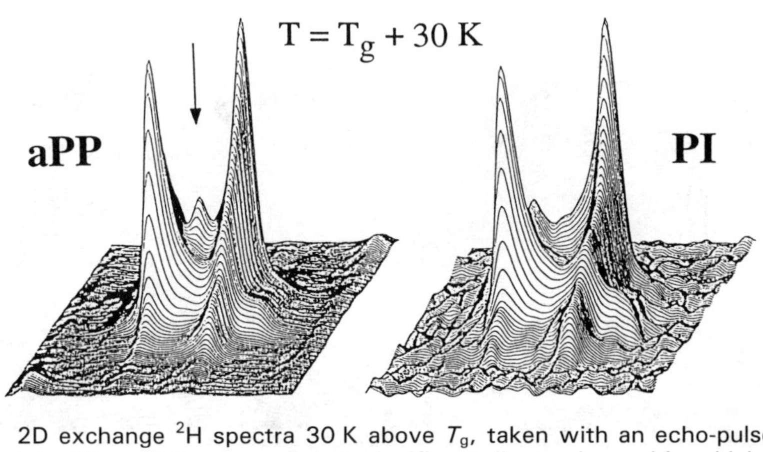

Figure 8.5 2D exchange ^2H spectra 30 K above T_g, taken with an echo-pulse delay of $\Delta = 40$ μs. (a) aPP exhibits a small but significant 'isotropic peak', which proves a heterogeneous distribution of correlation times. On the other hand, the asymmetry of the spectrum with respect to the diagonal sets an upper limit to the width of the distribution (cf. Figure 10.6). (b) PI shows no isotropic peak, which proves a narrower distribution of correlation times. Adapted from Schaefer and Spiess (1992).

values for the logarithmic-average correlation time and the width of the correlation-time distribution. The distribution has been found to be temperature-dependent, broadening as T_g is approached. From a practical point of view, it may be interesting to note that the narrow distribution observed in PI also sets a safe lower limit on the effects of temperature gradients on these 2D measurements performed with a nitrogen-gas flow for cooling the sample.

8.3 Exchange near T_g in inhomogeneous systems

The investigation of dynamics near the T_g by 2D exchange NMR is by no means restricted to standard homopolymers like PS or PI. This section will review the application of the technique to more complex organic systems. The exchange experiments concern the dynamics around T_g in a polymer blend, the reorientation of a low-molar-mass diluent in a blend, as well as the glass-transition dynamics in a discotic liquid crystal.

Figure 8.6 displays 2D exchange ^{31}P spectra (Zhang *et al.*, 1991) of a phosphate ester diluent (20 wt%) in a glassy PS–PXE blend with a T_g around 100 °C. The system is investigated far above the T_g (8 °C) of the pure diluent, therefore it is reasonable to analyse the situation in the context of glass-transition dynamics. In Figure 8.6(a) at 71 °C after a mixing time of 1 second no exchange intensity is observed (the spurious ridges extending from the maximum peak away from the rest of the spectrum are due

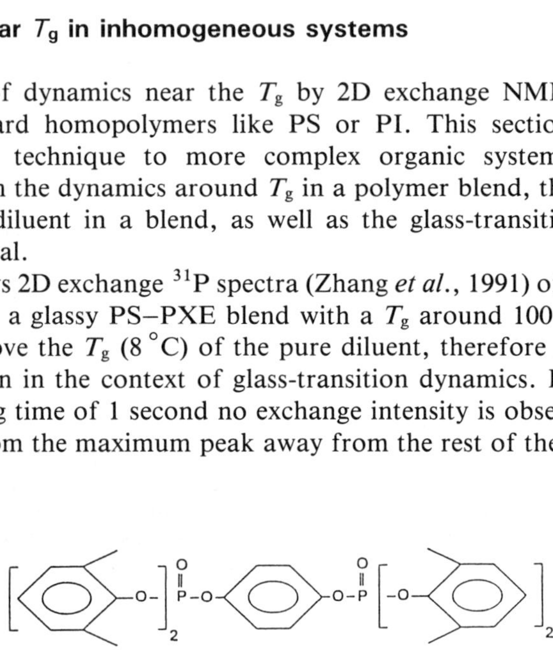

TXHQDP

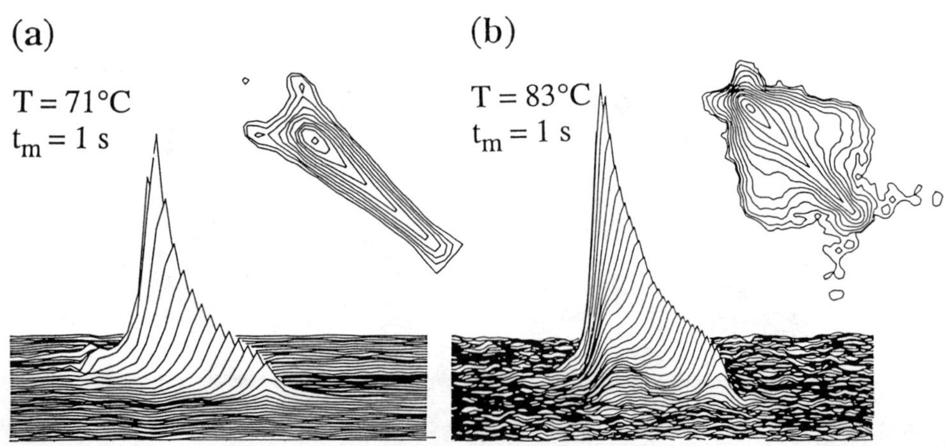

Figure 8.6 2D exchange ^{31}P spectra of a low-molar-mass diluent (20 wt%) in a PS-PXE blend; $t_m = 1$ s. (a) At 71 °C, no significant exchange is observed. (b) At 83 °C, about 20 K below the T_g of the system, motion of the diluent is observed by the exchange intensity in the 2D spectrum. Adapted from Zhang *et al.* (1991).

to strong Lorentzian line broadening). At 83 °C, that is, 17 K below the calorimetric glass transition of the mixed system, pronounced exchange intensity is observed after $t_m = 1$ s, but still a narrow diagonal portion of about 40% remains (Figure 8.6(b)). This indicates a bimodal distribution of correlation times: some diluent molecules reorient on a 1 s timescale, presumably those in contact with other diluent molecules, while the rest are immobilized by their contact to the polymer matrix. The appearance of exchange within mixing times of about 1 s is found to be related to the material's mechanical loss peak at a frequency of 1 Hz. This connection is indeed expected according to the discussion of Section 5.3.

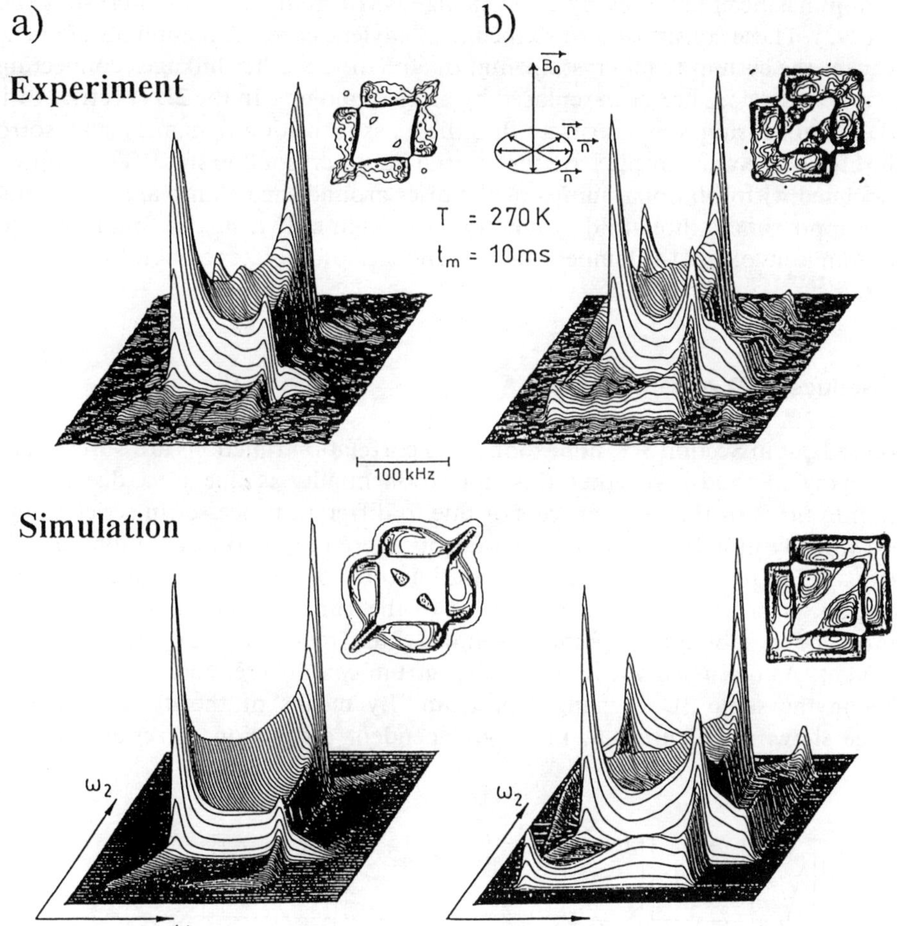

Figure 8.7 2D exchange spectra above T_g of a (low-molar-mass) discotic liquid-crystalline system. (a) Spectrum of an unoriented sample. (b) Spectrum of a macroscopically oriented sample. The indications of elliptical features in the contour plots show that large-angle jumps by ca. 60° or 120°, i.e. in a potential with three- or six-fold symmetry, occur together with small-angle processes. The simulations in the lower trace were obtained on the basis of a reorientation-angle distribution 30°-fwhm wide around the large central jump angles. Adapted from Leisen *et al.* (1992).

In 50 : 50 and 25 : 75 blends of PXE and PS, known to be homogeneous on a 5 nm scale, the glass-transition dynamics has been selectively observed in both components (Chin *et al.*, 1992). Working with 100% enriched $O^{13}CH_3$ groups in the PXE, an onset of mobility of the PXE below the calorimetric glass transition of the blend is detected. The authors observe indications of a trimodal distribution of correlation times. The resolution of the motionally narrowed ^{13}C chemical-shift spectrum of the methyl group is, however, limited. In addition, the mixing times have to be restricted to the range between 0.5 ms (inverse width of the powder spectrum) and 50 ms (spin-diffusion limit for this enriched system) which reduces the range of correlation times that can clearly be distinguished.

A somewhat exotic, but instructive, example of a glass transition with a single degree of freedom has been observed by 2D exchange NMR in discotic liquid-crystals (Leisen *et al.*, 1992). These consist of disc-shaped triphenylene cores with aliphatic side-chains. To prevent the system from crystallizing, one of the six ether linkages connecting the sidechains to the core has been replaced by an ester moiety. In the 2D spectra of Figure 8.7, the elliptical ridges demonstrate that the glass-transition dynamics is anisotropic, which reflects the anisotropic structure of the liquid-crystalline state. The ellipses can be associated with rotational jumps of the discs around the column axis in a potential with an approximate three-fold symmetry. Although only a single, rotational, degree of freedom is involved, the temperature dependence of the rates exhibits a typical WLF behaviour.

8.4 Reduced 4D experiment

As pointed out in Section 8.1, nonexponential correlation functions are observed above T_g by many methods. Whether this nonexponentiality is due to a distribution of correlation times of the same process or due to different processes in series cannot be probed directly in a 2D exchange experiment where only two orientations of a given moiety are correlated. To elucidate the nature of the nonexponential loss of correlation, information on the orientation of the same segment must be measured at three or more subsequent points in time. This requires a 3D or 4D exchange experiment. As discussed above, the 3D spectrum can differentiate between processes producing the same 2D exchange spectrum. By means of the 4D exchange pulse sequence shown in Figure 8.8, model-independent distinction of serial and parallel

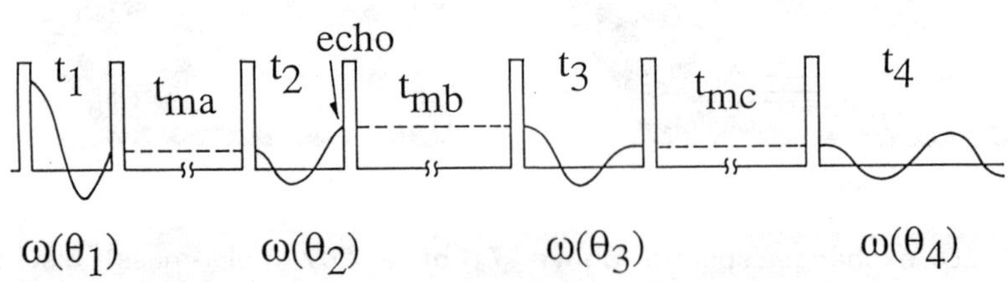

Figure 8.8 Pulse sequence for 4D exchange NMR. In the reduced 4D experiment, the evolution times t_1 and t_2 are held equal and constant. In the stimulated echo forming at $t_2 = t_1$, the magnetization from segments with $\tilde{\omega}_1 \cong \tilde{\omega}_2$ is refocused. Thus, the magnetization of slowly reorienting segments can be selected and subsequently analysed in terms of their exchange in a (ω_3, ω_4) 2D spectrum.

dynamics can be attained (Schmidt-Rohr and Spiess, 1991). In order to make the experiment feasible and to 'project out' the relevant information, the evolution times t_1 and t_2 are chosen to be of equal duration. At $t_2 = t_1$, a stimulated echo (see section 10.2.3) is produced by those components that have the same frequency before and after the first mixing time, t_{ma}, while the other isochromats interfere destructively. This

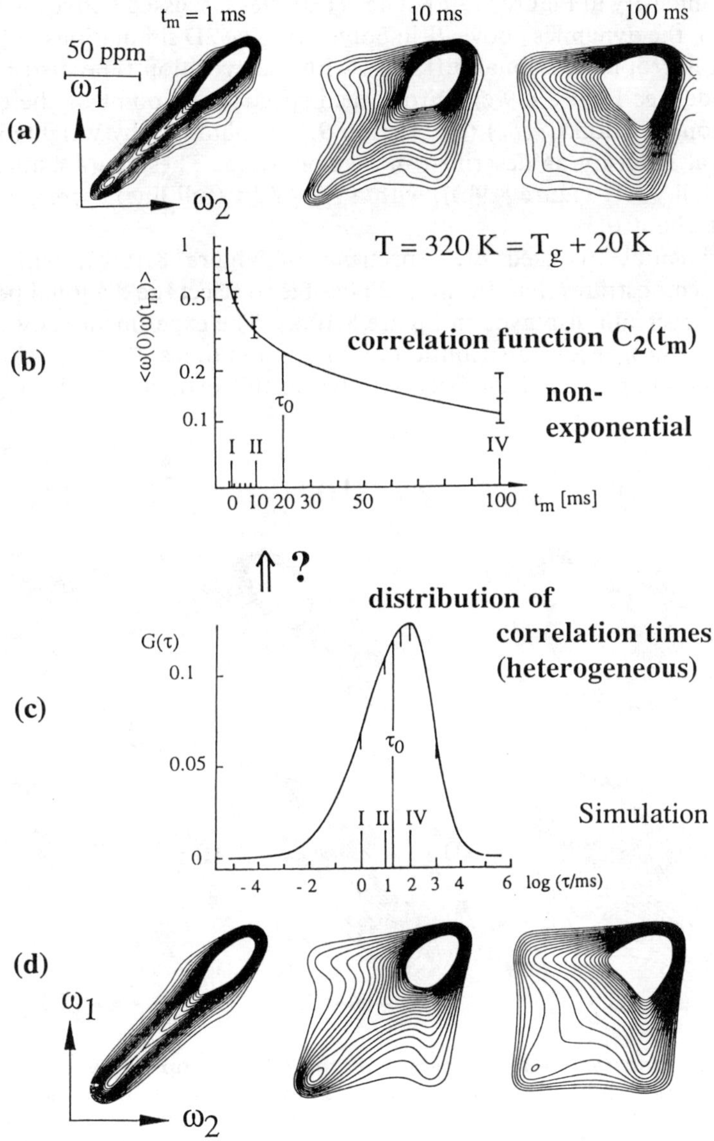

Figure 8.9 2D exchange ^{13}C NMR of the COO group of PVAc 20 K above T_g. (a) Series of 2D spectra. (b) Correlation-function data points obtained from the spectra, with KWW fit ($\beta_{ww} = 0.32$). (c) Distribution of correlation times corresponding to the KWW fit and used as the basis of the 2D simulations in (d). Adapted from Schmidt-Rohr and Spiess (1991).

results in a selection of the magnetization that corresponds to the diagonal of the 2D reference spectrum. The dynamical state of the selected species is analysed in terms of a 2D exchange spectrum with the mixing time t_{mc}.

For reference, Figure 8.9(a) displays a series of 2D spectra of the carbonyl carbon of PVAc (^{13}C in natural abundance) 20 K above T_g. The spectra exhibit the same diffusive behaviour as the ^2H 2D spectra shown in Section 8.2, with both slow and fast exchange processes as witnessed by the coexistence of pronounced diagonal as well as off-diagonal intensity in Figure 8.9(a). The ^2H studies discussed above have shown this to be typical of the dynamics above T_g in polymers. The 2D simulations in Figure 8.9(d) are based on isotropic rotational diffusion with a correlation time distribution about 3 decades wide (see Figure 8.9(c)). From each spectrum, a point of the orientational autocorrelation function $C_2(t_m)$ (cf. Section 9.1) is obtained by weighted integration of the spectral intensity as described in Section 9.1.5. These data exhibit a definite nonexponentiality (cf. Figure 8.9b)), with a KWW fit (full line) corresponding to a β parameter of about 0.3.

The experimental 'reduced 4D' spectrum of Figure 8.10(b), with $t_{ma} = t_{mb} = t_{mc} = 10$ ms, demonstrates that it is indeed possible to select the diagonal portion of the reference 2D spectrum displayed in Figure 8.10(a). The experiment shows directly the presence of a heterogeneous distribution of correlation times: there are slowly relaxing portions of the sample that have been selected in this experiment. They generate the

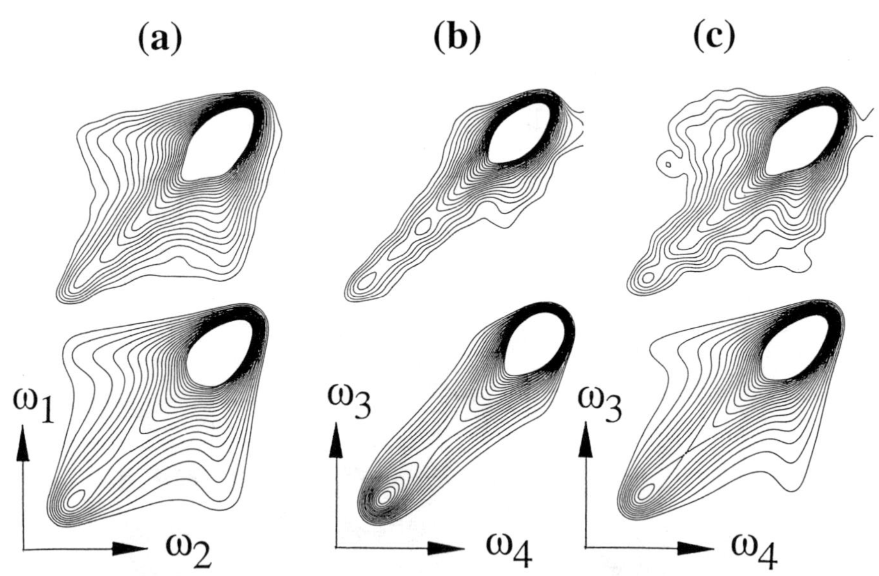

Figure 8.10 2D and reduced 4D spectra of the COO group of PVAc 20 K above T_g. Upper row: experimental spectra; lower row: simulations.
(a) 2D spectrum with $t_m = 10$ ms for reference. (b) Reduced 4D spectrum with $t_{ma} = t_{mb} = t_{mc} = 10$ ms. Only little exchange is observed: slow units have been selected. This proves that a heterogeneous distribution of correlation times causes the nonexponential loss of correlation, at least on the timescale of the logarithmic-average correlation time τ_0. (c) Similar to (b) but with $t_{mb} = 1000$ ms. Significantly more exchange is observed, indicating slow changes of the correlation times of the initially slow units. Adapted from Schmidt-Rohr and Spiess (1991).

long tail of the correlation curve, while its fast initial decay is due to the fast reorienting species that give rise to the off-diagonal intensity in the 2D spectrum of Figure 8.10(a).

This analysis concerns the timescale of $t_{ma} + t_{mb} + t_{mc} = 30$ ms probed in the spectrum of Figure 8.10(b), which is close to the logarithmic average τ_0. An extension of t_{mb} by two orders of magnitude, $t_{mb} = 1000$ ms, can elucidate whether the slow segments also remain slow for long times. The experimental spectrum of Figure 8.10(c) shows that indeed many segments have changed their correlation times: considerably more exchange intensity than in the spectrum for $t_{mb} = 10$ ms is observed. This can be attributed to fluctuations in the supercooled melt occurring on the timescale of the longest τ in the distribution of Figure 8.10(c). The spectrum of Figure 8.10(c) thus indicates a breakdown of the picture of a static distribution of correlation times on these long timescales. This also implies a return of the system to ergodicity. Note that these fluctuations do not contribute to the nonexponentiality, as they only relax the long-time tail of the correlation function; this actually brings the correlation function slightly closer to the mono-exponential form. The fluctuations may be responsible for the asymmetric shape of the correlation-time distributions, representing an effective cut-off for the wing at long τ values: the fluctuations involve structural rearrangements that allow segments to reorient even if the constraints by their original environment had resulted in a much longer correlation time.

It may be interesting to note that whether or not reorientations occur during the central mixing time t_{mb} is essentially of no effect on the spectra since the stimulated echo has refocused the phase differences for the relevant species; no labelling in terms of frequencies or orientations, but rather a distinction in terms of frequency differences, has been generated before t_{mb}. Rather than reorientations, changes of the correlation times during t_{mb} can be selectively observed.

The aPP ^{2}H spectrum 30 K above T_g shown in Figure 8.5(a) complement the results of the reduced 4D experiment on the heterogeneity of the distribution of correlation times. The narrow isotropic peak proves this heterogeneity in a different sample, for a different τ_0, and with respect to the short-τ wing of the distribution of correlation times.

8.5 Trans-gauche isomerizations above T_g

As discussed in Section 2.4.6, the isotropic chemical shift can be sensitive to the conformations close to a moiety under consideration, and to some extent also to the packing around the segment. These effects have been best characterized for aliphatic carbons, in particular methylene groups. With the isotropic chemical shifts of γ-trans/γ-trans (*tt*) and γ-gauche/γ-gauche (*gg*) conformations known from syndiotactic PP (at 48.6 ppm and 40 ppm, respectively), and of γ-trans/γ-gauche (*tg*) from the 3_1 helix of isotactic PP (44.5 ppm), amorphous atactic polypropylene is ideal for studying the distribution and dynamics of the conformations. Influences of the chain packing on the chemical shift are usually small compared with these conformational effects, inducing splittings or shifts in the range of below 1 ppm.

Figure 8.11 displays 1D spectra of aPP at 262 K and 250 K. At the lower temperature, 11 K above the calorimetric T_g, a distinct splitting is observed (Zemke *et al.*, 1991). The shifts correspond to the *tt* and *tg* conformations; *gg* components contribute only a minor shoulder hardly recognizable in the 1D spectrum. The splitting

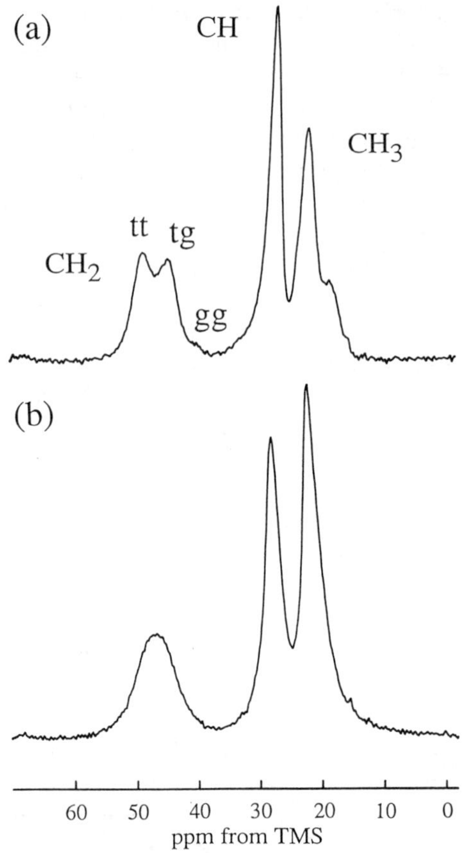

Figure 8.11 1D MAS spectra of aPP. (a) Spectrum taken at 11 K above T_g. The CH_2 region of the spectrum is spread over approximately 10 ppm, showing two maxima and a high-field shoulder. The dispersion is due to the γ-gauche effect (cf. Section 2.4.6). The maxima correspond to γ-trans/γ-trans and γ-trans/γ-gauche conformations, the shoulder to γ-gauche/γ-gauche. (b) 20 K above T_g, partial coalescence of the peaks is observed, indicating some exchange between the conformations, without providing information to what extent they exchange. Adapted from Zemke *et al.* (1991).

disappears at $T = T_g + 20$ K, but in contrast to the coalescence behaviour known in liquids, the peaks do not narrow above the coalescence temperature. This may lead to the idea that the averaging is incomplete, with a given segment not probing all conformations present. This conclusion is shown to be in error by 2D spectra taken at 250 K (11 K above T_g, Figure 8.12). After a mixing time of 500 ms, the exchange signals have very nearly reached the intensity of the diagonal signal. Since any slice parallel to ω_2 is nearly identical to the 1D spectrum, all conformations are apparently accessible from any start position. The broadening above the coalescence in the 1D spectra must therefore be traced back to interference of the motion with the high-power proton decoupling.

The timescale of the conformational changes appears to be similar to that of the reorientational dynamics observed in static exchange spectra. These findings may seem to suggest the description of the motion only in terms of trans and gauche

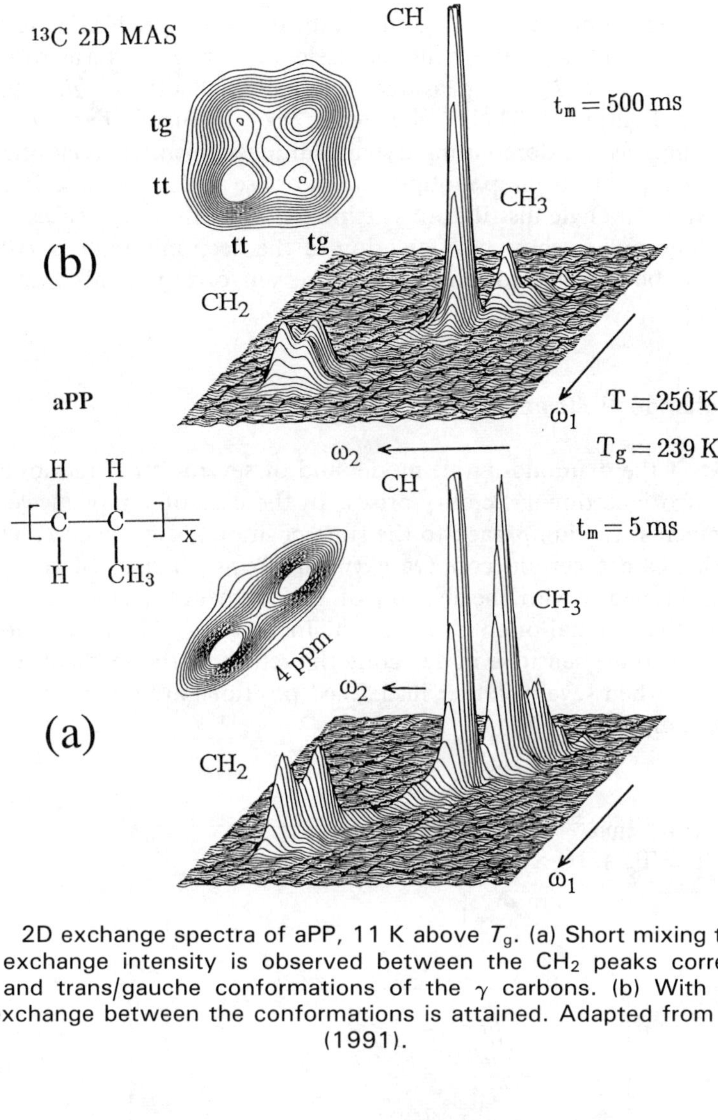

Figure 8.12 2D exchange spectra of aPP, 11 K above T_g. (a) Short mixing time of 5 ms; hardly any exchange intensity is observed between the CH_2 peaks corresponding to trans/trans and trans/gauche conformations of the γ carbons. (b) With $t_m = 500$ ms, nearly full exchange between the conformations is attained. Adapted from Zemke *et al.* (1991).

conformations, equivalent to dynamics on a diamond lattice (Monnerie and Gery, 1969). Experimental results proving that this is too simple a picture, that the potential is in fact not static, will be discussed in the following.

First of all, the lineshape along the diagonal for the spectrum with $t_m = 5$ ms shows that trans and gauche are the preferred, but by no means the only, accessible conformations. It demonstrates that the broadening of the signals observed in the 1D spectrum is not homogeneous. If it were homogeneous, the 2D spectrum of the methylene units would consist of two humps centred according to the maxima of the 1D spectrum with circular or star-shaped broadening into the 2D plane. Instead, the experimental 2D spectrum, with $t_m = 5$ ms, is extended along the diagonal with little homogeneous broadening as indicated by the small width perpendicular to the diagonal. This shows that the intensity between the maxima arises from conformations or environments other than those corresponding to the positions of the maxima.

Another strong indication against dynamics to be described predominantly in terms of trans-gauche transitions is given by the featureless exchange patterns of static 2D ^{2}H exchange spectra above T_g, as measured for aPP (Schaefer *et al.*, 1990) and for polyisoprene, cf. Figure 8.3. Their rather featureless, diffusive shapes clearly exclude well-defined jumps as the dominating dynamical mechanism. The elliptical exchange ridges which are typical for jump motions are not observed in the experimental spectra and the reorientation-angle distributions estimated from the 'edge slices' of measured spectra, as in Figure 8.3, show no indication of the deep minimum between 30° and 60° which would be expected for any jump process involving mainly trans and gauche conformations.

8.6 3D Exchange NMR above T_g

In the context of the diamond-lattice model and of several other motional processes, the question of orientational memory arises. In the case of jumps in well-structured static environments, the jump back to the start position always occurs. Therefore, its absence can indicate a certain cooperativity, involving changes of the environment which are simultaneous with the motion of the segment considered, such that the memory about the original orientation is lost. In Section 7.12, it was shown that 3D exchange NMR is quite sensitive in detecting the return of the segment to the original orientation even when several rather ill-defined positions are involved (as in the case of the helical jumps in PEO).

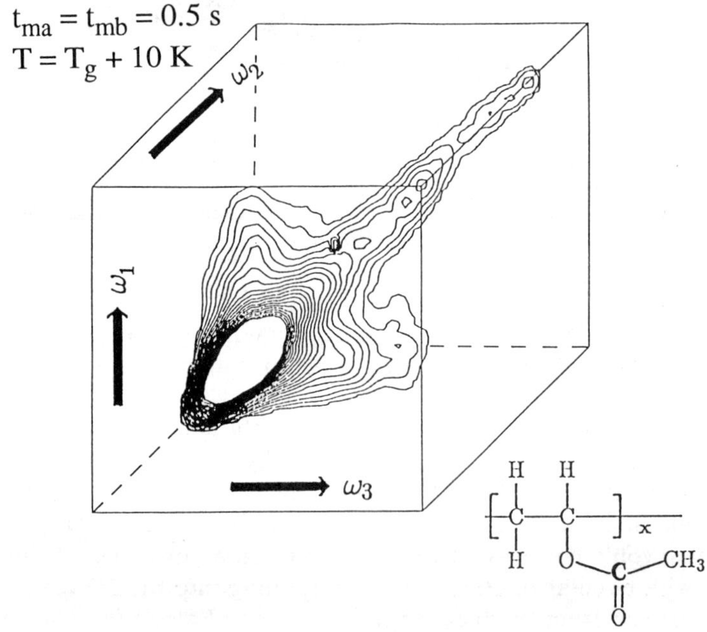

Figure 8.13 3D exchange ^{13}C spectrum of PVAc 10 K above T_g. The strong signal on the space diagonal indicates that many slow units with $\tilde{\omega}_1 \cong \tilde{\omega}_2 \cong \tilde{\omega}_3$ exist, which corroborates the result of the reduced 4D experiment.

Motions involving jumps between few positions exhibit diagonal intensity in a 2D spectrum due to jumps back to the start position (e.g. the helical jumps in Section 7.12). The diagonal in the 2D spectrum (see Figure 8.10(a)) of PVAc above T_g could be partially due to such jumps back to the start position, or just to units that have hardly moved at all. These contributions can be distinguished in the 3D spectrum. The example shown in Figure 8.13 exhibits a strong space-diagonal signal ($\tilde{\omega}_1 \cong \tilde{\omega}_2 \cong \tilde{\omega}_3$) from segments not changing their orientations; these immobile species thus account for the diagonal intensity in the 2D spectrum. The reduced 4D experiment discussed in Section 8.4 convincingly confirms this conclusion.

Consistently with this sufficient explanation of the diagonal in the 2D spectrum, signals of jumps back to the start position do not show up in the 3D spectrum of PVAc. They would produce intensity at $\tilde{\omega}_1 = \tilde{\omega}_3 \neq \tilde{\omega}_2$, that is, in the $\omega_1 = \omega_3$ diagonal plane, but off the space diagonal. No such intensity is apparent in the 3D spectrum of PVAc above T_g in Figure 8.14(a), while signals can clearly be detected in the $\omega_1 = \omega_3$ diagonal plane of 3D spectra involving jump processes, such as the helical jumps in PEO shown for comparison in Figure 8.14(b).

3D simulations based on a heterogeneous and a homogeneous distribution of correlation times of isotropic rotational diffusion are shown in Figure 9.4. In the heterogeneous case, which fits the experimental data better, 3D spectra for different correlation times have been superimposed, with the slow components producing a strong space-diagonal signal as observed in the experiment. The homogeneous model assumes reorientation of every segment with the reorientation-angle distribution

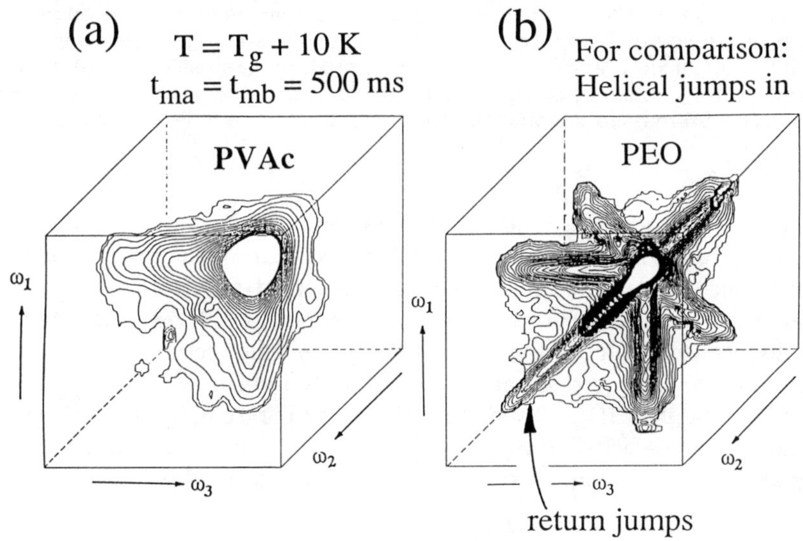

Figure 8.14 (a) View of the 3D spectrum of Figure 8.6(a) onto the ω_1–ω_3 side of the spectral cube, with short 3D diagonal spectrum (from upper right front to lower left back). No strong signal is observed in the $\omega_3 = \omega_1$ diagonal plane (apart from the 3D diagonal spectrum). This indicates that no significant orientational memory is present in this process. (b) For comparison, equivalent view of 3D exchange ^{13}C spectrum of PEO with exchange arising due to helical jumps of the chains in the crystallites (cf. Section 7.12). Here, the return-jump signals in the $\omega_3 = \omega_1$ plane are clearly detected, in spite of the somewhat disordered chain structure of PEO.

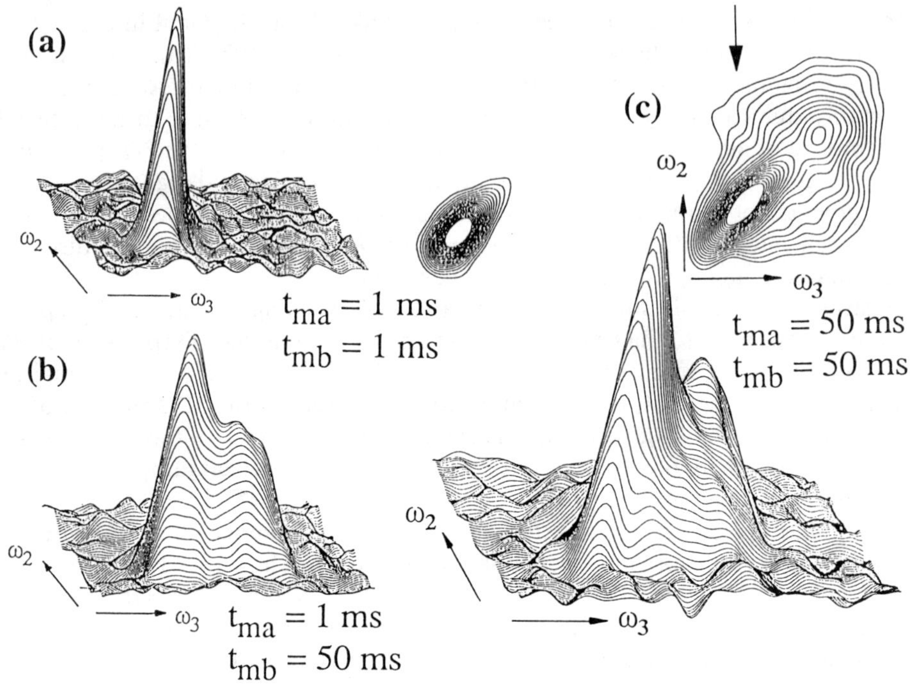

Figure 8.15 2D spectra corresponding to two-dimensional cross-sections through a 3D exchange ^{13}C spectrum of the CH_3 group in aPP at $T = 255$ K, about 15 K above T_g. The spectra were actually taken with selective ω_1-excitation near the low edge of the CH_3 powder pattern, followed by a mixing time t_{ma} and a 2D exchange experiment with mixing time t_{mb}. (a) Short mixing times $t_{ma} = t_{mb} = 1$ ms, proving clean selection of a narrow band of frequencies. (b) Exchange pattern for $t_{ma} = 1$ ms, $t_{mb} = 50$ ms. (c) Long mixing times $t_{ma} = t_{mb} = 50$ ms. No specific return-jump signal at $\omega_3 = \omega_1$ (arrow) is found.

$R(\beta; t_m)$ that fits the corresponding 2D spectrum; no fast and slow segments are distinguished. Here, the diagonal is more depleted than in the other model or in the experiment, since here the segments having reoriented only little during t_{ma} have a considerable chance of reorienting by large angles, just like all the other segments. This model is discussed with more theoretical background in Section 9.3.2.

Similar results as discussed so far for PVAc are found for aPP in a two-dimensional cross-section through a 3D ^{13}C spectrum shown in Figure 8.15(c). The spectrum was actually taken with selective excitation in ω_1. At the spectral position of the return-jump signal (indicated by an arrow) no strong intensity is observed.

8.7 Summary of results of multidimensional exchange NMR above T_g

- Local dynamics is detected over 5–7 orders of magnitude in 2D spectra (9 orders in combination with 1D NMR spectra, 12 decades if NMR T_1-relaxation data are included); it follows the $\tau(T)$ curves seen by other methods, and can be fitted by the WLF/VFT function.

- Liquid-like clusters in the literal sense, with reorientational rates exceeding 10 MHz, are not detected close to T_g (nevertheless, distributions of correlation times up to three decades wide are found).
- Orientation correlation functions can be extracted from 2D spectra and are nonexponential.
- The reduced 4D experiment establishes a heterogeneous distribution of correlation times as the basis of the nonexponential loss of correlation (nonergodicity on the time scale of τ_0). On the scale of the longest correlation times in the distribution, fluctuations are detected which change the correlation times (non-Markovian segmental dynamics).
- The widths of the distributions of correlation times are slightly temperature-dependent. They differ between materials, appearing to broaden with increasing T_g.
- 2D MAS ^{13}C spectra show conformational changes. In aPP, exchange between trans and gauche conformations is dominant.
- 2D ^2H spectra close to T_g show broad distributions of reorientation angles indicating ill-defined reorientational motions with strong diffusive components.
- In 3D ^{13}C spectra, no indications of orientational memory are found.

8.8 References

Angell, C. A. (1991). Relaxation in Liquids, Polymers and Plastic Crystals – Strong/Fragile. *J. Non-Cryst. Solids*, **131–133**, 13.

Bengtzelius, U., Goetze, W. and Sjoelander, A. J. (1984). Dynamics of Supercooled Liquids and the Glass Transition. *J. Phys. C.*, **17**, 5915.

Chin, Y. H., Zhang, C., Wang, P., Inglefield, P. T., Jones, A. A., Kambour, R. P., Bendler, J. T. and White, D. M. (1992). Glass Transition Dynamics in a Blend by 2D NMR. *Macromolecules*, **25**, 3031.

Davidson, D. W. and Cole, R. H. (1951). Dielectric Relaxation. *J. Chem. Phys.*, **19**, 1417.

Donth, E. -J. (1992). *Relaxation and Thermodynamics in Polymers: Glass Transition.* Akademie Verlag, Berlin.

Ferry, J. D. (1980). *Viscoelastic Properties of Polymers*. Wiley, New York.

Kaplan, J. I. and Garroway, A. N. (1982). Homogeneous and Inhomogeneous Distributions of τ_c-Lineshapes. *J. Magn. Reson.*, **49**, 464.

Lindsey, C. P. and Patterson, G. D. (1980). Comparison of WW and Cole-Davidson Functions. *J. Chem. Phys.*, **73**, 3348.

Leisen, J., Werth, M., Boeffel, C. and Spiess, H. W. (1992). Dynamics above T_g: NMR of a Glass-Forming Discotic LC. *J. Chem. Phys.*, **97**, 3749.

McCrum, N. G., Read, B. E. and Williams, G. (1967). *Anelastic and Dielectric Effects in Polymer Solids*. Wiley, New York.

Monnerie, L. and Gery, F. (1969). Mouvement d'une chaîne macromoléculaire: Méthode Monte-Carlo. *J. Chim. Phys.*, **66**, 1691.

Pschorn, U., Rössler, E., Sillescu, H., Kaufmann, S., Schaefer, D. and Spiess, H. W. (1991). Motions Above T_g of Polystyrene: 1D and 2D NMR. *Macromolecules*, **24**, 398.

Schaefer, D., Spiess, H. W., Suter, U. W. and Fleming, W. W. (1990). 2D NMR Studies of Ultraslow Motion: aPP vs. iPP. *Macromolecules*, **23**, 3431.

Schaefer, D. and Spiess, H. W. (1992). 2D Exchange NMR of Powders IV. Distribution $g(\tau)$ and Line Shapes in the Intermediate Dynamic Range. *J. Chem. Phys.*, **97**, 7944.

Schaefer, D. and Spiess H. W. (1993). Unpublished results.

Schmidt-Rohr, K. and Spiess, H. W. (1991). Nature of Nonexponential Loss of Correlation Above T_g. *Phys. Rev. Lett.*, **66**, 3020.

Schnauss, W., Fujara, F., Hartmann, K. and Sillescu, H. (1990). Nonexponential ^2H Spin-Lattice Relaxation. *Chem. Phys. Lett.*, **166**, 381.

Wefing, S., Kaufmann, S. and Spiess, H. W. (1988). 2D Exchange NMR of Powders II. Evolution of Two-Time Distributions. *J. Chem. Phys.*, **89**, 1234.

Williams, G., Cook, M. and Hains, P. J. (1972). Molecular Motion in Amorphous Polymers. *J. Chem. Soc. Trans. Faraday Soc.*, **68**, 1045.

Williams, G. and Watts, D. C. (1970). Nonsymmetrical Dielectric Relaxation from a Simple Decay Function. *Trans. Faraday Soc.*, **66**, 80.

Williams, M. L., Landel, R. F. and Ferry, J. D. (1955). Temperature Dependence of Relaxation in Amorphous Polymers and Glass-Forming Liquids. *J. Am. Chem. Soc.*, **77**, 3701.

Zemke, K., Chmelka, B., Schmidt-Rohr, K. and Spiess, H. W. (1991). Conformational Exchange near T_g: 2D ^{13}C NMR of aPP. *Macromolecules*, **24**, 6874.

Zhang, C., Wang, P., Jones, A. A., Inglefield, P. T. and Kambour, R. P. (1991). Diluent Motion in a Glassy Blend by 2D ^{31}P NMR. *Macromolecules*, **24**, 338.

CHAPTER NINE

Multidimensional Spectra, Correlation Functions and Stochastic Processes

Keywords: *correlation functions (c.f.): simple time c.f., orientational c.f., c.f. in NMR and dielectric relaxation, c.f. from 2D exchange spectra, space–time c.f. in neutron scattering and 2D exchange NMR; nonergodicity; joint/conditional/*a priori *probabilities, Markov processes; information content of 3D and 4D exchange spectra.*

9.1 2D exchange spectra and correlation functions

Exchange spectroscopy can be regarded as a special form of correlation spectroscopy. In fact, it is intuitively appealing, and correct, to call the 2D exchange spectrum a 'correlation diagram'. Does this mean that a 2D exchange NMR spectrum, or a set of 2D spectra, is a correlation function, or is it more? The answer to each of these questions is actually 'yes and no'! To clarify this, the definition and meaning of correlation functions will be reviewed. Several formulas for evaluating correlation functions will be compiled and compared. In order to make the treatment accessible to other experimentalists and stress the basic ideas, the mathematical description of these complex issues will be kept very simple, even to the point where the theoretical purist may consider it as incorrect. The crucial connection to the experiments will involve correlation functions in dielectric and NMR relaxation on the one hand, and in neutron scattering on the other. It will be pointed out how these different methods for the study of molecular dynamics probe different types of correlation functions, and how they relate to 2D exchange spectra.

9.1.1 Basics on thermal motions and correlation functions

Thermal molecular motion represents a stochastic process which causes certain observables to vary randomly in time (Figure 9.1). For instance, the position vectors $\vec{\mathbf{r}}(t)$ of the particles are the stochastic variables in the case of (translational) Brownian motion. Concerning the rotational parts of the motion, angles that describe the molecular orientations, or functions of these angles, such as $3\cos^2\theta(t)-1$, can be considered as the stochastic variables. The complicated functions $f(t)$, being nonidentical for all the different particles, are not suitable for a simple characterization of the molecular motion. A description in terms of joint or conditional probabilities of finding certain values of $f(t)$ at certain times, as outlined in Section 9.3, is suitable

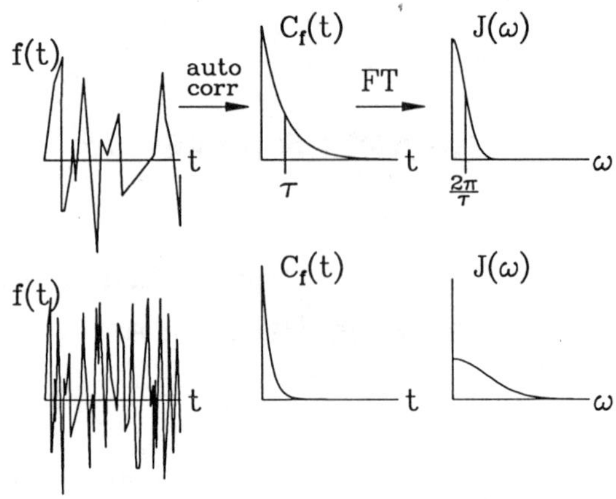

Figure 9.1 The upper diagram shows the stochastically varying function $f(t)$, its correlation function $C_f(t) = \langle f(0)f(t)\rangle / \langle f(0)^2 \rangle$, and the spectral density $J(\omega)$, which is the Fourier transform of $C_f(t)$. The lower diagram is the same as the upper one, but with a shorter correlation time τ.

for analysing various details of the stochastic process. The simplest and most commonly used characterization, however, is based on correlation functions:

$$C_f(t) \sim \langle f(0)f(t)\rangle. \tag{9.1}$$

Such a function probes the similarity or correlation between the stochastic variable f at time 0 and time t; if the motion is slow, this correlation will decrease only slowly with t whereas for a fast motion the correlation between $f(0)$ and $f(t)$ is lost rapidly (Figure 9.1). The angular brackets in equation (9.1) usually represent an ensemble average; various explicit expressions for these averaging brackets are described below. Due to the averaging, the correlation function of a stochastic process is a smoothly decaying function which is much easier to characterize than $f(t)$ itself. In order to eliminate irrelevant prefactors, $C_f(t)$ is normalized to fulfil $C_f(0) = 1$, and therefore the correlation expression (9.1) is usually divided by $\langle f^2(0)\rangle$.

Perhaps even more important than the simplification of the theoretical description is the fact that many experimental techniques determine averages of the type of equation (9.1), and others measure their Fourier transforms, which are commonly denoted as the 'spectral densities' of the motional processes. In particular, the autocorrelation functions relevant for relaxation techniques are usually time-dependent orientational autocorrelation functions $C_L(t)$ of order L:

$$C_L(t) = \langle P_L(\cos\theta(0))P_L(\cos\theta(t))\rangle(2L+1) \tag{9.2}$$

[using $\langle P_L^2(\cos\theta(0))\rangle = 1/(2L+1)$, see Equations (12.9) and (11.23). L denotes the order of the corresponding Legendre polynomial $P_L(\cos\theta)$ and $\theta(t)$ is the angle between some axis in the molecule at time t and some external axis (e.g. an external $\vec{\mathbf{B}}_0$ or electric field). NMR interactions, characterized by second-rank tensors, produce $\omega(\theta(t)) \cong \delta P_2(\cos\theta(t))$ (see equation (2.32)), therefore NMR relaxation generally probes a second-order orientational autocorrelation function $C_2(t)$. In contrast to this, 2D exchange NMR spectroscopy is not restricted to $C_2(t)$ but can be used to determine $C_L(t)$ of higher even orders L as well (Spiess, 1991). This will be shown in detail in

Section 9.1.5. Dielectric relaxation measurements with a harmonically varying field determine essentially the Fourier transform of the time derivative of $\langle \vec{\mu}(t)\cdot\vec{\mathbf{E}}\rangle \sim \langle \cos\beta(t)\rangle \sim C_1(t)$. It should be noted that dielectrics is classified as a relaxation technique here, even though dielectric diagrams that display $\varepsilon''(\omega)$ or $\varepsilon''(T)$ often look like 'spectra' with several peaks. In fact, $\varepsilon''(T)$ diagrams are similar to T_1 relaxation curves. The $\varepsilon''(\omega)$ curves are much different from resonance spectra as known, for instance, in IR or NMR spectroscopy. The dielectric 'spectrum' is essentially the imaginary part of the Fourier transform of the monotonically decreasing time derivative of the autocorrelation function (Cook *et al.*, 1970), instead of being the spectrum of an oscillating time signal as is the case for NMR. The peaks in the real part of the dielectric 'spectrum' are consequently all centred at $\omega = 0$, and of Lorentzian shape $\tau/(\omega^2 + \tau^{-2})$ for each exponential contribution to the autocorrelation function. The peaks in the imaginary part $\varepsilon''(\omega)$ of the Fourier transform are just the maxima of the dispersive 'wings' of the imaginary part $-\omega/(\omega^2 + \tau^{-2})$ of the Lorentzian, which occur at $|\omega| = \tau^{-1}$; see Figure 4.3.

In inelastic and quasi-elastic neutron scattering, the space–time Fourier transform of the scattering intensity $S(\vec{\mathbf{q}}, \omega)$ is the so-called van Hove correlation function (Lovesey, 1984), which can be expressed as:

$$G(\vec{\mathbf{r}}, t) = \langle \rho(\vec{\mathbf{0}}, 0)\rho(\vec{\mathbf{r}}, t)\rangle / \langle \rho^2(\vec{\mathbf{0}}, 0)\rangle \tag{9.3}$$

for liquids or amorphous solids (Springer, 1972). The relations of these various correlation functions and two-dimensional exchange spectra is the subject of the following sections.

9.1.2 Simple and space–time correlation functions

Simple time autocorrelation functions as already referred to in the previous section have the form:

$$C_f(t) = \langle f(0)f(t)\rangle / \langle f^2(0)\rangle. \tag{9.4}$$

Often, time autocorrelation functions are referred to simply as correlation functions or 'loss of correlation'. The averaging indicated by the pointed brackets will be specified in detail below.

Space–time correlation functions:

$$G_f(\vec{\mathbf{x}}, t) = \langle f(\vec{\mathbf{0}}, 0)f(\vec{\mathbf{x}}, t)\rangle / \langle f^2(\vec{\mathbf{0}}, 0)\rangle \tag{9.5}$$

involve both space and time together. $\vec{\mathbf{x}}$ is a general spatial coordinate meant to represent translational ($\vec{\mathbf{r}}$ or $|\vec{\mathbf{r}}|$) or rotational ((α, β, γ) or β) coordinates. The van Hove correlation function mentioned in the previous section is the best-known example of such a space–time correlation function.

The distinction between 'simple' time and 'space–time' correlation functions, which is rarely discussed in the literature, will turn out to be necessary in the context of characterizing the status of sets of 2D exchange spectra. Indeed, as will be shown, a set of 2D spectra contains more information on a dynamic process than a corresponding simple (orientational) time autocorrelation function. On the other hand, a set of 2D spectra is equivalent to an (orientational) space–time autocorrelation function.

As to the relation between time orientational autocorrelation functions $C_L(t)$ probed by NMR or dielectrics, and space–time correlation function $G(\vec{\mathbf{r}}, t)$ of inelastic neutron scattering, it may be interesting to note that $C_L(t)$ is not equivalent to $G(\vec{0}, t)$. Orientational correlation functions are determined by molecular rotations, and therefore the displacements $\vec{\mathbf{r}}$ are not equal to $\vec{\mathbf{0}}$ for the atoms concerned (Springer, 1972, p. 65). The lack of an explicit $\vec{\mathbf{r}}$ dependence in the autocorrelation functions $C_L(t)$ is not a result of setting the spatial component $\vec{\mathbf{r}}$ equal to $\vec{\mathbf{0}}$, but rather by integrating over the spatial coordinates: $C_L(t) = \int d\beta R(\beta, t) P_L(\cos\beta)$ (see equation (9.21c)).

The angular brackets in equations (9.4) and (9.5) represent averages, over time (see equation (9.11) or over a thermodynamic ensemble of systems. The formal definition of this classical average involves phase space with its points $\vec{\mathbf{\Gamma}}^N(t) = (\vec{\mathbf{r}}_1, \ldots, \vec{\mathbf{r}}_N, \vec{\mathbf{v}}_1, \ldots, \vec{\mathbf{v}}_N)$ made up of the position and velocity vectors of the n molecules of the system. The Liouville density $\rho(\vec{\mathbf{\Gamma}}^N(t), t)$ gives the probability of finding the system in the state $\vec{\mathbf{\Gamma}}^N(t)$. The (auto)correlation function of a stochastic variable f can thus be defined via the ensemble average (Landau and Lifshitz, 1980; Martynov, 1992):

$$\langle f(0) f(t) \rangle = \int f(\vec{\mathbf{\Gamma}}^N(t)) f(\vec{\mathbf{\Gamma}}^N(0)) \rho(\vec{\mathbf{\Gamma}}^N(0), 0,)\, d\vec{\mathbf{\Gamma}}^N(0) \tag{9.6}$$

where $\int d\vec{\mathbf{\Gamma}}^N(0)$ indicates the volume integration over phase-space at time 0. For single-particle orientation correlation functions, the formidable looking $6N$-dimensional phase-space integration of equation (9.6) can be reduced to a simple integral over an angle coordinate $\theta_m(t)$:

$$\langle f(0) f(t) \rangle = \int f(\theta_m(t)) f(\theta_m(0)) P(\theta_m(0))\, d\theta_m(0) \tag{9.7}$$

with the (time-independent) *a priori* probability density $P(\theta_m)$ of finding the segment, m at an angle θ_m. The index m refers to the considered molecule out of the N molecules of the system. Since we will mainly consider single-particle correlation functions in the following, we will soon drop the index of θ_m. It should be noted that the analysis of a single-particle function does not imply that the selected molecule does not interact with the others in the system. The multidimensional integral in (9.6) simplifies to become that of (9.7) because $f(\theta_m(t))$ depends only on θ_m so that all the other integrations, over θ_j with $j \neq m$, are lumped into $P(\theta_m(0))$.

The expression (9.7) still cannot be directly evaluated because the evolution from $\theta_m(0)$ to $\theta_m(t)$ is not specified. It is most conveniently described by the conditional-probability density $P(\theta_2, t \mid \theta_1, 0)$ of finding the segment at $\theta_m(t) = \theta_2$ provided it was at $\theta_m(0) = \theta_1$. Thus, we can rewrite equation (9.7) as:

$$\begin{aligned}\langle f(0) f(t) \rangle &= \int \left\{ \int f(\theta_2) P(\theta_2, t \mid \theta_1, 0)\, d\theta_2 \right\} f(\theta_1) P(\theta_1)\, d\theta_1 \\ &= \iint f(\theta_2) f(\theta_1) P(\theta_2, t; \theta_1, 0)\, d\theta_1\, d\theta_2 \end{aligned} \tag{9.8}$$

with the joint-probability density $P(\theta_2, t; \theta_1, 0) = P(\theta_2, t \mid \theta_1, 0) P(\theta_1)$.

In NMR spectroscopy, the frequency ω is measured, which is related to the angle θ by a functional relation $\omega(\theta)$ and its inverse $\theta(\omega)$. According to the principle of differential conservation of the integral (see Appendix E), the two-time probability densities for the angles and the frequencies are related by:

$$P(\theta_2, t; \theta_1, 0)\, \mathrm{d}\theta_1\, \mathrm{d}\theta_2 = S(\omega_2, t; \omega_1, 0)\, \mathrm{d}\omega_1\, \mathrm{d}\omega_2 \tag{9.9}$$

where $\omega_1 = \omega(\theta_1)$ and $\omega_2 = \omega(\theta_2)$. Inserting this in equation (9.8), we obtain:

$$\langle f(0) f(t) \rangle = \iint f(\theta(\omega_2)) f(\theta(\omega_1)) S(\omega_2, t; \omega_1, 0)\, \mathrm{d}\omega_1\, \mathrm{d}\omega_2. \tag{9.10}$$

Since $S(\omega_2, t; \omega_1, 0)$ can be identified with a 2D exchange spectrum, this relation (9.10) will be used repeatedly below.

Simple autocorrelation functions $C_f(t)$ characterize certain properties of a system of many particles by a single, integral function. Through the integral (9.10) it is related to the two-time distribution function $S(\omega_2, t; \omega_1, 0)$ which contains more details about the system or process than the simple autocorrelation function $C_f(t)$ itself, since many different $C_f(t)$ can be calculated from one distribution function $S(\omega_2, t; \omega_1, 0)$. Therefore, experiments yielding this distribution have a higher information content. Often, a two-time distribution is equivalent to a space–time autocorrelation function as introduced by equation (9.5).

To express the bracketed term in equations (9.4) and (9.5) explicitly, as an alternative to the ensemble average (9.6), a time or space–time average can be performed:

$$C_{\mathrm{f}}(t) = \lim_{t_f \to \infty} \frac{1}{t_f} \int_0^{t_f} f(t') f(t' + t)\, \mathrm{d}t' \tag{9.11}$$

$$G_{\mathrm{f}}(\vec{\mathbf{x}}, t) = \lim_{t_f \to \infty} \frac{1}{t_f} \int_0^{t_f} \mathrm{d}t' \lim_{V \to \infty} \frac{1}{V} \int_V \mathrm{d}^3 x'\, f(\vec{\mathbf{x}}', t') f(\vec{\mathbf{x}}' + \vec{\mathbf{x}}, t' + t). \tag{9.12}$$

The time average corresponds to the ensemble average if the system is ergodic (see Section 9.1.6).

9.1.3 Exchange NMR signals vs. correlation functions in neutron scattering

Inelastic and quasi-elastic neutron scattering comprise powerful techniques for investigating molecular motion, both translation and reorientation, with correlation times between 10^{-15} s and 10^{-7} s. Thus, they are complementary to NMR which detects reorientations in the range between 10^{-10} s (via T_1 relaxation) and 10^2 s (by exchange NMR). In the following, we will touch only on certain features of neutron scattering and assume some familiarity with its description in terms of correlation functions. For thorough discussions of this topic, the reader is referred to standard texts, for instance Lovesey (1984) or Springer (1972).

Exchange NMR as described here relies on single-particle interactions. In neutron scattering experiments, this corresponds to the single-particle or 'self' part of the scattering function, which is the sum over single-particle functions. It is measured in the incoherent scattering intensity, while for coherent scattering the more complicated 'distinct' part of the correlation and scattering functions with interferences from all pairs of particles is also present.

The analogies between 2D exchange NMR and quasielastic neutron scattering for the characterization of reorientational motions (Springer, 1972, Section 7.2) are found to be particularly close. However, as the treatment of neutron scattering which also incorporates translational motions is much more widely known, we will point out analogies of this theory with 2D exchange NMR.

The self part of the space–time van Hove correlation function $G(\vec{\mathbf{r}}, t)$, which is the inverse Fourier transform of the incoherent neutron scattering intensity $S_{\text{inc}}(\vec{\mathbf{q}}, \omega)$, is usually written in the classical approximation as:

$$G_s(\vec{\mathbf{r}}, t) = N^{-1} \left\langle \sum_{j=1}^{N} \delta(\vec{\mathbf{r}} - [\vec{\mathbf{r}}_j(t) - \vec{\mathbf{r}}_j(0)]\,) \right\rangle \tag{9.13a}$$

$$= \langle \delta(\vec{\mathbf{r}} - [\vec{\mathbf{r}}(t) - \vec{\mathbf{r}}(0)]\,) \rangle. \tag{9.13b}$$

This represents the probability that a point particle, characterized by the δ function, is found displaced by $\vec{\mathbf{r}}$ from its original position after having waited for a time t. $G_s(\vec{\mathbf{r}}, t)$ of equation (9.13b) can be characterized succinctly as the displacement-vector distribution, and $G_s(|\vec{\mathbf{r}}|; t)$ as the jump-distance distribution, after a time t. This is clearly analogous to the reorientation-angle or jump-angle distribution $R(\beta; t_m)$ after a mixing time t_m; $R(\beta; t_m)$ is in turn essentially equivalent to a 2D exchange spectrum (see Section 7.4). The equality of the sum of single-particle correlation functions, equation (9.13a), to one single-particle function, equation (9.13b), is often used tacitly, and we will employ it in the following, too. It holds if the N particles behave equivalently; the various configurations of the system are already contained in the correlation average.

Actually, the commonly cited form (9.13) for the classical van Hove correlation function does not obviously exhibit the form of a space–time correlation function as defined in equation (9.5). Nonetheless, equation (9.13) can be derived from the typical correlation function expression:

$$G(\vec{\mathbf{r}}, t) = \frac{1}{N} \left\langle \int \rho(\vec{\mathbf{r}}', 0)\rho(\vec{\mathbf{r}}' + \vec{\mathbf{r}}, t)\, \mathrm{d}^3 r' \right\rangle \tag{9.14}$$

with the spatial density of the point-like particles expressed as:

$$\rho(\vec{\mathbf{r}}, t) = N^{-1} \sum_j \delta(\vec{\mathbf{r}} - \vec{\mathbf{r}}_j(t)).$$

Quite often, instead of $G_s(\vec{\mathbf{r}}, t)$, its spatial Fourier transform, the intermediate scattering function, is analysed (for the following equation, recall the fundamental property of the δ function, $\int \delta(\vec{\mathbf{r}} - \vec{\mathbf{r}}_0) f(\vec{\mathbf{r}})\, \mathrm{d}^3 r = f(\vec{\mathbf{r}}_0)$):

$$I_s(\vec{\mathbf{q}}, t) = \mathrm{FT}_{\vec{\mathbf{r}}}(G) = \int \delta(\vec{\mathbf{r}} - \{\vec{\mathbf{r}}(t) - \vec{\mathbf{r}}(0)\}\,)\, \mathrm{e}^{-\mathrm{i}\vec{\mathbf{q}}\cdot\vec{\mathbf{r}}}\, \mathrm{d}^3 r$$

$$= \langle \mathrm{e}^{-\mathrm{i}\vec{\mathbf{q}}\cdot\vec{\mathbf{r}}(0)}\, \mathrm{e}^{\mathrm{i}\vec{\mathbf{q}}\cdot\vec{\mathbf{r}}(t)} \rangle. \tag{9.15}$$

This function clearly represents a time correlation function $\langle f^*(0) f(t) \rangle$ with $f(t) = \mathrm{e}^{\mathrm{i}\vec{\mathbf{q}}\cdot\vec{\mathbf{r}}(t)}$. Nevertheless, $I_s(\vec{\mathbf{q}}, t)$ is not a space–time correlation function, in contrast to its Fourier transform $G_s(\vec{\mathbf{r}}, t)$. It is obvious that in expression (9.15) the scattering vector $\vec{\mathbf{q}}$, unlike $\vec{\mathbf{r}}(t)$ in $G_s(\vec{\mathbf{r}}, t)$ of equation (9.13), is not a correlation variable taken at two different values. It is rather like a parameter, which is usually under the control of the experimentalist. As expected for time correlation functions, the $I_{\vec{\mathbf{q}}}(t) := I(\vec{\mathbf{q}}, t)$ functions are exponentials for simple relaxation processes. The fact that the space–time correlation function $G(\vec{\mathbf{r}}, t)$ can be obtained by Fourier transformation of the full $I(\vec{\mathbf{q}}, t)$ data set does not allow one to conclude that $I(\vec{\mathbf{q}}, t)$ is also a space–time correlation function.

The simple-time correlation functions $I_{\vec{\mathbf{q}}}(t)$ can be compared to the functions $C_L(t)$. Both the $\vec{\mathbf{q}}$ and the L dependences provide valuable information on the spatial nature of the molecular dynamics, translational/rotational and rotational, respectively. Similar to the Fourier synthesis of the space-and-time correlation function $G(\vec{\mathbf{r}}, t)$ from the simple time-correlation function $I_{\vec{\mathbf{q}}}(t)$, the full orientation-and-time correlation function $R_f(\beta, t)$ can be calculated by a 'Legendre synthesis':

$$R_f(\beta, t)/\sin\beta = \sum_{L=0}^{\infty} (2L+1)C_L(t)P_L(\cos\,\beta) \tag{9.16}$$

if the full set of the $C_L(t)$ is available (see equation (9.25) and Section 12.3.1).

$I(\vec{\mathbf{q}}, t)$ in equation (9.15) resembles the stimulated-echo decay function $I(t_e, t_m)$ (Spiess, 1980; Fujara *et al.*, 1986; Rössler, 1986) in the 2D NMR time signal $s(t_1, t_2; t_m)$ (see equation (10.3)):

$$I(t_e, t_m) := s(t_1 = t_e, t_2 = t_e; t_m) \sim \langle e^{-it_e\omega(0)}\, e^{it_e\omega(t_m)} \rangle \tag{9.17a}$$

$$I(t_e, t_m) \sim \iint d\omega_1\, d\omega_2\, e^{-i\omega_1 t_e}\, e^{i\omega_2 t_e}\, S(\omega_1, \omega_2; t_m). \tag{9.17b}$$

The negative sign in front of $i\omega_1 t_1$ can be generated by appropriate pulse phases; a term with $\exp(+i\omega_1 t_1)$ would not generate a stimulated echo (see Section 10.2.3). The connection (9.17b) between $I(t_e, t_m)$ and the 2D spectrum $S(\omega_1, \omega_2; t_m)$ is similar to the Fourier relationship between $I(\vec{\mathbf{q}}, t)$ and $G(\vec{\mathbf{r}}, t)$:

$$I(\vec{\mathbf{q}}, t) \sim \iint d\vec{\mathbf{r}}_1\, d\vec{\mathbf{r}}_2\, e^{-i\vec{\mathbf{q}}\cdot\vec{\mathbf{r}}_1}\, e^{i\vec{\mathbf{q}}\cdot\vec{\mathbf{r}}_2}\, G(\vec{\mathbf{r}}_2 - \vec{\mathbf{r}}_1; t). \tag{9.18}$$

Equations (9.17) and (9.18) indicate again the analogy between $S(\omega_1, \omega_2; t_m)$ and $G(\vec{\mathbf{r}}_2 - \vec{\mathbf{r}}_1; t_m)$.

The analogy of the final-state signal $I(t_e, t_m \to \infty)$ after the three-pulse sequence in ^2H NMR and the elastic incoherent structure factor $I(|\vec{\mathbf{q}}|, t \to \infty)$ in neutron scattering has been demonstrated in simulations and experiments by Fujara *et al.* (1986, 1988). Figure 9.2 portrays typical curves for the two methods in comparison. From the NMR point of view, it is worth noting that for obtaining $I(t_e, t_m)$ the stimulated-echo signal $s(t_1 = t_e, t_2 = t_e; t_m)$ measured experimentally is reduced by relaxation factors $\exp(-t_e/T_2)$ and $\exp(-t_m/T_1)$. These are eliminated through multiplication by $s(0, 0; \text{'0'})/s(0, 0; t_m) = \exp(t_m/T_1)$ and $1/s(t_e, t_e; \text{'0'}) = \exp(t_e/T_2)/s_0$, which also normalizes out the arbitrary intensity prefactor s_0 in the measured data:

$$I(t_e, t_m) = \left\{ s(t_e, t_e; t_m)\, \frac{1}{s(t_e, t_e; \text{'0'})}\, \frac{s(0, 0; \text{'0'})}{s(0, 0; t_m)} \right\}. \tag{9.19}$$

The T_1 elimination can actually not be performed for the 'alignment' signal $s_{\sin}(t_e, t_e; t_m \to \infty) = \langle \sin\tilde{\omega}_1 t_e\, \sin\tilde{\omega}_2 t_e \rangle$ because it vanishes for $t_e = 0$. This vanishing initial intensity also tends to render $I_{\sin}(t_e; t_m \to \infty)$ unreliable for small t_e values, because spurious signals due to pulse imperfections, otherwise negligible, can here become comparable to the real signal.

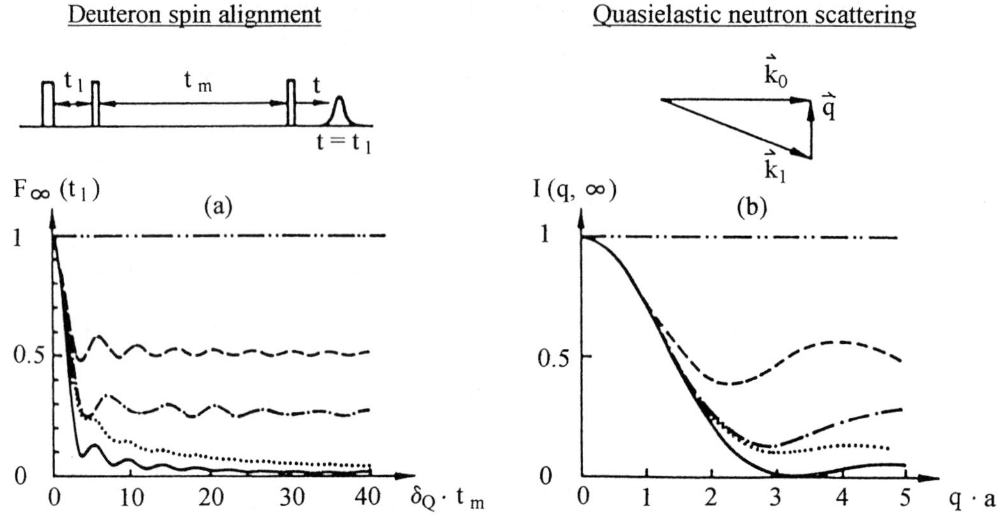

Figure 9.2 Analogy of quasielastic neutron scattering and the stimulated-echo decay in ^{2}H NMR for various motional models (-··· no motion, --- two-site jumps, -·- four-site jumps, ... rotational diffusion on a circle, —— isotropic rotational diffusion). The NMR data represent $\langle e^{-i\omega_1 t_e} e^{i\omega_2 t_e}\rangle$, while the single-particle incoherent neutron scattering function is $\langle e^{-i\vec{q}\cdot\vec{r}(0)} e^{i\vec{q}\cdot\vec{r}(t)}\rangle$. The final values for $t_e \to \infty$ or $q \to \infty$ are $1/N$, where N is the number of sites in the model. Adapted from Fujara *et al.* (1988).

9.1.4 Time-dependent orientational autocorrelation functions $C_L(t)$

The time-dependent orientational autocorrelation functions $C_L(t_m)$ of order L:

$$C_L(t_m) = \langle P_L(\cos\theta(t=0)) P_L(\cos\theta(t=t_m))\rangle (2L+1)$$

were introduced above. Here, $\theta(t)$ is the angle made at time t with some external axis (e.g. the $\vec{\mathbf{B}}_0$ field). An equivalent and useful formula for $C_L(t_m)$ is found by expressing the process in terms of the reorientational angle β relative to the start position ($\theta(t=0) = 0°$):

$$C_L(t_m) = \langle P_L(\cos(0°)) P_L(\cos\beta(t_m))\rangle / \langle P_L^2(\cos(0°))\rangle$$
$$= \langle P_L(\cos\beta(t_m))\rangle \tag{9.20}$$

(recall $\cos(0°) = 1$ and $P_L(1) = 1$).

Time correlation functions are suitable for summarizing the overall time dependence of stochastic processes. To discuss the relation between 2D exchange spectra and $C_2(t_m)$, we repeat in appropriate nomenclature the different expressions for $C_2(t)$ where $\omega(t)$ can be replaced by $\delta P_2(\cos\beta(t))$ (if the asymmetry parameter of the NMR interaction $\eta = 0$):

$$C_2(t_m) = \langle \omega(0)\omega(t_m)\rangle / \langle \omega^2(0)\rangle \tag{9.21a}$$

$$C_2(t_m) = \frac{1}{\delta^2} \iint \omega(\theta_2)\omega(\theta_1) P(\theta_2, t_m; \theta_1, 0)\, d\theta_1\, d\theta_2 \tag{9.21b}$$

$$C_2(t_m) = \frac{5}{\delta^2} \iint d\omega_1\, d\omega_2\, S(\omega_1, \omega_2; t_m)\omega_1\omega_2 \tag{9.21c}$$

$$C_2(t_m; t_s) = \frac{1}{\delta^2}\frac{1}{t_s}\int_0^{t_s} \omega(t)\omega(t+t_m)\, dt. \tag{9.21d}$$

The first expression, equation (9.21a), is more or less a memory-aid to be filled with a specific meaning by the following formulas. The ensemble average equation (9.21b) is not of much practical use when the two-time angle distribution $P(\theta_2, t; \theta_1, 0)$ is not known. In contrast, the two-time distribution $S(\omega_1,\omega_2; t_m)$ in equation (9.21c) can be measured; it is the 2D exchange spectrum (normalized to unit integral). The normalization factor $5/\delta^2$ in (9.21c) is obtained via:

$$\langle \omega^2(t=0)\rangle = \delta^2\langle \{P_2(\cos\theta)\}^2\rangle = \delta^2 \int_{-1}^{1} \{P_2(y)\}^2 \int_{-1}^{1} dy = \delta^2 \tfrac{1}{5}.$$

The time average in equation (9.21d) is equivalent to the ensemble average only if the system is ergodic. For solid systems, ergodicity is often reached only after extremely long times t_s (see Section 9.1.6).

9.1.5 How to determine $C_L(t)$ from 2D exchange spectra

For $I = \frac{1}{2}$ nuclei subject to a NMR interaction with vanishing asymmetry parameter $\eta = 0$, the value of the autocorrelation function can be determined by numerically integrating the experimental spectral intensity $S(\omega_1, \omega_2; t_m)$ weighted with the product of the frequency coordinates of the individual points:

$$C_2(t_m) = \frac{5}{\delta^2} \iint d\omega_1\, d\omega_2\, S(\omega_1, \omega_2; t_m)\omega_1\omega_2$$

(see equation (9.21c)). Thus, each spectrum directly yields an absolute value for $C_2(t_m)$. (It should be noted, however, that the treatment assumes the slow-motion approximation, i.e. $\tau > 0.1$ ms. On that timescale, the initial drop of the correlation function due to possible fast motions with $\tau < 1$ μs appears as instantaneous and cannot be directly observed because it is only reflected in the averaged interaction tensor.)

In fact, all higher C_L with even L can be obtained from the series of 2D spectra. For instance, C_4 is obtained after expressing the fourth Legendre polynomial $P_4(x)$ (see Table B.2) in terms of $P_2(x)$ according to:

$$\begin{aligned} P_4(x) &= \tfrac{1}{8}(35x^4 - 30x^2 + 3) \\ &= \tfrac{1}{18}(35\{P_2(x)\}^2 - 10P_2(x) - 7). \end{aligned} \tag{9.22a}$$

Substituting ω/δ for $P_2(x)$:

$$p_4(\omega) := \frac{1}{18}\left(35\left(\frac{\omega}{\delta}\right)^2 - 10\,\frac{\omega}{\delta} - 7\right) \tag{9.22b}$$

and with $9 = 2L + 1$ for $L = 4$, we obtain:

$$C_4(t_m) = 9 \iint d\omega_1\, d\omega_2\, S(\omega_1, \omega_2; t_m) p_4(\omega_1) p_4(\omega_2). \tag{9.22c}$$

In an analogous way, we find:

$$p_6(\omega) = \frac{1}{18}\left(77\left(\frac{\omega}{\delta}\right)^3 - 42\left(\frac{\omega}{\delta}\right)^2 - 21\,\frac{\omega}{\delta} + 4\right) \tag{9.23a}$$

$$C_6(t_m) = 13 \iint d\omega_1\, d\omega_2\, S(\omega_1, \omega_2; t_m) p_6(\omega_1) p_6(\omega_2). \tag{9.23b}$$

Another, even simpler, formula for obtaining these C_L, namely in terms of the reorientation-angle distribution $R(\beta; t_m)$, can be derived directly from:

$$C_L(t) = \langle P_L(\cos\beta(t)) \rangle. \tag{9.24}$$

$C_2(t_m)$, for instance, can be obtained directly from $R(\beta; t_m)$ if one specifies equation (9.24) for $L = 2$ as:

$$C_2(t_m) = \int_0^{90^\circ} d\beta\; R(\beta; t_m) \tfrac{1}{2}(3\cos^2\beta - 1). \tag{9.25a}$$

In fact, all C_L with even L can be calculated according to:

$$C_L(t_m) = \int_0^{90^\circ} d\beta\; R(\beta; t_m) P_L(\cos\beta). \tag{9.25b}$$

In practice, with a resolution of $\Delta\cos\beta$, reliable C_L values are obtained for $L \ll 1/\Delta\cos\beta$. The restriction to even L is due to $R(\beta; t_m)$ resulting from the full distribution $R_f(\beta)$ by symmetrization with respect to $\beta = 90^\circ$; while the $P_L(\cos\beta)$ for even L are symmetric with respect to 90° and are therefore unaffected by the symmetrization, those for odd L are antisymmetric with respect to 90°, so that the integral in (9.25b), extended over a range of 180°, vanishes.

From the perspective of the moment expansion as discussed in Sections 11.6 and 12.3.1, equation (9.25b) states that $C_L(t_m)$ is the Lth moment of $R_f(\beta, t_m)/\sin\beta$ for a given t_m. Therefore, the $C_L(t_m)$ can be obtained as the coefficients of a best fit of Legendre subspectra:

$$S_L(\omega_1, \omega_2; t_m) := \int_0^1 d\cos\beta\; S_\beta(\omega_1, \omega_2; t_m) P_L(\cos\beta) \tag{9.26}$$

to the experimental 2D exchange spectrum (see Sections 11.6 and 11.7). This approach could be applied to obtain $C_L(t_m)$ from ^{2}H spectra in spite of their overlapping transitions, as well as to spectra with $\eta \neq 0$, for which otherwise the analysis as outlined here is not directly applicable.

Based on the identity of the moments of $R_f(\beta, t_m)/\sin\beta$ with the $C_L(t_m)$, we can exploit relations known from the moment expansion. For instance, it is known that for an isotropic, angle-independent distribution, the moments with $L > 0$ all vanish; the same must hold for the C_L for the isotropic final state. Indeed, that corresponding reorientation-angle distribution $R_f(\beta, t_m)/\sin\beta = \frac{1}{2}$ is a constant.

If we consider the case of $\eta \neq 0$, the normalization prefactor for $\langle\omega(0)\omega(t_m)\rangle$, obtained by integrating $\omega^2(\theta, \phi)$ over θ and ϕ, reads $(5/\delta^2)(1/(1 + \eta^{2/3}))$. For small η,

$1/(1+\eta^2/3)$ is close to unity. For instance, it is 0.97 for $\eta = 0.3$, so that, up to this η value, the analysis outlined for $\eta = 0$ will be applicable to a good approximation.

The autocorrelation functions $C_L(t_m)$ can also be obtained from the 'edge slices' $S(\pm\delta, \omega; t_m)$ of ^{2}H spectra by inserting the relation $S(\delta, \omega; t_m)\,d\omega = R(\beta; t_m)\,d\beta$, equation (7.8b), into equation (9.25) for C_L. The following simple expressions result:

$$C_2(t_m) = \frac{1}{\delta}\int_{-\frac{1}{2}\delta}^{\delta} d\omega\; S(\pm\delta, \omega; t_m)\omega \tag{9.27a}$$

$$C_L(t_m) = \int_{-\frac{1}{2}\delta}^{\delta} d\omega\; S(\pm\delta, \omega; t_m)p_L(\omega) \tag{9.27b}$$

with the $p_4(\omega)$ and $p_6(\omega)$ functions of equations (9.22b) and (9.23a). These relations (9.27) represent an evaluation of the 'edge slice' directly in terms of frequencies, and are therefore relatively well-behaved in practical application, provided the 'edge slice' is available in sufficient quality.

9.1.6 Nonergodicity

Ergodicity plays a central role in statistical mechanics. A system is called ergodic on a time-scale of t_s if every subsystem probes all configurations accessible under the macroscopic boundary conditions within the time t_s. The classical, more restrictive, use of the term leaves out the specification of a timescale, and considers only the case $t_s \to \infty$. In an ergodic system, time average and ensemble average are equal. The more precise quasi-ergodic hypothesis assumes that in the course of time every subsystem comes arbitrarily close to any accessible configuration. While the gases and liquids usually considered in elementary statistical mechanics and thermodynamics are quasi-ergodic, solids can remain nonergodic on all practically relevant timescales. In particular, spontaneous symmetry-breaking at phase transitions leads to 'breaking' of ergodicity. For instance, a ferromagnet with its magnetization up and at sufficiently low temperature will 'never' fluctuate into the energetically equivalent configuration with its magnetization down.

In neutron scattering, residual correlation in the intermediate scattering function $I(\vec{\mathbf{q}}, t)$, according to:

$$f_{\vec{\mathbf{q}}} := \frac{I(\vec{\mathbf{q}}, t \to \infty\,[>10^{-6}\,\text{s}])}{I(\vec{\mathbf{q}}, t = 0)}$$

is denoted as the nonergodicity parameter (nonergodic on the time scale of 10^{-6} s). In this sense, all motions with an anisotropic final state appear as nonergodic.

9.1.7 Helical-jump correlation functions

In Section 7.12, it was mentioned that helical jumps can be approximated by a random walk in continuous time. The corresponding master equation quantifies the exchange between nearest neighbours in an elemental step:

$$\frac{dG_n(t)}{dt} = -2kG_n(t) + kG_{n+1}(t) + kG_{n-1}(t) \tag{9.28a}$$

where the indices label the positions along the helical chain. The rate is $k = \frac{1}{2}(1/\tau)$

where τ is the average residence time. The solution of (9.28a) with the initial condition $G_n(t) = \delta_{n,0}$ yields pair-correlation functions:

$$G_n(t) = I_{|n|}(2kt)\, \mathrm{e}^{-2kt} = \sum_{m=0}^{\infty} \frac{(kt)^{n+2m}}{m!(n+m)!}\, \mathrm{e}^{-2kt} \tag{9.28b}$$

where I_n is the nth-order Bessel function of imaginary argument (modified Bessel function) (Van Kampen, 1981). These functions exhibit a definitely nonexponential dependence on t_m (Figure 9.3). This is typical behaviour for a diffusive translational process. The corresponding orientational correlation function $C_L(t)$ reflecting the rotational part of the motion can be generated by a suitable superposition of the same pair correlation functions $G_n(t)$:

$$C_L(t) = \frac{1}{N} \sum_{n=1}^{N} P_L(\cos \beta_n) G_n(t) \tag{9.29}$$

where β_n specifies the orientation of the PAS $\vec{\mathbf{Z}}_P$ axis of the nth site. Although the $G_n(t)$ are nonexponential, $C_L(t)$ is a simple exponential (plus a constant). As an example, consider jumps on a 2_1 helix, in which all even-numbered sites are equivalent to site 0, while all odd-numbered sites are equivalent to site 1. This case must apparently be equivalent to jumps between two sites. Therefore, the exponential loss of correlation known to result for the two-site problem must be reproduced by the sum (9.29) over the helical jump pair-correlation functions $G_n(t)$. The mathematical basis for this are the addition theorems of Bessel functions which appear in the pair-correlation functions according to equation (9.28b).

The pair correlation functions $G_n(t)$ of the helical jumps are closely related to the van Hove correlation function $G_s(\vec{\mathbf{r}}, t)$. For a discrete random walk in continuous time, $G_n(t)$ represents the probability of finding a segment displaced by a distance $r = nr_0$

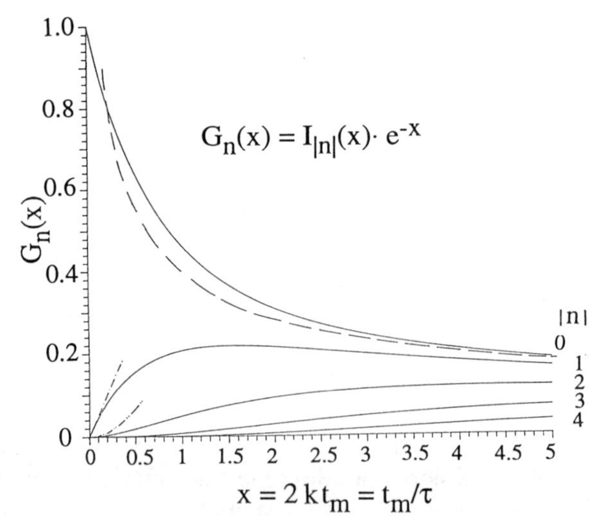

Figure 9.3 Solutions of the random walk in continuous time on a chain with the particle starting at $n = 0$. These functions provide the weighting factors for the subspectra in helical-jump simulations (see Section 7.12). Dashed curve: $\sim 1/\sqrt{t_m}$. The initial linear and quadratic t_m dependence for $n = 1$ and $n = 2$ is indicated by dashed-dotted curves.

from its original position after waiting for a time t, where r_0 is the distance between neighbouring sites. This is completely analogous to the interpretation of $G_s(r, t)$ as expressed in equation (9.13), so we can write $G_n(t) = G_s(nr_0, t)$.

9.2 Distributions of correlation times

Processes with a single correlation time are rarely found in partially or strongly disordered solids like semicrystalline or amorphous polymers. Rather, the time-dependent parts $C_t(t) := (C(t) - C_\infty)/(C(0) - C_\infty)$ of correlation functions $C(t)$ are, or seem to be, generated by a superposition of exponentials with different correlation times. This can be written as:

$$C_t(t) = \int_0^\infty e^{-t/\tau} g(\tau)\, d\tau \tag{9.30}$$

where $g(\tau)$ is the distribution of correlation times (Davidson and Cole, 1950; Connors, 1964). Such distributions have been shown to span several orders of magnitude for instance in the case of local motions in the glassy state. The reason for such a tremendous variation of τ can be understood on the basis of the strong dependence of τ on the energy barriers E_a according to Arrhenius' law:

$$\tau = \tau_0\, e^{-E_a/kT}. \tag{9.31}$$

Obviously, a logarithmic τ scale corresponds to a linear scale of E_a (if a weak variation of τ_0 can be assumed). On the basis of relations like equation (9.31) between τ, E_a, and T, the distributions of correlation times or even their partial integrals can be converted into the distribution of activation energies (Frauenfelder, 1984; Rössler *et al.*, 1990).

Concerning the shape of the distribution, different approaches have been taken. For a parametrization directly in terms of correlation times, a log-Gaussian distribution:

$$F_G(\ln \tau) = \frac{1}{\sigma_{\ln}\sqrt{2\pi}} \exp\left(- \frac{\{\ln(\tau/\tau_0)\}^2}{2\sigma_{\ln}^2} \right) \tag{9.32}$$

appears to be particularly simple. This may be converted into $g(\tau)$, as defined above in equation (9.30), by applying the principle of conservation of area of corresponding intervals (Appendix E):

$$g(\tau)\, d\tau = F(\ln \tau)\, d\,(\ln \tau) \Leftrightarrow g(\tau) = F(\ln \tau)\, \frac{d\,(\ln \tau)}{d\tau} = F(\ln \tau)\, \frac{1}{\tau}. \tag{9.33}$$

In addition, some authors define a function:

$$G(\tau) := F(\ln \tau) = \tau g(\tau). \tag{9.34}$$

The difference between this $G(\tau)$ and the actual linear distribution function $g(\tau)$ has sometimes not been considered in the literature.

Several different mean correlation times can be defined from the distributions. The most straightforward expression:

$$\langle \tau \rangle = \int_0^\infty \tau\, g(\tau)\, d\tau = \int_{-\infty}^{\infty} \tau\, F(\ln \tau)\, d\,(\ln \tau) \tag{9.35}$$

does not account well for the fact that the distribution spans orders of magnitude. Instead, a 'logarithmic average' τ_0 is introduced:

$$\ln \tau_0 := \langle \ln \tau \rangle = \int_{-\infty}^{\infty} \ln \tau\, F(\ln \tau)\, d\,(\ln \tau) = \int_0^\infty \ln \tau\, g(\tau)\, d\tau. \tag{9.36}$$

That is, $\ln(\tau_0)$ is the center of gravity of $F(\ln \tau)$.

The width of the distribution of correlation times can be specified as the halfwidth in decades (powers of ten, orders of magnitude):

$$\Delta_{\frac{1}{2}} := \log_{10} \tau^+ - \log_{10} \tau^- = \log_{10} \frac{\tau^+}{\tau^-} = \left(\ln \frac{\tau^+}{\tau^-}\right) \Big/ 2.303 \tag{9.37}$$

with τ^+ and τ^- being the τ values at half height of the maximum of $F(\ln \tau)$:

$$F(\ln \tau^{\pm}) = \tfrac{1}{2} \max\{F(\ln \tau)\}. \tag{9.38}$$

For the log-Gaussian distribution $F_G(\ln \tau)$, inserting the actual expression (9.32) into (9.38) produces $(\ln \tau^{\pm}/\tau_0) = \pm\, \sigma_{\ln}\sqrt{2 \ln 2}$, and employing the definition of $\Delta_{\frac{1}{2}}$ (9.37) we obtain as the width $\Delta_{\frac{1}{2}}^{G}$ of the log-Gaussian distribution:

$$\Delta_{\frac{1}{2}}^{G} = 2\sigma_{\ln}\sqrt{2 \ln 2}/2.303 = 1.02\sigma_{\ln}. \tag{9.39}$$

Instead of $g(\tau)$, often fitting functions for $C_t(t)$ are considered, since several methods yield data for $C_t(t)$ directly (or after Fourier transformation as in the case of dielectric relaxation). The stretched exponential (Kohlrausch–Williams–Watts, KWW) function $\exp(-\{t/\tau_{ww}\}^{\beta})$ has been found to fit the data quite universally. To obtain the corresponding $g(\tau)$ (Lindsey and Patterson, 1980), the integral equation (9.30) must be inverted. This corresponds to an inverse Laplace transformation, which is, however, not as straightforward to perform as a Fourier transformation. In particular for experimental $C_t(t)$ data with noise and a limited range of t values, special fit algorithms have been developed to obtain the corresponding $g(\tau)$.

From the $C_t(t)$ curve, the average correlation time can be defined as:

$$\langle \tau \rangle = \int_0^\infty C_t(t)\, dt \tag{9.40}$$

which is consistent with the average τ of equation (9.35) after inserting the superposition formula (9.30):

$$\int_0^\infty C_t(t)\, dt = \int_0^\infty g(\tau)\, d\tau \int_0^\infty e^{-t/\tau}\, dt = \int_0^\infty g(\tau)\tau\, d\tau \tag{9.41}$$

9.3 Stochastic processes and probabilities

As discussed in Section 9.1.1, stochastic processes produce random fluctuations of the microscopic variables (positions, orientations, etc.) of the physical system under consideration. Nonetheless, it is possible to treat the dynamics of the system in a relatively simple way by characterizing it in terms of probabilities (or probability densities). The correlation functions discussed in the previous section represent the most widely used examples that contain this probability approach. However, the integrals indicated by the pointed brackets remove much of the theoretically achievable information, in particular in the case of simple time correlation functions.

Before the advent of multidimensional exchange NMR, there was no experimental technique with access to these probability distributions for dynamics in disordered solids. Even the van Hove correlation function in neutron scattering, which has a similar quality to the 2D exchange NMR spectra, appears to occur as a basic theoretical concept rather than as the result of measurements and data analysis. Therefore, the superior information content of the space–time correlation functions (probability distributions) compared with simple time correlation functions has rarely been appreciated.

In the theory of stochastic processes (Oppenheim *et al.*, 1977; Van Kampen, 1981), one starts out with a more detailed approach, considering the probabilities that at certain points in time t_j certain values x_j of a certain stochasic variable (angle, position, etc.) are taken. These can be joint probabilities:

$$W(x_n, t_n; x_{n-1}, t_{n-1}; x_{n-2}, t_{n-2}; \ldots; x_0, t_0) \tag{9.42}$$

or conditional probabilities:

$$W(x_n, t_n \mid x_{n-1}, t_{n-1}; x_{n-2}, t_{n-2}; x_{n-3}, t_{n-3}; \ldots; x_0, t_0) \tag{9.43}$$

or *a priori* probabilities:

$$P(x_j, t_j). \tag{9.44}$$

The interpretation of these expressions can be given concisely by the following rules: $W(\ldots)$ or $P(\ldots)$ denotes 'the probability density that ...'; 'x_j, t_j' is to be read as 'the stochastical variable takes the value x_j at the time t_j' or more loosely as 'the system is in the state x_j at the time t_j'; each semicolon is read as 'and', while the bar '$|\ldots$' means 'under the condition that...' or 'provided that...'.

Let us consider simple examples taken from 2D NMR (Wefing and Spiess, 1988). The two-dimensional spectrum $S(\omega_2, t_m; \omega_1, 0)$ represents the probability of finding a segment with a frequency ω_1 at $t = 0$, and with the frequency ω_2 at $t = t_m$. This joint probability is the product of the probability that the segment is found at ω_1, the *a priori* probability $P(\omega_1)$, and the probability that the segment gets to ω_2 provided it has been at ω_1 a time t_m ago, which is the conditional probability $S(\omega_2, t_m \mid \omega_1, 0)$:

$$S(\omega_2, t_m; \omega_1, 0) = S(\omega_2, t_m \mid \omega_1, 0)P(\omega_1, 0). \tag{9.45}$$

9.3.1 Markov processes

In this section, we consider Markov processes (Oppenheim *et al.*, 1977), which are the simplest stochastic processes with finite correlation between successive events. The definition of a Markov process is the following relationship between conditional probabilities occurring in the process, written with increasing times from right to left ($t_j > t_{j-1}$):

$$W(x_n, t_n \mid x_{n-1}, t_{n-1}; x_{n-2}, t_{n-2}; x_{n-3}, t_{n-3}; \ldots; x_0, t_0) = W(x_n, t_n \mid x_{n-1}, t_{n-1}). \tag{9.46}$$

This states that the conditional probability at time t_n of finding the state x_n after an arbitrary succession ($t_j > t_{j-1}$) of preceding times depends only on the state at the last of these times. Note that this does not mean that the state at t_n is completely independent of states at earlier times, for instance of that at time t_{n-2}, as is to be shown now. The dependence on the state at t_{n-2} is in fact implicit in $W(x_n, t_n \mid x_{n-1}, t_{n-1})$, and $W(x_{n-1}, t_{n-1} \mid x_{n-2}, t_{n-2})$. A certain dependence of the state at t_n on that at t_{n-2} obviously exists, as equation (9.46) also implies that we have:

$$W(x_n, t_n \mid x_{n-2}, t_{n-2}; x_{n-3}, t_{n-3}; x_{n-4}, t_{n-4}; \ldots; x_0, t_0) = W(x_n, t_n \mid x_{n-2}, t_{n-2}). \quad (9.47)$$

The relationship between $W(x_n, t_n \mid x_{n-2}, t_{n-2})$ and $W(x_n, t_n \mid x_{n-1}, t_{n-1})$ in a Markov process is clear on the grounds of the basic relation (9.46): one obtains the probability $W(x_n, t_n \mid x_{n-2}, t_{n-2})$ of getting from state x_{n-2} at t_{n-2} to state x_n at t_n as the product of the probabilities of getting from (x_{n-2} at t_{n-2}) to (x_{n-1} at t_{n-1}) and from (x_{n-1} at t_{n-1}) to (x_n at t_n), integrated over all possible intermediate states x_{n-1} at fixed t_{n-1}:

$$W(x_n, t_n \mid x_{n-2}, t_{n-2}) = \int \mathrm{d}\tilde{x}\; W(x_n, t_n \mid \tilde{x}, t_{n-1}) W(\tilde{x}, t_{n-1} \mid x_{n-2}, t_{n-2}). \quad (9.48)$$

This is the Smoluchowski–Chapman–Kolmogorov equation (Oppenheim *et al.*, 1977).

For a Markov process, according to (9.46) the joint probability of having states x_j at times t_j can be factorized into two-time probabilities:

$$W(x_n, t_n; x_{n-1}, t_{n-1}; \ldots; x_1, t_1) = W(x_n, t_n \mid x_{n-1}, t_{n-1})$$
$$\times\; W(x_{n-1}, t_{n-1} \mid x_{n-2}, t_{n-2}) \ldots W(x_1, t_1 \mid x_0, t_0) P(x_0, t_0). \quad (9.49)$$

Markov processes are commonplace; examples include random walks and other diffusive processes. Non-Markovian behaviour, though more difficult to visualize, also occurs in nature. As an example, consider helical jumps of polymer chains in semicrystalline polymers (see Sections 7.12 and 7.14.2) with loop restraints taken into account, that is, the loops in the amorphous regions cannot be shortened below a certain length and are otherwise passive. In a crystallite, for a segment starting from position 0 at $t = 0$ the probability of finding it at the next-but-one position 2 after having detected it at the neighbouring position 1 is reduced due to possible strain in the shortened chain loop in the amorphous region. This represents a non-Markovian process since the probability of finding a segment at position 2 does not depend only on the probability of finding it at site 1 before and the transition probability from 1 to 2. It is slightly lower if the particle had come from site 0 than if it had started at site 2; this is obviously a non-Markovian influence.

To state this more formally, for a Markovian helical jump process it is required that:

$$W(2, t_2; 1, t_1; 0, t_0) = W(2, t_2 \mid 1, t_1) W(1, t_1 \mid 0, t_0) P(0, t_0)$$
$$= W(0, t_2 \mid 1, t_1) W(1, t_1 \mid 0, t_0) P(0, t_0) = W(0, t_2; 1, t_1; 0, t_0) \quad (9.50)$$

where we have used $W(n+1, t_2 \mid n, t_1) = W(n-1, t_2 \mid n, t_1)$ due to the equivalence of both directions along the chain. This means that jumps back to position 0 and jumps further to site 2 must be equally probable for a Markovian random walk of a chain segment. The process actually occurring is non-Markovian due to the preference of jumps back to position 0:

$$W(2, t_2; 1, t_1; 0, t_0) < W(0, t_2; 1, t_1; 0, t_0) \tag{9.51}$$

because of tightening of loops; this is in contradiction of the Markovian requirement (9.50).

Most natural Markov processes are stationary which means that only time differences play a role, such that for arbitrary t_{mb}:

$$W(\tilde{\theta}_2, t_2 \mid \tilde{\theta}_1, t_1) = W(\tilde{\theta}_2, t_2 + t_{\mathrm{mb}} \mid \tilde{\theta}_2, t_1 + t_{\mathrm{mb}}). \tag{9.52}$$

Often this is formulated by stating that 'no time origin is preferred', and the conditional probability is written as:

$$W(\tilde{\theta}_2, t_2 - t_1 \mid \tilde{\theta}_1) = W(\tilde{\theta}_2, t_2 \mid \tilde{\theta}_1, t_1). \tag{9.53}$$

In addition, for a stationary process the *a priori* probabilities are time-independent.

9.3.2 Information on stochastic processes from 3D exchange spectra

3D exchange spectra are three-time distributions:

$$S(\tilde{\omega}_3, t_{\mathrm{ma}} + t_{mb}; \tilde{\omega}_2, t_{\mathrm{ma}}; \tilde{\omega}_1, 0).$$

For the helical jumps with their discrete positions, a 3D exchange spectrum can directly yield three-time occupation-probability distributions like:

$$W(2, t_{\mathrm{c}}; 1, t_{\mathrm{b}}; 0, t_{\mathrm{a}}).$$

Thus, along the lines of the discussion in Section 7.10.2, one can detect non-Markovian behaviour (if it is sufficiently pronounced) as manifested by inequalities of the peak intensities along lines of ($\omega_1 = \text{const.}, \omega_2 = \text{const.}, \omega_3$).

In the context of 3D exchange NMR, we make the perhaps surprising observation that the dynamics in a system with a heterogeneous distribution of correlation times is a non-Markovian process, though made up of individual Markov processes. For such a reorientational process, the 3D orientation probability distribution of the diffusing axes violates the Markov decomposition condition (9.49) which should result in:

$$W(\tilde{\theta}_3, t_{\mathrm{ma}} + t_{\mathrm{mb}}; \tilde{\theta}_2, t_{\mathrm{ma}}; \tilde{\theta}_1, 0) \stackrel{!}{=} W(\tilde{\theta}_3, t_{\mathrm{ma}} + t_{\mathrm{mb}} \mid \tilde{\theta}_2, t_{\mathrm{ma}}) W(\tilde{\theta}_2, t_{\mathrm{ma}}; \tilde{\theta}_1, 0). \tag{9.54}$$

The overall 2D distributions are superpositions of the 2D distributions for the various correlation times τ:

$$W(\tilde{\theta}_3, t_{\mathrm{ma}} + t_{\mathrm{mb}} \mid \tilde{\theta}_2, t_{\mathrm{ma}}) = \int W(\tau; \tilde{\theta}_3, t_{\mathrm{ma}} + t_{\mathrm{mb}} \mid \tilde{\theta}_2, t_{\mathrm{ma}}) g(\tau)\, \mathrm{d}\tau \tag{9.55}$$

(note that τ, which had to be included in the list of arguments here, does not obey the rules given below equation (9.44)). The correct 3D distribution is a superposition of products like (9.54) for the different τ values:

$$\begin{aligned} W(\tilde{\theta}_3, t_{\mathrm{ma}} + t_{\mathrm{mb}}; \tilde{\theta}_2, t_{\mathrm{ma}}; \tilde{\theta}_1, 0) &= \int W(\tau; \tilde{\theta}_3, t_{\mathrm{ma}} + t_{\mathrm{mb}}; \tilde{\theta}_2, t_{\mathrm{ma}}; \tilde{\theta}_1, 0) g(\tau)\, \mathrm{d}\tau \\ &= \int W(\tau; \tilde{\theta}_3, t_{\mathrm{ma}} + t_{\mathrm{mb}} \mid \tilde{\theta}_2, t_{\mathrm{ma}}) W(\tau; \tilde{\theta}_2, t_{\mathrm{ma}}; \tilde{\theta}_1, 0) g(\tau)\, \mathrm{d}\tau. \end{aligned} \tag{9.56}$$

This integral of the products is indeed not equal to the product of the integrals:

$$\int W(\tau; \tilde{\theta}_3, t_{ma} + t_{mb} \mid \tilde{\theta}_2, t_{ma}) g(\tau)\, \mathrm{d}\tau \times \int W(\tau'; \tilde{\theta}_2, t_{ma}; \tilde{\theta}_1, 0) g(\tau')\, \mathrm{d}\tau'$$

which appears on the right-hand side of the expression (9.54). The invalidity of the Markovian condition (9.54) is exemplified by the differences between the spectra in Figures 9.4(a) and (b). The 3D spectrum in Figure 9.4(a) was simulated assuming the (overall) Markov condition (9.54) to hold, with the two dimensional $W(\tilde{\theta}_2, t_m; \tilde{\theta}_1, 0)$ angular distribution that would be obtained according to equation (9.55) from the same model and distribution of correlation times used to simulate the 3D spectrum in Figure 9.4(b). The model assumes isotropic rotational diffusion with a certain distribution of correlation times. For the spectrum of figure 9.4(b) 3D spectra for individual correlation times were summed up with weights given by the distribution of correlation times; this is exactly analogous to equation (9.56). These 3D spectra are definitely different, while the corresponding 2D spectra would be identical. This again demonstrates that the information content of a 3D exchange spectrum can be larger than that of the corresponding 2D spectrum.

Similarly, for inhomogeneously broadened spectra a larger information content of 3D spectra is revealed compared with 2D spectra: a 3D powder spectrum cannot be written as a product of 2D powder spectra $S(\omega_3; t_{mb}, \omega_2)$ and $S(\omega_2; t_{ma}, \omega_1)$ divided by the 1D powder spectrum $P(\omega_2)$:

$$S(\omega_3, t_{mb} + t_{ma}; \omega_2, t_{ma}; \omega_1, 0) \neq \overbrace{S(\omega_3, t_{mb}; \omega_2)/P(\omega_2)}^{S(\omega_3; t_{mb} \mid \omega_2)} \overbrace{S(\omega_2, t_{ma}; \omega_1)}^{S(\omega_2; t_{ma} \mid \omega_1) P(\omega_1)} \qquad (9.57)$$

as would be suggested by the decomposition relation (9.49) for probabilities in a Markovian process and relation (9.45) between joint, conditional and *a priori* probabilities. The inequality (9.57) becomes obvious when one considers the overall

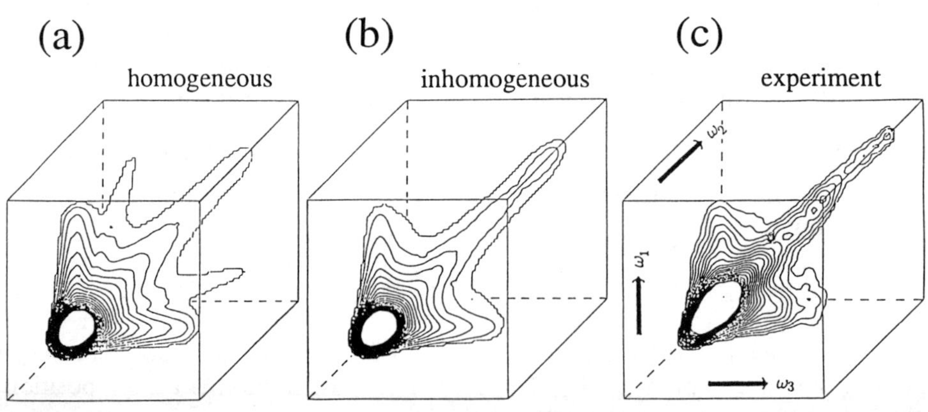

Figure 9.4 3D exchange spectra for a given reorientation-angle distribution originating from an isotropic diffusive process and consistent with the corresponding 2D exchange spectrum. (a) Markov process (homogeneous). (b) Isotropic rotational diffusion with a distribution of correlation times (inhomogeneous). (c) Experimental spectrum for PVAc 11 K above T_g (see Section 8.6); with its strong space-diagonal intensity, it resembles the simulated spectrum in (b).

intensity distribution of the two sides of the expression (9.57) for the simple case of two-site jumps. For the 3D spectrum on the left-hand side of relation (9.57), the spectral intensity was shown to be confined to three diagonal planes (see Section 7.10); this means that the intensity vanishes unless (at least) two frequency coordinates are equal ($\omega_m = \omega_n$). In contrast, for most values $\omega_1 \neq \omega_2$ and most values $\omega_3 \neq \omega_2$, the 2D spectra in the product on the right-hand side of (9.57) have nonvanishing intensity, including regions with $\omega_3 \neq \omega_1$, where consequently all three frequency coordinates are different. Thus, even for simple stationary Markov processes, the powder spectra do not exhibit the Markovian property, which would require an equality in relation (9.57). Nonetheless, considering certain peak ratios in a 3D spectrum, possible non-Markovian reorientation behaviour can be observed. This also means that it is not possible to identify molecular reorientation as Markovian or non-Markovian by inserting the spectral intensities of a series of 2D powder spectra with different mixing times into the Smoluchowski–Chapman–Kolmogorov equation (9.48).

Note that if there is a single line for each site and all lines are resolved, as may be the case in solution NMR or in single crystals, the spectral peak intensities directly reflect the conditional probabilities of transitions between sites and therefore relation (9.57) holds as an equality (if the process is Markovian).

9.3.3 Analysis of the reduced 4D experiment

The signal in the reduced 4D experiment can be written as

$$\iint d\omega_1\, d\omega_2 \cos(\omega_1 - \omega_2)t_1 S(\omega_4, t_{ma} + t_{mb} + t_{mc}; \omega_3, t_{ma} + t_{mb}; \omega_2, t_{ma}; \omega_1, 0). \quad (9.58)$$

For molecules with $\omega_1 = \omega_2$, we have $\cos(\omega_1 - \omega_2)t_1 = 1$, resulting in:

$$\iint d\omega_1\, d\omega_2\, S(\omega_4, t_{ma} + t_{mb} + t_{mc}; \omega_3, t_{ma} + t_{mb}; \omega_2, t_{ma}; \omega_1, 0) = S(\omega_4, t_{mc}; \omega_3). \quad (9.59)$$

Thus, the undistorted 2D spectrum of these segments results.

On the other hand, for fast segments (i.e. those with $t_{ma} > \tau$) and sufficiently large t_1, the $\cos(\omega_1 - \omega_2)t_1$ term oscillates fast as a function ω_1 and ω_2 compared with the variation of the full-exchange spectra $S(\omega_1, \omega_2; t_{ma}/\tau)$ and $S(\omega_3, \omega_2; t_{mb}/\tau)$. As a consequence, the positive and negative lobes of the cosine cancel approximately such that the integral (9.58) nearly vanishes.

Therefore, the behaviour of certain clusters of segments which all have similar correlation times can be considered. The reduced 4D spectrum in Figure 8.10(c) shows that the behaviour of the corresponding segments is not stationary. For instance, the probability of finding a segment at its start frequency, or orientation, decreases with the overall time t_{mb} after the selection pulse sequence:

$$W(\tilde{\theta}, t_{mc} + t_{ma} + 0 \mid \tilde{\theta}, t_{ma} + 0) > W(\tilde{\theta}, t_{mc} + t_{ma} + \tilde{t}_{mb} \mid \tilde{\theta}, t_{ma} + \tilde{t}_{mb}) \quad (9.60)$$

where we compare the probabilities with $t_{mb} \cong 0$ and a long $t_{mb} = \tilde{t}_{mb}$. Relation (9.60) obviously violates the condition of 'stationarity' (9.52). Thus, the reduced 4D experiment can detect a nonstationary process of certain sub-ensembles even in a quasi-equilibrium system. For the sample as a whole, the process is nevertheless stationary (Sillescu, 1971).

The information content in a series of 2D exchange spectra was shown to be larger than that of the corresponding simple time autocorrelation function, and equivalent to an orientational space–time autocorrelation function. The 3D spectrum, being a three-time distribution, contains information that cannot be directly retrieved from the 2D spectrum. The detailed information on the nature of the nonexponential loss of correlation obtained from the reduced 4D exchange spectrum is also not accessible in the corresponding 2D exchange spectrum. This shows that 3D and 4D spectra contain information that cannot be measured by techniques probing simple or space–time correlation functions.

9.4 The exchange matrix

Jump processes between discrete sites represent an important class of stochastic processes and can be treated with relative ease. Therefore, even continuous (e.g. diffusive) stochastic processes are made tractable by discretization with a sufficiently large number of sites. For a stationary Markov process, the time dependence of the occupation probabilities W_n of the sites $n = 1, \ldots, N$ by a particle having started from site 1 at $t = 0$ can be described by a simple master equation (Oppenheim *et al.*, 1977):

$$\frac{\mathrm{d}W_n}{\mathrm{d}t} = \sum_{m=1}^{N} \Pi_{nm} W_m \quad \text{or} \quad \frac{\mathrm{d}\vec{\mathbf{W}}}{\mathrm{d}t} = \mathbf{\Pi}\vec{\mathbf{W}}. \tag{9.61}$$

The probabilities $W_m(t)$ are recognized as the conditional probabilities $W(m, t \mid 1, 0)$ of finding the particle on site m at time t if it was at site 1 at time 0. On such a basis, the relation (9.61) can be extended into a matrix relation involving any site k as the start position:

$$\frac{\mathrm{d}W(n, t \mid k, 0)}{\mathrm{d}t} = \sum_{m=1}^{N} \Pi_{nm} W(m, t \mid k, 0) \quad \text{or} \quad \frac{\mathrm{d}\mathbf{W}}{\mathrm{d}t} = \mathbf{\Pi}\mathbf{W}. \tag{9.62}$$

The elements of the 'exchange matrix' Π_{nm} for $m \neq n$ are the probabilities of transitions from m to n (the fraction Π_{nm} of W_m is transferred to W_n per unit time). Note that the diagonal element Π_{mm} is not the 'transition probability from m to m'; it cannot be the probability of remaining at m because $\mathbf{\Pi}\vec{\mathbf{W}}$ is the *change* of $\vec{\mathbf{W}}$ with time according to equation (9.61). Rather, Π_{mm} is always $\leqslant 0$, and its absolute value is the total probability of transitions from site m to anywhere else. Therefore, $-\Pi_{mm} = \Sigma_{n(\neq m)} \Pi_{nm}$, which is equivalent to:

$$\sum_{n=1}^{N} \Pi_{nm} = 0. \tag{9.63}$$

For the solution of simple physical problems of the form (9.61) it is customary to make the ansatz:

$$\vec{\mathbf{W}}(t) = \mathrm{e}^{\pi_j t} \vec{\mathbf{r}}_j \tag{9.64}$$

transforming (9.61) into a principle-value problem (see Appendix D.2):

$$\pi_j \vec{\mathbf{r}}_j = \mathbf{\Pi} \vec{\mathbf{r}}_j \tag{9.65}$$

which can be solved if one finds a matrix **R** that diagonalizes **Π** (see Appendix D.3):

$$\pi_j \delta_{jk} = (\mathbf{R}^{-1}\mathbf{\Pi}\mathbf{R})_{jk}. \tag{9.66}$$

The rates π_j are the eigenvalues of **Π**, the $\vec{\mathbf{r}}_j$ are column vectors of the diagonalizing matrix **R**. **R** certainly exists if **Π** is symmetric. The final solution is found by adding solutions (9.64) for different j values with weighting factors appropriate to fulfil the initial conditions.

A similar but more elegant and more widely applicable approach, which includes the initial conditions automatically, consists in writing the formal solution of the master equation (9.61):

$$\vec{\mathbf{W}}(t) = \mathrm{e}^{\mathbf{\Pi}t}\vec{\mathbf{W}}(0). \tag{9.67}$$

As shown in Appendix D.4, the exponential matrix expression in (9.67) can be evaluated if a matrix **R** can be found that diagonalizes **Π**, $(\mathbf{R}^{-1}\mathbf{\Pi}\mathbf{R})_{jk} = \delta_{jk}\pi_j$, with the eigenvalues π_j of **Π**. According to equation (D.10), this matrix also diagonalizes $\mathrm{e}^{\mathbf{\Pi}t}$:

$$(\mathbf{R}^{-1}\,\mathrm{e}^{\mathbf{\Pi}t}\,\mathbf{R})_{jk} = \delta_{jk}\,\mathrm{e}^{\pi_j t} \tag{9.68}$$

so that all matrices are known for evaluating:

$$\vec{\mathbf{W}}(t) = \mathbf{R}(\mathbf{R}^{-1}\,\mathrm{e}^{\mathbf{\Pi}t}\,\mathbf{R})\mathbf{R}^{-1}\vec{\mathbf{W}}(0). \tag{9.69}$$

In order to diagonalize **Π**, it is helpful to transform it into a symmetric matrix on the basis of the principle of detailed balance; this will be explained in Section 10.3.

Since the probabilities $W_j(t)$ must be $\leqslant 1$ and usually decay to zero for long times, we expect the exponents π_j of $\mathrm{e}^{\pi_j t}$, which appear in the solution (9.69) due to (9.68), to be negative or null. The eigenvalue $\pi_1 = 0$ is always present since the time-invariance of the final state $\vec{\mathbf{W}}^\infty := \vec{\mathbf{W}}(t \to \infty)$ implies:

$$0 = \frac{\mathrm{d}\vec{\mathbf{W}}^\infty}{\mathrm{d}t} = \mathbf{\Pi}\vec{\mathbf{W}}^\infty \Rightarrow \mathbf{\Pi}\vec{\mathbf{W}}^\infty = \mathbf{0}\cdot\vec{\mathbf{W}}^\infty. \tag{9.70}$$

The exchange processes between isochromats that are observed in 2D exchange NMR can also be described directly in terms of an exchange-matrix master equation:

$$\frac{\mathrm{d}S(\omega_2, t_\mathrm{m} \mid \omega_1, 0)}{\mathrm{d}t_\mathrm{m}} = \sum_{\omega=1}^{N} \Pi_{\omega_2,\omega} S(\omega, t_\mathrm{m} \mid \omega_1, 0) \tag{9.71}$$

where exchange between N discrete frequencies is assumed, as in an N-site jump in a single crystallite. The t_m dependence of the exchange process is given by the solution of the master equation (9.71):

$$\mathbf{S}(t_\mathrm{m}) = \mathrm{e}^{\mathbf{\Pi}t_\mathrm{m}}\,\mathbf{S}(0) \tag{9.72}$$

which can be evaluated as described above. In the approach described here, the powder average over all crystallites is performed by adding the solutions (9.72) for all possible crystallite orientations.

For isotropic motional processes, the exchange matrix is often available in terms of θ (or $\cos\theta$) rather than ω (examples are given in Section 11.4). Then, it is more convenient to express all matrices in a space of discrete angles $\theta_1, \theta_2, \ldots, \theta_N$ to calculate (and later Fourier transform) 2D time signals like:

$$s(t_1, t_2; t_m) = \langle e^{i\tilde{\omega}_1 t_1} e^{i\tilde{\omega}_2 t_2} \rangle = \vec{1}\, e^{i\omega t_1}\, e^{\Pi t_m}\, e^{i\omega t_2}\, \mathbf{P}\vec{1} \tag{9.73}$$

where $\boldsymbol{\omega}$ is a diagonal matrix of frequencies, $\omega_{\theta_1, \theta_2} = \omega(\theta_1)\, \delta_{\theta_1, \theta_2}$, $\vec{1} = (1, 1, 1, \ldots, 1)$, and $\mathbf{P}$ is a diagonal matrix of *a priori* probabilities (see also Sections 2.4.7 and 10.3).

9.5 References

Davidson, D. W. and Cole, R. H. (1950). Dielectric Relaxation in Glycerine. *J. Chem. Phys.*, **18**, 1417.

Connors, T. M. (1964). Distributions of Correlation Times, and Relaxation. *J. Chem. Soc. Trans. Faraday Soc.*, **60**, 1574.

Cook, M., Watts, D. C. and Williams, G. (1970). Correlation Function Approach to Dielectric Behavior of Polymers. *J. Chem. Soc. Trans. Faraday Soc.*, **66**, 2503.

Frauenfelder, H. (1984). From Atoms to Biomolecules. *Helv. Phys. Acta*, **57**, 165.

Fujara, F., Wefing, S. and Spiess, H. W. (1986). Analogies Between Quasielastic n-Scattering and ^{2}H NMR Spin Alignment. *J. Chem. Phys.*, **84**, 4579.

Fujara, F., Petry, W., Schnauss, W. and Sillescu, H. (1988). Reorientation of Benzene in its Crystals: Analogy Between NMR and *n*-Scattering. *J. Chem. Phys.*, **89**, 1801.

Landau, L. D. and Lifshitz, E. M. (1980). *Theoretical Physics V. Statistical Physics, Part 1.* Pergamon Press, Oxford.

Lindsey, C. P. and Patterson, G. D. (1980). Comparison of KWW and Cole–Davidson Functions. *J. Chem. Phys.*, **73**, 3348.

Lovesey, S. W. (1984). *Theory of Neutron Scattering from Condensed Matter.* Clarendon Press, Oxford.

Martynov, G. A. (1992). *Fundamental Theory of Liquids. Method of Distribution Functions.* Adam Hilger, Bristol.

Oppenheim, I., Shuler, K. E. and Weiss, G. H. (1977). *Stochastic Processes in Chemical Physics: The Master Equation.* MIT Press, Cambridge, MA.

Rössler, E. (1986). 2D Exchange NMR Analyzed in the Time Domain. *Chem. Phys. Lett.*, **128**, 330.

Rössler, E., Taupitz, M., Börner, K., Schulz, M and Vieth, H. M. (1990). ^{2}H NMR: $g(E_a)$ of Guest Molecules in Disordered Systems. *J. Chem. Phys.*, **92**, 5847.

Sillescu, H. (1971). Theory of Molecular Reorientations: Magnetic Resonance Line Shapes. *J. Chem. Phys.*, **54**, 2110.

Spiess, H. W. (1980). ^{2}H Spin Alignment for Studying Ultraslow Motions in Solids. *J. Chem. Phys.*, **72**, 6755.

Spiess, H. W. (1991). 2D NMR: Elucidation of Molecular Dynamics in Complex Systems. *J. Non-Cryst. Solids*, **131–133**, 766.

Springer, T. (1972). *Quasielastic Neutron Scattering: Diffusive Motions in Solids and Liquids.* Springer, Berlin.

Van Kampen, N. G. (1981). *Stochastic Processes in Physics and Chemistry.* North-Holland, Amsterdam.

Wefing, S. and Spiess, H. W. (1988). 2D Exchange NMR of Powders. I. Two-Time Distributions. *J. Chem. Phys.*, **89**, 1219.

CHAPTER TEN

Time-Domain Signals for Multidimensional Spectra

Keywords: *pulse sequences for* $\mathrm{I}=\frac{1}{2}$ *and* $\mathrm{I}=1$, *stimulated echoes,* T_1 *relaxation effects; intermediate exchange regime; MAS phase problems, time reversal; practical hints for 2D spectroscopy.*

10.1 Introduction

In the preceding chapters, multidimensional exchange spectroscopy was discussed mainly from the viewpoint of the final spectra. Technical details on the acquisition of data in the time domain and the subsequent Fourier transformation were restricted to a minimum in order to keep the treatment concise. In this chapter, we want to fill that gap and discuss features of the two-dimensional time-domain signals and the data processing for generating the 2D spectra. These points are of interest mainly for the NMR spectroscopist who wants to perform 2D experiments or understand their details.

10.2 Pulse sequences for exchange spectroscopy

The basic pulse sequence of exchange spectroscopy consists of three pulses: the first is needed in order to obtain transverse magnetization; the second terminates the evolution time t_1 and defines the start of the mixing time. The third pulse terminates the mixing time by again creating transverse magnetization, or coherences that evolve into transverse magnetization. This three-pulse sequence can be extended by a pulse generating an echo before the actual detection in order to overcome the various dead-time problems. For mitigating effects of finite pulse lengths, it can also be useful to generate a spin echo at the start of the evolution period; this is particularly crucial in ^{2}H NMR, where the time signal decays fast due to the large width of the ^{2}H quadrupolar powder patterns.

We shall start the discussion with the simpler case of ^{13}C ($S=\frac{1}{2}$) and then comment on aspects of ^{2}H 2D exchange NMR in some detail.

10.2.1 Amplitude modulation for $S=\frac{1}{2}$

The principle of amplitude modulation of the detected signal by $\cos\tilde{\omega}_1 t_1$ or $\sin\tilde{\omega}_1 t_1$ can be most easily visualized for precessing ($S=\frac{1}{2}$) nuclei, see Figure 10.1. We consider one isochromat, precessing with a certain frequency $\tilde{\omega}_1$ during the evolution time. Only one component, e.g. M_x or M_y, of this magnetization can be stored along the z axis during

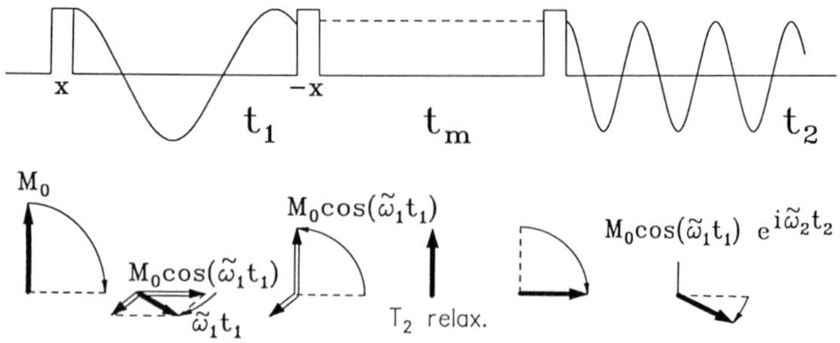

Figure 10.1 Pulse sequence, magnetization evolution, and schematic time signal for the 2D exchange experiment for $I = \frac{1}{2}$ nuclei. Of any precessing isochromat $M_0 \exp(i\tilde{\omega}_1 t_1)$ only the projection on one axis, $M_0 \cos \tilde{\omega}_1 t_1$ in the figure, is retained along the z direction during the mixing time. $M_0 \cos \tilde{\omega}_1 t_1$ is therefore the length of the isochromat vector after t_m, which then evolves as $\exp(i\tilde{\omega}_2 t_2)$.

the mixing time. If the excitation of the initial length M_0 of the isochromat is effected by an x pulse and the storage by a $-x$ pulse, the projection on the y axis at the end of the evolution period, $M_y = M_0 \cos \tilde{\omega}_1 t_1$, is stored along z during t_m, see Figure 10.1. The remaining transverse x component is quickly dephased by dipolar couplings, so that the length of the magnetization vector at the end of t_m is $M_0 \cos \tilde{\omega}_1 t_1$. After having been flipped from the z axis into the xy plane, it evolves as:

$$s^c(t_1, t_2) = \underbrace{(M_0 \cos \tilde{\omega}_1 t_1)}_{\text{amplitude}} e^{i\tilde{\omega}_2 t_2}. \tag{10.1}$$

If the second pulse is 90° out of phase with the first, that is, a y pulse in our case, the component $M_0 \sin \tilde{\omega}_1 t_1$ is stored, and as a function of t_1 and t_2, we obtain the signal of the 'sine dataset':

$$s^s(t_1, t_2) = M_0 \sin \tilde{\omega}_1 t_1 \, e^{i\tilde{\omega}_2 t_2}. \tag{10.2}$$

If the dipolar couplings are weak and if the T_2 time constant of (irreversible) relaxation is very long, the transverse magnetization may not have decayed within t_m. In this case, it can be removed by a phase cycle of pulse and receiver phases. For instance, if the sign of the second pulse is alternated, the undesired magnetization component that remains transverse during t_m is unaffected while the sign of the resulting z magnetization is inverted in alternate scans. If the detected signals are subtracted by also alternating the receiver phase by 180°, the undesired component is cancelled while the component that is along z during t_m adds up. (This phase cycle also mitigates T_1 relaxation effects, as will be discussed in Section 10.2.5.)

10.2.2 Pulse sequences for 2D exchange ^{13}C NMR

Figure 10.2 displays the simplest version of the pulse sequence actually employed for 2D exchange ^{13}C NMR in organic solids (Szeverenyi *et al.*, 1982). Dipolar decoupling (DD) of protons is usually indispensable during the evolution and detection periods, when the ^{13}C magnetization precesses in the xy plane. During the long mixing time, no dipolar decoupling is applied in general. (Only in certain MAS exchange

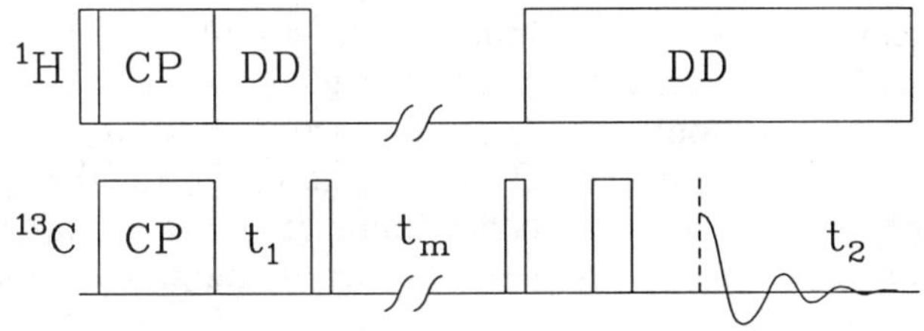

Figure 10.2 Pulse sequence actually used in ^{13}C 2D exchange NMR with cross polarization (CP), heteronuclear dipolar decoupling (DD), and a 180° pulse for generating a Hahn spin echo before detection.

experiments would it be desirable to suppress the ^{1}H–^{13}C couplings during t_m in order to quench ^{13}C spin diffusion, see Section 13.5; this is, however, reluctantly implemented because continuous high-power rf irradiation for a few seconds can ruin the equipment.) Hartmann–Hahn cross polarization (CP) with irradiation for about 2 ms is used in most experiments to obtain optimum signal strength. Nevertheless, sometimes single-pulse excitation of ^{13}C is also employed for selectively observing methyl groups or other fast-relaxing carbon sites.

For most actual experiments, a Hahn spin echo is generated before detection by means of a 180° pulse applied approximately 50 μs after the 90° pulse terminating the mixing time, (Figure 10.2). This permits dead-time free detection. For broad spectra (width > 10 kHz), phase distortions in the 2D plane have been found to be reduced by also generating a Hahn echo before the start of the evolution, with a pre-echo delay of less than 10 μs.

In some instances, it may be desirable to observe the unprotonated carbons selectively; the suppression of the protonated species is achieved by gated decoupling, that is, by switching the dipolar decoupling off for a certain time (Section 3.2).

With ^{13}C spins one usually has several options for solving the phase problem, namely to measure a single cosine dataset off-resonance, to measure a cosine-and-sine (hypercomplex) dataset, or to generate one dataset with TPPI in t_1 (see Section 4.6.1). The off-resonance technique has proven convenient in practice, involving only one pulse program and one dataset. However, in high fields and with very large chemical-shift anisotropies, of C≡N groups for instance, excitation problems may become so severe that measuring off resonance is not possible any more. Then, TPPI or the hypercomplex acquisition should be applied on resonance. Both approaches (TPPI and hypercomplex sampling) are nearly identical in practice; the only difference consists in the fact that t_1 is equal for the $\cos \tilde{\omega}_1 t_1$ and $\sin \tilde{\omega}_1 t_1$ hypercomplex datapoints, while t_1 is incremented between sine and cosine datapoints in the TPPI approach. For obtaining equivalent spectra in ω_1, the time between cosine (or between sine) data points must be chosen identical in both sampling modes (this is twice the dwell time t_{dw1} of the TPPI experiment).

10.2.3 Stimulated echoes in the 2D time signals

The most characteristic feature of 2D exchange time-domain data corresponding to inhomogeneously broadened spectra are the stimulated echoes which occur at $t_2 = t_1$. Note that the occurrence of these echoes must not prompt the spectroscopist to shift the start of data acquisition to the maximum of the stimulated echo at $t_2 = t_1$. Rather, the stimulated echo is just the maximum of the signal $\langle \cos \tilde{\omega} t_1 \exp(\mathrm{i}\omega t_2) \rangle$ for a fixed t_1 value, and this signal should be detected starting from $t_2 = 0$. The expression $\langle \cos \omega t_1 \exp(\mathrm{i}\omega t_2) \rangle$ derives from the 2D time signal (10.1) for the special case of no exchange, $\tilde{\omega}_1 = \tilde{\omega}_2 =: \omega$. The echo with its maximum at $t_2 = t_1 = t_e$ can be derived on the basis of the decomposition of the cosine into a pair of complex exponentials, equation (A.21):

$$\begin{aligned} \langle \cos \omega t_e \, \mathrm{e}^{\mathrm{i}\omega t_e} \rangle &= \langle \tfrac{1}{2} \{ \mathrm{e}^{-\mathrm{i}\omega t_e} + \mathrm{e}^{\mathrm{i}\omega t_e} \} \, \mathrm{e}^{\mathrm{i}\omega t_e} \rangle \\ &= \langle \tfrac{1}{2} \{ 1 + \mathrm{e}^{\mathrm{i}\omega 2 t_e} \} \rangle . \end{aligned} \tag{10.3}$$

The '1' represents the echo maximum, where independent of ω all isochromats contribute equally to the signal. Considering the prefactor of $\frac{1}{2}$, we find that the stimulated echo has only 50% of the intensity of the corresponding FID or Hahn echo (Figure 10.3). The term $\langle \frac{1}{2} \exp(\mathrm{i}\omega(t_1 + t_2)) \rangle$ represents the contribution of the 'virtual echo' with an (unobservable) maximum at $t_2 = -t_1$. Equivalently, $\langle \frac{1}{2} \exp(\mathrm{i}\omega(t_1 + t_2)) \rangle$ can also be viewed as the left-hand side tail of the stimulated echo, reflected back at $t_2 = 0$.

Stimulated echoes occur also for $\langle \cos \omega t_1 \cos \omega t_2 \rangle$ and $\langle \sin \omega t_1 \sin \omega t_2 \rangle$, due to:

$$\begin{aligned} \langle \cos \omega t_e \cos \omega t_e \rangle &= \tfrac{1}{2} \langle 1 + \cos(\omega 2 t_e) \rangle \\ \langle \sin \omega t_e \sin \omega t_e \rangle &= \tfrac{1}{2} \langle 1 - \cos(\omega 2 t_e) \rangle \end{aligned} \tag{10.4}$$

(see equations (A.7) and (A.8)).

Since the echo height $s(t_e, t_e)$ is related to the diagonal fraction of the spectrum, its t_e-dependence can be used to probe the decrease of the spectral diagonal in favour of exchange intensity. The observation of the decay of the stimulated echo can actually be exploited in practice to find the appropriate mixing time and temperature for an exchange experiment, in particular for studies of dynamics above T_g where the exchange intensity is strong. Contributions of T_2 relaxation to the echo decay can be corrected by comparing with the corresponding signals without exchange (short t_m). T_1 corrections are possible by referencing the measured echo heights to the $t_e = 0$ signal, which is free of exchange effects and thus selectively measures the T_1 decay during t_m. In addition to this qualitative time-domain testing for exchange in the 2D spectrum one can actually extract some quantitative information from the echo decay (Spiess, 1980; Fujara *et al.*, 1986; Rössler *et al.*, 1986) (see also Section 9.1.3). Concerning its strong

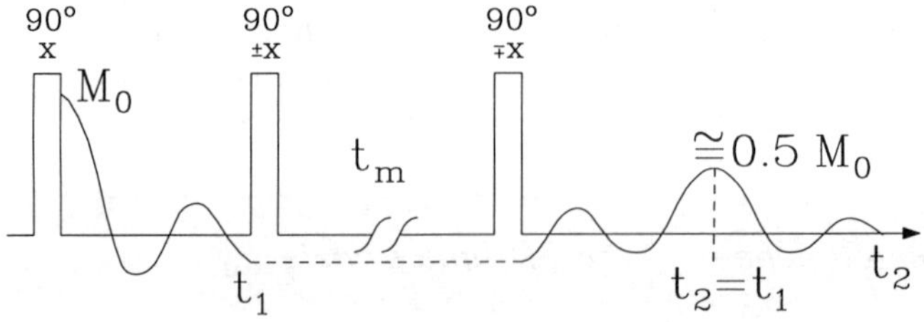

Figure 10.3 Stimulated echo occurring at $t_2 = t_1$ in a three-pulse sequence from z magnetization during t_m. Even without frequency changes during t_m, in the limit of large t_1 the echo height is only 50% of the height of the single-pulse time signal.

points, it is somewhat complementary to the full 2D spectrum. In particular, the echo decay is sensitive to small-angle motions: imagine a slow, very restricted (to $\pm 2^\circ$) rotational diffusion which broadens the diagonal of the spectrum by about 50% (this value depends of course on the homogeneous linebroadening); the exchange-induced spectral broadening would probably remain unnoticed, while in the time domain after sufficiently long times t_1 and t_2 it reduces the echo amplitude by a large, well detectable factor.

In some experiments, it is of interest to rationalize the actual relationship between the decay of the stimulated echo as a function of t_1 and the intensity distribution in the 2D spectrum. As was discussed above, the time signal $\langle \exp(-i\omega t_1)\exp(i\omega t_2)\rangle$ with its stimulated echoes represents the inverse Fourier transform of a 'diagonal' spectrum (which is actually located on the antidiagonal, where $(-\omega_1) = \omega_2$, but has the same shape as the diagonal spectrum obtained from the time signal $\langle \exp(i\omega t_1)\exp(i\omega t_2)\rangle$ which would have 'stimulated echoes' at $t_2 = -t_1$). In the general case of possibly unequal frequencies $\tilde{\omega}_1$ and $\tilde{\omega}_2$, the 'echo intensity' $s(t_e, t_e) = \langle \exp(-i\tilde{\omega}_1 t_e)\exp(i\tilde{\omega}_2 t_e)\rangle$ is related to the spectrum $S(\omega_1, \omega_2)$ by:

$$s(t_e, t_e) = w_{t_e}(t_1, t_2)\, s(t_1, t_2) \overset{FT}{\leftrightarrow} W_{t_e}(\omega_1, \omega_2) * S(\omega_1, \omega_2)\,\frac{1}{4\pi^2} \tag{10.5}$$

according to the convolution theorem (4.10), where the 'echo-picking' function $w_{t_e}(t_1, t_2)$ and its Fourier transform $W_{t_e}(\omega_1, \omega_2)$ are given by:

$$w_{t_e}(t_1, t_2) = \delta(t_1 - t_e)\,\delta(t_2 - t_e) \overset{FT}{\leftrightarrow} W_{t_e}(\omega_1, \omega_2) = e^{-i\omega_1 t_e}\, e^{-i\omega_2 t_e}. \tag{10.6}$$

$W_{t_e}(\omega_1, \omega_2)$ is obviously a 'plane wave':

$$W_{t_e}(\omega_1, \omega_2) = e^{-i(t_e, t_e)(\omega_1, \omega_2)} \tag{10.7}$$

with wavelength $1/t_e$ and 'crests' of an amplitude of unity parallel to the antidiagonal $(-\omega, \omega)$. One crest is always on the antidiagonal itself, for all t_e values. Thus, for a purely (anti-)diagonal spectrum, this crest of height 1 always multiplies the spectrum in the convolution integral (10.5), which is therefore independent of the parameter t_e. Off-diagonal exchange intensity, on the other hand, contributes less in the convolution integral in equation (10.5) when the wavelength $1/t_e$ becomes smaller. This corresponds to the situation described above, where a fast decay of the echo intensity with increasing $t_1 = t_e$ was seen to indicate strong exchange intensity in the spectrum.

10.2.4 2D exchange pulse sequence for ^{2}H quadrupolar coupling

For static exchange spectra of ^{2}H with its spin $I = 1$, the evolution of the density operator under the quadrupolar interaction is more complicated than the precession of the spin-$\frac{1}{2}$ magnetization. Nevertheless, how a signal:

$$\cos(\tilde{\omega}_1 t_1)\cos(\tilde{\omega}_2 t_2) \tag{10.8}$$

can be obtained is quite simply derived from the one-pulse evolution of the density operator as described in Section 2.5.5. The signal (10.8) can actually be generated by means of the usual three-pulse sequence of x, $-x$, x 90° pulses. Its effect can be understood fully in terms of the detectable magnetization of a given isochromat: the y magnetization obtained by the x pulse on Boltzmann z magnetization M_0 is modulated by $\cos(\omega_{Q,1} t_1)$ according to equation (2.89). This magnetization of length

$M_0 \cos(\omega_{Q,1} t_1)$ is flipped to the z axis by the $-x$ pulse. After t_m it is rotated back onto the y axis to evolve as $\cos(\omega_{Q,2} t_2)$. Thus, the detected magnetization is $M_0 \cos(\omega_{Q,1} t_1) \cos(\omega_{Q,2} t_2)$ and produces the cosine-signal (10.8). For general pulse flip angles χ_1, χ_2 and χ_3, only the projections $\sin \chi_k$ of the magnetization are retained or detected (e.g. $\sin \chi_k = 1$ for $\chi_k = 90^\circ$), so that the detected signal is proportional to:

$$M_0 \sin \chi_1 \sin \chi_2 \sin \chi_3 \cos(\omega_{Q,1} t_1) \cos(\omega_{Q,2} t_2). \tag{10.9}$$

In addition to the cosine signal, the signal:

$$\sin(\tilde{\omega}_1 t_1) \sin(\tilde{\omega}_2 t_2) \tag{10.10}$$

is required for quadrature (i.e. sign discrimination) in ω_1. In order to obtain a signal modulated by $\sin(\omega_{Q,1} t)$, the two first pulses will have to be 90° out of phase (Spiess, 1980; Schmidt *et al.*, 1986, 1988). For calculating the effects of the pulse sequence, we will rely heavily on Tables C.2 and C.3 which contain the effects of the quadrupolar coupling and of pulses on every possible term in the density operator for $I = 1$. $\hat{H}_Q$ acting on initial y magnetization produces:

$$\hat{I}_y \xrightarrow{e^{-i\hat{H}_Q t_1}} \hat{I}_y \cos(\omega_{Q,1} t_1) + (\hat{T}_{21} - \hat{T}_{2-1}) \sin(\omega_{Q,1} t_1). \tag{10.11}$$

Subject to a y pulse, the first term, representing magnetization along the $\vec{\mathbf{B}}_1$ field, remains unchanged along the y axis and decays in the ensuing long mixing time due to dipolar couplings of the ^{2}H spins or by T_2 relaxation. A y pulse of flip angle χ_2 transforms the remaining term in the density matrix (10.11) according to:

$$(\hat{T}_{21} - \hat{T}_{2-1}) \sin(\omega_{Q,1} t_1) \xrightarrow{e^{i\chi_2 \hat{I}_y}} -\sqrt{\tfrac{3}{2}}\hat{T}_{20} \sin(2\chi_2) + (\hat{T}_{21} - \hat{T}_{2-1})a + (\hat{T}_{22} + \hat{T}_{2-2})b. \tag{10.12}$$

The coefficients a and b, which can be read from Table C.3, are of no further relevance since their $\hat{T}$ terms, corresponding to off-diagonal density-matrix elements (see Section 3.12), relax with T_2-like time constants. By the x pulse of flip angle χ_3 after the mixing time, the surviving $\hat{T}_{20}$ term is transformed according to:

$$\hat{T}_{20} \xrightarrow{e^{i\chi_3 \hat{I}_x}} i\sqrt{\tfrac{3}{8}}(\hat{T}_{21} + \hat{T}_{2-1}) \sin(2\chi_3) + \hat{T}_{20}\tfrac{1}{2}(3\cos^2\chi_3 - 1) - \sqrt{\tfrac{3}{8}}(\hat{T}_{22} + \hat{T}_{2-2}) \sin^2\chi_3. \tag{10.13}$$

Under the quadrupole coupling, the terms $\hat{T}_{20}$ and $(\hat{T}_{22} + \hat{T}_{2-2})$ are invariant so that they do not transform into detectable magnetization and need not be considered for the NMR signal. Detectable magnetization arises only from:

$$(\hat{T}_{21} + \hat{T}_{2-1}) \xrightarrow{e^{-i\hat{H}_Q t_2}} (\hat{T}_{21} + \hat{T}_{2-1}) \cos(\omega_{Q,2} t_2) + i\hat{I}_x \sin(\omega_{Q,2} t_2). \tag{10.14}$$

The final magnetization is thus $\hat{I}_x$, that is, along the flip axis of the third pulse, and the signal is proportional to:

$$M_0 \sin \chi_1 \sin(\omega_{Q,1} t_1)\{-\sqrt{\tfrac{3}{2}} \sin(2\chi_2)\} i^2 \sqrt{\tfrac{3}{8}} \sin(2\chi_3) \sin(\omega_{Q,2} t_2)$$
$$= M_0 \tfrac{3}{4} \sin \chi_1 \sin(2\chi_2) \sin(2\chi_3) \sin(\omega_{Q,1} t_1) \sin(\omega_{Q,2} t_2) \tag{10.15}$$

where we have also included the factor $M_0 \sin \chi_1$ for the effect of the flip angle χ_1 of the first pulse.

To overcome the receiver dead time problem, echo detection of the signal after the mixing time is required, which extends the sequence by the echo pulse; this yields the four-pulse sequence shown in Figure 10.4. For the cosine signal, the standard quadrupole echo can be generated by a fourth pulse that is 90° out of phase with the

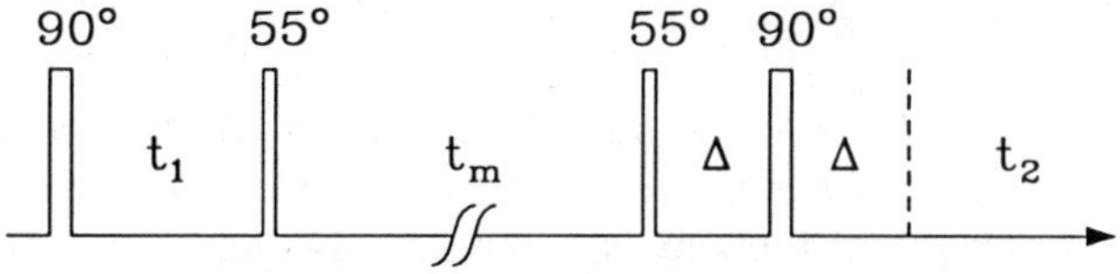

Figure 10.4 Four-pulse sequence for ^2H 2D exchange NMR. The fourth pulse serves to generate an echo for deadtime-free detection.

third pulse. For the sine signal (10.15), the problem is more involved. In order to keep the treatment simple, we assume that $\chi_3 = 54.74^\circ$ so that the $\hat{T}_{20}$ term vanishes in relation (10.13) and $\sin(2\chi_3) = \sqrt{\frac{2}{3}}$. For the fourth pulse, we choose the appropriate flip angle of $\chi_4 = 90^\circ$ right away, too. This pulse is applied at a time Δ after the third and generates a density operator:

$$\begin{gathered}\sqrt{\tfrac{2}{3}}(\hat{T}_{21} + \hat{T}_{2-1})\cos(\omega_{Q,2}\Delta) + \sqrt{\tfrac{2}{3}}\,\mathrm{i}\hat{I}_x \sin(\omega_{Q,2}\Delta) + \sqrt{\tfrac{3}{8}}(\hat{T}_{22} + \hat{T}_{2-2})\tfrac{2}{3} \\ \Big\downarrow\, e^{\mathrm{i}(\pi/2)\hat{I}_x} \\ \sqrt{\tfrac{2}{3}}(\hat{T}_{21} + \hat{T}_{2-1})\cos(\omega_{Q,2}\Delta)(-1) + \sqrt{\tfrac{2}{3}}\,\mathrm{i}\hat{I}_x \sin(\omega_{Q,2}\Delta) + \hat{T}_{20}c + (\hat{T}_{22} + \hat{T}_{2-2})d.\end{gathered} \tag{10.16}$$

The $\hat{T}_{20}$ and $(\hat{T}_{22} + \hat{T}_{2-2})$ terms are again without relevance for the detected signal. The quadrupolar coupling is refocused, since after another delay of Δ, the normal time evolution is resumed (only the sign of the signal becomes negative):

$$\begin{gathered}-(\hat{T}_{21} + \hat{T}_{2-1})\cos(\omega_{Q,2}\Delta) + \mathrm{i}\hat{I}_x \sin(\omega_{Q,2}\Delta) \\ \Big\downarrow\, e^{-\mathrm{i}\hat{H}_Q(\Delta + t_2)} \\ -(\hat{T}_{21} + \hat{T}_{2-1})\cos(\omega_{Q,2}t_2) - \mathrm{i}\hat{I}_x \sin(\omega_{Q,2}t_2)\end{gathered} \tag{10.17}$$

(prove by actually inserting

$$\begin{aligned}(\hat{T}_{21} + \hat{T}_{2-1}) &\xrightarrow{e^{-\mathrm{i}\hat{H}_Q\Delta'}} (\hat{T}_{21} + \hat{T}_{2-1})\cos(\omega_Q\Delta') + \mathrm{i}\hat{I}_x \sin(\omega_Q\Delta') \\ \hat{I}_x &\xrightarrow{e^{-\mathrm{i}\hat{H}_Q\Delta'}} \hat{I}_x \cos(\omega_Q\Delta') + \mathrm{i}(\hat{T}_{21} + \hat{T}_{2-1})\sin(\omega_Q\Delta')\end{aligned} \tag{10.18}$$

with $\Delta' = \Delta + t_2$, and the trigonometric addition theorems (A.5) and (A.6)). Thus, the desired signal is refocused at a time Δ after the fourth pulse (actually, only for $t_1 = 0$ in the cosine dataset does an echo maximum occur at this point; nevertheless, the detection time t_2 must always be started at this time (concerning the stimulated echoes see Section 10.2.3)).

At the end of the data processing, the 2D spectra obtained from the cosine and sine time signals have to be added up to obtain a spectrum with sign discrimination in ω_1 (see Section 4.6.1). At this point, the flip-angle dependent prefactors sin χ_k etc. in the signals (10.9) and (10.15) should be equal. As stated above, the third pulse is conveniently chosen as a 55° pulse and the fourth pulse should be a 90° pulse. The prefactors in (10.9) and (10.15) are equal if the second pulse is also chosen as a 55° pulse; their common value is then $\frac{2}{3}$. The shortening of the second and the third pulse is also favourable for making the pulse-excitation function broader (see Section 2.6.5).

In the cosine signal, it has to be taken into account that for $\chi_3 \neq 90^\circ$ there remains z magnetization $M_0 \sin\chi_1 \sin\chi_2 \cos\chi_3$ on which the fourth pulse acts. This effect is actually known from the normal solid-echo pulse sequence with the first pulse shortened (see Section 2.6.4); the undesired signal components can be removed by alternating the phase of the last pulse (the fourth pulse in the 2D experiment).

The addition parameter for the two datasets is exactly equal to unity only in theory. In practice, the intensities of the two datasets are usually somewhat different due to differences in the longitudinal relaxation behaviour during t_m: for the cosine dataset, the density operator during t_m is $\hat{I}_z$, 'Zeeman order', relaxing with a longitudinal relaxation time T_{1z}; for the sine dataset, the density operator during t_m is $\hat{T}_{20}$, 'quadrupolar order', relaxing with T_{1Q}, which is unequal to T_{1z} in general. Therefore, it is important to measure the two datasets independently in order to have the option of introducing a scaling parameter for correcting the effect of the difference in the relaxation times, and to be able to test it iteratively on the spectra. In the t_1-TPPI method, the scaling factor can only be applied to the time-domain data, such that a change of the scaling factor requires repeating the full data processing. To find a good value for the scaling factor is important: relatively small differences between the two datasets can lead to considerable artifacts since the combination of sine and cosine data involves subtraction of signals on the antidiagonal that are large compared with the exchange intensity.

The sine signal must be measured for two reasons: measuring only the cosine dataset in the off-resonance approach (see Section 4.6.1) would aggravate the pulse-excitation problem by an intolerable factor of two; in addition, it is found that the single cosine dataset has the desired structure only if the second and the third pulse are exact 54.74° pulses.

10.2.5 T_1 *relaxation during the mixing time*

In order to investigate exchange processes with small exchange rates, it is desirable to employ relatively long mixing times. In their course, however, T_1 relaxation can occur, generating the so-called 'axial' peaks in the spectrum. These artifacts can be eliminated by a simple phase cycle to be described in the following. Though it is quite widely known, doubts have recently been cast on its efficiency (Kenwright and Packer, 1990), and actually its mode of operation has rarely been described in the literature. For an intuitive understanding, a specific case is to be considered first. The signal is assumed to consist of two components: one, M_{slow}, with $T_1 \gg t_m$, the other, M_{fast}, with $T_1 \ll t_m$, as found, for instance, in samples with rigid crystalline, and highly mobile amorphous fractions. The 90° pulse at the beginning of the mixing time is phase-cycled such that it flips the magnetization $M_{slow}(t_1) + M_{fast}(t_1)$ (more precisely, its relevant component) to the z and $(-z)$ direction in alternating scans. At the end of the mixing time, M_{fast} has returned to its equilibrium M_{fast}^{∞} in both cases. The magnetizations at the end of the mixing time are therefore:

$$M^{+z}(t_1 + t_m) = M_{slow}(t_1) + M_{fast}^{\infty} \tag{10.19a}$$

$$M^{-z}(t_1 + t_m) = -M_{slow}(t_1) + M_{fast}^{\infty}. \tag{10.19b}$$

In their difference:

$$M^{+z}(t_1 + t_m) - M^{-z}(t_1 + t_m) = 2M_{slow}(t_1) \tag{10.20}$$

which is easily obtained by alternating either the phase of the pulse after the mixing time, or the receiver phase, M_{fast} is eliminated as desired.

To motivate a more thorough analysis of the general case, it should be mentioned that the change of the magnetization due to T_1 relaxation in the mixing time varies with its deviation from the equilibrium value (Figure 10.5). Therefore, in general it is not possible to split the signal $M(t_1 + t_m)$ into $\pm M(t_1) + \Delta M$ and repeat the argument given above. With the phase-cycled pulse as above, one has instead:

$$M^{+z}(t_1 + t_m) = M(t_1) + \{M^\infty - M(t_1)\}\left(1 - \exp\left(-\frac{t_m}{T_1}\right)\right) \qquad (10.21a)$$

which expresses that the magnetization, starting at $M(t_1)$, relaxes as $1 - \exp(-t_m/T_1)$ towards the final value M^∞. Similarly, starting at $-M(t_1)$:

$$M^{-z}(t_1 + t_m) = -M(t_1) + \{M^\infty - (-M(t_1))\}\left(1 - \exp\left(-\frac{t_m}{T_1}\right)\right). \qquad (10.21b)$$

In the difference:

$$M^{+z}(t_1 + t_m) - M^{-z}(t_1 + t_m) = 2M(t_1)\exp\left(-\frac{t_m}{T_1}\right) \qquad (10.22)$$

all undesired M^∞ terms are eliminated, and $M(t_1)$ is simply scaled by a relaxation factor $\exp(-t_m/T_1)$ which is independent of t_1. In Section 13.4.2, T_1 relaxation during t_m will be discussed specifically related to ^{1}H spin diffusion.

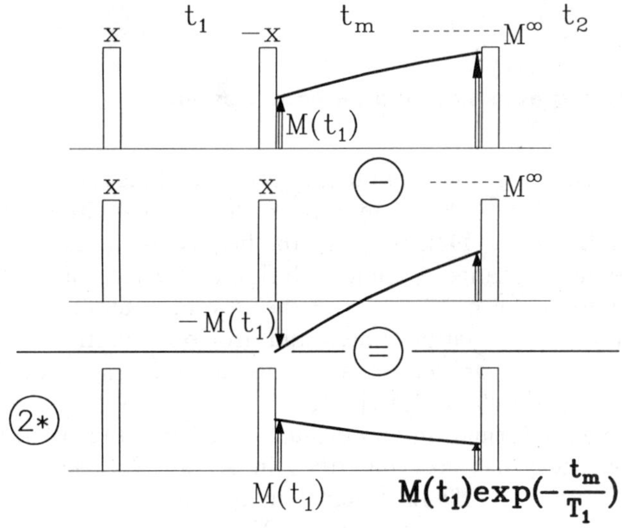

Figure 10.5 Visualization of the standard '$\pm z$' phase cycle for T_1-artifact cancellation. Inversion of the phase of the second pulse inverts the relevant magnetization component $M(t_1)$ on the z axis. In both cases, the relaxation is determined by the difference of $M(t_1)$ and the equilibrium value M^∞, which would lead to an undesired t_1-independent signal component $M^\infty \exp(i\tilde{\omega}_2 t_2)$. Taking the difference of the two signals eliminates M^∞ (bottom trace).

10.2.6 Limits to the mixing times

The duration of the mixing time is an important parameter in exchange experiments. The upper limit can be dictated by:

- T_1 relaxation during t_m
- the increase of the measuring times proportional to $t_0 + t_m$
- spin diffusion undesired in reorientation-exchange experiments (visible after mixing times of a few hundreds of milliseconds in typical ^2H, and after several seconds in natural-abundance ^{13}C exchange experiments).

The lower limit for t_m is given by the requirements of:

- destruction of transverse magnetization components during t_m
- $t_m > \text{maximum}(t_1 + t_2)$ because otherwise the assumption that the dynamics is essentially confined to the mixing time is not justified.

In ^{13}C NMR, motions with correlation times between 0.1 ms and 1 μs, or equivalently with rates between 10 kHz and 1 MHz are often difficult to investigate for practical reasons. Cross polarization becomes inefficient due to $T_{1\rho}$ relaxation during the spin lock and linebroadening occurs due to a breakdown of the proton decoupling; the coherent averaging on which this decoupling relies according to Section 3.4.1 is achieved only with a static (or fast-limit quasistatic) dipolar Hamiltonian. It breaks down when the interaction becomes time-dependent on the timescale of one averaging cycle, which is a 2π pulse for the commonly used constant irradiation on the protons.

10.3 Motion during evolution and detection periods

Lifting the restriction to the slow motion limit can extend the range of correlation times accessible to exchange 2D ^2H NMR by three orders of magnitude (Kaufmann *et al.*, 1990; Schaefer and Spiess, 1992). Some of the beauty of the exchange experiment is lost when the frequency changes during evolution and detection occur, but still some interesting information can be obtained in the case of complicated motions. With a broad distribution of correlation times and isotropic motions, the intermediate motional regime applies for certain segments even if $\tau_0 > 1$ ms and should be considered for a valid simulation of spectra.

The exchange-matrix formalism and evaluation of exchange of frequencies during the free induction decay has been described in Section 9.4. Analogous to equation (9.73), the (four-pulse) ^2H 2D time signal in the intermediate motional regime is obtained as a vector–matrix–vector product (Kaufmann *et al.*, 1990):

$$s(t_1, t_2; t_m) = \vec{\mathbf{1}}^{\mathrm{T}} \underbrace{e^{(\mathbf{\Pi} + i\omega)t_2}}_{\text{detect}} \underbrace{e^{(\mathbf{\Pi} + i\omega)\Delta}}_{\text{refocus}} \underbrace{e^{(\mathbf{\Pi} - i\omega)\Delta}}_{\text{refocus}} \underbrace{e^{\mathbf{\Pi} t_m}}_{\text{mix}} \underbrace{e^{(\mathbf{\Pi} + i\omega)t_1}}_{\text{evol}} \underbrace{\mathbf{P}\vec{\mathbf{1}}}_{a\ priori} \qquad (10.23)$$

where T_1 relaxation during t_m has been neglected. The $\vec{\mathcal{M}}(0)$ vector, which consists of the *a priori* probabilities P_j, has been written as a matrix–vector product $\mathbf{P}\vec{1}$ with the diagonal matrix:

$$\mathbf{P} = \begin{bmatrix} P_1 & 0 & \dots & 0 \\ 0 & P_2 & \dots & 0 \\ \vdots & \vdots & \ddots & \vdots \\ 0 & 0 & \dots & P_N \end{bmatrix} \tag{10.24}$$

which will be of use in the following.

The exponential matrix expressions in formula (10.23) can usually be transformed into simple matrices by diagonalization of $\mathbf{\Pi}$ and $\mathbf{\Pi} + \mathrm{i}\boldsymbol{\omega}$ as described in Appendix D.4. If $\mathbf{\Pi}$ can be transformed into a symmetric matrix, the diagonalizations become more convenient. On the basis of the principle of detailed balance, such a transformation is possible (Oppenheim *et al.*, 1977, p. 71). Detailed balance means that the number of units going from state k to state j per unit time is equal to the number making the opposite transition. These are the products of the transition probabilities, which are the exchange matrix elements $\mathbf{\Pi}_{kj}$ and $\mathbf{\Pi}_{jk}$, respectively, multiplied by the occupation numbers of the initial states or the corresponding *a priori* probabilities P_j. Thus:

$$\Pi_{kj} P_j = \Pi_{jk} P_k \tag{10.25}$$

and after multiplication of both sides by $\sqrt{P_k}^{-1}\sqrt{P_j}^{-1}$ (i.e. division by $\sqrt{P_k}\sqrt{P_j}$), we obtain a symmetrized matrix $\mathbf{\Pi}^{\mathrm{s}}$ with:

$$\Pi^{\mathrm{s}}_{kj} := \sqrt{P_k}^{-1}\Pi_{kj}\sqrt{P_j} = \sqrt{P_j}^{-1}\Pi_{jk}\sqrt{P_k} = \Pi^{\mathrm{s}}_{jk}. \tag{10.26}$$

On this basis, the diagonalization and calculation of the time signal (10.23) can usually (exceptions are discussed by Kaufmann *et al.*, 1990) be performed along the lines outlined in Appendix D.4. With the matrix:

$$\sqrt{\mathbf{P}} := \begin{bmatrix} \sqrt{P_1} & 0 & \dots & 0 \\ 0 & \sqrt{P_2} & \dots & 0 \\ \vdots & \vdots & \ddots & \vdots \\ 0 & 0 & \dots & \sqrt{P_N} \end{bmatrix} \tag{10.27}$$

and on the basis of equation (10.26) and the matrix relation (D.9) we can write:

$$\mathrm{e}^{(\mathbf{\Pi} + \mathrm{i}\omega)t} = \sqrt{\mathbf{P}}\ \mathrm{e}^{\sqrt{\mathbf{P}}^{-1}(\mathbf{\Pi} + \mathrm{i}\omega)t\sqrt{\mathbf{P}}}\ \sqrt{\mathbf{P}}^{-1} = \sqrt{\mathbf{P}}\ \mathrm{e}^{(\mathbf{\Pi}_{\mathrm{s}} + \mathrm{i}\omega)t}\ \sqrt{\mathbf{P}}^{-1} \tag{10.28}$$

and thus the time signal (10.23) can be rewritten as:

$$s(t_1, t_2; t_{\mathrm{m}}) = \vec{\mathbf{1}}^{\mathrm{T}}\sqrt{\mathbf{P}}\ \mathrm{e}^{(\mathbf{\Pi}_{\mathrm{s}} + \mathrm{i}\omega)t_2}\ \mathrm{e}^{(\mathbf{\Pi}_{\mathrm{s}} + \mathrm{i}\omega)\Delta}\ \mathrm{e}^{(\mathbf{\Pi}_{\mathrm{s}} - \mathrm{i}\omega)\Delta}\ \mathrm{e}^{\mathbf{\Pi}_{\mathrm{s}} t_{\mathrm{m}}}\ \mathrm{e}^{(\mathbf{\Pi}_{\mathrm{s}} + \mathrm{i}\omega)t_1}\ \sqrt{\mathbf{P}}\vec{\mathbf{1}}. \tag{10.29}$$

Asymmetry of the spectral intensity with respect to the diagonal (ω, ω) is one prominent feature of spectra taken with a long echo-delay Δ in the intermediate-exchange region. To analyse the appearance of this asymmetry, we note that the symmetry of the spectrum and of the time-domain signal is equivalent:

$$S(\omega_1, \omega_2) = S(\omega_2, \omega_1) \iff s(t_1, t_2) = s(t_2, t_1). \tag{10.30}$$

Before proving the asymmetry of equation (10.29) upon interchange of t_1 and t_2, for comparison we want to derive the symmetry of time-signals in the slow-exchange regime:

$$s_{\text{slow}}(t_1, t_2; t_m) = \vec{\mathbf{1}}^{\text{T}}\sqrt{\mathbf{P}}\underset{\text{detect}}{e^{i\omega t_2}}\ \underset{\text{refocus}}{e^{i\omega\Delta}}\ e^{-i\omega\Delta}\ \underset{\text{mix}}{e^{\mathbf{\Pi}_s t_m}}\ \underset{\text{evol}}{e^{i\omega t_1}}\ \sqrt{\mathbf{P}}\vec{\mathbf{1}}$$

$$= \vec{\mathbf{1}}^{\text{T}}\sqrt{\mathbf{P}}\ e^{i\omega t_2}\ e^{\mathbf{\Pi}_s t_m}\ e^{i\omega t_1}\ \sqrt{\mathbf{P}}\vec{\mathbf{1}} \tag{10.31}$$

and of time-signals in the intermediate case but without the fourth pulse:

$$s_{\text{3-pulse}}(t_1, t_2; t_m) = \vec{\mathbf{1}}^{\text{T}}\sqrt{\mathbf{P}}\ \underset{\text{detect}}{e^{(\mathbf{\Pi}_s + i\omega)t_2}}\ \underset{\text{mix}}{e^{\mathbf{\Pi}_s t_m}}\ \underset{\text{evol}}{e^{(\mathbf{\Pi}_s + i\omega)t_1}}\ \sqrt{\mathbf{P}}\vec{\mathbf{1}}. \tag{10.32}$$

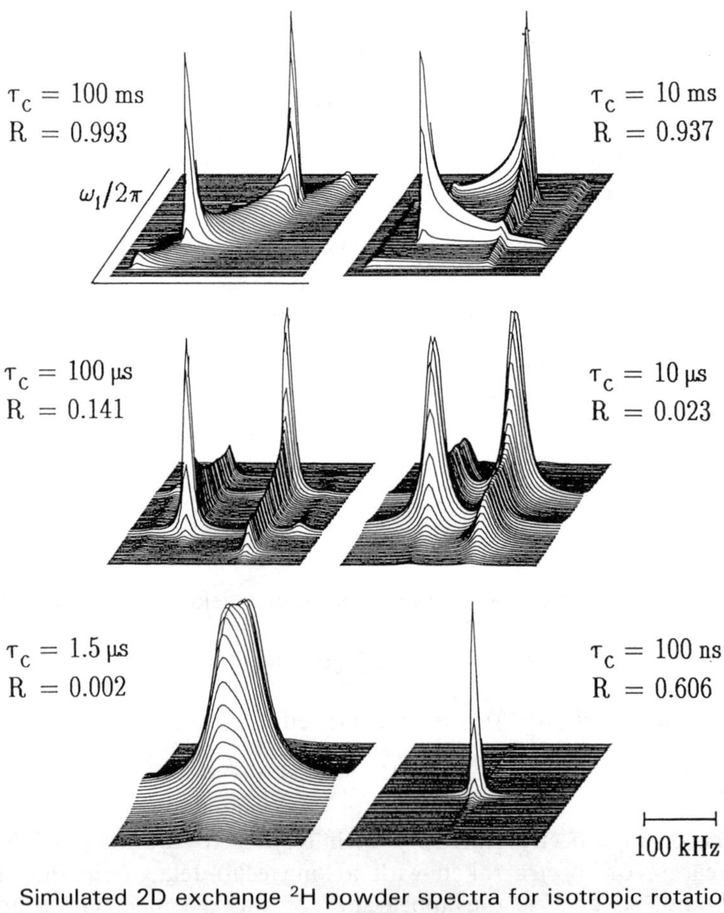

Figure 10.6 Simulated 2D exchange ^{2}H powder spectra for isotropic rotational diffusion in the intermediate motional regime, assuming a four-pulse experiment. The single correlation times of the diffusion process and the intensity reduction factors are given next to the spectra. In the spectra with $\tau_c < 1$ ms, note the asymmetry with respect to the diagonal, which is due to the fourth pulse. Adapted from Kaufmann *et al.* (1990).

For that purpose, consider that for a general matrix **A** and vectors $\vec{1} = (1, 1, \ldots, 1)^T$, **A** can be replaced by $\mathbf{A}^T$ in the product:

$$\vec{1}^T\mathbf{A}\vec{1} = \sum_{n=1}^{N}\sum_{m=1}^{N} A_{mn} = \sum_{n=1}^{N}\sum_{m=1}^{N} A_{nm} = \vec{1}^T\mathbf{A}^T\vec{1} \tag{10.33}$$

consider further that $(\mathbf{ABCD})^T = \mathbf{D}^T\mathbf{C}^T\mathbf{B}^T\mathbf{A}^T$, and that $\mathbf{B}^T = \mathbf{B}$ if the matrix **B** is symmetric. These relations can be applied to rewrite the signals (10.31) and (10.32) identically as:

$$S_{\text{slow}}(t_1, t_2; t_m) = \vec{1}^T\sqrt{\mathbf{P}}\, e^{i\omega t_1}\, e^{\mathbf{\Pi}_s t_m}\, e^{i\omega t_2}\, \sqrt{\mathbf{P}}\vec{1} \tag{10.34}$$

$$s_{3\text{-pulse}}(t_1, t_2; t_m) = \vec{1}\sqrt{\mathbf{P}}\, e^{(\mathbf{\Pi}_s + i\omega)t_1}\, e^{\mathbf{\Pi}_s t_m}\, e^{(\mathbf{\Pi}_s + i\omega)t_2}\, \sqrt{\mathbf{P}}\vec{1}. \tag{10.35}$$

If t_1 and t_2 are interchanged in expression (10.34), the resulting formula for $S_{\text{slow}}(t_2, t_1; t_m)$ is identical to $S_{\text{slow}}(t_1, t_2 \colon t_m)$ of equation (10.31), and analogously $s_{3\text{-pulse}}(t_2, t_1; t_m) = s_{3\text{-pulse}}(t_1, t_2; t_m)$ is found. For the full signal (10.29), the same procedure produces:

$$s(t_2, t_1; t_m) = \vec{1}^T\sqrt{\mathbf{P}}\, e^{(\mathbf{\Pi}_s + i\omega)t_2}\, e^{\mathbf{\Pi}_s t_m} \underbrace{e^{(\mathbf{\Pi}_s - i\omega)\Delta}\, e^{(\mathbf{\Pi}_s + i\omega)\Delta}}\, e^{(\mathbf{\Pi}_s + i\omega)t_1}\sqrt{\mathbf{P}}\vec{1}. \tag{10.36}$$

Compared to $s(t_1, t_2; t_m)$ in (10.29), the position of the term $e^{(\mathbf{\Pi}_s - i\omega)\Delta}\, e^{(\mathbf{\Pi}_s + i\omega)\Delta}$ is different, and since the matrices $\mathbf{\Pi}_s + i\boldsymbol{\omega}$, $\mathbf{\Pi}_s - i\boldsymbol{\omega}$, and $\mathbf{\Pi}_s$ no do not commute, this term cannot be interchanged with $e^{\mathbf{\Pi}_s t_m}$. The asymmetry becomes particularly pronounced for diffusive processes (Schaefer and Spiess, 1992), as shown by the spectra in Figure 10.6.

Since the interpretation of the spectral intensity as a probability density is not valid in the intermediate-exchange region, there can be regions of negative signal in the spectrum (the integral must, however, always be positive). Figure 10.7 shows an example for a simple two-site exchange between sharp lines.

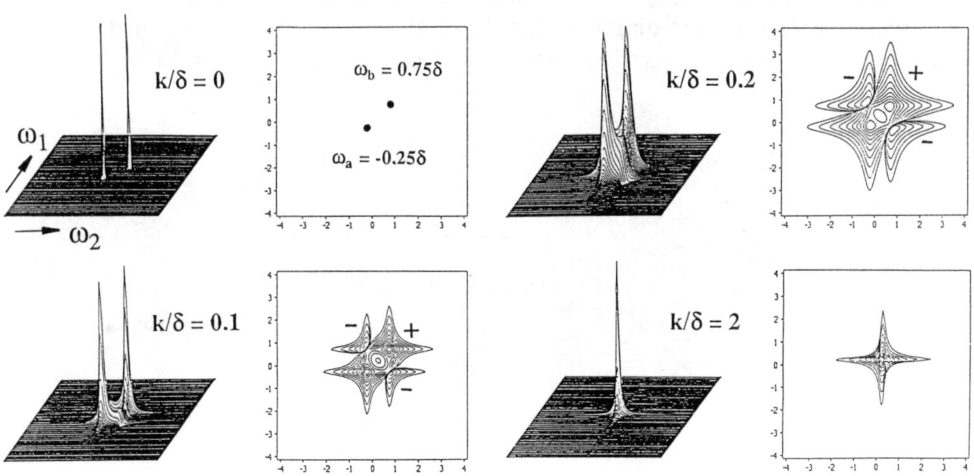

Figure 10.7 Simulated 2D exchange two-line spectra for a two-site jump in the intermediate regime, for $t_m = 0$. Although the mixing time is zero, off-diagonal intensity is observed. Note the dispersive and negative signal components (the latter are marked by a minus sign in the contour plots) which arise although the data processing produces an absorptive and positive spectrum in the slow and fast exchange limits. The rates k are given in terms of the frequency difference δ.

10.4 2D MAS exchange time signals

In Chapter 7, the various MAS exchange experiments were discussed only briefly. In this section, we shall give a detailed analysis of the corresponding MAS time signals.

10.4.1 Time signals in the sideband-exchange experiment

The MAS sideband exchange experiment was introduced in Section 7.11. It represents a compromise between resolution and orientational information: the spectral overlap is partially removed by splitting the powder patterns into sidebands while the anisotropic interaction is partially retained in the sidebands. In this section, the generation of the time signals required for obtaining absorptive 2D sideband patterns (Hagemeyer *et al.*, 1989) will be outlined.

If we apply the exchange pulse sequences shown in Figure 10.8 under MAS conditions, for a single segment we obtain signals:

$$g_{\substack{\mathrm{Re}\\ \mathrm{Im}}}(t_1, t_2) = \frac{\mathrm{Re}}{\mathrm{Im}}\{f_1^*(\gamma_1)f_1(\gamma_1 + \omega_r t_1)\}f_2^*(\gamma_2 + \omega_r t_1 + \omega_r t_m)f_2(\gamma_2 + \omega_r t_1 + \omega_r t_m + \omega_r t_2). \tag{10.37}$$

The imaginary part of $g(t_1, t_2)$ is measured by the 'sine' pulse sequence, that is, with the second pulse 90° out of phase with the first. By coadding the signals of the two pulse sequences with the receiver phases differing by 90° to effect the multiplication of the imaginary part by i, $g_{\mathrm{Re}} + \mathrm{i}g_{\mathrm{Im}} = g$ can be measured as a single dataset with one (phase-cycled) pulse program.

The indices ($_1$) and ($_2$) of the f functions and the angle γ are necessary because a molecular reorientation during the mixing time can change the angles α, β and γ so that the f functions are altered. Thus, we have defined $f_1(\gamma_1) := f(\alpha_1, \beta_1, \gamma_1)$ and $f_2(\gamma_2) := f(\alpha_2, \beta_2, \gamma_2)$.

In order to obtain no off-diagonal signal just from the reorientation of the rotor in the mixing time, the term $\omega_r t_m$ is eliminated from the f functions by choosing the

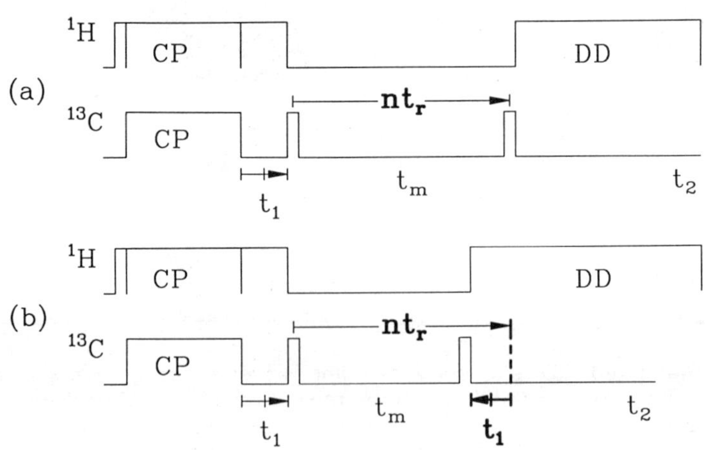

Figure 10.8 Pulse sequences for the MAS sideband-exchange 2D experiment. (a) Exchange pulse sequence with rotor-synchronized mixing time $t_m = nt_r$. (b) Timings for obtaining 't_1-time-reversed' data, which are necessary for generating absorptive spectra.

duration of the mixing time as an integer multiple of the rotor period:

$$t_m = nt_r \Rightarrow \omega_r t_m = n \times 2\pi \tag{10.38}$$

(de Jong *et al.*, 1984; Kentgens *et al.*, 1987). Because the f functions are periodic with 2π, $\omega_r t_m = n \times 2\pi$ can be dropped from the argument. Without a molecular exchange process, the rotor phase now continues where it had been before the mixing time, in general:

$$g(t_1, t_2) = f_1^*(\gamma_1) f_1(\gamma_1 + \omega_r t_1) f_2^*(\gamma_2 + \omega_r t_1) f_2(\gamma_2 + \omega_r t_1 + \omega_r t_2). \tag{10.39}$$

If no exchange occurs, we have $\gamma_1 = \gamma_2$ and $f_1 = f_2 = f$, so that due to $f(\gamma)f^*(\gamma) = 1$ (see equation (3.86)) the signal reduces to:

$$g^{\mathrm{ne}}(t_1, t_2) = f^*(\gamma) f(\gamma + \omega_r(t_1 + t_2)) \tag{10.40}$$

yielding single-segment $I^{\mathrm{ne}}_{M,N}(\alpha, \beta, \gamma)$ sideband intensities:

$$I^{\mathrm{ne}}_{M,N} = \int_0^{t_r} \frac{dt_1}{t_r} \int_0^{t_r} \frac{dt_2}{t_r}\, e^{-i\omega_r M t_1}\, e^{-i\omega_r N t_2}\, g^{\mathrm{ne}}(t_1, t_2)$$

$$\overset{(\tilde{\gamma} = \gamma + \omega_r(t_1 + t_2))}{=} f^*(\gamma)\, e^{i\gamma N} \underbrace{\int_0^{t_r} \frac{dt_1}{t_r}\, e^{-i\omega_r(Mt_1 - Nt_1)}}_{\delta_{M,N}} \int_0^{2\pi} \frac{d\tilde{\gamma}}{2\pi}\, e^{-iN\tilde{\gamma}}\, f(\tilde{\gamma})$$

$$= \delta_{M,N}\, e^{i\gamma N} \underbrace{f^*(\gamma) \int_0^{2\pi} \frac{d\tilde{\gamma}}{2\pi}\, e^{-iN\tilde{\gamma}} f(\tilde{\gamma})}_{I_N(\alpha, \beta, \gamma)}. \tag{10.41}$$

Thus, in the case without exchange, sidebands appear only where $M = N$, that is, on the diagonal, with the intensities $I_N(\alpha, \beta, \gamma)$ of the corresponding 1D MAS spectrum (see equation (3.83)).

Even in the case of exchange, the $I_{M,N}$ are real for a powder (but not necessarily positive). This can be understood on the basis of the decomposition of the signal in terms of contributions from mirror-image segment pairs (see Section 3.9.2). In the case of the exchange experiment it is important to realize that PAS in the mirror-image segment has the orientations $(-\alpha_1, \beta_1, -\gamma_1)$ and $(-\alpha_2, \beta_2, -\gamma_2)$ before and after the reorientation, so that the α and γ angles are reversed between the mirror-image segments in both f_1 and f_2. The pair signal thus reads:

$$\begin{aligned} g_p(t_1, t_2) = {} & f_1^*(\alpha_1, \beta_1, \gamma_1) f_1(\alpha_1, \beta_1, \gamma_1 + \omega_r t_1) \\ & f_2^*(\alpha_2, \beta_2, \gamma_2 + \omega_r t_1) \\ & \times f_2(\alpha_2, \beta_2, \gamma_2 + \omega_r t_1 + \omega_r t_2) \\ & + f_1^*(-\alpha_1, \beta_1, -\gamma_1) \\ & \times f_1(-\alpha_1, \beta_1, -\gamma_1 + \omega_r t_1) \\ & f_2^*(-\alpha_2, \beta_2, -\gamma_2 + \omega_r t_1) \\ & \times f_2(-\alpha_2, \beta_2, -\gamma_2 + \omega_r t_1 + \omega_r t_2). \end{aligned} \tag{10.42}$$

Writing down the complex conjugate of $g_p(t_1, t_2)$ and replacing t_1 and t_2 by $(-t_1)$ and $(-t_2)$, after application of $f(-\alpha, \beta, -\gamma) = f^*(\alpha, \beta, \gamma)$ (equation (3.88)) it is immediately shown that $g_p(t_1, t_2) = g_p^*(-t_1, -t_2)$, and consequently the same applies for the superposition of all pair signals in the powder, $\bar{g}(t_1, t_2) = \bar{g}^*(-t_1, -t_2)$, so that an absorptive spectrum results.

10.4.2 MAS 'time-reversed' signal

The full time signal for a powder in the 2D MAS exchange experiment can be written as:

$$\bar{s}(t_1, t_2) = \sum_M \sum_N I_{M,N}\, e^{it_1(M\omega_r + \omega_{iso})}\, e^{it_2(N\omega_r + \omega_{iso})}\, e^{-(t_1 + t_2)/T_2} \tag{10.43}$$

where we have included the isotropic shift ω_{iso} and T_2 relaxation; the treatment in the preceding section has shown that the $I_{M,N}$ coefficients are real.

In Section 4.6.2 it was pointed out that for obtaining absorptive peaks in experimental two-dimensional spectra, real Fourier components $I_{M,N}$ of the periodic time signal without T_2 decay and without isotropic shift are a necessary, but not a sufficient, condition. The phase problem occurring with double Fourier transformation as discussed in Section 4.6.2 must still be solved. Measuring off-resonance as described in Section 4.6.1 is not helpful, as we cannot measure $\langle\cos(\tilde{\omega}_1 t_1)\exp(i\tilde{\omega}_2 t_2)\}$ separately in the MAS experiment: the 'sine' and 'cosine' pulse sequences yield the terms given in equation (10.37), where t_1 is seen to play a role after the mixing time, outside of the amplitude modulation. Only when combined do the two datasets produce the product (10.43) of exponentials with the real prefactors (the double Fourier transform of $\mathrm{Re}\{s(t_1, t_2)\}$ does not even yield a diagonal spectrum in the case of no exchange).

Therefore, one has to produce the signal with t_1 'time-reversed' (except for the irreversible T_2 relaxation):

$$\bar{g}^{tr}(t_1, t_2) = \sum_M \sum_N I_{M,N}\, e^{-t_1 i(M\omega_r + \omega_{iso})}\, e^{it_2(N\omega_r + \omega_{iso})}\, e^{-(1/T_2)(t_1 + t_2)} \tag{10.44}$$

to create cosine and sine datasets $\bar{g}^c = (\bar{g} + \bar{g}^{tr})/2$ and $\bar{g}_s = (\bar{g} - \bar{g}^{tr})/2$:

$$\bar{g}_s^c(t_1, t_2) = \sum_M \sum_N I_{M,N} \begin{matrix}\cos\\ \sin\end{matrix}\{t_1(M\omega_r + \omega_{iso})\}\, e^{it_2(N\omega_r + \omega_{iso})}\, e^{-(1/T_2)(t_1 + t_2)} \tag{10.45}$$

from which the purely absorptive spectra can be generated as described in Section 4.6.1. The time signal with t_1 time reversal is obviously (see equation (10.37)):

$$g^{tr}(t_1, t_2) = f_1^*(\gamma_1) f_1(\gamma_1 - \omega_r t_1) f_2^*(\gamma_2 - \omega_r t_1) f_2(\gamma_2 - \omega_r t_1 + \omega_r t_2). \tag{10.46}$$

Time reversal in MAS was first considered by Munowitz and Griffin (1983). In the case of the exchange experiment, it was recognized by Hagemeyer *et al.* (1989) that a time-reversed signal can effectively be generated by choosing:

$$t_m = nt_r - t_1 \tag{10.47}$$

and substituting:

$$\gamma_n' = \gamma_n + \omega_r t_1$$

If this is inserted into the fundamental time-signal (10.37), we obtain:

$$g^{tr}_{\substack{Re\\Im}} = \begin{matrix}\mathrm{Re}\\ \mathrm{Im}\end{matrix}\{f_1^*(\overbrace{\gamma_1' - \omega_r t_1}^{\gamma}) f_1(\gamma_1')\} f_2^*(\gamma_2' - \omega_r t_1) f_2(\gamma_2' - \omega_r t_1 + \omega_r t_2). \tag{10.48}$$

This is nearly the desired expression (10.44), except that we want the complex conjugate of the term in the curly brackets, which is easily obtained, on the basis of $z^* = \mathrm{Re}(z) - \mathrm{i}\ \mathrm{Im}(z)$ from the two measured components:

$$g^{\mathrm{tr}} = g^{\mathrm{tr}}_{\mathrm{Re}} - \mathrm{i} g^{\mathrm{tr}}_{\mathrm{Im}} = f_1(\gamma_1' - \omega_r t_1) f_1^*(\gamma_1') f_2^*(\gamma_2' - \omega_r t_1) f_2(\gamma_2' - \omega_r t_1 + \omega_r t_2). \quad (10.49)$$

This is the t_1-time-reversed signal (10.46), except for γ_n being replaced by γ_n'. That is no problem in the case of powder samples: for each slice, the powder-averaged time signal is not affected by the constant rotor-phase shift of $\omega_r t_1$, because the f functions are periodic and the integral over a full period of γ is therefore unaffected by a shift of the phase. Every slice being unchanged by $\gamma_n' \to \gamma_n$, we can write γ_n instead of γ_n' in (10.49) for the complete signal $g^{\mathrm{tr}}(t_1, t_2)$.

The variation of the mixing time involved in the time-reversal condition $t_m = nt_r - t_1$ (10.47) can always be restricted to less than one rotor period (by choosing n appropriately) and has therefore no relevant effects on the dynamics in the mixing time.

Figure 10.9 demonstrates the increased resolution obtained in the absorptive version of the 2D MAS sideband-exchange experiment (Figure 10.9(a)), comparing it with a corresponding magnitude spectrum (Figure 10.9(b)).

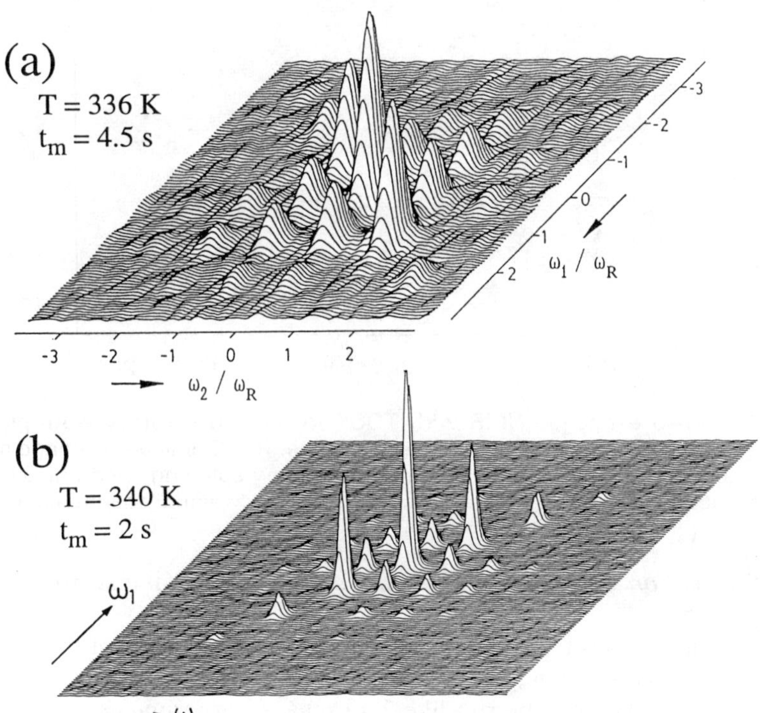

Figure 10.9 2D MAS sideband-exchange powder spectra of POM near 340 K with mixing times of several seconds. (a) Resolution-enhanced absolute-mode (nonabsorptive) spectrum. Adapted from Kentgens *et al.* (1987). (b) An absorptive spectrum, obtained by including the t_1-time-reversed data, exhibits superior resolution. In comparing the spectra, note that the diagonals in the two representations are oriented differently. Adapted from Hagemeyer *et al.* (1989).

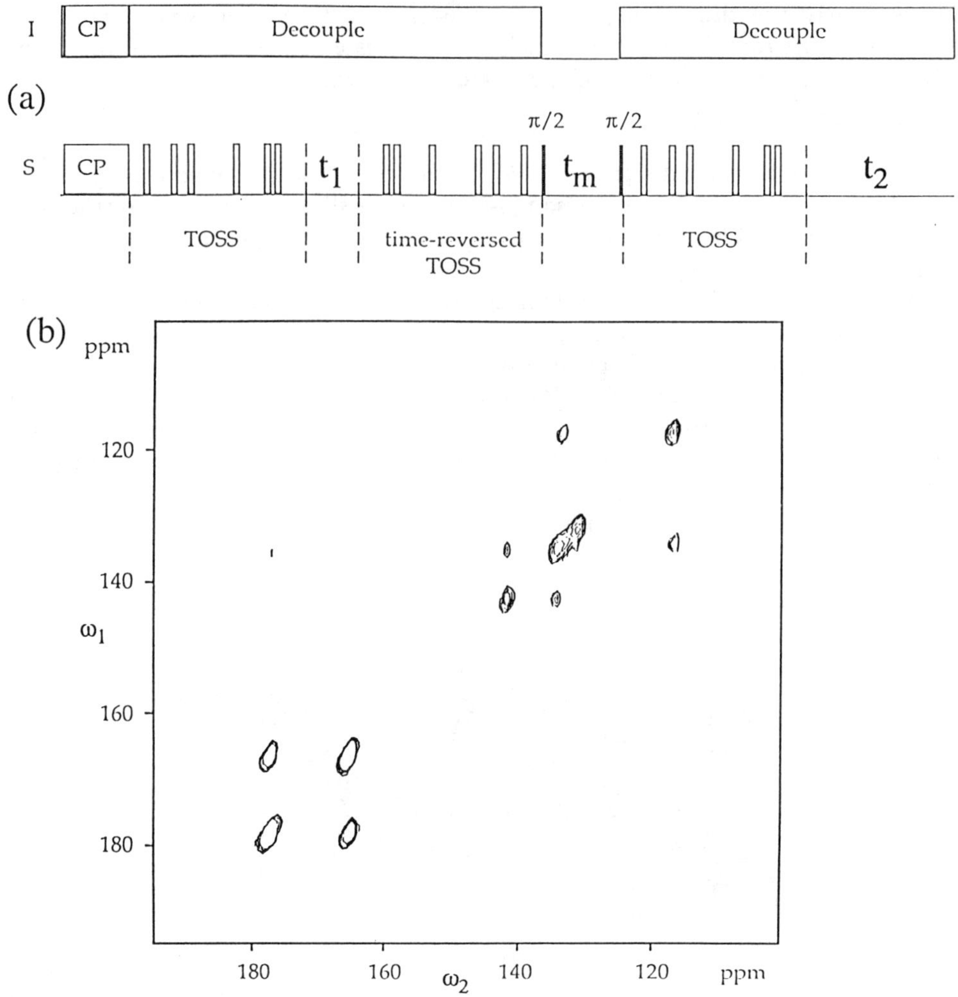

Figure 10.10 MAS exchange NMR with TOSS–t_1–deTOSS, totally isotropic evolution, mixing time, and TOSS detection. (a) Pulse sequence. (b) Chemical-exchange spectrum of tropolone (see Figure 7.28) taken with a relatively slow spinning speed of 3 kHz at a ^{13}C Larmor frequency of 75.47 MHz. Adapted from Geen and Bodenhausen (1982).

10.4.3 Correlation of isotropic chemical shifts with totally isotropic evolution

If magic-angle spinning is fast enough to remove all visible effects of the chemical-shift anisotropy, chemical-exchange 2D experiments (Jeener *et al.*, 1979) in solids (Szeverenyi *et al.*, 1982) can be run like 2D exchange experiments on static samples. In contrast, the 2D measurement of chemical exchange is not straightforward in MAS spectra that exhibit considerable sidebands of the chemical-shift anisotropy. In today's and tomorrow's B_0 fields, for aromatic and carbonyl carbons the 2D spectra are in fact choked with sidebands if the spinning speed is not higher than 10 kHz. These sidebands have dispersive components that overlap with the isotropic-exchange peaks of interest, which are often of much lower intensity. The 2D exchange experiment described in the previous sections can remove the sidebands induced by the reorientation of the sample in the mixing time, so that off-diagonal intensity appears only if exchange occurs: chemical exchange produces cross peaks between different sideband patterns, reorientational exchange within the sidebands of a site only. This experiment also provides information on the reorientation of the chemical-shift tensor in a chemical-exchange process. However, the sidebands complicate the exchange pattern and the spreading of intensity on all the sidebands decreases the overall signal-to-noise ratio.

The first approach to the challenge of removing all sidebands consisted in chemical-shift scaling (see Section 3.4.4) during t_1 to make the rotation frequency exceed the scaled anisotropies (Harbison *et al.*, 1985). This technique was applied to detect slow flips of phenylene rings with asymmetric substituents that make the interchanging sites nonequivalent in terms of their isotropic shifts so that chemical-exchange cross peaks between them can be detected. However, the scaling experiment has severe drawbacks: it requires many ^{13}C pulses and is therefore highly susceptible to experimental imperfections. Also, the scaling results in some reduction of the optimum achievable resolution. If the scaling factor is not sufficiently small, the anisotropy during t_1 is not fully eliminated.

All these problems are avoided in the more versatile method of totally-isotropic evolution by means of TOSS–t_1–deTOSS (see section 3.9.4) as indicated in Figure 10.10(a) (Geen and Bodenhausen, 1992). In the detection, TOSS or simple MAS can be employed without any of the special MAS phase problems. Figure 10.10(b) displays the application to the chemical-exchange process in tropolone at an intermediate spinning speed with TOSS detection.

In contrast to actual fast-spinning spectra, after TOSS–t_1–deTOSS neither with TOSS nor with MAS detection are the exchange peaks at $(\tilde{\omega}_A, \tilde{\omega}_B)$ and $(\tilde{\omega}_B, \tilde{\omega}_A)$ necessarily of equal intensity, that is to say, the spectra can be asymmetric with respect to the diagonal: this is due to the intensity reductions in the TOSS detection as well as for the centreband of the MAS spectrum. As the simplest case, consider exchange between two sites A and B that occur in a 1 : 1 ratio but exhibit different shift tensors. If the exchange probability is p, the crosspeak intensities are $pF_A(\alpha, \beta)$ and $pF_B(\alpha, \beta)$, since the intensity in the spectrum depends only on the anisotropy in the detection period; the anisotropic evolution has been completely refocused. The asymmetry is not apparent in the spectrum of Figure 10.10(b); this is probably due to the similarity of the tensor orientations of the exchanging sites.

10.5 Practical hints for 2D spectroscopy of polymers

For obtaining multidimensional spectra of high quality, certain points have to be considered in addition to the requirements of 1D NMR. A strong signal from the units of interest is always of great importance, with a good filling factor (of the rf coil) as one essential factor. If possible, block samples filling coil or rotor completely should be used, which contain typically three times as much material as corresponding powder samples even after strong manual compression. Three times less signal means measuring times nine times longer to get the the same signal-to-noise ratio. Many polymer powders can be turned into compact samples thermoplastically, by heating (with a hot-air dryer in the simplest case) and mechanical compression into a tube of suitable diameter. As in industrial compression moulding, pressure is necessary due to the high viscosity of most polymer melts.

As good phasing and a flat baseline are particularly important for 2D spectra, echo detection with left-shifts of the first points should be employed, and the phase should be properly adjusted in the time domain (receiver phase; see Section 4.5). These settings are most appropriately ensured by checking the spectrum obtained after running the 2D pulse program for the cosine dataset with $t_1 \cong 0$.

How far to measure out in the t_1 dimension is a crucial question, requiring some experience. It is the maximum t_1 value that counts for the resolution and smoothing of the spectrum, not the number of t_1 increments; this is directly related to the 'dwell time theorem', Section 4.9.1. To obtain good resolution, one must measure out far, into a region where little signal may be recognized in the time domain. On the other hand, if $t_1 > T_2$, there is a double penalty: one measures more noise and no signal in the time domain, increasing the noise level in the spectrum, and in doing so one wastes the additional measuring time needed to acquire the data for those t_1 values (see also Section 4.9.3). As a compromise, the signal for larger t_1 values can be measured with fewer scans, and scaled up accordingly before the data evaluation. Thus the resolution is available if needed, but not at the expense of so much measuring time. Scaling the measured signal to its correct height pulls the noise up, too. This is not very harmful, however, as these slices will not contribute much to the spectrum after having been subject to the Gaussian damping applied before the FTs.

Steps of data acquisition and evaluation

- Optimize the useful signal.
- Adjust irradiation frequencies: exactly on-resonance for ^2H, fully off-resonance for simple ^{13}C. The ^1H irradiation should be set on resonance for pure ^{13}C experiments as well as for SLF or WISE, but off resonance for ^1H evolution under multipulse irradiation (MREV-8, etc.).
- Adjust receiver phase and echo delay to obtain, for $t_1 = 0$ or smallest possible t_1, an absorptive 1D spectrum in ω_2 with minimal phasing.
- Observe echo decays as a function of t_m to test for exchange, perform time-domain checks of proper evolution behaviour (e.g. t_1 modulation in SLF experiments), decide on duration of mixing and/or cross polarization times.
- Set dwell times to have the spectrum well inside the spectral widths. In exchange experiments, to obtain symmetric spectral data arrays set $t_{dw1} = 2^n t_{dw2}$; $2^n = 1$ is typical for double TPPI, $2^n = 2$ for hypercomplex data (in particular, for ^2H exchange experiments), and $2^n = 4$ for ^{13}C off-resonance measurements (see Figure 10.11) in order to have plenty of baseline to avoid baseline or phase-distortion problems as listed in Section 4.5.3.
- Decide on maximum t_1 required, and on the duration of the experiment (number of t_1 increments, times number of scans per t_1 increment, times the sum of repetition delay and mixing time).
- Start the experiment, check the parameters.
- After the measurement is finished, start the data treatment on the cosine data set, with left shifts, Gaussian damping (for smoothing or apodization, Section 4.8.2) and possibly resolution enhancement in t_2 (multiplication with e^{+t_2/τ_2}, $\tau_2 \geqslant T_2^*$).
- After zero filling to obtain the desired number of ω_2 spectral points (e.g. 512 as in Figure 10.11), perform an FT over t_2.

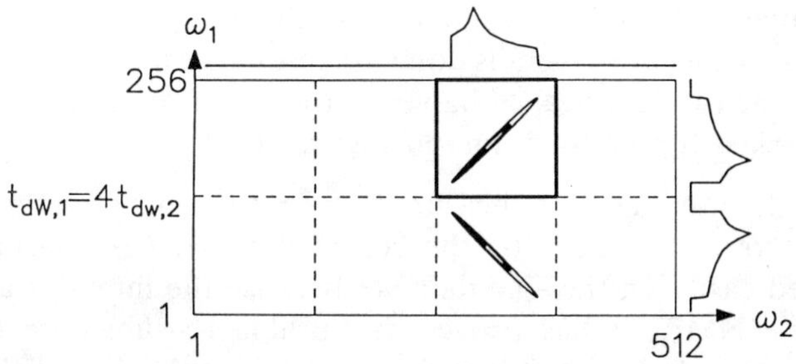

Figure 10.11 Typical format of 2D exchange ^{13}C spectra in an off-resonance measurement. Thick square frame: final 2D spectrum.

- Phasing in ω_2 must be checked on the $t_1 = 0$ slice; avoid linear phasing in ω_2. Finally, the same phasing is applied to all slices.
- Cut off the outer parts of the ω_2 range to leave only the region that actually contains signal in the $t_1 = 0$ slice.
- Damp (apodize) in t_1; check the result on the t_1 slice at highest peak in ω_2. For exchange spectra to be symmetrically broadened, use equal damping constants in absolute time units for both t_1 and t_2.
- Eliminate the dispersive contribution D_2 before FT over t_1 (not necessary for signals which are fully periodic in t_1 or t_2, or if in at least one dimension the signal covers both positive and negative time).

  For cosine data sets (e.g. in off-resonance measurements), and for TPPI in t_1:

  Set the imaginary part of the data to zero.

  For sine data sets $\langle \sin \tilde{\omega}_1 t_1 \exp(i\tilde{\omega}_2 t_2) \rangle$ (i.e. $\langle \sin \tilde{\omega}_1 t_1 (A_2 + iD_2) \rangle$ after the t_2-FT):

  Set the imaginary part of the data to zero, and then interchange the real part and the (zero in the) imaginary part, to obtain $\langle i \sin(\tilde{\omega}_1 t_1) A_2 \rangle$ yielding an absorptive real part after the t_1-FT.

  For $\langle \sin \tilde{\omega}_1 t_1 \sin \tilde{\omega}_2 t_2 \rangle$:

  Set the real part of the data to zero.

  For hypercomplex data (i.e. $\langle \cos \tilde{\omega}_1 t_1 (A_2 + iD_2) \rangle$, $\langle \sin \tilde{\omega}_1 t_1 (A_2 + iD_2) \rangle$ after the t_2-FT):

  Combine the real part of the cosine component, and the real part of the sine component as real and imaginary parts of the remaining complex data, $\langle (\cos\tilde{\omega}_1 t_1 + i \sin \tilde{\omega}_1 t_1) A_2 \rangle$.

- After zero filling to obtain the desired number of ω_1 spectral points (e.g. 256 as in Figure 10.11), perform an FT over t_1.
- Phase in ω_1; it is convenient to use the ω_2 slice containing the highest peak of the 2D spectrum. Avoid constant phasing here since the amplitude modulation in t_1, such as $\cos(\tilde{\omega}_1 t_1)$, is independent of the receiver phase. Some linear ω_1 phasing with respect to $\omega_1 = 0$ is appropriate. The same ω_1 phasing is applied to all slices.

- Cut the ω_1 range; for obtaining symmetric exchange spectra consider a sketch like Figure 10.11. If the measurement is done off resonance as indicated in Figure 10.11, cut off one of the two identical ω_1 halves of the spectrum, to retain only the square data array marked by the thick line in Figure 10.11.

If a separate sine data set has been measured, the processing is performed (essentially) with the parameters determined for the cosine data set (ω_2 phasing, for instance, cannot be checked easily on the sine data set because the intensity at $t_1 = 0$ is zero).

In ^{2}H exchange NMR, it has proved favourable to substitute the linear phase correction by t_1-extrapolation as well as t_1- and t_2- interpolation of the data (Schmidt *et al.*, 1988). In the simplest case, a small (integer) number of additional data points is generated at $t_1 \cong 0$. In order to be able to perform fine tuning of the t_1 extrapolation, not just in steps of a full point or dwell time, the whole data field is shifted by interpolation.

10.6 References

de Jong, A. F., Kentgens, A. P. M. and Veeman, W. S. (1984). 2D Exchange NMR in Rotating Solids: Study of Slow Molecular Reorientations. *Chem. Phys. Lett.*, **109**, 337.

Fujara, F., Wefing, S. and Spiess, H. W. (1986). Analogies Between Quasielastic *n*-Scattering and ^{2}H NMR Spin Alignment. *J. Chem. Phys.*, **84**, 4579.

Geen, H. and Bodenhausen, G. (1992). Chemical Exchange NMR with Sideband Suppression. *J. Chem. Phys.*, **97**, 2928.

Hagemeyer, A., Schmidt-Rohr, K. and Spiess, H. W. (1989). 2D NMR for Studying Molecular Order and Dynamics. *Adv. Magn. Reson.*, **13**, 85.

Harbison, G. S., Raleigh, D. P., Herzfeld, J. and Griffin, R. G. (1985). 2D Exchange NMR in Rotating Solids. *J. Magn. Reson.*, **64**, 284.

Jeener, J., Meier, B. H., Bachmann, P. and Ernst, R. R. (1979). Investigation of Exchange Processes by 2D NMR. *J. Chem. Phys.*, **71**, 4546.

Kaufmann, S., Wefing, S., Schaefer, D. and Spiess, H. W. (1990). 2D Exchange NMR of Powders: Transition to Motional Averaging. *J. Chem. Phys.*, **93**, 197.

Kentgens, A. P. M., de Boer, E. and Veeman, W. S. (1987). Ultraslow Motions in Crystalline POM. Complete Elucidation by 2D NMR. *J. Chem. Phys.*, **87**, 6859.

Kenwright, A. M. and Packer, K. J. (1990). T_1 Cancellation in Goldman–Shen Experiments. *Chem. Phys. Lett.*, **173**, 471.

Munowitz, M. G. and Griffin, R. G. (1983). 2D NMR in rotating solids: Time reversal in dip-shift spectra. *J. Chem. Phys.*, **78**, 613.

Oppenheim, I., Shuler, K. E. and Weiss, G. H. (1977). *Stochastic Processes in Chemical Physics: The Master Equation*. MIT Press, Cambridge, MA.

Rössler, E. (1986). 2D Exchange NMR Analyzed in the Time Domain. *Chem. Phys. Lett.*, **128**, 330.

Schaefer, D. and Spiess, H. W. (1992). 2D Exchange NMR of Powders IV: Distribution $g(\tau)$ and Intermediate Dynamic Regime. *J. Chem. Phys.*, **97**, 7944.

Schmidt, C., Wefing, S., Blümich, B. and Spiess, H. W. (1986). Determination of Rotational Angles from 2D NMR of Powders. *Chem. Phys. Lett.*, **130**, 84.

Schmidt, C., Blümich, B. and Spiess, H. W. (1988). ^{2}H 2D Exchange NMR in Solids. *J. Magn. Reson.* **79**, 269.

Spiess, H. W. (1980). ^{2}H Spin Alignment for Studying Ultraslow Motions in Solids. *J. Chem. Phys.*, **72**, 6755.

Szeverenyi, N. M., Sullivan, M. J. and Maciel, G. E. (1982). Observation of Spin-Exchange by 2D ^{13}C CP MAS. *J. Magn. Reson.*, **47**, 462.

CHAPTER ELEVEN

Multidimensional Exchange Spectra: Simulations, Models, Angle Distributions

Keywords: *simulation approaches; reorientation models, exchange matrices, random jumps, rotational diffusion;* R*(β) distributions, inverse problems and best fits, reconstructions, Legendre expansions, direct expansions.*

11.1 Motivation

A considerable amount of information on slow reorientational processes can be extracted model-independent from multidimensional exchange spectra. Nevertheless, to corroborate those conclusions and to determine details of the reorientation geometry, special simulations are required, in particular with $\eta \neq 0$, where for powder samples a direct inversion yielding the relevant jump angles is not possible.

Even inversion or best-fit procedures to extract the reorientation-angle distribution $R(\beta)$ from an exchange spectrum require simulated subspectra (although in principle, by means of the DECODER experiment described in Section 12.7, these powder 2D subspectra could be determined experimentally if $\eta = 0$). Calculations for important motional models, in particular for diffusive motions, are necessary to provide the $R(\beta)$ distributions to compare with the experimental data. Simulations on the basis of such models are widely used in the data analysis. They require much less effort than the full inversion procedures which are sensitive to partial imperfections of the experimental spectra.

11.2 Basic technicalities of spectral simulations

Most of the simulations of 1D or 2D spectra governed by orientation-dependent interactions require a numerical integration over many orientations. In a computer program, this can be implemented simply in terms of loops over the relevant angles (Θ^k, Φ^m) with sufficiently small increments ($<3°$). The corresponding contributions to $S(\omega_1(\Theta^k, \Phi^m), \omega_2(\Theta^k, \Phi^m))$ are weighted with the size of the surface element that the angles stand for, given as $\mathrm{d}\cos\Theta := \cos(\Theta^k) - \cos(\Theta^{k+1})$. Formally, such a numerical integration can be written as:

$$S(\omega_1, \omega_2) = \iint \mathrm{d}\Phi\ \mathrm{d}\cos\Theta\, P(\Theta, \Phi)\ \delta(\omega_1 - \omega(\theta_1(\Theta, \Phi), \phi_1(\Theta, \Phi))$$
$$\times\, \delta(\omega_2 - \omega(\theta_2(\Theta, \Phi), \phi_2(\Theta, \Phi))$$

(see Appendix E). For the actual calculations, various approaches can be taken. Conceptually, the simplest procedure is the calculation directly in frequency space, sorting a contribution from (Θ^k, Φ^m) into the 'frequency channel':

$$(n_{\omega 1}, m_{\omega 2}) = (\mathrm{nint}\{\omega(\theta_1(\Theta^k, \Phi^m), \phi_1(\Theta^k, \Phi^m))\}, \mathrm{nint}\{\omega(\theta_2(\Theta^k, \Phi^m), \phi_2(\Theta^k, \Phi^m))\})$$

where the 'nint' function rounds to the nearest integer. Summation of the surface-element-weighted contributions within each 'frequency channel' approximates the integration in equation (11.1). The calculation time is proportional to the product of the numbers of steps in Θ^k and Φ^m. Many small steps are required to reduce the numerical noise which can occur due to the finite step size for the angles, or due to the rounding of the frequencies. The latter problem can be avoided by calculating time domain signals of the type:

$$g(t_1, t_2) = \iint \mathrm{d}\Phi \ \mathrm{d} \cos \Theta P(\Theta, \Phi) \ \mathrm{e}^{\mathrm{i}t_1\omega\{\theta_1(\Theta, \Phi), \phi_1(\Theta, \Phi)\}} \ \mathrm{e}^{\mathrm{i}t_2\omega\{\theta_2(\Theta, \Phi), \phi_2(\Theta, \Phi)\}} \quad (11.2)$$

using the frequencies without rounding. The integral (11.2) is in practice again approximated by a sum. It must be calculated point by point in terms of t_1 and t_2, which is time-consuming.

Ruggedness of a spectrum calculated in the frequency domain with finite angular steps is particularly large in spectral regions where the frequency is most sensitive to the integration angles (Θ, Φ). As a simple example, consider a 1D spectrum spread over 128 frequency channels, with $\eta = 0$ and $\Theta = \theta$. Here, $\Theta = 45^\circ$ contributes to channel 64, $\Theta = 46^\circ$ to channel 62 ($128 \cos^2(46^\circ) = 61.766$). Thus, with an angular increment of 1°, a deep dip will remain at frequency channel 63. The use of more spectral points, say 1024 instead of 128, can help to reduce the effect of frequency-rounding errors in 1D spectra, producing 'stick spectra' which after appropriate convolution are essentially equivalent to the spectra obtained from the time signals. In multidimensional simulations, however, this approach would result in prohibitively large data fields (typically 1024^n for n-dimensional spectra).

11.3 2D powder spectra for axially symmetric coupling ($\eta = 0$)

A discrete jump with a fixed reorientation angle β probed by a uniaxial nuclear interaction ($\eta = 0$) presents the simplest situation for the calculation of an exchange 2D powder pattern. In addition, a set of such spectra $S(\omega_1, \omega_2, \beta)$ for a definite reorientation angle β can actually be useful for the simulation with any motional model: for $\eta = 0$, any 2D powder spectrum can be obtained by superposition of the spectra $S(\omega_1, \omega_2, \beta)$ weighted with the reorientation-angle distributions $R(\beta)$ for the particular process (see Section 7.4). In this approach, once a full set of subspectra $S(\omega_1, \omega_2, \beta)$ has been generated (1° steps in β are usually sufficient), the problem of simulating 2D spectra for any reorientational process (in the slow-exchange limit) reduces to determining $R(\beta)$.

For these $\eta = 0$ exchange powder spectra, analytical expressions can be calculated. A slice at a given $\tilde{\omega}_1$ corresponds to the 1D spectrum of a conical distribution: The C—^{2}H bonds are located on a cone with apex angle 2β around the initial direction with its polar angle $\theta_1(\tilde{\omega}_1)$ relative to $\vec{\mathbf{B}}_0$ (see Figure 7.6). Taking the initial C—^{2}H bond orientation as a temporary z axis, according to the three-axes relation (A.14) we find:

$$\cos \theta_2 = \cos \beta \cos \theta_1 + \sin \beta \sin \theta_1 \cos \psi \quad (11.3)$$

where ψ varies around the cone. From the constant density of axes on the cone, $P(\cos\theta_1, \psi) = p_0$, we can calculate the searched cross-section $P(\cos\theta_1, \omega_2)$ or $S(\tilde{\omega}_1, \omega_2)_1$ on the basis of the differential conservation of the integral (Appendix E), most conveniently in two steps, $\psi \to \cos\theta_2 \to \omega_2$:

$$P(\cos\theta_1, \cos\theta_2)\,|\,\mathrm{d}\cos\theta_2\,| = P(\cos\theta_1, \psi)\,|\,\mathrm{d}\psi\,|$$

$$\Rightarrow P(\cos\theta_1, \cos\theta_2) = \frac{P(\cos\theta_1, \psi)}{|\,\mathrm{d}\cos\theta_2/\mathrm{d}\psi\,|} = \frac{p_0}{\sin\beta\sin\theta_1\sin\psi(\theta_2)}$$

$$= p_0 / \sqrt{\sin^2\beta\sin^2\theta_1 - \{\cos\theta_2 - \cos\beta\cos\theta_1\}^2} \tag{11.4a}$$

where we have expressed $\sin\psi(\theta_2) = \sqrt{1 - \cos^2\psi(\theta_2)}$ through $\cos\psi(\theta_2)$ as obtained from equation (11.3). The distribution of equation (11.4a) has singularities at $\cos\theta_2 = \cos(\theta_1 \pm \beta)$ and vanishes outside the $\cos\theta_2$ range enclosed by the singularities (for a visualization of the related $P(\cos\theta_1, \theta_2)$ distribution, see Figures 7.6 and 12.3). The second step is the transformation from $\cos\theta_2$ to ω_2, where $\omega_2 = \delta\frac{1}{2}(3\cos^2\theta_2 - 1)$ is inverted to $\cos\theta_2 = \pm\sqrt{\frac{2}{3}(\omega_2/\delta + \frac{1}{2})}$:

$$S(\cos\theta_1, \omega_2)\,|\,\mathrm{d}\omega_2\,| = P(\cos\theta_1, \cos\theta_2)\,|\,\mathrm{d}\cos\theta_2\,|$$

$$S(\cos\theta_1, \omega_2) = p_0 / \sqrt{6\delta(\omega_2 + \tfrac{1}{2}\delta)}\,\sqrt{\sin^2\beta\sin^2\theta_1 - \{\pm\sqrt{\tfrac{2}{3}(\omega_2/\delta + \tfrac{1}{2})} - \cos\beta\cos\theta_1\}^2} \tag{11.4b}$$

using $\mathrm{d}\cos\theta_2/\mathrm{d}\omega_2 = 1/\sqrt{6\delta(\omega_2 + \frac{1}{2}\delta)}$, which is the Pake spectrum (2.47). The spectral intensity vanishes in the ω_2 regions for which the argument under either root in the denominator of (11.4b) is negative. The '$\pm$' sign in the denominator depends on the sign of $\cos\theta_2$, which changes at $\theta_2 = \pi/2$: if the cosine is positive, $\cos\theta_2 = +\sqrt{\frac{2}{3}(\omega_2/\delta + \frac{1}{2})}$, while in the negative lobe of the cosine function, $\cos\theta_2 = -\sqrt{\frac{2}{3}(\omega_2/\delta + \frac{1}{2})}$. The partial spectra resulting with either sign must be summed up to produce the total spectrum.

The cross-section spectrum (11.4b) has up to three singularities, occurring at the ω_2 values for which the denominator is zero. They correspond to the two crossings through the elliptical ridge at $\omega_2 = \delta\frac{1}{2}(3\cos^2(\theta_1 \pm \beta) - 1)$ (check the appearance of the singularities by inserting these ω_2 values into equation (11.4b)), and to the '90° ridge' at $\omega_2 = -\frac{1}{2}\delta = \omega(\theta_2 = 90°)$, where the first square-root in the denominator of equation (11.4b) vanishes. These prominent features of the 2D powder spectrum for $\eta = 0$ were rationalized in a simple manner in Section 7.3. $S(\omega_1, \omega_2)$ is obtained from $S(\cos\theta_1, \omega_2)$ according to $S(\omega_1, \omega_2)\,|\,\mathrm{d}\omega_1\,| = S(\cos\theta_1, \omega_2)\,|\,\mathrm{d}\cos\theta_1\,|$ analogous to the procedure in equation (11.4b). This amounts to only a few substitutions in equation (11.4b): it is multiplied by a Pake spectrum $\mathrm{d}\cos\theta_1/\mathrm{d}\omega_1 = 1/\sqrt{6\delta(\omega_1 + \frac{1}{2}\delta)}$, and $\cos\theta_1$ is substituted by $\sqrt{\frac{2}{3}(\omega_1/\delta + \frac{1}{2})}$. As a simple alternative to the analytical calculation, the 2D spectra for $\eta = 0$ can also be generated according to the numerical procedure for arbitrary η which is outlined in Section 11.5.1.

11.4 Reorientation-angle distributions for various models

As was discussed above, 2D spectra for $\eta = 0$ can be calculated directly from the reorientation-angle distribution $R(\beta)$. To simulate the 2D spectra for a certain motional model, it is therefore sufficient to determine $R(\beta, t_m)$. For several important models, the procedure for obtaining the reorientation-angle distribution is outlined here.

11.4.1 Random jumps

A standard model of reorientation in disordered materials is the random-jump model. A possible picture of its physical meaning is the following. The system is rigid except for places visited by roaming density fluctuations dilating the structure so much that isotropic reorientations occur; after the fluctuation has gone by, the structure is immobile again. The reorientation angle distribution $R(\beta; t_m)$ thus consists of a peak at $\beta = 0^\circ$, from those segments not reached by a fluctuation within t_m, and the broad $\sin\beta$ contribution from those isotropically redistributed by the motion. As the probability per unit time for a bond to be reached by a fluctuation is naturally assumed to be a constant $1/\tau$, the $\beta = 0$ component decreases as $\exp(-t_m/\tau)$ and the exchange rises as $1 - \exp(-t_m/\tau)$. 2D spectra can thus be easily generated as a weighted superposition of a diagonal spectrum and a spectrum with full (isotropic) exchange (Figure 11.1). A distribution of correlation times $g(\tau)$ in this model obviously changes only the time dependence of the decrease of diagonal spectrum:

$$w_{dia}(t_m) = \int d\tau\, g(\tau)\, e^{-t_m/\tau} \tag{11.5a}$$

and the increase of the isotropic exchange:

$$w_{exch}(t_m) = 1 - w_{dia}(t_m). \tag{11.5b}$$

The exchange matrix for this process is obtained after rendering $\cos\theta$ discrete, for instance in $N = 100$ intervals. The probability per unit time of reaching one $\cos\theta$ surface element from any other is $k = \tau^{-1}/N$):

$$\mathbf{\Pi} = \begin{bmatrix} -(N-1)k & k & \cdots & k \\ k & -(N-1)k & \cdots & k \\ \vdots & \vdots & \ddots & \vdots \\ k & k & & -(N-1)k \end{bmatrix}.$$

11.4.2 Uniaxial rotational diffusion

Another simple model is the diffusion of a bond (a vector) on a cone with an apex angle of 2χ. The process represents simply a one-dimensional diffusion with periodic, or equivalently reflecting, boundary conditions. Starting at the beginning of the mixing time from a position $\psi = 0$ on the cone, the solution $P(\psi, t_m)$ is just a Gaussian with maximum at $\psi = 0$, of single-sided width $\sigma = \sqrt{2Dt_m}$, wrapping around on the circle so that its values at $\psi + n \times 360^\circ$ are added up. The symmetry with respect to $\psi = 0^\circ$ allows to restrict the range of ψ to $0 \leqslant \psi \leqslant 180^\circ$. The reorientation-angle distribution

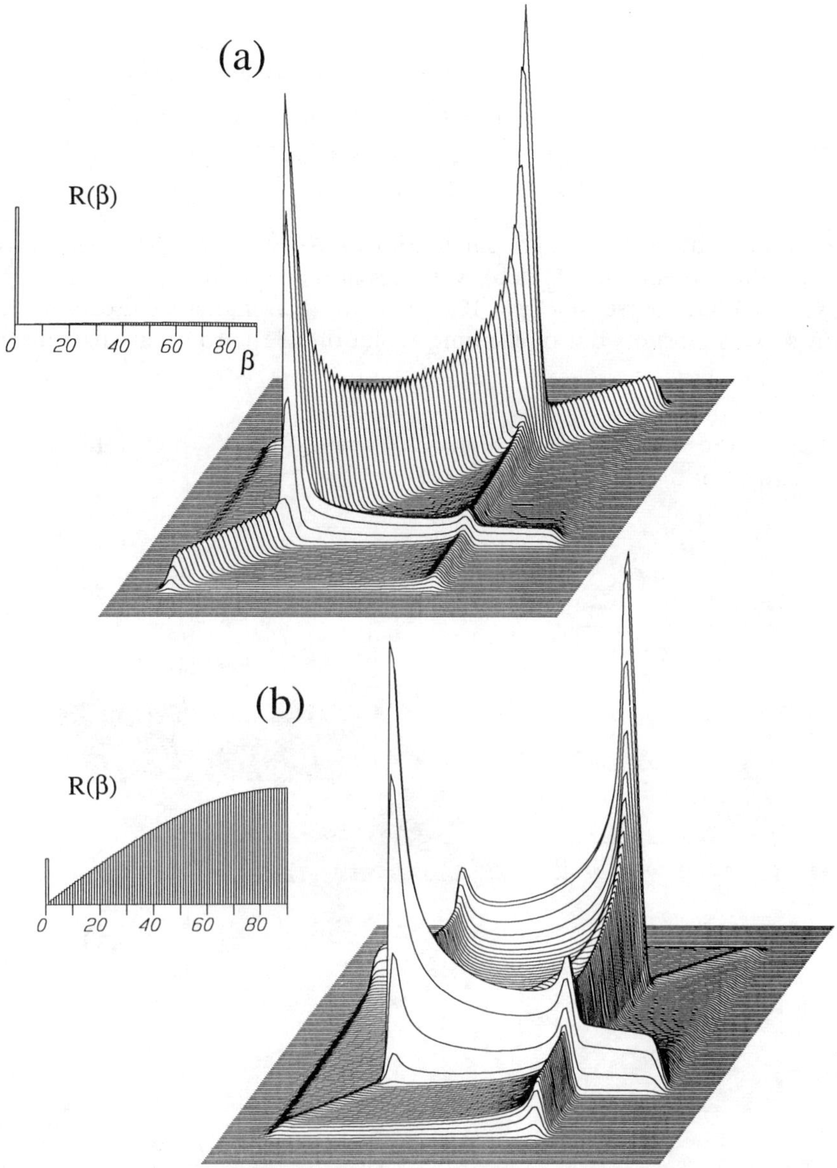

Figure 11.1 ^{2}H 2D exchange spectra resulting from the random-jump model, as a function of t_m/τ_c. The underlying reorientation-angle distributions $R(\beta)$ are shown as inserts. The spectra are simply weighted superpositions of diagonal and full-exchange patterns. (a) $t_m/\tau_c = 1$; (b) $t_m/\tau_c = 5$.

$R_f(\beta; t_m)$ is obtained from the rotation-angle distribution $P(\psi; t_m)$ through the usual conservation of area of corresponding intervals (Appendix E):

$$R_f(\beta; t_m)\ d\beta = P(\psi; t_m)\ d\psi$$

$$\Rightarrow R_f(\beta; t_m) = P(\psi(\beta); t_m)\ 1/\left(\frac{d\beta}{d\psi}\right) \qquad (11.6a)$$

with

$$\cos\,\beta = \cos^2\chi + \sin^2\chi\,\cos\,\psi$$

(see (A.14)). Using d $[\cos(\beta(\psi))]/\mathrm{d}\psi = (-\sin\,\beta)\mathrm{d}\beta/\mathrm{d}\psi$, we obtain:

$$\frac{\mathrm{d}\beta}{\mathrm{d}\psi} = \sin^2\chi\,\frac{\sin\,\psi}{\sin\,\beta} = 2\,\frac{\sin(\frac{1}{2}\beta)}{\sin\,\beta}\,\sin\,\chi\,\sqrt{1 - \frac{\sin^2(\frac{1}{2}\beta)}{\sin^2\chi}} \tag{11.6b}$$

to insert into equation (11.6a). A singularity in $R_\mathrm{f}(\beta; t_\mathrm{m})$ occurs where $\mathrm{d}\beta/\mathrm{d}\psi = 0$, which, according to equation (11.6b), vanishes at $\psi = 180^\circ$ or equivalently for $\beta = 2\chi$ (unless $\chi = 90^\circ$). Of course, if $P(\psi = 180^\circ; t_\mathrm{m}) = 0$, this singularity is cancelled. Figure 11.2 shows two snapshots out of the time evolution of the rotation-angle distribution

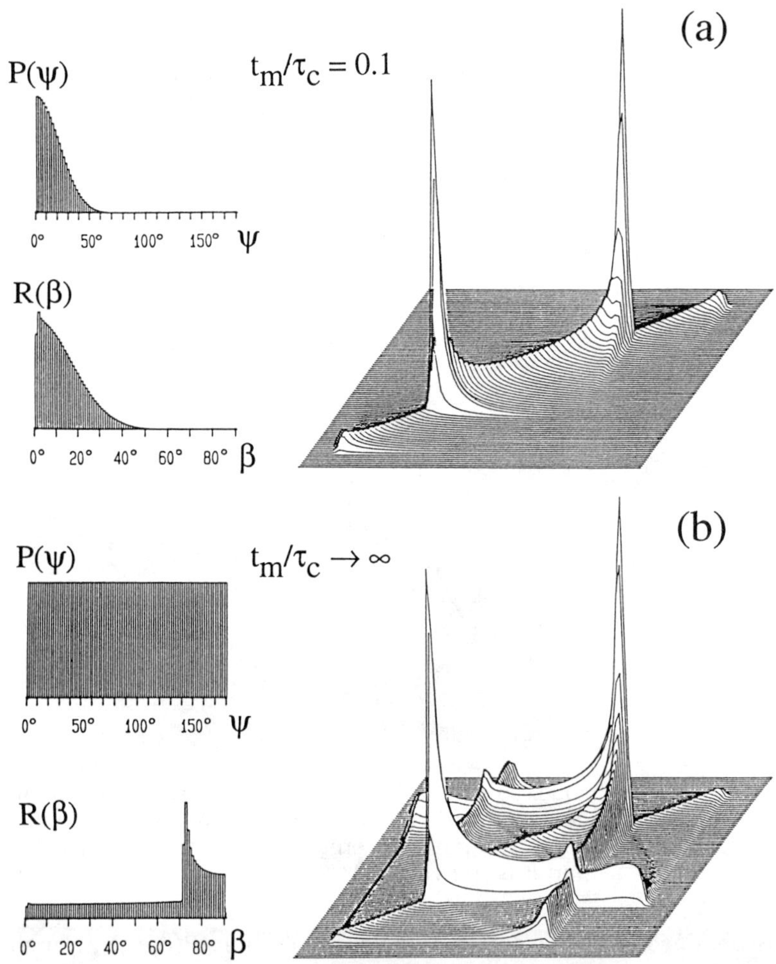

Figure 11.2 ^{2}H 2D exchange spectra for uniaxial rotational diffusion with the rotation-angle distributions $P(\psi; t_\mathrm{m}/\tau_\mathrm{c})$ and the reorientation-angle distributions $R(\beta; t_\mathrm{m}/\tau_\mathrm{c})$ shown as inserts (a) $t_\mathrm{m}/\tau_\mathrm{c} = 0.1$; (b) $t_\mathrm{m}/\tau_\mathrm{c} \rightarrow \infty$. Note the singularity in $R(\beta, t_\mathrm{m}/\tau_\mathrm{c} \rightarrow \infty)$, which arises due to the differential conservation of the integral at the point where $\mathrm{d}\beta/\mathrm{d}\psi = 0$. The corresponding elliptical ridge is observed faintly in the 2D spectrum.

$P(\psi; t_m)$, the corresponding reduced RADs $R(\beta; t_m) = R_f(\beta; t_m) + R_f(180^\circ - \beta; t_m)$ and the ^{2}H 2D exchange spectra.

11.4.3 Isotropic rotational diffusion

Isotropic rotational diffusion describes diffusion of a point (tip of a unit vector, C—^{2}H bond) on the unit sphere. In the calculation of the corresponding reorientation angle distribution, the angular grid is most conveniently chosen such that the vector starts from the north pole at $t_m = 0$. Then, due to the isotropy there is no Φ dependence of the probability $P(\beta, \Phi; t_m)$ of finding the vector at (β, Φ) after a time t_m. The RAD is simply $R_f(\beta, t_m) = \sin\beta . P(\beta, \Phi; t_m)$. The rotational-diffusion equation reads:

$$\frac{\partial P(\beta, \Phi, t)}{\partial t} = D\,\hat{\Delta}_\beta P(\beta, \Phi, t) \quad \hat{\Delta}_\beta = \frac{1}{\sin\beta}\frac{\partial}{\partial\beta}\left(\sin\beta\frac{\partial}{\partial\beta}\right). \tag{11.7}$$

This Laplacean $\hat{\Delta}_\beta$ in terms of β is familiar from the solution of the Schrödinger equation with a central potential (the hydrogen atom). For practical purposes, equation (11.7) can be solved in a discrete form (Sillescu, 1971; Wefing *et al.*, 1988) with N angular values spaced by an angular increment $\Delta = 180^\circ/N$, making the following substitutions:

$$\sin\beta\left.\frac{\partial P}{\partial\beta}\right|_{\beta_m} \rightarrow \sin\beta_m \frac{P(\beta_m + \frac{1}{2}\Delta, \Phi, t) - P(\beta_m - \frac{1}{2}\Delta, \Phi, t)}{\Delta} \tag{11.8a}$$

$$\frac{\partial}{\partial\beta}\left.\left(\sin\beta\frac{\partial P}{\partial\beta}\right)\right|_{\beta_m} \rightarrow \frac{1}{\Delta}\left\{\sin(\beta_m + \tfrac{1}{2}\Delta)\frac{P(\beta_m + \Delta, \Phi, t) - P(\beta_m, \Phi, t)}{\Delta} - \sin(\beta_m - \tfrac{1}{2}\Delta)\frac{P(\beta_m, \Phi, t) - P(\beta_m - \Delta, \Phi, t)}{\Delta}\right\} \tag{11.8b}$$

with $\beta_m = (m - \frac{1}{2})\Delta$. Setting:

$$P_n(t) := R(\beta_n, t) = P(\beta_n, \Phi, t)\sin\beta_n$$

$$\Rightarrow \frac{\partial P_m(t)}{\partial t} = \frac{\partial P(\beta_m, \Phi, t)}{\partial t}\sin\beta_m \tag{11.9}$$

the diffusion equation (11.7) with equations (11.8) and (11.9) inserted turns into a discrete master equation $\partial P_m/\partial t = \Sigma_n \Pi_{mn} P_n$, with a tridiagonal exchange matrix $\mathbf{\Pi}$ made up of the nonvanishing elements:

$$\Pi_{m,m\pm1} = \frac{D}{\Delta^2}\frac{\sin(\beta_m \pm \frac{1}{2}\Delta)}{\sin(\beta_{m\pm1})} \tag{11.10a}$$

$$\Rightarrow \Pi_{m\pm1,m} = \frac{D}{\Delta^2}\frac{\sin(\beta_m \pm \frac{1}{2}\Delta)}{\sin\beta_m} \tag{11.10b}$$

$$\Pi_{m,m} = -2\frac{D}{\Delta^2}\cos(\tfrac{1}{2}\Delta) \cong -2\frac{D}{\Delta^2} \tag{11.10c}$$

(all indices m or $m \pm 1$ are to be restricted to the range from 1 to N). The relations (11.10) obviously describe jumps between nearest neighbours, typical of a diffusive

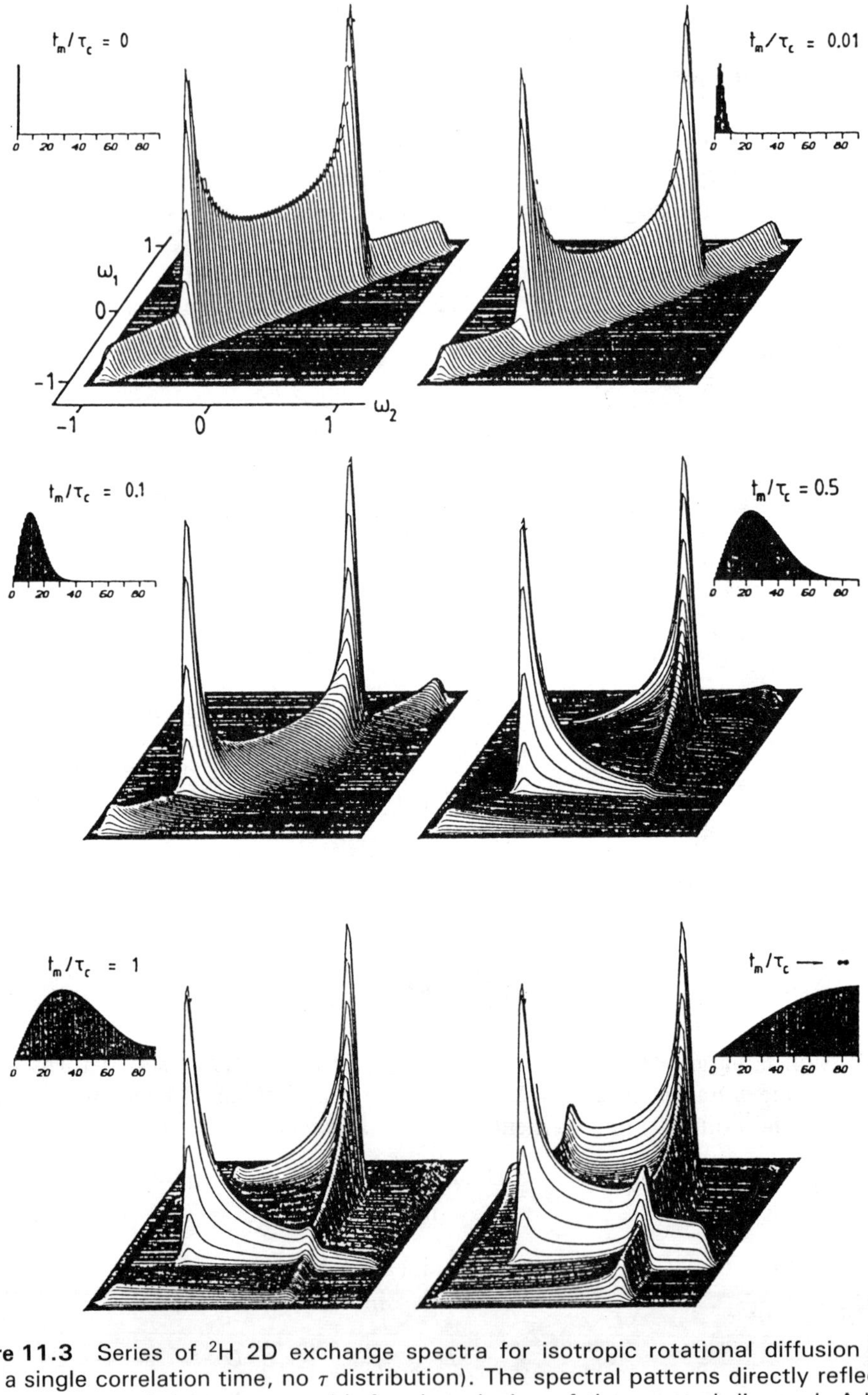

Figure 11.3 Series of ^{2}H 2D exchange spectra for isotropic rotational diffusion (each with a single correlation time, no τ distribution). The spectral patterns directly reflect the diffusive nature of the process, with fast broadening of the spectral diagonal. Adapted from Wefing *et al.* (1988).

process. The requirement (9.63) of a vanishing sum over a column, $\Sigma_m \Pi_{mn} = 0$, is automatically fulfilled, even for the first and last columns of the exchange matrix.

To obtain the jump-angle distribution $R(n\Delta\beta; t) = P_n(t)$, with $P_n(0) = \delta_{n1}$ representing the start at the north pole, according to Section 9.4 the exchange matrix must be diagonalized. Figure 11.3 displays the time-evolution of $R(\beta; t_m/\tau_c)$ for isotropic rotational diffusion with a single correlation time. The diffusive broadening of the reorientation-angle distribution is directly reflected in the broadening of the diagonal of the corresponding ^2H 2D exchange spectra.

11.5 Spectra for non-axially-symmetric coupling ($\eta > 0$)

For couplings with nonvanishing asymmetry parameter ($\eta > 0$), the spectral simulations are more complex than for $\eta = 0$ since the spectrum depends on three reorientation angles instead of a single one in the case $\eta = 0$ (see Section 7.5). In the most general case, a three-dimensional distribution in terms of the three reorientation angles would be required to generate a spectrum from a correspondingly immense number of subspectra. Therefore, other approaches must be taken.

11.5.1 Simulation for discrete jumps

In this section, we shall describe a useful approach for the simulation of 2D exchange spectra for discrete jumps. The simulation of DECODER spectra, or of dip-shift correlation spectra for static samples, can be performed along the same lines. In the latter case (see Section 6.7.3), in the ω_1 dimension the $\boldsymbol{\sigma}$ matrix is simply replaced by the $\frac{1}{2}\mathbf{D}$ matrix with its PAS diagonal elements δ_D and $-\frac{1}{2}\delta_D(1 \pm \eta_D)$.

For calculating the anisotropic frequencies, we exploit the simplifications by the transition into the unified PAS as discussed in Section 7.5. According to equation (2.28), the frequency can be calculated as:

$$\omega_1 = \vec{\mathbf{b}}_0^{PAS} \begin{bmatrix} \sigma_x & & \\ & \sigma_y & \\ & & \sigma_z \end{bmatrix} \vec{\mathbf{b}}_0^{PAS} \tag{11.11}$$

with the unit vector $\vec{\mathbf{b}}_0^{PAS} := \vec{\mathbf{B}}_0^{PAS}/B_0$ and the frequency given in ppm of the Larmor frequency (i.e. the absolute frequency is obtained after multiplying relation (11.11) by $(-\omega_0)$). The interaction tensor after a reorientation by the Euler angles (α, β, γ) is expressed in the original PAS according to:

$$\boldsymbol{\sigma}_2 = \mathbf{R}(\alpha, \beta, \gamma) \begin{bmatrix} \sigma_x & & \\ & \sigma_y & \\ & & \sigma_z \end{bmatrix} \mathbf{R}^{-1}(\alpha, \beta, \gamma) \tag{11.12}$$

(see equation (B.18)). The frequency after the jump is thus given by:

$$\omega_2 = \vec{\mathbf{b}}_0^{PAS} \boldsymbol{\sigma}_2 \vec{\mathbf{b}}_0^{PAS}. \tag{11.13}$$

For simulation of a powder spectrum, the unit vector:

$$\vec{\mathbf{b}}_0^{\mathrm{PAS}} = \begin{bmatrix} \cos\Phi \sin\Theta \\ \sin\Phi \sin\Theta \\ \cos\Theta \end{bmatrix} \tag{11.14}$$

is rotated through all orientations of a hemisphere by systematically varying Θ and Φ; either angle covers the range between 0 and 180° in steps of about 1°. Alternatively, Φ can be stepped through 360° and Θ only through 90°. Each such (Θ, Φ) orientation makes a contribution of $|\Delta \cos\Theta| = |\cos\Theta_i - \cos\Theta_{i-1}|$ to the spectral intensity at (ω_1, ω_2).

Often, it is useful to evaluate the orientation-dependent frequency not directly in the unified PAS, but rather in a molecular frame (rotated but fixed with respect to the PAS) in which the motional process can be characterized more easily. In the case of reorientations around a given axis, such as for two-site jumps or for helical jumps, the rotation axis is chosen as the z axis of the molecular frame MF. For evaluating the equivalent of equation (11.12):

$$\omega_j = \vec{\mathbf{b}}_0^{\mathrm{MF}} \boldsymbol{\sigma}_j^{\mathrm{MF}} \vec{\mathbf{b}}_0^{\mathrm{MF}} \tag{11.15}$$

the interaction tensors must be rotated into the MF:

$$\boldsymbol{\sigma}_1^{\mathrm{MF}} = \mathbf{R}^{-1}(\alpha_{\mathrm{P}}, \beta_{\mathrm{P}}, 0) \begin{bmatrix} \sigma_x & & \\ & \sigma_y & \\ & & \sigma_z \end{bmatrix} \mathbf{R}(\alpha_{\mathrm{P}}, \beta_{\mathrm{P}}, 0) \tag{11.16a}$$

$$\boldsymbol{\sigma}_2^{\mathrm{MF}} = \mathbf{R}^{-1}(\alpha_{\mathrm{P}}, \beta_{\mathrm{P}}, \psi) \begin{bmatrix} \sigma_x & & \\ & \sigma_y & \\ & & \sigma_z \end{bmatrix} \mathbf{R}(\alpha_{\mathrm{P}}, \beta_{\mathrm{P}}, \psi). \tag{11.16b}$$

The Euler angles α_{P} and β_{P} are the polar coordinates of the rotation axis in the PAS, while ψ denotes the rotation angle around the molecular axis (e.g. $\psi = 180^\circ$ for a two-site jump, $\psi \cong n \times 200^\circ$ for the helical jumps in POM, and $\psi \cong n \times 103^\circ$ in PEO).

For uniaxially oriented samples, in the molecular frame the orientational properties can be taken into account very conveniently. While the integral over Φ is performed as before, the contributions for different Θ are weighted by the orientation distribution $u(\Theta)$, which is centred on the average angle Θ_0 between the sample director and the $\vec{\mathbf{B}}_0$ field. If the distribution is sufficiently narrow, the Θ integration range can easily be restricted; this can reduce the calculation time considerably.

11.5.2 Simulation for random jumps ($\eta > 0$)

A random-jump 2D exchange spectrum was shown in Section 11.4.1 to consist just of a diagonal powder spectrum and the exchange 2D pattern for completely isotropic reorientation, with weighting factors (11.5). The spectrum for completely isotropic reorientation is simply made up of 1D powder patterns $S(\omega_1)$ and $S(\omega_2)$ in cross sections parallel to either axis, see Figure 11.4: after the mixing time, the segments are isotropically distributed and thus yield powder patterns $S(\omega_2)$ in ω_2 for every ω_1 slice.

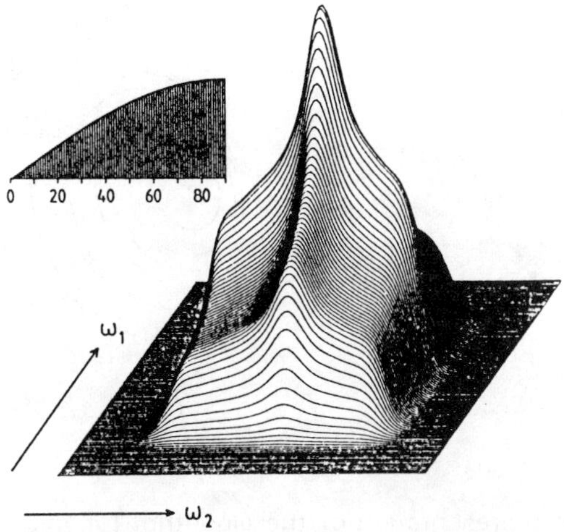

Figure 11.4 Simulated spectrum for $\eta = 0.82$ and complete isotropic exchange. Each cross-section parallel to either coordinate axis has the shape of the 1D powder pattern.

The height of these slices is determined by $S(\omega_1)$, the *a priori* probability of finding a frequency ω_1, which is itself a powder pattern. Consequently, the final-state spectrum is given by:

$$S_{\mathrm{iso}}(\omega_1, \omega_2; t_{\mathrm{m}} \to \infty) = S(\omega_1)S(\omega_2). \tag{11.17}$$

The cross-section along the diagonal is consequently proportional to $S^2(\omega)$, the square of the 1D powder pattern. This enhances the spectral maximum (at ω_y).

11.5.3 Simulation for isotropic rotational diffusion ($\eta > 0$)

Isotropic rotational diffusion represents an important motional model. With $\eta \neq 0$, the direct approach of considering the reorientations of the molecular segments in the laboratory frame leads to complicated situations, because in addition to the diffusive motion of the z axis of the PAS, rotation around this axis would also have to be taken into account. Therefore, it is advantageous to view the reorientational process from the unified PAS (see Section 7.5), where it can be described simply as an isotropic rotational diffusion of the $\vec{\mathbf{B}}_0$ axis. Using the reorientation-angle distribution $R(\beta, t_{\mathrm{m}})$ of an axis as calculated in Section 11.4.3, a pair of initial (θ_1, ϕ_1) and final (θ_2, ϕ_2) $\vec{\mathbf{B}}_0$ field orientations, separated by β, with:

$$\cos\beta = \cos\theta_1 \cos\theta_2 + \sin\theta_1 \sin\theta_2 \cos(\phi_1 - \phi_2) \tag{11.18}$$

(see (A.14)) contributes an intensity $\{R(\beta; t_{\mathrm{m}})/\sin\beta\} \sin\theta_1 \sin\theta_2$ at $(\omega(\theta_1, \phi_1), \omega(\theta_2, \phi_2))$. The scanning of the angles (θ_1, ϕ_1), (θ_2, ϕ_2) and the determination of β is indicated in Figure 11.5. Spectra simulated on the basis of these relations are shown in Section 8.4.

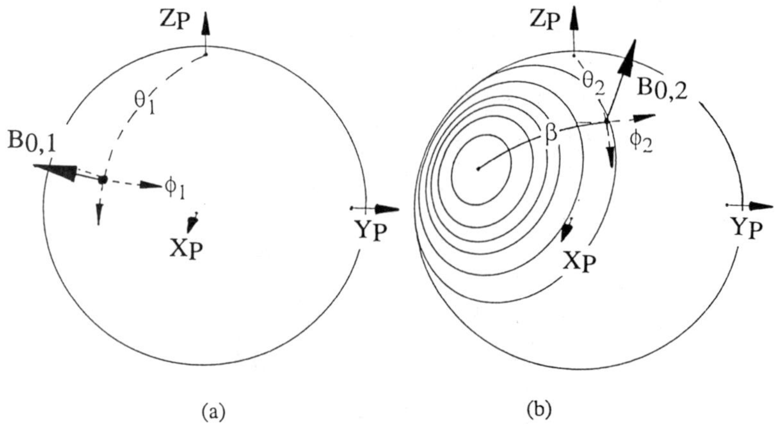

Figure 11.5 Schematic representation of the algorithm for the simulation of isotropic-rotational-diffusion 2D spectra with $\eta > 0$. (a) In the unified PAS, the initial $\vec{\mathbf{B}}_0$ direction $\vec{\mathbf{B}}_{0,1}$ is varied through θ_1 and ϕ_1 and the corresponding evolution-time frequency $\omega_1 = \omega(\theta_1, \phi_1)$ determined. (b) For each $\vec{\mathbf{B}}_{0,1}$ orientation, the detection-time orientation $\vec{\mathbf{B}}_{0,2}$ of the field in the PAS is varied over the whole unit sphere. The angular distance β between $\vec{\mathbf{B}}_{0,1}$ and $\vec{\mathbf{B}}_{0,2}$ determines the weight of the corresponding signal at the spectral point $(\omega(\theta_1, \phi_1), \omega(\theta_2, \phi_2))$. The usually bell-shaped weighting function $R(\beta; t_m)/\sin\beta$ is indicated by concentric circles.

11.6 Model-independent extraction of angular distributions from NMR spectra

Angular distributions are the ultimate results to be extracted from many 1D and 2D NMR spectra. This applies to reorientation-angle distributions discussed in this chapter as well as to orientation distributions that will be of interest in Chapter 12. In this section, different approaches to attain these distributions will be compared:

(i) direct reconstruction
(ii) direct expansion (best fit of delta-function subspectra)
(iii) moment expansion (best fit of Legendre-polynomial-weighted spectra).

The discussion to follow now will be restricted to the case of one-dimensional distributions $R(\beta)$, which is particularly simple and therefore best suited to make the principles clear. For the moment expansion, the full treatment in terms of Wigner functions will be outlined in Section 12.3.3.

Direct reconstruction (i) is the most convincing method of obtaining the distribution, but is applicable only to experiments amenable in this respect. It requires that the spectrum contains a direct image of the angular distribution, that is, a single spectral peak for each angular peak, and no coincidence of spectral peaks. The distribution can then be reconstructed directly from the spectral intensity, which needs only to be corrected for the frequency-to-angle distortion according to the principle of conservation of the area of corresponding intervals ('$S(\omega)\,d\omega = R(\beta)\,d\beta$', see Appendix E). This does not involve any data fitting at all, and is a relatively simple and robust procedure. Direct reconstruction is possible with a number of experiments in the case of interactions with $\eta = 0$: the one-dimensional orientation distribution $u(\theta)$ of $\vec{\mathbf{Z}}_P$ axes relative to $\vec{\mathbf{B}}_0$ can be obtained from 1D spectra for $I = \frac{1}{2}$; the two-dimensional orientation distribution of $\vec{\mathbf{Z}}_P$ axes can be extracted from 2D DECODER spectra to some extent, and fully, even for ^{2}H with its two 'transitions', from the 3D DECODER; and the reorientation angle distribution can be reconstructed from the 'edge slice' $S(\delta, \omega_2)$ of 2D exchange spectra.

Approach (ii) involves the expansion of a given experimental spectrum $S^{ex}(\vec{\omega})$ in terms of subspectra $S^{th}(\vec{\omega}; \beta)$ for individual β values:

$$S^{ex}(\vec{\omega}) = \int_{0^\circ}^{90^\circ} S^{th}(\vec{\omega}; \beta) R(\beta)\, d\beta. \tag{11.19}$$

(Here and in the following, the range of β is restricted to the interval between 0° and 90°, since the second-rank tensor nature does not allow us to distinguish between β and 180° − β.) A discrete version of equation (11.19) displays its character as a linear equation more clearly:

$$S_\omega^{ex} = \sum_\beta S_{\omega\beta}^{th} R_\beta. \tag{11.20}$$

This can also be expressed as a matrix equation:

$$\vec{\mathbf{S}}^{ex} = \mathbf{S}^{th} \vec{\mathbf{R}}. \tag{11.21}$$

Such a direct-expansion procedure has been applied by Hagemeyer *et al.* (1990) to experimental 2D exchange spectra for extracting the reorientation-angle distribution $R(\beta)$ (Figure 11.6).

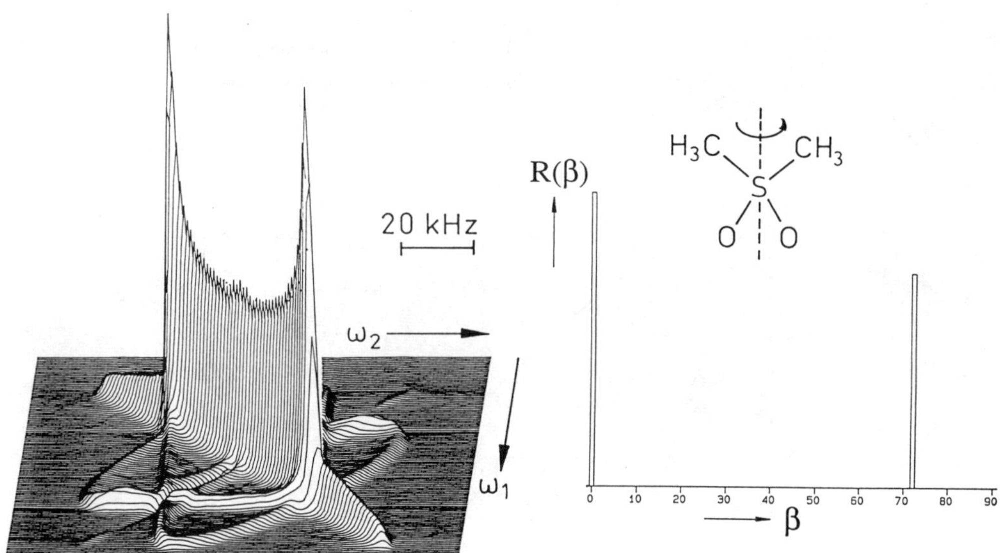

Figure 11.6 Left: ^{2}H 2D exchange spectrum of dimethylsulfone $(CH_3)_2SO_2$ undergoing two-site jumps. Right: reorientation-angle distribution obtained by a best-fit of slices out of the spectrum. The reorientation angle of 72° found by the best-fit algorithm is equivalent to 180° − 72° = 108°, which is the angle between the C_3 axes of the exchanging CH_3 groups in this molecule. Adapted from Hagemeyer *et al.* (1990).

The moment expansion, approach (iii), is traditionally applied in the analysis of orientation distributions of ordered materials. In fact, several classical methods (such as birefringence or dichroism) for characterizing orientation yield directly the ensemble averages of $P_2(\cos\Theta)$ or $P_4(\cos\Theta)$, which correspond to orientational moments $\langle P_2\rangle$ and $\langle P_4\rangle$. The second moment $\langle P_2\rangle$ is also convenient for roughly characterizing the width and the centre of the orientation distribution. These points will be discussed in more detail in Section 12.3.1. The moment expansion involves the expansion of the angular distribution in terms of the Legendre polynomials $P_L(\cos\beta)$:

$$\tilde{R}(\beta) = \sum_L (2L+1)\langle P_L\rangle P_L(\cos\beta). \tag{11.22}$$

The $P_L(\cos\beta)$ for $L = 2, 4, 6, 8$ are plotted in Figure 11.7. Each order parameter $\langle P_L\rangle$ characterizes the weight with which the corresponding $P_L(\cos\beta)$ contributes to $\tilde{R}(\beta)$:

$$\langle P_L\rangle = \int_0^1 \mathrm{d}\cos\beta\, \tilde{R}(\beta)P_L(\cos\beta) = \int_{-1}^1 \mathrm{d}\cos\beta\, \tilde{R}(\beta)P_L(\cos\beta)\Big/\int_{-1}^1 \mathrm{d}\cos\beta. \tag{11.23}$$

In equation (11.22), the expansion is written for $\tilde{R}(\beta) = R(\beta)/\sin\beta$ to be consistent with common conventions. $\tilde{R}(\alpha, \beta)$ is the probability density of finding an orientation with the polar angles (α, β), or a corresponding reorientation in the case of exchange NMR. The conventions expect that for an isotropic distribution, defined by $\tilde{R}(\beta) = 1$ which results in $R(\beta) \sim \sin\beta$, all higher moments vanish, $\langle P_L\rangle = \delta_{L,0}$; that holds only if the distribution is a constant, like $\tilde{R}(\beta) = 1$, and is not true for $R(\beta) \sim \sin\beta$.

The expansion (11.22) of $\tilde{R}(\beta)$ can be inserted into equation (11.19), which produces:

$$S^{\mathrm{ex}}(\vec{\omega}) = \sum_L (2L+1)\langle P_L\rangle \underbrace{\int_{0^\circ}^{90^\circ} S^{\mathrm{th}}(\vec{\omega};\beta)P_L(\cos\beta)\,\mathrm{d}\beta}_{S_L^{\mathrm{th}}(\vec{\omega})}. \tag{11.24}$$

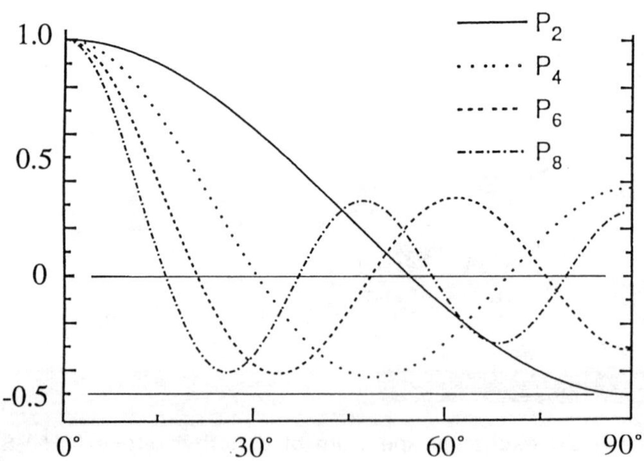

Figure 11.7 Legendre polynomials $P_L(\cos\beta)$ for $L = 2, 4, 6, 8$. Note the decreasing widths around $\beta = 0$ (approximately proportional to $2/L$) and the increase of the number $L/2$ of zero-crossings.

This means that the experimental spectrum can be expressed as a linear superposition:

$$S^{\mathrm{ex}}(\vec{\omega}) = \sum_L (2L+1)\langle P_{\mathrm{L}}\rangle S_{\mathrm{L}}^{\mathrm{th}}(\vec{\omega}) \tag{11.25}$$

of P_{L}-weighted spectra:

$$S_{\mathrm{L}}^{\mathrm{th}}(\vec{\omega}) := \int_{0^\circ}^{90^\circ} \mathrm{d}\beta\, S^{\mathrm{th}}(\vec{\omega};\beta) P_{\mathrm{L}}(\cos\beta). \tag{11.26}$$

(Contrary to the nomenclature used originally (Hentschel *et al.*, 1978), we do not refer to these spectra as subspectra since they do not correspond to realistic sub-ensembles of the sample; this is in turn due to the negative portions of the $P_{\mathrm{L}}(\cos\beta)$ 'distributions'.) Relation (11.25) has a form suitable for a best-fit procedure with a relatively restricted number of coefficients $\langle P_{\mathrm{L}}\rangle$. Writing relation (11.25) in a vector–matrix form (with $(\vec{\mathbf{P}})_{\mathrm{L}} = (2L+1)\langle P_{\mathrm{L}}\rangle$, $(\vec{\mathbf{S}}^{\mathrm{ex}})_\omega = S^{\mathrm{ex}}(\vec{\omega})$ and $(\mathbf{S}_{\mathrm{th}})_{\mathrm{L}\omega} := S_{\mathrm{L}}^{\mathrm{th}}(\vec{\omega})$):

$$\vec{\mathbf{S}}^{\mathrm{ex}} = \mathbf{S}_{\mathrm{th}}\vec{\mathbf{P}} \tag{11.27}$$

we have a linear equation that we essentially have to invert to obtain the desired coefficients $\langle P_{\mathrm{L}}\rangle$:

$$\vec{\mathbf{P}} = \mathbf{S}_{\mathrm{th}}^{-1}\vec{\mathbf{S}}^{\mathrm{ex}}. \tag{11.28}$$

This procedure is discussed in more detail in the following section.

The direct-expansion approach (ii) and the moment-expansion approach (iii) are complementary in many respects. The former is based on the superposition of separate contributions from each angle. To show the analogy to the moment expansion, the direct expansion and the corresponding subspectra can be written in terms of $\delta(\beta'-\beta)$ functions (instead of $P_{\mathrm{L}}(\cos\beta)$):

$$\begin{aligned} R(\beta) &= \int \mathrm{d}\beta' R(\beta')\,\delta(\beta'-\beta) \text{ (see equation (11.22))} \\ S_{\beta'}^{\mathrm{th}}(\vec{\omega}) &= \int \mathrm{d}\beta\, S^{\mathrm{th}}(\vec{\omega};\beta)\,\delta(\beta-\beta') \text{ (see equation (11.26))} \end{aligned} \tag{11.29}$$

and the fitting coefficients, corresponding to the $\langle P_{\mathrm{L}}\rangle$, are the $R(\beta')$ values themselves. Being based on a δ-function expansion, this approach is particularly useful and promising for distributions consisting of a few narrow peaks.

In contrast to that, the Legendre moment expansion has the isotropic distribution as its starting point, since the lowest Legendre polynomial is a constant, $P_0(\cos\beta) = 1$. Therefore, it is better suited for characterizing broad distributions. On the other hand, analysis of an $R(\beta)$ distribution with two sharp peaks, as exemplified by the reorientation-angle distribution for discrete jumps, is hopeless by a moment expansion.

In some respects, the moment expansion incorporates more *a priori* information than the direct expansion, for instance on the position of the centre of the distribution. On the one hand, this knowledge is easily incorporated here by choosing the position of the maximum as $\beta = 0$, while the choice of the origin does not affect the direct expansion. On the other hand, without such information the moment expansion does not produce the desired results. If the peak of the distribution occurs at $\beta \neq 0$, say, at 50°, the second moment $\langle P_2\rangle$, weighting $P_2(\cos\beta)$ which has no maximum whatsoever near $\beta = 50^\circ$, does not at all reflect the width of the distribution around its maximum. In fact, many terms in the moment expansion are needed to reproduce just the approximate position and width of the peak.

Compared with the moment expansion, where rarely more than 6 coefficients ($\langle P_2 \rangle$), $\langle P_4 \rangle$, $\langle P_6 \rangle$, $\langle P_8 \rangle$, $\langle P_{10} \rangle$, $\langle P_{12} \rangle$) can be extracted from experimental spectra, the direct-expansion approach is more ambitious. It involves determining 90 coefficients if the resolution is 1° for a distribution extending over 90°, and still 18 values if the resolution is 'only' 5°. So if the direct expansion can be inverted, more detailed information is available. On the other hand, if the distribution is broad and the resolution of the spectra restricted, or if the spectra possess imperfections, the less demanding moment expansion produces better results.

The quality of the fit by the series (11.22) or (11.25) truncated at a certain $L_{\max}$ requires some closer scrutiny. In the best case, we have a measured distribution function, or a spectrum from which the distribution function could be reconstructed. Then, no matter where we truncate the expansion series (11.22), the best fit gives the same values for the coefficients included in the series (if we disregard noise and experimental imperfections); this is a consequence of the orthogonality of the Legendre polynomials. Nevertheless, the distribution corresponding to such a best fit does not represent a good fit to the actual distribution unless the distribution is broad or many moments are included.

As a consequence, if for some technique the spectrum of a highly oriented sample is fitted well by subspectra for few moments, this actually means that the respective technique has a low angular resolution. Sharp features in the orientation distribution are reproduced exclusively by the Legendre polynomials of high order.

For all fit procedures, problems arise at the singularities (sharp maxima) in the spectra: at those points the frequencies are independent of the angle to the first approximation so that the resolution is low. In addition, small mismatches of experimental and theoretically assumed linewidths produce large differences between experimental and theoretical spectra, without being of great importance for the actual quality of the fit.

11.7 Linear least-squares fits

As discussed above, the experimental spectral data combined into a vector $\vec{\mathbf{S}}^{\text{ex}}$ can often be written as a superposition of theoretical (sub)spectra combined into a matrix $\mathbf{S}_{\text{th}}$ and weighted with the coefficients R_k (e.g. $R_L = \langle P_L \rangle$ or $R_\beta = R(\beta)$). This relation can be written in a compact way as a linear matrix equation:

$$\vec{\mathbf{S}}^{\text{ex}} = \mathbf{S}_{\text{th}} \vec{\mathbf{R}}. \tag{11.30}$$

It is our purpose to obtain the $\vec{\mathbf{R}}$ coefficient vector solely from the experimental data $\vec{\mathbf{S}}^{\text{ex}}$ and the theoretical subspectra $\mathbf{S}_{\text{th}}$. Multiplying both sides of (11.30) with the inverse matrix $\mathbf{S}_{\text{th}}^{-1}$, assuming that it exists, we obtain $\mathbf{S}_{\text{th}}^{-1}\vec{\mathbf{S}}^{\text{ex}} = \mathbf{S}_{\text{th}}^{-1}\mathbf{S}_{\text{th}}\vec{\mathbf{R}} = \vec{\mathbf{R}}$. Thus, equation (11.30) can indeed be inverted to yield $\vec{\mathbf{R}}$:

$$\vec{\mathbf{R}} = \mathbf{S}_{\text{th}}^{-1} \vec{\mathbf{S}}^{\text{ex}}. \tag{11.31}$$

However, the applicability of this inversion formula depends on the dimensions of $\vec{\mathbf{R}}$ and $\vec{\mathbf{S}}^{\text{ex}}$. When there are fewer elements of $\vec{\mathbf{S}}^{\text{ex}}$ (measured data points) than elements of $\vec{\mathbf{R}}$ (fitting coefficients), the problem is underdetermined: there are infinitely many solutions for $\vec{\mathbf{R}}$ and the correct one cannot be distinguished from the rest, and no left-hand side inverse matrix exists.

In practice, one often encounters the more favourable case that there are more data points S_j^{ex} than elements R_k of the angular distribution; the problem is over-determined. We look for an $\vec{\mathbf{R}}$ coefficient vector fulfilling:

$$\vec{\mathbf{S}}^{ex} = \mathbf{S}_{th}\vec{\mathbf{R}}. \tag{11.32}$$

With noise and imperfections of the experimental data, no $\vec{\mathbf{R}}$ exists that exactly fulfils (11.32), but approximations are possible. There are many left-hand-side inverse matrices $\mathbf{S}_L^{-1}$ of the asymmetric matrix $\mathbf{S}_{th}$. If relation (11.32) is just assumed to hold, it results in $\mathbf{S}_L^{-1}\vec{\mathbf{S}}^{ex} = \mathbf{S}_L^{-1}\mathbf{S}_{th}\vec{\mathbf{R}} = \vec{\mathbf{R}}$, so that the matrices $\mathbf{S}_L^{-1}$ provide suggestions for $\vec{\mathbf{R}}$ on the basis of the experimental data $\vec{\mathbf{S}}^{ex}$. Among these $\vec{\mathbf{R}}$ vectors, we search for the one producing the 'best' (linear least squares) fit $\vec{\mathbf{R}}_{min}$ for $\vec{\mathbf{R}}$. It is given by:

$$\vec{\mathbf{R}}_{min} = (\mathbf{S}_{th}^T\mathbf{S}_{th})^{-1}\mathbf{S}_{th}^T\ \vec{\mathbf{S}}^{ex}. \tag{11.33}$$

Before deriving this useful formula, let us give heuristic arguments for understanding its structure: a symmetric and thus (possibly) uniquely invertible matrix is produced by multiplying $\mathbf{S}_{th}^T$ and $\mathbf{S}_{th}$. This product matrix must again be multiplied by $\mathbf{S}_{th}^T$ in order to reproduce equation (11.31) when $\mathbf{S}_{th}$ is symmetric and therefore $(\mathbf{S}_{th}^T\mathbf{S}_{th})^{-1}\mathbf{S}_{th}^T = (\mathbf{S}_{th})^{-1}(\mathbf{S}_{th}^T)^{-1}\mathbf{S}_{th}^T = \mathbf{S}_{th}^{-1}$.

The $\vec{\mathbf{R}}_{min}$ vector given in equation (11.33) is the $\vec{\mathbf{R}}$ with the minimal squared Euclidean distance:

$$\mathscr{R}(\vec{\mathbf{R}}) := |\,\vec{\mathbf{S}}^{ex} - \mathbf{S}_{th}\vec{\mathbf{R}}\,|^2. \tag{11.34}$$

To find the minimum of $\mathscr{R}(\vec{\mathbf{R}})$, the null point of its derivatives with respect to the R_i has to be determined. Therefore, let us consider the problem written in terms of coordinates:

$$\mathscr{R}(\vec{\mathbf{R}}) = \sum_j \Big(S_j^{ex} - \sum_k S_{jk}^{th}R_k\Big)^2$$

$$\Rightarrow \frac{\partial\mathscr{R}}{\partial R_i} = \sum_j 2\left(S_j^{ex} - \sum_k S_{jk}^{th}R_k\right)(-S_{ji}^{th}) \stackrel{!}{=} 0 \text{ (for } R_k = R_k^{min})$$

$$\Rightarrow \sum_{j,k} S_{ji}^{th}S_{jk}^{th}R_k^{min} = \sum_j S_{ji}^{th}S_j^{ex} \Leftrightarrow (\mathbf{S}_{th}^T\mathbf{S}_{th})\vec{\mathbf{R}}_{min} = \mathbf{S}_{th}^T\vec{\mathbf{S}}^{ex}.$$

This last matrix expression is equivalent to equation (11.33) if $(\mathbf{S}_{th}^T\mathbf{S}_{th})$ is invertible (multiply by $(\mathbf{S}_{th}^T\mathbf{S}_{th})^{-1}$).

The geometric interpretation (Figure 11.8(a)) of this linear least-squares fit is as follows: $\vec{\mathbf{S}}(\vec{\mathbf{R}}) = \mathbf{S}_{th}\vec{\mathbf{R}}$ defines a (hyper)plane in $\vec{\mathbf{S}}$ space, and $\mathbf{S}_{th}\vec{\mathbf{R}}_{min}$ is the point in this plane with the minimum distance to the point $\vec{\mathbf{S}}^{ex}$.

In many cases, for instance with the direct inversion, one has the additional information that the coefficients R_k must not be negative, $R_k \geqslant 0$. This constraint or

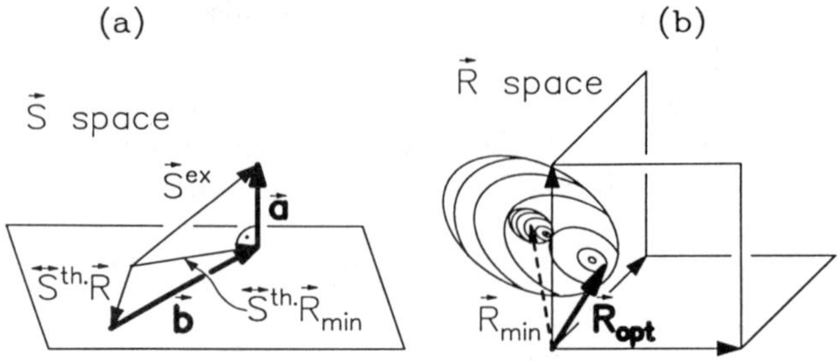

Figure 11.8 (a) Schematic representation of the determination of the least-squares fit vector $\vec{\mathbf{R}}_{\min}$ by minimization of the S-space distance of the experimental data $\vec{\mathbf{S}}^{ex}$ and the (hyper)plane of the possible fit vectors. The vectors $\vec{\mathbf{a}}$ and $\vec{\mathbf{b}}$ are referred to in the text. (b) Schematic representation of the determination of $\vec{\mathbf{R}}_{opt}$ on the basis of the requirement that its components must be nonnegative.

'prejudice' can be essential for finding a reliable set of coefficients $\vec{\mathbf{R}}_{opt}$. For the further analysis, $\mathscr{R}(\vec{\mathbf{R}})$ is rewritten:

$$\begin{aligned}\mathscr{R}(\vec{\mathbf{R}}) &= |\,\vec{\mathbf{S}}^{ex} - \mathbf{S}_{th}\vec{\mathbf{R}}\,|^2 = |\,\vec{\mathbf{S}}^{ex} - \mathbf{S}_{th}(\vec{\mathbf{R}} - \vec{\mathbf{R}}_{min} + \vec{\mathbf{R}}_{min})\,|^2 \\ &= \underbrace{|\,\vec{\mathbf{S}}^{ex} - \mathbf{S}_{th}\vec{\mathbf{R}}_{min}\,|^2}_{\vec{\mathbf{a}}} + \underbrace{|\,\mathbf{S}_{th}(\vec{\mathbf{R}} - \vec{\mathbf{R}}_{min})\,|^2}_{\vec{\mathbf{b}}} \\ &= \mathscr{R}_{min} + |\,\mathbf{S}_{th}(\vec{\mathbf{R}} - \vec{\mathbf{R}}_{min})\,|^2. \end{aligned} \qquad (11.35)$$

In the third step, $\vec{\mathbf{a}}.\vec{\mathbf{b}} = 0$ has been used, where $\vec{\mathbf{a}} = \vec{\mathbf{S}}^{ex} - \mathbf{S}_{th}\vec{\mathbf{R}}_{min}$ and $\vec{\mathbf{b}} = \mathbf{S}_{th}(\vec{\mathbf{R}} - \vec{\mathbf{R}}_{min})$. This can be proved after inserting $\vec{\mathbf{R}}_{min}$ of equation (11.33) into the expressions for $\vec{\mathbf{a}}$ and $\vec{\mathbf{b}}$, or by considering that $\vec{\mathbf{a}} = \vec{\mathbf{S}}^{ex} - \mathbf{S}_{th}\vec{\mathbf{R}}_{min}$ is the minimum-distance vector from the $\mathbf{S}_{th}\vec{\mathbf{R}}$ plane to $\vec{\mathbf{S}}^{ex}$ (see Figure 11.8(a)), and therefore perpendicular to the plane. The $\vec{\mathbf{S}}$-vector $\vec{\mathbf{b}} = \mathbf{S}_{th}(\vec{\mathbf{R}} - \vec{\mathbf{R}}_{min})$ is in the plane and thus perpendicular to $\vec{\mathbf{a}}$.

We now search for the minimal $\mathscr{R}$, $\mathscr{R}(\vec{\mathbf{R}}_{opt})$, consistent with the constraint that $\vec{\mathbf{R}}$ must be located in the 'positive hyperoctant' of $\vec{\mathbf{R}}$ space, $R_i \geqslant 0$. To see how that goal can be attained, it is helpful to have a geometrical interpretation of equation (11.35). An ellipsoid in $\vec{\mathbf{R}}$ space (not to be mixed up with the $\vec{\mathbf{S}}$ space that contains the (hyper)plane discussed above) is defined by all vectors $\vec{\mathbf{R}}$ with the same $|\,\mathbf{S}_{th}(\vec{\mathbf{R}} - \vec{\mathbf{R}}_{min})\,|^2 = \text{constant}$, since $|\,\mathbf{S}_{th}(\vec{\mathbf{R}} - \vec{\mathbf{R}}_{min})\,|^2 = (\vec{\mathbf{R}} - \vec{\mathbf{R}}_{min})^T\mathbf{S}_{th}^T\mathbf{S}_{th}(\vec{\mathbf{R}} - \vec{\mathbf{R}}_{min})$ is a bilinear form (see Appendix D.3). The point where such an ellipsoid first touches the positive hyperoctant of $R_k \geqslant 0$ (Figure 11.8(b)) represents the set $\vec{\mathbf{R}}_{opt}$ of nonnegative coefficients that produce the smallest deviations from the experimental data. This point can be searched in an appropriate algorithm.

11.8 References

Hagemeyer, A., Brombacher, L., Schmidt-Rohr, K. and Spiess, H. W. (1990). Reconstruction of Angular Distributions from 2D NMR Spectra. *Chem. Phys. Lett.*, **167**, 583.

Hentschel, R., Schlitter, J., Sillescu, H. and Spiess, H. W. (1978). Orientational distributions from NMR and ESR line shapes. *J. Chem. Phys.*, **68**, 56.

Sillescu, H. (1971). Theory of Molecular Reorientations: Magnetic Resonance Line Shapes. *J. Chem. Phys.*, **54**, 2110.

Wefing, S., Kaufmann, S. and Spiess, H. W. (1988). 2D Exchange NMR of Powders II: Evolution of Two-Time Distributions. *J. Chem. Phys.*, **89**, 1234.

CHAPTER TWELVE

Determination of Order in Polymers by Multidimensional NMR

Keywords: *orientation distributions, moment expansion, uniaxiality, transverse isotropy, biaxiality; sync-MAS 2D, DECODER, analogies to WAXS; orientation distributions in LCs, fibres, thin films.*

12.1 Introduction

Orientational order plays an important role in solid polymers. It is often induced by the industrial processing, for example in fibres and injection- or compression-moulded parts. In polymers with liquid-crystalline properties of the melt or solution, the anisotropies generated by the flow pattern are particularly pronounced. In order to improve the mechanical properties of polymer fibres or films, the degree of orientation is intentionally enhanced by drawing. At the same time, anisotropy of mechanical properties can result in low tolerance to unfavourably directed loads. In many liquid-crystalline polymers (see Section 5.6), in the mesophase near the transition to the isotropic phase, electric or magnetic fields can induce macroscopic orientational order.

The behaviour of solid polymers under cold drawing is particularly complex (Woodward, 1989, Chapter X). Several methods, including ^{2}H and ^{13}C NMR, prove that at high strain the chains and not the lamellar crystallites are oriented parallel to the direction of the applied strain. The spherulites (see Section 5.2) are destroyed in the process; the lamellae, however, remain intact according to on-line X-ray measurements (Peterlin, 1979). It appears that a visualization in terms of fuzzy boards (the crystalline lamellae with amorphous loops) connected by many threads (the tie molecules) is quite appropriate. NMR shows that at least parts of the material in the noncrystalline regions are still quite isotropically mobile, while the interfacial layer seems to be anisotropically mobile.

The complete description of orientational order of polymers requires a specification of the orientation-distribution function of every relevant molecular moiety in terms of three angles that relate the segmental orientation to the macroscopic sample directions (such as the draw direction and the normal in film samples). However, in many cases the characterization can be simplified. Often, the sample exhibits macroscopic uniaxiality which reduces the number of relevant angles. In addition, in many polymers some or all segments are fixed relative to one another by intra- or intermolecular forces so that the whole system can be described by the orientation distribution of a chain axis, or of the unit cell in crystalline material. If the orientation distribution is broad, it is sufficient to characterize it in terms of few order parameters (see Section 11.6).

Experimentally, a variety of techniques for studying orientational order are available. Several of these (e.g. birefringence measurements) can only determine one or a few order parameters. This means that the angular resolution is restricted since quite different distributions can have identical second moments. Also, with those methods it is often not clear which are the molecular units whose orientation distribution is reflected in the experimental data. In NMR, due to its well defined local interactions, this problem does not arise. In wide-angle X-ray scattering (WAXS), orientation distributions of crystallites can be fully determined by means of the 'pole-figure analysis' (Baltá-Calleja and Vonk, 1989) (see also Section 12.7.3), which yields the orientation distributions of sets of crystallite planes. It should be noted, however, that for obtaining the desired orientation distributions of the chain axes, this X-ray-scattering information on the crystallite planes must be transformed via the known or assumed structure. Similar complications will be seen to arise in obtaining the chain-axes distribution from the PAS orientation distributions measured by NMR.

12.2 Angles and coordinate systems

The analysis of orientation distributions can be performed in terms of various angular variables. These are all interrelated, but some of them are more suitable from the NMR-spectroscopic viewpoint, while others describe the distribution of the structurally most relevant axes in the sample. For reference, we define all these angles and coordinate systems together in this section, in part repeating previous definitions for the convenience of the reader. The relations of coordinate systems and angles are portrayed schematically in Figure 12.1.

(i) *Laboratory frame, LF*. The z axis $\vec{\mathbf{Z}}_L$ of the laboratory frame is defined by the $\vec{\mathbf{B}}_0$ field.

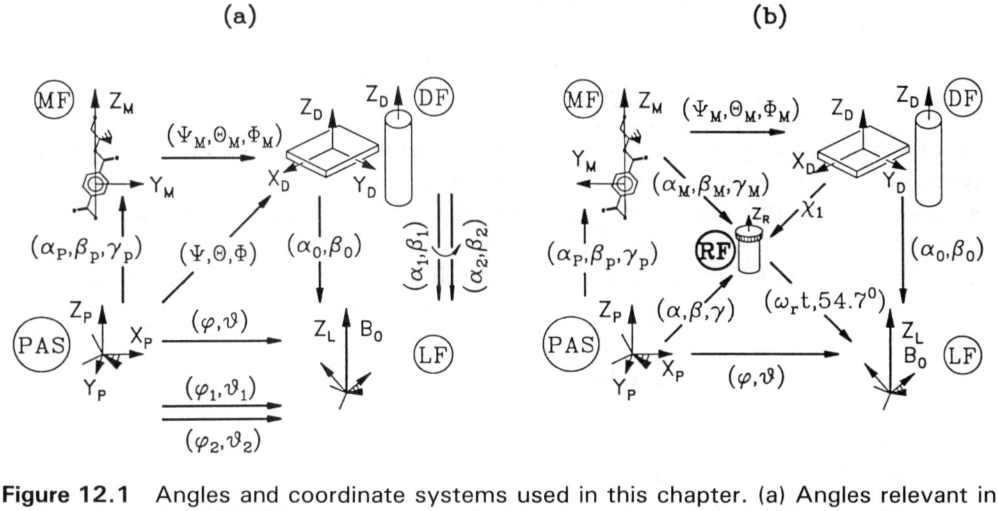

Figure 12.1 Angles and coordinate systems used in this chapter. (a) Angles relevant in static 1D and DECODER experiments. (b) Angles relevant in MAS experiments. In particular, the angles relating the rotor frame RF to the other coordinate systems are included in the figure.

(ii) *Sample-fixed director frame, DF.* The primary order direction of the sample is denoted as the sample director and usually chosen as the z axis $\vec{\mathbf{Z}}_{\mathrm{D}}$ of the DF.
In 1D experiments, β_0 denotes the experimentally set, fixed angle between the sample director and $\vec{\mathbf{B}}_0$. In the DECODER 2D experiment described below, it proves useful to denote the orientation of the $\vec{\mathbf{B}}_0$ field during the kth period (evolution, detection) of the experiment by the polar coordinates (β_k, α_k) in the DF; the flip angle between two of these $\vec{\mathbf{B}}_0$ orientations is denoted by β_{f}. The direction $\vec{\mathbf{F}}$ of the flip axis, which is sample-fixed, plays an important role in the DECODER experiment. If $\vec{\mathbf{F}}$ is perpendicular to $\vec{\mathbf{B}}_0$, the plane of orientations perpendicular to $\vec{\mathbf{F}}$, denoted as the field plane since it contains $\vec{\mathbf{B}}_0$, is of particular relevance. In the field plane, Γ denotes the angle between a given direction and a reference direction.

(iii) *Molecular frame, MF.* The molecular frame is fixed in the molecular repeat unit; usually, the local chain axis is chosen as the z axis $\vec{\mathbf{Z}}_{\mathrm{M}}$ of the MF.
The Euler angles $(\Psi_{\mathrm{M}}, \Theta_{\mathrm{M}}, \Phi_{\mathrm{M}})$ denote the orientation of a given molecular frame in the sample-fixed director frame; these are usually the structurally most meaningful angles for describing the molecular orientation.

(iv) *Principal-axes system, PAS* (see Sections 2.4.2 and D.3). (θ, ϕ) are the polar coordinates of the $\vec{\mathbf{B}}_0$ field in the PAS, so θ is the angle between the $\vec{\mathbf{Z}}_{\mathrm{P}}$ axis and $\vec{\mathbf{B}}_0$. (Ψ, Θ, Φ) are the Euler angles describing the orientation of a given PAS in the sample-fixed DF. The orientation of the PAS in the molecular frame MF is specified by the Euler angles $(\alpha_{\mathrm{P}}, \beta_{\mathrm{P}}, \gamma_{\mathrm{P}})$. We can set $\gamma_{\mathrm{P}} = 0$ without restricting the generality of the treatment: in the transformation PAS $\rightarrow$ MF $\rightarrow$ DF, both γ_{P} and Ψ_{M} stand for a rotation around the $\vec{\mathbf{Z}}_{\mathrm{M}}$ axis; therefore, only the sum $\gamma_{\mathrm{P}} + \Psi_{\mathrm{M}}$ is relevant so that γ_{P} can be included in Ψ_{M} with a suitably redefined null point.

(v) *Rotor-fixed frame, RF.* The RF is a sample-fixed frame whose z axis $\vec{\mathbf{Z}}_{\mathrm{R}}$ is the rotor axis.
χ_1 is the angle between the sample director $\vec{\mathbf{Z}}_{\mathrm{D}}$ and the rotor axis $\vec{\mathbf{Z}}_{\mathrm{R}}$. The angles (α, β, γ) describe the relative orientations of the PAS to the RF. The analogous angles $(\alpha_{\mathrm{M}}, \beta_{\mathrm{M}}, \gamma_{\mathrm{M}})$ relate the MF to the RF. In particular, the angles γ and γ_{M} correspond to rotations around the rotor axis. The polar coordinates of the $\vec{\mathbf{B}}_0$ field in the rotor frame are $(54.7^\circ, \omega_{\mathrm{r}} t)$.

The various angles given in Figure 12.1 are interrelated. In particular, simple and important relations exist in each 'triangle' of relative orientations or transformations between three coordinate systems, generally labelled A, B, and C, in Figure 12.1: for a transformation A $\rightarrow$ C by $(\alpha_{\mathrm{AC}}, \beta_{\mathrm{AC}}, \gamma_{\mathrm{AC}})$ which is equivalent to two successive transformations A $\rightarrow$ B by $(\alpha_{\mathrm{AB}}, \beta_{\mathrm{AB}}, \gamma_{\mathrm{AB}})$ and B $\rightarrow$ C by $(\alpha_{\mathrm{BC}}, \beta_{\mathrm{BC}}, \gamma_{\mathrm{BC}})$ we have:

$$\cos \beta_{\mathrm{AC}} = \cos \beta_{\mathrm{AB}} \cos \beta_{\mathrm{BC}} + \sin \beta_{\mathrm{AB}} \sin \beta_{\mathrm{BC}} \cos(\alpha_{\mathrm{BC}} + \gamma_{\mathrm{AB}} - 180^\circ) \qquad (12.1)$$

according to the 'three-axes relation' (A.14). To apply that formula, we have considered the z axes of systems A and C in system B; according to equation (B.8), their polar coordinates are $(\beta_{\mathrm{AB}}, 180^\circ - \gamma_{\mathrm{AB}})$ and $(\beta_{\mathrm{BC}}, \alpha_{\mathrm{BC}})$, respectively.

Applying this relation (12.1) to the transformations PAS $\rightarrow$ MF $\rightarrow$ DF, we find:

$$\cos \Theta = \cos \beta_{\mathrm{P}} \cos \Theta_{\mathrm{M}} + \sin \beta_{\mathrm{P}} \sin \Theta_{\mathrm{M}} \cos(\gamma_{\mathrm{P}} + \Psi_{\mathrm{M}}). \qquad (12.2)$$

In the transformation MF $\rightarrow$ PAS $\rightarrow$ DF, the 'short cut' MF $\rightarrow$ DF is described by

$(\Psi_M, \Theta_M, \Phi_M)$, the transformation MF → PAS by $(-\gamma_P, -\beta_P, -\alpha_P)$, and PAS → DF by (Ψ, Θ, Φ), so that equation (12.1) produces:

$$\cos \Theta_M = \cos \beta_P \cos \Theta + \sin \beta_P \sin \Theta \cos(\Psi - \alpha_P). \tag{12.3}$$

Similarly, the transformation PAS → DF → MF yields:

$$\cos \beta_P = \cos \Theta \cos \Theta_M + \sin \Theta \sin \Theta_M \cos(\Phi - \Phi_M). \tag{12.4}$$

For the other 'triangles' in Figure 12.1 we find:

$$\cos \theta_k = \cos \Theta \cos \beta_k - \sin \Theta \sin \beta_k \cos(\Phi + \alpha_k) \tag{12.5}$$

which is relevant in the DECODER experiments, and:

$$\cos \theta = \cos \beta \cos(54.7^\circ) - \sin \beta \sin(54.7^\circ) \cos(\gamma + \omega_r t) \tag{12.6}$$

$$\cos \Theta_M = \cos \chi_1 \cos \beta_M + \sin \chi_1 \sin \beta_M \cos(\gamma_M) \tag{12.7}$$

which play a role in MAS experiments.

12.3 Moment expansions

To extract information on molecular orientation distributions from experimental data, various approaches can be taken, as was outlined in Section 11.6. Of these, the most widely known, though theoretically complex, technique is the Legendre moment expansion. This approach is to be discussed in more detail in the following.

12.3.1 Moment expansions for a uniaxial system

In the beginning, we assume that the system exhibits transverse isotropy (macroscopic and microscopic uniaxiality), so that the orientation distributions can be specified in terms of a single angular variable Θ_M between the macroscopic director $\vec{\mathbf{Z}}_D$ and the unique molecular axis $\vec{\mathbf{Z}}_M$. This distribution can be expanded in terms of the Legendre polynomials $P_L(\cos \Theta_M)$ which represent a complete basis set of functions over the $\cos \Theta_M$-interval [0...1]:

$$\tilde{R}(\Theta_M) = \sum_L (2L+1)\langle P_L \rangle P_L(\cos \Theta_M). \tag{12.8}$$

The range $0 \leqslant \cos \Theta_M \leqslant 1$ corresponds to the angle range $0^\circ \leqslant \Theta_M \leqslant 90^\circ$, which, as discussed in Section 11.6, is sufficient for the analysis of NMR second-rank-tensor orientations.

The coefficients $\langle P_L \rangle$ in (12.8) are the order parameters. Since the $P_L(\cos \Theta_M)$ functions are orthogonal:

$$\int_0^1 P_{L'}(\cos \Theta_M) P_L(\cos \Theta_M)\, \mathrm{d} \cos \Theta_M = \delta_{L'L}/(2L+1). \tag{12.9}$$

Each order parameter $\langle P_L \rangle$ characterizes the weight with which the corresponding $P_L(\cos \Theta_M)$ contributes to $\tilde{R}(\Theta_M)$:

$$\int_0^1 \tilde{R}(\Theta_M) P_L(\cos \Theta_M) \, d \cos \Theta_M$$

$$\overset{(12.8)}{=} \int_0^1 \sum_{L'} (2L' + 1)\langle P_{L'} \rangle P_{L'}(\cos \Theta_M) P_L(\cos \Theta_M) \, d \cos \Theta_M \overset{(12.9)}{=} \langle P_L \rangle. \quad (12.10)$$

Since we are interested only in the shape of $\tilde{R}(\Theta_M)$, we have to eliminate effects of arbitrary prefactors by normalizing $\tilde{R}(\Theta_M)$ such that the following integral of $\tilde{R}(\Theta_M)$ is unity:

$$\langle P_0 \rangle = \int_0^{90°} d\Theta_M \sin \Theta_M \, \tilde{R}(\Theta_M) = 1. \quad (12.11)$$

This $\langle P_0 \rangle$ is commonly considered as representing the isotropic part of the distribution, though it is unity even if, strictly speaking, the distribution has no isotropic part (for instance when the distribution is zero over a certain angular range). The order parameters $\langle P_L \rangle$ for $L > 0$ contain the information about the orientation distribution. $\langle P_2 \rangle$ in particular characterizes the width of the distribution, if the distribution is centred around $\Theta_M = 0$: the smaller the $\langle P_2 \rangle$ value, the larger the width of $\tilde{R}(\Theta_M)$. The moment expansion is not convergent for distributions containing isolated δ functions (very sharp peaks): if, for instance, the distribution is $\delta(\Theta_M)$, that is, a very sharp peak at $\Theta_M = 0$, the coefficients do not decrease, rather $\langle P_L \rangle = 1$ for all L.

In addition to the order parameters, moments P_{L00} have been used in the past (Hentschel *et al.*, 1978), with:

$$R(\Theta_M) = \sum_{L=0}^{\infty} P_{L00} P_L(\cos \Theta_M). \quad (12.12)$$

In our frame of conventions, the relation between the moments and the order parameters is simply $P_{L00} = (2L + 1)\langle P_L \rangle$ (for an additional factor of $8\pi^2$ see below). However, when people talk about the 'second moment' of the orientation distribution, they usually refer to $\frac{1}{2}\langle 3 \cos^2\theta - 1 \rangle$, which is the second order parameter in the more general nomenclature outlined here. Since the difference consists only in the factor $(2L + 1)$, we will also indulge in the more convenient and widespread usage of the term 'moment'. (If we introduced orthonormal Legendre polynomials $\tilde{P}_L(\cos \Theta_M) := (2L + 1)^{\frac{1}{2}} P_L(\cos \Theta_M)$, the difference between the weighting factors (moments) in equation (12.8) and the weighted averages (order parameters) in equation (12.12) would be eliminated. However, $\tilde{P}_2(\cos \Theta_M) = \sqrt{5}\,\frac{1}{2}(3 \cos^2\Theta_M - 1)$ is somewhat unorthodox and would only produce confusion. The spherical harmonics Y_{Lm}, such as $Y_{2,0} = \sqrt{5}\,\frac{1}{2}(3 \cos^2\Theta_M - 1)/(4\pi)^{\frac{1}{2}}$, actually contain the factor $(2L + 1)^{\frac{1}{2}}$.)

In comparing with other texts, it should be noted that we have eliminated factors $8\pi^2$, which otherwise make many relations lengthy and more difficult to memorize. We have included this factor into the definition of the average $\langle\rangle$ (see equation (12.10)). We restrict the integral to the relevant variable θ and to the interval [0...1] over which the Legendre polynomials are generally considered. This can also be viewed as representing an average taken over the surface of the unit sphere, instead of the integral taken over that surface. The factor $8\pi^2$, being common to all order parameters, represents nothing but a certain normalization of $\tilde{R}(\Theta_M)$. For a moment expansion, $\tilde{R}(\Theta_M)$ must always be normalized, for instance according to equation (12.11), and the normalization conditions determine whether $8\pi^2$ appears or not. If $\langle P_0 \rangle = 1$ is ensured, the normalization of all the other order parameters will be consistent with the usual conventions.

A few basic properties of the moments in the expansion are to be mentioned here. With $\langle P_0 \rangle = 1$, the other moments range between 1 and -0.5 (for $L > 2$, the minimal values of $P_L(x)$ are slightly above -0.5). If all moments for $L > 0$ are zero, the distribution is isotropic; if all are equal to 1, the uniaxial orientation is perfect with $\tilde{R}(\Theta_M)$ consisting of a single peak at $\Theta_M = 0°$. Often, the distribution can be approximated by Gaussian centred at $\Theta_M = 0°$:

$$\tilde{R}(\Theta_M) = N \exp(-\sin^2\Theta_M / 2\sigma^2_{\Theta M})$$

(In contrast to the literature (Spiess, 1984), we choose $\sigma^2_{\Theta M}$ instead of $\sin^2(\sigma_{\Theta M})$ as the parameter determining the width. Since $\sin(\sigma_{\Theta M}) \leqslant 1$, it does not allow the width of the distribution to be increased beyond a certain limit. For sufficiently small widths $\sigma_{\Theta M} < 30°$, the values of $\sin(\sigma_{\Theta M})$ and $\sigma_{\Theta M}$ are actually quite similar.) For the Gaussian distribution, all moments are positive, as can be shown by actually evaluating the integrals of equation (12.10). The values of the moments decrease monotonically with

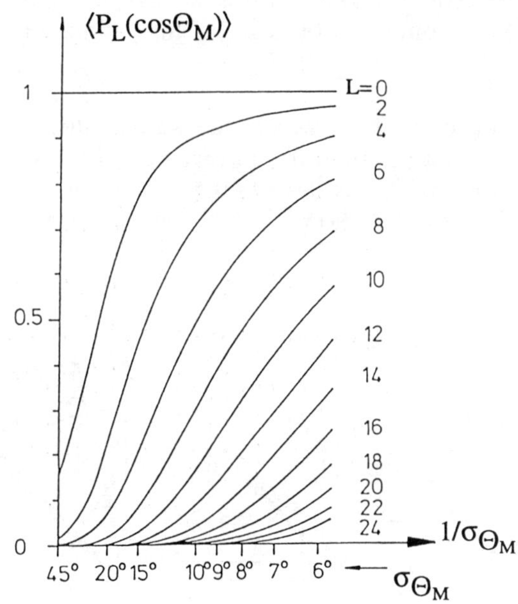

Figure 12.2 Dependence of order parameters on the width $\sigma_{\Theta M}$ of a Gaussian orientation distribution. For this specific type of distribution, the order parameters decrease with increasing L. Note the nonlinearity of the $1/\sigma_{\Theta M}$ angle scale; on a linear $\sigma_{\Theta M}$ scale, the drastic increase of the higher order parameters with decreasing $\sigma_{\Theta M}$ would become more apparent, but the values would not be easy to read off the diagram. For comparing widths of distributions as given in different publications, it may be important to remember that the full width at half maximum of the Gaussian is $2.35\sigma_{\Theta M}$. Adapted from Hentschel *et al.* (1981).

increasing order L. Figure 12.2 shows the order parameters of Gaussian distributions plotted as a function of $1/\sigma_{\Theta_M}$.

An important property of the moment expansion based on Legendre polynomials is the direct relation of moments involving different frames of reference. Consider, for instance, polymer chain axes $\vec{\mathbf{Z}}_M$ distributed uniaxially around the sample director $\vec{\mathbf{Z}}_D$, with the angle Θ_M between $\vec{\mathbf{Z}}_D$ and an individual $\vec{\mathbf{Z}}_M$. The $\vec{\mathbf{Z}}_P$ axes of the NMR interaction, typically the C—^{2}H bonds, make an angle β_P with the chain axis $\vec{\mathbf{Z}}_M$ (e.g. $\beta_P = 90^\circ$ for an all-trans chain) if there is no conformational disorder. In a uniaxially oriented sample, both the distribution $\tilde{R}(\Theta_M)$ of the chain axes and the $\tilde{R}^\Theta(\Theta)$ of the C—^{2}H bonds are uniaxial around the $\vec{\mathbf{Z}}_D$ axis, but they are not identical. Due to the cylindrically symmetric distribution around $\vec{\mathbf{Z}}_D$, each fixed chain orientation $\delta(\Theta_M)$ produces a cone of C—^{2}H bonds, which results in a broadened distribution $p(\Theta;\Theta_M)$ in Θ with (integrable) singularities at its ends as indicated in Figure 12.3. The actual functional form of this $p(\Theta;\Theta_M)$ can be calculated as for the conical distribution (11.4), where $\Theta_M \triangleq \theta_1, \Theta \triangleq \theta_2$, and $\beta_P \triangleq \beta$. While $\tilde{R}^\Theta(\Theta)$ is obtained from $\tilde{R}(\Theta_M)$ simply by the superposition:

$$\tilde{R}^\Theta(\Theta) = \int_0^1 \tilde{R}(\Theta_M)p(\Theta;\Theta_M)\ \mathrm{d}\cos\Theta_M \tag{12.13}$$

of the distributions $p(\Theta;\Theta_M)$, it is not trivial to invert relation (12.13) to obtain $\tilde{R}(\Theta_M)$, the searched chain-axes distribution, out of the measured principal-axes distribution $\tilde{R}^\Theta(\Theta)$. Nevertheless, such an inversion is possible because the moments of $\tilde{R}(\Theta_M)$ and $\tilde{R}^\Theta(\Theta)$ are connected by a surprisingly simple relation:

$$\langle P_L(\cos\Theta)\rangle = P_L(\cos\beta_P)\langle P_L(\cos\Theta_M)\rangle. \tag{12.14a}$$

(To test the relation and to check on which side we have to write Θ or Θ_M, consider a very narrow Θ_M distribution around 0°, so that $\langle P_2(\cos\Theta_M)\rangle = 1$, and a structure with $\beta_P = 90^\circ$, resulting in $P_2(\cos\beta_P) = -0.5$. Then the C–^{2}H axes are all at $\Theta = \beta_P = 90^\circ$, such that $\langle P_2(\cos\Theta)\rangle = -0.5$; this is also the result of relation

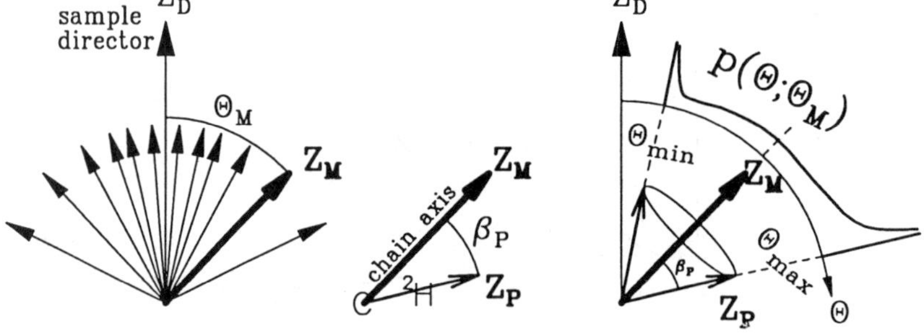

Figure 12.3 (Left) Distribution of chain axes $\vec{\mathbf{Z}}_M$ around the sample director $\vec{\mathbf{Z}}_D$. (Middle) One $\vec{\mathbf{Z}}_M$ singled out. With the $\vec{\mathbf{Z}}_P$ axis (e.g. C—^{2}H bond), it makes an angle β_P determined by the molecular structure. (Right) Broad $\vec{\mathbf{Z}}_P$ distribution $p(\Theta;\Theta_M)$ for the one selected $\vec{\mathbf{Z}}_M$. The broadening of $p(\Theta;\Theta_M)$ arises because $\vec{\mathbf{Z}}_M$ is not parallel to $\vec{\mathbf{Z}}_D$ and $\vec{\mathbf{Z}}_P$ not parallel to $\vec{\mathbf{Z}}_M$. The full $\vec{\mathbf{Z}}_p$ distribution $\tilde{R}^\Theta(\Theta)$ is an intricate combination of the Θ_M distribution $\tilde{R}(\Theta_M)$ indicated by the $\vec{\mathbf{Z}}_M$ arrows on the left and the broadened $p(\Theta;\Theta_M)$ distributions. Nevertheless, a conversion of the measured $\tilde{R}^\Theta(\Theta)$ into the corresponding $\tilde{R}(\Theta_M)$ distribution can be achieved based on the moment-relationship given in the text.

(12.14a).) The derivation of relation (12.14a) is based on the spherical-harmonics addition theorem (B.31), subject to the uniaxial average $\langle\,\rangle$, which does not affect β_P or γ_P:

$$\langle P_L(\cos\Theta)\rangle = \frac{4\pi}{2L+1}\sum_{m=-L}^{L}(-1)^m\, Y_{Lm}(\beta_P, -\gamma_P)\,\langle Y_{Lm}(\Theta_M, -\Psi_M)\rangle. \quad (12.14b)$$

Due to the uniaxiality, $\langle Y_{Lm}(\Theta_M, -\Psi_M)\rangle$ vanishes unless $m = 0$, so that only the $Y_{L0} \sim P_L$ remain.

Relation (12.14a) makes it possible to transform a measured distribution of C—^{2}H bonds $\tilde{R}^{\Theta}(\Theta)$ into the desired distribution of chain axes $\tilde{R}(\Theta_M)$. First, the moments of $\tilde{R}^{\Theta}(\Theta)$ are easily calculated by integration according to equation (12.10) (with $\tilde{R}^{\Theta}(\Theta)$ inserted in the place of $\tilde{R}(\Theta_M)$]. Then they are transformed into the moments of $\tilde{R}(\Theta_M)$ according to relation (12.14a). Out of its moments, $\tilde{R}(\Theta_M)$ can be easily built up according to equation (12.8).

The transformation of the moments, equation (12.14), is valid for arbitrary η. However, with $\eta \neq 0$ and $\beta_P \neq 0$ it is not very useful for determining the $\langle P_L(\cos\Theta_M)\rangle$, since the $\langle P_L(\cos\Theta)\rangle$ themselves are accessible only via an involved fit of the NMR spectra. In this fit, since the distribution in Ψ does not exhibit uniaxiality, other order parameters $\langle D^L_{m0}(\Psi, \Theta, 0)\rangle$ of the PAS orientation distribution, which do not vanish according to equation (B.31), must be taken into account, too.

In 1D NMR spectra for $\eta = 0$ interactions, the lineshape reflects the moments $\langle P_L(\cos\theta)\rangle$ of $\vec{\mathbf{Z}}_P$ axes (C—^{2}H bonds) relative to the $\mathbf{B}_0$ field rather than the moments with respect to the sample order axis $\vec{\mathbf{Z}}_D$. With an angle β_0 between $\mathbf{B}_0$ and $\vec{\mathbf{Z}}_D$, by double application of equation (12.14a) we find:

$$\langle P_L(\cos\theta)\rangle = P_L(\cos\beta_0) P_L(\cos\beta_P)\langle P_L(\cos\Theta_M)\rangle \quad (12.15)$$

(check the arrangement of angles again as outlined below relation (12.14a), e.g. for $\beta_P = 0^\circ$ and a sample orientation with $\beta_0 = 90^\circ$). The $P_L(\cos\beta_0)$ $P_L(\cos\beta_P)$ terms in this expression can be considered as 'transformation factors' which take into account the deviations of the $\vec{\mathbf{Z}}_L$ axis $\mathbf{B}_0$ from the orientationally more relevant $\vec{\mathbf{Z}}_D$ director axis, and of the $\vec{\mathbf{Z}}_P$ PAS axis from the more relevant $\vec{\mathbf{Z}}_M$ chain axis.

12.3.2 P_L-weighted spectra for the moment expansion

So far, the expansion of orientation distributions in terms of Legendre polynomials has been considered. In this section, it will be discussed how NMR spectra can be expanded in terms of P_L-weighted spectra that correspond to the order parameters.

$S(\vec{\omega})$ belonging to a certain distribution $\tilde{R}(\Theta_M)$ can be expressed as a linear superposition:

$$S(\vec{\omega}) = \int_0^1 S(\vec{\omega}; \Theta_M)\tilde{R}(\Theta_M)\, \mathrm{d}\cos\Theta_M \quad (12.16)$$

where a subspectrum $S(\vec{\omega}; \Theta_M)$ is the spectrum that corresponds to a sub-ensemble with an angular distribution consisting of a single peak at Θ_M.

Now the expansion (12.8) $\tilde{R}(\Theta_M) = \Sigma_L(2L+1)\langle P_L\rangle P_L(\cos\Theta_M)$ is inserted into equation (12.16):

$$S(\vec{\omega}) = \int_0^1 S(\vec{\omega};\Theta_M)\overbrace{\sum_L (2L+1)\langle P_L\rangle P_L(\cos\Theta_M)}^{\tilde{R}(\Theta_M)}\,\mathrm{d}\cos\Theta_M$$

$$= \sum_L \langle P_L\rangle \underbrace{(2L+1)\int_0^1 S(\vec{\omega};\Theta_M)P_L(\cos\Theta_M)\,\mathrm{d}\cos\Theta_M}_{S_L(\omega)}$$

That tells us that a given spectrum can be expressed as a linear superposition:

$$S(\vec{\omega}) = \sum_L \langle P_L(\cos\Theta_M)\rangle S_L(\vec{\omega}) \tag{12.17}$$

of the spectra $S_L(\vec{\omega})$ defined by:

$$S_L(\vec{\omega}) := (2L+1)\int_0^1 \mathrm{d}\cos\Theta_M\, S(\vec{\omega});\Theta_M)P_L(\cos\Theta_M). \tag{12.18}$$

At this point it might be interesting to recall that a $\langle P_L(\cos\Theta_M)\rangle$ for a given L is just a number and does not depend on Θ_M explicitly. Still, it makes sense to indicate Θ_M in $\langle P_L(\cos\Theta_M)\rangle$ because different order parameters result depending on whether the orientation is parameterized in terms of Θ_M, Θ or θ (see equations (12.14a) and (12.15)).

According to equation (12.17), the spectrum $S(\vec{\omega})$ can be expressed in terms of few $\langle P_L\rangle$-weighted spectra $S_L(\vec{\omega})$ if the distribution of orientations can be characterized by few order parameters. The procedure for determining the coefficients $\langle P_L\rangle$ in the linear superposition (12.17) has been described in Section 11.7.

The subspectra $S(\vec{\omega};\Theta_M)$ are even with respect to $\Theta_M = 90^\circ$, that is, $S(\vec{\omega};\Theta_M) = S(\vec{\omega};180^\circ - \Theta_M)$, due to the second-rank tensor character of NMR interactions. The $P_L(\cos(\Theta_M))$ are odd with respect to 90° for odd L, and even with respect to 90° for even L. Therefore, the $S_L(\vec{\omega})$ given by the integral of equation (12.18) vanish for odd L, while the integral can be restricted to $0 \leqslant \cos\Theta_M \leqslant 1$, or equivalently $0^\circ \leqslant \Theta_M \leqslant 90^\circ$, for the $S_L(\vec{\omega})$ spectra with even L.

Above, it was shown that moment expansions can be considered for distributions of MF or PAS orientations in the sample or relative to $\vec{\mathbf{B}}_0$, which we expressed in terms of different angles Θ_M, Θ or θ. Quite like the order parameters, the P_L-weighted spectra belonging to the different orientation distributions are related in a simple manner.

As was mentioned above, for sites with $\eta = 0$ an expansion can also be performed in terms of Θ. It is completely analogous to the treatment given for Θ_M, and thus produces:

$$S(\vec{\omega}) = \sum_L S_L^{\Theta}(\vec{\omega})\langle P_L(\cos\Theta)\rangle \tag{12.19}$$

$$S_L^{\Theta}(\vec{\omega}) = (2L+1)\int_0^1 S(\vec{\omega},\Theta)\, P_L(\cos\Theta)\,\mathrm{d}\cos\Theta. \tag{12.20}$$

Inserting $\langle P_L(\cos\Theta)\rangle = P_L(\cos\beta_P)\langle P_L(\cos\Theta_M)\rangle$, equation (12.14a), into equation (12.19) we find:

$$S(\vec{\omega}) = \sum_L S_L^{\Theta}(\vec{\omega}) P_L(\cos\beta_P)\langle P_L(\cos\Theta_M)\rangle.$$

Comparison with equation (12.17) yields:

$$S_L(\vec{\omega}) = P_L(\cos\beta_P)\, S_L^{\Theta}(\vec{\omega}). \tag{12.21}$$

Note that here, compared with $\langle P_L(\cos\Theta)\rangle = P_L(\cos\beta_P)\langle P_L(\cos\Theta_M)\rangle$, equation (12.14a), the $P_L(\cos\beta_P)$ factor has 'changed sides', appearing in front of the Θ expression while in the moment relation (12.14a) it appears before the Θ_M term.

Similarly, the expansion in terms of $P_L(\cos\theta)$ (Hentschel *et al.*, 1978) leads to P_L-weighted spectra $S_L^{\theta}(\omega)$ with:

$$S(\omega) = \sum_L S_L^{\theta}(\omega)\langle P_L(\cos\theta)\rangle. \tag{12.22}$$

The weighting can be expressed in analogy to equation (12.18), and $S(\tilde{\omega};\theta)$, the spectrum for a single θ, is actually known to be a single peak $\delta(\tilde{\omega}-\omega(\theta))$. Thus:

$$\begin{aligned} S_L^{\theta}(\tilde{\omega}) &= (2L+1)\int_0^1 \delta(\tilde{\omega}-\omega(\theta))P_L(\cos\theta)\,\mathrm{d}\cos\theta \\ &= (2L+1)P_L\{\cos\theta(\tilde{\omega})\}\,\frac{\mathrm{d}\omega}{\mathrm{d}\cos\theta} \\ &= (2L+1)P_L\{\sqrt{\tfrac{2}{3}(\tilde{\omega}/\delta+\tfrac{1}{2})}\}\; S_{\mathrm{Pake}}(\tilde{\omega}) \end{aligned} \tag{12.23}$$

where for evaluating the δ function we have used relation (E.14) combined with (E.6), and the Pake spectrum is given by equation (2.47). The relation (12.23) between the spectra $S_L^{\theta}(\omega)$, the Legendre polynomial P_L, and the Pake spectrum is visualized in Figure 12.4. For instance, $S_2^{\theta}(\omega)$ passes through zero at $\omega(54.7^{\circ})$. Note that the $S_L^{\theta}(\omega)$ are independent of both the sample orientation and the relative orientation of C—^{2}H bonds with the $\vec{\mathbf{Z}}_M$ axis.

Inserting the relation (12.15) between $\langle P_L(\cos\theta)\rangle$ and $\langle P_L(\cos\Theta_M)\rangle$ into the expansion of the spectrum equation (12.22), we obtain:

$$S(\omega) = \sum_L S_L^{\theta}(\omega)P_L(\cos\beta_0)P_L(\cos\beta_P)\langle P_L(\cos\Theta_M)\rangle \tag{12.24}$$

which upon comparison with the spectral expansion (12.17) in terms of $S_L(\omega)$ and $\langle P_L(\cos\Theta_M)\rangle$ yields the following instructive relation between the P_L-weighted spectra for $\eta = 0$:

$$S_L(\omega) = P_L(\cos\beta_0)P_L(\cos\beta_P)S_L^{\theta}(\omega). \tag{12.25}$$

12.3.3 Full moment expansion for non-axial systems

Up to this point, we have restricted the discussion to uniaxial systems with transverse isotropy. An expansion of the corresponding orientation-distribution function $\tilde{R}(\Theta_M)$ can be performed in terms of the Legendre polynomials $P_L(\cos\Theta_M)$. The general case

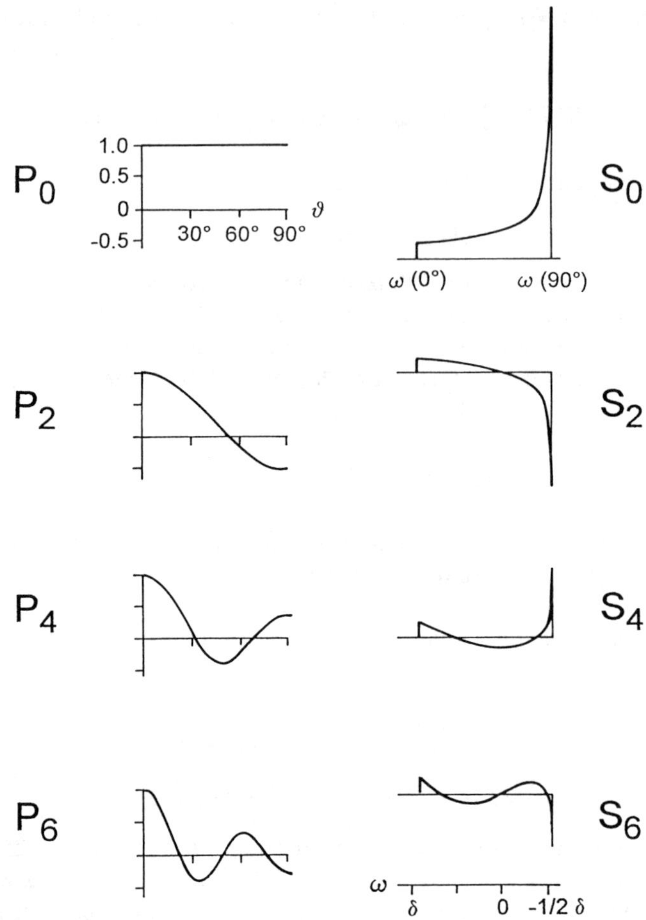

Figure 12.4 $P_L(\cos\theta)$-weighted spectra for $\eta = 0$. Note that the features of the $P_L(\cos\theta)$, in particular their zero crossings, are directly reflected in the spectra S_L. Adapted from Henschel *et al.* (1978).

is based on the same principles but is naturally more complex: an orientation distribution function $R(\Psi, \Theta, \Phi)$ is needed to describe the probability (density) of finding a coordinate system at an orientation given in terms of the three Euler angles (Ψ, Θ, Φ) relative to the system of reference. The suitable set of basis functions here are the Wigner rotation matrices $D^L_{mn}(\Psi, \Theta, \Phi)$ (or Wigner functions, as they are indeed a set of functions); these are discussed in some detail in Appendix B.5). The expansion reads:

$$R(\Psi, \Theta, \Phi) = \sum_{L=0}^{\infty} \sum_{m=-L}^{L} \sum_{n=-L}^{L} P_{Lmn}\, D^L_{mn}(\Psi, \Theta, \Phi). \tag{12.26}$$

The Wigner functions are orthogonal:

$$\langle D^L_{mn}(\Psi, \Theta, \Phi) D^{L'}_{m'n'}(\Psi, \Theta, \Phi) \rangle = \delta_{LL'}\, \delta_{mm'}\, \delta_{nn'}/(2L+1) \tag{12.27}$$

$$\langle\rangle := \frac{1}{8\pi^2} \int_0^{2\pi} \int_{-1}^{1} \int_0^{2\pi} \mathrm{d}\Psi \; \mathrm{d}\cos\Theta \; \mathrm{d}\Phi. \tag{12.28}$$

The combination of (12.26) and (12.27) shows that the moment P_{Lmn} is essentially the average of the orientation-distribution function weighted with the D^L_{mn} function:

$$P_{Lmn} = (2L+1)\langle D^L_{mn}(\Psi, \Theta, \Phi)\rangle \tag{12.29}$$

(the proof is analogous to equation (12.10)). The $\langle D^L_{mn}(\Psi, \Theta, \Phi)\rangle$ are denoted as the order parameters.

Symmetries decrease the number of relevant order parameters considerably. Due to the second-rank-tensor behaviour of the NMR interactions, only the order parameters for even L contribute to the NMR spectra and can be determined from those spectra. In most polymers, we have orthorhombic symmetry both macroscopically and microscopically (i.e. the structures exhibit reflection symmetry with respect to three orthogonal planes). This leads to:

$$\langle D^L_{mn}\rangle = \langle D^{|L|}_{|m|,|n|}\rangle \tag{12.30}$$

and establishes that the $\langle D^L_{mn}\rangle$ are real.

If the orientations exhibit cylindrical symmetry with respect to the sample director, the distribution is independent of Φ and can be described by the $\langle D^L_{m0}\rangle$: Φ is related to the second index (n) according to equation (B.27). In the case of microscopic cylindrical symmetry, there is no dependence on Ψ so that only the $\langle D^L_{0n}\rangle$ contribute in the moment expansion. For transverse isotropy, that is, microscopic and macroscopic uniaxiality, only the $\langle D^L_{00}\rangle = \langle P_L\rangle$ remain, since $D^L_{00}(\Psi, \Theta, \Phi) = P_L(\cos\Theta)$ (see equation (B.29)).

12.4 Synchronized-MAS 2D experiment: site-resolved orientation measurement

For samples with some degree of macroscopic orientational order, there is a dependence of the MAS sideband spectrum, as shown in Figure 12.5, on the rotor phase γ at the start of the signal detection (if the sample is arranged in the rotor such that it does not exhibit cylindrical symmetry around the rotor axis). This orientation dependence can be exploited to analyse the degree of orientation in the material: the higher the degree of order of a sample, the stronger the γ dependence of the spectra.

In one-dimensional rotor-synchronized spectra, however, mixing of absorptive and dispersive lineshapes occurs (Maricq and Waugh, 1979) as seen in Figure 12.5, with the dispersive components often broadening the lines in an intolerable fashion. Therefore, the orientation dependence of rotor-synchronized MAS spectra could only be fully exploited after it was discovered that a Fourier transformation against the rotor phase $\gamma_1 = \omega_r t_1$, applied to 2D datasets with $\gamma_1 = \omega_r t_1$ regularly incremented through 360° (see the pulse sequence in Figure 12.6), under most circumstances produces absorptive lines along ω_2 for each fixed ω_1 value. Another major advantage of this 2D experiment over 1D 'MAS rotation patterns' is the selective extraction of deviations from cylindrical symmetry around the rotor axis in terms of the spectral intensity with $\omega_1 \neq 0$. Therefore, the technique is particularly useful for samples with relatively low degrees of order, which are quite difficult to study in detail with other NMR methods. In isotropic samples, there is no initial-rotor-phase dependence so that no modulation occurs; consequently, the spectral intensity is confined to the slice at $\omega_1 = 0$, as seen in Figure 12.7(a). The higher the degree of order perpendicular to the rotation axis,

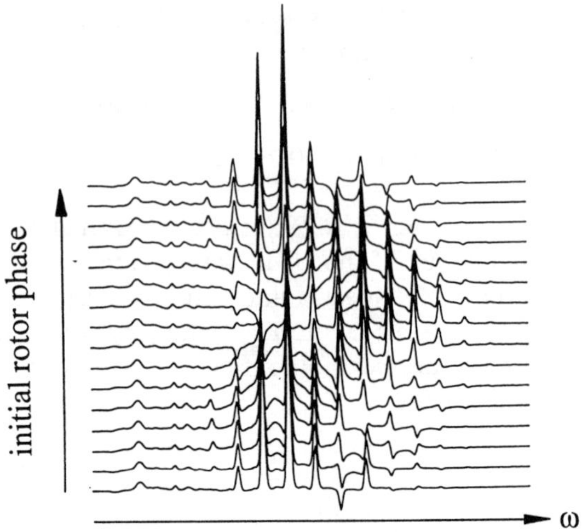

Figure 12.5 Dependence of the MAS sideband pattern of a highly oriented uniaxial material on the initial rotor phase $\omega_r t_1$ at the start of the acquisition. Dispersive lines with broad wings are recognized in many of the spectra. These are eliminated by Fourier transformation over t_1 (see the spectrum in Figure 12.7(c)).

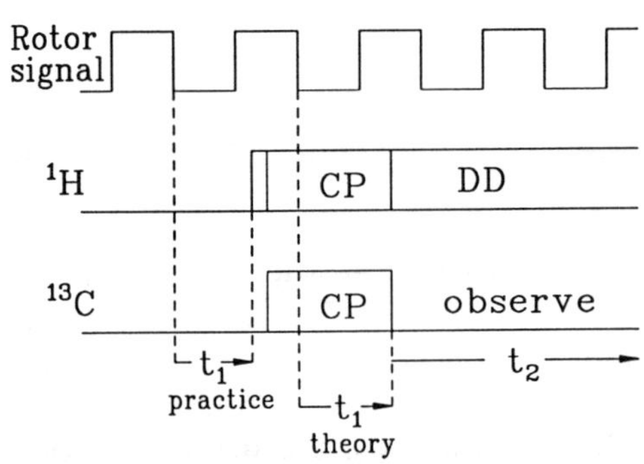

Figure 12.6 Pulse sequence of the sync-MAS 2D experiment: CP-MAS with rotor-synchronized initial phase. Two different t_1 definitions are indicated in the figure. In theory, $\omega_r t_1$ is the rotor phase at the start of the precession of the ^{13}C magnetization, i.e. right at the end of the cross-polarization pulses. In practice, it is risky to wait for the trigger pulse during CP irradiation. In addition, incrementing the CP time would lead to signal-intensity changes. Due to the exact periodicity of the rotor phase and because of the reproducible duration of the ^1H 90° pulse and the long CP pulses, the t_1 delay can also be inserted before the ^1H 90° pulse.

the stronger the modulation as a function of γ, and the higher the orders of the harmonics (sidebands in ω_1) that show significant intensities; an example is shown in Figure 12.7(c).

12.4.1 Molecular orientation and sidebands in the sync-MAS 2D spectrum

For the pulse sequence of Figure 12.6, in terms of the f functions as introduced in

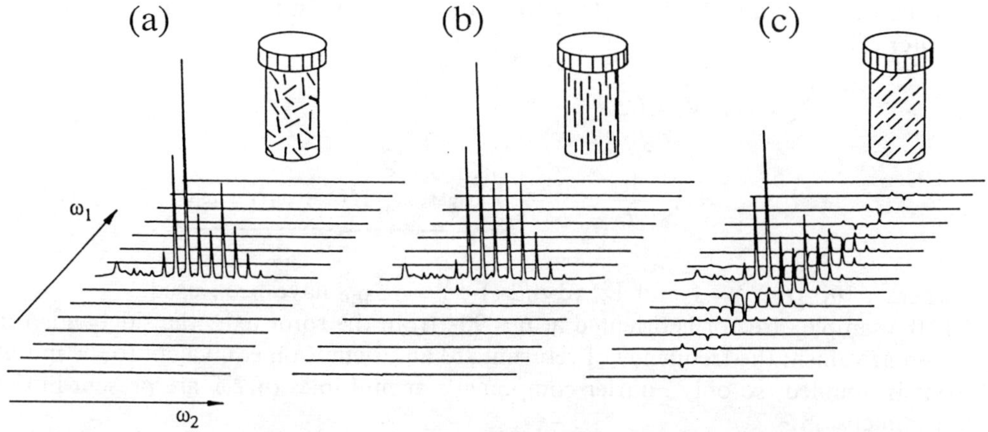

Figure 12.7 Orientation dependence of the spectral-intensity distribution in sync-MAS 2D spectra. For an isotropic sample (a), or a uniaxial sample with the unique axis along the rotor axis (b), the intensity is confined to the $\omega_1 = 0$ slice. (c) If the preferential orientation of the sample is not collinear with the rotor axis, intensity appears also at $\omega_1 \neq 0$. The further the intensity extends in the ω_1 direction, the higher the degree of orientation of the sample.

Section 3.9 the time signal for a single segment or crystallite, with initial phase $\gamma + \omega_r t_1$, is given as:

$$g(t_1, t_2) = f^*(\gamma + \omega_r t_1) f(\gamma + \omega_r t_1 + \omega_r t_2) \tag{12.31}$$

according to equation (3.82) (the isotropic chemical shift is disregarded here). The sideband intensities are produced by FT over $g(t_1, t_2)$:

$$I_{M,N} = \int_0^{t_r} \frac{dt_1}{t_r} e^{-iM\omega_r t_1} f^*(\gamma + \omega_r t_1) \int_0^{t_r} \frac{dt_2}{t_r} e^{-iN\omega_r t_2} f(\gamma + \omega_r t_1 + \omega_r t_2) \tag{12.32}$$

and with $\gamma_1 = \gamma + \omega_r t_1, \gamma_2 = \gamma + \omega_r t_1 + \omega_r t_2 = \gamma_1 + \omega_r t_2 \Rightarrow -N\omega_r t_2 = -N\gamma_2 + N\gamma_1$:

$$\begin{aligned} I_{M,N}(\alpha, \beta, \gamma) &= e^{iM\gamma} \int_0^{2\pi} \frac{d\gamma_1}{2\pi} e^{-i(M-N)\gamma_1} f^*(\gamma_1) \int_0^{2\pi} \frac{d\gamma_2}{2\pi} e^{-iN\gamma_2} f(\gamma_2) \\ &= e^{iM\gamma} F^*_{N-M}(\alpha, \beta) F_N(\alpha, \beta). \end{aligned} \tag{12.33}$$

These sideband intensities have the following properties:

- The centre slice at $M = 0$ is always real (absorptive) and positive. In fact, it is equal to the unsynchronized 1D spectrum of the oriented sample:

$$I_{0,N} = e^0 F^*_{N-0}(\alpha, \beta) F_N(\alpha, \beta) = |F_N(\alpha, \beta)|^2. \tag{12.34}$$

- The total intensity for each ω_1 (i.e. M) slice vanishes, except for the central ($M = 0$) slice. This can be simply derived in the framework of the theory of Fourier series, where the integral in the frequency domain is known to be equal to the first point in the time domain, $\Sigma_N I_N = g(t_2 = 0)$ in the 1D case. In two dimensions, we can write:

$$\sum_N I_{M,N} = \text{`}g_M(t_2 = 0)\text{'}$$

$$= \int_0^{t_r} \frac{dt_1}{t_r} e^{-iM\omega_r t_1} \underbrace{f^*(\gamma + \omega_r t_1) f(\gamma + \omega_r t_1 + \omega_r 0)}_{=1} = \delta_{M,0} \quad (12.35)$$

where $f^*(\gamma')f(\gamma') = 1$ and $\int_0^{2\pi} (d\gamma'/2\pi)\, e^{iM\gamma'} = \delta_{M,0}$ have been used.

- If the sample director is oriented at $\beta = 90°$ from the rotor axis, the sidebands for even M vanish: the frequency of returning to an orientation equivalent to the initial one is doubled, so only Fourier components at multiples of $2\omega_r$ are present in the ω_1 dimension.
- According to Section (4.6.2), for the 2D spectra to be absorptive, the sideband intensities $\bar{I}_{M,N}$ summed over all orientations present in the sample must be real or be easily phased to be real. This condition is not fulfilled for most single crystals (see Figure 12.8(a) and the following discussion). According to basic Fourier theory (see also Section 4.6.2), real $\bar{I}_{M,N}$ are equivalent to invariance of the time signal $\bar{g}(t_1, t_2)$ under simultaneous time reversal of t_1 and t_2 together with complex conjugation:

$$\bar{I}_{M,N} = \bar{I}^*_{M,N} \Leftrightarrow \bar{g}(t_1, t_2) = \bar{g}^*(-t_1, -t_2). \quad (12.36)$$

Before deriving the general condition for absorptive sidebands, we want to consider the instructive case of a single crystallite. Then, absorptive spectra require $g(t_1, t_2) \overset{!}{=} g^*(-t_1, -t_2)$, and with the signal $g(t_1, t_2)$ of equation (12.31) this reads:

$$f^*(\gamma + \omega_r t_1) f(\gamma + \omega_r t_1 + \omega_r t_2) \overset{!}{=} f(\gamma - \omega_r t_1) f^*(\gamma - \omega_r t_1 - \omega_r t_2). \quad (12.37)$$

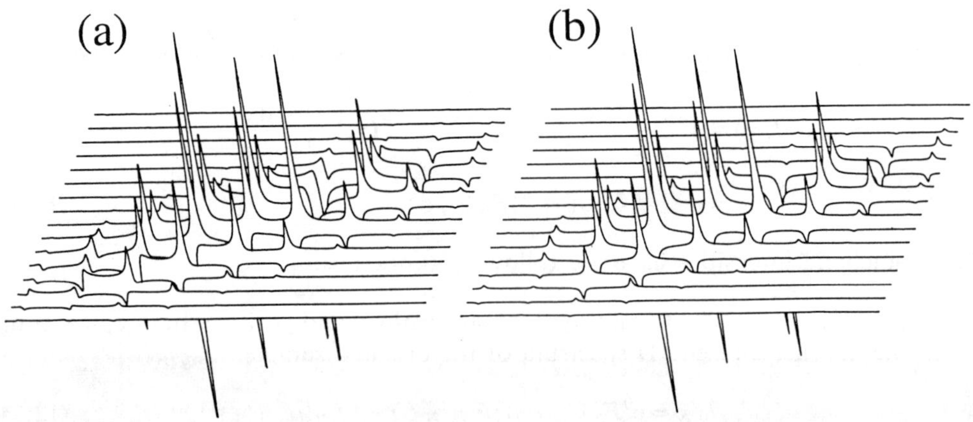

Figure 12.8 Simulated sync-MAS 2D spectra with $\delta/\omega_r = 2$ and $\eta = 0.6$. (a) Dispersive sidebands are clearly visible for a single site in a single crystal, $\alpha = 25°$, $\beta = 40°$ and $\gamma = 0°$ here. (b) These dispersive components cancel if the spectrum for the mirror-image orientation $\alpha = -25°$, $\beta = 40°$ and $\gamma = -0°$ is added to the spectrum in (a).

We can only fulfil this condition if $\gamma = 0$. This corresponds to a shift of the time origin by γ/ω_r which can be corrected by a linear phase correction $\exp(\mathrm{i}M\gamma)$ of the spectrum. Now we demand $f(-\omega_r t) \stackrel{!}{=} f^*(\omega_r t)$ in order to fulfil relation (12.37). On the basis of:

$$\mathrm{e}^{\mathrm{i}\Phi(-t)} \stackrel{(3.81)}{=} f(-\omega_r t) \stackrel{!}{=} f^*(\omega_r t) \stackrel{(3.81)}{=} \mathrm{e}^{-\mathrm{i}\Phi(t)} \tag{12.38}$$

this means that we require $\Phi(-t) \stackrel{!}{=} -\Phi(t)$. To check under which conditions this holds, we analyse formula (3.79) for $\gamma = 0$:

$$\Phi(t) = \frac{1}{\omega_r}\{C_1 \sin \omega_r t + \tfrac{1}{2} C_2 \sin(2\omega_r t) - S_1 \cos \omega_r t + \tfrac{1}{2} S_2 \cos(2\omega_r t)\}.$$

The phase $\Phi(t)$ changes sign upon the substitution $t \rightarrow -t$ only if the S_1, S_2 coefficients vanish (recall $\cos(-\omega_r t) = \cos \omega_r t$, $\sin(-\omega_r t) = -\sin \omega_r t$). S_1 and S_2 are proportional to η and to $\sin 2\alpha$, so we conclude that the $I_{M,N}$ are real if $\eta = 0$, or if $\alpha = n \times 90^\circ$ in the transformation PAS $\rightarrow$ RF. It should be noted, however, that for the general case of a single crystal (e.g. $\alpha = 25^\circ$, $\beta = 40^\circ$, $\eta = 0.6$, as in Figure 12.8(a)) the sidebands have dispersive components that cannot be eliminated by simple phase corrections.

For a sample containing two or a few more crystallites with different orientational angles β_j and γ_j, the total spectrum also cannot be properly phased even if η is zero for each interaction tensor: the sidebands of each crystallite require phasing by $\exp(\mathrm{i}\gamma_j M)$, and these phasing factors are different due to the difference in γ_j. (The difference in β_j must be demanded in this context to ensure that the dispersive parts do not cancel on the basis of the mirror-image pair relation after a phase correction $\exp\{\mathrm{i}\tfrac{1}{2}(\gamma_1 + \gamma_2)M\}$, as discussed in the following and shown in Figure 12.8(b).)

In partially ordered samples with continuous distributions of orientations, the condition (12.36), $\bar{g}(t_1, t_2) = \bar{g}^*(-t_1, -t_2)$, for absorptive sidebands is usually fulfilled. This can be shown on the basis of the decomposition of the signal in terms of contributions from mirror-image segment pairs as in Section 3.9.2. In that context, reflection symmetry is required with respect to the zx plane of the rotor frame. On the other hand, in the case of uniaxially oriented samples a reflection symmetry exists only with respect to the 'director plane', defined as the plane containing the sample director and the rotor axis. Both planes rotate with ω_r with the sample. We first assume that both planes coincide. This can be achieved by starting the evolution time t_1 when the director is in the plane that contains the $\vec{\mathbf{B}}_0$ field and the rotor axis: this '$\mathbf{B}_0$ plane' (not to be mixed up with the field plane of Sections 12.2 and 12.7) is the rotor-frame zx plane at $t = 0$ (see Section 3.9.2). Now we can apply the mirror-image pair signal argument (see equation (3.91)):

$$\begin{aligned} g_p(t_1, t_2) = {} & f^*(\alpha, \beta, \gamma + \omega_r t_1) f(\alpha, \beta, \gamma + \omega_r t_1 + \omega_r t_2) \\ & + f^*(-\alpha, \beta, -\gamma + \omega_r t_1) f(-\alpha, \beta, -\gamma + \omega_r t_1 + \omega_r t_2). \end{aligned} \tag{12.39}$$

If we write down the complex conjugate of $g_p(t_1, t_2)$ and replace t_1 and t_2 by $(-t_1)$ and $(-t_2)$, using $f(-\alpha, \beta, -\gamma) = f^*(\alpha, \beta, \gamma)$ (see equation (3.88)), the equality $g_p(t_1, t_2) = g_p^*(-t_1, -t_2)$ is found. Consequently, the same applies for the superposition of all pair signals in the sample, $\bar{g}(t_1, t_2) = \bar{g}^*(-t_1, -t_2)$, so that real $\bar{I}_{M,N}$ result.

Now we can relax the condition that the director plane must start in the $\vec{\mathbf{B}}_0$ plane. If the angle between these planes is γ_0, this corresponds to a change of all previous γ values by γ_0. According to the Fourier shift theorem (4.15), which is applicable to this periodic signal, this shift by γ_0 changes the sideband intensities $I_{M,N}$ only by a phase factor $\exp(\mathrm{i}\gamma_0 M)$. This can be compensated by linear phase correction in the $M(\omega_1)$ dimension (an ambiguity occurring in this phasing is discussed below). Note that this factor $\exp(\mathrm{i}\gamma_0 M)$ is only vaguely related to the term $\exp(\mathrm{i}\gamma M)$ in the single-crystallite signal (12.33). Even though in the uniaxial sample there are many different values of γ present, the total signal exhibits only one common phase factor with the single-valued γ_0.

- If the sideband coefficients after linear phase correction in M, $\bar{\bar{I}}_{M,N} := \mathrm{e}^{\mathrm{i}\gamma_0 M}\bar{I}_{M,N}$, are real, we have a symmetry relation between the sidebands (M, N) and $(-M, N-M)$:

$$\bar{\bar{I}}_{M,N} \overset{(\bar{\bar{I}}_{\text{real}})}{=} \bar{\bar{I}}^{*}_{M,N} = F_{N-M}F^{*}_{N} = \underbrace{F^{*}_{N}}_{N'-M'} \; \underbrace{F_{N-M}}_{N'} \overset{(12.33)}{=} \bar{\bar{I}}_{\underbrace{-M}_{M'},\underbrace{N-M}_{N'}}. \tag{12.40}$$

- In practice an ambiguity arises in the linear phase correction required to eliminate the factor $\mathrm{e}^{\mathrm{i}M\gamma_0}$. Often, the correct γ_0 is not known, therefore one looks for a phase angle ϕ that makes the peak shapes absorptive when a phase correction $\mathrm{e}^{-\mathrm{i}M\phi}$ is applied to the experimental spectrum. In addition to the correct solution $\phi = \gamma_0$, absorptive peaks are produced also by multiplication with:

$$(-1)^M \, \mathrm{e}^{-\mathrm{i}M\gamma_0} = \mathrm{e}^{-\mathrm{i}M(\gamma_0 + 180^\circ)} \qquad (\text{i.e. } \phi = \gamma_0 + 180^\circ)$$

which inverts the intensities in every odd-M slice. The correct solution $\phi = \gamma_0$ is *a priori* not preferred: the signs of the sideband intensities are not specified, because both positive and negative peak intensities $\bar{\bar{I}}_{M,N}$ can occur for $M \neq 0$.

This ambiguity has to be kept in mind when performing and analysing sync-MAS 2D experiments. In fact, it is not too difficult to arrange the sample orientation and the triggering such that it is known that $\gamma_0 \cong 0^\circ$. Then, the ambiguity is eliminated. A reliable choice between the phase values can also be made if signals from different sites are resolved whose relative orientations are fixed by chemical bonds. The order parameters of the molecular axis in the DF must be similar for these sites, which can serve as a criterion for finding the correct one of the two possible γ_0 values.

- The angle χ_1 between the rotor axis and the sample director $\vec{\mathbf{Z}}_{\mathrm{D}}$ strongly influences the sync-MAS 2D spectra. In terms of sample preparation, $\chi_1 = 0^\circ$ with the sample director along the rotor axis would usually be most convenient. It is useless, however, since it does not produce a γ-dependent modulation for uniaxial samples. The configuration with $\chi_1 = 90^\circ$ (sample director perpendicular to the rotor axis) would still be relatively easy to prepare but as noted above, every second slice in the spectral order dimension ω_1 vanishes in that case. The practical dilemma is obvious: for the spectroscopy, the sample configuration in the rotor has to be as asymmetric as possible, while the stability of the mechanical rotation requires a configuration that is as rotationally symmetric as possible.

For evaluating the orientational information contained in the 2D spectra, an analysis directly in terms of individual Θ values, equivalent to a δ-function expansion as discussed in Section 11.6, is not possible. A Legendre (or Y_{Lm}) moment expansion, however, suits the data very well as will be shown here.

12.4.2 Analysis for systems with transverse isotropy

For systems with transverse isotropy, Harbison *et al.* (1987) employ the following moments and P_L-weighted spectra in their analysis:

$$S^{ex}(\vec{\omega}) = \sum_L S_L^{MF}(\vec{\omega})\langle P_L(\cos\Theta_M)\rangle \tag{12.41a}$$

$$S_L^{MF}(\vec{\omega}) = (2L+1)\int S^{MF}(\vec{\omega};\Theta_M)P_L(\cos\Theta_M)\,d\cos\Theta_M. \tag{12.41b}$$

The $S_L^{MF}(\vec{\omega})$ depend on the sample orientation and the relative orientation of the PAS and the MF (more precisely, the $\vec{\mathbf{Z}}_M$ axis) as characterized by the angles α_P and β_P. (This distinguishes them from the orientation-independent P_L-weighted spectra (12.23) for $\eta = 0$. There, the effects of the sample (β_0) and tensor (β_P) orientation could be factored out on the basis of $\langle P_L(\cos\theta)\rangle = P_L(\cos\beta_0)P_L(\cos\beta_P)\ \langle P_L(\cos\Theta_M)\rangle$, relation (12.15).)

How well the moment-analysis approach suits the spectral features of the 2D sync-MAS experiment can be appreciated from Figure 12.9. It displays the relationship between the location of the signal intensity in the 2D spectrum and the order L of the Legendre polynomial: the P_L-weighted spectra $I_{L,m,n}$ for $P_L(\cos\Theta_M)$ contain intensity only for $|m| \leqslant L$ (Harbison *et al.*, 1987). (Note that compared, for instance, with equation (12.33), the indices of $I_{L,m,n}$ have been changed from M, N to m, n to avoid confusion with the index M for the molecular-frame angle Θ_M.) To rationalize the m and L dependence of the intensity, we have to write down the expression for 2D sideband intensities, equation (12.33), with a distribution $p_M(\Theta_M) = P_L(\cos\Theta_M)$:

$$I_{L,m,n} = \iiint d\alpha_M\, d\cos\beta_M\, d\gamma_M\ P_L[\cos\Theta_M(\alpha_M,\beta_M,\gamma_M)]$$
$$\times\, e^{im\gamma_M}F^*_{n-m}(\alpha_M,\beta_M)F_n(\alpha_M,\beta_M). \tag{12.42}$$

Here, we have exploited the fact that, as mentioned in Section 3.9.1, the formal expressions by f or F functions are analogous whether given in terms of angles (α,β,γ) relating PAS and RF, or in terms of angles that relate any other molecule-fixed system to the RF, such as $(\alpha_M,\beta_M,\gamma_M)$ which connect MF and RF. The angle Θ_M can be expressed in terms of $(\alpha_M,\beta_M,\gamma_M)$ according to equation (12.7):

$$\cos\Theta_M(\alpha_M,\beta_M,\gamma_M) = \cos\chi_1\cos\beta_M + \sin\chi_1\sin\beta_M\cos\gamma_M \tag{12.43}$$

where χ_1 is the angle between $\vec{\mathbf{Z}}_D$ and the rotor axis. In equation (12.42), the integral over γ_M represents a Fourier transformation of $P_L[\cos\Theta_M(\alpha_M,\beta_M,\gamma_M)]$. $P_L(x)$ is a polynomial with x^L as its highest power:

$$P_L(x) = \sum_{k=0}^{L} a_k x^k. \tag{12.44}$$

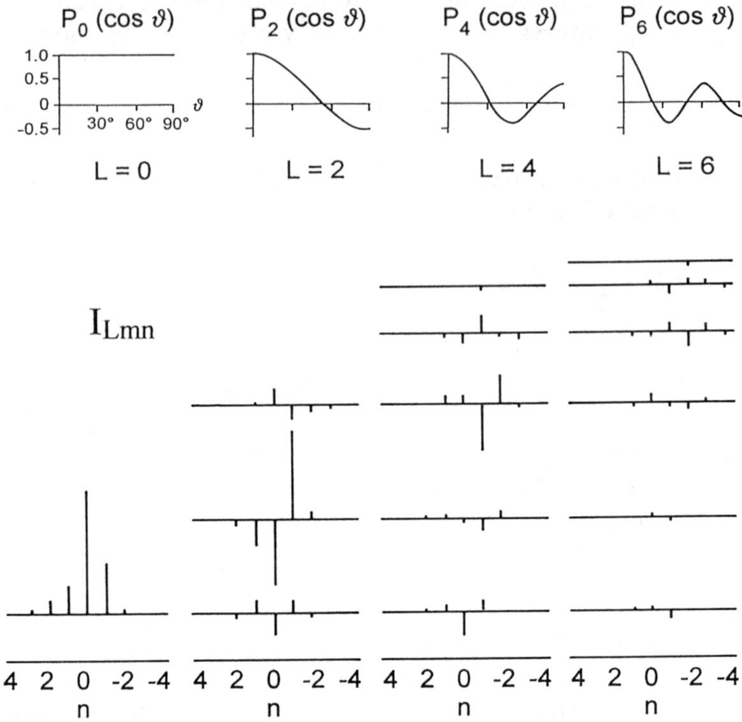

Figure 12.9 Sync-MAS 2D P_L-weighted spectra. Note that the spectral intensity is identically zero for ω_1 slices with $|m| > L$. This provides for a direct qualitative relationship between the order parameters and the spectral intensity distribution: the further the experimental intensity extends to slices with large m, the larger the contribution of higher order parameters. Adapted from Harbison *et al.* (1987).

Taking into account relation (12.43) for $\cos \Theta_M(\alpha_M, \beta_M, \gamma_M)$, we can write:

$$P_L[\cos \Theta_M(\alpha_M, \beta_M, \gamma_M)] = \sum_{k=0}^{L} \tilde{a}_k(\cos \gamma_M)^k = \sum_{k=0}^{L} b_k \cos(k\gamma_M) \qquad (12.45)$$

(for a proof of the last step from $(\cos \gamma_M)^k$ to $\cos(k\gamma_M)$ see formula (A.29)). Thus, we obtain:

$$\int_0^{2\pi} P_L[\cos \Theta_M(\alpha_M, \beta_M, \gamma_M)] \exp(im\gamma_M)\, d\gamma_M = \sum_{k=0}^{L} b_k \tfrac{1}{2}(\delta_{k,m} + \delta_{k,-m}) \qquad (12.46)$$

so that indeed signal appears only for $|m| \leqslant L$. To summarize, significant intensity in a slice of order m means that moments $\langle P_L \rangle$ with $L \geqslant |m|$ must contribute significantly.

The reverse conclusion is not always valid: a slice of order m can have low intensity even if the corresponding moment $\langle P_{L=m}(\cos \Theta_M) \rangle$ is large. This can occur because the intensity of a P_L-weighted spectrum $S_L^{MF}(\vec{\omega})$ can vanish or at least become very small for certain tensor orientations in the MF. For these cases, the coefficient $\langle P_{L=m}(\cos \Theta_M) \rangle$ in the sum (12.41a) cannot be determined at all, or it exhibits a large error margin. The occurrence of vanishing $S_L^{MF}(\vec{\omega})$ can be demonstrated easily in the case of $\eta = 0$, where $S_L^{MF}(\vec{\omega}) = P_L(\cos \beta_P) S_L^{\Theta}(\vec{\omega})$ according to equation (12.21).

Obviously, the spectrum $S_L^{MF}(\vec{\omega})$ vanishes if $P_L(\cos\beta_P)=0$, that is, if the angle β_P between the $\vec{Z}_P$ axis and the chain direction $\vec{Z}_D$ is at a null point of the Lth Legendre polynomial. For instance, for $\beta_P = 54.74°$, $S_2^{MF}(\vec{\omega})$ vanishes and thus $\langle P_2(\cos\Theta_M)\rangle$ cannot be determined. In fact, for all β_P values except for 0° and 90°, at some L value such a 'blind spot' occurs. The fit procedure will crash or produce an arbitrary coefficient if β_P is at the null point of a $P_L(\cos\beta_P)$ in the fit; the subspectra are not linearly independent when one of them is identically zero. In practice, β_P will never hit such a null point exactly. Still, the closer β_P approaches a null point, the less stable the fit will become so that the $\langle P_L(\cos\Theta_M)\rangle$ coefficient for the critical L will be very unreliable.

Finally, it should be recognized that a small or even vanishing $\langle P_2(\cos\Theta_M)\rangle$, reflected in relatively low intensity of $m=\pm1, \pm2$ slices, does not necessarily imply an unoriented sample, but can arise if the orientational distribution is not centred around 0° but, for instance, a conical distribution with half-apex angle close to 54.7°.

12.4.3 Macroscopically uniaxial systems without transverse isotropy

In the discussion of the preceding section, presence of transverse isotropy was assumed. The first requirement for this property, namely macroscopic uniaxiality, is readily assured in fibres and extruded samples. The second condition for transverse isotropy is the existence of a molecular axis around which the interaction tensors are distributed with cylindrical symmetry. If this condition is not fulfilled or if such an assumption is to be avoided, the orientation distribution must be considered in terms of two microscopic angles, such as Θ_M and Ψ_M. The macroscopic uniaxiality requires that the distribution exhibits cylindrical symmetry in the third angle, which is Φ_M in the present case. In terms of order parameters, this means that the $\langle D_{m0}^L(\Psi_M,\Theta_M,\Phi_M)\rangle \sim \langle Y_{Lm}(\Theta_M,-\Psi_M)\rangle$ (see equation (B.27)) do not vanish.

Alternatively, the distribution can be expressed in terms of Θ and Ψ, since the system is also uniaxial in Φ. This uniaxiality in Φ is not completely trivial, but it can be rationalized: if the chain axes are distributed cylindrically symmetric around $\vec{Z}_D$, the $\vec{Z}_P$ axes are also distributed with cylindrical symmetry, regardless of the conformation of the chain or the tensor orientations. This can be formally derived from the Wigner-matrix addition theorem (B.30), where for uniaxiality in Φ_M, only the following terms remain:

$$\begin{aligned}\langle D_{m0}^L(\Psi,\Theta,\Phi)\rangle &= \sum_{n'=-L}^{L} D_{mn'}^L(\alpha_P,\beta_P,\gamma_P)\langle D_{n'0}^L(\Psi_M,\Theta_M,\Phi_M)\rangle \\ &= \langle Y_{Lm}(\Theta,-\Psi)\rangle \\ &= \sum_{n'=-L}^{L} D_{mn'}^L(\alpha_P,\beta_P,\gamma_P)\,\langle Y_{Ln'}(\Theta_M,-\Psi_M)\rangle. \end{aligned} \tag{12.47}$$

The fact that all $\langle D_{mn}^L(\Psi,\Theta,\Phi)\rangle$ with $n\neq0$ vanish proves the cylindrical symmetry of the distribution in Φ. The expansion reads:

$$S^{ex}(\vec{\omega}) = \sum_{L=0}^{\infty}\sum_{m=-L}^{L} S_{Lm}^{PAS}(\vec{\omega})\langle Y_{Lm}(\Theta,\Psi)\rangle \tag{12.48}$$

$$S_{Lm}^{PAS}(\vec{\omega}) = \iint S^{PAS}(\vec{\omega};\Theta,\Psi)\; Y_{Lm}(\Theta,\Psi)\; d\Psi\; d\cos\Theta. \tag{12.49}$$

The analysis in terms of Θ and Ψ has the considerable advantage of being independent of the tensor-orientation angles α_P and β_P which are often not known precisely. On the other hand, a certain price must be paid for the increased information content obtained from this analysis: the increased number of fitting coefficients $S_{Lm}^{PAS}(\vec{\omega})$ require more spectral information and thus the number of sidebands and the reliability of their intensities must be greater than for the simpler case of systems with transverse isotropy. For the interpretation of the resulting distribution it is helpful to realize that (Θ, Ψ) can be considered as the polar coordinates of the macroscopic $\vec{\mathbf{Z}}_D$ axis in the PAS. (This makes it obvious that this type of analysis is not useful in the case of vanishing η: for $\eta = 0$, different Ψ values, or $\vec{\mathbf{X}}_P$ and $\vec{\mathbf{Y}}_P$ axes, are by no means distinguishable.) In the case of transverse isotropy, the contour lines of the distribution in Θ and Ψ are circles if the system is sufficiently highly oriented. The common centre of these circles indicates the orientation of the unique molecular axis (e.g. the chain axis) in the PAS and can thus serve to determine α_P and β_P. Note that analogous information is contained in DECODER 2D spectra (see below) and in certain static 1D spectra, where α_P and β_P can be determined if the system is highly oriented and known to be uniaxial with respect to a given macroscopic order axis. When the chemical-shift PAS orientation in the segment is known, the experiments designed to probe orientational order can actually yield valuable structural information, providing the relative orientation(s) of the segments and the chain axis. An example is given below for oriented DNA.

In the sync-MAS 2D experiment, the experimental values for the order parameters are obtained as the coefficients of the best fit of the subspectra S_L or S_{Lm} to the experimental spectrum S^{ex}, as outlined in Section 11.7.

12.4.4 Sync-MAS 2D studies of oriented polymers

PET fibres were studied by Harbison *et al.* (1987) employing the synchronized MAS 2D experiment. For the four sites distinguishable in terms of the isotropic chemical shift, order parameters of the principal axes systems relative to the sample director were determined. Originally, the $\langle P_L(\cos\Theta)\rangle$ values resulting from the analysis for the OCH_2 and the carboxyl groups were hard to rationalize; the results have become consistent in the light of recent, more precise determinations of the tensor orientations (Schmidt-Rohr *et al.*, 1993; Chmelka *et al.*, 1993), which have corrected the originally assumed orientations of Harbison *et al.* (1987) of the $\vec{\mathbf{Z}}_P$ axes for the carboxyl and OCH_2 groups by 13° and −19°, respectively. The resulting chain order parameters $\langle P_L(\cos\Theta_M)\rangle$, which can be calculated by $\langle P_L(\cos\Theta_M)\rangle = \langle P_L(\cos\Theta)\rangle / P_L(\cos\beta_P)$, with $\beta_P = 8°$ for the OCH_2 groups and $\beta_P = 18°$ for the carboxyls, are

Table 12.1 *Order parameters measured for PET fibres by sync-MAS 2D NMR and DECODER 2D NMR*

		P_2	P_4	P_6	P_8	P_{10}
sync-MAS	OCH_2	0.5(5 ± 2)	0.5(5 ± 3)			
sync-MAS	COO	0.5(1 ± 5)	0.4(1 ± 6)	0.(55 ± 15)		
DECODER	OCH_2	0.6(0 ± 5)	0.4(7 ± 5)	0.3(7 ± 5)	0.26	0.17

given in Table 12.1 and compared with results of recent 2D DECODER measurements on PET fibres. The values agree quite well, with an overall $\langle P_2(\cos\Theta_M)\rangle = 0.55 \pm 0.04$ and $\langle P_4(\cos\Theta_M)\rangle = 0.48 \pm 0.05$. It should be noted that the small ratio of P_2/P_4 for all these data consistently shows that the distribution is not a simple Gaussian, where $\langle P_2(\cos\Theta_M)\rangle = 0.55 \pm 0.05$ would be accompanied by $\langle P_4(\cos\Theta_M)\rangle = 0.17 \pm 0.05$. In fact, the distribution $p_M(\Theta_M)$ reconstructed from the DECODER experiments below (see figure 12.25) proves that the distribution is at least bimodal, with a highly oriented and a disordered component.

For the aromatic carbons, distributions of the $\vec{Z}_D$ axis in the PAS, as displayed in Figure 12.10, were generated from the data. The microscopic cylindrical symmetry is broken for the protonated phenylene sites in this system (sample and spectrum) since the two pairs of protonated sites exhibit different orientations relative to the chain axis, while they are spectrally not directly distinguishable in this experiment. This circumstance required an analysis in terms of Θ and Ψ (while the sample is still uniaxial and not biaxial). As mentioned above, in these plots the contour lines must be circular if the tensor orientation is unique. Thus, the plots make it possible to detect the

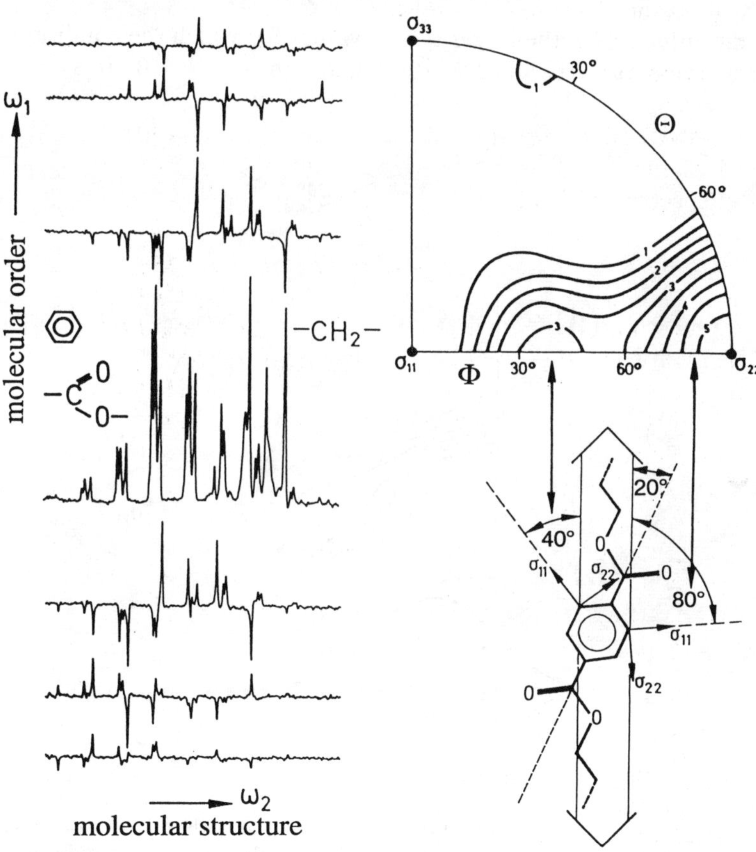

Figure 12.10 Two-dimensional distribution of the director $\vec{Z}_D$ axis in the PAS of the protonated aromatics in PET fibres, as obtained from the sync-MAS 2D data shown to the left. Two maxima, corresponding to the two orientationally inequivalent protonated sites on the phenylene rings of PET, are clearly detected. Adapted from Harbison *et al.* (1987).

presence of various tensor orientations (i.e. α_P and β_P values). Various tensor orientations for the same set of sidebands can be due to a variety of conformations or to a spectral superposition of sites with different dispositions relative to the molecular (chain) axis $\vec{\mathbf{Z}}_M$. The two maxima in Figure 12.10, at $\Psi \simeq 40^\circ$ and at $\Psi \approx 80$–90° correspond to the two different α_P values of the two orientationally distinguishable protonated aromatics. This agrees well with the values of $60^\circ + 20^\circ$ and $60^\circ - 20^\circ$ expected according to the X-ray structure of crystalline PET (Hall, 1984).

Special caution is required in the interpretation of the maxima of the distribution near the boundaries of the octant of the sphere as plotted in Figures 12.10 and 12.11(a). The distribution shown represents the superposition of the distribution of $\vec{\mathbf{Z}}_D$ in all eight octants; therefore, there is a spurious shift of the maximum towards the boundaries due to the portions of the distribution that are 'folded back' from the neighbouring octants. This circumstance is portrayed for a one-dimensional case in Figure 12.11(b) (this could represent a relevant cross-section through the 2D plot of Figure 12.11(a)): the measured intensity at $\Psi = 0^\circ$ is higher than at the real maximum position of $p(\Psi_0)$ if

$$2p(0^\circ) > p(\Psi_0) + p(2\Psi_0). \tag{12.50}$$

The wider the curve $p(\Psi)$, the larger the Ψ_0 values for which the condition (12.50) still holds. For a Gaussian of fwhm $\Delta\Psi$, the maximum is at $\Psi = 0^\circ$ if $\Psi_0 < 0.44\Delta\Psi$. For

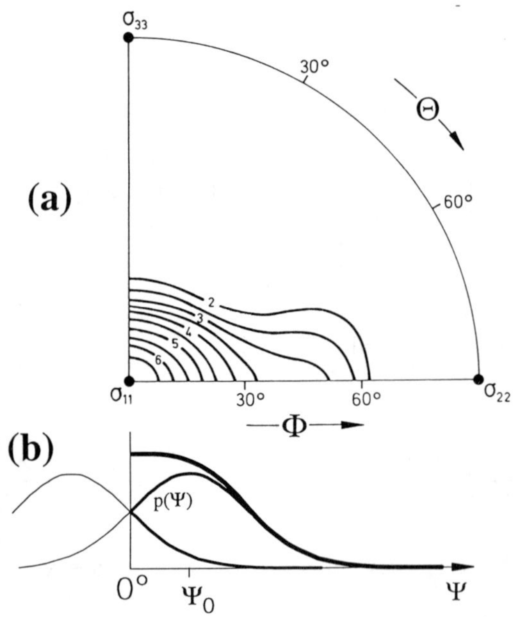

Figure 12.11 (a) Orientation distribution of the director $\vec{\mathbf{Z}}_D$ axis in the PAS of the unprotonated aromatic sites of PET. The maximum of the plotted distribution is centred around the σ_{11} direction, but this need not correspond exactly to the true maximum of the distribution, which is expected near $\Phi = 20^\circ$. In (b), the spurious maximum-shift caused by 'folding back' of the distribution at the limits of an octant of the unit sphere of orientations is shown for a schematic 1D cross-section. The thinner lines indicate the true distribution functions, the thick line the measured distribution. Adapted from Harbison *et al.* (1987).

larger Ψ_0, the maximum is still shifted close to $\Psi = 0$, so that analysis with limited resolution still detects a maximum at $\Psi = 0$. This is probably the reason for the maximum at $\Psi = 0$ for the distribution obtained from the signals of the unprotonated carbons (Figure 12.11(a)). The true maximum must be shifted by 60° relative to the maxima for the protonated sites according to the structure of the phenylene ring and its well-established tensor orientations. This would locate it at $\Psi_0 = 20°$ ($80° - 60°$, $|40° - 60°|$), which seems not to contradict the contour plot in Figure 12.11(a) according to the arguments just given.

Other highly oriented fibres have also been investigated by means of the sync-MAS 2D experiment in the same fashion as PET: for instance, polyethylene (Harbison and Spiess, 1986) and PPTA (Wilhelm *et al.*, 1994). The order parameters found were consistent with those obtained by wide-angle X-ray scattering.

The sync-MAS 2D experiment has been applied to polymers with liquid-crystalline side groups, where the connection between the benzophenone mesogens and the main chains is provided by flexible spacers typically consisting of six CH_2 units (Boeffel and Spiess, 1989). Here, orientational order is very different for the various segments in a repeat unit. Therefore, the site-specific information from the sync-MAS 2D experiment is particularly useful. The spectra show that the orientational order is highest in the mesogens ($P_2 \cong 0.94$, $P_4 \cong 0.77$, $P_6 \cong 0.67$), somewhat lower in the spacers ($P_2 \cong 0.66$, $P_4 \cong 0.44$), and drops to a small value for the main chains ($P_2 \cong 0.1$, $P_4 \cong 0.05$). These results are consistent with ^{2}H NMR lineshape studies of the same systems (Boeffel and Spiess, 1989) and partially confirm the 'spacer concept' which assumes that the aliphatic portion of the side groups provides a decoupling of the main chains from the ordering of the mesogens.

From sync-MAS 2D spectra of oriented DNA in films of wet-spun fibres (Tang *et al.*, 1989), information on the conformation in the solid state could be obtained. Figure 12.12 shows the distribution of the sample director in the PAS as obtained from the sync-MAS 2D data on ^{31}P in the phosphodiester group of the DNA. As outlined above, the maxima of the distribution indicate the orientation of the molecular (chain) director relative to the PAS and thus to the molecular moiety, which in this case is the

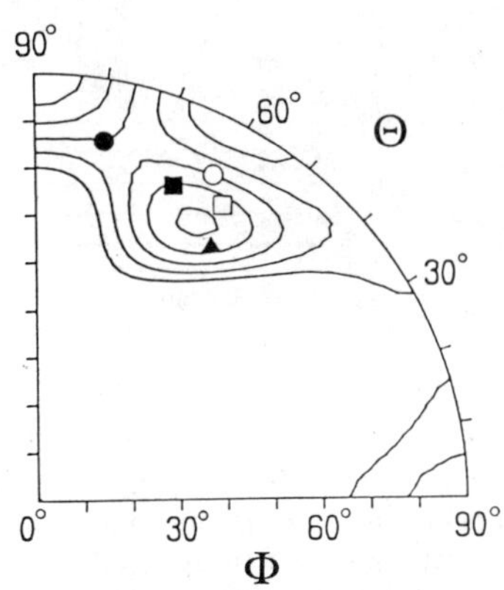

Figure 12.12 Oriented DNA: distribution of director orientations in the principal-axis system, obtained from sync-MAS ^{31}P data. The points indicate conformations of various forms of A-DNA determined by X-ray diffraction. Adapted from Tang *et al.* (1989).

DNA phosphate group. The width of the patterns can be due to macroscopic (fibre) or microscopic (conformational) disorder, and some contribution from the truncation of the Legendre series is also possible. Tang *et al.* (1989) consider conformational and orientational disorder as the major reason. The points with the symbols represent the conformations of A-DNA according to diffraction studies. It should be noted that, since there is no overlap problem in the ^{31}P spectrum with its single site, 'rotation patterns' of a static sample can provide essentially the same information as the MAS experiment; the results of the two approaches actually agree quite well. With a DECODER 2D experiment (see Section 12.7), this analysis of the static lineshape could be further refined.

12.5 3D spectra with sideband-separated orientation spectra

Considering sync-MAS 2D spectra as in Figures 12.10 and 12.13(a), it is apparent that overlap of various sideband patterns is a problem in the sync-MAS 2D experiment even though the anisotropic patterns are split up into sidebands. Sideband patterns overlap in spite of appreciable differences in their isotropic chemical shifts, in particular in polymeric systems where the sidebands are broadened by conformational and packing effects (see Section 2.4.6). Therefore, a 3D experiment is desirable that includes a 2D technique separating the sideband patterns by the isotropic shifts. Two such 2D experiments have been described in Section 6.5. For both of them, prefixing the rotor-phase evolution represents no problem: unlike TOSS, these techniques do not rely on

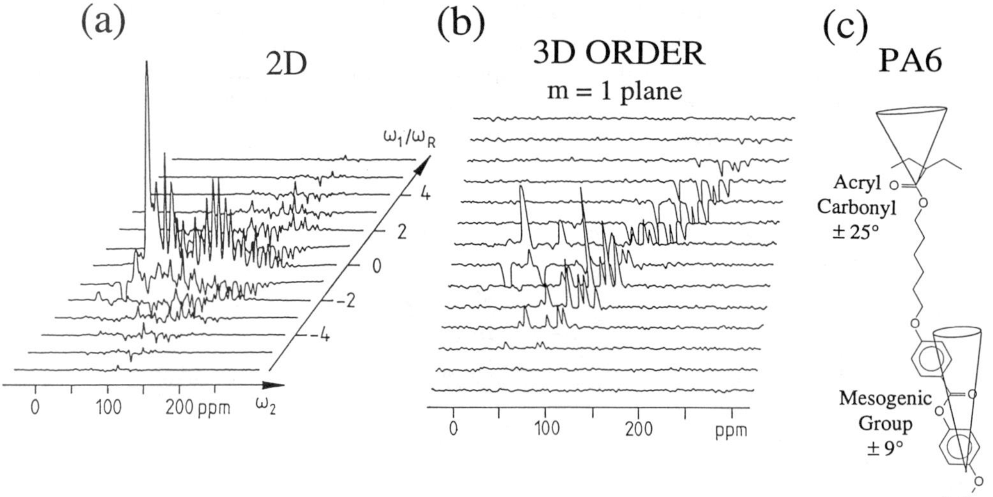

Figure 12.13 Sync-MAS 2D and 3D experiments applied to a liquid-crystalline side-group polymer PA6. (a) 2D spectrum taken at a spinning speed of 1.5 kHz. Sidebands of various orders and different isotropic shifts overlap heavily. (b) In the $m = 1$ plane from the 3D spectrum, the overlap of the sideband patterns is removed by spreading them out according to the sideband orders. (c) Schematic representation of structure and degrees of orientation of mesogens and spacer of PA6 as obtained from the multidimensional sync-MAS spectra. Adapted from Titman *et al.* (1993).

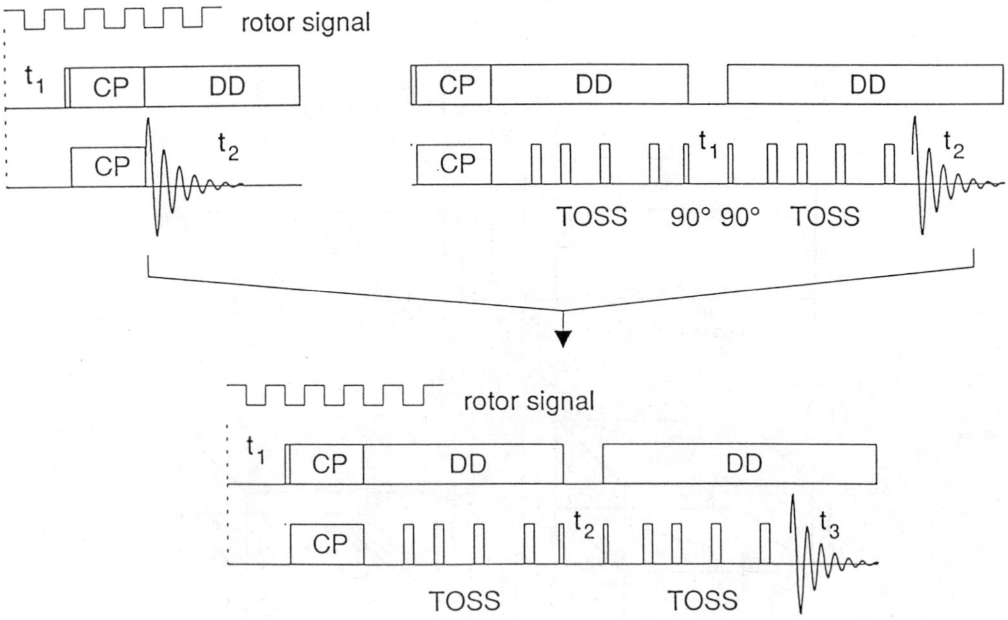

Figure 12.14 Pulse sequence of the sync-MAS 3D experiment which combines the sync-MAS 2D experiment with separation of sidebands by their order (see Section 6.5.3) to remove the overlap of sideband patterns that occurs in many sync-MAS 2D spectra. Adapted from Titman *et al.* (1993).

the integral over the rotor phase γ. Of the two experiments, the sideband-order separation experiment appears to be the better choice due to the smaller number of evolution increments which keeps the 3D datafiles manageable (Titman *et al.*, 1993). The increased clarity of the spectrum is appreciated by comparing the crowded 2D spectrum in Figure 12.13(a) with a plane out of the 3D dataset, Figure 12.13(b). The combination of the pulse sequences is indicated in Figure 12.14. A full view of such a 3D spectrum is shown in Figure 1.6(b). The data evaluation for the signals thus separated can be performed exactly as for the sync-MAS 2D experiment described at length in the previous section. The experiment can be run in a day or even overnight, and the results obtained indicate a high degree of reliability and wide applicability. For instance, order parameters in PPTA fibers were consistently found for carboxyl and unprotonated-aromatic sites (Wilhelm *et al.*, 1994). For polymers with liquid-crystalline side-chains, the results of the sync-MAS 2D experiment have been confirmed in detail (Titman *et al.*, 1993).

12.6 3D correlation of order and motion

The sync-MAS 2D experiment can be combined with the MAS 2D exchange experiment (see Sections 7.11 and 10.4) to yield a 3D experiment (Yang *et al.*, 1990). The detection period of the former 2D experiment is merged with the evolution period of the latter, see Figure 12.15(a). The theory consists of a combination of the properties of both 2D experiments.

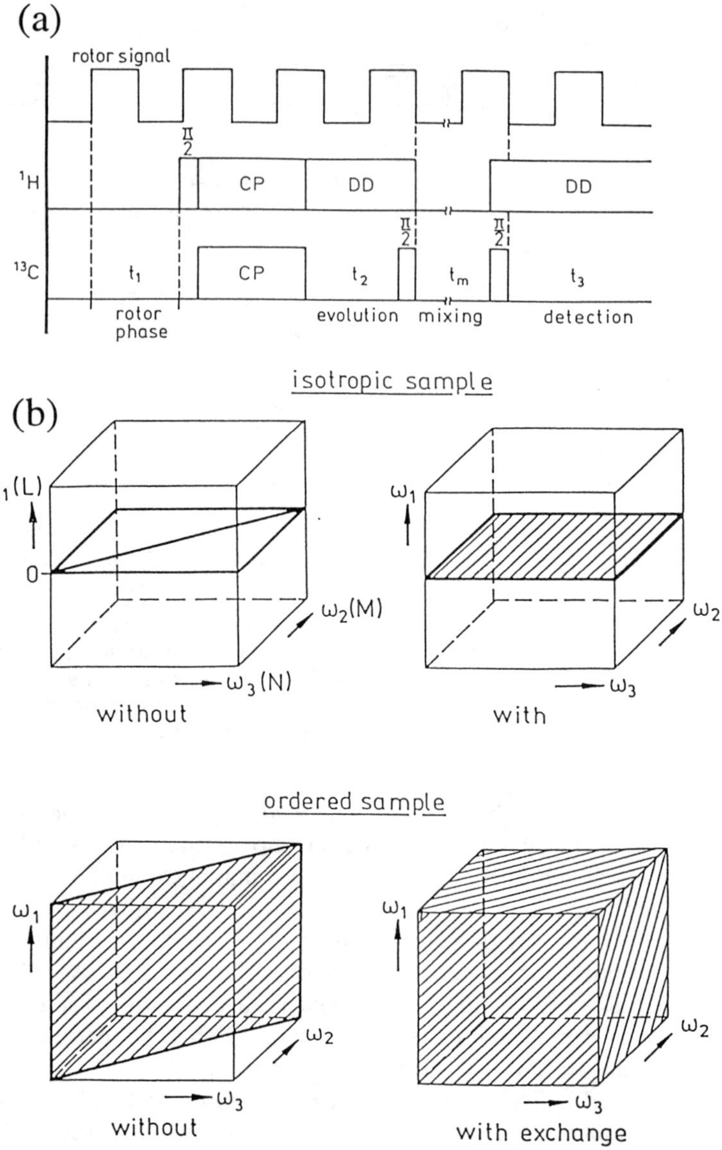

Figure 12.15 (a) Pulse-sequence of the sync-MAS-exchange (correlation of order and motion, 'CORD') 3D experiment, which combines the sync-MAS 2D experiment with the sideband-exchange experiment. (b) Typical intensity distributions in CORD 3D spectra: without exchange, $\omega_2 = \omega_3$; without order, $\omega_1 = 0$. Adapted from Yang *et al.* (1990).

The main features of 3D intensity distributions in this experiment are shown schematically in Figure 12.15(b). They are obvious from the constituent 2D experiments: for powder samples, the intensity in the order dimension ω_1 is restricted to the $L = 0$ slice. Without exchange, the signal is confined to the diagonal plane $\omega_2 = \omega_3$ ($M = N$). The application of this experiment to highly oriented POM is demonstrated in Figure 12.16.

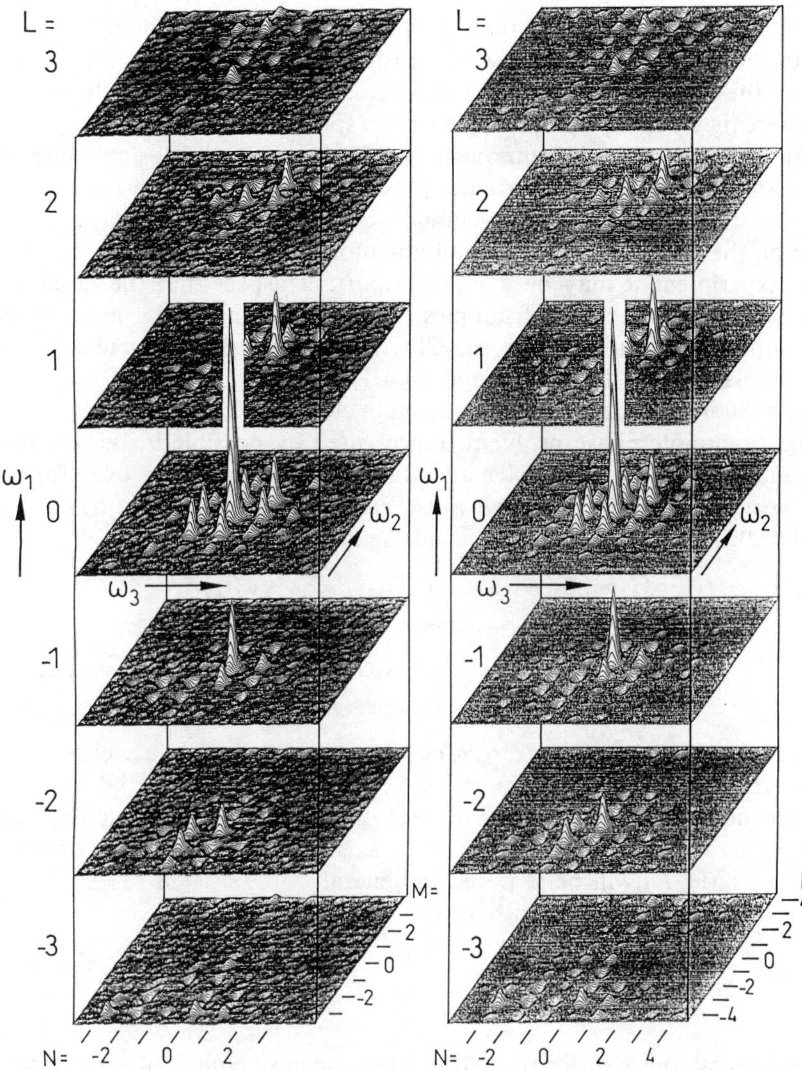

Figure 12.16 CORD 3D spectrum of oriented POM. The spectrum shows a large number of positive and negative peaks, which should make the analysis reliable; on the other hand, effects of static and dynamic disorder are mixed in the peaks, which makes it more difficult to analyse deviations from idealizing models. Adapted from Yang *et al.* (1990).

The number of sidebands in this experiment is considerably larger than in the 2D MAS exchange experiment. In the $\omega_1 = 0$ slice, dynamical information is found without effects of the orientation, so that the motional model can be checked independently of the orientational properties. The basic information on the degree of order can be extracted from the projection onto the ω_1–ω_3 plane (integral over the ω_2 direction). Finally, both dynamical and orientational parameters can be refined on the basis of all sidebands in the 3D cube.

Though in this 3D experiment both order and motion are probed, it is actually difficult to extract information on the dependence of the exchange process on the degree of orientation. Although the signal from the highly oriented material is fairly isolated in the outer slices of the ω_1 dimension, the contributions from less oriented portions of the sample are not separately accessible. They reside in the region around $\omega_1 = 0$, where they overlap completely with the spectrum of the oriented portions (note that in this experiment, each component contributes with its full signal intensity in the central ω_1 slices, regardless of how far it extends into the $\omega_1 \neq 0$ region). In a static 2D or reduced 4D experiment, 'enrichment' of one component with its characteristic lineshape on the diagonal would actually be more directly detectable. In fact, for the 3D MAS experiment it may be a more important aspect that the large number of exchange sidebands for oriented samples, with both positive and negative intensities, provides more information than the 2D MAS sideband-exchange experiment for powder samples (see Sections 7.11 and 10.4.1).

From the theoretical point of view, the main challenge in the 3D experiment consists in solving the double phase problem (represented by possible imaginary portions of $I_{L,M,N}$), and by mixing of dispersive and absorptive 3D lineshapes even for purely real $I_{L,M,N}$). For $I_{L,M,N}$ to be real, we must have $g(t_1, t_2, t_3) = g^*(-t_1, -t_2, -t_3)$ (see Section 4.6.2). Mirror-image segment pair signals $g_p(t_1, t_2, t_3)$ with:

$$\begin{aligned} g_p(t_1, t_2, t_3) = {} & f_a^*(\alpha_a, \beta_a, \gamma_a + \omega_r t_1) f_a(\alpha_a, \beta_a, \gamma_a + \omega_r t_2) \\ & f_b^*(\alpha_b, \beta_b, \gamma_b + \omega_r t_2) \times f_b(\alpha_b, \beta_b, \gamma_b + \omega_r t_2 + \omega_r t_3) \\ & + f_a^*(-\alpha_a, \beta_a, -\gamma_a + \omega_r t_1) \times f_a(-\alpha_a, \beta_a, -\gamma_a + \omega_r t_2) \\ & f_b^*(-\alpha_b, \beta_b, -\gamma_b + \omega_r t_2) \times f_b(-\alpha_b, \beta_b, -\gamma_b + \omega_r t_2 + \omega_r t_3) \end{aligned}$$

can again be shown to obey $g_p(t_1, t_2, t_3) = g_p^*(-t_1, -t_2, -t_3)$ (see Section 3.9.2). As in the rotor-synchronized 2D experiment of Section 12.4.2, in uniaxial samples mirror-image segment pairs are necessarily present only with respect to the reflection plane containing the sample director and the rotor axis. Therefore, a phase correction of the $I_{L,M,N}$ with $\exp(i\gamma_0 L)$ will be required in general.

The 3D time signal can be written as:

$$\bar{s}_s^c(t_1, t_2, t_3) = \sum_L \sum_M \sum_N I_{L,M,N}\, e^{iL\omega_r t_1} \begin{matrix} \cos(\{M\omega_r + \omega_{iso}\} t_2) \\ \sin(\{M\omega_r + \omega_{iso}\} t_2) \end{matrix}\, e^{it_3(N\omega_r + \omega_{iso})}\, e^{-(t_2 + t_2)/T_2}. \tag{12.51}$$

As the first steps, the Fourier transformations against t_3 and t_1 are performed. The resulting data are phased first in ω_3, and then in ω_1 to compensate the $e^{iL\gamma_0}$ rotor-phase factor arising from the signal in equation (12.51). The phase values are determined on the data of the ($t_2 = 0$) plane of the cosine dataset. The next processing step involves combining the real parts of cosine and sine datasets as the real and imaginary parts, respectively, of the signal which is then Fourier transformed against t_2 (see Section 4.6). If the spectrum has not been phased properly in ω_3 and particularly in ω_1, the coefficients multiplying the $t_2 - \omega_3$ signal are complex and the dispersive components ($D(\omega_3 - N\omega_r)$ in our specific case) are not eliminated, as explained in Section 4.6.2.

12.7 The DECODER technique

On the basis of the angular dependence of the frequency, NMR measurements of molecular order can be performed in nonrotating samples. The DECODER technique extends this approach to two and three dimensions; the multiple frequency measurement enhances the information content of the spectra in such a way that angular distributions can be directly reconstructed from the spectral intensity. How that can be achieved will be outlined in this section.

12.7.1 Introduction and motivation

Before the advent of the sync-MAS 2D experiment, ^2H 1D spectroscopy on static samples had established itself as a widely applicable NMR technique for analysing molecular orientations in partially ordered materials (Spiess, 1984, 1985). However, a direct reconstruction (see Section 11.6) of the orientational distribution of C—^2H bonds in the sample is practically impossible from 1D ^2H spectra. The determination of the orientational distribution from 1D ^2H spectra has instead been based on (eye-ball) fits of a moment-expansion series or of suitable assumed distributions (usually Gaussians) (Hentschel *et al.*, 1981). With a series of spectra for various sample orientations, the results of this procedure are quite reliable if the orientational behaviour is sufficiently simple. However, in principle NMR can provide information on the orientational order with fewer assumptions. Among other things, one can overcome the two major problems in 1D ^2H NMR that prevent a more direct approach in the analysis: (i) the overlap of the spectral patterns of two 'transitions' in the easily accessible central region of the spectra; and (ii) the extremely poor angular resolution at the $\omega(\theta = 90^\circ)$ singularity and at the $\omega(\theta = 0^\circ)$ end of the spectral intensity range. This latter problem results in instabilities for best-fit or reconstruction routines, since in those critical regions, the frequency is angle-independent in the first approximation, $(d\omega/d\theta)|_{0^\circ} = (d\omega/d\theta)|_{90^\circ} = 0$, which makes the transformation of ω into θ unreliable.

For natural-abundance ^{13}C NMR in static samples, spectral overlap from various carbon sites represents the most serious problem. In oriented samples, it can be mitigated at certain orientations (e.g. with $\vec{\mathbf{Z}}_D \parallel \vec{\mathbf{B}}_0$) where the spectral patterns are compressed as much as possible; however, only at the price of decreased angular resolution.

12.7.2 Relation between orientational and spectral features

A solution for the mentioned problems along with qualitatively improved information on the orientation distribution can be achieved by measuring the orientation-dependent frequency two or three times for each individual segment by means of 2D or 3D spectroscopy with a sample flip between the evolution and detection periods (Henrichs, 1987). The corresponding pulse sequence is shown in Figure 12.17. This technique and the corresponding reconstruction of the orientation distribution from the spectral intensity has recently been termed DECODER (direction exchange with correlation for orientation–distribution evaluation and reconstruction). The 2D correlation approach provides spreading-out of signals in the 2D spectral plane that overlap in the 1D spectrum.

The spectral spreading described in the last paragraph applies for the spectral patterns of each individual site. In addition, overlapping spectral patterns can be separated relatively well by means of the multidimensional approach. This can be achieved by aligning the main order direction along the $\vec{\mathbf{B}}_0$ field in evolution or detection, which yields the narrowest patterns (see Section 2.4.3), while the other

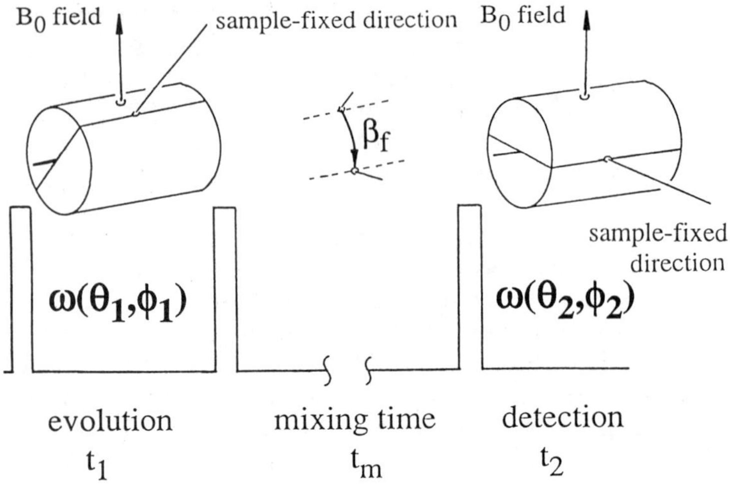

Figure 12.17 Principle of the 2D DECODER experiment. The orientation-dependent frequencies of a given segment at two different sample orientations are correlated in a two-dimensional spectrum. The pulse sequence is identical to that of the generic ^{13}C 2D exchange experiment, except for the macroscopic flip of the sample during the mixing time (and its return to the initial orientation during the recycle delay). Adapted from Schmidt-Rohr *et al.* (1992).

spectral dimension provides a wide spectral spread and correspondingly good resolution. This can be useful even in ^2H NMR to separate the two 'transitions' in a sufficiently highly oriented sample (see Figure 12.24). As will be demonstrated below for PET fibres, 2D DECODER spectra taken with this configuration provide not only separation of 1D spectra, but actually yield the full two-dimensional information which is not available even from non-overlapping 1D spectra.

The 2D sample-flip experiment was first applied by Henrichs (1987) to characterize the orientation distributions in biaxially drawn industrial PET films. Spectral separation of OCH_2-, aromatic, and carboxyl patterns was achieved and the data could be analysed in terms of orientational moments. The biaxiality of the films results in a considerable number of nonvanishing moments $\langle D^L_{mn} \rangle$ even for small L. In the study, it was found that the planes of the phenylene rings are nearly parallel to the plane of the film. The chain axes are also in the plane of the films, and slightly more aligned along the transverse than along the machine direction.

In the moment analysis, however, it is not apparent that the orientation distribution is reflected nearly point by point in the spectral intensity distribution. For understanding how the orientation distribution can be decoded from the DECODER spectra, we first consider the relatively simple case of $\eta = 0$. The analysis is performed in a suitable sample-fixed system, which involves some mental transformations from the laboratory frame (see Figure 12.18(a)). However, the relevant orientation distribution of the $\vec{Z}_P$ axes (C—^2H bonds) is static in the sample and can therefore be described most conveniently in a sample-fixed system, where usually the macroscopic director is chosen as the z axis $\vec{Z}_D$ (director frame DF).

Figure 12.18(b) indicates how a peak at a certain point in the spectral plane is related to very few symmetry-related orientations, represented as points on the unit sphere of

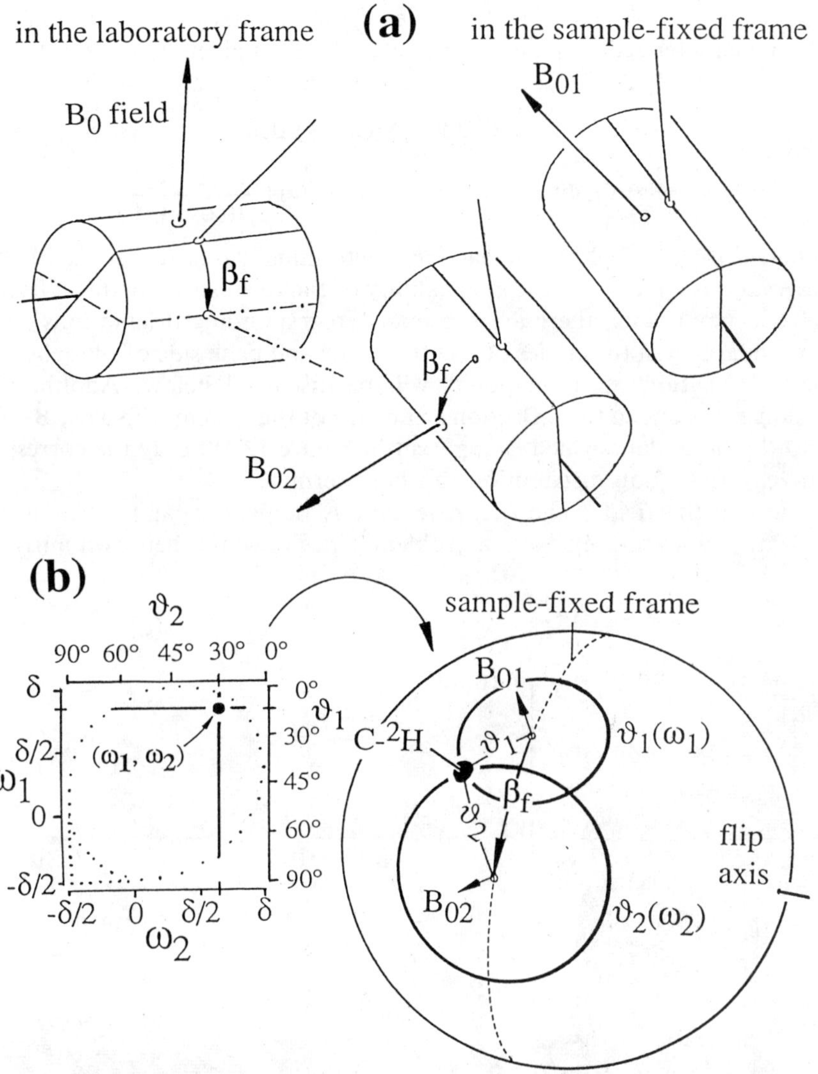

Figure 12.18 Principle of the reconstruction of the orientation distribution from a DECODER 2D spectrum (for $\eta = 0$). (a) Transition from the laboratory frame into a sample-fixed frame. In a sample-fixed frame, the $\vec{\mathbf{B}}_0$ field appears to reorient between two directions $\vec{\mathbf{B}}_{0,1}$ and $\vec{\mathbf{B}}_{0,2}$. (b) The frequency coordinates ω_1 and ω_2 of a given signal $S(\omega_1, \omega_2)$ define two cones in the sample-fixed frame (circles on the unit sphere) on which the corresponding $\vec{\mathbf{Z}}_P$ axis (C—^{2}H bond) must be located. The circles' intersection points thus pin down the few possible $\vec{\mathbf{Z}}_P$ orientations. Adapted from Schmidt-Rohr *et al.* (1992).

orientations in the sample-fixed frame. On the hemisphere shown, each of the two frequency coordinates $\tilde{\omega}_i$ specifies a circle around the $\mathbf{B}_{0i}$ orientation corresponding to a cone with an apex angle of $2\theta(\tilde{\omega}_i)$. The intersection points of these circles specify the orientations that can contribute to the spectral intensity at $(\tilde{\omega}_1, \tilde{\omega}_2)$.

From this point-by-point relationship between the frequency points and the orientations, it is clear that a direct reconstruction of the orientation distribution from the spectral intensity is possible: the value of the orientation distribution at a given orientation characterized by polar coordinates (Θ, Φ) is given by the spectral intensity at the corresponding $(\tilde{\omega}_1, \tilde{\omega}_2)$, corrected for the transformation of the intervals:

$$P(\Theta, \Phi)\ \mathrm{d}\cos\Theta\ \mathrm{d}\Phi \overset{!}{=} S(\omega_1, \omega_2)\ \mathrm{d}\omega_1\ \mathrm{d}\omega_2$$

$$\Rightarrow P(\Theta, \Phi) = S[\omega_1(\Theta, \Phi), \omega_2(\Theta, \Phi)]\ \frac{\partial(\omega_1, \omega_2)}{\partial(\cos\Theta, \Phi)} \tag{12.52}$$

where $\partial(\omega_1, \omega_2)/\partial(\cos\Theta, \Phi)$ is the Jacobian determinant (see Appendix E). However, a few reservations concerning the applicability of the reconstruction must be made: for certain frequency points, there are too many corresponding orientations since the θ_i values are so large that the circles on the front side and back side of the sphere overlap. The spectral location of those points will be discussed below. Additional loss of information arises due to the reflection symmetry of the system (flip axis, $\vec{\mathbf{B}}_0$ field) with respect to the field plane which is apparent in Figure 12.18; only the correspondingly symmetrized orientation distribution can be determined.

For points in the field plane (i.e. orientations perpendicular to the flip axis) this symmetrization does not represent a problem, since they are their own mirror images.

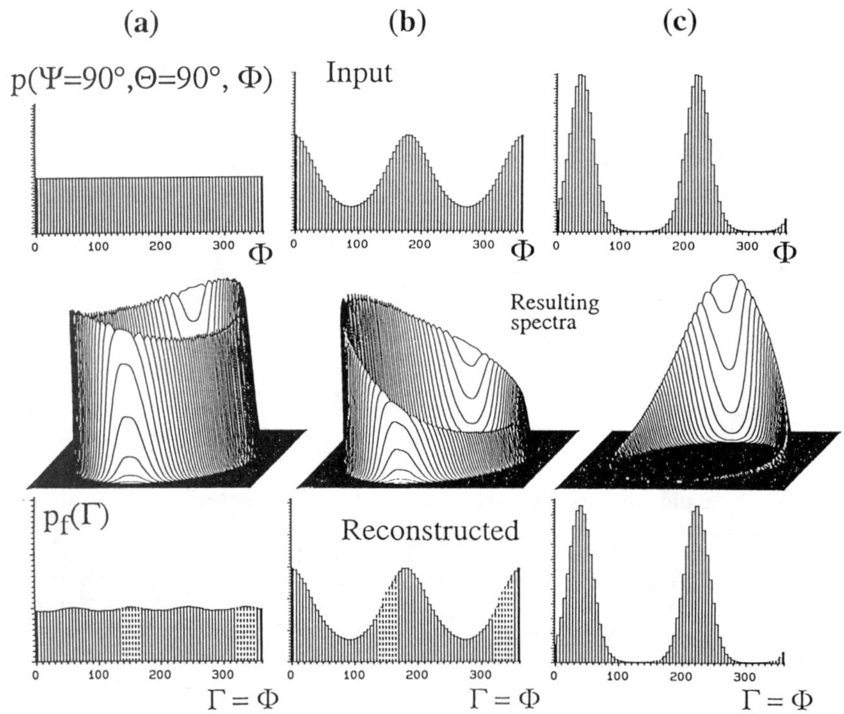

Figure 12.19 Demonstration of the reliability of the reconstruction of the orientation distribution in the field plane from simulated DECODER spectra with $\eta = 0.28$ and $\beta_f = 40°$. The distributions (top row) which were used to simulate DECODER spectra are directly reflected in the spectral intensities on the nearly circular ellipse (middle row) and are reconstructed faithfully from these spectra (bottom row). The hatched areas indicate regions where the reconstruction is not unique (see text). Adapted from Chmelka *et al.* (1993).

The one-dimensional slice through the orientation distribution along the field-plane meridian can be reconstructed with particular reliability, as is demonstrated in Figure 12.19. In the spectrum, the intensity from the field-plane orientations appears on an ellipse whose Lissajou representation (see Appendix F):

$$\begin{bmatrix}\omega_1\\ \omega_2\end{bmatrix}_f = \tfrac{1}{2}\delta\begin{bmatrix}3\cos^2(\Gamma-\beta_1)-1\\ 3\cos^2(\Gamma-\beta_2)-1\end{bmatrix} = \tfrac{3}{2}\delta\begin{bmatrix}\cos(2\Gamma')\\ \cos\{2\Gamma'-2\beta_f\}\end{bmatrix} + \begin{bmatrix}\tfrac{1}{4}\delta\\ \tfrac{1}{4}\delta\end{bmatrix} \tag{12.53}$$

($\beta_f = \beta_2 - \beta_1, \Gamma' = \Gamma - \beta_1$) can be derived quite easily by calculating the scalar products $\cos\theta_1$ and $\cos\theta_2$ between the $\vec{\mathbf{Z}}_P$ axes ($\sin\Gamma, 0, \cos\Gamma$) in the field plane and the $\vec{\mathbf{B}}_0$ fields ($\sin\beta_1, 0, \cos\beta_1$) and ($\sin\beta_2, 0, \cos\beta_2$). The ratio of the semi-axes a and b of this ellipse is directly related to the flip angle β_f according to:

$$|\tan\beta_f| = \frac{b}{a}. \tag{12.54}$$

For $\beta_f = 45^\circ$, the ellipse becomes a circle with the favourable property of constant angular resolution in terms of Γ and no spectral-intensity compression along the circle.

Before discussing the spectral features for various sample-orientation configurations, we want to specify in which portions of the 2D frequency plane the spectral intensity can at all occur for a given flip angle β_f. This is most easily found by considering the intensity distribution in powder spectra: in isotropic samples, all possible tensor orientations are present. The powder spectra for various η values and flip angles are displayed in Figure 12.20. Three elliptical features are prominent in these spectra. For each of these ellipses, the ratio of the semi-axes a and b is again characterized by $|\tan\beta_f| = b/a$. For $\eta = 0$, the smaller ellipse contracts into the singularity at $(-\tfrac{1}{2}\delta, -\tfrac{1}{2}\delta)$, while the two larger ellipses degenerate into one. This is the ellipse of the field-plane orientations. In particular, it contains the points at the lower ends of the spectrum, at $\omega_1 = \omega(\theta_1 = 0) = \delta$ and at $\omega_2 = \omega(\theta_2 = 0) = \delta$; these correspond to $\vec{\mathbf{Z}}_P$ axes along $\vec{\mathbf{B}}_{01}$ (for $\theta_1 = 0$) and along $\vec{\mathbf{B}}_{02}$ (for $\theta_2 = 0$), which are indeed located in the field plane (see Figure 12.18(b)).

In general, for $\eta = 0$ the orientation distribution cannot be reconstructed for points outside the field-plane ellipse since the frequency points between the 90° ridges and the ellipse correspond to two unrelated quartets of orientations as discussed above. In the special case of a $\beta_f = 90^\circ$ flip, however, the symmetry in the experiment is increased so that the intersection points are symmetrically arranged with respect to each of the two $\vec{\mathbf{B}}_0$ field orientations: the plane perpendicular to the $\vec{\mathbf{B}}_0$ field in the evolution period is also a symmetry plane in the detection, since it contains the $\vec{\mathbf{B}}_0$ field then. Therefore, from the full intensity range in the $\beta_f = 90^\circ$ spectrum, a contracted orientation distribution can be reconstructed, but at the cost of an additional degeneracy: the reconstruction can cover only one octant of the unit sphere, in which the distribution from the eight octants of the sphere are superimposed. The three apices of the octant are marked by the flip axis and the two $\vec{\mathbf{B}}_0$ field orientations.

An ($\eta = 0$) 3D DECODER spectrum for an isotropic sample flipped twice around a single axis is a single, concave 'sheet' which is infinitely thin (apart from homogeneous broadening). The first two frequencies either determine the third frequency uniquely or correspond to only two ω_3 values. At the $\omega_j = \omega(90^\circ)$ planes, the sheet bends backwards so that a structure results which resembles a wide icecream-cone (Figure 12.21) with the closed tip at the '90° corner' ($\omega(90^\circ), \omega(90^\circ), \omega(90^\circ)$) and a broad aperture with 'lips' producing the elliptical ridges in the three 2D projections. The

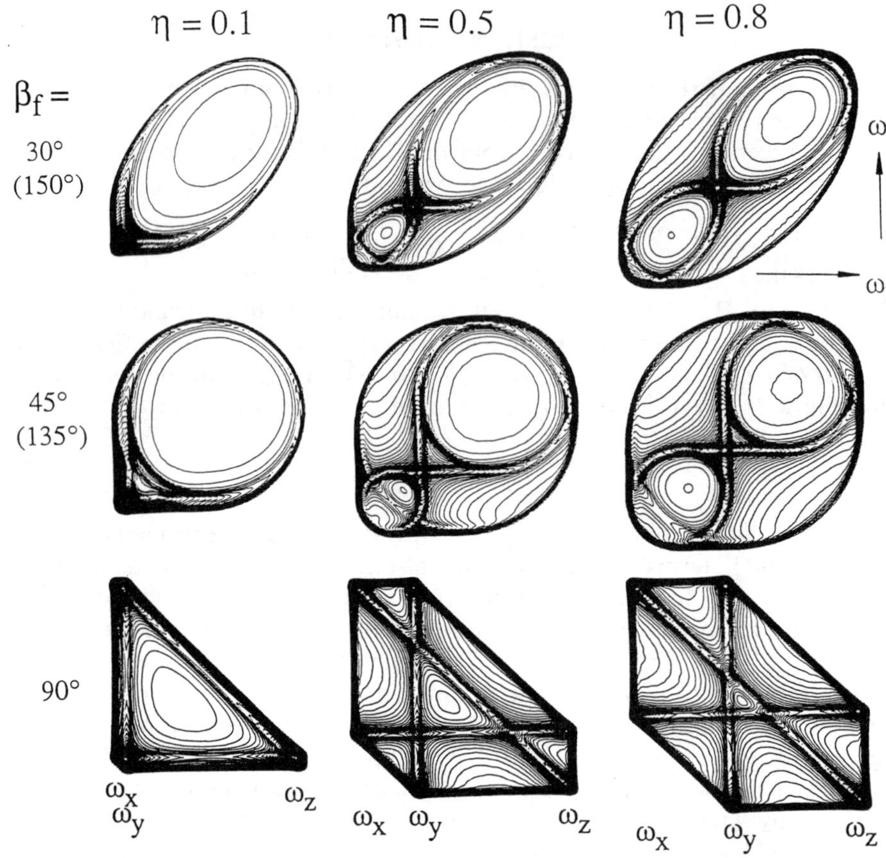

Figure 12.20 Simulated 2D DECODER patterns for unoriented samples ('powders') as a function of η and the flip angle β_f. Elliptical patterns with aspect ratios characteristic of β_f are observed in all the spectra. The three ellipses observed in the general case degenerate into a single one for $\eta = 0$. Adapted from Chmelka *et al.* (1993).

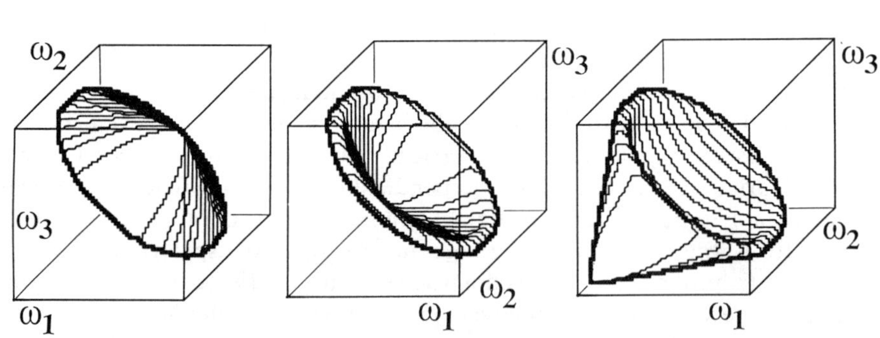

Figure 12.21 Simulated 3D DECODER powder spectrum for $\eta = 0$ and two successive 45° sample flips. The plots show various views of the surface to which the spectral intensity is confined. (Here, the contour lines indicate the depths at which this surface is found, while in the 3D plot of Figure 1.6(a) they indicate maximum intensities along rays traced perpendicular to the plane of the paper.) The surface is a wide distorted cone with its tip at $(-\frac{1}{2}\delta, -\frac{1}{2}\delta, -\frac{1}{2}\delta)$, which corresponds to $\vec{\mathbf{Z}}_P$ axes oriented along the sample-flip axis. The projections onto the (ω_1, ω_2) and (ω_2, ω_3) planes are 2D powder spectra for $\beta_f = 45^\circ$, each with a circular ridge pattern (see Figure 12.20), while the (ω_1, ω_3) projection is a 2D powder pattern for $\beta_f = 90^\circ$ with a triangular ridge pattern. The partial overlap of signals from inequivalent orientations in 2D DECODER spectra is removed in the 3D spectrum by separating these orientations ('opening the cone') in the third dimension.

overlap from the pairs of unrelated quartets of orientations in the 2D spectrum is removed, and the variation of the angular sensitivity in the spectrum is reduced (except at the 90° corner, which corresponds to the flip-axis direction).

For $\eta \neq 0$, the third Euler angle Ψ must be included in the analysis. As shown by several examples in Figure 12.22, it leads to a considerable complication of the 2D DECODER spectra. The corresponding $\eta = 0$ powder spectra in the top row in Figure 12.22 can serve as guides for the interpretation of the $\eta \neq 0$ patterns for various Ψ values, since these must degenerate into the $\eta = 0$ pattern upon reduction of η. Each

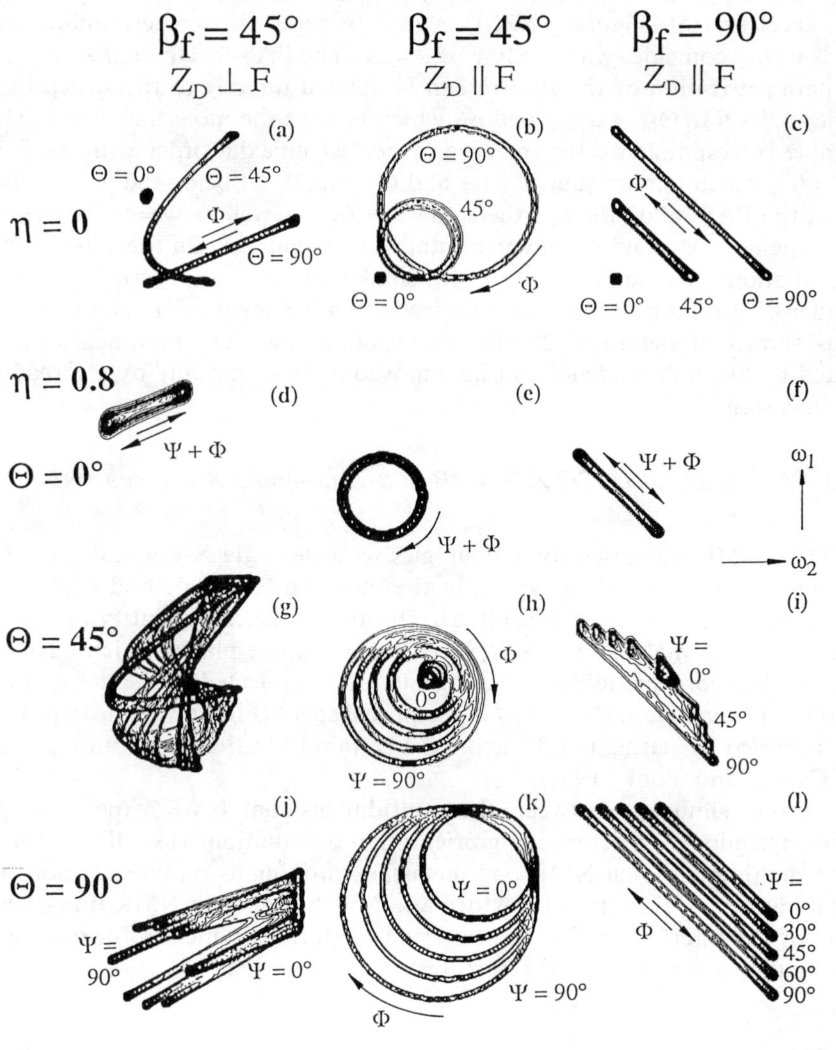

Figure 12.22 Typical 2D DECODER patterns for highly oriented samples which are uniaxial around the $\vec{Z}_p$ axis (constant Φ distribution). The geometry of the flip is defined above each column. Top row: spectra for $\eta = 0$. Below: $\eta = 0.8$. Within each column, the overall similarity of the patterns for equal Θ is apparent. With $\eta = 0$, the spectra for all Ψ values are identical; with $\eta > 0$, the spectral patterns depend on Ψ as shown by superimposing spectra for Ψ values of 0°, 30°, 45°, 60° and 90°.

of the spectra shown in Figure 12.22 is calculated for a fixed value of Θ, with a cylindrical distribution in Φ and $\Psi = 0^\circ, 30^\circ, 45^\circ, 60^\circ, 90^\circ$ as a parameter to indicate the spread in frequency produced by this variable. For each Ψ value, a spectral ridge results along which Φ varies. Typical flip angles β and sample configurations have been chosen, namely with $\vec{\mathbf{Z}}_D$ along or perpendicular to the flip axis $\vec{\mathbf{F}}$. The spectra can be understood by considering the distribution of $\vec{\mathbf{B}}_0$ fields in the unified PAS, with lines of constant frequency marked on a unit sphere (see Figure 2.7). The spectra of Figures 12.22(e) and (f), for instance, correspond to $\vec{\mathbf{B}}_0$ fields along the equator of this sphere, with the frequencies varying between σ_x and σ_y, while the spectra of Figures 12.22(k) and (l) correspond to meridians with the frequencies in the range between σ_z at the north pole and $\frac{1}{2}\{-\sigma_z + (\sigma_x - \sigma_y)\cos(2\Psi)\}$ on the equator. The strong intensity of the $\Psi = 0^\circ$ spectrum of Figure 12.22(i) arises because the corresponding circle of $\vec{\mathbf{B}}_0$ fields nearly coincides with the line of $\omega = \sigma_y$. The PAS-to-DF angles (Ψ, Θ, Φ) used in the parametrization of the spectra can be related directly to the structurally more relevant angles $\Phi_M, \Theta_M, \Psi_M, \beta_P$ and α_P which involve the molecular frame MF: there is a simple correspondence for the case of perfect uniaxial order with $\vec{\mathbf{Z}}_M \| \vec{\mathbf{Z}}_D$, that is, $\Theta_M = 0^\circ$, which implies that $\Psi = \alpha_P$ and $\Theta = \beta_P(\Theta, \Psi)$ being the polar coordinates of $\vec{\mathbf{Z}}_D$ in the PAS, and (β_P, α_P) the same for $\vec{\mathbf{Z}}_M$, as well as $\Phi = \Psi_M + \Phi_M$, since all of these angles Φ, Ψ_M and Φ_M involve rotations around $\vec{\mathbf{Z}}_D$. On this basis, for narrow Θ_M distributions around $\Theta_M = 0^\circ$ spectra similar to those of Figure 12.22 are found. A broader distribution in Θ_M naturally leads to a broadening of each of the spectral patterns shown in Figure 12.22. By extension to the 3D experiment, the overlap generated by the Ψ dependence can be removed at least partially by a spread into the third dimension.

12.7.3 Analogies of DECODER NMR and wide-angle X-ray scattering

DECODER NMR possesses some analogies to wide-angle X-ray scattering (WAXS) pole-figure analysis, which is currently the most powerful method for determining orientational distributions of crystallites. This technique can quantify the distribution $p(\Theta, \Phi)$ of the normal vectors $\vec{\mathbf{G}}(\Theta, \Phi)$ of sets of lattice planes which have the same spacing d. The corresponding scattering intensity is selected by scanning the 'sphere of position' of diameter $|\vec{\mathbf{q}}| = 2\pi/d$ in reciprocal space (Figure 12.23). In practice, this can be achieved by setting the detector to a certain reflection and rotating the sample (Baltá-Calleja and Vonk, 1989).

The formal similarity between the multidimensional NMR experiment and the scattering technique for determining orientation distributions is similar to the relation between multidimensional NMR and inelastic scattering as outlined in Section 9.1.3. The multidimensional Fourier transformation of the coherent NMR time-signal takes the role of the interference in the scattering amplitude, which is formally a Fourier transformation over the spatial structure:

WAXS:

$$I_G(\vec{\mathbf{q}}) = \iint d\Theta\, d\Phi\, p(\Theta, \Phi) \left| \int \underbrace{e^{i\vec{\mathbf{G}}(\Theta,\Phi)\cdot\vec{\mathbf{r}}}}_{\text{structure}}\, \underbrace{e^{-i\vec{\mathbf{q}}\cdot\vec{\mathbf{r}}}\, d\vec{\mathbf{r}}}_{\text{interference}} \right|^2 \tag{12.55}$$

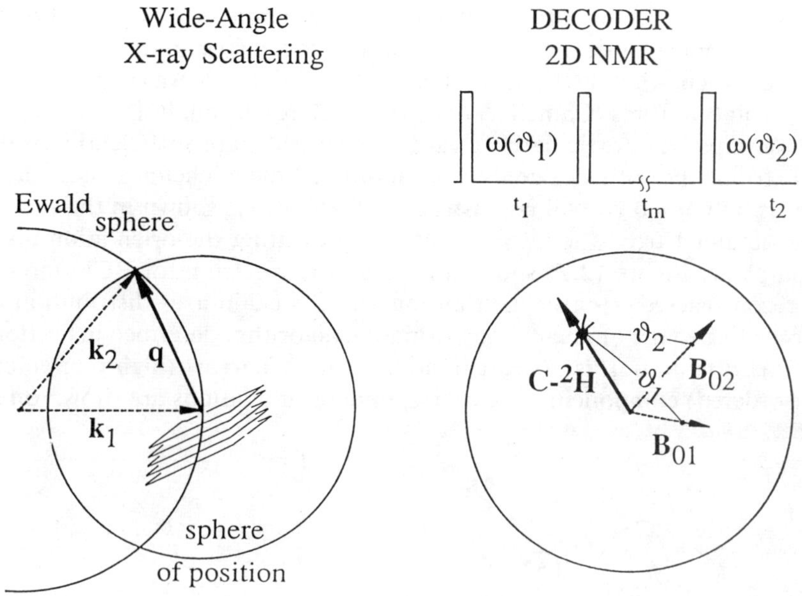

Figure 12.23 Analogy between DECODER and WAXS (wide-angle X-ray scattering) pole-figure analysis. The measurement of the $\vec{\mathbf{k}}$ vectors before and after the scattering yields the orientation distribution of the normals of certain sets of crystal-lattice planes. The measurement of the NMR frequencies before and after the sample flip provides direct information on the orientation distribution of $\vec{\mathbf{Z}}_P$ axes. Adapted from Schmidt-Rohr *et al.* (1992).

DECODER:

$$S_\beta(\vec{\omega}) = \iint \mathrm{d}\Theta\ \mathrm{d}\Phi\ p(\Theta,\Phi) \int \underbrace{\mathrm{e}^{\mathrm{i}\vec{\omega}_\beta(\Theta,\Phi)\cdot\vec{\mathbf{t}}}}_{\text{precession}}\ \underbrace{\mathrm{e}^{-\mathrm{i}\vec{\omega}\cdot\vec{\mathbf{t}}}}_{\text{FT}}\ \mathrm{d}\vec{\mathbf{t}} \tag{12.56}$$

with $\vec{\omega} := (\omega_1, \omega_2)$, $\vec{\mathbf{t}} := (t_1, t_2)$ and $\vec{\omega}\cdot\vec{\mathbf{t}} = \omega_1 t_1 + \omega_2 t_2$. Similarly, $\vec{\omega}_\beta(\Theta,\Phi) := \{\tilde{\omega}_1(\Theta,\Phi;\beta_1,\alpha_1),\ \tilde{\omega}_2(\Theta,\Phi;\beta_2,\alpha_2)\}$ contains the precession frequencies during evolution and detection. The Fourier transformations of the complex exponentials in (12.55) and (12.56) produce δ peaks:

$$I_G(\vec{\mathbf{q}}) = \iint \mathrm{d}\Theta\ \mathrm{d}\Phi\ p(\Theta,\Phi)\ \delta[\vec{\mathbf{G}}(\Theta,\Phi) - \vec{\mathbf{q}}] \tag{12.57}$$

$$S_\beta(\vec{\omega}) = \iint \mathrm{d}\Theta\ \mathrm{d}\Phi\ p(\Theta,\Phi)\ \delta[\vec{\omega}_\beta(\Theta,\Phi) - \vec{\omega}]\,. \tag{12.58}$$

These expressions show the close analogy of the techniques. For further discussion of the signals in (12.58) see Schmidt-Rohr *et al.* (1992).

12.7.4 Applications of the DECODER experiment

After the Legendre-moment expansion for ^{13}C spectra of PET films performed by Henrichs (1987), the 'decoding' of the orientation distribution was first tested on ^{2}H

NMR spectra of highly oriented deuterated PE (Schmidt-Rohr *et al.*, 1992). Figure 12.24 shows a typical experimental spectrum with the corresponding simulation and a reconstructed orientation distribution. DECODER ^{13}C NMR was applied to industrial PET fibres and thin films (Chmelka *et al.*, 1993). Here, the analysis of the signals from the OCH_2 groups was particularly rewarding since they are sufficiently upfield to be separated from the other signals. In addition, their chemical-shift asymmetry parameter is only $\eta = 0.28$. All reconstructed distributions shown in the figures below have been obtained from the OCH_2 signals, representing the orientation distribution in the field plane. Figure 12.25 shows a DECODER spectrum for PET fibres together with the reconstructed $\vec{\mathbf{Z}}_P$ axes distribution and the chain-axes distribution obtained from it according to the moment-transformation algorithm described in Section 12.3.1. The $p_M(\Theta_M)$ distribution is at least bimodal, with a narrow (highly oriented) and a broad (disordered) component. The corresponding simulations are shown on the right in Figure 12.25.

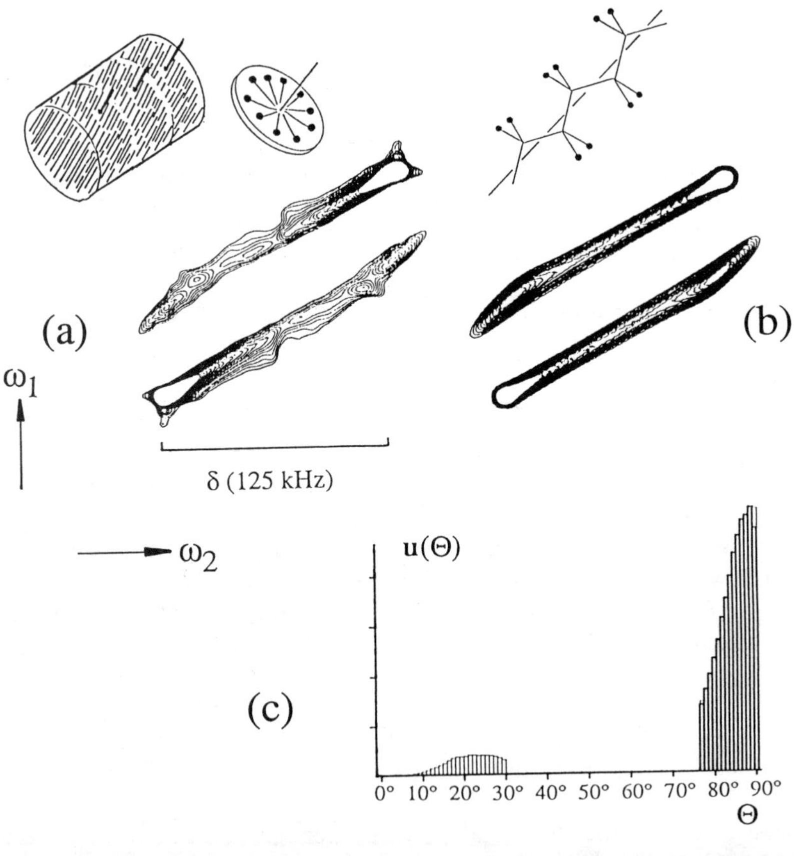

Figure 12.24 (a) 2D DECODER ^2H spectrum of highly oriented deuterated PE with $\beta_f = 16^\circ$. Two narrow straight stripes are observed, centrosymmetrically arranged with respect to the origin of the spectrum, corresponding to the two transitions of the quadrupolar ^2H spectrum. The small widths of these spectral intensity patterns directly reflect a narrow orientation distribution of the C—^2H bonds. (b) Corresponding simulation, assuming an 8° width (fwhm) of the Θ distribution. (c) Reconstructed orientational distribution. Adapted from Schmidt-Rohr *et al.* (1992).

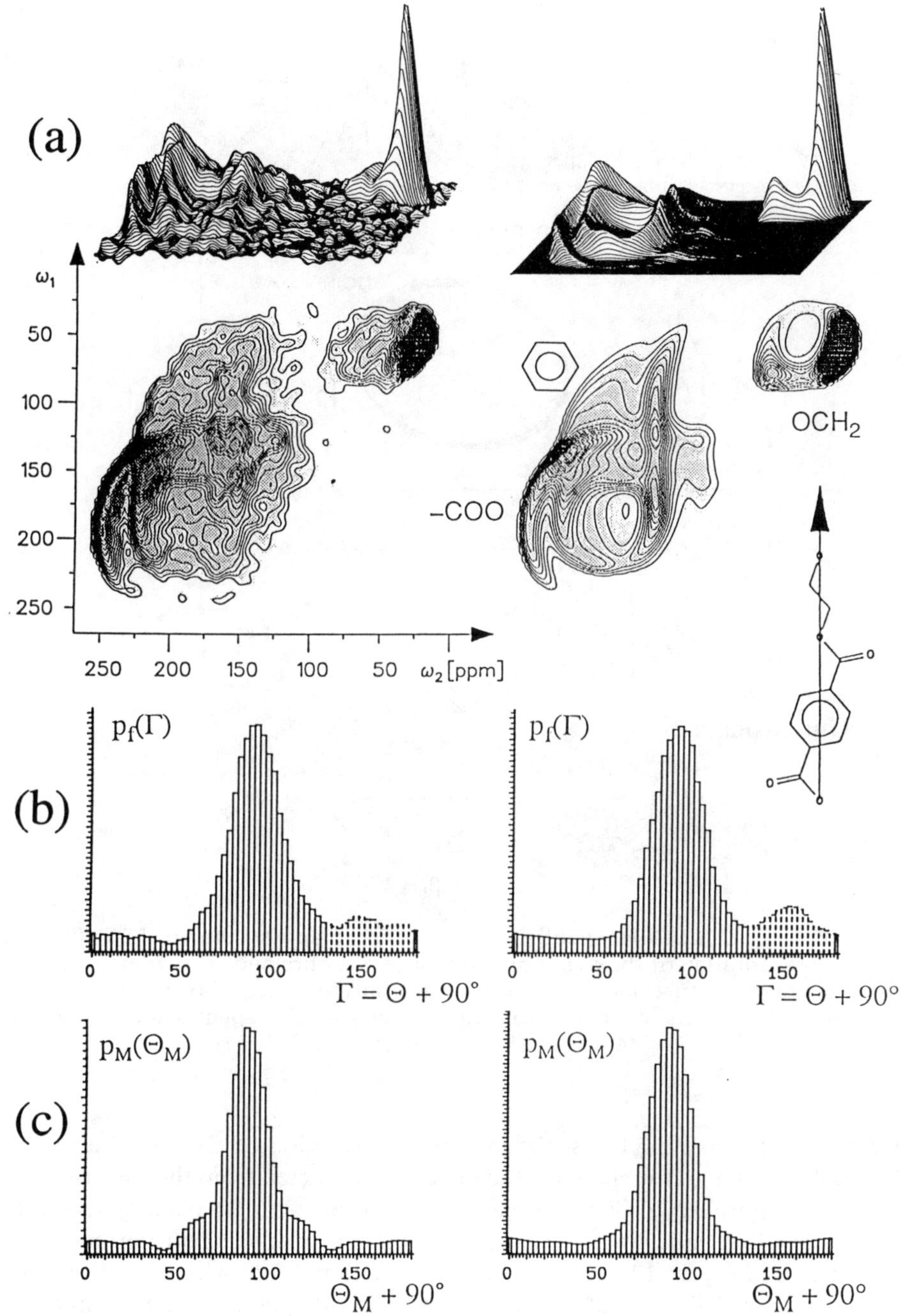

Figure 12.25 (a) ^{13}C DECODER spectrum of industrial PET fibres taken with a flip angle of 42°. (b) Reconstructed field-plane distribution of the Z_P axes of the OCH_2 groups. (c) Chain-axes orientation distribution calculated from the distribution in (b) by means of the moment-transformation based on equation (12.14a). The distribution is at least bimodal, with a narrow component (fwhm of 24°, standard deviation of 10°) and a very broad component. Corresponding simulated spectrum and distributions are shown on the right. Adapted from Chmelka *et al.* (1993).

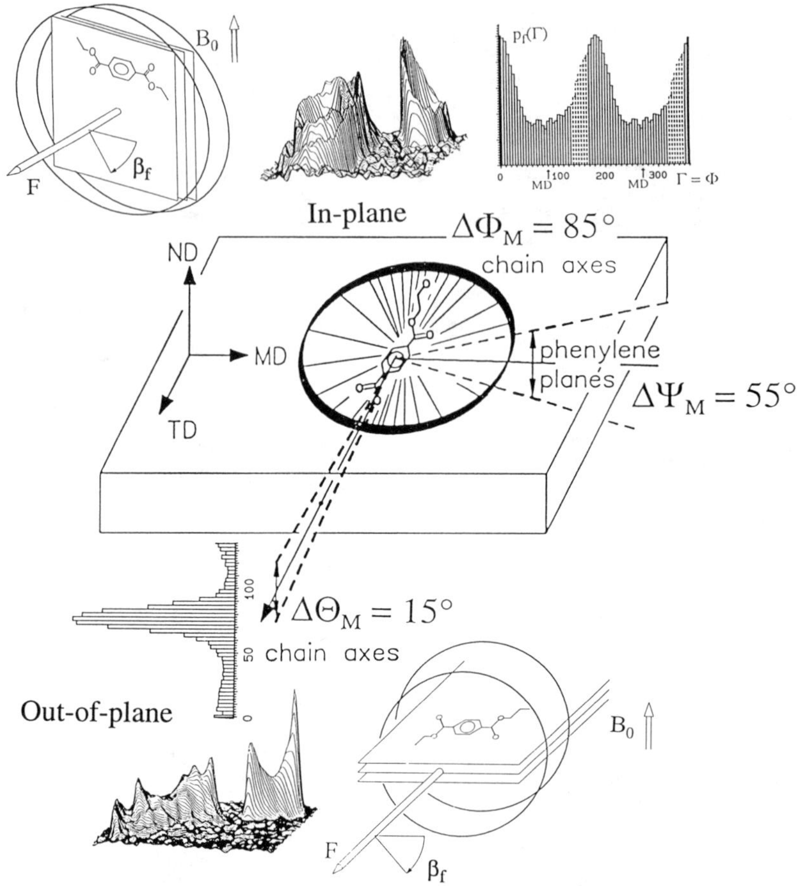

Figure 12.26 ^{13}C DECODER spectra of industrial biaxially drawn PET films with schematic representation of the orientation distributions obtained. Top: in-plane spectrum and reconstructed distribution of chain axes. Bottom: Out-of-plane spectrum and reconstructed distribution of chain axes. Simulations of the spectra are shown in the literature (Chmelka *et al.*, 1993).

For biaxially drawn PET films, the in-plane distribution of OCH_2 $\vec{\mathbf{Z}}_P$ axes (which are essentially parallel to the chain axes) can be reconstructed from the spectrum shown at the top of Figure 12.26. This in-plane distribution with its maximum along the transverse draw direction is much broader than the out-of-plane distribution of the $\vec{\mathbf{Z}}_P$ axes also shown in Figure 12.26, obtained from the same films with a different sample configuration. Parallelization of the phenylene and COO planes with a fwhm of about 55° could also be deduced from these spectra. The in-plane distribution reconstructed for samples taken from the edge of the biaxially drawn PET sheet in the production line is narrower than for a sample from the centre of the sheet. In addition, the preferential chain direction in the edge-sheet sample is rotated by about 35° from the transverse direction. More details, including a transformation of the distributions into order parameters, are given in the original literature (Chmelka *et al.*, 1993).

12.8 References

Baltá-Calleja, F. J. and Vonk, C. G. (1989). *X-ray Scattering of Synthetic Polymers*. Elsevier, Amsterdam.
Boeffel, C. and Spiess, H. W. (1989). NMR for Studying Order and Motion in LC Side Group Polymers. In C. B. McArdle (ed.), *Side Chain LC Polymers*. Blackie, London.
Chmelka, B. F., Schmidt-Rohr, K. and Spiess, H. W. (1993). Orientation Distributions in PET Films and Fibers from DECODER NMR. *Macromolecules*, **26**, 2282.
Hall, I. H. (1984). *Structure of Crystalline Polymers*. Elsevier Applied Science, London.
Harbison, G. S. and Spiess, H. W. (1986). Partially Oriented Solids: Characterization by 2D MAS NMR. *Chem. Phys. Lett.*, **124**, 128.
Harbison, G. S., Vogt, V.-D. and Spiess, H. W. (1987). Partially Oriented Solids: Characterization by 2D MAS NMR. *J. Chem. Phys.*, **86**, 1206.
Henrichs, P. M., (1987). Molecular Orientation and Structure with ^{13}C NMR: A Study of Biaxial Films of PET. *Macromolecules*, **20**, 2099.
Hentschel, R., Schlitter, J., Sillescu, H. and Spiess, H. W. (1978). Orientational Distributions from NMR and ESR Line Shapes. *J. Chem. Phys.*, **68**, 56.
Hentschel, R., Sillescu, H. and Spiess, H. W. (1981). Orientational Distribution of Polymers Studied by ^{2}H NM. *Polymer.*, **22**, 1516.
Maricq, M. M. and Waugh, J. S. (1979). NMR in Rotating Solids. *J. Chem. Phys.*, **70**, 3300.
Peterlin, A. (1979). In A. Ciferri and I. M. Ward (eds), *Ultra-High Modulus Polymers*, Chapter 10. Applied Science Publishers, London.
Schmidt-Rohr, K., Hehn, M., Schaefer, D. and Spiess, H. W. (1992). 2D NMR for Characterizing Orientation Distributions, Analogy to X-Ray. *J. Chem. Phys.*, **97**, 2247.
Schmidt-Rohr, K., Wilhelm, M., Johansson, A. and Spiess, H. W. (1993). Shift Tensors in Methylene Groups. *Magn. Reson. in Chem.*, **31**, 352.
Spiess, H. W. (1984). NMR in Oriented Polymers. In I. M. Ward (ed.), *Developments in Oriented Polymers*. Applied Science Publishers, London.
Spiess, H. W. (1985). 2H NMR for Studying Mobility and Orientation in Polymers. *Adv. Polym. Sci.*, **66**, 23.
Tang, P., Santos, R. A. and Harbison, G. S. (1989). 2D Solid-State NMR: Conformation of Oriented DNA, *Adv. Magn. Reson.*, **13**, 225.
Titman, J. J., Féaux de Lacroix, S. and Spiess, H. W. (1993). Structure and Order in Solids by 3D MAS NMR. *J. Chem. Phys.*, **98**, 3816.
Wilhelm, M., Féaux de Lacroix, S., Titman, J. J., Schmidt-Rohr, K. and Spiess, H. W. (1994). Orientation in PPTA by ^{13}C NMR, *Acta Polymerica*.
Woodward, A. E. (1989). *Atlas of Polymer Morphology*. Hanser Verlag, München.
Yang, Y., Hagemeyer, A., Zemke, K. and Spiess, H. W. (1990). 3D MAS NMR: Molecular Structure, Order, and Dynamics. *J. Chem. Phys.*, **93**, 7740.

CHAPTER THIRTEEN

Domain Sizes and Internuclear Distances from Spin Diffusion and Dipolar Couplings

Keywords: *exchange by ^{1}H spin diffusion, 2D CRAMPS vs. 1D chemical-shift filter, ^{13}C detection; diffusion simulations, initial rates, domain sizes, model morphologies,* D$(\vec{r})$ *in lattice calculation; analogies to SAXS; polymer blends, copolymers, microphase separation, nano-heterogeneities; ^{13}C spin diffusion: natural, 'rf driven', 'rotor driven', rotational resonance; low-abundance short-range problem, local orientational order.*

13.1 Introduction

The characterization of a polymeric material in terms of its structure on a nanometer to micrometer scale represents an important aspect of polymer analytics in the solid state.

Through magnetization diffusion, which is mediated by dipolar couplings, solid-state NMR permits to probe distances up to 200 nm. (Note that by contrast in NMR spectroscopy, the localized nature of quadrupolar and chemical-shift interactions is essential for the analysis in terms of chemical moieties, in particular in disordered systems, e.g. amorphous polymers.) 2D and 1D exchange experiments allow for a combination of spectroscopic information with the dipolar multi-step magnetization transfer (see Sections 3.5–3.7) termed 'spin diffusion' by Bloembergen (1949).

The characteristics of the dipolar magnetization transfer depend somewhat on the spin species concerned, in particular on its abundance and on the strength of the dipolar couplings compared with other interactions (see Section 3.7). We shall discuss ^{1}H and ^{13}C spin diffusion processes as the most relevant in polymeric systems. In one or the other case, most of the features relevant with other isotopes, like ^{29}Si, ^{2}H (Schajor *et al.*, 1980), or ^{19}F, are addressed implicitly.

For ^{1}H spin diffusion in typical organic polymers, the proton–proton dipolar coupling is by far the dominating interaction, which makes the process very efficient. By contrast, in unmodified ('proton driven') ^{13}C spin diffusion, the essential $^{13}C-^{13}C$ couplings (15 Hz for the typical distance of 0.8 nm) are dominated by the $^{1}H-^{13}C$ couplings (~20 000 Hz) mixing with $^{1}H-^{1}H$ couplings (~40 000 Hz), as well as ^{13}C chemical-shift differences (~1000–20 000 Hz), because all those Hamiltonians do not commute. To make the system accessible to coherent averaging techniques, in experiments aimed at speeding up ^{13}C spin diffusion, $^{1}H-^{13}C$ couplings and chemical-shift difference effects are eliminated for the ^{13}C sites that are to communicate with one another.

In the light of modern NMR theory and experiments, it is found that the term 'spin diffusion' does not provide a particularly appropriate description of the physical basis of the process. It is based on reversible coherent dipolar evolution (Zhang *et al.*, 1992) (see also Section 3.5.3) and thus does not represent a diffusion process in the classical sense. 'Dipolar magnetization transfer' might be a more correct description. However, experimental data for protons usually show an initial $\sqrt{t_m}$ dependence typical for a diffusive process. It comes about due to the large number of different couplings. Since the expression 'spin diffusion' is well established we shall continue to use it here.

13.2 ^{1}H spin diffusion

In organic polymers, ^{1}H spin diffusion proceeds quite swiftly in the dense network of strong ^{1}H–^{1}H dipolar couplings, which justifies a continuum description of the diffusion process (see Section 3.5.1). The time dependence of the diffusion process of non-equilibrium distributions of z magnetization contains information on the domain sizes in heterogeneous materials: in structures with small domains, the magnetization equilibrates faster than in a system consisting of large domains. The relations between space and time evolution for a diffusive process are analysed in Section 13.3. With a diffusivity of about 1 nm^2/ms (see Section 3.5.1) and mixing times of more than a second, ^{1}H spin diffusion can probe domain diameters of up to 200 nm. The lower limit is about 0.5 nm. More precisely speaking, spin diffusion probes 'typical smallest diameters': smallest diameters since the equilibrium occurs via the shortest paths, but they must be typical, because a few particularly thin domains will not dominate the measured curve which represents all domains. In the range below 5 nm, the 'domains' may often be more correctly denoted as heterogeneities. However, spin diffusion is equally efficient for domains and heterogeneities; in the extreme case, even the diameter of the polymer chains or segments which were initially polarized is reflected in the spin-diffusion behaviour. To reach equilibrium (to within 10%) in a large monomer unit requires more than 500 μs; such an effect is readily detected in the experiment.

13.2.1 ^{1}H experiments: principles and 1D spectra

Spin-diffusion experiments are typical exchange experiments, consisting of an evolution or selection period, a mixing time t_m, and a detection period. For spin diffusion to occur, a spatially inhomogeneous distribution of z magnetization must be generated (see Figure 13.1). In the most favourable case, the magnetization of one component is selected, while that of the rest is suppressed ('saturated'). The exchange process during t_m consists in the diffusion out of the source region into its surrounding which is initially devoid of magnetization. Then, the distribution of integral magnetization on the components of the sample is monitored in the NMR spectrum. For small domains, the equilibration of magnetization will occur fast, while in very large structures, the magnetization from the source can only slowly penetrate into relatively thin boundary layers of the other domains. Consequently, the t_m dependence of the magnetization exchange, reflected in the intensities in the spectrum taken after the mixing time, yields direct information on domain sizes.

The selection of ^{1}H magnetization can be based on differences in the decay of the transverse proton magnetization, which itself is due to differences in the dipolar

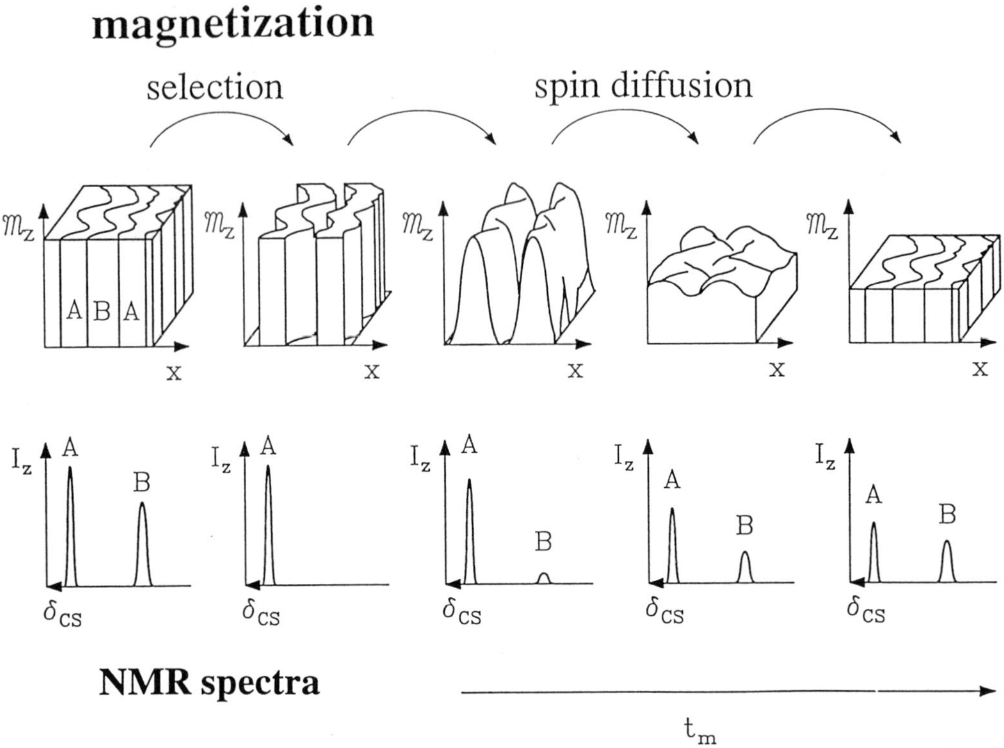

Figure 13.1 Basic structure of spin-diffusion exchange experiments: after generating a gradient of magnetization by selection of magnetization in component A, spin diffusion occurs during the mixing time t_m and is detected through the growth of the signal of component B. Adapted from Clauss *et al.* (1993).

couplings. This is the basis of the Goldman–Shen experiment (Goldman and Shen, 1966) sketched in Figure 13.2. Starting with Boltzmann z magnetization, the first of three pulses flips the ^1H magnetization to the xy plane; after a certain time, typically 30 μs, the magnetization of the rigid components has been dephased by their strong dipolar couplings, so that only ^1H magnetization from the mobile regions in the sample remains. A second pulse flips this magnetization back to the z axis, where it is stored for the ensuing mixing time during which ^1H spin diffusion to the rigid regions will occur. The third pulse terminates the mixing time, and the magnetization is detected to monitor how the spin diffusion has proceeded. The magnetization description given here represents a somewhat simplistic description of the evolution of the density operator in the multiparticle dipolar interactions. In fact, multiple-quantum coherences evolving during the mixing time produce spectral distortions (Packer and Pope, 1983). However, in sufficiently long mixing times these coherences are suppressed by transverse relaxation.

Other selection techniques are based on differences in relaxation times, such as $T_{1\rho}$. This approach is beset with the problem of spin-diffusion during the selection, since the dipolar couplings are not removed during the spin lock of the selection period (see Section 3.4.2). Therefore, short effective mixing times are not accessible and consequently materials with small domains cannot be investigated with this approach.

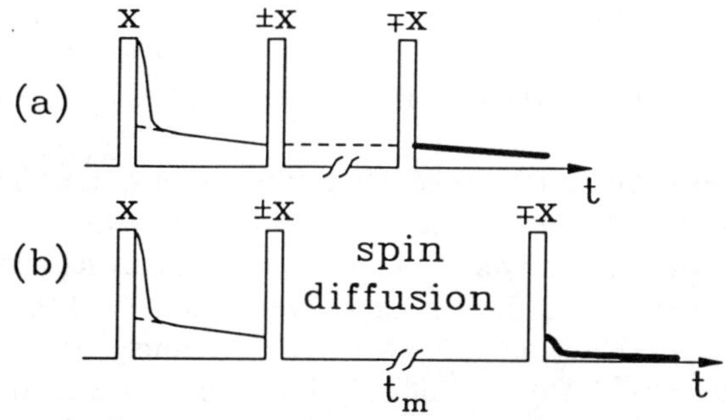

Figure 13.2 Principle of the Goldman–Shen experiment, the simplest ^{1}H spin diffusion experiment, which is applicable to systems consisting of both rigid and mobile regions. (a) Selection of magnetization in the mobile regions is achieved through the fast dephasing of the proton magnetization in the rigid regions by their strong dipolar couplings. The remaining ^{1}H magnetization in the mobile regions is flipped to the *z* axis and can then (b) diffuse to the rigid regions, which is monitored in the signal acquired after the mixing time.

^{1}H T_1 times are generally not suitable for selection in spin-diffusion experiments: if the T_1 times for different components are sufficiently different for a selection, this means that spin diffusion does not equilibrate the magnetization within T_1; this in turn means that the diffusion does not proceed very far even within the longest mixing times, since these are themselves limited by the T_1 relaxation times.

The selection of ^{1}H magnetization can also be achieved on the basis of differences in chemical shifts, in particular under CRAMPS conditions. Such 1D selective-inversion or chemical-shift filter experiments are closely related to the 2D exchange approach and will be discussed with it in the following section.

13.2.2 The CRAMPS 2D exchange experiment

With good CRAMPS available, it is quite straightforward to perform a CRAMPS 2D exchange experiment, with cross-peaks indicating the coupling of certain sites by ^{1}H spin diffusion. Technically, it is to be noted that the simple off-resonance solution of the phase problem (see Section 4.6.1) can be applied: that is, only a cosine dataset must be measured, since the 1D CRAMPS measurement must be performed off-resonance anyway (see Section 3.3.4).

Figure 1.7 displays 2D ^{1}H spin-diffusion CRAMPS spectra obtained by Caravatti *et al.* (1985), with $t_m = 100$ ms, of two blends of polystyrene (PS) and polyvinyl methylether (PVME) cast from different solvents. Resulting from unavoidable spin diffusion within each component, intramolecular exchange peaks appear in both spectra. Intermolecular cross-peaks between the signals of the aromatic protons of PS and the —OCH_3 protons of PVME are observed only in the blend cast from toluene. This proves that mixing in this blend occurs on a much smaller length scale than in the material cast from chloroform. This is in agreement with the results of other techniques. Nonetheless, for concluding whether the mixing occurs on a 'molecular' scale, comparable to the size of the monomer units, the mixing-time dependence of the exchange intensity has to be measured and analysed. In a similar PS/PVME blend cast from toluene and measured at a similar temperature, spin diffusion was found to be essentially completed after $t_m = 20$ ms, but not yet at 1 ms, and a WISE experiment showed the motional state of the two components, PS and PVME, to be quite different. These observations indicate nano-heterogeneities in the blend (Schmidt-Rohr *et al.*, 1992).

As to the requirements for observing the inter-component spin diffusion unambiguously in CRAMPS exchange experiments, it is apparent that each of the constituents must possess a relatively isolated proton line of its own. In the case of PS and PVME, the aromatic band derives exclusively from the PS, while the line in the centre of the spectrum is due to the PVME alone, namely, the OCH_3 group.

As CRAMPS spectra typically exhibit only few lines, for measuring the full mixing-time dependence of the spin diffusion process, it can be advantageous to perform selective 1D experiments. Under conditions of strong dipolar couplings, this goal will not be attained by simply irradiating a long soft pulse. Rather, frequency-selective CRAMPS methods are required. The technique introduced by Caravatti *et al.* (1986a) is a combination of the principle of the DANTE sequence (see Section 3.8), with the requirements of CRAMPS. It is quite selective, but is also demanding as it requires multipulse sequences with some pulses extended by small-flip angles. The second approach (Schmidt-Rohr *et al.*, 1990), termed 'chemical-shift filter', is based on standard CRAMPS and is conceptually quite simple (see Section 3.8), separating signals directly on the basis of their different precession behaviour, see Figure 3.12. This approach has proven to be sufficient for many problems in solid polymers (Schmidt-Rohr *et al.*, 1990; Clauss *et al.*, 1993).

13.2.3 ^1H spin diffusion with ^{13}C detection

The main problem in ^1H spin diffusion experiments is the low resolution of ^1H spectra. In the experiment described in the previous section, it is required that each component exhibits a peak without significant overlap from the signals of the other component. Only on that basis can it be established where the magnetization came from and where it went; in the PS/PVME spectrum of Figure 1.7, for instance, the aromatic signal for PS and the OCH_3 peak of PVME fulfil this condition. For the detection period, it is actually sufficient to achieve the separation in the highly resolved ^{13}C MAS spectrum if the ^1H magnetization is transferred to ^{13}C spins by cross polarization. The intensity of the ^{13}C signals created in this way is proportional to the ^1H magnetization in the proximity of the given ^{13}C spin. These considerations show that for NMR experiments, the picture of each nuclear spin carrying a small amount of magnetization provides a valid description. It is more useful than the widespread notion that each spin is either 'up' or 'down', which is the correct approach for measurements of single spins, but not useful for NMR experiments on more than 10^{16} spins. Conventional NMR measures only the averages over the huge number of similar structures, and these magnetization expectation values can indeed be 'a little up', 'null', or 'a little down'.

As an example for the selective transfer, Figure 13.3 displays ^1H spectra after application of the chemical shift filter and the corresponding ^{13}C spectra, for a heterogeneous blend of PS and PMMA. Selection of proton magnetization of the PS yields the clean PS ^{13}C spectrum, and selection of OCH_3 protons yields only PMMA ^{13}C peaks. That the CH_2 signals are also present in the ^{13}C spectra is due to ^1H spin diffusion during the cross polarization time of 0.4 ms (see Section 3.4.3). The further

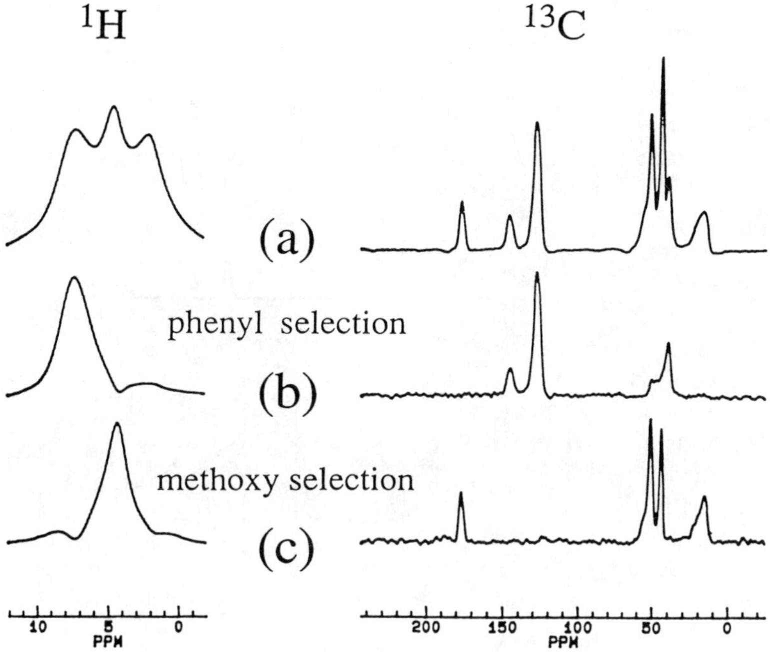

Figure 13.3 ^1H chemical-shift filter with ^1H and ^{13}C detection for a heterogeneous blend of PMMA and PS. (a) Without selection. (b) Selection of the magnetization of the aromatic protons in the PS. The ^{13}C spectrum of PS is observed selectively. (c) Selection of the OCH_3 protons of PMMA, yielding the ^{13}C spectrum of PMMA. Adapted from Clauss *et al.* (1993).

detection of ^1H spin diffusion between PS and PMMA in such ^{13}C spectra is shown in Figure 13.7(a).

13.2.4 2D vs. selective 1D ^1H–^{13}C spectra

In Section 7.2, it was outlined that selective 1D spectra can be equivalent to slices (cross-sections) through 2D spectra. In this section, we shall analyse this relationship for the chemical-shift filter with CP-^{13}C-MAS detection and the heteronuclear ^1H-CRAMPS/^{13}C-MAS 2D experiment (section 6.7.2), taking the effect of homogeneous line-broadening particularly into account.

Figure 13.4 indicates the relationship between these experiments. It shows that the 1D method is advantageous for homogeneously broadened, partially overlapping (^1H) lines in the ω_1 dimension or selection period. In the corresponding 2D spectrum, the overlap of the ^1H lines, which results in wings extending into the ^{13}C slice of the other lines (Figure 13.4 bottom), makes a quantitative analysis more difficult. In the 1D experiment, the other lines are completely eliminated (the broadening was assumed to be homogeneous!), such that there are no disturbing residual signals. This can be termed, somewhat frivolously, 'fooling Fourier'.

Quite generally, in terms of data handling and storage requirements, 1D experiments are certainly more convenient. With the advent of even faster computers and larger storage media, and with convenient, comprehensive or flexible 2D evaluation programs

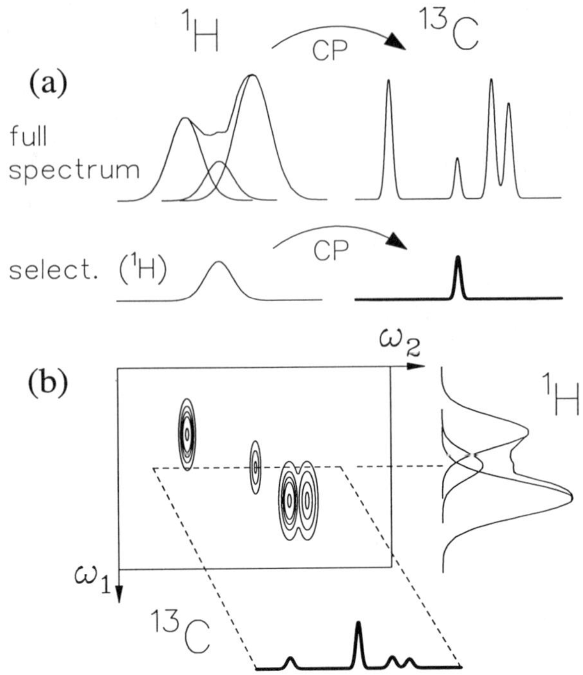

Figure 13.4 2D vs. selective-1D isotropic-shift ^1H–^{13}C correlation experiments in the presence of ^1H line-broadening. (a) In the 1D experiment, the central line in the ^1H spectrum is selected and used as the magnetization source for the ^{13}C spectrum. The undesired lines can be suppressed completely in this 1D experiment. (b) In the 1D cross-section through the 2D spectrum at the maximum of the central ^1H peak (fat trace), the wings of the other ^1H bands produce undesired peaks, i.e. they are suppressed only incompletely.

becoming routinely available, this argument will become less and less important. Nonetheless, large series of data are handled and evaluated more easily in terms of 1D spectra. When it comes to measuring the mixing-time dependence of exchange signals, as is required for spin-diffusion experiments, 1D experiments are definitely more convenient; on the other hand, for establishing all the connectivities between the peaks in a spectrum, a 2D experiment is the method of choice.

An important advantage of 2D spectra consists in the greater redundancy of information and clear criteria for possible artifacts. For exchange spectra, one of the most important indicators of quality is the symmetry of the 2D spectrum with respect to the diagonal; another consists in the fact that no exchange signal must appear outside the range of the intensity on the diagonal. In the 1D experiment, a drift of spectrometer parameters producing fake exchange peaks is much more difficult to detect than in the 2D spectrum.

13.2.5 WISE 2D experiments with ^1H spin diffusion

^{13}C detection of ^1H spin diffusion can also be achieved by incorporating ^1H spin diffusion into the WISE 2D experiment (see Section 6.6.6). Using the pulse sequence of Figure 13.5(a), the ^1H magnetization at the end of the ^1H evolution is flipped to

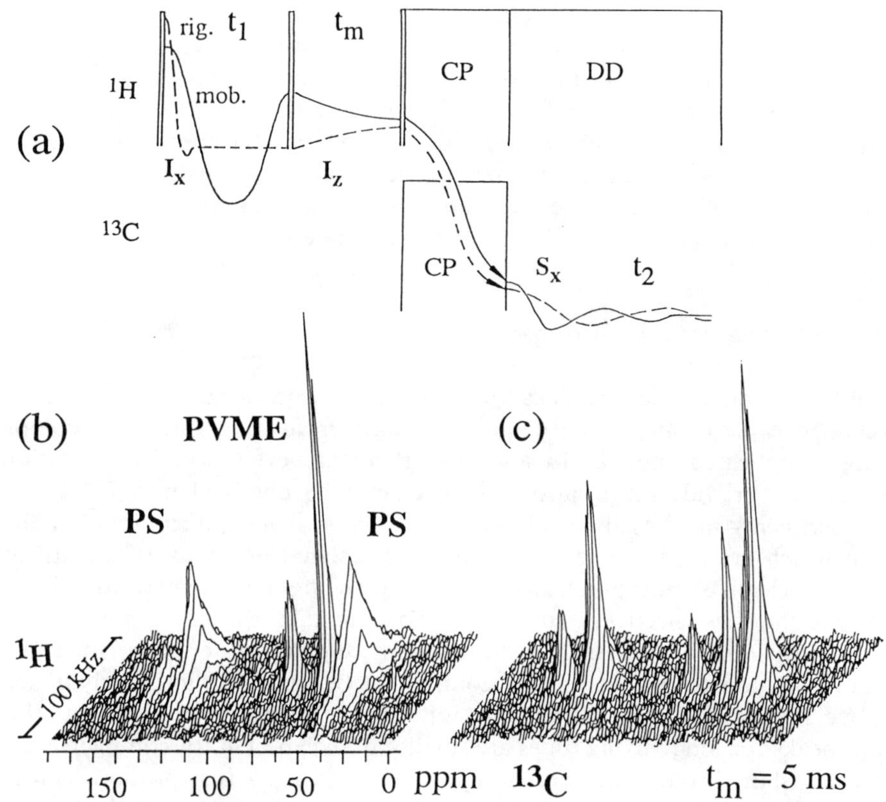

Figure 13.5 WISE spectroscopy with ^{1}H spin diffusion. (a) Pulse sequence with ^{1}H evolution, ^{1}H spin diffusion during t_m, cross polarization and ^{13}C signal detection. (b) and (c) Application to a 50:50 PS/PVME blend cast from toluene which is homogeneous by DSC and electron-microscopy criteria. (b) Reference spectrum with a minimum effective mixing time of 0.2 ms (due to CP). Narrow PVME and broad PS ^{1}H lines are observed, which proves dynamical heterogeneities correlated with the chemical structure. (c) After a mixing time of only 5 ms, the silhouette of the spectrum is completely changed, because the original lineshape differences in the ^{1}H dimension have been eliminated by the ^{1}H spin diffusion. This fast equilibration shows that the dynamical heterogeneities are only about 3 nm in diameter. Adapted from Schmidt-Rohr *et al.* (1992).

the z axis by a 90° pulse, subject to spin diffusion during a mixing time, and flipped back into the transverse plane by a second 90° pulse to be immediately subject to the cross-polarization locking field. The effect of the spin diffusion on the ^{1}H magnetization is again detected through the amplitude modulation of the ^{13}C magnetization generated by the cross polarization. In particular, complete equilibration of the ^{1}H magnetization in the system is reflected in equal lineshapes of all lines in the ω_1 (^{1}H wideline) dimension: complete equilibration means that at the end of t_m, all ^{1}H magnetization components vary with t_1 in the same way. In the cross polarization, they produce identical amplitude modulations of all ^{13}C signals, which after FT against t_1 result in identical ^{1}H lineshapes.

An example of the utility of the WISE experiment with spin diffusion has been provided for a 'homogeneous' blend of PVME and PS (similar to the material analysed in the experiments of Section 13.2.2) (Schmidt-Rohr *et al.*, 1992). As shown in Figure 13.5(b), a 'normal' WISE experiment with a very short mixing time exhibits PVME ^{1}H lines which are much narrower than for the PS. This means that the PVME is more mobile. In a second WISE experiment with a mixing time of 5 ms, these lineshape differences are found to have been equilibrated by the ^{1}H spin diffusion (Figure 13.5(c)). This tells us that the magnetization has equilibrated within only 5 ms over the dynamic and chemical heterogeneities, which means that these are nano-heterogeneities, with diameters of 3.5 ± 1.5 nm estimated from the spin-diffusion time dependence.

13.2.6 Selective saturation transfer

Selective 1D exchange experiments can provide the same information as spin-diffusion 2D exchange experiments, albeit requiring longer measuring times for completely analysing complicated spectra. In addition, there are even selective-saturation 1D techniques that provide information which cannot be obtained in a 2D approach. These experiments are based on periods of selective inversion/saturation and spin diffusion which are repeated several times before signal detection (Caravatti *et al.*, 1986b). This selective saturation transfer procedure eliminates magnetization in spatial proximity to the sites selectively suppressed. To rationalize that this can provide more information than a 2D experiment, consider a three-phase system consisting of large regions of pure component A, of pure component B, and of an intimately mixed AB phase. In a 2D spin-diffusion experiment, for such a system it would be found that the exchange peaks for long mixing times are smaller than expected for complete exchange (where every slice through the spectrum parallel to either axis is proportional to the normal 1D spectrum). However, the exchange peaks are smaller in the same way for various compositions: it is indistinguishable whether exchange from A to B is not efficient because the magnetization cannot diffuse out of the cores of very large A domains or because it cannot efficiently penetrate into huge B domains (note that the relative amounts of A, B and AB phases, i.e. three numbers, are not uniquely determined by the two peaks, one for A and one for B, in the 1D spectrum). In the selective-saturation experiment the signal contributions from the mixed phase are eliminated together with those of the selectively irradiated species, so that the remaining signals directly represent the pure-phase fraction of the second component.

Selective-saturation transfer spin-diffusion experiments were first performed by Caravatti *et al.* (1986b). They applied the technique to blends of PS and PVME cast from toluene ('homogeneous' blend) and from chloroform (heterogeneous mixture). Figure 13.6 shows typical datasets for selective irradiation on the OCH_3 lines of the PVME, as a function of the number of cycles of selective saturation and spin diffusion. Due to loss by spin diffusion to the PS, the magnetization decays much faster in the homogeneous than in the heterogeneous blend. In an ideal experiment, the fraction of pure-phase polystyrene could be directly determined from the ratio of the PS peak after sufficiently long selective PVME saturation to that without irradiation. In practice, however, due to the relatively strong signal decay caused by the residual dipolar couplings or because of imperfect selectivity, the spin diffusion effects are actually recognized only upon comparison with the decay in the heterogeneous blend. These exhibit different mobilities, dipolar couplings and decay behaviours from the components of the nano-heterogeneous blend, which make the interpretation of the

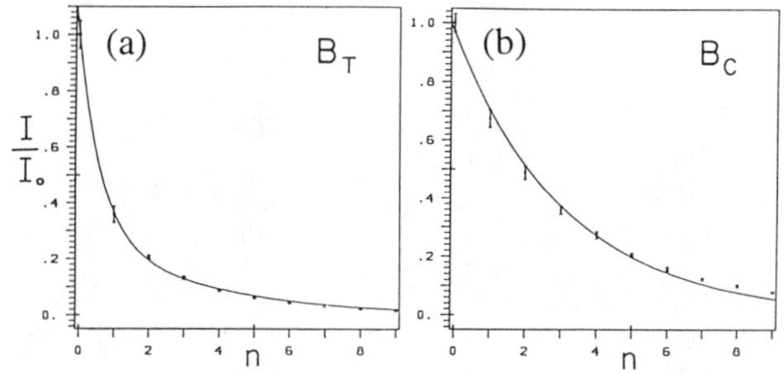

Figure 13.6 Selective-saturation transfer in 50:50 PS/PVME blends cast (a) from toluene (homogeneous) and (b) from chloroform (inhomogeneous) (Caravatti *et al.*, 1986b). The plots show the normalized intensity of the OCH_3 lines of PVME as a function of the number of selective-saturation transfer cycles, with the PS signals being selectively saturated, and mixing times of 10 ms for the transfer. In (a) the decay is faster since, in this intimately mixed blend, spin diffusion occurs from the PVME to the PS where the magnetization is destroyed. Adapted from Caravatti *et al.* (1986).

data more difficult. But this problem does not beset the method in general, as it does not arise in the more common case of homogeneous mixtures of two rigid components. As an alternative, normalization by the signal with the same number of selective saturation but no spin diffusion (minimal mixing times) could account for the trivial signal reduction.

For the analysis of the phase composition, other NMR methods should also be taken into account. The presence of a pure PVME phase more than 70 K above its T_g, for instance, should be detectable with high sensitivity simply in terms of narrow PVME peaks in the single-pulse proton spectrum; in a similar PS/PVME system, the most highly mobile components still exhibit a considerable linebroadening and also contain a 40% phenyl signal, which casts doubts on the presence of large regions of pure PVME.

13.2.7 Short review of ^1H spin-diffusion studies

Due to the relevance of the structural information that can be obtained from ^1H spin-diffusion experiments, and since these experiments can be performed with standard equipment, a relatively large and constantly growing number of ^1H spin-diffusion studies of multi-component polymer systems have been reported in the literature. Many of these are based on selection in terms of differences in mobility, either through the reduction of dipolar couplings or via differences in $T_{1\rho}$ relaxation times; others exploit chemical-shift differences. For detection, ^1H wideline (Cheung and Gerstein, 1981; Packer *et al.*, 1984; Colquhoun and Packer, 1987); CRAMPS (Caravatti *et al.*, 1985, 1986b), and ^{13}C CP/MAS spectroscopy (Schmidt-Rohr *et al.*, 1990; Clauss *et al.*, 1993) have been employed. The systems studied comprise polyurethanes (Assink, 1978), semicrystalline polymers in the bulk (Cheung and Gerstein, 1981; Packer *et al.*, 1984; Colquhoun and Packer, 1987; Kimura *et al.*, 1992) and as fibres (Havens and VanderHart, 1984, 1985), polymer blends (Caravatti *et al.*, 1985, 1986b; Schmidt-Rohr *et al.*, 1990), and diblock-copolymers (Clauss *et al.*, 1993). The evaluation is usually based on spin-diffusion simulations, though in favourable cases comparison of

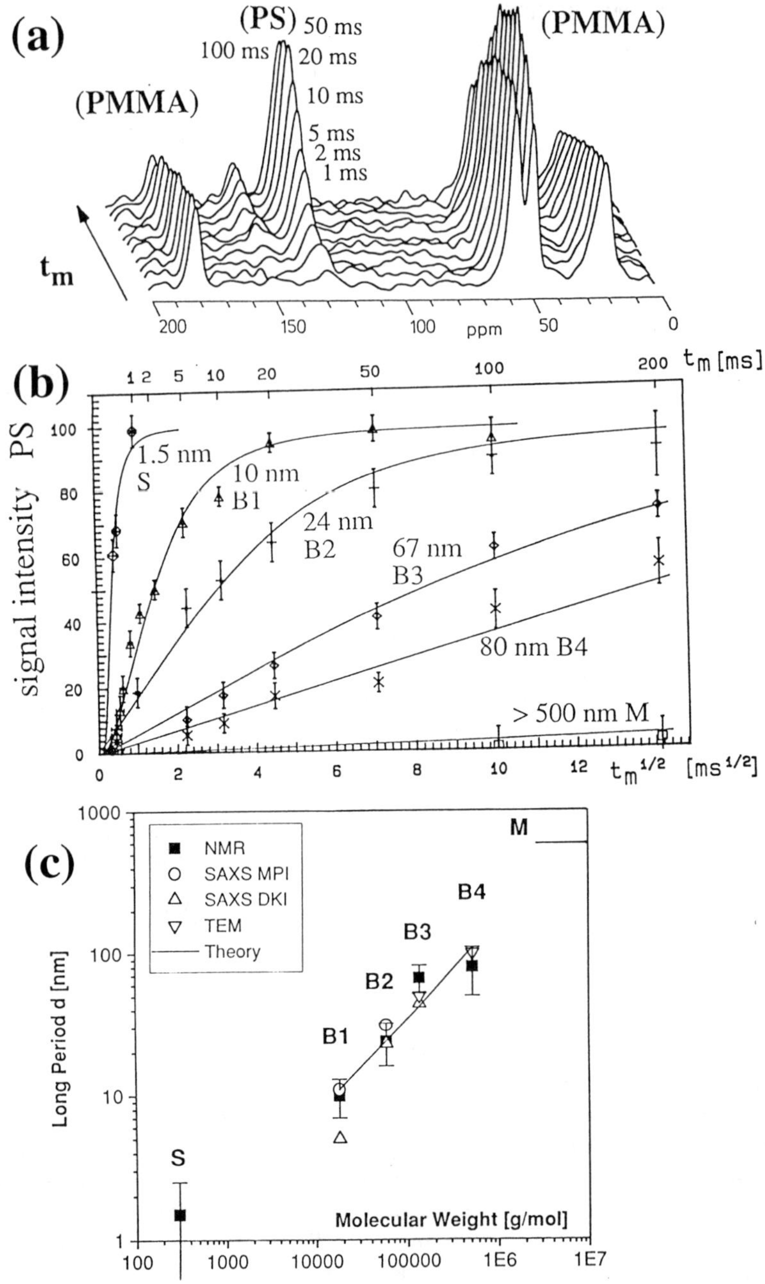

Figure 13.7 Determination of the spin diffusion coefficient and the molecular-weight dependence of domain sizes in PS-PMMA systems (block copolymers B1–B4 of various molecular weights, a statistical copolymer S, and a blend M). (a) ^{13}C CP-MAS TOSS spectra of a PS-*b*-PMMA block copolymer after ^{1}H chemical-shift filter, selecting the magnetization of the OCH_3 protons of PMMA, followed by a mixing time and CP to ^{13}C. The mixing times increase from front to back. Spin diffusion to the PS is apparent by the increase of the initially suppressed PS signals. (b) Mixing-time dependence of the strongest PS peaks for the investigated systems. Long periods as obtained from the fits with a diffusivity of 0.8 nm^2/ms are given next to the curves and shown in (c) as a function of the molecular weight, together with small-angle X-ray scattering (SAXS) and transmission electron microscopy (TEM) data. Adapted from Clauss *et al.* (1993).

intercomponent and intracomponent spin diffusion can provide direct distance information (Schmidt-Rohr *et al.*, 1990); in that study, homogeneous mixing on a 1 nm scale could be proved by showing that sites less than 1 nm away from the selected methyl groups in poly(ether imide) chains were reached by the methyl-proton magnetization in the same time as were sites on the blend partner, an aromatic poly(ether ketone).

In the spin-diffusion simulations, the value of the spin-diffusion coefficient in polymers has to be determined. Two studies have focused on this problem (Kimura *et al.*, 1992; Clauss *et al.*, 1993); both find diffusivities of about 0.7 nm^2/ms consistent with values previously derived from certain theoretical considerations (Cheung and Gerstein, 1981; Cheung, 1981), but they are considerably larger than the value of 0.01 nm^2/ms estimated by Abragam (1961). (Actually, Abragam gives a D value of 10^{-13} without units, but from the context it is clear that these should be the cgs units cm^2/s. It is to be noted that $D = 0.2\ nm^2/ms$, much closer to the experimental value, is obtained from Abragam's formula $D = Wa^2$ if $a = 2$ Å is use and typical values are inserted into Abragam's formula for W (Abragam, 1961, Chapter 5B), e.g. $W = 2\pi \times 25$ kHz/30.) In one of the experimental efforts to determine D (Kimura *et al.*, 1992), numerical simulations were employed which take into account relaxation and for the first time a spatial dependence of the diffusivity. Nevertheless, several shortcomings restrict the significance of the diffusivity values obtained in that study: the T_1 relaxation effects are dominant over the spin-diffusion, the initial magnetizations states are not very far from equilibrium, and due to pronounced differences in mobility of the various phases in the investigated polyethylene samples, several different diffusivities had to be used as fit parameters. In the second study (Clauss *et al.*, 1993), a set of PS-PMMA block-copolymer samples characterized by small-angle scattering and electron microscopy was studied by means of the chemical-shift filter with spin diffusion and ^{13}C CP/MAS detection. Figure 13.7(a) shows a representative series of such spectra; note the initial complete suppression of the PS peaks and their gradual increase with t_m. In Figure 13.7(b), the time dependences of the largest PS peak are plotted for the various samples that were studied. Figure 13.7(c) portrays a plot of the molecular-weight dependence of the long periods, where values obtained from SAXS and electron microscopy are compared with those resulting from the NMR data for a diffusivity of 0.8 nm^2/ms.

13.3 Analysis of ^1H spin-diffusion data

In order to obtain information on domain sizes from ^1H spin diffusion data, it is necessary to analyse the time-dependence of the magnetization exchange. In this section, several procedures for extracting such information will be outlined. The spin diffusion data are usually given in terms of the integral z magnetization in one component as obtained from the spectral intensity of a corresponding NMR signal. In most cases, the rise of intensity from the component initially devoid of magnetization is considered, a sorption curve (Crank, 1975), as it is particularly simple to interpret and involves large relative changes of signal intensities. Various approaches will be presented: a general initial-rate relationship, a rigorous solution for lamellar structures, related approximations for more complicated disordered morphologies, and a versatile lattice-calculation approach for arbitrary distributions of diffusivity values and initial z magnetization.

In our treatment, we start from the diffusion equation for z magnetization $M(\vec{\mathbf{r}}, t_{\mathrm{m}})$:

$$\frac{\partial M(\vec{\mathbf{r}}, t_{\mathrm{m}})}{\partial t_{\mathrm{m}}} = \vec{\nabla} \cdot \{D(\vec{\mathbf{r}}) \vec{\nabla} M(\vec{\mathbf{r}}, t_{\mathrm{m}})\}$$

$$= \frac{\partial}{\partial x}\left\{D(\vec{\mathbf{r}}) \frac{\partial}{\partial x} M(\vec{\mathbf{r}}, t_{\mathrm{m}})\right\} + \frac{\partial}{\partial y}\left\{D(\vec{\mathbf{r}}) \frac{\partial}{\partial y} M(\vec{\mathbf{r}}, t_{\mathrm{m}})\right\} + \frac{\partial}{\partial z}\left\{D(\vec{\mathbf{r}}) \frac{\partial}{\partial z} M(\vec{\mathbf{r}}, t_{\mathrm{m}})\right\} \quad (13.1)$$

justified in Section 3.5.1. The assumptions concerning the spatial dependence of the diffusivity $D(\vec{\mathbf{r}})$ will be specified in the individual subsections.

13.3.1 *Initial rate, interfacial area and domain sizes*

Between the initial rate of the spin diffusion behaviour and the interfacial area per volume, simple relations exist that can be exploited for estimating domain sizes. This approach can also be exploited for deriving what types of averages over a distribution of domain sizes are actually measured in a spin diffusion experiment. To keep the treatment simple, we assume equal ^{1}H spin densities in all regions of the sample. The analysis of the more general case without this assumption can be found in the literature (Clauss *et al.*, 1993).

We consider an arbitrary two-phase structure, with the magnetization initially homogeneously distributed in phase A and zero in the B regions. The mixing-time dependence of the integral magnetization in B is to be considered, with the complete-exchange intensity $I_{\mathrm{B}}(t_{\mathrm{m}} \rightarrow \infty)$ as the 100% level. For sufficiently short times, everywhere along the A–B interface we have the same dependence of the magnetization density $M(x_\perp, t_{\mathrm{m}})$ on the spatial coordinate $x_\perp$ perpendicular to the interface:

$$I_{\mathrm{B}}(t_{\mathrm{m}}) := \int_{\mathrm{B}} M(\vec{\mathbf{r}}, t_{\mathrm{m}})\, \mathrm{d}^3 r \cong S^{\mathrm{tot}}_{\mathrm{A-B}} \int_0^\infty M(x_\perp, t_{\mathrm{m}})\, \mathrm{d}x_\perp \quad (13.2)$$

with the total area of all interfaces given by $S^{\mathrm{tot}}_{\mathrm{A-B}}$. For spatially constant diffusivity D, the magnetization profile perpendicular to the interface is given by:

$$M(x_\perp, t_{\mathrm{m}}) \simeq M_0/2 \ \mathrm{erfc}(x_\perp/\sqrt{4Dt_{\mathrm{m}}}) \quad (13.3)$$

if t_{m} is short enough for effects of the finite size of the domains to be negligible. In fact, equation (13.3) represents the solution of the diffusion equation (13.1) for two semi-infinite regions (Crank, 1975). The complement of the error function, erfc(x), is defined in Table 13.1; in fact we only need its integral between 0 and ∞, which is $1/\sqrt{\pi}$ according to Table 13.1. Thus, we can evaluate equations (13.2) and (13.3):

$$I_{\mathrm{B}}(t_{\mathrm{m}}) = S^{\mathrm{tot}}_{\mathrm{A-B}} M_0 \sqrt{Dt_{\mathrm{m}}/\pi}. \quad (13.4)$$

The complete-exchange intensity $I_{\mathrm{B}}(t_{\mathrm{m}} \rightarrow \infty)$ is given by:

$$I_{\mathrm{B}}(t_{\mathrm{m}} \rightarrow \infty) = f_{\mathrm{B}} I^{\mathrm{tot}} = f_{\mathrm{B}} M_0 f_{\mathrm{A}} V^{\mathrm{tot}} \quad (13.5)$$

using the volume fractions $f_{\mathrm{B}} := V^{\mathrm{tot}}_{\mathrm{B}}/V^{\mathrm{tot}}$, $f_{\mathrm{A}} := V^{\mathrm{tot}}_{\mathrm{A}}/V^{\mathrm{tot}}$, and the relation $I^{\mathrm{tot}} = M_0 V^{\mathrm{tot}}_{\mathrm{A}}$ between the total integral intensity I^{tot} and the overall volume $V^{\mathrm{tot}}_{\mathrm{A}}$ of domains of phase A. Combination of equations (13.4) and (13.5) yields:

$$\frac{I_{\mathrm{B}}(t_{\mathrm{m}})}{I_{\mathrm{B}}(t_{\mathrm{m}} \rightarrow \infty)} = \sqrt{Dt_{\mathrm{m}}/\pi} \ \frac{1}{f_{\mathrm{A}} f_{\mathrm{B}}} \frac{S^{\mathrm{tot}}_{\mathrm{A-B}}}{V^{\mathrm{tot}}} - O(\sqrt{t_{\mathrm{m}}}^{\,2}). \quad (13.6)$$

Table 13.1 *The error function erf(x) and related functions*

$$\mathrm{erf}(x) := \frac{2}{\sqrt{\pi}} \int_0^x \exp(-\tilde{x}^2)\, d\tilde{x}$$

$\mathrm{erf}(-\infty) = -1$ $\quad$ $\mathrm{erf}(0) = 0$ $\quad$ $\mathrm{erf}(\infty) = 1$ $\quad$ $\mathrm{erf}(-x) = -\mathrm{erf}(x)$

$$\mathrm{erfc}(x) := 1 - \mathrm{erf}(x) = \frac{2}{\sqrt{\pi}} \int_x^\infty \exp(-\tilde{x}^2)\, d\tilde{x}$$

$\mathrm{erfc}(-\infty) = 2$ $\quad$ $\mathrm{erfc}(0) = 1$ $\quad$ $\mathrm{erfc}(\infty) = 0$ $\quad$ $\mathrm{erfc}(-x) = 1 + \mathrm{erf}(x) = 2 - \mathrm{erfc}(x)$

$$\mathrm{ierfc}(x) := \int_x^\infty \mathrm{erfc}(\tilde{x})\, d\tilde{x} = \frac{1}{\sqrt{\pi}} \exp(-x^2) - x\, \mathrm{erfc}(x)$$

$x < -3$: $\quad$ $\mathrm{ierfc}(x) \cong -2x$ $\quad$ $\mathrm{ierfc}(0) = 1/\sqrt{\pi}$ $\quad$ $\mathrm{ierfc}(\infty) = 0$

where $O(\sqrt{t_m})^2$ denotes a term quadratic in $\sqrt{t_m}$. S^{tot}_{A-B}/V^{tot} can be translated into domain sizes d, since generally $d \sim (S^{tot}_{A-B}/V^{tot})^{-1}$. For a periodic lamellar structure consisting of alternating A and B sheets of thicknesses d_A and d_B, respectively, we have $V = (d_A + d_B)S^{tot}_{A-B}/\frac{1}{2}$ (factor $\frac{1}{2}$ due to the two surfaces per lamella). For A-phase cylinders of volume fraction f_A and diameter d_A in a matrix of phase B, we have $(Vf_A)/S_{A-B} = L_c\pi(\frac{1}{2}d_A)^2/(L_c 2\pi\frac{1}{2}d_A) = \frac{1}{2}d_A/2$; the same result is also found for cylinders with quadratic base. For A-phase spheres (or cubes) of diameter d_A in a B-matrix, $(Vf_A)/S_{A-B} = 4\pi/3(\frac{1}{2}d_A)^3/(4\pi(\frac{1}{2}d_A)^2) = \frac{1}{2}d_A/3$. Introducing the 'dimensionality' ε as the number of orthogonal directions relevant for the diffusion process (lamellae: $\varepsilon = 1$, cylinders: $\varepsilon = 2$, spheres/cubes: $\varepsilon = 3$) the geometric relations can be summarized as:

$$d_A/2 = \varepsilon f_A (S^{tot}_{A-B}/V^{tot})^{-1}. \tag{13.7}$$

For extracting the domain sizes in a two-component system out of the initial rate of the $I_B(\sqrt{t_m})$ graph, draw a straight line through the points in the initial linear range and extend it to determine the t_m value t^s_m of this line's intersection with the $I = 100\%$ (i.e. $I = I_B(t_m \to \infty)$) value (see Figure 13.8). At this point, the left-hand side of equation (13.6) is equal to 1, so that with (13.7) we have:

$$d_A = \frac{\varepsilon}{f_B}\sqrt{4Dt^s_m/\pi} \qquad (f_B = 1 - f_A) \tag{13.8}$$

showing that common formulae like $d \simeq \sqrt{\langle x^2 \rangle} = \sqrt{2Dt^s_m}$ or $d = \sqrt{4Dt^s_m/3}$ are only unnecessarily crude approximations, yielding domain sizes which are typically too small by a factor of 4.

These results can be generalized easily. For instance, if $D_A \neq D_B$, the magnetization profile in phase B is given by:

$$M_B(x_\perp, t_m) \cong M_0 \frac{\sqrt{D_A}}{\sqrt{D_A} + \sqrt{D_B}} \mathrm{erfc}(x_\perp/\sqrt{4D_B t_m}) \tag{13.9}$$

(Crank, 1975, p. 39), which can be integrated to yield:

$$I_B(t_m) \cong S^{tot}_{A-B} M_0\, 2\frac{\sqrt{D_A D_B}}{\sqrt{D_A} + \sqrt{D_B}}\sqrt{t_m/\pi}. \tag{13.10}$$

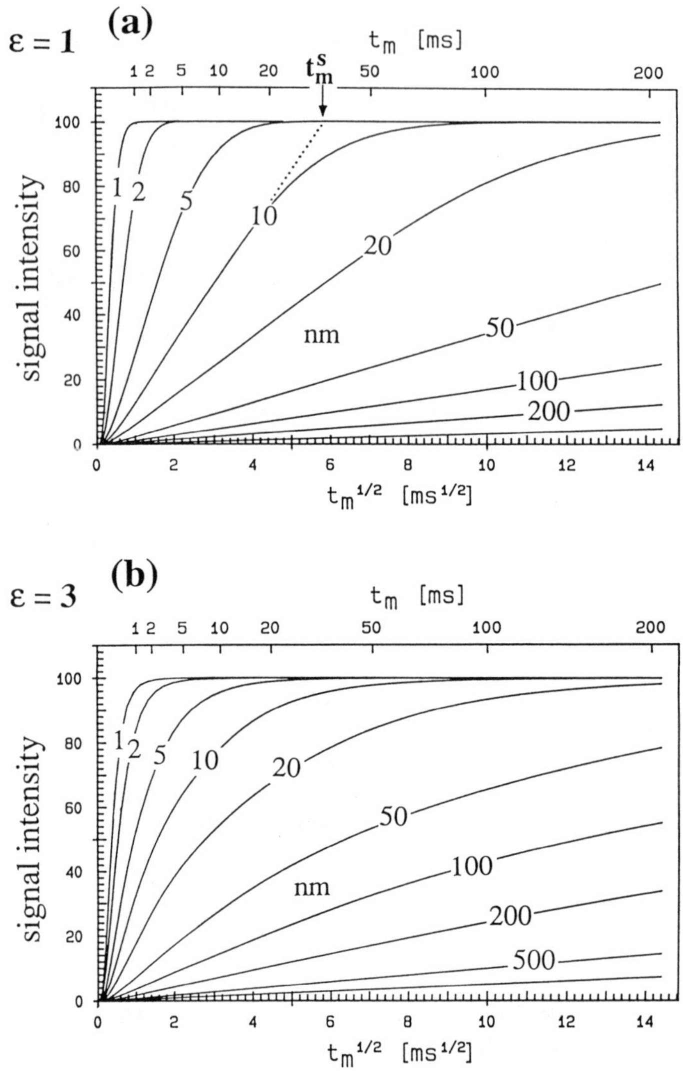

Figure 13.8 Simulated spin-diffusion curves of a 50:50 system, $D_A = D_B = 0.6\ nm^2/ms$, for $\varepsilon = 1$ (lamellae) and $\varepsilon = 3$ (cubes, spheres). The domain size d_A in nanometres is given next to each curve. The extrapolation of the initial linear regime to 100% for determining t_m^s is indicated for one of the curves. Adapted from Clauss *et al.* (1993).

Comparison with equation (13.4) shows that most of the equations given above for constant diffusivity can still be used after replacing $\sqrt{D}$ with $\sqrt{D_{eff}}$ as defined by:

$$\sqrt{D_{eff}} := \frac{\sqrt{D_A D_B}}{(\sqrt{D_A} + \sqrt{D_B})/2} \tag{13.11}$$

which is the geometric average of D_A and D_B divided by the arithmetic average of their square roots.

With a distribution of domain sizes, on the basis of:

$$S_{A-B}^{tot}/V^{tot} = \langle S_{A-B}\rangle/\langle V\rangle$$

the initial rate is determined by an effective domain size d_A^{eff} that takes the place of d_A:

$$\begin{aligned} &\text{lamellae:} && d_A^{eff} = \langle d_A\rangle = 2f_A\langle V\rangle/\langle S_{A-B}\rangle \\ &\text{cylinders:} && d_A^{eff} = \frac{\langle d_A^2\rangle}{\langle d_A\rangle} = 4f_A\langle V\rangle/\langle S_{A-B}\rangle \\ &\text{spheres:} && d_A^{eff} = \frac{\langle d_A^3\rangle}{\langle d_A^2\rangle} = 6f_A\langle V\rangle/\langle S_{A-B}\rangle. \end{aligned} \quad (13.12)$$

Beyond the initial-rate information, precise $I_B(t_m)$ data also contain hints as to the dimensionality of the system, in particular at intermediate t_m values (Figure 13.8). In addition, a bimodal distribution, with small and large domains, would be detectable as a superposition of a fast and a slow signal rise.

The simple and useful initial-rate approach breaks down if the system is not a two-component system with a sharp interface. Then, a complete modelling of the diffusion behaviour is required. Such full solutions, which also permit the exploitation of the information contained in the larger t_m values, are the subject of the following section.

13.3.2 Analytical solutions

We look for solutions of the diffusion equation (13.1) for constant diffusivity $D(\vec{\mathbf{r}}) = D$:

$$\frac{dM(\vec{\mathbf{r}}, t)}{dt} = D\Delta M(\vec{\mathbf{r}}, t) \quad (13.13)$$

in simple model morphologies containing the essential features of the morphology in the sample.

In the previous section, it was seen that a classification of the morphologies in terms of their dimensionality is useful. The simplest structure is a periodic lamellar morphology, with only the direction perpendicular to the surface of the lamellae being relevant for the diffusion process. Consequently, only a one-dimensional diffusion equation must be solved. The solution can be derived from scratch just on the basis of the knowledge that an initial spatial δ-function distribution develops into a Gaussian of constant area, broadening proportional to $\sqrt{2Dt_m}$.

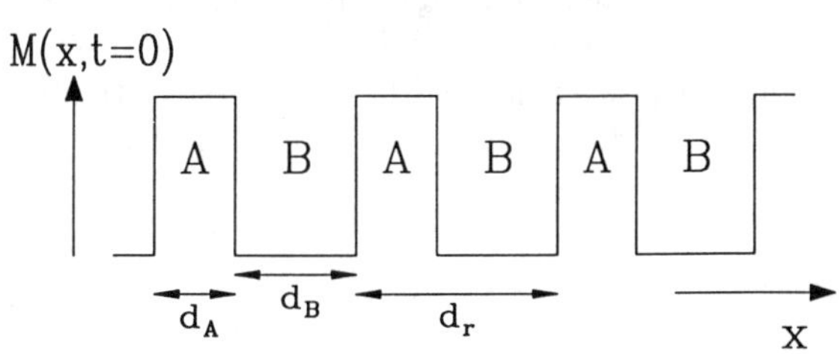

Figure 13.9 Domain size definitions for spin-diffusion simulations in a regular lamellar system of two phases A and B.

The initial distribution of magnetization, a periodic array of boxes of width d_A and height M_0 (Figure 13.9), can be conceived as generated by a dense array (superposition) of spatial δ functions. The magnetization starting from the lamella of width d_A and centred around $x = 0$, with:

$$M_{n=0}(x, t_m = 0) = M_0 \int_{-\frac{1}{2}d_A}^{\frac{1}{2}d_A} dx' \, \delta(x' - x) \tag{13.14}$$

evolves as

$$M_{n=0}(x, t_m) = \frac{M_0}{\sqrt{4\pi D t_m}} \int_{-\frac{1}{2}d_A}^{\frac{1}{2}d_A} dx' \exp(-(x' - x)^2/4Dt_m). \tag{13.15}$$

For the nth lamella, centred around nd_r, with the repetition period of the structure $d_r = d_A + d_B$, the integration limits are $nd_r - \frac{1}{2}d_A$ and $nd_r + \frac{1}{2}d_A$, so that the integral yields:

$$M_n(x, t_m) = \frac{M_0}{2}\left\{\operatorname{erfc}\left(\frac{nd_r - \frac{1}{2}d_A - x}{\sqrt{4Dt_m}}\right) - \operatorname{erfc}\left(\frac{nd_r + \frac{1}{2}d_A - x}{\sqrt{4Dt_m}}\right)\right\} \tag{13.16}$$

erfc(x) is the complement [$1 - \operatorname{erf}(x)$] of the error function erf(x), which is the integral of the Gaussian function; see Table 13.1. (We write equation (13.16) in terms of erfc(x) instead of using erf(x) for sake of simplicity in evaluating equation (13.17).)

The behaviour of the magnetization in all detection (B) regions is equivalent, therefore it is sufficient to consider a single one. We choose the one extending from $\frac{1}{2}d_A$ to $\frac{1}{2}d_A + d_B$:

$$\begin{aligned} I_B(t_m) &= \int_{\frac{1}{2}d_A}^{\frac{1}{2}d_A + d_B} dx \quad S \sum_{n=-N}^{N} M_n(x, t_m) \\ &= \frac{M_0}{2} S \sum_{n=1}^{N} \sqrt{4Dt_m}\left\{\operatorname{ierfc}\left(\frac{nd_r - d_A}{\sqrt{4Dt_m}}\right) - \operatorname{ierfc}\left(\frac{nd_r}{\sqrt{4Dt_m}}\right)\right. \\ &\quad \left. - \operatorname{ierfc}\left(\frac{(n-1)d_r}{\sqrt{4Dt_m}}\right) + \operatorname{ierfc}\left(\frac{(n-1)d_r + d_A}{\sqrt{4Dt_m}}\right)\right\} \end{aligned} \tag{13.17}$$

with $d_r = d_A + d_B$, and exploiting the symmetry of the structure about $x = \frac{1}{2}d_r$ to restrict the sum to $n \geqslant 1$. The integral of the error function complement, ierfc, can be calculated according to Table 13.1 from the error function, which is usually available in numerical libraries.

Only the first few terms in the virtually infinite sum (13.7) contribute significantly for most practical purposes. Each term represents the contribution of a lamella a distance nd_r away from the considered detection region. Therefore, the terms with large n become relevant only after very long times. Those with $n > n_c$ can be approximated as a semi-infinite region with initial magnetization density $M_0 d_A / d_r$, separated by a distance $x_c = n_c d_r$ from the detection region with its extension d_B. With the solution $\operatorname{erfc}(x/\sqrt{4Dt_m})$ for a semi-infinite source, we have:

$$\begin{aligned} I_{B, n > n_c} &\simeq M_0 \frac{d_A}{d_r} S \int_{x_c}^{x_c + d_B} dx \operatorname{erfc}\left(\frac{x}{\sqrt{4Dt_m}}\right) \\ &= M_0 S \frac{d_A}{d_r} \sqrt{4Dt_m}\left\{\operatorname{ierfc}\left(\frac{x_c}{\sqrt{4Dt_m}}\right) - \operatorname{ierfc}\left(\frac{x_c + d_B}{\sqrt{4Dt_m}}\right)\right\}. \end{aligned} \tag{13.18}$$

With the terms for $n = 0, \ldots, 4$ in (13.17) and the correction term (13.18) for $n_c = 4$, good results are obtained for arbitrary t_m.

These relations can easily be extended. For instance, a smearing-out of the initial distribution by a width of u_a can be incorporated by replacing $\sqrt{4Dt_m}$ in all relations by $\sqrt{4Dt_m + 2u_a^2}$. For a three-phase system with an interphase of thickness d_i, belonging neither to the initially magnetized nor to the detection region, $d_r = d_A + d_B + 2d_i$ and the final result:

$$I_B(t_m) = \frac{M_0}{2} S \sum_{n=1}^{N} \sqrt{4Dt_m} \left\{ \operatorname{ierfc}\left(\frac{nd_r - d_A - d_i}{\sqrt{4Dt_m}}\right) - \operatorname{ierfc}\left(\frac{nd_r - d_i}{\sqrt{4Dt_m}}\right) - \operatorname{ierfc}\left(\frac{(n-1)d_r + d_i}{\sqrt{4Dt_m}}\right) + \operatorname{ierfc}\left(\frac{(n-1)d_r + d_A + d_i}{\sqrt{4Dt_m}}\right) \right\}. \tag{13.19}$$

The decrease of the integral magnetization in region A, $I_A(t_m)$, can be calculated in an analogous fashion, but is also easily determined from $I_B(t_m)$ via the conservation of magnetization $I_A(t_m) + I_B(t_m) = d_A S M_0$.

For dimensionalities $\varepsilon = 2$ and $\varepsilon = 3$, that is, cylinders and spheres/cubes in a matrix, the arrangement of domains is more complicated. Figure 13.10(a) shows a typical structure, whose magnetization distribution can be viewed as the superposition of the magnetization from all individual sources, Figure 13.10(b). Figure 13.10(c) shows how each of these individual structures can be approximated by a pure A (source) region, surrounded by a pure B region according to the given f_B/f_A volume fraction, and a more distant environment consisting of both A and B in the ratio f_B/f_A.

For square-base cylinders and for cubes, the structures simplified as shown in Figure 13.10(d) are products of (equal) 1D structures along two or three orthogonal directions. Consequently, the magnetization distribution can be written as a product of 1D solutions of the diffusion equation:

$$M(\vec{r}, t_m) = M_0 \prod_{k=1}^{\varepsilon} \tfrac{1}{2} \left\{ \operatorname{erfc}\left(\frac{-\frac{1}{2}d_A - r_k}{\sqrt{4Dt_m}}\right) - \operatorname{erfc}\left(\frac{\frac{1}{2}d_A - r_k}{\sqrt{4Dt_m}}\right) \right\}. \tag{13.20}$$

The integral intensity in component B is found as:

$$I_B(t_m) = I(\tfrac{1}{2}d_r; t_m) - I(\tfrac{1}{2}d_r - d_B; t_m) + (I_0 - I(\tfrac{1}{2}d_r; t_m))f_B \tag{13.21}$$

with

$$\begin{aligned} I(s; t_m) &:= aM_0 \prod_{k=1}^{\varepsilon} \int_{-s}^{s} dr_k \, \tfrac{1}{2} \left\{ \operatorname{erfc}\left(\frac{-\frac{1}{2}d_A - r_k}{\sqrt{4Dt_m}}\right) - \operatorname{erfc}\left(\frac{\frac{1}{2}d_A - r_k}{\sqrt{4Dt_m}}\right) \right\} \\ &= aM_0 \left[\sqrt{4Dt_m} \tfrac{1}{2} \left\{ \operatorname{ierfc}\left(\frac{-\frac{1}{2}d_A + s}{\sqrt{4Dt_m}}\right) - \operatorname{ierfc}\left(\frac{-\frac{1}{2}d_A - s}{\sqrt{4Dt_m}}\right) - \operatorname{ierfc}\left(\frac{\frac{1}{2}d_A + s}{\sqrt{4Dt_m}}\right) + \operatorname{ierfc}\left(\frac{\frac{1}{2}d_A - s}{\sqrt{4Dt_m}}\right) \right\} \right]^{\varepsilon} \end{aligned} \tag{13.22}$$

and $I_0 = aM_0(d_a)^{\varepsilon}$.

For simulating diffusion from the matrix into the cylinders or cubes, one can make use of 'Babinet's principle' (Baltá-Calleja and Vonk, 1989) for (spin) diffusion: for an

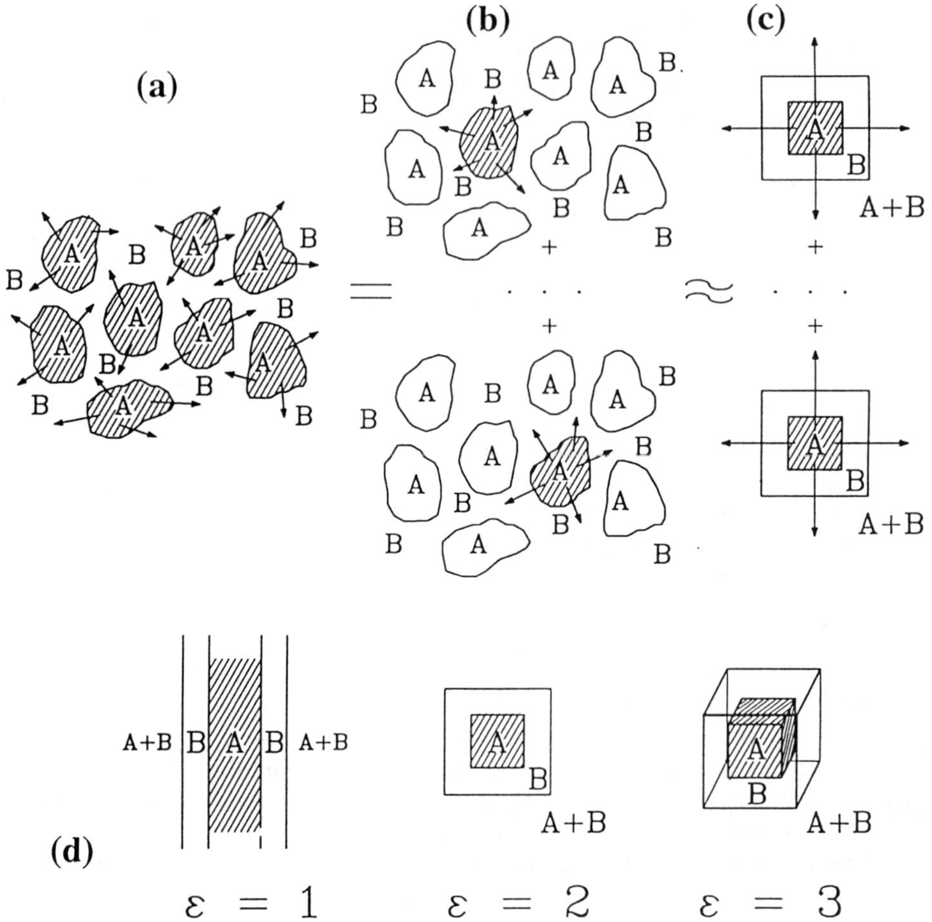

Figure 13.10 Visualization of the approximation for structures with dimensionalities $\varepsilon = 2$ or $\varepsilon = 3$ for spin-diffusion calculations. Due to the linearity of the diffusion equation, the distribution of magnetization from *N* domains of component A (a) can be represented as the superposition of *N* structures in which only one of the A domains contains magnetization (b). These can on average be substituted by the structure shown in (c), with an immediate pure-B environment of the A domain, and the rest of the structure made up of a mixture of A and B with the correct f_A:f_B ratio. (d) Structures simplified along these lines for dimensionalities $\varepsilon = 1$, 2 and 3. Adapted from Clauss *et al.* (1993).

arbitrary system made up of two components I and II as sketched in Figure 13.11, the integral diffusion behaviour for diffusion from I into II, case (i), is equal to that of diffusion from II into I, case (ii). This is analogous to the equality of the intensities of the two cross peaks, for I → II and II → I, in a corresponding 2D exchange spectrum.

For a proof, convince yourself that $\widetilde{M}(\vec{\mathbf{x}}, t_m) = M_0 - M(\vec{\mathbf{x}}, t_m)$ is also a solution of the diffusion equation. $\widetilde{M}(\vec{\mathbf{x}}, 0)$ fulfils the desired 'complementary' initial condition, case (ii). The asserted equal integral diffusion behaviour thus reads:

$$\int_{\mathrm{II}} M(\vec{\mathbf{x}}, t_m)\, \mathrm{d}^3x = \int_{\mathrm{I}} \widetilde{M}(\vec{\mathbf{x}}, t_m)\, \mathrm{d}^3x \tag{13.23}$$

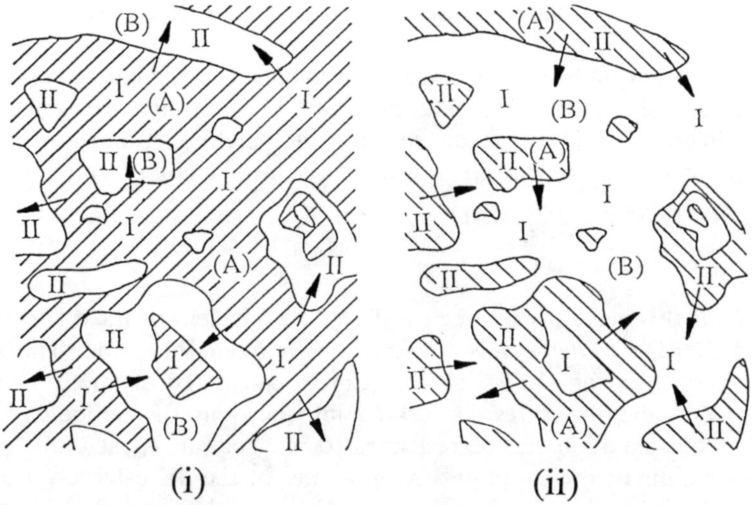

Figure 13.11 Complementary initial conditions (Babinet's principle) in spin diffusion. The increase of the integral magnetization in the non-hatched regions, which are initially free of magnetization, is exactly the same in both cases, though shapes and volume fractions of the sources (hatched) are quite different in the structures.

with the integrals encompassing all regions of the corresponding component. This relation is easily proved on the basis of conservation of magnetization ('c.o.m.'):

$$\int_{\mathrm{I}} M_0 \, \mathrm{d}^3x = \int_{\mathrm{sample}} M(\vec{\mathbf{x}}, 0) \, \mathrm{d}^3x \overset{\text{(c.o.m.)}}{=} \int_{\mathrm{sample}} M(\vec{\mathbf{x}}, t_{\mathrm{m}}) \, \mathrm{d}^3x$$

$$= \int_{\mathrm{I}} M(\vec{\mathbf{x}}, t_{\mathrm{m}}) \, \mathrm{d}^3x + \int_{\mathrm{II}} M(\vec{\mathbf{x}}, t_{\mathrm{m}}) \, \mathrm{d}^3x.$$

In the last step, $\int_{\mathrm{sample}} = \int_{\mathrm{I}} + \int_{\mathrm{II}}$ has been used. Considering the very first and the last expression in this string of equalities, one finds indeed that:

$$\int_{\mathrm{II}} M(\vec{\mathbf{x}}, t_{\mathrm{m}}) \, \mathrm{d}^3x = \int_{\mathrm{I}} (M_0 - M(\vec{\mathbf{x}}, t_{\mathrm{m}})) \, \mathrm{d}^3x \overset{\text{(per def.)}}{=} \int_{\mathrm{I}} \tilde{M}(\vec{\mathbf{x}}, t_{\mathrm{m}}) \, \mathrm{d}^3x.$$

To indicate that this relation need not appear as trivial, consider a similar case where not only the initial magnetization, but also the diffusivities, are interchanged between I and II. Then, though the initial rate is unchanged according to equation (13.10), the further behaviour can be changed (and is indeed changed, as shown by lattice simulations discussed in the following section).

13.3.3 Lattice calculations of diffusion

In this section, the treatment of structures with spatially dependent diffusivity, arbitrary initial conditions, etc., in terms of the discrete equivalent of the diffusion equation will be outlined. The elementary step of this 'diffusion on a lattice' is simple and intuitively appealing. The approach is completely deterministic: it has nothing to do with Monte Carlo calculations of a random walk, which are also much more time consuming.

For the magnetization transfer between protons in a solid, a lattice approach is even somewhat more realistic than a continuum theory. Per unit time, every nuclear magnet gives a certain fraction of its magnetization to its neighbours and in turn takes that fraction of their magnetization. In the case of a linear chain of spins this results in:

$$\frac{\Delta M_j}{\Delta t} = -2\Omega M_j + \Omega M_{j+1} + \Omega M_{j-1}. \tag{13.24}$$

In Section 3.5.1 this was shown to be the discrete equivalent of the diffusion equation (13.1), with $D = \Omega a^2$, where a is the spacing between the spins. The space–time evolution of arbitrary initial distributions can be calculated on the basis of formula (13.24). For periodic structures, by taking into account the periodicity conditions $M_{N+1} = M_1$ the calculation can be restricted to one spatial repeat unit.

Spatial variations of the coupling strengths, i.e. of the diffusivities, are taken into account by permitting the transition rate Ω to be spatially dependent. Transitions to the right (^{r}) and to the left (^{l}) must now be distinguished for the chain of spins:

$$\frac{\Delta M_j}{\Delta t} = -(\Omega_j^{\mathrm{l}} + \Omega_j^{\mathrm{r}})M_j + \Omega_{j+1}^{\mathrm{l}} M_{j+1} + \Omega_{j-1}^{\mathrm{r}} M_{j-1}. \tag{13.25}$$

Conservation of magnetization is fulfilled because M_j loses in ΔM_j exactly what M_{j+1} and M_{j-1} gain from M_j in ΔM_{j+1} and ΔM_{j-1}.

Being due to the same coupling between same pair of spins, the transition rates from spin j to spin $j+1$ and from $j+1$ to j must be equal:

$$\Omega_j^{\mathrm{r}} = \Omega_{j+1}^{\mathrm{l}}.$$

This condition also resolves possible uncertainties at the interface between regions with different diffusivities.

Different coupling strengths can result from differences in the proton densities of the components, but more often from partial averaging of dipolar couplings by molecular motions. This case is met quite often in practice, since soft components are used for increasing the impact strength and other mechanical properties of polymers.

Couplings to next-nearest neighbours (at distance $2a$, the coupling strength reduced by 2^{-3}) can easily be taken into account in the lattice calculation:

$$\frac{\Delta M_j}{\Delta t} = -2\Omega M_j + \Omega M_{j+1} + \Omega M_{j-1} + \tfrac{1}{8}(-2\Omega M_j + \Omega M_{j+2} + \Omega M_{j-2}). \tag{13.26}$$

The calculation can be extended to diffusion in two and three dimensions, either by setting up a corresponding 2D or 3D lattice, or with approximations in terms of products of 1D solutions as described in Section 13.3.2. Both approaches produce very similar results.

13.3.4 ^{1}H spin diffusion compared with small-angle scattering

The information content in ^{1}H spin diffusion experiments resembles that of small angle scattering (for an introduction see for instance, Baltá-Calleja and Vonk (1989)). With either technique:

- one typical dimension is probed
- the specific surface can be determined
- Babinet's inversion principle applies
- an interfacial transition layer causes a deviation from a linear initial behaviour
- with large domains, data correction (desmearing, T_1-correction) becomes necessary

However, there are also obvious differences, insofar as in well-ordered systems SAXS is more precise, while on the other hand spin diffusion experiments do not require regular structures. In addition, the difference in the generation of contrast also makes the techniques complementary in some instances.

13.4 ^{1}H spin diffusion data at long mixing times

For determining sizes of large domains by spin diffusion experiments, measurements with long mixing times are necessary. These are limited by T_1 relaxation of the ^{1}H magnetization during the mixing times. In many cases, useful data can be obtained for mixing times up to and beyond T_1 by the phase cycle alternating the magnetization between $+z$ and $-z$, as commonly used in exchange experiments (see Section 10.2.5) and a correction of the data by e^{t_m/T_1}. The conditions under which this procedure is valid will be analysed in Section 13.4.1. Even if reliable data are available only for the initial linear range of the spin-diffusion curve, domain sizes can be estimated if the complete-exchange intensity $I_B(t_m \to \infty)$ can be determined. This value can often be obtained using some additional information, such as the spectrum without selection or the overall composition of the sample, as will be outlined in Section 13.4.2.

13.4.1 Effects of T_1 relaxation

The phase cycle for mitigating the effects of T_1 relaxation during t_m in exchange experiments as analysed in Section 10.2.5 has been widely used successfully. Nevertheless, it has been asserted (Kenwright and Packer, 1990) that:

(i) the phase cycle could not separate T_1 and spin diffusion effects, and
(ii) even in a two-component system without spin diffusion, relaxation under the said phase cycle produced effects like spin diffusion.

In the following, the second assertion is shown to be incorrect for suitably designed experiments, and the limited validity of the first to be pointed out.

The basis of the alleviation of the T_1 effects is the simple phase cycle of the pulse before the mixing time as described in Section 10.2.5. The relevant component of the magnetization present at the end of the evolution time is stored alternately along the $+z$ and $-z$ directions. In the detection, the $-z$ signal is effectively subtracted from the $+z$ one. Without exchange during t_m, it was shown in Section 10.2.5 that the resulting difference is:

$$M_z^d(t_m) := M_z^+(t_m) - M_z^-(t_m) = 2M_0 \exp\left(-\frac{t_m}{T_1}\right) \tag{13.27}$$

consequently in a 1D experiment with complete initial suppression of one component, or in a 2D experiment, without spin diffusion the exchange signal remains zero even in the presence of T_1 relaxation. This refutes assertion (ii).

The evolution of z magnetization under both spin diffusion and T_1 relaxation is governed by:

$$\frac{(\mathrm{d}M_z(\vec{\mathbf{r}}, t)}{\mathrm{d}t} = D\Delta M_z(\vec{\mathbf{r}}, t) - \frac{M_z(\vec{\mathbf{r}}, t) - M^{\infty}}{T_1(\vec{\mathbf{r}})}. \tag{13.28}$$

Due to the linearity of this differential equation, the equation for $M_z^{\mathrm{d}}(\vec{\mathbf{r}}, t_{\mathrm{m}}) = M_z^+(\vec{\mathbf{r}}, t_{\mathrm{m}}) - M_z^-(\vec{\mathbf{r}}, t_{\mathrm{m}})$ is given as the difference of the equations for M_z^+ and M_z^-, with M^{∞} being eliminated regardless of the spatial dependence of $T_1(\vec{\mathbf{r}})$:

$$\frac{(\mathrm{d}M_z^{\mathrm{d}}(\vec{\mathbf{r}}, t)}{\mathrm{d}t} = D\Delta M_z^{\mathrm{d}}(\vec{\mathbf{r}}, t) - \frac{M_z^{\mathrm{d}}(\vec{\mathbf{r}}, t)}{T_1(\vec{\mathbf{r}})}. \tag{13.29}$$

If T_1 is independent of $\vec{\mathbf{r}}$, this equation (13.29) is solved by the product:

$$M_z^{\mathrm{d}}(\vec{\mathbf{r}}, t_{\mathrm{m}}) = M_{z\,\mathrm{diffus}}^{\mathrm{d}}(\vec{\mathbf{r}}, t_{\mathrm{m}}) \exp\left(-\frac{t_{\mathrm{m}}}{T_1}\right) \tag{13.30}$$

$M_{z\,\mathrm{diffus}}^{\mathrm{d}}(\vec{\mathbf{r}}, t_{\mathrm{m}})$ being the solution without T_1 relaxation (prove by inserting solution (13.30) into equation (13.29)). Contrary to assertion (i), the effect of homogeneous T_1 relaxation can thus be simply corrected by multiplication of the data by $\exp(+t_{\mathrm{m}}/T_1)$, with a value of T_1 easily obtained in a measurement without selection of magnetization.

If T_1 exhibits a strong spatial dependence, assertion (i) is generally correct since only the data for mixing times much shorter than the shortest T_1 can be evaluated easily. However, since it is usually possible to determine the complete-exchange intensities as described in Section 13.4.2, the data for short mixing times can be put on the right scale and thus allow for an estimate of the domain sizes.

If exchange by spin diffusion becomes only appreciable at long mixing times, contrary to assertion (ii) it can still be distinguished qualitatively from T_1 relaxation if one component was completely suppressed initially and the z-alternation phase cycle is used: under those conditions, reappearance of the initially suppressed signal can only be due to spin diffusion. For a quantitative analysis of data in this regime, one can take the effect of T_1 relaxation into account in a lattice calculation:

$$\frac{\Delta M_j}{\Delta t} = -\left(\Omega_j^{\mathrm{l}} + \Omega_j^{\mathrm{r}} + \left(\frac{1}{T_1}\right)_j\right) M_j + \Omega_{j+1}^{\mathrm{l}} M_{j+1} + \Omega_{j-1}^{\mathrm{r}} M_{j-1}. \tag{13.31}$$

This is the discrete form of equation (13.29), with a spatially dependent $D(\vec{\mathbf{r}})$ also considered. Note that due to the use of the z-alternation phase cycle, the unknown M^{∞} values play no role. The phase cycle also has a further beneficial effect: after a clean selection, often only a relatively small fraction of the overall signal is left, and therefore a spectrum taken without the z-alternation would be dominated by the contributions from the T_1-relaxing bulk of the sample. This would mask the small spin diffusion effects and thus make their analysis unreliable.

13.4.2 How to obtain complete-exchange intensities

For experimental spin diffusion data from systems with large domains, the determination of complete-exchange intensity values $I_B(t_m \to \infty)$ can be very important to put the data for the accessible t_m values on the correct scale in a spin-diffusion plot such as those of Figure 13.8; without such a 100% value, the intensity scaling would be arbitrary for measured data that do not reach a saturation level at the longest accessible mixing times. Depending on the type of spectrum and additional knowledge about the system, different approaches can be taken to determine the complete-exchange value; three of them will be described in this section.

In ^1H multipulse spectra, on the basis of constant spectral area $A(t_m)$ (T_1 relaxation can be disregarded here), we have $A(t_m \cong 0) = A(t_m \to \infty)$. The unfiltered reference spectrum, with total area A_u, and a peak height (or area) h_u for the relevant component, is a scaled image of the equilibrium spectrum ($t_m \to \infty$), that is, $h_u / A_u = h(t_m \to \infty) / A(t_m \to \infty)$. Thus, the required peak height (or area):

$$h(t_m \to \infty) = h_u \frac{A(t_m \cong 0)}{A_u} \tag{13.32}$$

is expressed in terms of directly accessible and reliably measurable values, without specific assumptions on composition or phase structure.

Final values for ^{13}C spectra of a material with relatively rigid and well detectable components which nevertheless have unknown CP efficiencies are easily determined for a two-phase system of known overall composition (proton fractions f_A and f_B), but otherwise arbitrary domain structure. The unfiltered reference spectrum is then to be scaled by a factor x_{scal} such that signal of the source component (A) in this scaled spectrum is only a fraction f_A of that of the filtered spectrum ($t_m \cong 0$):

$$x_{scal} h_A^u = f_A h_A(t_m \cong 0). \tag{13.33}$$

Again due to the identical shapes of the $t_m \to \infty$ and the unfiltered spectrum, one has:

$$h_B(t_m \to \infty) = x_{scal} h_B^u \overset{(13.33)}{=} f_A \frac{h_A(t_m \cong 0)}{h_A^u} h_B^u. \tag{13.34}$$

If it is preferable to obtain experimental data points without reference to the sample composition, one can determine directly the ratio $q(t_m)$ of source and detection peak heights (or areas):

$$q(t_m) = h_B(t_m) / h_A(t_m) \tag{13.35}$$

from the spin-diffusion spectra, and the final-state ratio:

$$q(t_m \to \infty) = h_B(t_m \to \infty) / h_A(t_m \to \infty) = h_B^u / h_A^u \tag{13.36}$$

from the reference spectrum without selection. To determine the domain sizes, theoretical curves are fitted directly to the $q(t_m)$ data by calculating both $h_B(t_m)$ and $h_A(t_m)$, which are on a well-defined scale and little affected by T_1 relaxation at sufficiently short times. If the intrinsic T_1 values are sufficiently similar for components A and B, the ratio (13.35) even makes a T_1 correction unnecessary.

13.5 ^{13}C spin diffusion

^{13}C spin diffusion is of interest mainly because of the superior resolution of ^{13}C MAS spectra, compared with ^1H solid-state spectra even under the best multiple-pulse conditions. The considerable number of unprotonated ^{13}C sites that are structurally important, such as CO groups, also enhances the value of the ^{13}C spectra. The higher resolution results in a better distinction of components in a composite system, and a better chance for identification of presumed specific interactions of polymers in homogeneous blends.

13.5.1 Low-abundance problem in short-range structural analysis

The low natural abundance of ^{13}C reduces the usefulness of ^{13}C spin diffusion experiments for detecting preferential proximities of certain carbon sites, which would be relevant in the study of spatial structures or as indicators of specific interactions. This will be illustrated for a simple example. Imagine that due to specific interactions, two particular and distinguishable carbon sites on different molecules are always close in space. In a 100% ^{13}C enriched system, due to the proximity of the interacting sites, an immediate rise of the exchange signal between these signals to approximately 50% of the original diagonal signal would be observed, which is more than the final ($t_m \to \infty$) value if other sites are present. By contrast, with ^{13}C in only 1.1% abundance, for a ^{13}C in one of the sites, the probability is only 1.1% of having a ^{13}C in its partner site. Therefore, the portion of the exchange signal that rises fast and thus indicates the specific interaction represents only 1.1% of the original diagonal signal. The further rise to the final state value, which is the same fraction of the diagonal signal as in the enriched system, can yield only information on coarser structures.

13.5.2 2D exchange powder patterns

^{31}P, ^{13}C, ^{15}N or ^{29}Si spin diffusion 2D exchange spectroscopy can be employed to probe local structure. More precisely, from the exchange patterns of the anisotropically broadened powder spectrum relative orientations of the molecular moieties containing the nuclei under observation can be determined. Due to its local nature, the technique is not restricted to crystalline materials. This makes it particularly valuable for probing the structure of amorphous substances. For efficient spin diffusion and sufficiently small-scale information (see the preceding section), an isotopic abundance of more than about 20% is desirable for the species to be observed.

One of the first examples involved 90% ^{13}C-labelled zinc acetate dihydrate (Henrichs and Linder, 1984). After a mixing time of 0.5 seconds, a final-state spin diffusion spectrum was obtained. The pattern with a diagonal ridge and 90° ridges reflects the two distinguishable sites in the sample. The reorientation angle of the $\vec{\mathbf{Z}}_P$ axis in the simulation amounts to $\beta \cong 90^\circ$ (the corresponding ridge perpendicular to the diagonal was not resolved due to the strong broadening of the spectrum, presumably resulting from ^{13}C$-^{13}$C dipolar couplings). The analysis of these spectra with exchange by spin diffusion between sites in equivalent chemical moieties is similar to that of the exchange spectra for molecular reorientation discussed in Chapter 7. The principal values and the relative orientations of the principal-axes systems of the sites between which the magnetization exchanges determine the exchange patterns. In addition, however, modulations of these patterns will usually arise due to the angular dependence of the dipolar exchange rate: reduced intensity will be observed for those regions of the exchange patterns that correspond to ^{13}C$-^{13}$C internuclear vectors close to the magic angle with respect to $\vec{\mathbf{B}}_0$. For exchange between different moieties (e.g. CH_2 and COO), due to the different principal values, the spectral intensity appears shifted compared with typical exchange spectra as discussed in Chapter 7; for instance, if the exchanging powder patterns do not overlap, the corresponding exchange intensity is completely separated from the spectral diagonal (nevertheless still symmetric with respect to the diagonal, because two patterns occur). This latter case, however, has rarely been relevant in experimental spectra because most systems investigated contained only one relevant site.

In unlabelled polyethylene, Edzes and Bernards (1984) and VanderHart (1987) observed relatively strong ^{13}C exchange intensity after mixing times of only 10 seconds. Figure 13.12(a) shows a 2D spectrum with a straight ridge which starts on the spectral diagonal at the point $(\omega_{33}, \omega_{33})$, and runs to the point $(\omega_{11}, \omega_{22})$. It corresponds to a rotation of the chemical-shift tensor by $\psi = 90^\circ$ (see Section 7.7 and the simulated spectrum Figure 13.12(b)) around the $\vec{\mathbf{Z}}_P$ (i.e. ω_{33}) axis, which is also the chain axis in PE. This corresponds exactly to the relative orientation of certain neighbouring chains in PE crystallites (see Figure 5.1(b)), suggesting that the exchange is basically due to ^{13}C spin diffusion. Its observed temperature dependence (VanderHart, 1987) and the relatively high efficiency at and above room temperature is probably related to the chain-diffusion process described in Section 7.14.2: the relative displacements of the chains bring many pairs of ^{13}C nuclei closer together for a certain period in the mixing time (note that it is not the nearly invariant sum $\Sigma_{ij} 1/|\vec{\mathbf{r}}_i - \vec{\mathbf{r}}_j|^3$ that is relevant for the exchange intensity but rather the number of pairs of nuclei that have experienced a strong coupling for a certain period during the mixing time).

Tycko and Dabbagh (1991) have found an interesting application of the spin-diffusion 2D exchange experiment to the study of local order in ^{13}C-labelled amorphous methanol. In the crystalline α modification that was measured for comparison, an elliptical ridge for a well-defined relative tensor orientation in neighbouring molecules is found (Figure 13.13(a)). In the β modification, hardly any exchange is detected (Figure 13.13(b)) since the orientations are nearly indistinguishable for the NMR interaction. In a glassy sample obtained by quenching the methanol to 103 K, a diffusive exchange pattern is found (Figure 13.13(c)), which

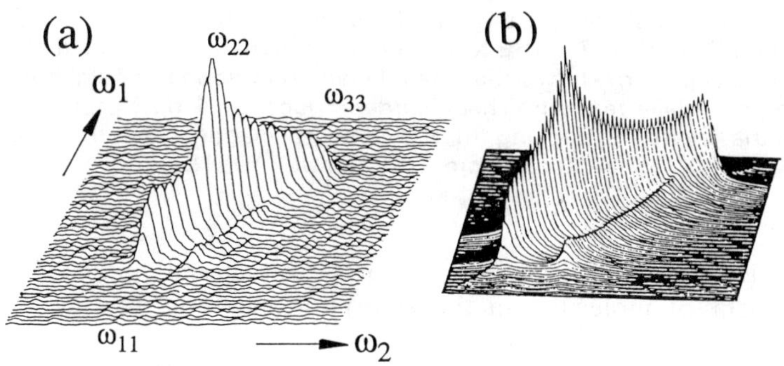

Figure 13.12 (a) ^{13}C exchange 2D spectrum of polyethylene obtained with $t_m = 10$ s. A straight exchange ridge is observed, starting on the diagonal at the ω_{33} end of the spectrum and ending at the point $(\omega_{11}, \omega_{22})$. It indicates exchange between ^{13}C-shift-tensor orientations rotated by 90° around the ω_{33} direction. This is consistent with the simulation (b) assuming spin exchange between methylene units in neighbouring chains in the crystallites. Adapted from Edzes *et al.* (1984).

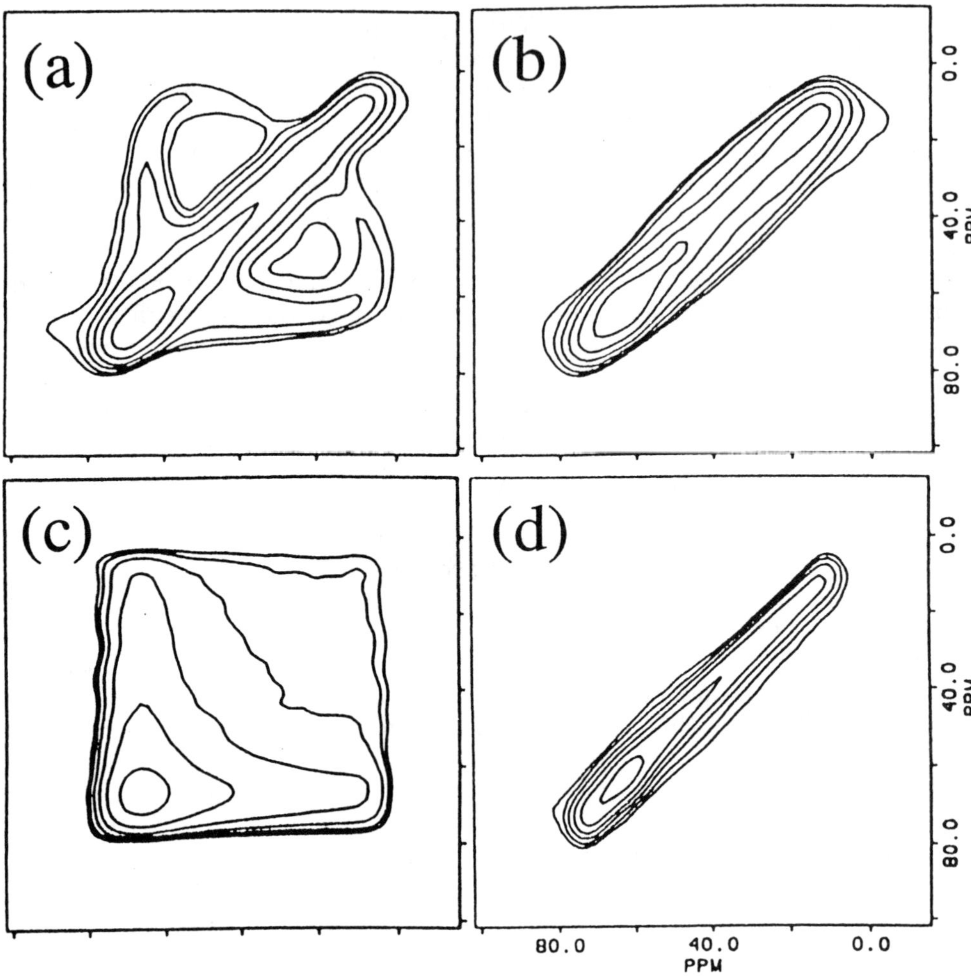

Figure 13.13 ^{13}C spin exchange spectra of ^{13}C-labelled solid methanol, taken with $t_m = 1$ s (Tycko and Dabbagh, 1991). (a) Crystalline α modification at $T = 133$ K. (b) Crystalline β modification at $T = 168$ K. (c) Amorphous material obtained by quenching to 103 K. The spectral pattern shows that the relative orientations of neighbouring $^{13}CH_2OH$ molecules are essentially random. This disordered local packing is obviously completely different from the crystalline packing. (d) Experimental diagonal spectrum for comparison. Adapted from Tycko *et al.* (1991).

proves local orientational disorder in the amorphous material, and no similarity of the local arrangement of molecules in the amorphous material to that of one of the crystalline forms.

Robyr *et al.* (1991) used the spin exchange between ^{13}COO-labelled benzoic-acid dimers to determine the relative orientation of the ^{13}C chemical-shift tensors in the two magnetically inequivalent dimers as a function of temperature. Since the orientation of the chemical-shift tensors in each dimer, which is equal for both corboxyl groups in the dimer (Robyr *et al.*, 1991), depends on the averaged position of the two hydrogen atoms, from the relative orientations of the two dimer's shift tensors an asymmetry of 0.42 ± 0.13 kJ/mol in the hydrogen-bond double-minimum potential could be determined.

13.5.3 ^{13}C spin exchange between resolved sites

Instead of observing spin diffusion between different orientations, it is also possible to study the spin exchange between different isotropic chemical shifts or the sharp peaks in oriented samples (including single crystals). Some of the earliest such spin-diffusion investigations on ^{13}C in natural abundance were focused on adamantane (Bronniman *et al.*, 1983) and mixtures of adamantane and another simple aliphatic molecule (2,2,3,3-tetramethylbutane) (Caravatti *et al.*, 1982). The spin diffusion is exceptionally efficient for these plastic crystals due to their relatively small chemical-shift differences and motionally narrowed dipolar lineshapes providing efficient line overlap. Nevertheless, these materials represent an interesting testing ground for the ^{13}C spin-exchange theory.

In rigid organic materials, after sufficiently long mixing times of 10 to 100 seconds, 'natural' or 'proton driven' spin diffusion produces appreciable cross peaks (~10% of the diagonal peaks) (Szeverenyi *et al.*, 1982) between signals from segments that are less than about 1 nm apart. If the S/N is good enough, even after only a few seconds cross peaks from spin diffusion become visible; an example is provided by experiments on amino acids and peptides by Frey and Opella (1984). However, in many unlabelled systems of simple structure, the spin-diffusion approach does not yield much information of interest, mainly because of the short-range problem laid out in Section 13.5.1. Nevertheless, for multicomponent materials, large molecules, or samples containing different crystalline modifications, relevant structural information can be gained. In labelled materials, more information can be obtained, in particular if the ^{13}C nuclei are introduced into one or even two specific sites.

As an example, Figure 13.14 displays a ^{15}N spin-diffusion exchange spectrum from oriented fd viral coat protein which was ^{15}N-labelled in its four valine residues (Cross *et al.*, 1983). From the cross peaks observed in the spectrum, information on the spatial proximity of these residues is obtained, which is consistent with the peak assignment: cross peaks are observed between valine residues 29 and 30, as well as 30 and 31, but not between the more distant residue 33 and the others.

In a polymer blend with at least one component enriched in ^{13}C, it is quite straightforward to study spin diffusion in a 2D MAS exchange experiment. Linder *et al.* (1985) demonstrated this for PET/PC blends and showed that these are mixed at distances of 4.5 Å to 6 Å. If significant spinning sidebands are present, to avoid phase problems it is advisable to apply the TOSS–t_1–deTOSS evolution as described in Sections 3.9.4 and 7.14.1. While the intimacy of mixing can be investigated even for natural-abundance ^{13}C by rf-driven spin diffusion, according to the discussion given in Section 13.5.1 specific interactions can hardly be detected without labelling, while this should be possible at least in principle for blends with one suitably enriched component.

Roy *et al.* (1987) investigated the ^{13}C spin exchange from a labelled position in a low-molar mass additive (di-*n*-butyl phthalate, DBP) to a polycarbonate matrix. At 10% DBP, where the additive acts as an antiplasticizer, the exchange rate to the quaternary carbon is smaller than at 25% DBP plasticizer concentration. This is interpreted as the effect of a specific interaction of the plastifier with the aromatic region of the polymer.

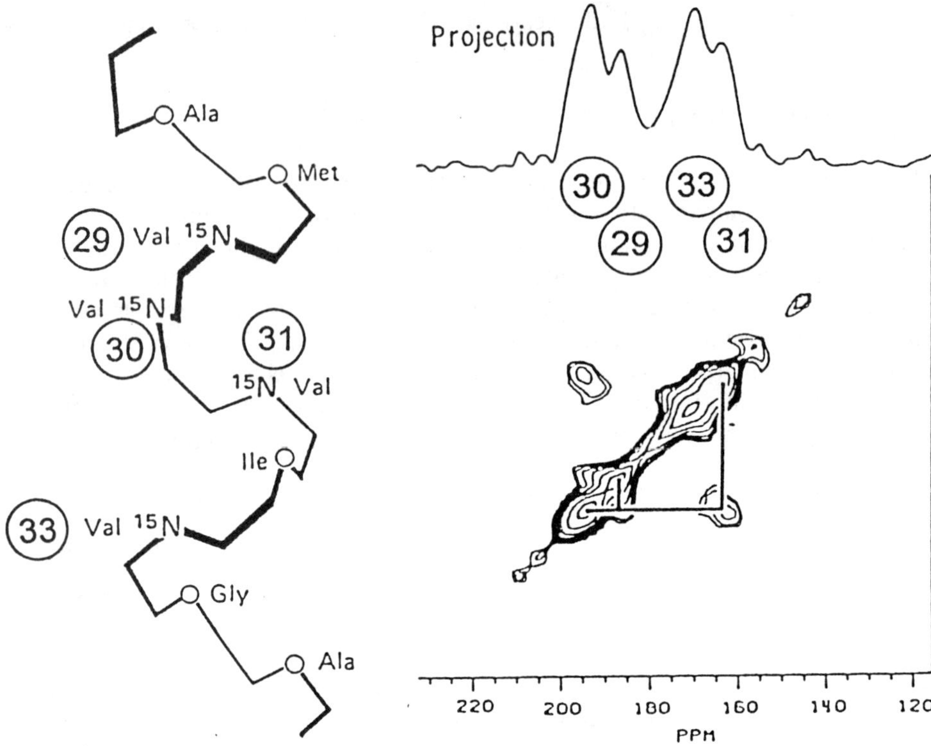

Figure 13.14 ^{15}N spin exchange in selectively ^{15}N labelled, oriented fd-virus coat protein. Left: helical backbone conformation with the labelled valine residues (29, 30, 31 and 33) indicated. Right: 2D ^{15}N spin exchange spectrum with cross peaks observed between valine residues 29 and 30, as well as 30 and 31, but not between the more distant residue 33 and the others. Adapted from Cross *et al.* (1983).

However, such a change is expected even for an unspecific distribution of the labelled additive because of the difference in the concentration of ^{13}C in the two samples: the smaller the concentrations of the label, the smaller the exchange rate (nearly zero in the limit of no ^{13}C labelling). This suggests that the reported similarity between the rates of exchange from the label to the PC phenylene carbons for the two concentrations may be fortuitous. It could be due to differences in the phenylene flip dynamics. The intensity of the evaluated peaks is actually quite small and possibly biased by baseline distortions from the large signals in the spectrum.

13.5.4 Speeding-up of ^{13}C spin diffusion

Natural, proton-driven spin diffusion is so slow that it can often not be observed due to problems with T_1 relaxation during t_m or prohibitively long measuring times due to excessively long mixing times. On the other hand, the $^{13}C-^{13}C$ couplings are typically 15 Hz so that in principle, if all other interactions had been removed, substantial transfer could be achieved within a time of only $1/(15\ \text{Hz}) \cong 70$ ms. Therefore, it is appealing to consider speeding up ^{13}C spin diffusion by manipulating the interactions in the spirit of the average-Hamiltonian theory outlined in Section 3.3. In order to be able to manipulate the ^{13}C Hamiltonian, it is necessary to remove the dominant ^{1}H dipolar effects first. However, if not 'disturbed' by $^{1}H-^{13}C$ dipolar couplings, magic-angle spinning as well as ^{13}C chemical-shift differences introduce time dependencies into the system that can average out the $^{13}C-^{13}C$ dipolar couplings (for the chemical-shift differences, the time dependence arises after transition into the interaction representation that transforms their immediate action away). To eliminate these effects that quench ^{13}C spin diffusion, different approaches have been taken by Meier, Ernst and their co-workers. 'Rotor-driven' spin diffusion (Colombo *et al.*, 1988) relies on clever combination of the time dependencies of MAS and chemical shift difference to cancel each other. This is achieved by resonant matching of the chemical shift frequency difference $\omega_0\Delta\sigma$ of two ^{13}C signals and the spinning frequency ω_r. 'Radio-frequency driven' spin diffusion (Robyr *et al.*, 1989) is based on averaging the chemical shift to zero by rf pulses, in the simplest case by a spin-lock field during t_m (see Figure 13.15(a)); under these conditions, however, nothing prevents the $^{13}C-^{13}C$ couplings from being averaged away by MAS. Therefore, the sample must be static or rotate off the magic angle during the mixing time. Both approaches require high-power irradiation during t_m, which limits the mixing time to less than 1 second on standard spectrometer systems. Additionally, in rf-driven spin diffusion experiments under the locking field during t_m, $T_{1\rho}$ decay of the ^{13}C magnetization occurs.

13.5.5 Radio-frequency-driven spin diffusion

In this approach, both the chemical shift and the $^{13}C-^{1}H$ couplings are eliminated by spin-locking the ^{13}C magnetization in the *xy* plane during the mixing time. As derived

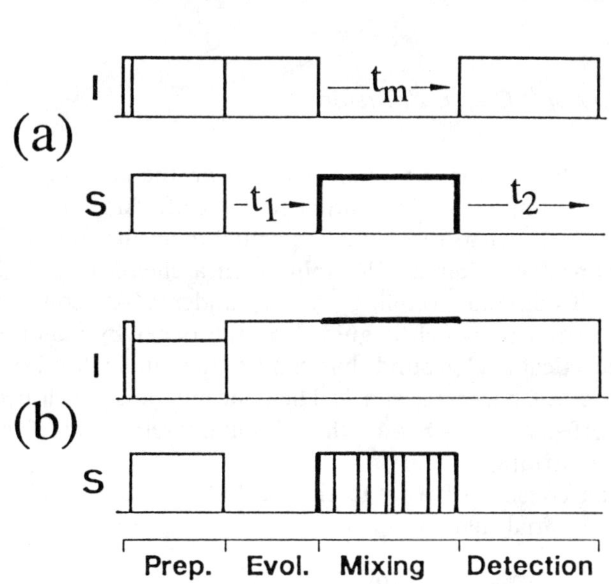

Figure 13.15 Pulse sequences for rf-driven ^{13}C spin diffusion. After cross-polarization and ^{1}H-decoupled evolution, in the mixing time t_m the magnetization is spin-locked to average to zero the chemical-shift difference and the heteronuclear dipolar Hamiltonian. (a) Simple locking field. (b) WALTZ-17 sequence for broad-band averaging of the chemical shifts and high-power ^{1}H decoupling during t_m. Adapted from Robyr *et al.* (1989).

in Section 3.4.2, under spin-locking conditions the chemical-shift and heteronuclear dipolar couplings are averaged to zero, and the $^{13}C-^{13}C$ dipolar coupling, scaled by $-\frac{1}{2}$, is the only remaining interaction. This interaction, typically amounting to 15 Hz, would be averaged out by MAS; therefore, the sample must be static or spun off the magic angle during t_m.

In practice, it was found by Robyr *et al.* (1989) that it is decisive to achieve more complete heteronuclear decoupling by irradiation on the ^1H frequency simultaneous to the ^{13}C spin lock (see Figure 13.15(b)). Of course, this must be done with a power level sufficiently far off the Hartmann–Hahn condition in order to prevent cross polarization during the mixing time. In order to average out the chemical-shift differences to the necessary extent, Robyr *et al.* (1989) employed the WALTZ-17 sequence which is particularly suited for that purpose. It is windowless and consists of 90°, 180°, 270° and 360° pulses, taking only the phases $+x$ or $-x$. After 16 such pulses, an α° pulse is added to provide an effective field.

At this point it can be recognized easily that, contrary to some statements in the literature, it is not the square of ω_D^2 in equations for the transfer probability $P_{if}(t)$ (see Section 3.7) that prevents MAS from averaging out the weak $^{13}C-^{13}C$ dipolar couplings ω_D on which ^{13}C spin diffusion relies. A term with ω_D^2 (e.g. in equation (3.48)) occurs also when the $^{13}C-^{13}C$ coupling is the only relevant interaction, and in that case MAS quenches spin diffusion immediately. The actual reason why the $^{13}C-^{13}C$ dipolar couplings survive MAS in proton-driven ^{13}C spin diffusion is the 'protection' by the stronger $^{13}C-^1H$ and $^1H-^1H$ dipolar couplings that are not spun out (see Section 3.9.5). Due to the noncommutation of the S–S and I–S couplings as well as of the I–S and the I–I couplings, all these Hamiltonians have cross-terms with one another so that MAS cannot effectively refocus (and thus remove) the effects of the S–S couplings.

13.5.6 Rotor-driven ^{13}C spin diffusion

The fact that rotor-driven spin diffusion is based on the compensation of harmonic time-dependences of sample-rotation and chemical-shift differences has already been pointed out above. In order to recognize the actual quantum-mechanical basis of this explanation, we consider a pair of ^{13}C spins with a chemical-shift difference which exceeds the $^{13}C-^{13}C$ dipolar coupling by far, under MAS conditions and proton decoupling. The treatment, which is guided by the paper by Colombo *et al.* (1988), requires some theoretical background, but can be fully understood on the basis of the relations introduced in Chapters 2 and 3. The concepts thus developed are also useful for rationalizing effects of MAS on other dipolar interactions, for instance in the context of cross polarization.

The Hamiltonian consists of a large chemical shift and a small $^{13}C-^{13}C$ dipolar term, acting on the initial density operator:

$$\hat{\rho}(0) = c_a\hat{S}_z^a + c_b\hat{S}_z^b \tag{13.37}$$

where the coefficients must be unequal, $c_a \neq c_b$, for spin exchange to be observable (a and b are the labels for the two spins). Since the chemical-shift sum Hamiltonian $\hat{H}_\Sigma(t) = \{\omega_a(t) + \omega_b(t)\}\frac{1}{2}(\hat{S}_z^a + \hat{S}_z^b)$ commutes with $\hat{\rho}(0)$ of relation (13.37) and all the other Hamiltonians (due to the Zeeman truncation), it is irrelevant for the time evolution. The chemical-shift difference Hamiltonian:

$$\hat{H}_{\Delta}(t) = \tfrac{1}{2}(\hat{S}_z^{\mathrm{a}} - \hat{S}_z^{\mathrm{b}})\{\omega_{\mathrm{a}}(t) - \omega_{\mathrm{b}}(t)\} \tag{13.38}$$

remains very important, however. ($\hat{H}_{\Delta}(t)$ and $\hat{H}_{\Sigma}(t)$ together indeed make up the total chemical-shift Hamiltonian $\hat{H}_{\Delta}(t) + \hat{H}_{\Sigma}(t) = \omega_{\mathrm{a}}(t)\hat{S}_z^{\mathrm{a}} + \omega_{\mathrm{b}}(t)\hat{S}_z^{\mathrm{b}} = \hat{H}_{\mathrm{cs}}$.)

The relevant portion of the homonuclear dipolar Hamiltonian is:

$$\hat{H}_{S,S}(t) = \omega_{\mathrm{D}}(t)(\hat{S}_x^{\mathrm{a}}\hat{S}_x^{\mathrm{b}} + \hat{S}_y^{\mathrm{a}}\hat{S}_y^{\mathrm{b}}) \tag{13.39}$$

according to Section 3.5.1. As demonstrated in Section 3.3.4, the circumstances with a strong Hamiltonian $\hat{H}_{\Delta}$ and a weak Hamiltonian $\hat{H}_{S,S}$ call for an interaction representation in terms of the strong Hamiltonian $\hat{H}_{\Delta}$, with:

$$\begin{aligned}\hat{U}(t) &= \exp\left\{-\mathrm{i}\int_0^t \hat{H}_{\Delta}(t')\,\mathrm{d}t'\right\}\\ &= \exp\left\{-\mathrm{i}\tfrac{1}{2}(\hat{S}_z^{\mathrm{a}} - \hat{S}_z^{\mathrm{b}})\int_0^t [\omega_{\mathrm{a}}(t') - \omega_{\mathrm{b}}(t')]\,\mathrm{d}t'\right\}\\ &= \exp\{-\mathrm{i}(\hat{S}_z^{\mathrm{a}} - \hat{S}_z^{\mathrm{b}})\tfrac{1}{2}\Phi_{\Delta}(t)\}\end{aligned} \tag{13.40}$$

where the time-dependence and self-commutation of $\hat{H}_{\Delta}$ has been taken into account. In the last step, the phase angle $\Phi_{\Delta}(t) := \Phi_{\mathrm{a}}(t) - \Phi_{\mathrm{a}}(0) - \Phi_{\mathrm{b}}(t) + \Phi_{\mathrm{b}}(0)$ for the chemical-shift-difference evolution under MAS was introduced. The explicit expression for $\Phi(t)$ in terms of the parameter δ and η, as well as the Euler angles α and β, could be looked up in Section 3.9.1. According to Section 3.3.4, the dipolar coupling Hamiltonian in the interaction frame is:

$$\hat{H}_{S,S}^{\prime}(t) = \hat{U}^{-1}(t)\hat{H}_{S,S}(t)\hat{U}(t). \tag{13.41}$$

The propagator $\hat{U}(t)$ of equation (13.40) represents rotations about the z axis of the spin operators, but with opposite directions for a and b; this means that $\hat{S}_x^{\mathrm{a}}$ in the dipolar Hamiltonian (13.39) must be replaced by $\{\hat{S}_x^{\mathrm{a}}\cos[\tfrac{1}{2}\Phi_{\Delta}(t)] - \hat{S}_y^{\mathrm{a}}\sin[\tfrac{1}{2}\Phi_{\Delta}(t)]\}$, $\hat{S}_x^{\mathrm{b}}$ by $\{\hat{S}_x^{\mathrm{b}}\cos[\tfrac{1}{2}\Phi_{\Delta}(t)] + \hat{S}_y^{\mathrm{b}}\sin[\tfrac{1}{2}\Phi_{\Delta}(t)]\}$, and analogously for $\hat{S}_y^{\mathrm{a}}$ and $\hat{S}_y^{\mathrm{b}}$. After multiplication of these terms and application of trigonometric relations of Appendix A this produces:

$$\hat{H}_{S,S}^{\prime}(t) = \omega_{\mathrm{D}}(t)\{(\hat{S}_x^{\mathrm{a}}\hat{S}_x^{\mathrm{b}} + \hat{S}_y^{\mathrm{a}}\hat{S}_y^{\mathrm{b}})\cos[\Phi_{\Delta}(t)] + (\hat{S}_x^{\mathrm{a}}\hat{S}_y^{\mathrm{b}} - \hat{S}_y^{\mathrm{a}}\hat{S}_x^{\mathrm{b}})\sin[\Phi_{\Delta}(t)]\}. \tag{13.42}$$

For obtaining the time average (i.e. the secular terms) of $\hat{H}_{S,S}^{\prime}(t)$, we have to specify the expression for $\omega_{\mathrm{D}}(t)$:

$$\omega_{\mathrm{D}}(t) = C_1^{\mathrm{D}}\cos(\gamma^{\mathrm{D}} + \omega_{\mathrm{r}}t) + C_2^{\mathrm{D}}\cos(2\gamma^{\mathrm{D}} + 2\omega_{\mathrm{r}}t) \tag{13.43}$$

obtained from equation (3.74) in the case of $\eta = 0$ (which applies for spin-pair dipolar coupling). Combining this with the Fourier expansion:

$$\exp\{-\mathrm{i}\Phi_{\Delta}(t)\} = \mathrm{e}^{-\mathrm{i}\Delta_{\mathrm{iso}}t}\sum_{N=-\infty}^{\infty} I_N\,\mathrm{e}^{\mathrm{i}N\omega_{\mathrm{r}}t} \tag{13.44}$$

and using the simple reformulation:

$$C_1^{\mathrm{D}} \cos(\gamma^{\mathrm{D}} + \omega_{\mathrm{r}}t) = \tfrac{1}{2} \sum_{M=-1,1} C_1^{\mathrm{D}}\, \mathrm{e}^{\mathrm{i}(M\gamma^{\mathrm{D}} + M\omega_{\mathrm{r}}t)}$$

as well as an analogous relation for the $\cos(2\gamma^{\mathrm{D}} + 2\omega_{\mathrm{r}}t)$ term in $\omega_{\mathrm{D}}(t)$ of equation (13.43), we can write:

$$\omega_{\mathrm{D}}(t) \exp\{-\mathrm{i}\Phi_{\Delta}(t)\} = \tfrac{1}{2} \sum_{M=\pm 1, \pm 2} \sum_{N=-\infty}^{\infty} C_{|M|}^{\mathrm{D}} I_N\, \mathrm{e}^{\mathrm{i}M\gamma^{\mathrm{D}}}\, \mathrm{e}^{\mathrm{i}\{(N+M)\omega_{\mathrm{r}} - \Delta_{\mathrm{iso}}\}t}. \quad (13.45)$$

The real and imaginary parts of this expression are proportional to the cosine and sine terms respectively, of $\hat{H}'_{S,S}(t)$ in equation (13.42). Time-independent terms in (13.45) occur for:

$$(N+M)\omega_{\mathrm{r}} = \Delta_{\mathrm{iso}} \quad (13.46)$$

that is, when the isotropic-shift difference Δ_{iso} is an integer multiple $(N+M)\omega_{\mathrm{r}} = K\omega_{\mathrm{r}}$ of the rotation frequency ω_{r}. In that case, the average consists of the terms with $N = K - M$ in equation (13.45):

$$\overline{\omega_{\mathrm{D}}(t)\exp\{-\mathrm{i}\Phi_{\Delta}(t)\}} = \tfrac{1}{2} \sum_{M=\pm 1, \pm 2} C_{|M|}^{\mathrm{D}} I_{K-M}\, \mathrm{e}^{\mathrm{i}M\gamma^{\mathrm{D}}}. \quad (13.47)$$

With the time-independent $\overline{\hat{H}'_{S,S}(t)}$ obtained by inserting the real and imaginary parts of the average (13.47) into the dipolar Hamiltonian (13.42), we can calculate the time-evolution according to equation (2.75), using the following commutators (see equation (3.39)):

$$[S_x^{\mathrm{a}}S_x^{\mathrm{b}} + S_y^{\mathrm{a}}S_y^{\mathrm{b}}, S_y^{\mathrm{a}}S_x^{\mathrm{b}} - S_x^{\mathrm{a}}S_y^{\mathrm{b}}] = \mathrm{i}\tfrac{1}{2}(S_z^{\mathrm{a}} - S_z^{\mathrm{b}})$$

$$[\tfrac{1}{2}(S_z^{\mathrm{a}} - S_z^{\mathrm{b}}), S_x^{\mathrm{a}}S_x^{\mathrm{b}} + S_y^{\mathrm{a}}S_y^{\mathrm{b}}] = \mathrm{i}(S_x^{\mathrm{a}}S_y^{\mathrm{b}} - S_y^{\mathrm{a}}S_x^{\mathrm{b}})$$

$$[S_y^{\mathrm{a}}S_x^{\mathrm{b}} - S_x^{\mathrm{a}}S_y^{\mathrm{b}}, \tfrac{1}{2}(S_z^{\mathrm{a}} - S_z^{\mathrm{b}})] = \mathrm{i}(S_x^{\mathrm{a}}S_x^{\mathrm{b}} + S_y^{\mathrm{a}}S_y^{\mathrm{b}}).$$

As shown in Section 3.5.2, the result is an oscillation of the z magnetization between spin a and spin b with the frequency:

$$\overline{\omega_{\mathrm{D}}} = \tfrac{1}{2} \sum_{M=\pm 1, \pm 2} C_{|M|}^{\mathrm{D}}\, | I_{K-M} |. \quad (13.48)$$

While in this expression the $C_{|M|}^{\mathrm{D}}$ coefficients are independent of the rotation frequency, the I_N can be considered as the sideband intensities of the chemical-shift difference tensor and thus depend on ω_{r}. Nevertheless, even in the fast-rotation limit where only the centreband intensity I_0 remains, rotor-driven spin-diffusion can work, but according to equation (13.46) only for $(0+M)\omega_{\mathrm{r}} = \Delta_{\mathrm{iso}}$, $M = \pm 1, \pm 2$. The effective coupling frequency (13.48) in this case (i.e. with $K = M$ and $I_0 = 1$) is $\omega_{\mathrm{D}}^{\mathrm{f}} = \tfrac{1}{2}C_{|K|}^{\mathrm{D}}$. As an heuristic argument, it can be said that the coupling is not spun out because the chemical-shift difference $\Delta_{\mathrm{iso}} \geqslant \omega_{\mathrm{r}}$ is not affected by fast spinning.

Having derived a quasi-static behaviour with an effective coupling frequency $\overline{\omega_{\mathrm{D}}}$ for the two-spin system, the treatment can be generalized to a multispin system as shown in Section 3.7. A certain complication arises because there is a significant fraction of small isolated clusters of ^{13}C spins which show some oscillatory behaviour, while among the rest the magnetization will behave more or less diffusively.

Figure 13.16 shows one of the first applications of rotor-driven spin diffusion. By means of the DANTE sequence (Section 3.8), one line in the MAS spectrum of threonine is selectively inverted. The rotation frequency ω_{r} is set such that $n\omega_{\mathrm{r}}$ is equal

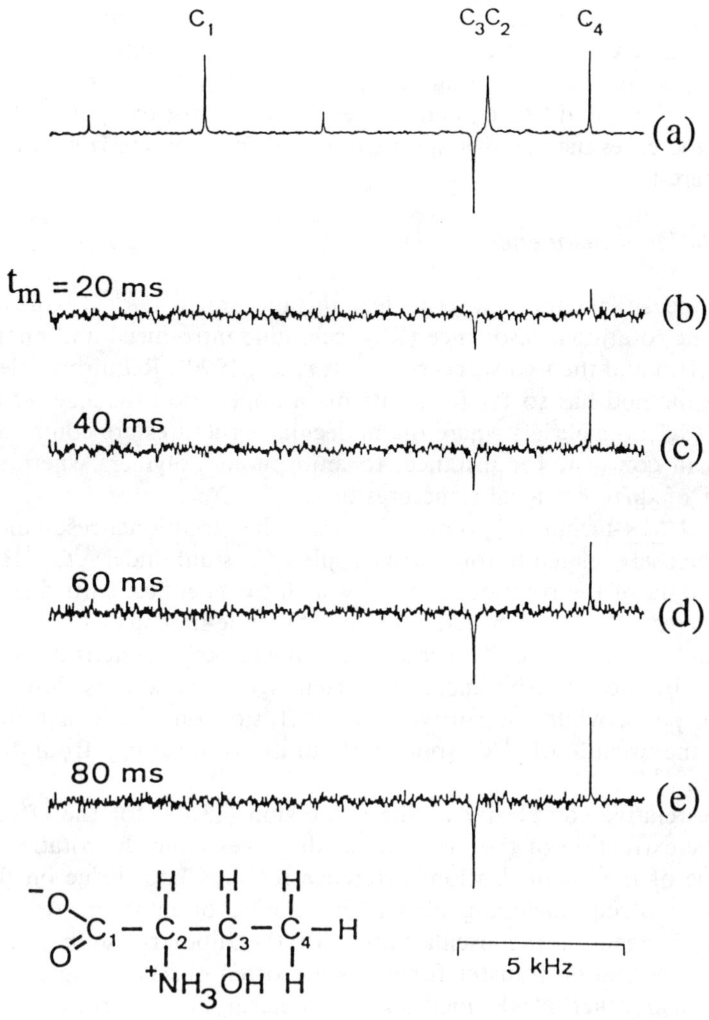

Figure 13.16 Rotor-driven spin diffusion of threonine at a spinning speed of 3511 kHz which matches the shift difference between the C_3 and C_4 resonances (Colombo *et al.*, 1988). (a) Selective inversion of the C_3 resonance after application of the DANTE sequence (see Section 3.8). (b)–(e) Difference spectra (1024 scans), scaled up by a factor of 10, for different t_m values show spin diffusion between C_3 and C_4 resonances which is speeded up by a factor of 200 compared with proton-driven spin diffusion. Adapted from Colombo *et al.* (1988).

to the chemical-shift difference between the inverted line and another line of interest. The decrease of the intensity of that line is monitored as a function of the mixing time; Figures 13.16(b)–(e) show the difference spectra, obtained by subtracting the spectrum with decoupling during the mixing time from the analogous spectrum without such decoupling, where only the slow ^{1}H-driven spin diffusion occurs.

The technique of rotor-driven spin diffusion has certain limitations. It is hardly applicable if the splitting between the lines of interest is only a few hundred Hz, since this case would require correspondingly slow spinning. In addition, due to the sharpness of the rotational-resonance condition only a fraction of the inhomogeneously broadened MAS lines found in many polymeric materials can participate in the rotor-driven spin diffusion, so that the process will not be very efficient in these materials. That the spin diffusion can be measured only for one pair of lines at a time considerably increases the measuring time required for characterizing a heterogeneous system of interest.

13.5.7 Rotational resonance

Very precise measurements of internuclear distances of $^{13}C-^{13}C$ pairs in solids are achieved by the rotational-resonance (R^2) technique introduced and analysed mainly by Levitt, Griffin and their co-workers (Levitt *et al.*, 1990). Requiring selective double labelling, the method has so far found its main application (Spencer *et al.*, 1991) to non-fluid biological materials where the molecular structures are complex but usually well-defined, in contrast, for instance, to amorphous polymers where apparently a large number of different local structures exist.

From the NMR-technical point of view, the rotational-resonance exchange experiment is closely related to rotor-driven spin diffusion: under $^{13}C-^{1}H$ decoupling, through matching of the rotation frequency with the chemical-shift difference of the species of interest (i.e. the labelled sites in R^2), time-independent $^{13}C-^{13}C$ dipolar couplings result which naturally produce a transfer of magnetization between the carbon spins. In the R^2 experiment, the exchange behaviour is dominated by the $^{13}C-^{13}C$ spin pairs, while rotor-driven spin diffusion on ^{13}C in natural-abundance depends on the wealth of ^{13}C spin configurations resulting from the statistical distribution of ^{13}C nuclei.

Due to the relative complexity of the expression (13.48) for the effective dipolar frequency, the extraction of the internuclear distances from the rotational-resonance t_m-dependence of the magnetization difference requires knowledge on the chemical-shift tensors involved, including also their relative orientation. As an additional complication, instead of the oscillations with the dipolar-coupling frequency ω_D, monotonic magnetization-transfer functions are often observed experimentally.

There are also other NMR methods for determining internuclear distances in $^{13}C-^{13}C$ pairs. One example is CPGM (Carr–Purcell–Gill–Meiboom) dipolar spectroscopy, where a train of 180° ^{13}C pulses, under $^{13}C-^{1}H$ decoupling and stroboscopic signal detection, refocuses the chemical shift, while leaving the $^{13}C-^{13}C$ interactions unaffected. In the homonuclear dipolar Pake pattern that results, the distance of the singularities is a sensitive measure of the internuclear distance, without other parameters entering the analysis. However, in complex systems as investigated by Griffin and co-workers, the peaks of the labelled sites are often small even in the MAS spectrum, in fact of lower intensity than other, natural abundance, signals. This makes it difficult to detect the dipolar-coupling effects under experimental conditions other than fast MAS as employed for the R^2 technique.

Figure 13.17 displays R^2 data obtained for a doubly labelled peptide fragment of the noncrystalline amyloid-forming protein of Alzheimer's disease (Spencer *et al.*, 1991). The plotted decrease of the magnetization difference as a function of the mixing time after selective inversion of one of the lines is significantly faster than expected for the commonly assumed structure of an ideal antiparallel β sheet. Instead of the expected distance of 4.74 Å, the experimental data yield 4.0 Å, which strongly suggests an

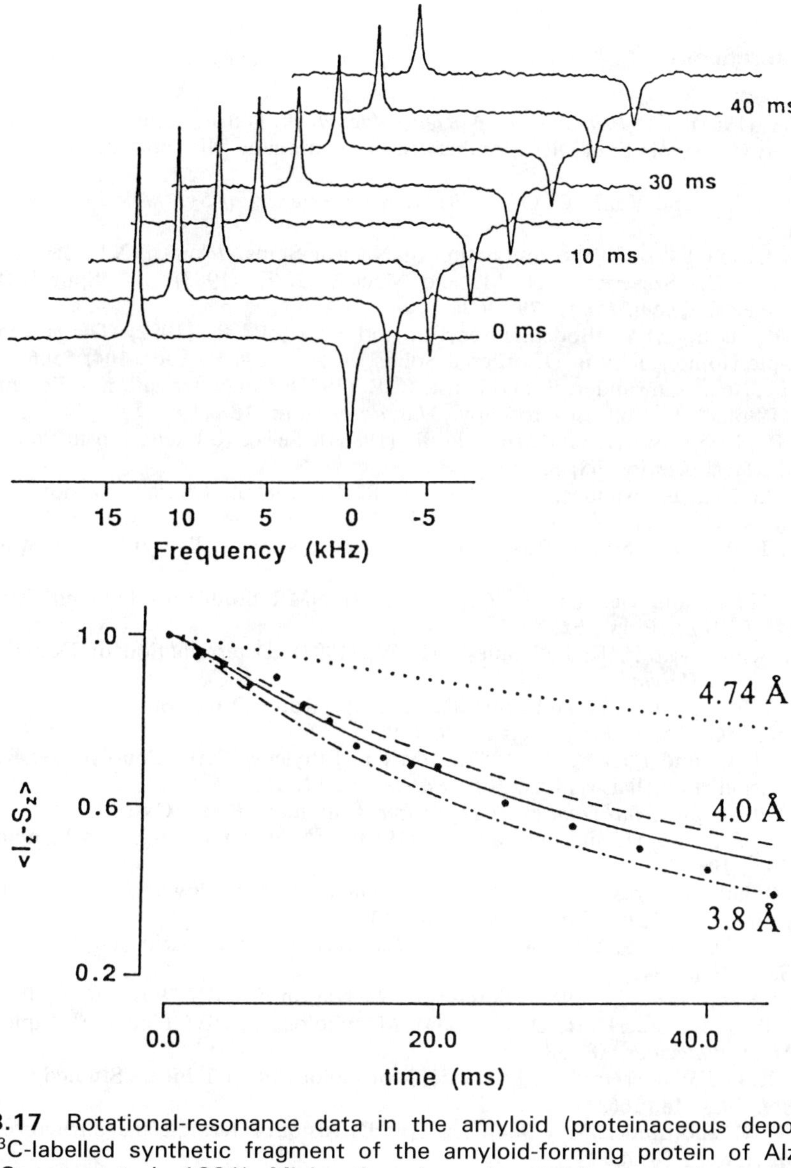

Figure 13.17 Rotational-resonance data in the amyloid (proteinaceous deposit) of a doubly ^{13}C-labelled synthetic fragment of the amyloid-forming protein of Alzheimer's disease (Spencer *et al.*, 1991). Mixing-time dependence, after selective inversion at a rotation frequency of 12.487 kHz, of the resonances of the labelled α carbon of a glycine residue and the labelled carbonyl in the neighbouring glycine, which are separated by four bonds. The distance obtained from the fit of the rotational-resonance data is 4.0 ± 0.2 Å, while 4.74 Å are expected for an ideal antiparallelized β strand. This indicates an unusual conformation of this peptide, which may be related to the high stability of the amyloid. Adapted from Spencer *et al.* (1991).

unusual peptide conformation which may be linked to the (fatal) stability of the amyloid deposits.

13.6 References

Abragam, A. (1961). *The Principles of Nuclear Magnetism*. Oxford University Press, London.

Assink, R. A. (1978). ^{1}H Spin Diffusion between Polyurethane Microphases. *Macromolecules*, **11**, 1233.

Baltá-Calleja, F. J. and Vonk, C. G. (1989). *X-ray Scattering of Synthetic Polymers*. Elsevier, Amsterdam.

Bloembergen, N. (1949). On the Interaction of Nuclear Spins. *Physica*, **XV**, 386.

Bronniman, C. E., Szeverenyi, N. M. and Maciel, G. E. (1983). ^{13}C Spin Diffusion of Adamantane. *J. Chem. Phys.*, **79**, 3694.

Caravatti, P., Deli, J. A., Bodenhausen, G. and Ernst, R. R. (1982). Direct Evidence of Microscopic Homogeneity in Disordered Solids. *J. Am. Chem. Soc.*, **104**, 5506.

Caravatti, P., Neuenschwander, P. and Ernst, R. R. (1985). Characterization of Polymer Blends by 2D ^{1}H Spin Diffusion Spectroscopy. *Macromolecules*, **18**, 119.

Caravatti, P., Levitt, M. H. and Ernst, R. R. (1986a). Selective Excitation in Solid-State ^{1}H NMR. *J. Magn. Reson.*, **68**, 323.

Caravatti, P., Neuenschwander, P. and Ernst, R. R. (1986b). Characterization of Polymer Blends by Selective ^{1}H Spin-Diffusion NMR. *Macromolecules*, **19**, 1889.

Cheung, T. T. P. (1981). Spin Diffusion in NMR in Solids. *Phys. Rev. B: Condens. Matter* **23**, 1404.

Cheung, T. T. P. and Gerstein, B. C. (1981). ^{1}H NMR Studies of Domain Structures in Polymers. *J. Appl. Phys.*, **52**, 5517.

Clauss, J., Schmidt-Rohr, K. and Spiess, H. W. (1993). Determination of Domain Sizes in Polymers. *Acta Polymer.*, **44**, 1.

Colombo, M. G., Meier, B. H. and Ernst, R. R. (1988). Rotor-Driven Spin Diffusion in Natural-Abundance ^{13}C. *Chem. Phys. Lett.*, **146**, 189.

Colquhoun, I. C. and Packer, K. (1987). NMR in Ethylene/Olefin Copolymers: Relaxation, Spin-Diffusion, Lamellar Widths. *Brit. Polym. J.*, **19**, 151.

Crank, J. (1975). *The Mathematics of Diffusion*. Clarendon Press, Oxford.

Cross, T. A., Frey, M. H. and Opella, S. J. (1983). ^{15}N Spin Exchange in a Protein. *J. Am. Chem. Soc.*, **105**, 7471.

Edzes, H. T. and Bernards, J. P. C. (1984). 2D NMR in Static Powders: Interchain ^{13}C Spin Exchange in PE. *J. Am. Chem. Soc.*, **106**, 1515.

Frey, M. H. and Opella, S. J. (1984). ^{13}C Spin Exchange in Amino Acids and Peptides. *J. Am. Chem. Soc.*, **106**, 4942.

Goldman, M. and Shen, L. (1966). Spin–Spin Relaxation in LaF_3. *Phys. Rev.*, **144**, 321.

Havens, J. R. and VanderHart, D. L. (1984). Morphology of PET Fibers: Multiple-Pulse ^{1}H NMR. *Macromolecules*, **18**, 589.

Havens, J. R. and VanderHart, D. L. (1985). Morphology of PET Fibers Studied by ^{1}H NMR. *Macromolecules*, **18**, 1663.

Henrichs, P. M. and Linder, M. (1984). ^{13}C Spin Diffusion: Intermolecular Structure in Solids. *J. Magn. Reson.*, **58**, 458.

Kenwright, A. M. and Packer, K. J. (1990). On T_1 Cancellation Schemes in Goldman–Shen Experiments. *Chem. Phys. Lett.*, **173**, 471.

Kimura, T., Neki, K., Tamura, N., Horii, F., Nakagawa, M. and Odani, H. (1992). ^{13}C NMR Study of ^{1}H Spin Diffusion and T_1 in Polymers. *Polymer*, **33**, 493.

Levitt, M. H., Raleigh, D. P., Creuzet, F. and Griffin, R. G. (1990). Theory of Homonuclear Spin Pair Systems in Rotating Solids, **92**, 6347.

Linder, M., Henrichs, P. M., Hewitt, J. M. and Massa, D. J. (1985). ^{13}C Spin Diffusion: Miscibility of Polymer Blends. *J. Chem. Phys.*, **82**, 1585.

Packer, K. J. and Pope, J. M. (1983). Distortion of NMR Spectra by Multiple Quantum Coherence Effects from Goldman–Shen Sequences. *J. Magn. Reson.*, **55**, 378.

Packer, K. J., Pope, J. M., Yeung, R. R. and Cudby, M. E. A. (1984). Morphology, ^{1}H Spectra and Relaxation in Polyolefins. *J. Polym. Sci., Polym. Phys. Ed.*, **22**, 589.

Robyr, P., Meier, B. H. and Ernst, R. R. (1989). RF-Driven Spin Diffusion in Solids. *Chem. Phys. Lett.*, **162**, 417.

Robyr, P., Meier, B. H. and Ernst, R. R. (1991). Tensor Correlation by 2D Spin-Diffusion Powder NMR: H-Bond in Benzoic Acid. *Chem. Phys. Lett.*, **187**, 471.

Roy, A. K., Inglefield, P. T., Shibata, J. H. and Jones, A. A. (1987). Intermolecular Structure in an Antiplasticized Glass by NMR. *Macromolecules*, **1434**, 20.

Schajor, W., Pislewski, N., Zimmermann, H. and Haberlen, U. (1980). Spin Diffusion and T_1 of ^{2}H in Solids. *Chem. Phys. Lett.*, **76**, 409.

Schmidt-Rohr, K., Clauss, J., Blümich, B. and Spiess, H. W. (1990). Miscibility of Polymer Blends. ^{1}H Spin Diffusion, ^{13}C NMR Detection. *Magn. Reson. in Chem.*, **28**, 3.

Schmidt-Rohr, K., Clauss, J. and Spiess, H. W. (1992). Correlation of Structure, Mobility, and Morphology by 2D WISE NMR. *Macromolecules*, **25**, 3273.

Spencer, R. G. S., Halverson, K. J., Auger, M., McDermott, A. E., Griffin, R. G. and Lansbury, P. T. (1991). Application of Solid-State NMR to the Determination of Protein Structure. *Biochemistry*, **30**, 10382.

Szeverenyi, N. M., Sullivan, M. J. and Maciel, G. E. (1982). ^{13}C Spin Exchange by 2D FT ^{13}C CP/MAS. *J. Magn. Reson.*, **47**, 462.

Tycko, R. and Dabbagh, G. (1991). Molecular Orientational Ordering in 3 Forms of Solid Methanol. *J. Am. Chem. Soc.*, **113**, 3592.

VanderHart, D. L. (1987). Spin Exchange for Natural-Abundance ^{13}C. *J. Magn. Reson.*, **72**, 13.

Zhang, S., Meier, B. H. and Ernst, R. R. (1992). Polarization Echoes in NMR. *Phys. Rev. Lett.*, **69**, 2149.

Appendices

In these appendices, we compile various mathematical topics formulated from the perspective of NMR applications. In particular, basic relations and topics that recur in different contexts in solid-state NMR have been assembled here, thus avoiding awkward cross-references between various chapters. Most of these issues are well known—or at least they are expected to be so! Therefore, they represent the basis for understanding many NMR papers or texts. The search for these topics in mathematical texts is often tedious since they are usually formulated in a different notation and the relations important for NMR are 'hidden' by other subjects and by detailed proofs. It is hoped that these appendices not only make it easier to follow the derivations in this book but also represent useful references for everyday NMR work and studies.

Appendix A Trigonometric and geometric relations, complex numbers

Trigonometric relations are needed for deriving and exploiting the angle-dependent frequencies, as well as for the analysis of multidimensional solid-state NMR spectra and of molecular geometries.

A.1 Simple trigonometry

$$\sin^2\theta + \cos^2\theta = 1 \qquad \sin^2\theta = 1 - \cos^2\theta \tag{A.1}$$

$$\cos 2\theta = \cos^2\theta - \sin^2\theta \qquad \sin^2\theta = \cos^2\theta - \cos 2\theta \tag{A.2}$$

$$\sin^2\theta = \tfrac{1}{2}(1 - \cos 2\theta) \tag{A.3}$$

$$\cos^2\theta = \tfrac{1}{2}(1 + \cos 2\theta) \tag{A.4}$$

$$\cos(\beta \pm \theta) = \cos\beta\cos\theta \mp \sin\beta\sin\theta \tag{A.5}$$

$$\sin(\beta \pm \theta) = \sin\beta\cos\theta \pm \cos\beta\sin\theta \tag{A.6}$$

$$\cos\beta\cos\theta = \tfrac{1}{2}\cos(\beta - \theta) + \tfrac{1}{2}\cos(\beta + \theta) \tag{A.7}$$

$$\sin\beta\sin\theta = \tfrac{1}{2}\cos(\beta - \theta) - \tfrac{1}{2}\cos(\beta + \theta) \tag{A.8}$$

$$\cos(-\beta) = \cos\beta \tag{A.9}$$

$$\sin(-\beta) = -\sin\beta \tag{A.10}$$

Transition from Cartesian to polar coordinates:

$$\begin{aligned} \vec{\mathbf{a}} &:= (a_1, a_2, a_3) \\ |\vec{\mathbf{a}}| &= \sqrt{a_1^2 + a_2^2 + a_3^2} \end{aligned} \tag{A.11}$$

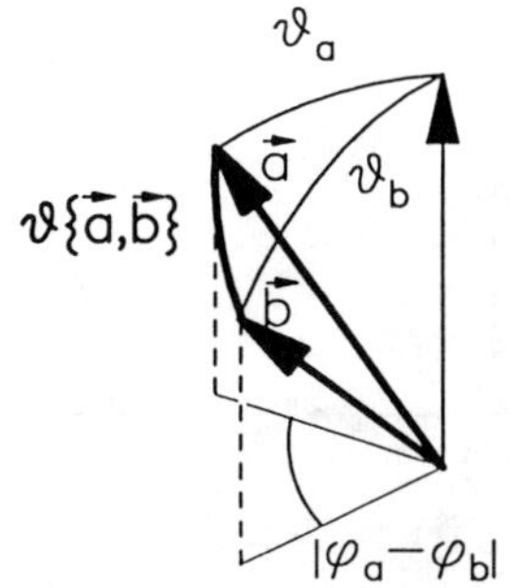

Figure A.1 How to find the angle $\theta\{\vec{\mathbf{a}}, \vec{\mathbf{b}}\}$ between two axes $\vec{\mathbf{a}}$ and $\vec{\mathbf{b}}$ given in terms of their polar coordinates (θ_a, ϕ_a) and (θ_b, ϕ_b) in a coordinate system of which only the z axis needs to be specified.

$$a_1 = |\vec{\mathbf{a}}| \cos \phi_a \sin \theta_a$$
$$a_2 = |\vec{\mathbf{a}}| \sin \phi_a \sin \theta_a \tag{A.12}$$
$$a_3 = |\vec{\mathbf{a}}| \cos \theta_a.$$

Scalar product of vectors $\vec{\mathbf{a}}$ and $\vec{\mathbf{b}}$:

$$\vec{\mathbf{a}}\cdot\vec{\mathbf{b}} := a_1 b_1 + a_2 b_2 + a_3 b_3 = |\vec{\mathbf{a}}|\,|\vec{\mathbf{b}}| \cos(\theta\{\vec{\mathbf{a}}, \vec{\mathbf{b}}\}). \tag{A.13}$$

Relative angle $\theta\{\vec{\mathbf{a}}, \vec{\mathbf{b}}\}$, see Figure A.1, between vectors $\vec{\mathbf{a}}$ and $\vec{\mathbf{b}}$ (obtained from the scalar product (A.13) with the coordinates (A.12), also using (A.5)):

$$\cos(\theta\{\vec{\mathbf{a}}, \vec{\mathbf{b}}\}) = \cos(\phi_a - \phi_b) \sin \theta_a \sin \theta_b + \cos \theta_a \cos \theta_b. \tag{A.14}$$

This relation is used many times in the text. Typically, three axes are considered and the angle between axes 1 and 3 is to be determined from the angles between 1 and 2, and between 2 and 3. For applying equation (A.14), axis 2 is chosen as the z axis of the coordinate system and axes 1 and 3 correspond to the vectors $\vec{\mathbf{a}}$ and $\vec{\mathbf{b}}$, respectively.

A.2 Complex numbers and complex exponentials

The complex number:

$$z = \mathrm{Re}(z) + \mathrm{i}\ \mathrm{Im}(z) \tag{A.15}$$

can be displayed in terms of a two-dimensional vector $[\mathrm{Re}(z), \mathrm{Im}(z)]$ (see Figure A.2). The imaginary unit has the following properties often used in the text:

$$\mathrm{i}^2 = -1 \Rightarrow \frac{1}{\mathrm{i}} = -\mathrm{i}.$$

The complex conjugate z^* is defined by:

$$z^* := \mathrm{Re}(z) - \mathrm{i}\ \mathrm{Im}(z). \tag{A.16}$$

If z is a more complicated expression, it is useful to realize that z^* can be obtained by replacing all i in z by $(-\mathrm{i})$.

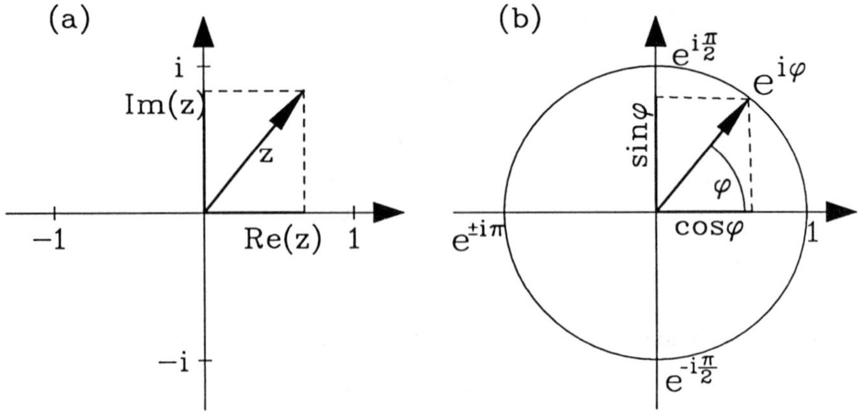

Figure A.2 (a) Vector representation of a complex number z. (b) Vector representation of a complex exponential $\exp(i\phi)$.

From relations (A.15) and (A.16), it follows that:

$$z \text{ is a real number} \Leftrightarrow z = \mathrm{Re}(z) \Leftrightarrow z = z^*. \tag{A.17}$$

The absolute value $|z|$ is defined by:

$$|z|^2 := \{\mathrm{Re}(z)\}^2 + \{\mathrm{Im}(z)\}^2 \tag{A.18}$$

$$|z|^2 = zz^*. \tag{A.19}$$

The complex exponential $e^{i\omega t}$ or $\exp(i\omega t)$ provides virtually indispensable simplifications of many calculations that involve periodic processes, and in particular for Fourier transformations (Chapter 4). It is related to the cosine and sine functions according to Euler's formula:

$$e^{i\omega t} = \cos(\omega t) + i\,\sin(\omega t) \tag{A.20}$$

$$\cos(\omega t) = \tfrac{1}{2}\,(e^{i\omega t} + e^{-i\omega t}) \tag{A.21}$$

$$\sin(\omega t) = \frac{1}{2i}\,(e^{i\omega t} - e^{-i\omega t}). \tag{A.22}$$

As mentioned below equation (A.16), the complex conjugate is obtained by inverting the sign of i:

$$(e^{i\omega t})^* = e^{-i\omega t}. \tag{A.23}$$

According to (A.19) and (A.23):

$$|e^{i\omega t}| = 1. \tag{A.24}$$

The following series expansion can serve as the actual definition of the exponential function:

$$e^{i\omega t} = \sum_{k=0}^{\infty} \frac{(i\omega t)^k}{k!} = 1 + i\omega t - \tfrac{1}{2}\omega^2 t^2 + \ldots \tag{A.25}$$

The related series expansions of cosine and sine are the basis of (A.20)–(A.22):

$$\cos \omega t = \mathrm{Re}(e^{i\omega t}) = \sum_{m=0}^{\infty} \frac{(i\omega t)^{2m}}{(2m)!} = 1 - \tfrac{1}{2}\omega^2 t^2 + \ldots \tag{A.26}$$

$$\sin \omega t = \mathrm{Im}(e^{i\omega t}) = \frac{1}{i} \sum_{m=0}^{\infty} \frac{(i\omega t)^{2m+1}}{(2m+1)!} = \omega t - \ldots \,. \tag{A.27}$$

A.3 Miscellaneous

Decomposition of $(\cos\gamma)^k$ into $\cos(m\gamma)$ is required in Section 12.4.2. Examples for small k are:

$$(\cos\gamma)^2 = \tfrac{1}{2} + \tfrac{1}{2}\cos(2\gamma) \qquad \text{(see (A.4))} \tag{A.28a}$$

$$(\cos\gamma)^3 = \tfrac{3}{4}\cos\gamma + \tfrac{1}{4}\cos(3\gamma) \qquad \text{(see (A.4) and (A.7))} \tag{A.28b}$$

$$(\cos\gamma)^4 = \tfrac{3}{8} + \tfrac{1}{2}\cos(2\gamma) + \tfrac{1}{8}\cos(4\gamma). \tag{A.28c}$$

No sine terms occur, and in the $\cos(m\gamma)$ terms the maximum value for m is obviously equal to k. Therefore, we expect:

$$(\cos\gamma)^k = \sum_{m=0}^{k} c_m \cos(m\gamma). \tag{A.29}$$

An indication of proof by complete induction follows. As shown in (A.28), for small k relation (A.29) is valid:

$$\text{if } (\cos\gamma)^{k-1} = \sum_{m=0}^{k-1} \tilde{c}_m \cos(m\gamma), \text{ we find}$$

$$(\cos\gamma)^k = (\cos\gamma)^{k-1}\cos\gamma = \sum_{m=0}^{k-1} \tilde{c}_m \cos(m\gamma)\cos\gamma$$

$$\overset{(A.7)}{=} \tilde{c}_0 \cos\gamma + \sum_{m=1}^{k-1} \tilde{c}_m\{\tfrac{1}{2}\cos(m\gamma-\gamma) + \tfrac{1}{2}\cos(m\gamma+\gamma)\} = \sum_{m=0}^{k} c_m \cos(m\gamma)$$

where every other coefficient vanishes, and for the nonvanishing coefficients we have $c_0 = \tfrac{1}{2}\tilde{c}_1$, $c_1 = \tilde{c}_0 + \tfrac{1}{2}\tilde{c}_2$ and $c_m = \tfrac{1}{2}(\tilde{c}_{m-1} + \tilde{c}_{m+1})$ otherwise, up to $c_k = \tfrac{1}{2}\tilde{c}_{k-1}$. These relations can be verified in the examples (A.28).

The sum of cosine and sine terms of identical argument α is a cosine or sine function of the same argument α, but with a phase-shift and with a different amplitude:

$$\cos\alpha + \sin\alpha = \sqrt{2}\cos(\alpha - 45^\circ) = \sqrt{2}\sin(\alpha + 45^\circ) \tag{A.30}$$

$$C\cos\alpha + S\sin\alpha = \sqrt{C^2 + S^2}\cos(\alpha - \phi_0) \qquad \tan(\phi_0) = \frac{S}{C}. \tag{A.31}$$

Proof:

$$C\cos\alpha + S\sin\alpha \overset{(A.21/22)}{=} \tfrac{1}{2}(C - iS)\, e^{i\alpha} + \tfrac{1}{2}(C + iS)\, e^{-i\alpha}$$

$$= \tfrac{1}{2}A\, e^{-i\phi_0} e^{i\alpha} + \tfrac{1}{2}A\, e^{i\phi_0} e^{-i\alpha} \overset{(A.21)}{=} A\cos(\alpha - \phi_0)$$

where $C = A\,\mathrm{Re}(e^{i\phi_0}) = A\cos\phi_0$ and $S = A\,\mathrm{Im}(e^{i\phi_0}) = A\sin\phi_0$ so that $C^2 + S^2 = A^2$ and $S/C = \sin\phi_0/\cos\phi_0 = \tan\phi_0$.

Appendix B Rotations and Euler angles

Keywords: *active vs. passive rotations, mathematically positive rotations, Euler angle determination, Euler rotation matrix, rotations of spin operators, Cartesian and spherical tensors, Wigner rotation matrices, addition theorems.*

Many NMR experiments and analyses involve rotations of one kind or another. Examples include the transition to the rotating frame, pulse rotations, sample rotation techniques (MAS, OMAS, DAS, DOR), and rotations of molecular segments.

B.1 Active vs. passive rotations

The rotations discussed here involve transitions from coordinates before the rotation (unprimed), to coordinates afterwards (primed). On a fundamental level, it can be important to distinguish between active and passive rotations (Figure B.1):

Active rotation

Rotation of vectors in a fixed coordinate system produces new coordinates r_i' of a vector:

$$r_i' = \sum_{j=1}^{3} R_{ij} r_j \qquad (\vec{\mathbf{r}}' = \mathbf{R}\vec{\mathbf{r}}). \tag{B.1}$$

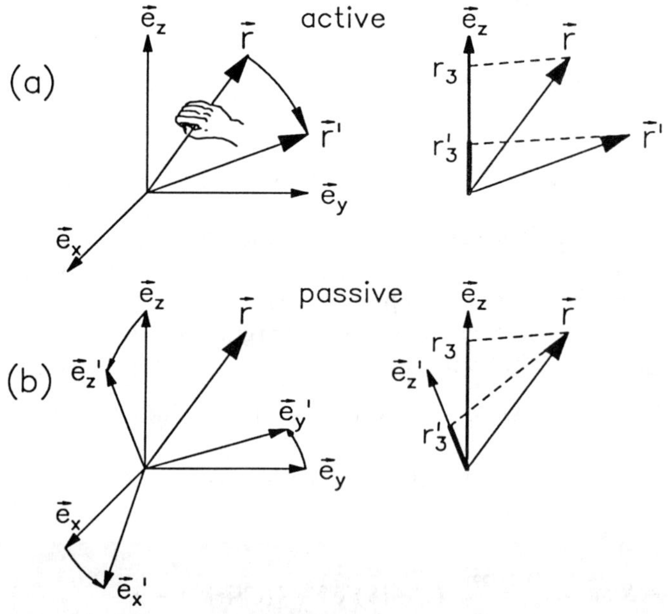

Figure B.1 Active vs. passive rotations. (a) An active rotation rotates the vector in a fixed coordinate system. (b) In a passive rotation, the vector is fixed in space, while the coordinate system is rotated. The change of the coordinates is indicated on the right of the figures.

For a matrix **T**, the invariance of the scalar $\vec{\mathbf{r}}'^{\mathrm{T}}\mathbf{T}'\vec{\mathbf{r}}' \overset{!}{=} \vec{\mathbf{r}}^{\mathrm{T}}\mathbf{T}\vec{\mathbf{r}}$ requires:

$$\mathbf{T}' = \mathbf{RTR}^{-1}. \tag{B.2}$$

Passive rotation

Rotation of coordinate system axes $\vec{\mathbf{e}}'_\alpha = \mathbf{R}\vec{\mathbf{e}}_\alpha$ produces new coordinates of a vector:

$$r'_i = \sum R^{-1}_{ij} r_j \qquad (\vec{\mathbf{r}}' = \mathbf{R}^{-1}\,\vec{\mathbf{r}}). \tag{B.3}$$

For a matrix **T**, invariance $\vec{\mathbf{r}}'^{\mathrm{T}}\mathbf{T}'\vec{\mathbf{r}}' \overset{!}{=} \vec{\mathbf{r}}^{\mathrm{T}}\mathbf{T}\vec{\mathbf{r}}$ requires:

$$\mathbf{T}' = \mathbf{R}^{-1}\mathbf{TR}. \tag{B.4}$$

The R_{ij} are the components of the rotation matrix **R**, for which:

$$\det \mathbf{R} = 1 \qquad \mathbf{R}^{-1} = \mathbf{R}^{\mathrm{T}}. \tag{B.5}$$

For instance, for a rotation in the mathematically positive sense around the z axis, the matrix reads:

$$\mathbf{R}_z(\phi) = \begin{bmatrix} \cos\phi & -\sin\phi & 0 \\ \sin\phi & \cos\phi & 0 \\ 0 & 0 & 1 \end{bmatrix}. \tag{B.6}$$

This can be tested by application of this rotation matrix, say, for $\phi = 90^\circ$ to a vector, for instance, along the x axis:

$$\begin{bmatrix} 0 & -1 & 0 \\ 1 & 0 & 0 \\ 0 & 0 & 1 \end{bmatrix} \begin{bmatrix} 1 \\ 0 \\ 0 \end{bmatrix} = \begin{bmatrix} 0 \\ 1 \\ 0 \end{bmatrix}$$

and comparison of the resulting rotation to the y axis with the expected result obtained by employing the definition of a mathematically positive rotation:

When the thumb of the right hand points along the rotation vector, the fingers indicate the sense of rotation (Figure B.2). (B.7)

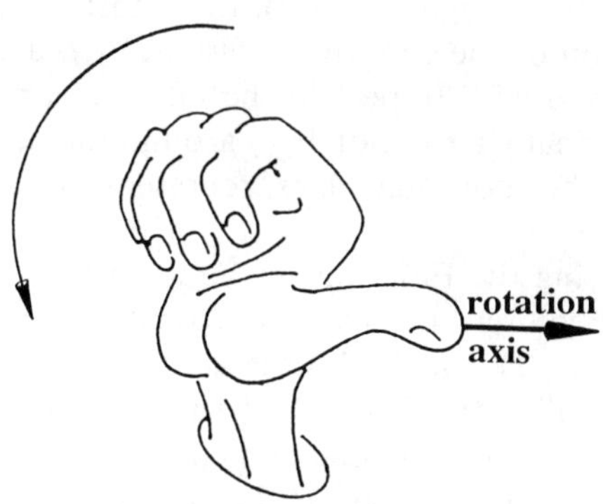

Figure B.2 Definition of right-handed (mathematically positive) rotations. With the thumb of the right hand pointing along the rotation-axis vector, the fingers indicate the direction of rotation.

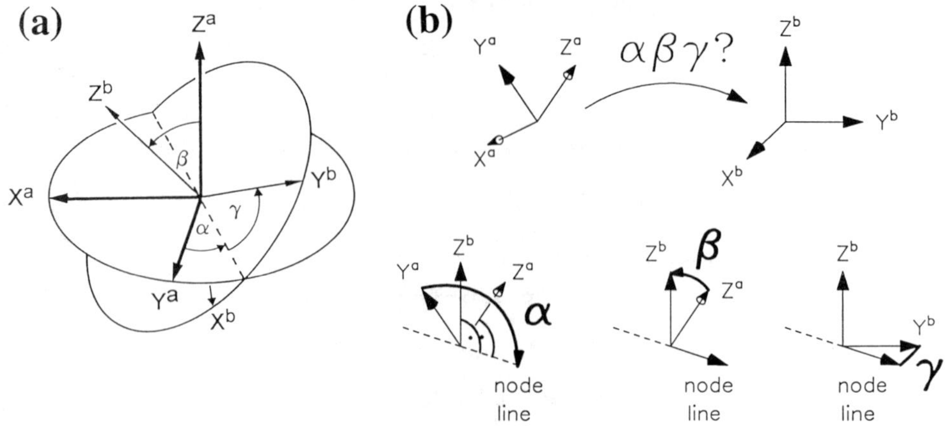

Figure B.3 (a) Definition of the Euler angles (α, β, γ) in the Rose convention. (b) How to find the Euler angles between a given pair of coordinate systems, labelled a and b. α is the angle of rotation of the Y^a axis, around the Z^a axis, onto the node line, which is perpendicular to both the Z^a and Z^b axes. β is the angle between the Z^a and Z^b axes. γ is the angle between the node line and the Y^b axis.

Note that for a passive rotation, such a positive rotation of the coordinate system induces a mathematically negative rotation of (the coordinates of) a vector $\vec{\mathbf{r}}$.

Usually, we consider active rotations. It is to be noted, however, that the standard definition of the Euler angles, as also given below, involves a rotation of the coordinate system, which is a passive rotation according to the above.

B.2 Euler angles and rotation matrices

It is customary and convenient to use triplets of Euler angles (α, β, γ) to specify rotations, as well as relative orientations of two coordinate systems. However, their definition is not unique and numerous conventions are used in the literature. We follow the convention of Rose (1957) (Figure B.3(a)): the transition from system a to system b is described by a rotation of the coordinate system by α around the original z axis, followed by a rotation by β with respect to the intermediary y axis (the 'node line', see next paragraph), and finally a rotation by γ around the new z axis. This definition, involving the rotation of the coordinate axes, represents a passive rotation as defined above.

Procedures for identifying the Euler angles between two given coordinate systems (indices ^{a} and ^{b}), as required in many practical applications, are represented graphically in Figure B.3(b). The angle β is the angle between the two Z axes; α is the rotation angle required to align the Y^a-axis perpendicular to the Z^b-axis when system (^{a}) is rotated around the Z^a axis. The line perpendicular to both z axes, along which the Y^a axis is thus intermediately aligned, is the node line of intersection of the X^aY^a and X^bY^b planes. There are two possible directions for the rotated Y^a axis along the node line (α and $360^\circ - \alpha$); the correct one is determined by the requirement of $0 \leqslant \beta \leqslant 180^\circ$ (positive β-rotation around the rotated Y^a axis). γ is the angle required to rotate the node line around the Z^b axis onto the Y^b axis, as indicated in Figure B.3.

An alternative method for determining (α, β, γ) exploits the following facts:

$$(\beta, \alpha) \text{ are the polar coordinates of the } Z^{\mathrm{b}} \text{ axis in frame } a \tag{B.8}$$

$$(\beta, 180^\circ - \gamma) \text{ are the polar coordinates of the } Z^{\mathrm{a}} \text{ axis in frame } b. \tag{B.9}$$

The Euler rotation matrix is obtained by considering the sequence of the three passive rotations (as specified in equation (B.3)) according to the definition of the Euler angles:

$$\vec{\mathbf{r}}^{\mathrm{b}} = \mathbf{R}_z^{-1}(\gamma)\{\mathbf{R}_y^{-1}(\beta)(\mathbf{R}_z^{-1}(\alpha)\,\vec{\mathbf{r}}^{\mathrm{a}})\}. \tag{B.10}$$

The overall matrix $\mathbf{R}$ also characterizes a passive rotation:

$$\vec{\mathbf{r}}^{\mathrm{b}} = \mathbf{R}^{-1}(\alpha, \beta, \gamma)\,\vec{\mathbf{r}}^{\mathrm{a}}. \tag{B.11}$$

Combining (B.10) and (B.11), we obtain:

$$\begin{aligned} \mathbf{R}^{-1}(\alpha, \beta, \gamma) &= \mathbf{R}_z^{-1}(\gamma)\mathbf{R}_y^{-1}(\beta)\mathbf{R}_z^{-1}(\alpha) \\ \Rightarrow \mathbf{R}(\alpha, \beta, \gamma) &= \mathbf{R}_z(\alpha)\mathbf{R}_y(\beta)\mathbf{R}_z(\gamma). \end{aligned} \tag{B.12}$$

In order to be able to treat the Euler rotation like the more familiar active rotations, it is customary to introduce more or less tacitly a 'pseudo-active' Euler rotation matrix $\tilde{\mathbf{R}}(\alpha, \beta, \gamma) := \mathbf{R}^{-1}(\alpha, \beta, \gamma)$:

$$\vec{\mathbf{r}}^{\mathrm{b}} \stackrel{!}{=} \tilde{\mathbf{R}}(\alpha, \beta, \gamma)\,\vec{\mathbf{r}}^{\mathrm{a}} \qquad \mathbf{T}^{\mathrm{b}} \stackrel{!}{=} \tilde{\mathbf{R}}(\alpha, \beta, \gamma)\mathbf{T}^{\mathrm{a}}\tilde{\mathbf{R}}^{-1}(\alpha, \beta, \gamma)$$

$$\tilde{\mathbf{R}}(\alpha, \beta, \gamma) = \mathbf{R}^{-1}(\alpha, \beta, \gamma) = \mathbf{R}^{\mathrm{T}}(\alpha, \beta, \gamma) \stackrel{(\mathrm{B}.12)}{=} \mathbf{R}_z^{\mathrm{T}}(\gamma)\mathbf{R}_y^{\mathrm{T}}(\beta)\mathbf{R}_z^{\mathrm{T}}(\alpha) \tag{B.13}$$

$$\tilde{\mathbf{R}}(\alpha, \beta, \gamma) = \begin{bmatrix} c\gamma & s\gamma & 0 \\ -s\gamma & c\gamma & 0 \\ 0 & 0 & 1 \end{bmatrix} \begin{bmatrix} c\beta & 0 & -s\beta \\ 0 & 1 & 0 \\ s\beta & 0 & c\beta \end{bmatrix} \begin{bmatrix} c\alpha & s\alpha & 0 \\ -s\alpha & c\alpha & 0 \\ 0 & 0 & 1 \end{bmatrix}$$

with $c\alpha := \cos\alpha$, $s\alpha := \sin\alpha$, $c\beta := \cos\beta$, etc. This yields:

$$\tilde{\mathbf{R}}(\alpha, \beta, \gamma) = \begin{bmatrix} c\alpha \cdot c\beta \cdot c\gamma - s\alpha \cdot s\gamma & s\alpha \cdot c\beta \cdot c\gamma + c\alpha \cdot s\gamma & -s\beta \cdot c\gamma \\ -c\alpha \cdot c\beta \cdot s\gamma - s\alpha \cdot c\gamma & -s\alpha \cdot c\beta \cdot s\gamma + c\alpha \cdot c\gamma & s\beta \cdot s\gamma \\ c\alpha \cdot s\beta & s\alpha \cdot s\beta & c\beta \end{bmatrix}. \tag{B.14}$$

Note that the angle arguments still refer to a passive Euler rotation, while the matrix $\tilde{\mathbf{R}}$ performs an active rotation! In the literature, $\mathbf{R}$ and $\tilde{\mathbf{R}}$ are sometimes not properly distinguished. Note that this distinction is necessary due to the fact that the definition of the Euler angles clearly involves passive rotations, while active rotations are more commonly used. It is *not* related to the possibility of defining the Euler rotations with respect either to the rotated axes or to space-fixed axes (Tinkham, 1964; Mehring, 1983), as in both cases the coordinate axes are rotated, that is, the rotations are passive. As an example, consider $\alpha = 0$, $\gamma = 0$: rotated-axes and fixed-axes rotations are identical (equal to $\mathbf{R}_y(\beta)$), while $\tilde{\mathbf{R}}$ and $\mathbf{R}$ are not the same:

$$\tilde{\mathbf{R}}(0, \beta, 0) = \mathbf{R}(0, -\beta, 0) = \mathbf{R}(180^\circ, \beta, 180^\circ) \neq \mathbf{R}(0, \beta, 0).$$

Around the coordinate axes, active positive (i.e. passive negative) rotations are performed with the following values for the Euler angles (note again that the angle arguments of $\tilde{\mathbf{R}}$ refer to a passive Euler rotation, while the matrix $\tilde{\mathbf{R}}$ performs an active rotation):

$$y \text{ axis:} \quad \mathbf{R}_y(\chi) = \tilde{\mathbf{R}}(0, -\chi, 0) = \tilde{\mathbf{R}}(180^\circ, \chi, 180^\circ) \tag{B.15a}$$

$$z \text{ axis:} \quad \mathbf{R}_z(\chi) = \tilde{\mathbf{R}}(\alpha, 0, \gamma) \text{ with any } \alpha + \gamma = -\chi(+360^\circ) \tag{B.15b}$$

$$x \text{ axis:} \quad \mathbf{R}_x(\chi) = \tilde{\mathbf{R}}(-90^\circ, -\chi, 90^\circ) = \tilde{\mathbf{R}}(90^\circ, \chi, 270^\circ). \tag{B.15c}$$

Around an axis with the polar coordinates (β, α) an active rotation by an angle χ is performed using the rotation matrix:

$$\mathbf{R}_{\beta,\alpha}(\chi) = \tilde{\mathbf{R}}(\sigma, -\beta, -\alpha)\tilde{\mathbf{R}}(\alpha, \beta, -\chi). \tag{B.15d}$$

Generally, the Euler angles of the inverse rotation are determined by:

$$\tilde{\mathbf{R}}(\alpha, \beta, \gamma) = \mathbf{R}^{-1}(\alpha, \beta, \gamma) = \mathbf{R}(-\gamma, -\beta, -\alpha) = \mathbf{R}(180^\circ - \gamma, \beta, 180^\circ - \alpha). \tag{B.16}$$

In solid-state NMR, the Euler angles are often used to characterize the relative orientation of second-rank interaction tensors (more precisely, of their respective principal-axes systems). Even for a given interaction tensor, the PAS can change its orientation due to a rotation of the molecular segment to which it is related. After an Euler rotation by (α, β, γ) that transforms system A into system B, the matrix representation of the interaction tensor $\boldsymbol{\sigma}_A$ in the PAS B corresponds to a passive rotation:

$$(\boldsymbol{\sigma}_A)_B = \mathbf{R}^{-1}\boldsymbol{\sigma}_A\mathbf{R} = \tilde{\mathbf{R}}\boldsymbol{\sigma}_A\tilde{\mathbf{R}}^{-1} \tag{B.17}$$

while expressing $\boldsymbol{\sigma}_B$ in the PAS A is like an active rotation:

$$(\boldsymbol{\sigma}_B)_A = \mathbf{R}\boldsymbol{\sigma}_A\mathbf{R}^{-1} = \tilde{\mathbf{R}}^{-1}\boldsymbol{\sigma}_A\tilde{\mathbf{R}}. \tag{B.18}$$

B.3 Rotation of functions

While the rotation of vectors, particularly spatial vectors $\vec{\mathbf{r}}$, and of matrix representations of Cartesian tensors is performed by a rotation matrix $\mathbf{R}(\alpha, \beta, \gamma)$, the rotation of (wave) functions and operators is effected by a rotation operator $\hat{P}(\alpha, \beta, \gamma)$. They are related by the requirement that:

$$\psi'(\vec{\mathbf{r}}') \overset{!}{=} \psi(\vec{\mathbf{r}}) \qquad \hat{P}\psi(\mathbf{R}\vec{\mathbf{r}}) := \psi'(\vec{\mathbf{r}}')$$
$$\Rightarrow \hat{P}(\alpha, \beta, \gamma)\psi(\vec{\mathbf{r}}) = \psi(\mathbf{R}^{-1}\vec{\mathbf{r}}).$$

The rotated function $\psi'(\vec{\mathbf{r}}') = \hat{P}\psi(\vec{\mathbf{r}}')$ is equal to the original function ψ taken at the original spatial position $\vec{\mathbf{r}} = \mathbf{R}^{-1}\vec{\mathbf{r}}'$ ($\mathbf{R}$ is taken to describe an active rotation here). In summary, we have obtained an expression for the action of $\hat{P}$ on ψ:

$$\hat{P}\psi(\vec{\mathbf{r}}) = \psi(R^{-1}\vec{\mathbf{r}}). \tag{B.19}$$

On this basis, it can be derived in quantum mechanics that an active rotation around an axis $\vec{\mathbf{e}}$ ($|\vec{\mathbf{e}}| = 1$) by an angle ϕ is performed by means of:

$$\hat{P}_L(\vec{\mathbf{e}}, \phi) = \exp(-i\phi\,\vec{\mathbf{e}} \cdot \hat{\vec{\mathbf{L}}}) \tag{B.20}$$

where $\hat{\vec{\mathbf{L}}}$ is the vector operator of the total angular momentum. The components of $\hat{\vec{\mathbf{L}}}$

do not commute with $\hat{r}_j$, while for the components of the nuclear-spin operator $\hat{\vec{\mathbf{I}}}$ alone the commutators vanish, $[\hat{I}_i, \hat{r}_j] = 0$.

Operators are consistently transformed according to:

$$\hat{T}' = \hat{P}\hat{T}\hat{P}^{-1} \tag{B.21}$$

as this makes the expectation values independent of the coordinate system:

$$\langle \psi'(\vec{\mathbf{r}}) \,|\, \hat{T}' \,|\, \psi'(\vec{\mathbf{r}}) \rangle = \langle \psi(\vec{\mathbf{r}}) \,|\, \hat{T} \,|\, \psi(\vec{\mathbf{r}}) \rangle.$$

B.4 Rotations in spin space

Rotations of spin operators are caused by the interaction of spin operators with magnetic fields, such as the Larmor precession around $\vec{\mathbf{B}}_0$ or pulse flip by a resonant $\vec{\mathbf{B}}_1$ field, on the basis of $\hat{H} = -\hat{\vec{\mathbf{I}}} \cdot \gamma\vec{\mathbf{B}}$ and $\hat{\rho}(t) = \exp(-\mathrm{i}\hat{H}t)\hat{\rho}(0)\exp(\mathrm{i}\hat{H}t)$. That these operators $\exp(-\mathrm{i}\hat{\vec{\mathbf{I}}} \cdot \gamma\vec{\mathbf{B}}t)$ actually represent rotations can be concluded directly from evaluating the operator relation (2.75) which yields:

$$\mathrm{e}^{-\mathrm{i}\phi\hat{I}_y}\hat{I}_z\,\mathrm{e}^{\mathrm{i}\phi\hat{I}_y} = \hat{I}_z\cos\phi + \hat{I}_x\sin\phi \tag{B.22a}$$

$$\mathrm{e}^{-\mathrm{i}\phi\hat{I}_x}\hat{I}_z\,\mathrm{e}^{\mathrm{i}\phi\hat{I}_x} = \hat{I}_z\cos\phi - \hat{I}_y\sin\phi \tag{B.22b}$$

$$\mathrm{e}^{-\mathrm{i}\phi\hat{I}_z}\hat{I}_x\,\mathrm{e}^{\mathrm{i}\phi\hat{I}_z} = \hat{I}_x\cos\phi + \hat{I}_y\sin\phi \tag{B.22c}$$

etc.

This obviously describes active right-handed rotations of spin operators by angles ϕ around the α axis of the $\hat{I}_\alpha$ operator in the exponential operators.

Note that the signs in (B.22) are determined not by any conventions other than the basic commutation relations ($[\hat{I}_x, \hat{I}_y] = \mathrm{i}\hat{I}_z$ and its cyclic permutations). To check these signs, it is sufficient to employ the linear approximations of equation (B.22). For instance, for (B.22a) we find:

$$(\hat{1} - \mathrm{i}\phi\hat{I}_y)\hat{I}_z(\hat{1} + \mathrm{i}\phi\hat{I}_y) = \hat{I}_z - \mathrm{i}\phi\underbrace{(\hat{I}_y\hat{I}_z - \hat{I}_z\hat{I}_y)}_{[\hat{I}_y,\,\hat{I}_z] = i\hat{I}_x} + \ \ldots = \hat{I}_z + \phi\hat{I}_x \ldots$$

where ϕ is the initial approximation for $\sin\phi$.

The relations (B.22) show that an operator $\hat{P}_\alpha^+(\phi) := \exp(-\mathrm{i}\phi\hat{I}_\alpha)$ $(\alpha = x, y, z)$ effects an active, mathematically positive rotation (indicated by the superscript $^+$) of spin operators by an angle ϕ around the axis $\vec{\mathbf{e}}_\alpha$ (this should be checked according to the right-hand rule (B.7)). Utilization of this concept (use your right hand) makes long tables of relations like (B.22) unnecessary.

As discussed in Section 2.6.2, two different conventions for the rotation directions of rf pulses are used in NMR, therefore care must be taken in comparing results. The relations (B.22) are of course independent of any convention.

B.5 Irreducible tensors and Wigner rotation matrices

The theoretical treatment of rotations is often facilitated by formulating the problem in terms of irreducible spherical tensors and Wigner rotation matrixes. Particularly important in solid-state NMR are irreducible tensors $\mathbf{R}_2$ and tensor operators $\hat{\mathbf{T}}_2$ of rank 2. Each such $\mathbf{R}_2$ is made up of a set of elements $\mathbf{R}_{2m}$ $(-2 \leqslant m \leqslant 2)$ which can be constructed from a symmetric traceless Cartesian 3×3 matrix $\mathbf{R}$ according to the relations given in Table B.1 (Spiess, 1978).

R_{1m} is equivalent to the antisymmetric part of a matrix, or to a vector. The trace of a matrix, which is rotationally invariant and thus a scalar can be interpreted as a 'tensor' of rank zero, R_{00}.

Many applications of spherical tensors in solid state NMR start from the relation for the Hamiltonian:

$$\hat{H} = \hat{\mathbf{I}}^{\mathrm{T}}\mathbf{R}\hat{\mathbf{J}} = \sum_{L=0}^{2} \sum_{m=-L}^{L} (-1)^m R_{Lm}\hat{T}_{L(-m)} \tag{B.23}$$

between a Cartesian bilinear form (i.e. linear in both $\hat{\mathbf{I}}$ and $\hat{\mathbf{J}}$) and the 'scalar product' of two spherical tensors, $\mathbf{R}$ and $\hat{\mathbf{T}}$ (see Appendix G). This represents a separation into the spatial part R_{Lm}, constructed from $\mathbf{R}$ according to Table B.l, and the spin-operator part $\hat{T}_{Lm}$, constructed in an analogous fashion from the dyadic-product matrix with elements $(\hat{\mathbf{J}}^* \hat{\mathbf{I}})_{kn} := \hat{J}_k \hat{I}_n$.

Rotations are often required for a transformation of properties, conveniently described in one system, into a system of reference that has other advantages. Here, the important transformation property of a spherical tensor $\hat{\mathbf{T}}_L$ comes into play: a tensor element $(\hat{T}_{Lm})'$ in one system can be expressed in terms of the set of $\hat{T}_{Lm'}$ in the other system, and the transformation only involves tensors of equal rank L:

$$(\hat{T}_{Lm})' := \hat{P}(\alpha, \beta, \gamma)\hat{T}_{Lm}\, \hat{P}^{-1}(\alpha, \beta, \gamma) \tag{B.24}$$

$$(\hat{T}_{Lm})' = \sum_{m'=-L}^{L} \hat{T}_{Lm'}\, D^L_{m'm}(\alpha, \beta, \gamma) \tag{B.25}$$

(compare with rotation of a Cartesian vector $r_i' = \Sigma_{j=1}^3 r_j R_{ij}(\alpha, \beta, \gamma)$).

The coefficients $D^L_{m'm}(\alpha, \beta, \gamma)$ are the Wigner rotation matrices, with the Euler angles (α, β, γ) specifying the coordinate transformation. The transformation properties (B.25) can actually be used as the defining condition for irreducible spherical tensor operators.

The dependence of the $D^L_{m'm}(\alpha, \beta, \gamma)$ on α and γ is simply $\mathrm{e}^{-im'\alpha}\,\mathrm{e}^{-im\gamma}$:

$$D^L_{m'm}(\alpha, \beta, \gamma) = d^L_{m'm}(\beta)\,\mathrm{e}^{-im'\alpha}\,\mathrm{e}^{-im\gamma}. \tag{B.26}$$

The nontrivial dependence on β is described by the 'reduced rotation matrix' elements $d^L_{m'm}(\beta)$ which are listed in Table B.2. In most cases, only a limited set of these elements is relevant: if there is no dependence on γ, or no dependence on α, only the elements:

$$D^L_{m0}(\alpha, \beta, \gamma) = Y_{Lm}(\beta, -\alpha)\left(\frac{4\pi}{2L+1}\right)^{\frac{1}{2}} \tag{B.27}$$

Table B.1

$R_{0,0} = \frac{1}{3}(R_{xx} + R_{yy} + R_{zz})$	
$R_{1,0} = \dfrac{-\mathrm{i}}{\sqrt{2}}(R_{xy} - R_{yx})$	$R_{1,\pm 1} = -\frac{1}{2}\{(R_{zx} - R_{xz}) \pm \mathrm{i}(R_{zy} - R_{yz})\}$
$R_{2,0} = \sqrt{\frac{3}{2}}(R_{zz} - R_{0,0})$	$R_{2,\pm 1} = \mp \frac{1}{2}\{(R_{zx} + R_{xz}) \pm \mathrm{i}(R_{zy} + R_{yz})\}$
	$R_{2,\pm 2} = \frac{1}{2}\{(R_{xx} - R_{yy}) \pm i(R_{xy} + R_{yx})\}$

Table B.2 *Legendre polynomials* $P_L(\cos\beta)$*, spherical harmonics.* $Y_{Lm}(\beta,\gamma)$*, and β-dependent factors* $d^{(2)}_{mn}(\beta)$ *of the Wigner functions* $D^{(2)}_{mn}(\alpha,\beta,\gamma)$

$P_0(\cos\beta) = 1$

$P_1(\cos\beta) = \cos\beta$

$P_2(\cos\beta) = \frac{1}{2}(3\cos^2\beta - 1)$

$P_3(\cos\beta) = \frac{1}{2}(5\cos^3\beta - 3\cos\beta)$

$P_4(\cos\beta) = \frac{1}{8}(35\cos^4\beta - 30\cos^2\beta + 3)$

$P_6(\cos\beta) = \frac{1}{16}(231\cos^6\beta - 315\cos^4\beta + 105\cos^2\beta - 5)$

$$Y_{0,0} = 1\left(\frac{1}{4\pi}\right)^{\frac{1}{2}}$$

$$Y_{1,0} = \cos\beta\left(\frac{3}{4\pi}\right)^{\frac{1}{2}} \qquad Y_{1,\pm1} = \mp\sin\beta\, e^{\pm i\gamma}\left(\frac{3}{8\pi}\right)^{\frac{1}{2}}$$

$$Y_{2,0} = \tfrac{1}{2}(3\cos^2\beta - 1)\left(\frac{5}{4\pi}\right)^{\frac{1}{2}} \qquad Y_{2,\pm1} = \mp\sin\beta\cos\beta\, e^{\pm i\gamma}\left(\frac{15}{8\pi}\right)^{\frac{1}{2}}$$

$$Y_{2,\pm2} = \sin^2\beta\, e^{\pm 2i\gamma}\left(\frac{15}{32\pi}\right)^{\frac{1}{2}}$$

	d^2_{x2}	d^2_{x1}	d^2_{x0}
d^2_{2x}	$\frac{1}{4}(1+\cos\beta)^2$	$-\frac{1}{2}(1+\cos\beta)\sin\beta$	$\sqrt{\frac{3}{8}}\sin^2\beta$
d^2_{1x}	$\frac{1}{2}(1+\cos\beta)\sin\beta$	$\frac{1}{2}(\cos\beta-1)+\cos^2\beta$	$-\sqrt{\frac{3}{8}}\sin 2\beta$
d^2_{0x}	$\sqrt{\frac{3}{8}}\sin^2\beta$	$\sqrt{\frac{3}{8}}\sin 2\beta$	$\frac{1}{2}(3\cos^2\beta-1)$
d^2_{-1x}	$\frac{1}{2}(1-\cos\beta)\sin\beta$	$\frac{1}{2}(1+\cos\beta)-\cos^2\beta$	$\sqrt{\frac{3}{8}}\sin 2\beta$
d^2_{-2x}	$\frac{1}{4}(1-\cos\beta)^2$	$\frac{1}{2}(1-\cos\beta)\sin\beta$	$\sqrt{\frac{3}{8}}\sin^2\beta$

	d^2_{x-1}	d^2_{x-2}
d^2_{2x}	$-\frac{1}{2}(1-\cos\beta)\sin\beta$	$\frac{1}{4}(1-\cos\beta)^2$
d^2_{1x}	$\frac{1}{2}(1+\cos\beta)-\cos^2\beta$	$-\frac{1}{2}(1-\cos\beta)\sin\beta$
d^2_{0x}	$-\sqrt{\frac{3}{8}}\sin 2\beta$	$\sqrt{\frac{3}{8}}\sin^2\beta$
d^2_{-1x}	$\frac{1}{2}(\cos\beta-1)+\cos^2\beta$	$-\frac{1}{2}(1+\cos\beta)\sin\beta$
d^2_{-2x}	$\frac{1}{2}(1+\cos\beta)\sin\beta$	$\frac{1}{4}(1+\cos\beta)^2$

Relations between the $d^2_{m,n}$

Of the 25 elements, only 9 (a, ..., i) are independent. The table is symmetric with respect to the diagonal from the lower left to the upper right corner. Relative to the other diagonal, only the absolute values exhibit a symmetry.

m \ n	2	1	0	−1	−2
2	a	$-b$	c	$-d$	e
1	b	f	$-g$	h	$-d$
0	c	g	i	$-g$	c
−1	d	h	g	f	$-b$
−2	e	d	c	b	a

$a = \frac{1}{4}(1+\cos\beta)^2$

$b = \frac{1}{2}(1+\cos\beta)\sin\beta$

$c = \sqrt{(\frac{3}{8})}\sin^2\beta$

$d = \frac{1}{2}(1-\cos\beta)\sin\beta$

$e = \frac{1}{4}(1-\cos\beta)^2$

$f = \frac{1}{2}(\cos\beta-1)+\cos^2\beta$

$g = \sqrt{(\frac{3}{8})}\sin 2\beta$

$h = \frac{1}{2}(1+\cos\beta)-\cos^2\beta$

$i = \frac{1}{2}(3\cos^2\beta-1)$

or

$$D^L_{0m}(\alpha, \beta, \gamma) = (-1)^m Y_{Lm}(\beta, -\gamma)\left(\frac{4\pi}{2L+1}\right)^{\frac{1}{2}} \tag{B.28}$$

are retained, respectively (Zare, 1988). If only the angle β between the two z axes determines the physical behaviour, then only:

$$D^L_{00}(\alpha, \beta, \gamma) = P_{\mathrm{L}}(\cos\beta) \tag{B.29}$$

is of interest.

A useful relation of the Wigner matrices is their 'addition theorem'. It relates the Wigner matrices of two successive rotations A → B and B → C to the Wigner matrices of the overall rotation A → C:

$$D^L_{n',n}(\Omega_{\mathrm{AC}}) = \sum_{m=-L}^{L} D^L_{n',m}(\Omega_{\mathrm{AB}})\, D^L_{m,n}(\Omega_{\mathrm{BC}}). \tag{B.30}$$

The Euler angles have been combined into $\Omega_{\mathrm{AC}} = (\alpha_{\mathrm{AC}}, \beta_{\mathrm{AC}}, \gamma_{\mathrm{AC}})$ etc. This expression (B.30) resembles relation (B.25), but the indices are interchanged. It can be proved by applying equation (B.25) for each of the two successive rotations as well as for the equivalent product rotation and comparing the coefficients.

For $n' = 0$ and/or $n = 0$, according to (B.28)–(B.29) the Wigner matrices can be replaced by spherical harmonics or Legendre polynomials, so that expression (B.30) reduces to the spherical-harmonics addition theorem:

$$P_{\mathrm{L}}(\cos\beta_{\mathrm{AC}}) = \frac{4\pi}{2L+1}\sum_{m=-L}^{L}(-1)^m Y_{Lm}(\beta_{\mathrm{AB}}, -\gamma_{\mathrm{AB}})\, Y_{Lm}(\beta_{\mathrm{BC}}, -\alpha_{\mathrm{BC}}) \tag{B.31}$$

which may be familiar from the multipole expansion in electrodynamics (Jackson, 1975) in the form:

$$P_{\mathrm{L}}(\cos\beta_{\mathrm{AC}}) = \frac{4\pi}{2L+1}\sum_{m=-L}^{L} Y_{Lm}(\beta_{\mathrm{BA}}, \alpha_{\mathrm{BA}})\, Y^*_{Lm}(\beta_{\mathrm{BC}}, \alpha_{\mathrm{BC}})$$

which is directly related to (B.31) considering the Euler angles for the B → A rotation, $\alpha_{\mathrm{BA}} = 180^\circ - \gamma_{\mathrm{AB}}$ and $\beta_{\mathrm{BA}} = \beta_{\mathrm{AB}}$, as well as $Y_{Lm}(\beta, 180^\circ - \alpha) = (-1)^m Y_{Lm}(\beta, \alpha)$, and $Y^*_{Lm}(\beta_{\mathrm{BC}}, \alpha_{\mathrm{BC}}) = Y_{Lm}(\beta_{\mathrm{BC}}, -\alpha_{\mathrm{BC}})$. For $L = 1$, from (B.31) we obtain:

$$\cos\beta_{\mathrm{AC}} = \cos\beta_{\mathrm{AB}}\cos\beta_{\mathrm{BC}} - \sin\beta_{\mathrm{AB}}\sin\beta_{\mathrm{BC}}\cos(\gamma_{\mathrm{AB}} + \alpha_{\mathrm{BC}})$$

which corresponds to equation (A.14).

The explicit $D^L_{m',m}(\alpha, \beta, \gamma)$ are needed, for instance, in the calculation of order parameters of complete orientation distributions (see Section 12.3.3). They can be obtained from the general formula (Rose, 1957; Wigner, 1959):

$$D^L_{m'm}(\alpha, \beta, \gamma) = \mathrm{e}^{-im'\alpha}\,\mathrm{e}^{-im\gamma}\sum_k \frac{(-1)^k\{(L+m)!(L-m)!(L+m')!(L-m')!\}^{\frac{1}{2}}}{k!(L+m-k)!(L-m'-k)!(k+m'-m)!}$$
$$\times\left(\cos\frac{\beta}{2}\right)^{2L-2k-m'+m}\left(\cos\frac{\beta}{2}\right)^{2k+m'-m}. \tag{B.32}$$

The sum runs over all k values for which the denominator does not vanish, that is, from $k_{\mathrm{min}} = \max(0, m - m')$ to $k_{\mathrm{max}} = \min(L+m, L-m')$. The formula (B.32) may appear unwieldy, but it is in fact quite easily implemented on a computer.

Appendix C Evolution of ^{2}H density-operator terms under pulses and quadrupole coupling

Here we want to summarize relations facilitating the description of the time evolution of $\hat{\rho}$ for an $I = 1$ nucleus with a small quadrupole moment, such as deuterons, in high field. For $\hat{\rho}$ given in terms of spin operators and spherical tensor operators (Table C.1), the action of rf pulses and the quadrupolar interaction is summarized in Tables C.2 and C.3.

The results for the quadrupolar interaction can be obtained from the time-evolution formula (2.75) and its simplified version (2.77). The necessary commutators can be obtained by Clebsch–Gordan decomposition of products of spherical tensor operators (Spiess, 1978) (which is analogous to the coupling of spins in atomic physics). Alternatively, the commutators can be derived or checked by means of the 3×3 matrix representations of the $\hat{I}_\alpha$ operators for $I = 1$ (see equation (2.86)).

The action of pulses on the $\hat{T}_{2m}$ can be obtained according to equation (B.25), which describes the transformation of the $\hat{T}_{2m}$ under a rotation specified by the Euler angles (α, β, γ). The necessary Wigner rotation matrix elements are given in Table B.2. The operator $e^{i\chi\hat{I}_y}$ corresponds to a left-handed (mathematically negative) active rotation by χ around the y axis and is characterized by the Euler angles $(0, \chi, 0)$; $e^{i\chi\hat{I}_x}$ corresponds to $(-90^\circ, \chi, 90^\circ)$. Note that according to equation (B.25) these rotations leave the index L of the $\hat{T}_{Lm}$ operators invariant.

Table C.1 *Cartesian spin operators and spherical tensor operators*

$\hat{T}_{10} = \hat{I}_z$	$\sqrt{\tfrac{2}{3}}\hat{T}_{20} = \tfrac{1}{3}\{3\hat{I}_z^2 - \hat{\mathbf{I}}^2\}$
$\hat{T}_{11} = -(\hat{I}_x + i\hat{I}_y)/\sqrt{2}$	$\hat{T}_{2\pm1} = (\hat{T}_{10}\hat{T}_{1\pm1} + \hat{T}_{1\pm1}\hat{T}_{10})/\sqrt{2}$
$\hat{T}_{1-1} = (\hat{I}_x - i\hat{I}_y)/\sqrt{2}$	$\hat{T}_{2\pm2} = (\hat{T}_{1\pm1})^2$

Table C.2 *Effect of the quadrupolar coupling* $\omega_Q\sqrt{\tfrac{2}{3}}\hat{T}_{20}$ on spherical tensor operators

$\hat{\rho}$	$\exp(-i\sqrt{\tfrac{2}{3}}\hat{T}_{20}\omega_Q t)\hat{\rho}\exp(i\sqrt{\tfrac{2}{3}}\hat{T}_{20}\omega_Q t$	
$\hat{I}_z$	$\hat{I}_z$	(invariant)
$\hat{T}_{20}$	$\hat{T}_{20}$	(invariant)
$(\hat{T}_{22} \pm \hat{T}_{2-2})$	$(\hat{T}_{22} \pm \hat{T}_{2-2})$	(invariant)
$\hat{I}_x$	$\hat{I}_x\cos(\omega_Q t) + i(\hat{T}_{21} + \hat{T}_{2-1})\sin(\omega_Q t)$	
$(\hat{T}_{21} + \hat{T}_{2-1})$	$(\hat{T}_{21} + \hat{T}_{2-1})\cos(\omega_Q t) + i\hat{I}_x\sin(\omega_Q t)$	
$\hat{I}_y$	$\hat{I}_y\cos(\omega_Q t) + (\hat{T}_{21} - \hat{T}_{2-1})\sin(\omega_Q t)$	
$(\hat{T}_{21} - \hat{T}_{2-1})$	$(\hat{T}_{21} - \hat{T}_{2-1})\cos(\omega_Q t) - \hat{I}_y\sin(\omega_Q t)$	

Table C.3 *Effect of rotations (rf pulses) on spherical tensor operators*

$\hat{\rho}$	$\exp(-i\chi\hat{I}_x)\hat{\rho}\exp(i\chi\hat{I}_x)$	$\exp(-i\chi\hat{I}_y)\hat{\rho}\exp(i\chi\hat{I}_y)$
$\hat{I}_z$	$\hat{I}_z \cos\chi - \hat{I}_y \sin\chi$	$\hat{I}_z \cos\chi + \hat{I}_x \sin\chi$
$\hat{I}_x$	$\hat{I}_x$	$\hat{I}_x \cos\chi - \hat{I}_z \sin\chi$
$\hat{I}_y$	$\hat{I}_y \cos\chi + \hat{I}_z \sin\chi$	$\hat{I}_y$
$\hat{T}_{20}$	$\hat{T}_{20}\,\frac{1}{2}(3\cos^2\chi - 1)$ $-i\sqrt{\frac{3}{8}}(\hat{T}_{21} + \hat{T}_{2-1})\sin(2\chi)$ $-\sqrt{\frac{3}{8}}(\hat{T}_{22} + \hat{T}_{2-2})\sin^2\chi$	$\hat{T}_{20}\,\frac{1}{2}(3\cos^2\chi - 1)$ $-\sqrt{\frac{3}{8}}(\hat{T}_{21} - \hat{T}_{2-1})\sin(2\chi)$ $+\sqrt{\frac{3}{8}}(\hat{T}_{22} + \hat{T}_{2-2})\sin^2\chi$
$(\hat{T}_{21} + \hat{T}_{2-1})$	$-i\sqrt{\frac{3}{8}}\hat{T}_{20}\sin(2\chi)$ $+(\hat{T}_{21} + \hat{T}_{2-1})\cos(2\chi)$ $-i\frac{1}{2}(\hat{T}_{22} + \hat{T}_{2-2})\sin(2\chi)$	$(\hat{T}_{21} + \hat{T}_{2-1})\cos\chi - (\hat{T}_{22} - \hat{T}_{2-2})\sin\chi$
$(\hat{T}_{21} - \hat{T}_{2-1})$	$(\hat{T}_{21} - \hat{T}_{2-1})\cos\chi$ $-i(\hat{T}_{22} - \hat{T}_{2-2})\sin\chi$	$\sqrt{\frac{3}{2}}\hat{T}_{20}\sin(2\chi)$ $+(\hat{T}_{21} - \hat{T}_{2-1})\cos(2\chi)$ $-\frac{1}{2}(\hat{T}_{22} + \hat{T}_{2-2})\sin(2\chi)$
$(\hat{T}_{22} + \hat{T}_{2-2})$	$-\sqrt{\frac{3}{2}}\hat{T}_{20}\sin^2\chi$ $-i\frac{1}{2}(\hat{T}_{21} + \hat{T}_{2-1})\sin(2\chi)$ $+\frac{1}{2}(\hat{T}_{22} + \hat{T}_{2-2})(\cos^2\chi + 1)$	$\sqrt{\frac{3}{2}}\hat{T}_{20}\sin^2\chi$ $+\frac{1}{2}(\hat{T}_{21} - \hat{T}_{2-1})\sin(2\chi)$ $+\frac{1}{2}(\hat{T}_{22} + \hat{T}_{2-2})(\cos^2\chi + 1)$
$(\hat{T}_{22} - \hat{T}_{2-2})$	$(\hat{T}_{22} - \hat{T}_{2-2})\cos\chi$ $-i(\hat{T}_{21} - \hat{T}_{2-1})\sin\chi$	$(\hat{T}_{22} - \hat{T}_{2-2})\cos\chi + (\hat{T}_{21} + \hat{T}_{2-1})\sin\chi$

Appendix D Useful tools in matrix algebra

Keywords: *eigenvalue problem, diagonalization, PAS, exponential matrices.*

D.1 Introduction

The interactions in solid-state NMR are characterized by second-rank tensors that can be represented by 3×3 matrices. Consequently, certain relations from matrix algebra are indispensable for many calculations and visualizations. Important concepts like diagonalization, the principal-axes system, etc. will be concisely introduced, also pointing out their mutual relations.

D.2 The principal-value problem

The principal-value or eigenvalue problem for a given matrix **A** involves finding eigenvectors $\vec{\mathbf{r}}_i$ and eigenvalues (principal values) a_i such that:

$$\mathbf{A}\vec{\mathbf{r}}_n \overset{!}{=} a_n\vec{\mathbf{r}}_n. \tag{D.1}$$

This means that multiplication by **A** does not change the direction of an eigenvector $\vec{\mathbf{r}}_n$. (Note that the index n does not indicate vector components here, but rather labels the eigenvectors and the corresponding eigenvalues.) If two eigenvalues a_n and a_m are different (nondegenerate), the corresponding eigenvectors $\vec{\mathbf{r}}_n$ and $\vec{\mathbf{r}}_m$ are orthogonal. In the most general case, the eigenvalues can be complex numbers. However, the N eigenvalues of a Hermitian $N \times N$ matrix $\mathbf{A}$ are real numbers ($\mathbf{A}$ is Hermitian if $\mathbf{A} = (\mathbf{A}^{\mathrm{T}})^*$):

$$\text{(D.1)} \Rightarrow a_n^* \vec{\mathbf{r}}_n^* = \mathbf{A}^* \vec{\mathbf{r}}_n^*$$

$$\Downarrow$$

$$a_n^* \vec{\mathbf{r}}_n \cdot \vec{\mathbf{r}}_n^* = \vec{\mathbf{r}}_n \cdot (\mathbf{A}^* \vec{\mathbf{r}}_n^*) = (\mathbf{A}^{*\mathrm{T}}\, \vec{\mathbf{r}}_n) \cdot \vec{\mathbf{r}}_n^* \overset{\text{(Herm.)}}{=} (\mathbf{A}\vec{\mathbf{r}}_n) \cdot \vec{\mathbf{r}}_n^* \overset{\text{(D.1)}}{=} a_n \vec{\mathbf{r}}_n \cdot \vec{\mathbf{r}}_n^*$$

$$\Rightarrow a_n^* = a_n$$

where we have used $\vec{\mathbf{b}} \cdot (\mathbf{A}\vec{\mathbf{c}}) = \Sigma_{ij} b_i A_{ij} c_j = \Sigma_{cj} (\mathbf{A}^{\mathrm{T}})_{ji} b_i = (\mathbf{A}^{\mathrm{T}}\vec{\mathbf{b}}) \cdot \vec{\mathbf{c}}$. Hermitian operators play an important role in quantum mechanics, where all observables must be associated with Hermitian operators so that their eigenvalues, the results of individual measurements, are real. In particular, the eigenvalues are real for symmetric real matrices ($\mathbf{A} = \mathbf{A}^{\mathrm{T}}$), which can be considered a special class of Hermitian matrices. The eigenvectors $\vec{\mathbf{r}}_n$ for different eigenvalues a_n are orthogonal, and if the eigenvalues are degenerate, the eigenvectors can be chosen to be orthogonal.

Many algorithms for finding the eigenvalues or both eigenvectors and eigenvalues of a given matrix are available in numerical libraries. For small matrices, the eigenvalues can be determined, for instance, as the roots (null points) of the characteristic polynomial:

$$c(x) := \det(\mathbf{A} - x\mathbf{1}) \overset{!}{=} 0. \tag{D.2}$$

This derives directly from the basic equation (D.1) of the eigenvalue problem which can be rewritten as $(\mathbf{A} - a_n\mathbf{1})\vec{\mathbf{r}}_n \overset{!}{=} \vec{\mathbf{0}}$. The $\vec{\mathbf{0}}$ can only result if the matrix by which $\vec{\mathbf{r}}_n$ is multiplied has a vanishing determinant. This is the requirement expressed in equation (D.2).

D.3 Diagonalization and the principal-axes system

In solid-state NMR, the principal values and the principal-axes systems (PASs) of second-rank tensors are of great relevance. It will be shown now that in the PAS, the matrix representation of the tensor is diagonal, and the principal values are its (diagonal) elements; thus finding the diagonal representation is equivalent to solving the principal-value problem. The simplification resulting from the diagonal representation in the PAS is also helpful for solving other problems in matrix calculus; an important example is treated in Section D.4.

The representation of a tensor in terms of a matrix $\mathbf{A}$ in one system changes to $\tilde{\mathbf{A}}$ upon transition into a different coordinate system. If the corresponding passive rotation is characterized by a rotation matrix $\mathbf{R}$, according to relation (B.4) the transformation is effected according to:

$$\tilde{\mathbf{A}} = \mathbf{R}^{-1}\mathbf{A}\mathbf{R}. \tag{D.3}$$

The coordinate system in which the matrix representation $\mathbf{A}^{\mathrm{D}}$ is diagonal, $(\mathbf{A}^{\mathrm{D}})_{nm} = \delta_{nm} a_n$, is the principal-axes system (PAS). In order to obtain the diagonal elements a_m, we have to fulfil:

$$\mathbf{A}^{\mathrm{D}} \overset{!}{=} \mathbf{R}_{\mathrm{D}}^{-1}\mathbf{A}\mathbf{R}_{\mathrm{D}} \tag{D.4}$$

$$\mathbf{R}_\mathrm{D}\mathbf{A}^\mathrm{D} \stackrel{!}{=} \mathbf{A}\mathbf{R}_\mathrm{D} \quad \begin{bmatrix} \vec{\mathbf{r}}_1 & \vec{\mathbf{r}}_2 & \vec{\mathbf{r}}_3 \end{bmatrix} \begin{bmatrix} a_1 & 0 & 0 \\ 0 & a_2 & 0 \\ 0 & 0 & a_3 \end{bmatrix} \stackrel{!}{=} \begin{bmatrix} \mathbf{A} \end{bmatrix} \begin{bmatrix} \vec{\mathbf{r}}_1 & \vec{\mathbf{r}}_2 & \vec{\mathbf{r}}_3 \end{bmatrix}$$

$$\Longleftrightarrow \vec{\mathbf{r}}_n a_n \stackrel{!}{=} \mathbf{A}\vec{\mathbf{r}}_n. \tag{D.5}$$

This is a principal-value problem (D.1) as discussed in the previous section. The diagonal elements a_n of $\mathbf{A}^\mathrm{D}$ are the principal values of the matrix $\mathbf{A}$. The eigenvectors $\vec{\mathbf{r}}_n$ are the column vectors of the rotation matrix $\mathbf{R}_\mathrm{D}$ which diagonalizes $\mathbf{A}$. These $\vec{\mathbf{r}}_n$ define the directions of the principal axes expressed in the coordinate system in which the matrix $\mathbf{A}$ was specified. For instance, due to $\vec{\mathbf{r}}_n \cdot \vec{\mathbf{r}}_m = \delta_{nm}$, we have $\mathbf{R}_\mathrm{D}^{-1}\,\vec{\mathbf{r}}_1 = \mathbf{R}_\mathrm{D}^\mathrm{T}\,\vec{\mathbf{r}}_1 = (1,0,0)^\mathrm{T}$. This means that $\mathbf{R}_\mathrm{D}^{-1}$ rotates the $(1,0,0)$ axis onto $\vec{\mathbf{r}}_1$; that is to say, $\vec{\mathbf{r}}_1$ is along the x axis of the PAS.

It is customary to 'visualize' a second-rank tensor $\mathbf{A}$ in terms of an ellipsoid which represents the set of vectors $\vec{\mathbf{x}}$ fulfilling:

$$\vec{\mathbf{x}}^\mathrm{T}\mathbf{A}\vec{\mathbf{x}} \stackrel{!}{=} 1. \tag{D.6}$$

As the '1' is a scalar, equation (D.6) is rotationally invariant and can be evaluated in any coordinate system. In the PAS, due to $(\mathbf{A}^\mathrm{D})_{nm} = \delta_{nm}\, a_m$, the product of equation (D.6) simplifies to yield:

$$a_1 x_1^2 + a_2\, x_2^2 + a_3\, x_3^2 = \frac{x_1^2}{a_1^{-1}} + \frac{x_2^2}{a_2^{-1}} + \frac{x_3^2}{a_3^{-1}} \stackrel{!}{=} 1. \tag{D.7}$$

This is the equation of an ellipsoid (see Appendix F). Its semi-axes are parallel to the principal axes and their lengths are $1/\sqrt{a_m}$. Alternatively, $\mathbf{A}$ could be characterized by the ellipsoid equation $(\mathbf{A}\,\vec{\mathbf{x}})^\mathrm{T}(\mathbf{A}\mathbf{x}) = \vec{\mathbf{x}}^\mathrm{T}\mathbf{A}^\mathrm{T}\mathbf{A}\,\vec{\mathbf{x}} = \vec{\mathbf{x}}^\mathrm{T}\mathbf{A}^2\,\vec{\mathbf{x}} = 1$. In this case, the directions of the semi-axes are unchanged compared to those of (D.7), but their lengths are now equal to $1/a_m$.

D.4 How to simplify exponential matrices by diagonalization

Exponential matrix expressions $\exp(\mathbf{A}t)$ appear several times in the analysis of time dependencies: examples include the formal solutions of the master equation (9.62) (with $\mathbf{A} = \mathbf{\Pi}$) and of the von Neumann equation (2.71) in a matrix representation (2.82), where $\mathbf{A} = -\mathrm{i}\mathbf{H}$. A method for making these kinds of expressions tractable will be described here.

The exponential of a matrix $\mathbf{D}$ can be evaluated in a simple fashion if $\mathbf{D}$ is diagonal. Matrix multiplication immediately shows that all $\mathbf{D}^n$ are also diagonal matrices with $(D^N)_{nn} = (D_{nn})^N$. Thus, the exponential of a diagonal matrix is equal to a diagonal matrix of exponentials:

$$\exp(\mathbf{D}) := \sum_{N=0}^{\infty} \frac{\mathbf{D}^N}{N!} = \begin{bmatrix} \exp(D_{11}) & 0 & \dots & 0 \\ 0 & \exp(D_{22}) & \dots & 0 \\ \vdots & \vdots & \ddots & \vdots \\ 0 & 0 & \dots & \exp(D_{ii}) \end{bmatrix}. \tag{D.8}$$

This suggests that diagonalization of $\mathbf{A}$ can make $\exp(\mathbf{A}t)$ tractable. If a matrix $\mathbf{R}_\mathrm{D}$ diagonalizes $\mathbf{A}$, the product matrix $(\mathbf{R}_\mathrm{D}\mathbf{A}\mathbf{R}_\mathrm{D}^{-1})$ is diagonal and its diagonal elements are the eigenvalues a_i of $\mathbf{A}$ (see equations (D.4) and (D.5)):

$$(\mathbf{R}_\mathrm{D}^{-1}\mathbf{A}\mathbf{R}_\mathrm{D})_{nm} = \delta_{nm} a_n.$$

According to (D.8), the exponential $\exp(\mathbf{R}_\mathrm{D}^{-1}\mathbf{A}\mathbf{R}_\mathrm{D})$ of this diagonal matrix is again a simple diagonal matrix, with $\exp(a_n)$ as its diagonal elements.

In addition, there is the relation:

$$\exp(\mathbf{R}^{-1}\mathbf{A}\mathbf{R}) = \mathbf{R}^{-1}\exp(\mathbf{A})\mathbf{R} \tag{D.9}$$

which is valid for any $\mathbf{A}$ and invertible $\mathbf{R}$.

[*Proof*: $(\mathbf{R}^{-1}\mathbf{A}\mathbf{R})^N = \mathbf{R}^{-1}\mathbf{A}\overbrace{\mathbf{R}\mathbf{R}^{-1}}^{1}\mathbf{A}\mathbf{R}\ldots\mathbf{R}^{-1}\mathbf{A}\mathbf{R} = \mathbf{R}^{-1}\mathbf{A}^N\mathbf{R}$

$$\Rightarrow \sum_{N=0}^{\infty}\frac{(\mathbf{R}^{-1}\mathbf{A}\mathbf{R})^N}{N!} = \sum_{N=0}^{\infty}\mathbf{R}^{-1}\frac{\mathbf{A}^N}{N!}\mathbf{R} = \mathbf{R}^{-1}\left(\sum_{N=0}^{\infty}\frac{\mathbf{A}^N}{N!}\right)\mathbf{R} \iff \text{(D.9)}.$$

So we obtain:

$$\mathbf{R}_\mathrm{D}^{-1}\exp(\mathbf{A})\mathbf{R}_\mathrm{D} = \exp(\mathbf{R}_\mathrm{D}^{-1}\mathbf{A}\mathbf{R}_\mathrm{D}) = \begin{bmatrix} \exp(a_1) & 0 & \ldots & 0 \\ 0 & \exp(a_2) & \ldots & 0 \\ \vdots & \vdots & \ddots & \vdots \\ 0 & 0 & \ldots & \exp(a_i) \end{bmatrix}$$

where a_m is the mth eigenvalue of $\mathbf{A}$. This leads directly to:

$$\exp(\mathbf{A}t) = \overbrace{\mathbf{R}_\mathrm{D}\mathbf{R}_\mathrm{D}^{-1}}^{1}\underbrace{\exp(\mathbf{A}t)\,\overbrace{\mathbf{R}_\mathrm{D}\mathbf{R}_\mathrm{D}^{-1}}^{1}}$$

$$= \mathbf{R}_\mathrm{D}\begin{bmatrix} \exp(a_1 t) & 0 & \ldots & 0 \\ 0 & \exp(a_2 t) & \ldots & 0 \\ \vdots & \vdots & \ddots & \vdots \\ 0 & 0 & \ldots & \exp(a_i t) \end{bmatrix}\mathbf{R}_\mathrm{D}^{-1}. \tag{D.10}$$

The exponential matrix expression $\exp(\mathbf{A}t)$ can thus be expressed in terms of the eigenvalues a_j and the eigenvector-matrix $\mathbf{R}_D$ of the matrix $\mathbf{A}$.

Appendix E Differential conservation of the integral and the Jacobian determinant

The transformation between variables, for instance from $P(\theta)$ to $S(\omega)$, with a nonlinear relationship $\omega(\theta)$ is required quite often in solid-state NMR and other fields. For one-dimensional functions, examples are the transition from θ to

$\omega(\theta) = \delta\frac{1}{2}(3\cos^2\theta - 1)$ or from τ to $\log\tau$. Often, it is physically unreasonable to work with the direct identification of the functions:

$$S(\omega) \stackrel{???}{=} P(\theta(\omega))$$

when the area under the functional curves rather than the height of the functions is the conserved quantity. Therefore, we demand the integral intensity of corresponding intervals in θ and ω to be equal:

$$\int_{\omega(\theta_1)}^{\omega(\theta_2)} S(\omega)\,\mathrm{d}\omega \stackrel{!}{=} \int_{\theta_1}^{\theta_2} P(\theta)\,\mathrm{d}\theta \tag{E.1}$$

(Figure E.1). Since this must apply for arbitrary θ_1 and θ_2, expression (E.1) represents an infinite number of relations. However, these can be summarized in the short infinitesimal formulation:

$$S(\omega)\,\mathrm{d}\omega \stackrel{!}{=} P(\theta)\,\mathrm{d}\theta. \tag{E.2}$$

This can be termed 'the principle of conservation of area of corresponding intervals' or 'the principle of differential conservation of the integral'. By virtue of Leibniz' suggestive notation of the derivative and the integral, 'dividing' by $\mathrm{d}\omega$ directly provides:

$$S(\omega) = P(\theta(\omega))\,\frac{\mathrm{d}\theta}{\mathrm{d}\omega} \tag{E.3}$$

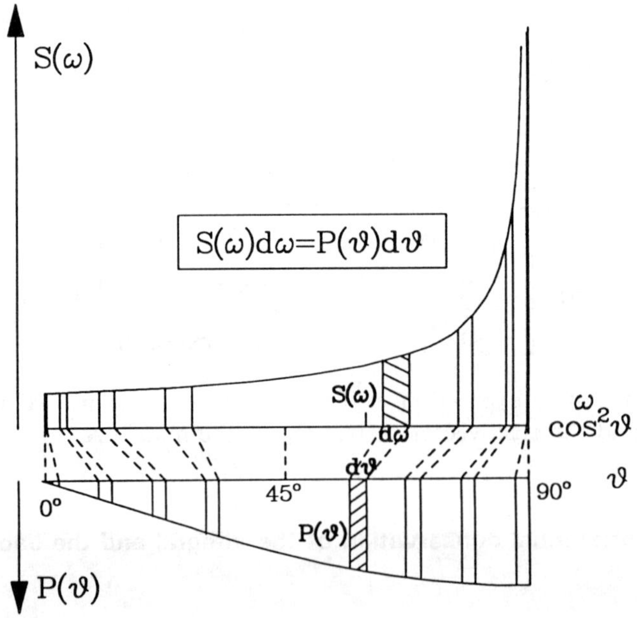

Figure E.1 Visualization of the principle of differential conservation of the integral of $P(\theta)$ and $S(\omega)$ over corresponding intervals $[\theta \ldots \theta + \mathrm{d}\theta]$ and $[\omega(\theta) \ldots \omega(\theta + \mathrm{d}\theta)]$. The area $P(\theta)\,\mathrm{d}\theta$ must be equal to the area $s(\omega)$. The chosen example shows the generation of a (single-transition) Pake pattern $S(\omega)$ from $\omega(\theta)/\delta = \frac{1}{2}(3\cos^2\theta - 1)$ and $P(\theta) = \sin\theta$.

which fulfils equation (E.1), producing a well-known integration theorem for implicit functions (insert (E.3) into (E.1)):

$$\int_{\omega(\theta_1)}^{\omega(\theta_2)} P(\theta(\omega)) \frac{d\theta}{d\omega} \, d\omega = \int_{\theta_1}^{\theta_2} P(\theta) \, d\theta.$$

If the inversion $\theta(\omega)$ of $\omega(\theta)$ is not unique (i.e. two or more θ values correspond to one ω value), the θ range has to be divided into uniquely invertible segments of $\omega(\theta)$ and the partial spectra resulting from (E.3) are added up in the end. Avoiding the problems of the inversion of $\omega(\theta)$ to some extent, it can sometimes be more convenient to perform a 'division by dθ' on equation (E.2) to obtain:

$$S(\omega) \frac{d\omega}{d\theta} = P(\theta) \Rightarrow S(\omega) = P(\theta) \Big/ \frac{d\omega}{d\theta}. \tag{E.4}$$

Nevertheless, finally θ has to be expressed in terms of ω, that is, $\theta(\omega)$ must be determined.

If ω decreases with increasing θ, such that $\omega(\theta_1) > \omega(\theta_2)$ it is necessary to take the absolute values of the corresponding intervals:

$$S(\omega) \, | d\omega | = P(\theta) \, | d\theta | \tag{E.5}$$

$$S(\omega) = P(\theta) \left| \frac{d\theta}{d\omega} \right|. \tag{E.6}$$

As a purist, one would also insist on using different symbols for the variable ω and the function $\omega(\theta)$, as well as for θ and $\theta(\omega)$. However, since this does not facilitate the understanding we have chosen to label 'the frequency' generally as ω and 'the angle' as θ.

For functions of several variables, the principal idea of the transformation is the same as for the one-dimensional functions, while the actual mathematics is a little more involved. The best-known example is the transformation between Cartesian coordinates $\vec{\mathbf{x}} = (x_1, x_2, x_3)$ and polar coordinates $(\Omega_1, \Omega_2, \Omega_3) = (r, \theta, \phi)$:

$$P(x_1, x_2, x_3) \, dx_1 \, dx_2 \, dx_3 \stackrel{!}{=} \tilde{P}(r, \theta, \phi) \, dr \, d\theta \, d\phi. \tag{E.7}$$

The result is known to be:

$$P(\vec{\mathbf{x}}) \, dx_1 \, dx_2 \, dx_3 = P[\vec{\mathbf{x}}(r, \theta, \phi)] \; r^2 \sin\theta \, dr \, d\theta \, d\phi \tag{E.8}$$

$$\Rightarrow \tilde{P}(r, \theta, \phi) = P[\vec{\mathbf{x}}(r, \theta, \phi)] r^2 \sin\theta. \tag{E.9}$$

The factor $r^2 \sin\theta$ quantifies the distortion of the volume element dr dθ dϕ relative to the corresponding volume element dx_1 dx_2 dx_3. In analogy to the correction factor dω/dθ, we now look for a derivative of the Cartesian coordinates x_i with respect to the polar coordinates Ω_j. Such a derivative is the gradient tensor $\partial x_i / \partial \Omega_j$. However, we need a scalar factor in the integration. It is provided by the determinant of the gradient tensor, the so-called Jacobian (determinant):

$$\frac{\partial(x_1, x_2, x_3)}{\partial(r, \theta, \phi)} := \begin{vmatrix} \dfrac{\partial x_1}{\partial r} & \dfrac{\partial x_2}{\partial r} & \dfrac{\partial x_3}{\partial r} \\[2ex] \dfrac{\partial x_1}{\partial \theta} & \dfrac{\partial x_2}{\partial \theta} & \dfrac{\partial x_3}{\partial \theta} \\[2ex] \dfrac{\partial x_1}{\partial \phi} & \dfrac{\partial x_2}{\partial \phi} & \dfrac{\partial x_3}{\partial \phi} \end{vmatrix} = r^2 \sin\theta. \tag{E.10}$$

In 2D NMR, for the transformation from orientations (Θ, Φ) to frequencies $(\tilde{\omega}_1, \tilde{\omega}_2)$ in:

$$S(\tilde{\omega}_1, \tilde{\omega}_2) = P(\Theta, \Phi) \bigg/ \frac{\partial(\tilde{\omega}_1, \tilde{\omega}_2)}{\partial(\Theta, \Phi)} \tag{E.11}$$

we have to analyse the Jacobian:

$$\frac{\partial(\tilde{\omega}_1, \tilde{\omega}_2)}{\partial(\Theta, \Phi)} := \begin{vmatrix} \dfrac{\partial \omega_1}{\partial \Theta} & \dfrac{\partial \tilde{\omega}_2}{\partial \Theta} \\ \dfrac{\partial \tilde{\omega}_1}{\partial \Phi} & \dfrac{\partial \tilde{\omega}_2}{\partial \Phi} \end{vmatrix} = \frac{\partial \tilde{\omega}_1}{\partial \Theta} \frac{\partial \tilde{\omega}_2}{\partial \Phi} - \frac{\partial \tilde{\omega}_1}{\partial \Phi} \frac{\partial \tilde{\omega}_2}{\partial \Theta}. \tag{E.12}$$

Considering the gradient vectors (columns of the gradient tensor):

$$\operatorname{grad}_{\Theta, \Phi}(\tilde{\omega}_i) = \begin{bmatrix} \dfrac{\partial \tilde{\omega}_i}{\partial \Theta} \\ \dfrac{\partial \tilde{\omega}_i}{\partial \Phi} \end{bmatrix} \tag{E.13}$$

we can write the Jacobian as the absolute value of the two-dimensional vector product of the gradient vectors:

$$\frac{\partial(\tilde{\omega}_1, \tilde{\omega}_2)}{\partial(\Theta, \Phi)} = |\operatorname{grad}_{\Theta,\Phi}(\tilde{\omega}_1) \times \operatorname{grad}_{\Theta, \Phi}(\tilde{\omega}_2)|. \tag{E.14}$$

This proves useful for interpreting the null-points of the Jacobian geometrically in terms of lines of constant frequency (see Section 6.7.3).

An alternative, frequently used formulation of the nonlinear transformation of variables as described so far makes use of a delta function:

$$S(\tilde{\omega}) = \int P(\theta)\, \delta(\omega(\theta) - \tilde{\omega})\, \mathrm{d}\theta. \tag{E.15}$$

The delta function relates θ and $\tilde{\omega}$ via $\omega(\theta)$ with the correct integral weight. (Here we felt forced to distinguish the variable $\tilde{\omega}$ and the function $\omega(\theta)$ since they are directly confronted.)

The δ-function formulation formally circumvents the problems encountered above if $\omega(\theta)$ is not uniquely invertible. In practical calculations, however, the delta-function expression does not provide any new possibilities. The integral is solved by substitution $\omega' = \omega(\theta)$, $\mathrm{d}\theta = (\mathrm{d}\theta/\mathrm{d}\omega')\, \mathrm{d}\omega'$, and then the delta function removes the integral:

$$S(\tilde{\omega}) = \int P(\theta(\omega'))\, \delta(\omega' - \tilde{\omega})\, \frac{\mathrm{d}\theta}{\mathrm{d}\omega'}\, \mathrm{d}\omega' = P(\theta(\tilde{\omega}))\, \frac{\mathrm{d}\theta}{\mathrm{d}\omega}\bigg|_{\tilde{\omega}}. \tag{E.16}$$

The resulting relationship between $S(\tilde{\omega})$ and $P(\theta)$ is identical to equation (E.3) above.

The multidimensional δ-function formulation is completely analogous to the one-dimensional relation (E.15):

$$\tilde{S}(\vec{\boldsymbol{\omega}}') = \int P(\vec{\boldsymbol{\Omega}})\, \delta(\vec{\boldsymbol{\omega}}(\vec{\boldsymbol{\Omega}}) - \vec{\boldsymbol{\omega}}')\, \mathrm{d}\vec{\boldsymbol{\Omega}}. \tag{E.17}$$

Appendix F Ellipses, ellipsoids and the Lissajou representation

Elliptical ridges are prominent features in many 2D experiments on powder samples. To prove that these features are indeed elliptical, some mathematical background is required. This is to be outlined here. In passing, we shall also mention a few features of ellipsoids in three-dimensional space.

An ellipse consists of the points (x, y) that fulfil the equation:

$$\frac{x^2}{a^2} + \frac{y^2}{b^2} = 1. \tag{F.1}$$

a and b are the semi-axes of the ellipse as indicated in Figure F.1(a). Ellipsoids in three dimensions are described by:

$$\frac{x^2}{a^2} + \frac{y^2}{b^2} + \frac{z^2}{c^2} = 1. \tag{F.2}$$

For $\vec{\mathbf{r}} = (x, y, z)^{\mathrm{T}}$, this can also be expressed as:

$$(x, y, z)\begin{bmatrix} a^{-2} & 0 & 0 \\ 0 & b^{-2} & 0 \\ 0 & 0 & c^{-2} \end{bmatrix}\begin{pmatrix} x \\ y \\ z \end{pmatrix} = 1. \tag{F.3}$$

This relationship is sometimes exploited for graphical representations of Cartesian tensors of second rank (3×3 matrices) with eigenvalues $1/a$, $1/b$ and $1/c$ (see Section D.3).

The Lissajou representation of ellipses is a parameter representation of the coordinates: the points of the ellipse are given by:

$$\begin{pmatrix} x' \\ y' \end{pmatrix} = \begin{pmatrix} \cos\phi \\ \cos(\phi - 2\beta) \end{pmatrix} \tag{F.4}$$

where ϕ is the parameter that must be varied from $0°$ to $360°$ to trace out the full

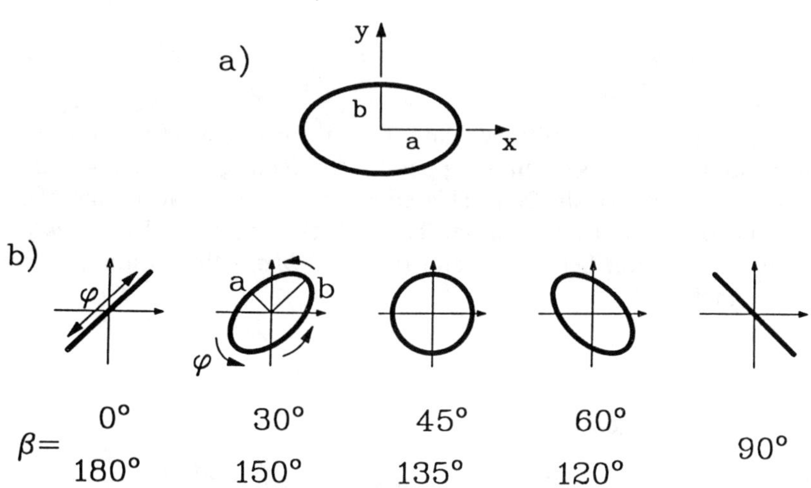

Figure F.1 (a) Ellipse $(x^2/a^2) + (y^2/b^2) = 1$ with semiaxes a and b. (b) Lissajou representations, equation (F.4), of ellipses with different eccentricities.

ellipse, and the fixed parameter β characterizes the shape of the ellipse. That equation (F.4) indeed produces an ellipse that fulfils equation (F.1) will be shown below. Such an elliptical Lissajou figure can be generated on an oscilloscope by sweeping the beam like $\cos\phi$ in the horizontal direction and with the same input signal but phase-shifted by 2β in the vertical direction. In contrast to the ellipse equation (F.1), in the Lissajou representation the semi-axes of the ellipse make angles of 45° with the coordinate axes (see Figure F.1(b)).

To gain an intuitive understanding of the Lissajou equation (F.4), consider a few special cases. For $\beta = 45°$, the coordinates:

$$\begin{pmatrix} x' \\ y' \end{pmatrix} = \begin{pmatrix} \cos\phi \\ \cos(\phi - 90°) \end{pmatrix} = \begin{pmatrix} \cos\phi \\ \sin\phi \end{pmatrix} \tag{F.5}$$

describe a circle; for $\beta = 0°$ and $\beta = 90°$, the ellipses degenerate into straight lines, with slope of $+1$ and -1, respectively:

$$\beta = 0°: \begin{pmatrix} x' \\ y' \end{pmatrix} = \cos\phi \begin{pmatrix} 1 \\ 1 \end{pmatrix} \tag{F.6}$$

$$\beta = 90°: \begin{pmatrix} x' \\ y' \end{pmatrix} = \begin{pmatrix} \cos\phi \\ \cos(\phi - 180°) \end{pmatrix} = \cos\phi \begin{pmatrix} 1 \\ -1 \end{pmatrix}. \tag{F.7}$$

To compare the Lissajou representation with the ellipse equation (F.1) the coordinate system must be rotated by 45°:

$$\begin{pmatrix} x \\ y \end{pmatrix} = \begin{bmatrix} \cos 45° & \sin 45° \\ -\sin 45° & \cos 45° \end{bmatrix} \begin{pmatrix} \cos\phi \\ \cos(\phi - 2\beta) \end{pmatrix} = \frac{1}{\sqrt{2}} \begin{pmatrix} \cos\phi + \cos(\phi - 2\beta) \\ -\cos\phi + \cos(\phi - 2\beta) \end{pmatrix} \tag{F.8}$$

($\cos 45° = \sin 45° = 1/\sqrt{2}$). From the plots in Figure F.l(b), we can guess that the lengths of the semi-axes are:

$$a = |\sqrt{2}\cos\beta| \qquad b = |\sqrt{2}\sin\beta|. \tag{F.9}$$

Inserting these equations (F.8) and (F.9) into the ellipse equation $x^2/a^2 + y^2/b^2 = 1$, we find:

$$\frac{\frac{1}{2}(\cos\phi + \cos\{\phi - 2\beta\})^2}{2\cos^2\beta} + \frac{\frac{1}{2}(-\cos\phi + \cos\{\phi - 2\beta\})^2}{2\sin^2\beta} \stackrel{!}{=} 1. \tag{F.10}$$

That this equality is true for arbitrary ϕ and β can be proved after multiple applications of trigonometric relations as given in Appendix A (yielding, for instance, $a^2 + b^2 = 2$, $a^2 - b^2 = 2\cos 2\beta$, $a^2b^2 = \sin^2 2\beta$). This equality (F.10) confirms that the Lissajou representation does describe an ellipse. Through the relations (F.9) thus proved we obtain the useful relation between β and the ratio b/a which is independent of any length scaling factor:

$$|\tan\beta| = \frac{b}{a}. \tag{F.11}$$

Finally, it should be noted that both:

$$\begin{pmatrix} x' \\ y' \end{pmatrix} = \begin{pmatrix} \cos(2\phi') \\ \cos(2\phi' - 2\beta) \end{pmatrix} \text{ and } \begin{pmatrix} x' \\ y' \end{pmatrix} = \begin{pmatrix} \cos\phi \\ \cos(\phi - 2\beta) \end{pmatrix} \tag{F.12}$$

produce the same ellipse; the former expression is usually encountered in 2D NMR.

Generally, the same ellipse results if the same term $f(\phi')$ appears at the place of ϕ in both $\cos\phi$ and $\cos(\phi - 2\beta)$; to trace out the full ellipse, the parameter $f(\phi')$ must cover a range of 360° or more.

Appendix G Truncation of Hamiltonians

Truncation denotes the phenomenon that a weak interaction $\hat{H}_1$ in the presence of a stronger interaction $\hat{H}_0$ is effectively reduced to certain components of $\hat{H}_1$ that commute with $\hat{H}_0$. The best-known example is the averaging-out of the local fields' nonsecular components by the Zeeman interaction (see Section 2.4.1). Another example is 'second averaging' of the Hamiltonians remaining after the first averaging according to average-Hamiltonian theory.

A visualization of the truncation effect is provided by considering the simultaneous action of a strong field $\vec{\mathbf{B}}_0$, and a weak (static) field $\vec{\mathbf{B}}_1$ with components B_{1a} along $\vec{\mathbf{B}}_0$ and B_{1p} perpendicular to $\vec{\mathbf{B}}_0$. The length of the resultant field vector $\vec{\mathbf{B}}_{tot} = \vec{\mathbf{B}}_0 + \vec{\mathbf{B}}_1$ is given by:

$$|\vec{\mathbf{B}}_{tot}| = (B_0 + B_{1a})\sqrt{1 + \left(\frac{B_{1p}}{B_0 + B_{1a}}\right)^2} \tag{G.1}$$

(Pythagorean theorem: $(B_0 + B_{1a})^2 + B_{1p}^2 = B_{tot}^2$). For $B_{1p}/B_0 \ll 1$, this relation (G.1) reduces to:

$$|\vec{\mathbf{B}}_{tot}| \cong (B_0 + B_{1a}). \tag{G.2}$$

Obviously, only the projection $\vec{\mathbf{B}}_{1a}$ of $\vec{\mathbf{B}}_1$ onto $\vec{\mathbf{B}}_0$ is relevant to first order.

A more general treatment of truncation considers an interaction representation (see Section 2.5.8) that transforms away the immediate effects of $\hat{H}_0$ in the time evolution of the density operator:

$$\hat{\rho}'(t) := e^{i\hat{H}_0 t}\hat{\rho}(t)\, e^{-i\hat{H}_0 t} \tag{G.3a}$$

$$\hat{H}_1'(t) := e^{i\hat{H}_0 t}\hat{H}_1\, e^{-i\hat{H}_0 t} \tag{G.3b}$$

$$\dot{\hat{\rho}}'(t) = -i\,[\hat{H}_1'(t), \hat{\rho}'(t)]. \tag{G.3c}$$

In the transformed von Neumann equation of motion, equation (G.3c), the Hamiltonian $\hat{H}_0$ does not appear directly any more. This can be viewed as a transition into a frame moving such that those time-dependencies appear as static. In this interaction frame, components of $\hat{H}_1$ that do not commute with $\hat{H}_0$ appear as time-dependent in $\hat{H}_1'(t)$ according to equation (G.3b). Since $\hat{H}_0$ is much stronger than $\hat{H}_1$, the harmonic oscillations imposed on these components of $\hat{H}_1$ by $\hat{H}_0$ exceed the time-dependencies caused by $\hat{H}_1$ itself. Therefore, the effect of those 'perpendicular' components is averaged to zero. Since the time-averaged, effective Hamiltonian:

$$\hat{H}_1^0 = \overline{\exp(i\hat{H}_0 t)\hat{H}_1\,\exp(-i\hat{H}_0 t)} \tag{G.4}$$

is again time-independent, the time evolution in the interaction frame can be approximated by the solution (2.73) of the von Neumann equation for time-independent Hamiltonians:

$$\hat{\rho}'(t) \cong e^{-i\hat{H}_1^0 t}\,\hat{\rho}'(0)\,e^{i\hat{H}_1^0 t}. \tag{G.5}$$

(This approximation can be formally justified as the application of the Magnus expansion, see Section 3.3.3, to the time-dependent Hamiltonian $H_1'(t)$). Combining $\hat{\rho}'(t)$ with the relation (G.3a) between $\hat{\rho}'(t)$ and $\hat{\rho}(t)$, in particular $\hat{\rho}'(0) = e^0\,\hat{\rho}(0)e^{-0} = \hat{\rho}(0)$, we obtain:

$$\hat{\rho}(t) \cong e^{-i\hat{H}_0 t}\, e^{-i\hat{H}_1^0 t}\, \hat{\rho}(0)\, e^{i\hat{H}_1^0 t}\, e^{i\hat{H}_0 t}. \tag{G.6}$$

Comparison of this equation (G.6) with the exact solution (2.73):

$$\hat{\rho}(t) = e^{-i(\hat{H}_0 + \hat{H}_1)t}\, \hat{\rho}(0)\, e^{i(\hat{H}_0 + \hat{H}_1)t}$$

proves that averaging in the interaction representation amounts to replacing:

$$e^{-i(\hat{H}_0 + \hat{H}_1)t} \cong e^{-i\hat{H}_0 t}\, e^{-i\hat{H}_1^0 t}. \tag{G.7}$$

The Hamiltonian $\hat{H}_1^0$ represents the time average (G.4) of $\hat{H}_1$ in the $\hat{H}_0$-interaction frame. Since $\hat{H}_0$ and $\hat{H}_1^0$ commute, the propagators on the right-hand side of (G.7) can be interchanged, or the exponents can be combined to stress the equivalence to equation (G.2).

Equation (G.7) with $\hat{H}_0 = \omega_0 \hat{I}_z$ represents the full description of Zeeman truncation. In contrast, the truncation procedure described in Section 2.4.1 is somewhat simplistic. It is based on splitting the Hamiltonian into two parts, one that commutes with $\omega_0 \hat{I}_z$ and one that does not. This, however, is not a unique prescription: consider $\hat{H}_1 = \hat{H}_1^c + \hat{H}_1^n$ with an $\hat{I}_z$-commuting part $\hat{H}_1^c$ and a noncommuting part $\hat{H}_1^n$. Then, by adding a commuting operator $\hat{C}$ to $\hat{H}_1^c$, we form another $\hat{I}_z$-commuting operator $\hat{H}_1^{c\prime} = \hat{H}_1^c + \hat{C}$, and subtracting $\hat{C}$ from $\hat{H}_1^n$, a noncommuting $\hat{H}_1^{n\prime} = \hat{H}_1^n - \hat{C}$. These primed operators represent another way of splitting $\hat{H}_1$ into $\hat{I}_z$-commuting and $\hat{I}_z$-noncommuting parts, since $\hat{H}_1 = \hat{H}_1^{c\prime} + \hat{H}_1^{n\prime}$. In simple cases, the correct way of splitting $\hat{H}_1$ is intuitively obvious. In the general case, however, the procedure based on equation (G.7) (or some equivalent of it) has to be applied.

Overall, the principle (G.7) of separating a strong and a weak Hamiltonian which act simultaneously is applied on three different levels in the NMR theory: for high-field truncation, for calculation of the average-Hamiltonian during rf irradiation (see Section 3.3.2), and for 'second averaging' (Section 3.3.4).

To find the secular terms of the homonuclear dipolar Hamiltonian $\hat{H}_D^{II}$ and of the quadrupolar Hamiltonian $\hat{H}_Q$ (see Section 2.4.1), the formulation of the bilinear Hamiltonian in terms of spherical tensors (Spiess, 1978; Mehring, 1983) (see Appendix B.5) can prove useful:

$$\hat{H}_1 = \sum_k \sum_{L=0}^{2} \sum_{m=-L}^{L} (-1)^m (R_{L,m}\, \hat{T}_{L,-m})_k \tag{G.8}$$

(the index k numbers spins or spin pairs). The transition into the rotating frame imposes a rotation around the z axis, which corresponds to the Euler angles $(0, 0, -\omega_0 t)$. Table B.2 shows that:

$$D_{m',m}(0, 0, -\omega_0 t) = \delta_{m',m}\, e^{im\omega_0 t} \tag{G.9}$$

so that the rotation formula for spherical tensor operators (B.25) yields:

$$\hat{H}_1' = \sum_k \sum_{L=0}^{2} \sum_{m=-L}^{L} (-1)^m (R_{L,m}\, \hat{T}_{L,-m})_k\, e^{-im\omega_0 t}. \tag{G.10}$$

This shows that the time-independent (secular) elements of $\hat{H}_1'$ are those for which $m = 0$: $R_{00}\hat{T}_{00} + R_{10}\hat{T}_{10} + R_{20}\hat{T}_{20}$. Of these, only:

$$R_{20}\hat{T}_{20} = R_{zz}\tfrac{1}{2}(3\hat{I}_z^i\hat{I}_z^j - \hat{\mathbf{I}}^i \cdot \hat{\mathbf{I}}^j) \tag{G.11}$$

(see Tables B.l and C.1) remains for quadrupolar and dipolar couplings because the isotropic and antisymmetric parts of their spatial terms vanish: $R_{00} \sim \Sigma_{\alpha=1}^{3} R_{\alpha\alpha} = 0$, and $R_{10} \sim R_{xy} - R_{yx} = 0$. For dipolar couplings, this follows from the explicit form of equation (2.14). For the quadrupolar interaction, the field gradient tensor $\mathbf{V}$ ($\triangleq\mathbf{R}$) is symmetric due to $V_{\alpha\beta} = \partial^2 V/\partial r_\alpha \partial r_\beta = \partial^2 V/\partial r_\beta \partial r_\alpha = V_{\beta\alpha}$. The quadrupolar scalar portion $R_{00}\hat{T}_{00} = \frac{1}{3}\,\mathrm{tr}(\mathbf{V})\,\hat{\mathbf{I}}^2$ produces no effect in the NMR spectra: for the given I, the factor $\hat{\mathbf{I}}^2$ shifts the levels of all $\hat{I}_z$ eigenfunctions by the same amount, such that the energy-level differences remain unchanged. Equivalently, one can state that the matrix representation in the basis of $\hat{I}_z$ eigenfunctions, $R_{00}\mathbf{T}_{00} = \frac{1}{3}\,\mathrm{tr}(\mathbf{V})I(I+1)\mathbf{1}$, is proportional to the unit matrix $\mathbf{1}$ and therefore commutes with all other operators. As a consequence, only the traceless portion of $\mathbf{V}$ is spectroscopically relevant.

References

Jackson, J. D. (1975). *Classical Electrodynamics*, Chapter 3. Wiley, New York.

Mehring, M. (1975). *High Resolution NMR of Solids*. Springer, New York.

Rose, M. E. (1957). *Elementary Theory of Angular Momentum*. Wiley, New York.

Spiess, H. W. (1978). Dynamic NMR Spectroscopy. *NMR Basic Principles and Progress*, **15**, 55.

Tinkham, M. (1964). *Group Theory and Quantum Mechanics*. McGraw-Hill, New York.

Wigner, E. P. (1959). *Group Theory & Application to the Quantum Mechanics of Atomic Spectra*. Academic Press, New York.

Zare, R. N. (1988). *Angular Momentum: Spatial Aspects in Chemistry and Physics*, Chapter 3. Wiley, New York.

Index